APPLIED COLLEGE ALGEBRA AND TRIGONOMETRY

Third Edition

Linda Davis

APPLIED COLLEGE ALGEBRA AND TRIGONOMETRY

Third Edition

Linda Davis

Prentice Hall

Upper Saddle River, New Jersey
Columbus, Ohio

Library of Congress Cataloging-in-Publication Data

Davis, Linda,
   Applied college algebra and trigonometry / Linda Davis.—3rd ed.
     p. cm.
   Includes bibliographical references and index.
   ISBN 0-13-093893-9
   1. Algebra.   2. Trigonometry.   I. Title.

QA154.3 .D38 2003
512′.13—dc21
                            2002019606

Editor in Chief: Stephen Helba
Executive Editor: Frank Mortimer
Assistant Editor: Michelle Churma
Production Editor: Tricia L. Rawnsley
Design Coordinator: Diane Ernsberger
Cover Designer: Ali Mohrman
Cover art: Hiroshi Sakuramoto/Photonica
Production Manager: Brian Fox
Electronic Text Management: Karen L. Bretz
Marketing Manager: Tim Peyton

This book was set in Times Roman by The Clarinda Company. It was printed and bound by Courier Westford, Inc. The cover was printed by The Lehigh Press, Inc.

Pearson Education Ltd., *London*
Pearson Education Australia Pty. Limited, *Sydney*
Pearson Education Singapore Pte. Ltd.
Pearson Education North Asia Ltd., *Hong Kong*
Pearson Education Canada, Ltd., *Toronto*
Pearson Educación de Mexico, S. A. de C. V.
Pearson Education—Japan, *Tokyo*
Pearson Education Malaysia Pte. Ltd.
Pearson Education, *Upper Saddle River, New Jersey*

10 9 8 7 6 5 4 3 2 1
ISBN: 0-13-093893-9

To Carlos, my best friend and soul mate.

Preface

APPROACH

After years of watching technology students struggle with mathematics, my objective was to write a student-oriented textbook that would make learning mathematics easier. Of course, any textbook that makes learning easier for the student will also make teaching easier for the instructor. The features that make this book student-oriented include

- The easy-to-follow writing style
- A spiral approach to learning
- A balance between pure math and technical applications
- The boxed step-by-step procedures
- The functional use of second color and graphics to support student learning

Writing Style

Experienced teachers know what approach works best in their classes for getting concepts across to students. In this book, I have used the approaches that have worked best for me in promoting student understanding. I have employed a conversational style to communicate the material at a reading level appropriate to the intended audience and to make it easy for students to read and understand. Also, I use examples to illustrate and explain every key concept presented. Further, though offering many examples is vital to the success of students in mathematics, more vital still is that they have at their disposal all of the explanation needed to grasp the concepts when they are doing their homework and the instructor is not around. So I have incorporated into the text verbal nudges similar to those a teacher might use in class or a tutoring session:

- **Learning Hints** give students practical study tips.
- **Notes** point out trouble spots and help students overcome them.
- **Cautions** alert students to common errors and ways to avoid them.
- **Remember** notes reinforce previously learned concepts.
- **Calculator symbols** in the examples and exercises let students know that they should use their calculators for those particular problems.

■ **Application** problems from the technologies are keyed to their specific area by labels next to the problem number:

E	Electronics	**P**	Physics		**Bi**	Biology
M	Mechanical	**Ch**	Chemistry/Chemical Engineering			
B	Business	**A**	Architecture			
C	Civil	**CS**	Computer Science			

Spiral Approach

In the spiral approach, a concept is first introduced, then discussed again in subsequent chapters. For example, equations involving radicals are introduced in Chapter 3 and discussed again in Chapters 5, 6, and 7 in relation to material on quadratic equations, systems of equations, and solving higher degree equations, respectively. The spiral approach used in this text enhances student retention by

■ building on what already has been learned

■ challenging students to apply concepts that have been mastered to new problems

■ periodically and methodically reviewing what has gone before

Balanced Coverage

In a technical mathematics textbook, the balance between pure mathematics and technology is very important. It is essential that students not only understand the mathematical concepts but also be able to apply them to technical problems. Thus, my aim has been to stress topics required for a career in technology without sacrificing the mathematics on which these topics are based. Applications are an integral part of the chapters and presume little prior technical knowledge. This integration affords students the opportunity to learn about technology as they learn math. When math can be shown to relate to the "real world," students' motivation to understand it increases.

Step-by-Step Procedures

Students progress through chapters in a consistent format:

■ For key concepts, students are presented a clearly marked **Procedure Box** that gives them a step-by-step method for solving problems.

■ Within the chapters, there is a natural progression from simple to more difficult examples.

■ Each example includes a step-by-step procedure with accompanying explanation.

■ Rules and formulas are also featured in highlight boxes.

■ Within sections, subtopics are clearly labeled in the margins, making it easier for students to find those subjects they need to review.

This consistent method for problem solving significantly aids students in mastering the subject of technical mathematics.

Functional Use of Second Color and Graphics

We have taken great care to use color to aid student learning and understanding. The second color has been used to highlight steps taken in solving a problem. This allows students to see what has been done in each step easily. Additionally, we have included graphical representations of concepts where appropriate and where they will help students retain the material. For students to develop good problem-solving skills, they must learn to take a written problem, determine the important information, and formalize it into a mathematical problem to solve. In many instances, representing the information graphically helps in this process. The old adage "a picture is worth a thousand words" is very true when solving a mathematics problem.

ORGANIZATION

The organization of the text is also centered around a consistent format built on several key pedagogical features:

- Each chapter opens with an application problem related to the chapter material.
- A detailed solution to the application problem is given at the end of the chapter.
- Chapters also begin with a list of learning objectives that are keyed to sections in which relevant topics are covered.
- End-of-chapter tests are, in turn, keyed to the learning objectives—incorrect answers on the self-tests can be referenced to objectives and then to chapter sections for additional study.
- Every chapter concludes with a summary that serves as a handy guide for reviewing terms (page references provided) and formulas.
- Chapters also end with a review that provides a sufficient number of problems for students to strengthen their knowledge of the chapter material and that separates topics by section so students can pinpoint those problem sections.
- Every chapter also concludes with a group activity, which applies the concepts of the chapter to a "real world" problem. The purpose of the group activity is to introduce students to problem-solving situations they may encounter on the job and to improve written and verbal presentation of their results/conclusions.
- The graphing calculator is a powerful tool. It has been integrated into every situation where it is applicable. The output of the screen is also shown.

ACCURACY

Nothing is more frustrating to both teachers and students than inaccuracies in a mathematics textbook. Our goal has been to eliminate that frustration by putting the text—the examples, the answers in the back of the text, and the solutions manual—through rigorous accuracy checking. All told, the book went through six rounds of accuracy reviews by fifteen professors at crucial stages of its development.

APPENDIXES

The textbook includes three appendixes:

- Because Appendix A includes example problems and exercises, it can serve as Chapter 0 to review topics in arithmetic and algebra, including real numbers, laws of operations of real numbers, zero and orders of operations, exponents, scientific notation, roots and radicals, addition and subtraction of algebraic expressions, multiplication of polynomials, polynomial division, percentages, and significant digits and rounding. In response to instructor reviews, apprendixes have been added on units and unit conversion, calculator usage, and computer number systems.
- Appendix B provides a review of basic geometric terms and formulas.
- Appendix C contains tables of integrals and metric units and conversions, which can be used for quick reference.

Step-by-step examples illustrate every topic.

Annotations explain the mathematical steps in the problem-solving process.

Symbol signifies end of example.

Margin label.

Procedure box provides a step-by-step method for solving problems related to essential concepts.

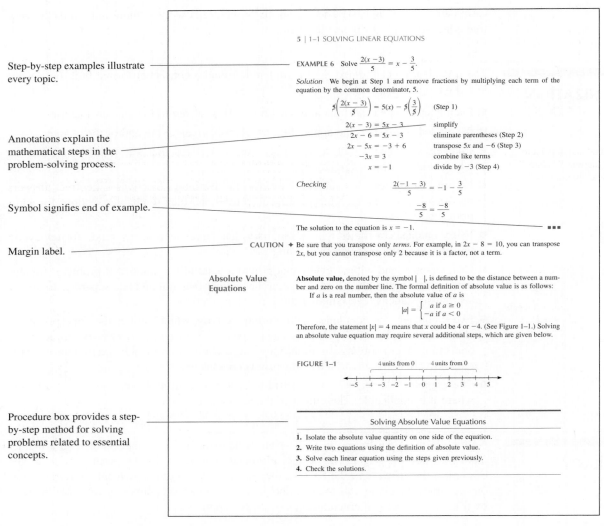

FIGURE 1
EXAMPLES OF STUDENT-ORIENTED FEATURES

Labels in the margin point to friendly "Cautions," "Hints," "Notes," and "Reminders."

Step-by-step examples.

"Bubbles" in color alert students to key steps in an informal manner.

Margin label.

---

91 | 3–1 INTEGRAL EXPONENTS

c. $\dfrac{x^8}{x^4} = x^{8-4} = x^4$ \hspace{2cm} Rule 3–4

d. $\dfrac{y^2}{y^8} = \dfrac{1}{y^{8-2}} = \dfrac{1}{y^6}$ \hspace{1.5cm} Rule 3–5

e. $\left(\dfrac{x^{-2}}{y^3}\right)^{-4} = \dfrac{(x^{-2})^{-4}}{(y^3)^{-4}} = \dfrac{x^8}{y^{-12}}$ \hspace{0.5cm} Rules 3–6 and 3–3

f. $4m^0 = 4(1) = 4$ \hspace{1.5cm} Rule 3–7

g. $(4m)^0 = 4^0 m^0 = 1$ \hspace{1cm} Rules 3–2 and 3–7 ▪▪▪

CAUTION ✦ Be careful in simplifying expressions containing zero exponents. In an expression such as $4m^0$, the zero exponent applies only to m. On the other hand, in an expression such as $(4m)^0$, the zero exponent applies to both the 4 and the m.

As a general rule, you should not leave zero or negative exponents in an algebraic expression. Rule 3–8 provides us with a method of eliminating negative exponents. To remove the negative exponents on a factor, move the factor across the fraction line. This technique is illustrated in the next example.

EXAMPLE 2 Simplify each expression using the Laws of Exponents. Express your answers with positive exponents only.

a. $y^4 \cdot y^{-9}$ \hspace{1cm} **b.** $(2x^{-3}y^2)^3$ \hspace{1cm} **c.** $\left(\dfrac{4}{x^3}\right)^{-2}$ \hspace{1cm} **d.** $\left(\dfrac{m^{-4}}{n^3}\right)^{-2}$

Solution

a. $y^4 \cdot y^{-9} = y^{4+(-9)} = y^{-5} = \dfrac{1}{y^5}$ \hspace{0.5cm} (change the sign of the exponent)

b. $(2x^{-3}y^2)^3 = (2)^3(x^{-3})^3(y^2)^3 = 8x^{-9}y^6 = \dfrac{8y^6}{x^9}$ \hspace{0.3cm} (change the sign of the exponent)

c. $\left(\dfrac{4}{x^3}\right)^{-2} = \dfrac{(4)^{-2}}{(x^3)^{-2}} = \dfrac{4^{-2}}{x^{-6}} = \dfrac{x^6}{4^2} = \dfrac{x^6}{16}$ \hspace{0.3cm} $\left(\dfrac{1}{x^{-6}} = x^6\right)$ $\left(4^{-2} = \dfrac{1}{4^2}\right)$

d. $\left(\dfrac{m^{-4}}{n^3}\right)^{-2} = \dfrac{(m^{-4})^{-2}}{(n^3)^{-2}} = \dfrac{m^8}{n^{-6}} = m^8n^6$ \hspace{1cm} ▪▪▪

NOTE ✦ There may be more than one correct method of simplifying an expression using the Laws of Exponents, as shown in the next example.

EXAMPLE 3 Simplify the following:

$$\left(\frac{-2a^3b^{-4}}{5a^{-5}b^{-6}}\right)^{-2}$$

Express the answer with positive exponents only.

FIGURE 1 (continued)
EXAMPLES OF STUDENT-ORIENTED FEATURES

SUGGESTIONS FOR USE

Applied College Algebra and Trigonometry can be used in an introductory algebra and trigonometry course for students in an engineering technology program. I have chosen to organize the text by grouping the algebra topics together and the trigonometry topics together. This sequencing of topics greatly contributes to the text's flexibility, as the chapters can be easily integrated. Tables 1 through 6 indicate suggested course schedules for two-quarter and two-semester sequences. Since some schools teach algebra and trigonometry as separate courses, I have included a schedule for each. I have also provided a schedule that includes coverage of Chapter 16, "Introduction to Statistics and Empirical Curve Fitting," and Chapter 17, "Sequences, Series, and the Binomial Theorem." In preparing these schedules, I have made these assumptions:

- the quarter courses are 5 hours per week for 10 weeks
- the semester courses are 3 hours per week for 18 weeks
- a test is given after each chapter, unless otherwise noted
- one day is spent on each section, unless otherwise noted

TABLE 1
Two Quarters (Algebra/Trig, excluding Chapters 16 & 17)

First Quarter (Algebra)		Second Quarter (Trig)	
Chapter	Number of days	Chapter	Number of days
1	5	9	5**
2	6	10	5**
3	5	11	6**
4	6	12	6
5	5	13	6**
6	6**	14	6**
7	5	15	9**
8	4		
		**Spend 2 days each on Sections 6–2, 6–3, 9–4, 10–4, 11–3, 11–4, 13–4, 14–4, 15–4, 15–5, 15–6.	

TABLE 2
Two Quarters (Algebra/Trig, including Chapters 16 & 17)

First Quarter (Algebra)		Second Quarter (Trig)	
Chapter	Number of days	Chapter	Number of days
1	5	9	4
2	6	10	4
3	5	11	4**
4	6	12	6
5	5	13	5
6	6*	14	6
7	5	15	6
8	4	16	4
		17	4**
*Spend 2 days each on 6–2, 6–3		**Combine Chapters 10 and 11 and Chapters 16 and 17 into one chapter test.	

TABLE 3
Two Semesters (Algebra/Trig, including Chapters 16 & 17)

First Semester (Algebra)		Second Semester (Trig)	
Chapter	Number of days	Chapter	Number of days
1	5	9	5**
2	8*	10	4
3	5*	11	6**
4	6	12	6
5	6*	13	5
6	7	14	5
7	5	15	6
8	4	16	4
		17	4
*Spend 2 days each on Sections 2–4, 2–5, and 5–5.		**Spend 2 days each on Sections 9–4, 11–3, and 11–4.	

TABLE 4
Two Quarters (Integrated, excluding Chapters 16 & 17)

First Quarter		Second Quarter	
Chapter	Number of days	Chapter	Number of days
1	5	11	6**
2	6	12	6
9	5*	7	5
3	5	8	4
4	6	14	7**
5	5	13	6**
6	6*	15	9**
10	4		
Spend 2 days on Section 6–2, 6–3 and 9–4.		**Spend 2 days each on Sections 11–3, 11–4, 13–4, 14–4, 14–5, 15–4, 15–5, 15–6.	

TABLE 5
Two Quarters (Integrated, including Chapters 16 & 17)

First Quarter		Second Quarter	
Chapter	Number of days	Chapter	Number of days
1	5	11	6**
2	6	12	6
9	5*	7	5
3	5	8	4
4	6	14	5
5	5	13	5
6	6*	15	6
10	4	16	4
		17	4
*Spend 2 days on Sections 6–2, 6–3 and 9–4.		Spend 2 days each on Sections 11–3 and 11–4.	

TABLE 6
Two Semesters (Integrated, including Chapters 16 & 17)

First Semester		Second Semester	
Chapter	Number of days	Chapter	Number of days
1	6*	11	6
2	7*	12	6
9	5*	7	5
3	5	8	4
4	6	14	5
5	6*	13	5
6	7*	15	6
10	4	16	4
		17	4
*Spend 2 days each on Sections 1–5, 9–4, 2–6, 6–2, 6–3, 6–4, and 5–5.			

ANCILLARIES

The entire package—textbook and supplementary materials—has been designed to give instructors all the support they need to teach technical mathematics efficiently and effectively:

- The **Instructor's Resource Manual** (0-13-093894-7) includes solutions to all even-numbered in-text problems, and contains additional worked-out example problems for instructors to use in class—these examples can be made into transparencies.

- The **Test Item File** (0-13-093896-3) provides over 1700 multiple-choice questions, which are organized by chapter and section.

- The **Test Manager** (0-13-093895-5) (hardware requirements: Windows 95, 98 or NT; Pentium processor or higher; 16 MB RAM, 32 MB recommended; 60 MB available hard disk space; VGA monitor or greater, screen resolution 800 × 600) utilizes the full set of questions of the printed test item file, is formatted for IBM or Macintosh, and provides full graphics and editing capabilities.

- The **Student Study Guide** (0-13-093897-1) contains worked-out solutions to selected odd-numbered problems in the text, gives students helpful study tips and additional explanation of material, and provides additional exercises and application problems.

- The **Companion Website** (*http://www.prenhall. com/davistech*) contains problems from each chapter, allowing students to practice the concepts and receive immediate feedback.

- The **Study Wizard** contains chapter objectives, a glossary of terms, and over 600 practice questions. The testing feature of this CD allows students to review problems as a practice activity or timed test. The feedback to the practice problems provides page number references to identify where the concept appears in the text. When students give incorrect answers, they are prompted with the correct answer, hints, and intermediate results to assist them with identifying and correcting mistakes.

ACKNOWLEDG-MENTS

Any textbook from its inception to completion involves a great many people willing to give of their time and talent. This textbook is no exception. I would like to acknowledge the contribution of the following colleagues in the development of this textbook:

Sandra Beken, Horry Georgetown Technical College

Michael G. Blitch, DeVry Institute of Technology—Decatur

Glenn R. Boston, Catawba Valley Community College

Robert J. Campbell, North Shore Community College

Ray E. Collings, Tri-County Technical College

Sabah Al-Hadad, California Polytechnic State University—San Luis Obispo

Harold Hauser, Mt. Hood Community College

Patricia L. Hirschy, Delaware Technical and Community College

Linda Isabel, DeVry Institute of Technology—Decatur

Baback A. Izadi, DeVry Institute of Technology—Columbus

Wendell Johnson, University of Akron

Dimitrios Kyriazopolous, DeVry Institute of Technology—Chicago

Michael Latina, Community College of Rhode Island

Lynn Mack, Piedmont Technical College

David L. Potter, ITT Technical Institute—Dayton

Larry Richard, DeVry Institute of Technology—Lombard

Ursula Rhoden, Nashville State Technical Institute

Fred Ueberschaer, DeVry Institute of Technology—Decatur

William Van Alstine, Aiken Technical College

Cecilia Vargas, DeVry Institute of Technology—North Brunswick

Kelly P. Wyatt, Umpqua Community College

Robert J. Zabek, Norwalk State Technical College

Accuracy checkers for the textbook and solutions manual:

John Bailey, Clark State Community College

Doug Cook, Owens Technical College

John L. Crawford, Lee College

Jane Downey, St. Cloud State University

Allan Edwards, West Virginia University—Parkersburg

J. Vernon Gwaltney, John Tyler Community College

John Heublein, Kansas Institute of Technology

Margaret Kimbell, Texas State Technical Institute—Waco

George Marketos, University of Cincinnati

Kylene Norman, Clark State Community College

Samuel F. Robinson, New Hampshire Technical College

John A. Seaton, Prince George's Community College

Molly Sumner, Pikes Peak Community College

Ed Turner, University of Cincinnati

Andrew S. Wargo, Bucks County Community College

In addition, I would especially like to thank Dr. Ravi Chandran for his work on the Test Item File, the Instructor's Resource Manual and the solutions, Lynn Mack for her work on Appendixes A and B; and Margie DeAngelo for her work on the accuracy check of the textbook. I would also like to thank James Page and Ben Kirk for the computer science problems. I would like to express my undying gratitude to Doug Cook for his work on the accuracy check of the textbook. His attention to detail and perseverance are greatly appreciated.

I would like to acknowledge the assistance of Michelle Churma, Tricia Rawnsley, and Frank Mortimer at Prentice Hall. Michelle was great to work with again. She was always very responsive to my questions and requests, and always encouraging and upbeat. As Production Editor, Tricia did a great job of coordinating the production process. She was able to coordinate the editing, production of proof pages, copyediting, and accuracy checking. I would also like to thank Steve Helba for initiating the process of producing a third edition of the book.

In conclusion, I would like to thank my husband Carlos. He helped me keep on schedule, and created the prototypes of new graphics for this edition. His assistance was invaluable and is appreciated greatly.

Contents

APPENDIXES

ANSWERS TO ODD-NUMBERED EXERCISES   739

INDEX   815

APPLIED COLLEGE ALGEBRA AND TRIGONOMETRY

Third Edition

Linda Davis

S

olving equations or inequalities of various types is one of the most important applications of algebra. To move a wrecked truck, two cables are attached to the truck and pulled by two winches. However, one of the winches breaks. To ensure that the cable from one winch can move the truck without breaking, you must solve the following equation $R = \sqrt{R_x^2 + R_y^2 + R_z^2}$ when $R_x = -2,100$ lb, $R_y = 2,300$ lb, and $R_z = 210$ lb. (The solution to this problem is given at the end of the chapter.)

Technical problems can be stated in the form of an equation or formula. However, many technical problems are stated verbally and must be translated into an equation or inequality before they can be solved. In this chapter we will begin our study of solving equations and inequalities with the linear type.

Learning Objectives

After you complete this chapter, you should be able to

1. Use the Principles of Equality to solve linear equations in one variable (Section 1–1).

2. Solve an equation involving absolute value (Section 1–1).

3. Solve literal equations for a specified variable (Section 1–2).

4. Solve linear inequalities in one variable and graph the solution (Section 1–3).

5. Solve an inequality involving absolute value and graph the solution (Section 1–3).

6. Transform a verbal statement of variation into a formula (Section 1–4).

7. Solve applied problems using variation (Section 1–4).

8. Apply the techniques of solving equations and inequalities learned in this chapter to technical problems (Section 1–5).

Chapter 1

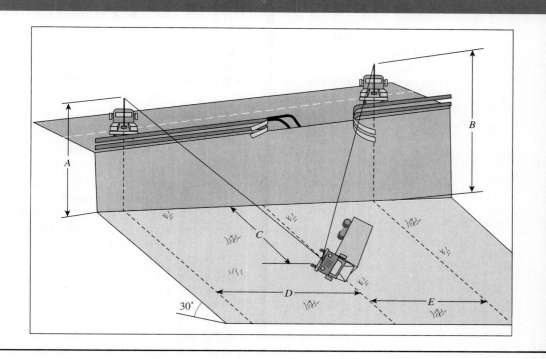

Solving Linear Equations and Inequalities

1–1

SOLVING LINEAR EQUATIONS

An **equation** is a statement that two expressions are equivalent or equal. In a linear equation in one variable, *the variable has an exponent of one.* (A *term* is an expression separated by an addition or subtraction symbol.) A **linear equation** is one that can be written in the form $ax + b = 0$, where a and b are real numbers and a is not zero. For example, both of the following are linear equations in one variable:

$$2x + 6(3x - 1) = 7$$

and

$$8x = 7 + 5x - 10$$

However, neither of the following is a linear equation:

$$6x^2 + 5x = 10$$

and

$$\frac{7}{x} + 3x - 8 = 0$$

An equation or **open statement** may be true for some values of the variable and false for other values. However, a **solution** or **root** of an equation is a number that produces a true statement when substituted for the variable. Two is the solution or root for the equation $x - 6 = -4$ because substituting two for x results in a true statement: $2 - 6 = -4$. In Chapter 4 we will discuss in detail the solution to an equation from a graphical approach. You will see that the solution or root is where the graph crosses the x axis. We will also discuss solutions to linear equations using a graphing calculator later in this section.

Since not all linear equations are easy to solve by trial-and-error substitutions, we must develop a systematic method of solving them. Before discussing a systematic procedure, however, let us look at the Principles of Equality, which serve as the basis for solving equations. The **Principles of Equality** state that if the same operation is performed on each side of an equation, an equivalent equation results. Thus, if you add, subtract, multiply, or divide both sides of an equation by the same quantity, you do not change the solution to the equation. (That is, you do not change the value of the unknown.) Another useful property in solving equations, called **transposition,** states that moving any *term* from one side of the equation to the opposite side and reversing its sign does not alter the equality. For example, $x + 4 = 0$ is equivalent to $x = -4$. Transposition produces the same result as adding or subtracting the same quantity to both sides of an equation ($x + 4 - 4 = 0 - 4$), but transposition is shorter and faster. A procedure for solving linear equations is given in the following table. An explanation of each step involved in this process is given in the examples.

Solving Linear Equations

1. Eliminate fractions by multiplying each term of the equation by the lowest common denominator. Simplify.
2. Eliminate parentheses using the Distributive Property. Simplify.
3. Using transposition, isolate all terms containing the variable on one side of the equation and all other terms on the opposite side of the equation. Simplify by combining like terms.
4. Divide both sides of the equation by the coefficient of the variable.
5. Check your solution by substituting into the original equation.

EXAMPLE 1   Solve $2x - 8 = 4$.

Solution   Since this equation does not contain fractions or parentheses, we can start the solution at Step 3. We have only one term containing an x, and it is on the left side of the equals sign. So we will isolate all x terms on the left and all terms without an x on the right. The choice of which side of the equals sign on which to isolate x terms is completely arbitrary. The resulting solution is the same either way. Isolating x terms on the left and all other terms on the right requires transposing -8 to the right side of the equals sign.

$$2x - 8 = 4$$
$$2x = 4 + 8 \qquad \text{transpose } -8$$
$$2x = 12 \qquad \text{combine like terms}$$

Next, we proceed to Step 4. The coefficient of the variable is 2; therefore, we divide both sides of the equation by 2.

$$\frac{2x}{2} = \frac{12}{2} \qquad \text{divide by 2}$$

$$x = 6 \qquad \text{simplify}$$

Checking
$$2x - 8 = 4$$
$$2(6) - 8 = 4$$
$$12 - 8 = 4$$
$$4 = 4$$

∎∎∎

EXAMPLE 2   Solve $3x - 9 = 4x + 5$.

Solution   We can start with Step 3 since this equation does not contain a fraction or parentheses. Isolating the x terms on the left side of the equals sign and all terms without an x on the right side requires that the -9 and $4x$ terms be moved. Using transposition, move -9 and $4x$ across the equals sign and reverse each sign.

$$3x - 9 = 4x + 5$$
$$3x - 4x = 5 + 9 \qquad \text{transpose } 4x \text{ and } -9 \text{ (Step 3)}$$
$$-x = 14 \qquad \text{combine like terms}$$
$$\frac{-x}{-1} = \frac{14}{-1} \qquad \text{divide by } -1 \text{ (Step 4)}$$

$$x = -14$$

Checking
$$3x - 9 = 4x + 5$$
$$3(-14) - 9 = 4(-14) + 5$$
$$-42 - 9 = -56 + 5$$
$$-51 = -51$$

∎∎∎

NOTE ◆ The coefficient of the variable must be $+1$. For example, $-x = 7$ is not an acceptable solution.

EXAMPLE 3 Solve $2(3x + 4) = 6 - (2x - 5)$.

Solution Since this equation does not contain fractions, we begin at Step 2 using the Distributive Property to remove the parentheses.

$$2(3x + 4) = 6 - (2x - 5) \qquad \text{(Step 2)}$$
$$6x + 8 = 6 - 2x + 5 \qquad \text{eliminate parentheses}$$
$$6x + 8 = 11 - 2x \qquad \text{combine like terms}$$
$$6x + 2x = 11 - 8 \qquad \text{transpose } -2x \text{ and } 8 \text{ (Step 3)}$$
$$8x = 3 \qquad \text{combine like terms}$$
$$\frac{8x}{8} = \frac{3}{8} \qquad \text{divide by 8 (Step 4)}$$

After checking, the solution is $x = 3/8$. ■■■

CAUTION ✦ Check your solution in the original equation. Later versions of the equation may contain errors.

EXAMPLE 4 Solve $2.1x + 6 = 7.3x - 4.8$.

Solution Since this equation does not contain fractions or parentheses, we begin with Step 3. To isolate the variable on the left side of the equation, transpose 6 and $7.3x$.

$$2.1x + 6 = 7.3x - 4.8$$
$$2.1x - 7.3x = -4.8 - 6 \qquad \text{transpose } 7.3x \text{ and 6 (Step 3)}$$
$$-5.2x = -10.8 \qquad \text{combine like terms}$$
$$x \approx 2 \qquad \text{divide by } -5.2 \text{ (Step 4)}$$
$$10.8 \boxed{+/-} \boxed{\div} 5.2 \boxed{+/-} \boxed{=} \longrightarrow 2.076923077$$

■■■

NOTE ✦ The $\boxed{+/-}$ key is used to change the sign of the previous entry.

EXAMPLE 5 Solve $\dfrac{3}{4}x - 2 = \dfrac{1}{3}x + \dfrac{1}{2}$.

Solution Begin by removing the fractions; that is, multiply each term by the lowest common denominator, 12.

$$\overset{3}{\cancel{12}}\left(\frac{3}{4}x\right) - 12(2) = \overset{4}{\cancel{12}}\left(\frac{1}{3}x\right) + \overset{6}{\cancel{12}}\left(\frac{1}{2}\right) \qquad \text{(Step 1)}$$
$$9x - 24 = 4x + 6 \qquad \text{simplify}$$
$$9x - 4x = 6 + 24 \qquad \text{transpose } 4x \text{ and } -24 \text{ (Step 3)}$$
$$5x = 30 \qquad \text{combine like terms}$$
$$x = 6 \qquad \text{divide by 5 (Step 4)}$$

■■■

EXAMPLE 6 Solve $\dfrac{2(x-3)}{5} = x - \dfrac{3}{5}$.

Solution We begin at Step 1 and remove fractions by multiplying each term of the equation by the common denominator, 5.

$$\cancel{5}\left(\frac{2(x-3)}{\cancel{5}}\right) = 5(x) - \cancel{5}\left(\frac{3}{\cancel{5}}\right) \quad \text{(Step 1)}$$

$$2(x-3) = 5x - 3 \qquad \text{simplify}$$

$$2x - 6 = 5x - 3 \qquad \text{eliminate parentheses (Step 2)}$$

$$2x - 5x = -3 + 6 \qquad \text{transpose } 5x \text{ and } -6 \text{ (Step 3)}$$

$$-3x = 3 \qquad \text{combine like terms}$$

$$x = -1 \qquad \text{divide by } -3 \text{ (Step 4)}$$

Checking

$$\frac{2(-1-3)}{5} = -1 - \frac{3}{5}$$

$$\frac{-8}{5} = \frac{-8}{5}$$

The solution to the equation is $x = -1$. ∎ ■ ■ ■

CAUTION ✦ Be sure that you transpose only *terms*. For example, in $2x - 8 = 10$, you can transpose $2x$, but you cannot transpose only 2 because it is a factor, not a term.

Absolute Value Equations

Absolute value, denoted by the symbol $|\ \ |$, is defined to be the distance between a number and zero on the number line. The formal definition of absolute value is as follows:

If a is a real number, then the absolute value of a is

$$|a| = \begin{cases} a \text{ if } a \geq 0 \\ -a \text{ if } a < 0 \end{cases}$$

Therefore, the statement $|x| = 4$ means that x could be 4 or -4. (See Figure 1–1.) Solving an absolute value equation may require several additional steps, which are given below.

FIGURE 1–1

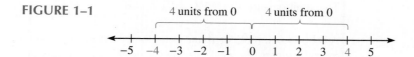

The solution to the equation is $x = -1$.

Solving Absolute Value Equations

1. Isolate the absolute value quantity on one side of the equation.
2. Write two equations using the definition of absolute value.
3. Solve each linear equation using the steps given previously.
4. Check the solutions.

EXAMPLE 7   Solve $|3x + 5| - 7 = 10$.

Solution   We begin with Step 1 by isolating the absolute value quantity on the left side of the equals sign by transposing -7.

$$|3x + 5| - 7 = 10$$
$$|3x + 5| = 10 + 7$$
$$|3x + 5| = 17$$

At Step 2 we write two equations using the definition of absolute value.

$$3x + 5 = 17 \qquad \text{or} \qquad 3x + 5 = -17$$

At Step 3 we solve each of these equations using the methods we developed earlier.

$$3x + 5 = 17 \qquad\qquad 3x + 5 = -17$$
$$3x = 17 - 5 \qquad\qquad 3x = -17 - 5$$
$$3x = 12 \qquad\qquad 3x = -22$$
$$x = 4 \qquad\qquad x = -\frac{22}{3}$$

The solutions are $x = 4$ and $-22/3$. ∎∎∎

EXAMPLE 8   Solve the following equation: $4|2x - 3| + 6 = 11$.

Solution   First, we isolate the absolute value quantity by transposing 6.

$$4|2x - 3| = 11 - 6$$
$$4|2x - 3| = 5$$

Normally, we would divide both sides of the equation by 4. However, in this case, a fraction results on the right side of the equation. Therefore, at this point, we will apply the definition of absolute value.

$$4(2x - 3) = 5 \qquad \text{or} \qquad 4(2x - 3) = -5$$
$$8x - 12 = 5 \qquad\qquad 8x - 12 = -5$$
$$8x = 17 \qquad\qquad 8x = 7$$
$$x = 17/8 \qquad\qquad x = 7/8$$

The solutions are $x = 17/8$ and $x = 7/8$. ∎∎∎

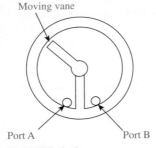

Moving vane

Port A   Port B

FIGURE 1–2

Applications

A robotic manipulator (hand) may be powered by a hydraulic fluid that is pumped by a rotary actuator. Figure 1–2 shows a single-action rotary actuator. If hydraulic fluid enters port A, the vane moves clockwise. If the fluid enters port B, the vane moves counterclockwise. The single-action rotary actuator can move through an angle of approximately 280°. The torque T developed by the actuator is given by

$$T = (P \cdot A \cdot Rc) - Tf$$

where P = fluid pressure, A = vane area, Rc = center radius of the vane, and Tf = friction torque.

EXAMPLE 9   Find the torque (in ft-lb) delivered to the rotary actuator of a robotic manipulator if the vane center radius is 1.10 in., the vane area is 0.780 in^2, the fluid pressure is 1,600 lb/in^2, and the friction torque is 18.0 in-lb.

Solution   From the information given, we know $P = 1{,}600$ lb/in^2, $A = 0.780$ in^2, $Rc = 1.10$ in., and $Tf = 18.0$ in-lb. Substituting these values gives

$$T = (1{,}600 \text{ lb/in}^2)(0.780 \text{ in}^2)(1.10 \text{ in.}) - (18.0 \text{ in-lb})$$

$$T = (1{,}600 \text{ lb})(0.780)(1.10 \text{ in.}) - (18.0 \text{ in-lb})$$

$$T = 1{,}354.8 \text{ in-lb}$$

$$T = 1{,}354.8 \text{ in-lb} \cdot \frac{1 \text{ ft}}{12 \text{ in.}} \quad \text{(converting to ft-lb)}$$

$$T = 113 \text{ ft-lb}$$

$$1600 \; \boxed{\times} \; 0.780 \; \boxed{\times} \; 1.10 \; \boxed{=} \; -18.0 \; \boxed{=} \; \boxed{\div} \; 12 \; \boxed{=} \; \longrightarrow \; 112.9000 \qquad \blacksquare\blacksquare\blacksquare$$

NOTE ✦ Units of measure (such as in^2 in the previous example) may be added, subtracted, multiplied, divided, or canceled just as other algebraic expressions. Refer to Appendix A–12 for a review of this process.

EXAMPLE 10   Both residential and commercial property have two values. The first value is the market value: the price for which the property could be sold. The second value is the assessed value used to calculate property taxes. The assessed value is usually smaller than the market value and is usually a percentage of the market value. The formula is

$$RB = P$$

where R = tax rate (can be expressed in property tax per hundred dollars of assessed value), B = assessed value, and P = amount of property taxes. Jose and Maria Garcia buy a house where the tax rate is \$10.80 per hundred of assessed value. Their house price is \$122,000 and their property taxes are \$4,578 a year. Find the assessed value of the house.

Solution   We begin by identifying what we know and what we must determine. From the problem statement, we know

$$R = \text{tax rate} = \$10.80 \text{ per } 100 = 10.8\% = 0.108$$
$$B = \text{must be determined}$$
$$P = \text{property tax} = \$4{,}578$$

Substituting these values into the equation $RB = P$ gives the following:

$$0.108B = 4{,}578$$

$$B = \frac{4{,}578}{0.108} \qquad \text{divide both sides by 0.108}$$

$$B = 42{,}389$$

The assessed value is approximately \$42,389. $\blacksquare\blacksquare\blacksquare$

LEARNING HINT ✦ If you have difficulty solving linear equations, write the steps for solution given in this section on an index card. Then use the index card when doing homework, and go through these steps one at a time.

1–1 EXERCISES

Solve each of the following equations. Treat the integers in these equations as exact numbers. Leave your solution in fractional form. When the equation has decimal coefficients, round to the correct number of significant digits.

1. $x + 6 = -11$

2. $x + 10 = 8$

3. $2x - 6 = -14$

4. $-3x + 4 = 19$

5. $6x - 7 + 4x = 8 - 5x + 10 + 5$

6. $10x - 5 + 6 - 2x = 3x - 9$

 7. $3.7x + 8.6 = 10.5$

8. $4.2 - 3.5x = 14.4$

9. $2(x - 6) = 6$

10. $3(x - 1) = 10$

11. $4(3x - 6) - 10x = 12$

12. $-2(4x - 7) + 5x = 7$

13. $\dfrac{x}{2} - 5 = 3$

14. $8 - \dfrac{x}{3} = 2$

15. $8.1x + 6.0 = 3.2x - 16.4$

16. $9.7x + 4.3 = 15.8 - 2.4x$

17. $\dfrac{3(-2x + 4)}{5} = 7$

18. $\dfrac{-5(x - 4)}{7} = -1$

19. $3(x - 1) - (2x + 6) = 10$

20. $4(2x - 1) - (3x + 2) = 8$

21. $7x - 5(2x + 3) = 4 - (2x + 6)$

22. $5x - 7(2x - 1) = 9 + 2(3 - x)$

23. $\dfrac{2(x - 2)}{3} + 5x = 10$

24. $\dfrac{4(3 - x)}{5} - 3x = -6$

25. $\dfrac{8x}{9} - 2 = \dfrac{-5x}{6} + \dfrac{1}{3}$

26. $\dfrac{x}{6} + \dfrac{3x}{4} + 1 = \dfrac{2x}{3} - \dfrac{5}{6}$

27. $\dfrac{x}{3} - \dfrac{1}{3} = \dfrac{x}{2} + 3$

28. $\dfrac{-2x}{3} + 2 = \dfrac{5x}{6} - \dfrac{5x}{2}$

29. $4(3x - 6) - 7x + 6 = -2(x - 6) + 4x$

30. $-9(x - 2) + 6 + 6x = 7x - 2(3x + 4)$

31. $-5(2x - 1) + 6 = 4(8 - x) - (3x + 2) + 5$

32. $3(2x - 4) + 16 = -5(x + 6) + (2 - 5x)$

33. $1.8x - 4.3 = 2.4(x - 3.0)$

34. $5.1 + 6.8x = 7.3(2.0x - 3.9)$

35. $\dfrac{3(x - 5)}{7} = 6 + 2x$

36. $\dfrac{-2(3x + 4)}{5} = -7 + 3x$

37. $\dfrac{x - 3}{4} + 6 = \dfrac{2x + 9}{3}$

38. $\dfrac{2x + 5}{6} + 1 = \dfrac{x - 3}{2}$

39. $\dfrac{3(4x - 6)}{2} - \dfrac{3}{4} = \dfrac{x + 6}{3}$

40. $\dfrac{2(2x + 3)}{3} - 2 = \dfrac{-(x + 4)}{2}$

41. $|4x - 6| = 10$

42. $|3x - 8| = 16$

43. $5|2x - 1| + 4 = 11$

44. $2|3x - 6| - 2 = 15$

45. $2|x - 5| + 7 = 14$

46. $4|3x + 1| - 7 = 16$

M 47. If the efficiency of an engine is 0.56 when the absolute initial temperature is 800 kelvins (800 K), the resulting equation is

$$0.56 = 1 - \frac{T}{800}$$

Find the absolute final temperature T.

P 48. In finding the resultant of two forces, a scientist must solve the following equation:

$$0.94A - (0.64)200 = 100$$

Find A.

B 49. The stockholders of Interstate Electro Optical expect a $2.1 million return on their investment of $132 million. What is the stockholders' rate of return? (Use $RB = P$ where $R = \%$ rate of return, $B =$ investment, and $P =$ return on investment.)

E 50. Kirchhoff's current law states that the sum of the currents flowing into a junction equals the sum of currents flowing out of the junction. For a particular circuit, the following equation results:

$$2 + \frac{v}{5} = 3$$

Solve for v.

P 51. A formula from physics is $v = v_0 + at$. Determine the value for t when $v = 73$ ft/s, $v_0 = 22$ ft/s, and $a = 8.0$ ft/s^2.

E 52. The voltage V in a certain circuit is given by the formula $V = IR$ where I is the current and R is the resistance. Determine the current in a circuit where the voltage is 110 V and the resistance is 15 Ω.

B 53. The yearly depreciation on a piece of business equipment is given by the formula

$$\text{Yearly depreciation} = \frac{\text{depreciable value}}{\text{expected life of equipment}}$$

Find the depreciable value of a computer system when its expected life is 5 years and the yearly depreciation is $3,500.

B 54. Using the formula given in Exercise 53, determine the yearly depreciation of an office copying machine if the expected life is 8 years and the depreciable value is $20,000.

P 55. The frequency of sound f' in hertz (Hz) is given by

$$f' = f\left(1 + \frac{v}{s}\right).$$ Solve for f' when $f = 400$ Hz, $v = 45$ mi/h, and $s = 760$ mi/h.

56. The total surface area A of a cylinder is given by $A = 2\pi rh + 2\pi r^2$. Solve for h if $r = 36$ ft and $A = 6{,}192\pi$.

C 57. The maximum length of a line on a map S is given by $S = \sqrt{8RE - E^2}$. Determine the maximum length of the line when $R = 0.38$ mm and $E = 0.18$ mm.

E 58. The total resistance R_T of two resistors connected in parallel is given by

$$\frac{1}{R_T} = \frac{1}{R_1} + \frac{1}{R_2}$$

Find R_T when $R_1 = 7.5 \ \Omega$ and $R_2 = 5.6 \ \Omega$.

B 59. An overnight delivery company has a fleet of vans. The annual operating cost of each van is $C = 0.32m + 3,750$ where m is the yearly mileage of the truck. If the owner's break-even point on cost is \$15,000, how many miles can each truck travel per year?

P 60. The distance d a rocket travels is given by $d = rt$ where r = rate and t = time. How far will a rocket traveling at 8.73 km/s travel in 9 s?

M 61. The kinetic energy of a moving body is given by K.E. = $mv^2/2$ where m is the mass and v is the velocity. Determine the kinetic energy of a truck whose mass is 200 slugs and whose velocity is 35 ft/s.

$$\left(Note:\ \text{slugs} = \frac{\text{lb}}{\text{ft/s}^2}. \right)$$

62. The volume V of a sphere is given by $V = 4\pi r^3/3$. Find the volume of a sphere if $r = 18$ ft.

63. The temperature in degrees Fahrenheit is given by $F = \frac{9}{5}C$ + 32. Find the temperature in degrees Fahrenheit corresponding to 35°C.

1–2

SOLVING LITERAL EQUATIONS

One of the most important applications of solving equations is evaluating and solving formulas, or literal equations. Many times in a science or technology application, you will need to rearrange a given formula to solve for a different variable and then evaluate that formula. In evaluating a formula, we substitute a numerical value for variables and calculate the resulting numerical value of another variable. For example, the perimeter of an equilateral triangle is given by $P = 3s$. We can evaluate the perimeter of different triangles by substituting the length of the sides for s. Solving a formula requires rearranging an equation in a different form. This section discusses both of these concepts.

Notation

A common practice in algebra is to *use the first letter of the word* to represent the *variable*. For example, we could use b to represent the base of a rectangle. However, in the case of the area of a trapezoid, there are two bases of different lengths, and the letter b cannot be used to represent both bases. To alleviate this problem, we can use subscripts. A **subscript** is a number or letter written below and to the right of a variable. In the case of the area of a trapezoid, the subscripts 1 and 2 could be used on the letter b to denote the two bases. Thus, the bases of the trapezoid could be represented by b_1 and b_2. Another method used to denote different quantities with the same letter is to use uppercase and lowercase letters. Thus, the bases of the trapezoid could also be represented as b and B.

Solving Literal Equations

To solve a formula or literal equation for a specific variable, you follow the same steps that were given in the previous section for linear equations. These steps are repeated below for your convenience.

Solving Literal Equations

1. Eliminate fractions by multiplying each term of the equation by the lowest common denominator. Simplify.

2. Eliminate parentheses in the equation using the Distributive Property. Simplify.

3. Using transposition, isolate all terms containing the variable on one side of the equation and all other terms on the opposite side of the equation. Simplify.

4. Divide both sides of the equation by the coefficient of the variable.

5. Check the solution.

EXAMPLE 1   Solve the following literal equation for h:

$$A = \frac{bh}{2} \qquad \text{area of a triangle}$$

Solution   First, we eliminate the fraction by multiplying each term by the common denominator, 2.

$$2(A) = 2\left(\frac{bh}{2}\right) \qquad \text{eliminate fractions (Step 1)}$$

$$2A = bh \qquad \text{simplify by removing parentheses (Step 2)}$$

Since the term containing h is isolated on the right side of the equals sign, we divide both sides of the equation by the coefficient of h, which is b.

$$\frac{2A}{b} = \frac{bh}{b} \qquad \text{divide by } b \text{ (Step 4)}$$

$$h = \frac{2A}{b}$$

Checking   Substitute into the equation just as we did in the previous section.

$$A = \frac{bh}{2}$$

$$A = \frac{b(2A/b)}{2}$$

$$A = A \qquad\qquad\qquad ∎∎∎$$

EXAMPLE 2   Solve the following for v_2:

$$m(v_2 - v_1) = Ft \qquad \text{change in momentum from physics}$$

Solution   We begin by using the Distributive Property to remove the parentheses.

$$m(v_2 - v_1) = Ft$$

$$mv_2 - mv_1 = Ft \qquad \text{eliminate parentheses (Step 2)}$$

$$mv_2 = Ft + mv_1 \qquad \text{transpose (Step 3)}$$

$$v_2 = \frac{Ft + mv_1}{m} \qquad \text{divide by } m \text{ (Step 4)}$$

Checking

$$m\left[\left(\frac{Ft + mv_1}{m}\right) - v_1\right] = Ft$$

$$Ft = Ft \qquad\qquad ∎∎∎$$

CAUTION ✦ Remember that like terms must have the same subscript and exponent. Therefore, we cannot combine v_1 and v_2 in Example 2.

EXAMPLE 3   Solve the following formula for b_2:

$$A = \frac{h(b_1 + b_2)}{2} \qquad \text{area of a trapezoid}$$

Solution

$$2(A) = 2(\frac{h(b_1 + b_2)}{2}) \qquad \text{eliminate fractions}$$

$$2A = h(b_1 + b_2) \qquad \text{simplify}$$

$$2A = hb_1 + hb_2 \qquad \text{eliminate parentheses}$$

$$2A - hb_1 = hb_2 \qquad \text{transpose}$$

$$\frac{2A - hb_1}{h} = b_2 \qquad \text{divide by } h$$

$$b_2 = \frac{2A - hb_1}{h}.$$

∎ ∎ ∎

CAUTION ✦ Be careful simplifying the last expression in Example 3. You can divide factors contained in both the numerator and the denominator of a fraction, but not terms. We cannot cancel the h's in the formula given in Example 3 because the h in the numerator is part of a term and is not a factor.

Evaluating Formulas

In evaluating a formula, we calculate the value of a specific variable, given numerical values for the remaining variables. In some instances, we may need to rearrange the formula and solve for a different variable. This process is illustrated in the next six examples.

EXAMPLE 4   The voltage across an inductance coil can be written as $V = IR + \dfrac{L(\Delta I)}{\Delta t}$. Solve for ΔI.

Solution   The quantities ΔI and Δt (read "delta I" and "delta t," respectively) are single quantities representing the change in I, current, and the change in t, time, respectively. We begin solving this formula by removing fractions.

$$\Delta t(V) = \Delta t(IR) + \Delta t\left[\frac{L(\Delta I)}{\Delta t}\right]$$

$$(\Delta t)V = \Delta t(IR) + L(\Delta I)$$

Next, we transpose $\Delta t(IR)$ to isolate the term containing ΔI on the right side of the equation.

$$V(\Delta t) - IR(\Delta t) = L(\Delta I)$$

Last, we divide the equation by the coefficient of ΔI, which is L.

$$\Delta I = \frac{V(\Delta t) - IR(\Delta t)}{L}$$

∎ ∎ ∎

EXAMPLE 5   The total distance s an object travels from some initial point s_0 at time zero to some point s_t at time t is given by $s = s_t - s_0$. If s represents the total distance traveled in time t, the formula is

$$s = vt + \frac{1}{2}at^2$$

where v is initial velocity, t is time, and a is acceleration. Find the acceleration of an object that travels 100 m in 8 s if the initial velocity is 10 m/s. Solve for a, then substitute numerical values.

Solution   We rearrange the formula to solve for *a* and substitute the numerical values for *s*, *v*, and *t*.

$$2(s) = 2(vt) + 2\left(\frac{1}{2}at^2\right) \qquad \text{eliminate fractions}$$

$$2s = 2vt + at^2 \qquad \text{simplify}$$

$$2s - 2vt = at^2 \qquad \text{transpose}$$

$$\frac{2s - 2vt}{t^2} = a \qquad \text{divide by } t^2$$

Substituting $s = 100$ m, $v = 10$ m/s, and $t = 8$ s into the formula gives

$$a = \frac{2(100 \text{ m}) - 2(10 \text{ m } \cancel{s})(8 \cancel{s})}{(8 \text{ s})^2}$$

$$a = 0.6 \text{ m/s}^2$$

$$2 \boxed{\times} 100 \boxed{-} 2 \boxed{\times} 10 \boxed{\times} 8 \boxed{=} \boxed{\div}$$

$$8 \boxed{x^2} \boxed{=} \longrightarrow 0.625 \qquad \blacksquare\blacksquare\blacksquare$$

EXAMPLE 6   In materials engineering a relationship exists between the stress and strain placed on an object and the deformation or elongation of the object. The modulus of elasticity ε is given by the equation:

$$\epsilon = \frac{PL}{Ae}$$

where ϵ = modulus of elasticity, P = total applied external axial load, L = length of the object, A = cross-sectional area of the object, and e = total axial deformation. Find the total axial deformation of a steel rod 20 in. long with a cross-sectional area of 1.77 in^2 if 25,000 lb of external axial load is applied and the modulus of elasticity is 30,000,000 psi.

Solution   First, we rearrange the formula to solve for *e*.

$$Ae(\epsilon) = \cancel{Ae}\left(\frac{PL}{\cancel{Ae}}\right) \qquad \text{eliminate fractions}$$

$$Ae\epsilon = PL \qquad \text{simplify}$$

$$\frac{Ae\epsilon}{A\epsilon} = \frac{PL}{A\epsilon} \qquad \text{divide by } A\epsilon$$

$$e = \frac{PL}{A\epsilon} \qquad \text{simplify}$$

Then, we substitute $L = 20$ in., $A = 1.77$ in^2, $P = 25,000$ lb, and $\epsilon = 30,000,000$ psi (psi = pounds/in^2).

$$e = \frac{(25,000 \cancel{\text{lb}})(20 \text{ in.})}{(1.77 \cancel{\text{in}^2})(30,000,000 \cancel{\text{lb/in}^2})}$$

$$e = 0.009 \text{ in.}$$

The steel rod would elongate 0.009 inches. $\qquad \blacksquare\blacksquare\blacksquare$

EXAMPLE 7   For two resistors r and R connected in series in a dc circuit of voltage V, the current I is given by

$$I = \frac{V}{(r + R)}$$

Find r if $I = 2$ amperes (A), $V = 12$ volts (V), and $R = 5$ ohms (Ω), the unit of resistance. Solve and substitute.

Solution   First, we solve the formula for r. We eliminate the fraction by multiplying each term by $(r + R)$.

$$(r + R)I = (r + R)\left(\frac{V}{r + R}\right)$$

$$I(r + R) = V \qquad\qquad \text{simplify}$$

$$Ir + IR = V \qquad\qquad \text{eliminate parentheses}$$

$$Ir = V - IR \qquad\qquad \text{transpose}$$

$$r = \frac{V - IR}{I} \qquad\qquad \text{divide by } I$$

Then we substitute $V = 12$ V, $I = 2$ A, and $R = 5$ Ω into the formula.

$$r = \frac{12 \text{ V} - 2 \text{ A}(5 \text{ }\Omega)}{2 \text{ A}}$$

Since $V = IR$, volts = amps $\cdot$ ohms.

$$r = \frac{12 \text{ A} \cdot \Omega - 10 \text{ A}\Omega}{2 \text{ A}}$$

$$r = 1 \text{ }\Omega$$

$$12 \boxed{-} 2 \boxed{\times} 5 \boxed{=} \boxed{\div} 2 \boxed{=} \longrightarrow 1$$

∎∎∎

EXAMPLE 8   The maximum deflection δ_{max} of a cantilever beam of length L with a uniformly distributed load W plus a concentrated load P at the free end (ϵ is modulus of elasticity and I is moment of inertia) is given by

$$\delta_{max} = \frac{8PL^3 + 3WL^3}{24\epsilon I}$$

Solve for P.

Solution
$$24\epsilon I(\delta_{max}) = 24\epsilon I \left(\frac{8PL^3 + 3WL^3}{24\epsilon I}\right) \qquad \text{remove fraction}$$

$$24\epsilon I\delta_{max} = 8PL^3 + 3WL^3 \qquad\qquad \text{simplify}$$

$$24\epsilon I\delta_{max} - 3WL^3 = 8PL^3 \qquad\qquad \text{transpose}$$

$$\frac{24\epsilon I\delta_{max} - 3WL^3}{8L^3} = P \qquad\qquad \text{divide}$$

∎∎∎

EXAMPLE 9   In determining profitability of his small business, Ned uses the relationships

$$\text{Markup} = \text{overhead} + \text{net profit}$$

and

$$\text{Cost} + \text{markup} = \text{selling price}$$

If Ned manufactures calculators that sell for $24.50, allows $4.75 for overhead expenses, and expects to make a net profit of $4.00, find Ned's cost to manufacture each calculator.

Solution First, we determine the markup.

$$M = OH + NP$$
$$M = 4.75 + 4.00$$
$$M = 8.75$$

Once we know the markup,

$$\text{Cost} + \text{markup} = \text{selling price}$$
$$C + M = S$$

Since we must find cost, we must solve the formula for C.

$$C + M = S$$
$$C = S - M$$

Substituting S = 24.50 and M = 8.75 gives

$$C = 24.50 - 8.75$$
$$C = 15.75$$

Each calculator costs Ned $15.75 to manufacture. ■ ■ ■

1–2 EXERCISES

Solve each formula for the specified variable.

1. $A = Lwh$ for h

2. $P = 2L + 2w$ for L

3. $S = 2\pi rh + 2\pi r^2$ for h

4. $L = \dfrac{1}{2}Ps$ for P

5. $E = mgh$ for m

6. $V = \dfrac{4}{3}\pi r^3$ for π

7. $I = kL(T - t)$ for t

8. $R = \dfrac{rL}{D^2}$ for L

9. $v^2 = v_0^2 + 2as$ for a

10. $P = \dfrac{fs}{t}$ for f

11. $T = \dfrac{D - d}{2}$ for D

12. $A = \dfrac{hb}{2}$ for b

13. $E = \dfrac{1}{2}mv^2$ for m

14. $V = \dfrac{1}{3}\pi r^2 h$ for h

15. $L = 2\pi rh$ for r

16. $d = \dfrac{fL}{f + w}$ for w

17. $F = \dfrac{9}{5}C + 32$ for C

18. $A = \dfrac{h(B + b)}{2}$ for B

19. $a = \dfrac{v - v_0}{t}$ for v_0

20. $s = vt + \dfrac{1}{2}at^2$ for v

21. $v_{av} = \dfrac{v_f + v_0}{2}$ for v_f

22. $P = \dfrac{N + 2}{D_0}$ for N

23. $L = 3.14(r_1 + r_2) + 2d$ for r_1

24. $P = \dfrac{1}{3}Nmv^2$ for m

25. $a = \dfrac{w_t - w_0}{t}$ for w_t

26. $mv_2 - mv_1 = Ft$ for v_1

27. $F = \dfrac{kq_1q_2}{r^2}$ for q_1

28. $M = \dfrac{P(C + L)}{T}$ for C

29. $y = mx + b$ for m

30. $h = \dfrac{k(t_2 - t_1)aT}{d}$ for t_2

31. $f = \dfrac{f_s u}{u + v_s}$ for f_s

32. $h = \dfrac{D - d}{2}$ for d

33. $a = \dfrac{2T}{(d_1 - d_2)g}$ for d_1

34. $\dfrac{P_1 V_1}{T_1} = \dfrac{P_2 V_2}{T_2}$ for T_2

35. $I = \dfrac{E}{R + r}$ for r

36. $s = \dfrac{H}{m(t_2 - t_1)}$ for t_1

37. $P = \dfrac{V_1(V_2 - V_1)}{gJ}$ for V_2

38. $m = \dfrac{y_2 - y_1}{x_2 - x_1}$ for y_1

39. $r = \dfrac{ab}{a + b + c}$ for c

40. $y - y_0 = m(x - x_0)$ for x

Solve for the required quantity by rearranging the formula (if necessary) and then substituting values. Assume integers are exact numbers. Round your answer to the correct number of significant digits.

P 41. The distance d that a free-falling object falls in t seconds is $d = gt^2/2$ where $g = 9.8 \text{ m/s}^2$, the acceleration due to gravity. How far will a cement block fall in 16 seconds?

E 42. The power P in watts (W) in a dc circuit is the product of the voltage V and the current I. What power is consumed by a radio requiring 110 V and 0.15 A?

B 43. Use the information in Example 7. An electronics store wants to sell a radio for $20.00 to meet the price of the competitor across the street. If overhead expenses are $6.00 and net profit is $2.40, determine the cost.

E 44. The resistance R of a copper wire changes with the temperature according to $R = R_1(1 + \alpha[t - t_1])$ where R_1 = initial resistance, α = coefficient of resistance, t = final temperature, and t_1 = initial temperature. Find the resistance R of a copper coil if the resistance is 750 Ω at 35°C and $\alpha = 0.00393$ at 10°C.

E 45. When two resistors, R_1 and R_2, are connected in parallel, the equivalent resistance R equals their product divided by their sum. If the two resistors are 15 Ω and 9 Ω, find the equivalent resistance.

A 46. The living room of a house requires 120 ft^2 of carpeting. Find the length of the room if the width is 8 ft. (*Note: A = LW.*)

B 47. The interest I earned on an investment equals the product of the amount invested P, the number of years invested t, and the interest rate r, expressed as a decimal. Mr. Jones earns $270 interest on an investment of $1,500 at 9%. How long did it take Mr. Jones to earn this interest?

P 48. A body is moving with an acceleration of 2 ft/s^2. How much time is required for its velocity to increase from 10 ft/s to 16 ft/s? (*Note: $a = (v - v_0)/t$.*)

B 49. A 90-day loan with interest at 8% was repaid with a check for $1,530.00. How much was the original loan (l)? $S = P(1 + rt)$ where S = maturity value (loan + interest), P = loan amount, r = interest rate in decimal form, and t = time in years.

M 50. Using Example 6, find the total axial deformation of a steel rod 34 in. long with a cross-sectional area of 1.96 in^2 if 18,000 lb of external axial load is applied and the modulus of elasticity is 30,000,000 psi.

A 51. A living room requires 20 yd^2 of carpeting. Find the width of the living room (in feet) if the length is 15 ft. ($A = LW$. Note the difference in the units of measure.)

B 52. At the end of a model year, a car dealer advertises that the list prices of last year's cars are reduced 12%. What was the original price of a car that is on sale for $16,280?

1–3

SOLVING LINEAR INEQUALITIES

An **inequality** is a statement that a certain quantity is not equal to a second quantity. The **solution** to an inequality consists of all real numbers that result in a true statement when substituted for the variable. Unlike the solution to a linear equation in one variable, the solution to a linear inequality is an infinite range of numbers. For example, the solution to the linear equation $x = 3$ means that three is the only number that can be substituted for x and yield a true statement. However, the solution to the inequality $x \geq 3$ means that substituting three or any number greater than three results in a true statement. Therefore, the solution to this inequality is the infinite range of numbers starting with three.

Properties of Inequalities

Properties similar to the Principles of Equality apply to inequalities. These properties can be divided into three categories: addition and subtraction, multiplication and division by a positive number, and multiplication and division by a negative number. These properties are summarized below.

Properties of Inequalities

Addition and Subtraction

If the same quantity is added to or subtracted from both sides of an inequality, an equivalent inequality results, thus leading to the same solution. For example,

$$7 > 1$$

is equivalent to

$$7 - 8 > 1 - 8 \qquad \text{subtract 8}$$
$$7 + 8 > 1 + 8 \qquad \text{add 8}$$

Multiplication and Division: Positive Number

If a positive number multiplies or divides both sides of an inequality, an equivalent inequality results. For example,

$$7 > 1$$

is equivalent to

$$3(7) > 3 \qquad \text{multiply by 3}$$

and to

$$\frac{7}{3} > \frac{1}{3} \qquad \text{divide by 3}$$

Multiplication and Division: Negative Number

If a negative number multiplies or divides both sides of an inequality, the direction of the inequality symbol must be reversed to produce an equivalent inequality. For example,

$$7 > 1$$

is equivalent to

$$-3(7) < -3 \qquad \text{multiply by } -3$$

and also to

$$-\frac{7}{3} < -\frac{1}{3} \qquad \text{divide by } -3$$

The process of solving a linear inequality uses the same steps that have been used throughout this chapter. *The only difference in solving an inequality is that the direction of the inequality symbol must be reversed if you multiply or divide both sides of the inequality by a negative number.* These steps are restated here for completeness.

Solving Linear Inequalities

1. Eliminate fractions by multiplying each term of the inequality by the lowest common denominator. Simplify.
2. Eliminate parentheses using the Distributive Property. Simplify.
3. Using transposition, isolate all terms containing the variable on one side of the inequality and all other terms on the opposite side of the inequality. Simplify.
4. Divide both sides of the inequality by the coefficient of the variable.
5. Check the solution.

EXAMPLE 1   Solve the inequality $6x + 5 > 2x - 15$.

Solution   We begin at Step 3 and isolate x terms on the left and all other terms on the right side of the inequality sign. We transpose $2x$ and 5.

$$6x + 5 > 2x - 15$$
$$6x - 2x > -15 - 5$$
$$4x > -20$$

Next, we perform Step 4 and divide both sides of the inequality by 4.

$$\frac{4x}{4} > -\frac{20}{4}$$
$$x > -5$$

Checking   Substitute any number greater than -5 into the original inequality. Let us use -2 for checking since it is greater than -5.

$$6x + 5 > 2x - 15$$
$$6(-2) + 5 > 2(-2) - 15$$
$$-12 + 5 > -4 - 15$$
$$-7 > -19 \text{ (true)}$$ ■ ■ ■

EXAMPLE 2   Solve the inequality $10 - 5x < 2x - 11$.

Solution   Since this inequality does not contain fractions or parentheses, we begin by isolating the variable.

$$10 - 5x < 2x - 11$$
$$-5x - 2x < -11 - 10 \qquad \text{transpose } 2x \text{ and } 10$$
$$-7x < -21 \qquad \text{combine like terms}$$
$$x > 3 \qquad \text{divide by } -7 \text{ and simplify}$$

Notice that the order of the inequality is reversed because we divided by -7.

Checking   Substitute any number from the solution set into the original inequality. Substituting $x = 6$ gives

$$10 - 5(6) < 2(6) - 11$$
$$-20 < 1 \quad \text{(true)}$$ ■ ■ ■

CAUTION ✦ The most common error in solving inequalities is neglecting to reverse the inequality symbol when multiplying or dividing by a negative number. Every time you multiply or divide, check the sign of the number to determine if reversing the inequality symbol is necessary.

EXAMPLE 3   A student must have at least an 80 average to earn a B in a math course. If the student has earned 76, 84, and 73 on the first three tests, what must he earn on the last test to earn a B, assuming all tests are equally weighted?

Solution   To calculate the student's average, we add the grades and divide by the number of grades. If $x =$ the grade on the last test, the average is given by

$$\frac{76 + 84 + 73 + x}{4}$$

Since the average must be 80 or above, the inequality becomes

$$\frac{76 + 84 + 73 + x}{4} \geq 80$$

Solving this inequality gives

$76 + 84 + 73 + x \geq 320$	eliminate fractions
$233 + x \geq 320$	combine like terms
$x \geq 320 - 233$	transpose
$x \geq 87$	combine like terms

The student must earn an 87 or above on the last test to earn a B in the course.　■ ■ ■

Graphing the Solution

In solving linear inequalities, it is often helpful to graph the solution to gain a visual representation on the number line. The graphical solution shows the starting point and the direction of the solution. To show that the starting number is included in the solution, place a dot at the number; to show that the starting number is not part of the solution, place a circle at the number. To show the direction of the solution, draw a line with an arrow pointing in the proper direction. Figure 1–3 illustrates this information.

FIGURE 1–3   The circle and arrow show that −1 is not included in the solution, but all numbers less than −1 are part of the solution.　　　The dot and arrow show that 2 is included in the solution as well as all numbers greater than 2.

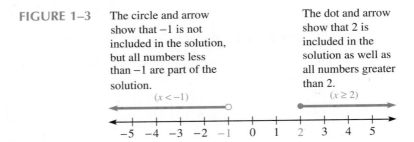

$(x < -1)$　　$(x \geq 2)$

LEARNING HINT ✦ In graphing the solution, you may confuse greater than and less than. To avoid this error, always isolate the variable on the left side of the inequality. Then the inequality symbol points in the direction of the arrow on the number line.

EXAMPLE 4   Solve the following inequality and graph the solution:

$$-3(2x - 3) + 15 \geq 2x + 4(x - 6)$$

Solution

$-3(2x - 3) + 15 \geq 2x + 4(x - 6)$	
$-6x + 9 + 15 \geq 2x + 4x - 24$	eliminate parentheses
$-6x + 24 \geq 6x - 24$	combine like terms
$-6x - 6x \geq -24 - 24$	transpose
$-12x \geq -48$	combine like terms
$x \leq 4$	divide by -12

Checking   Substitute a number from the range into the inequality. We substitute 3.

$$-3[2(3) - 3] + 15 \geq 2(3) + 4[(3) - 6]$$
$$-3(3) + 15 \geq 6 + 4(-3)$$
$$6 \geq -6 \quad \text{(true)}$$

The solution is $x \leq 4$. The graph of the solution set is a dot at 4 and a line with an arrow pointing to the left of 4. The graph of the solution is shown in Figure 1–4.

FIGURE 1–4

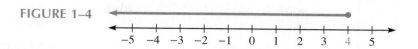

EXAMPLE 5 Solve the following inequality and graph the solution:

$$\frac{2(5x - 3)}{5} < \frac{1}{3} + \frac{x}{5}$$

Solution First, we eliminate the fractions by multiplying by the common denominator of 15.

$$15\left(\frac{2(5x - 3)}{5}\right) < 15\left(\frac{1}{3}\right) + 15\left(\frac{x}{5}\right)$$

$$
\begin{array}{ll}
6(5x - 3) < 5(1) + 3(x) & \text{simplify} \\
30x - 18 < 5 + 3x & \text{eliminate parentheses} \\
30x - 3x < 5 + 18 & \text{transpose } 3x \text{ and } -18 \\
27x < 23 & \text{combine like terms} \\
x < \dfrac{23}{27} & \text{divide by 27}
\end{array}
$$

The solution is $x < 23/27$. The graph of the solution set consists of a circle at 23/27 and a line with an arrow pointing to the left of 23/27. The solution is shown in Figure 1–5.

FIGURE 1–5

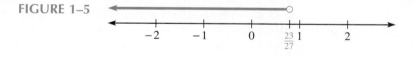

Applications

EXAMPLE 6 The strength s of a sheet of material sufficient to hold a certain weight is given by $s + 7 \geq 2s + 4$, where s is measured in lb/in^2. Find the range of s.

Solution Solving the inequality for s gives

$$
\begin{aligned}
s + 7 &\geq 2s + 4 \\
s - 2s &\geq 4 - 7 \\
-s &\geq -3 \\
s &\leq 3
\end{aligned}
$$

The material is able to withstand no more than 3 lb/in^2.

EXAMPLE 7 The lots in a subdivision development sell for \$32,000. Joan buys a lot expecting its value to increase \$4,500 each year. What is the minimum number of years Joan must wait for the property to be worth at least \$49,100?

Solution The inequality we must solve is

$$32{,}000 + 4{,}500t \geq 49{,}100$$

Solving the inequality gives

$$4{,}500t \geq 49{,}100 - 32{,}000 \qquad \text{transpose}$$
$$4{,}500t \geq 17{,}100 \qquad \text{simplify}$$
$$t \geq 3.8 \qquad \text{divide}$$

Joan must wait at least 3.8 years for the lot to be worth $49,100.   ■■■

EXAMPLE 8   For an investment of $750 to grow to more than $900 in 3 years, what must be the annual interest rate?

Solution   From the previous section, we know $S = P(1 + rt)$ is the formula relating maturity value, principal or initial investment, rate, and time. The maturity value must be greater than $900. Therefore $S > 900$ or

$$P(1 + rt) > 900$$

where $P = 750$ and $t = 3$. Substituting these values results in the following inequality.

$$750(1 + 3r) > 900$$
$$750 + 2{,}250r > 900$$
$$2{,}250r > 900 - 750$$
$$2{,}250r > 150$$
$$r > 0.0666$$

The annual interest rate must be more than $6\frac{2}{3}\%$.   ■■■

Absolute Value Inequalities

Earlier in this chapter we discussed solving equations involving absolute value. Now we extend those methods to solving inequalities involving absolute value. Recall from our previous discussion that absolute value results in two equations. An absolute value inequality also results in two inequalities. When a statement consists of two or more inequalities connected by "and" or "or," the statement is called a **compound inequality.** The expression $x > 2$ or $x < -1$ is an example of a compound inequality. Inequalities involving absolute value result in a compound inequality.

Solving an Absolute Value Inequality

If x represents a variable and $a \geq 0$, then

$$|x| \leq a \text{ is equivalent to } -a \leq x \leq a \ (x \leq a \text{ and } x \geq -a)$$
$$|x| \geq a \text{ is equivalent to } x \geq a \quad \text{or} \quad x \leq -a$$

These statements are also true for $|x| < a$ and $|x| > a$.

Let's examine the inequality $|x| > 4$. The solution consists of all numbers more than four units from zero on the number line. This set of numbers is shown in Figure 1–6. The

FIGURE 1–6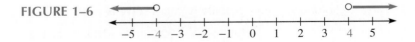

solution set can be written as $x > 4$ or $x < -4$. Therefore, an absolute value inequality containing the $>$ or $\geq$ symbol consists of two solutions connected by the word "or."

$$\text{solution:} \quad x < -4 \quad \text{or} \quad x > 4$$

On the other hand, the solution to an absolute value inequality such as $|x| < 2$ consists of all numbers less than two units from zero on the number line. This solution set is shown in Figure 1–7. The solution consists of all numbers $x < 2$ and $x > -2$. Therefore, an absolute value inequality containing the $<$ or $\leq$ symbol consists of two inequalities connected by the word "and." This type of solution can be expressed in a simpler notation as $-2 < x < 2$. When the solution consists of a line segment, the solution is written

$$\text{smallest number} < x < \text{largest number}$$
$$\text{or smallest number} \leq x \leq \text{largest number}$$
$$\text{solution:} \ -2 < x < 2$$

FIGURE 1–7

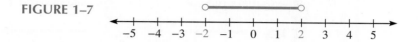

The next three examples illustrate solving absolute value inequalities.

EXAMPLE 9   Solve the inequality $|2x - 1| \leq 9$.

Solution   To solve an absolute value inequality, we follow the same steps used in solving equations involving absolute value, remembering that multiplication or division of the entire inequality by a negative number reverses the direction of the inequality. Since the absolute value quantity is isolated, we write the two inequalities that result from the definition of absolute value. Remember that the $\leq$ symbol results in two inequalities connected by "and."

$$
\begin{array}{lll}
2x - 1 \leq 9 & \text{and} & 2x - 1 \geq -9 \\
2x \leq 9 + 1 & & 2x \geq -8 \\
2x \leq 10 & & x \geq -4 \\
x \leq 5 & &
\end{array}
$$

Checking   Substituting -2 into the original inequality gives

$$|2(-2) - 1| \leq 9$$
$$|-5| \leq 9$$
$$5 \leq 9 \quad \text{(true)}$$

The solution is the set of numbers such that $x \geq -4$ and $x \leq 5$. The solution is written

$$\text{smallest number} \leq x \leq \text{largest number}$$
$$-4 \leq x \leq 5$$

and is shown in Figure 1–8.

FIGURE 1–8

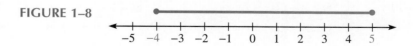

EXAMPLE 10 Solve $|5x + 3| > 7$.

Solution Since this absolute value inequality contains the $>$ symbol, we write two inequalities connected by "or."

$$\begin{aligned} 5x + 3 &> 7 \\ 5x &> 7 - 3 \\ 5x &> 4 \\ x &> 4/5 \end{aligned} \qquad \text{or} \qquad \begin{aligned} 5x + 3 &< -7 \\ 5x &< -10 \\ x &< -2 \end{aligned}$$

Checking To check the solution, we substitute a number from the solution set into the original inequality. Since the solution set consists of two rays, we substitute -3 and 2 into the inequality.

$$|5(-3) + 3| > 7 \qquad\qquad |5(2) + 3| > 7$$
$$12 > 7 \quad \text{(true)} \qquad\qquad 13 > 7 \quad \text{(true)}$$

The solution set is written $x > 4/5$ or $x < -2$ and is shown in Figure 1–9.

FIGURE 1–9

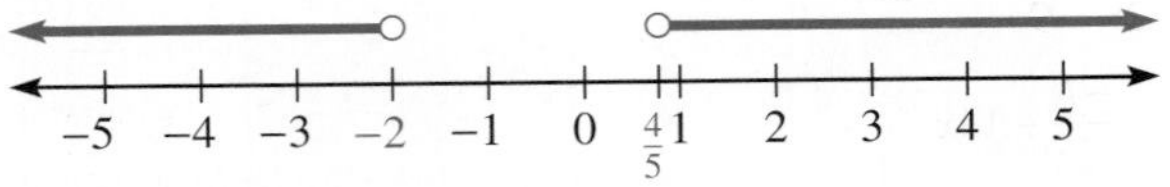

■■■

Application

EXAMPLE 11 All measuring devices are calibrated to within a plus or minus tolerance for error. Scales in the grocery are no exception. Suppose that you bought apples at \$1.29 per pound, and the scale is accurate to within ±0.03 pound. If the apples weighed 2.4 pounds and you were charged \$3.10, how much might you have been undercharged or overcharged due to the error in the scale?

Solution Let x represent the true weight of the apples. Since the scale is accurate to within 0.03 pound, the difference between the exact weight x and the scale weight 2.4 lb is less than or equal to 0.03 pound.

$$|x - 2.4| \le 0.03$$

Solving the resulting inequalities gives the following.

$$\begin{aligned} x - 2.4 &\le 0.03 \\ x &\le 0.03 + 2.4 \\ x &\le 2.43 \end{aligned} \qquad \text{and} \qquad \begin{aligned} x - 2.4 &\ge -0.03 \\ x &\ge -0.03 + 2.4 \\ x &\ge 2.37 \end{aligned}$$

This means your apples could have weighed between 2.37 and 2.43 pounds. The resulting cost is \$3.06 and \$3.13. So you were undercharged by as much as \$0.03 and overcharged by as much as \$0.04.

■■■

1–3 EXERCISES

Solve each of the following inequalities and graph the solution set on the number line.

1. $x + 5 \ge -8$

2. $7 \le 3 + x$

3. $2x - 6 \le 8$

4. $4x - 10 \le -14$

5. $2(-3x + 6) > -10$

6. $-5(3 - 2x) < 15$

7. $\frac{x}{3} > -4$

8. $\frac{3x}{5} < -6$

9. $2x + 3 - 7x < 9 - x + 6$

10. $8 - x + 4 > 6x - 10 + 1$

11. $5x - 3 \ge 7x + 7$

12. $3x + 17 < 5x + 3$

13. $-6 - 4x > -9x + 24$

14. $-7x \ge 35$

15. $-(2x + 6) + 8 < 2(2x - 5)$

16. $-4x + 17 + 8x < -5 - x + 2$

17. $\dfrac{3x}{7} + \dfrac{1}{4} < 1 + \dfrac{x}{2}$ **18.** $\dfrac{5x}{6} - \dfrac{2}{3} + x > \dfrac{1}{2} + \dfrac{2x}{3}$

19. $4(3 - 2x + 1) + 3x + 4 \geq 0$

20. $-2(5 + x) - 7 + 4x < -2x - 1$

21. $\dfrac{2x + 1}{4} < \dfrac{1}{2}$ **22.** $\dfrac{3x - 5}{6} > \dfrac{1}{3}$

23. $3x - \dfrac{1}{3} > \dfrac{x}{2} + 6$ **24.** $\dfrac{3}{4} - 5x + 2 < \dfrac{1}{2} - x + \dfrac{1}{4}$

25. $-8x + 7 - 3(x + 1) + 17 \leq 0$

26. $\dfrac{2(x + 1)}{3} > 6 - \dfrac{1}{4} + 3x$ **27.** $3\left(2x - \dfrac{1}{3}\right) \leq 0$

28. $\dfrac{7x + 1}{3} \geq \dfrac{2x - 1}{2}$ **29.** $\dfrac{x}{4} - \dfrac{x}{6} + \dfrac{x}{3} > \dfrac{x}{2}$

30. $\dfrac{2x}{3} - \dfrac{x}{9} + \dfrac{5x}{6} > \dfrac{x}{2} - \dfrac{4x}{9}$ **31.** $\dfrac{3(2x - 4)}{5} + \dfrac{1}{3} > \dfrac{-8x}{3}$

32. $\dfrac{-2(5 - 6x)}{2} + 3 < \dfrac{-7}{2} + 14x$

33. $2x - 8 + 5x \leq 3(x - 2) + 6$

34. $8(-2x + 1) - 6x > -7 + (2x - 1) - 4$

35. $\dfrac{3(4 - 2x)}{4} + 1 > \dfrac{-(x + 1)}{3} - \dfrac{5}{6}$

36. $\dfrac{2 + 5(x - 1)}{3} < \dfrac{3}{4}$ **37.** $\dfrac{4(x - 3) + (2x + 1)}{2} \geq \dfrac{1}{2}$

38. $\dfrac{5(x - 6) - (2x + 3)}{3} \geq 0$

39. $-3(2x - 1) + 6 + 2x < 7 - (2x - 1)$

40. $9 - 3(x - 5) + 6x \leq -4 + 2(4x - 7) + 6x$

41. $|3x + 4| < 8$ **42.** $|2x - 5| \geq 7$

43. $|4x - 1| + 3 > 16$ **44.** $5|x - 2| + 6 \leq 25$

45. $8|x - 3| < 21$ **46.** $3|7x - 2| - 5 \geq 19$

B 47. An intermediate car can be rented from Rent a Wreck for $250 per week while the same car can be rented from Cheap Cars for $175 per week plus 20 cents a mile. How many miles must you drive per week to make the rental fee for Cheap Cars more than that for Rent a Wreck?

M 48. The measurement m of an automobile engine part must satisfy a tolerance given by $|m - 6| \leq 0.05$, where m is measured in millimeters. Solve for m to determine the range of tolerance.

P 49. The in-flight time t of a projectile is given by $|4t - 9| < 9$. Find the interval of time in seconds that the object is in flight.

B 50. Using Example 10, find the undercharge and overcharge if the apples weighed 3.8 pounds, the scale error was ±0.1 pound, and you were charged $4.90.

B 51. A full size car can be rented from Uncle Buddy's Rentals for $275 per week with unlimited mileage. The same size car can be rented from Cheap Rentals for $200 per week plus $0.10 a mile. How many miles must you drive in a week to make Cheap Rentals cost more than Uncle Buddy's?

52. Judy begins a diet and exercise program designed to reduce her weight by 2 pounds per week. If she weighs 148 lbs at the beginning of her program, find the maximum number of weeks before her weight reaches (or falls below) 120 pounds.

CS 53. A company offers Internet access via satellite for a basic charge of $29.99 per month for the first 25 hours of online time. Each additional hour (or portion of an hour) costs $1.99. One subscriber to the service allocated $75 per month for online time. What is the maximum number of hours the subscriber can stay online and remain within the budgeted amount?

B 54. A semiconductor manufacturer knows that it will cost $65,000 for research and development of a new semiconductor chip and $2.50 per unit to manufacture. If x is the number of units sold, total cost $C = 65,000 + 2.50x$. The revenue from sales is expected to be $R = 3.20x - 15,000$. What level of sales will result in a profit? (When is $R > C$?)

E 55. A radio transmitter has a fuse that will blow if its power exceeds 4.50 A. However, the transmitter requires a minimum of 98.0 V to operate properly. If the total resistance in the circuit is a constant 33.0 Ω, what is the range of voltage at which the transmitter can operate?

1–4

VARIATION

Variation describes the relationship between variables. When two quantities are related so that their ratio is a constant, the variables are said to **vary directly** or to be **proportional.** If a car is traveling at 55 miles per hour, at the end of one hour the car has traveled 55 miles, at the end of two hours it has traveled 110 miles, and at the end of three hours it has traveled 165 miles. The equation $d = 55t$ describes this relationship between the distance traveled d and the time t. If we rearrange the equation, then $d/t = 55$. In other words, the ratio of the variables is a constant, called the **constant of variation.** We can describe this relationship by saying that distance varies directly with time. You should also notice from this example that as time increases, the distance traveled also increases. This kind of relationship is true for direct variation. In general, if we let x and y represent variables and k represent the constant of variation, then the formula for direct variation could be written as $y/x = k$ or as follows:

$$y = kx \quad \text{direct variation}$$

EXAMPLE 1 Hooke's law states that the force F required to stretch a spring is directly proportional to the change of length x of the spring. Write a formula for this relationship.

Solution Using k as the constant of variation, the formula is

$$F = kx \qquad \blacksquare\blacksquare\blacksquare$$

NOTE ✦ A variable may also vary with a second variable that has been raised to some power. In this case, the appropriate exponent is attached to the second variable.

EXAMPLE 2 When a ball rolls down an inclined plane, the distance d it travels is directly proportional to the square of the time t. Write a formula to represent this relationship.

Solution This relationship is a statement of direct variation where the distance varies with the square of the time. The formula representing this relationship is

$$d = kt^2 \qquad \blacksquare\blacksquare\blacksquare$$

NOTE ✦ At this point we do not know the value of k, the constant of variation, but it will be discussed later in this section.

Inverse Variation

Let us consider the possible dimensions for a rectangle when its area is fixed at 100 square feet. Several possibilities for the dimensions are 2 ft by 50 ft, 1 ft by 100 ft, and 4 ft by 25 ft. This relationship can be expressed as $LW = 100$. Two variables are said to **vary inversely** or to be **inversely proportional** if the product of the variables is a constant. In our example, L and W vary inversely because their product is a constant, and in this case 100 is the constant of variation. Notice from this example that as one variable increases (for example, length increases from 2 ft to 4 ft), the other variable decreases (for example, width decreases from 50 ft to 25 ft). If x and y represent the variables and k represents the constant of variation, the formula for inverse variation can be written as $y \cdot x = k$ or as follows:

$$y = \frac{k}{x} \quad \text{inverse variation}$$

For a graphical perspective of direct and inverse variation, look at Figures 1–10 and 1–11. For direct variation, shown in Figure 1–10, as x increases, y also increases. This means that as you move to the right along the x axis, move up until you touch the graph, and then left until you touch the y axis, the values of y get larger. For inverse variation, shown in Figure 1–11, as x increases, y decreases.

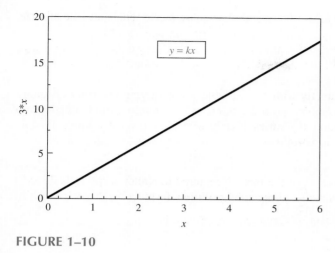

FIGURE 1–10

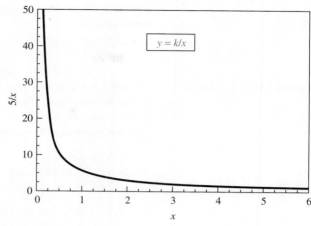

FIGURE 1–11

EXAMPLE 3   Write a formula for the following relationship: The pressure of a compressed gas varies inversely with the volume.

Solution   If P represents the pressure and V represents the volume, the formula is

$$P = \frac{k}{V}$$ ■ ■ ■

One variable can also be inversely proportional to the power of a second variable, as shown in the next example.

EXAMPLE 4   The illumination provided by a light source is inversely proportional to the square of the distance from the source. Write a formula to express this relationship.

Solution   If the letter I represents the illumination and d represents the distance, the formula for the inverse variation is

$$I = \frac{k}{d^2}$$ ■ ■ ■

Joint Variation

The volume of a cylinder changes with both the radius and the height of the cylinder. This example describes one variable whose value depends upon the value of two other variables. **Joint variation** occurs when one variable varies directly with several other variables. The general formula for joint variation with y varying jointly with x and z is as follows:

$$y = kxz \quad \text{joint variation}$$

EXAMPLE 5   The volume V of a cone varies jointly as the square of the radius r of the base and the altitude h. Write a formula to represent this relationship.

Solution   Since the volume varies jointly with the square of the radius and the altitude, the formula is

$$V = kr^2h$$

■ ■ ■

Combined Variation

The same variable can vary directly with one variable while varying inversely or jointly with other variables. We can combine several types of variation into a single relationship by using one constant of variation. If y varies directly with x and inversely with z, the formula for **combined variation** is as follows:

$$y = \frac{kx}{z} \quad \text{combined variation}$$

EXAMPLE 6   The safe load L for a horizontal beam supported at the ends varies jointly with the width w and the square of the depth d and inversely with the distance s between the supports. Write a formula to represent this relationship.

Solution   This is a combined variation problem because it involves both joint and inverse variation. The formula for the expression "L varies jointly with w and d^2" would be written as

$$L = k_1 wd^2$$

Similarly, the formula for "L varies inversely as s" would be written

$$L = \frac{k_2}{s}$$

Combining these two formulas using only one constant of variation, k, gives the formula for the verbal statement as

$$L = \frac{kwd^2}{s}$$

■ ■ ■

Solving Variation Problems

Variation is used to extrapolate or go beyond given information. For example, the perimeter of a regular polygon varies directly with the length of the side. If the perimeter of a polygon is 20 ft when the length of a side is 4 ft, then what is the perimeter when the length of a side is 7 ft? Solving a problem of this type requires a procedure such as the one outlined below.

Solving Variation Problems

1. Write a variation formula from the verbal statement.
2. Solve for k, the constant of variation, by substituting the complete set of values given for the variables.
3. Substitute the value of k into the variation formula.
4. Using the value of k and the remaining variables, solve for the value of the specified variable.

EXAMPLE 7 Boyle's law states that the volume V of a gas varies inversely as the pressure P of the gas. If the volume of the gas at a certain temperature is 75 in^3 when the pressure is 30 lb/in^2, find the volume when the pressure is 25 lb/in^2.

Solution We begin by writing a formula for the inverse variation described.

$$V = \frac{k}{P}$$

To solve for k, we must know values for V and P. We can have only one unknown to yield a unique solution, but the volume is 75 in^3 when the pressure is 30 lb/in^2. Therefore, substituting in values for $V = 75$ in^3 and $P = 30$ lb/in^2 gives

$$75 \text{ in}^3 = \frac{k}{30 \text{ lb/in}^2}$$

$$75 \text{ in}^3 \cdot 30 \frac{\text{lb}}{\text{in}^2} = k$$

$$2{,}250 \text{ lb} \cdot \text{in.} = k$$

Next, we substitute 2,250 lb · in. for k into the variation formula.

$$V = \frac{2{,}250 \text{ lb} \cdot \text{in.}}{P}$$

Finally, the verbal statement requires finding the volume when the pressure is 25 lb/in^2. Substituting in the value of $P = 25$ lb/in^2 and solving for V gives

$$V = \frac{2{,}250 \text{ lb} \cdot \text{in.}}{25 \text{ lb/in}^2}$$

$$V = 90 \text{ in}^3$$

Thus, the volume is 90 in^3 when the pressure is 25 lb/in^2. ■■■

EXAMPLE 8 The power in a circuit varies jointly as the resistance and the square of the current. If the power in a 10-Ω resistor is 3.6 W when the current is 0.60 A, what is the power in the resistor when the current is 0.80 A?

Solution Using P for power, R for resistance, and I for current, the formula is given by

$$P = kRI^2$$

To solve for k, substitute $P = 3.6$ W, $R = 10$ Ω, and $I = 0.60$ A.

$$3.6 = k(10)(0.60)^2$$

$$k = 1$$

$$P = RI^2$$

To find the power when the resistance is 10 Ω and the current is 0.80 A, substitute $R = 10$ and $I = 0.80$.

$$P = (10)(0.80)^2$$

Therefore, the power is 6.4 W. ■■■

EXAMPLE 9 Chuck's Pizza has found that the demand for its pizza is inversely proportional to its price. When the price for a pizza is $8.95, the demand is 50 pizzas per day. What will be the demand if the price is increased to $9.95?

Solution Let P = price and D = demand. Since they are inversely proportional, the equation is

$$D = \frac{k}{P}$$

We know that the demand is 50 when the price is \$8.95. Substituting these values and solving for k gives the following result.

$$50 = \frac{k}{8.95}$$

$$447.5 = k$$

The equation becomes

$$D = \frac{447.5}{P}$$

When the price is \$9.95, the demand is given by the following.

$$D = \frac{447.5}{9.95}$$

$$D = 45$$

Demand will decrease to 45 pizzas when the price is increased to \$9.95.

1–4 EXERCISES

Write a formula to express the following statements.

1. x varies directly with z.
2. m varies inversely with p^2.
3. a varies jointly with b and c.
4. m varies directly with p^2.
5. x varies directly with y and inversely with s.
6. d is inversely proportional to t^2 and directly proportional to s.
7. m varies jointly with n and p^3 and inversely with r.
8. x varies directly with y and inversely with z^3.
9. s varies jointly with t^2 and v.
10. a varies inversely with b^2 and jointly with c and d^3.

Set up a formula and solve for k, the constant of variation.

11. x varies directly as y^2, and $x = 10$ when $y = 2$.
12. m varies inversely with n, and $m = 5$ when $n = 2$.
13. a varies jointly with b and c^2, and $a = 30$ when $b = 5$ and $c = 1$.
14. s varies directly with a and inversely with t^2, and $s = 81$ when $a = 4$ and $t = 2$.
15. x varies directly with y^2 and inversely with z, and $x = 32$ when $y = 4$ and $z = 2$.

Solve the following variation problems.

16. x varies inversely with y^2. If $x = 2$ when $y = 3$, find x when $y = 4$.
17. y varies directly with t. If $y = 14$ when $t = 4$, then find y when $t = 16$.
18. m varies jointly with n and p. If $m = 30$ when $n = 1$ and $p = 3$, find m when $n = 3$ and $p = 4$.
19. a varies directly with b^2 and inversely with c. If $a = 4$ when $b = 4$ and $c = 7$, then find a when $b = 6$ and $c = 2$.
20. s varies inversely with t. If $s = 3$ when $t = 9$, find s when $t = 4$.
21. r varies directly as the square of s. If $r = 40$ when $s = 2$, then find r when $s = 3$.
22. z varies inversely as the square root of y. If $z = 1$ when $y = 9$, then find z when $y = 25$.
23. a varies jointly with b and the square of c. If $a = 72$ when $b = 9$ and $c = 2$, find b when $a = 36$ and $c = 3$.
24. d varies directly with t. If $d = 30$ when $t = 6$, then find t when $d = 55$.
25. m varies directly as n and inversely as p. If $m = 2$ when $n = 16$ and $p = 16$, then find p when $m = 16$ and $n = 25$.

P 26. The intensity of light varies inversely as the square of the distance from the source. If the intensity is 10 foot-candles at a distance of 2 ft, find the intensity when the distance is 7 ft.

P 27. Hooke's law states that the force needed to stretch a spring is directly proportional to the length of the stretch. Use Hooke's law to determine the length that a spring is stretched when a force of 5 newtons (N) is applied, if a force of 10 N stretched the spring 2.7 m.

E 28. The power in a resistance circuit varies jointly as the resistance and the square of the current. If a power of 100 W results from a 25-Ω resistance with a 2-A current, find the resistance when the power is 75 W and the current is 6 A.

29. The area of a circle varies directly with the square of the radius. The radius of a circle whose area is 12.56 cm² is 2 cm. Find the area of a circle whose radius is 5 cm.

P 30. The volume of a gas varies directly with the temperature and inversely with the pressure. If a volume of 500 cm³ results from a temperature of 273 K and pressure of 76 N/cm², find the volume if the temperature is 353 K and the pressure is 90 N/cm².

M 31. The strength of a rectangular beam is directly proportional to the square of the thickness. If a 2-in.-thick beam supports 1,000 lb, how much would a 5-in. beam support?

P 32. The acceleration a of an object is directly proportional to the difference between the final velocity v_f and the initial velocity v_0 and inversely proportional to the elapsed time t. Write a formula to express this relationship.

E 33. Ohm's law states that current in a circuit varies directly as the voltage and inversely as the resistance. If the current is 2 A, the voltage is 10 V, and the resistance is 5 Ω, what resistance is caused by a current of 7 A and a voltage of 3 V?

34. The volume of a cone varies jointly with the height and the square of the radius. If a volume of 12.56 m³ results from a cone of height 3 m and radius 2 m, what is the volume of a cone whose height is 2 m and radius is 5 m?

P 35. The potential energy of a stationary object is jointly proportional to the mass and elevation of the object. If the potential energy of a 5-kg object whose elevation is 3 m is 147 J, then find the elevation of an object if the potential energy is 100 J and its mass is 6 kg.

P 36. The kinetic energy of an object varies jointly with its mass and the square of its velocity. If the kinetic energy of a 10-kg object whose velocity is 8 m/s is 320 J, find the kinetic energy of an object whose mass is 20 kg and whose velocity is 6 m/s.

P 37. The period of a pendulum is directly proportional to the square root of its length and inversely proportional to the square root of the acceleration due to gravity. If the period is 1.57 s, the length is 2.0 ft, and the acceleration due to gravity is 32 ft/s², find the period of a pendulum 5.0 ft long.

E 38. The resistance of a wire varies directly as the length and inversely as the cross-sectional area of the wire. If the resistance of a wire 14 ft long with a cross-sectional area of 0.01 in² is 0.5 Ω, find the resistance of the same wire 20 ft long.

39. The distance an object falls is jointly proportional to the acceleration due to gravity and the square of time. If the acceleration due to gravity is 9.8 m/s², the time is 4 s, and the distance it falls is 78.4 m, how far does the object fall in 6 s?

40. The surface area of a sphere varies directly as the square of the radius. The surface area of a sphere with a radius of 8 in. is 804 in². Find the surface area of a sphere with a radius of 10 in.

P 41. The speed of a pulley is inversely proportional to the diameter of the pulley. Two pulleys are connected by a belt. One has a diameter of 5.00 in. and rotates at 1,350 revolutions per minute (rpm). How fast is the other pulley moving if its diameter is 8.25 in.?

A 42. The amount of heat lost through insulation is inversely proportional to the thickness of the insulation. If 3.0 in. of insulation loses 1,900 btu per hour, how much insulation is required to reduce heat loss to 1,200 btu per hour?

M 43. The speed of a gear is inversely proportional to the number of teeth in the gear. Two gears are meshed together. The smaller gear travels at 57 rpm and has 16 teeth. How many teeth does the larger gear have if it rotates at 48 rpm?

B 44. The cost of labor varies directly with the number of workers and number of days worked. If 16 people work 5 days for $9,600, how many days can 9 people work to earn $6,480?

P 45. Kepler's third law states that the time it takes a planet to revolve around the sun varies directly as the square root of the cube of the maximum radius of its orbit. If the maximum radius of the orbit of Earth is 93 million mi, how long does it take Mars to revolve around the sun if its maximum radius is 142 million mi?

P 46. The volume of a gas varies directly with temperature and inversely with pressure. If a volume of 125 in³ results from a temperature of 215° C and pressure of 43.0 lb/in², find the volume when the temperature is 270° C and the pressure is 75.0 lb/in².

B 47. The cost of publishing a magazine is $500 plus an amount directly proportional to the number of magazines published. If 20,000 magazines cost $35,000, how much will it cost to publish 55,000 magazines?

E 48. In a series circuit, the voltage drop across a resistor is directly proportional to the resistance. If the voltage drop across a 75-Ω resistor is 10 V, what is the voltage drop across a 30-Ω resistor?

1–5

APPLICATIONS

So far in this chapter, we have primarily solved equations and inequalities that are explicitly stated. However, most problems encountered in mathematics, science, and engineering job situations are stated verbally. Therefore, it is necessary to transform a verbal statement into an equation that can be solved. Solving problems of this type is essential for the technician but often most difficult to master. The following steps should help in formulating an equation from a verbal statement.

Solving Verbal Problems

1. Read the entire verbal statement carefully. If possible, draw a diagram or chart.
2. Select a letter to represent one of the unknown quantities. Many times the question at the end of the verbal statement will give a clue about the unknown quantity.
3. If there are several unknown quantities, they must all be expressed in terms of the letter selected in Step 2.
4. Write an equation from the verbal statement.
5. Solve the equation and relate the solution to the variable specified in the verbal statement.
6. Check the solution to make sure the conditions of the verbal statement are satisfied.

EXAMPLE 1 A technician earns $12.40 an hour for the first 40 hours and $18.60 for each hour over 40. How many hours of overtime must the technician work to earn $719.20 in a week?

Solution Since the verbal statement asks for overtime hours, let x = number of overtime hours. To write the equation, we know that weekly pay is the sum of salary from regular hours and salary from overtime.

$$\text{pay for regular hours} + \text{pay for overtime hours} = \text{total salary}$$
$$12.40(40) \quad + \quad 18.60x \quad = 719.20$$

Solving the equation gives

$$496.00 + 18.60x = 719.20$$
$$18.60x = 719.20 - 496.00$$
$$18.60x = 223.20$$
$$x = 12$$

The technician must work 12 hours of overtime to earn $719.20

$$719.20 \;\boxed{-}\; 496.00 \;\boxed{=}\; \boxed{\div}\; 18.60 \;\boxed{=}\; \longrightarrow 12$$

Checking

$$\text{technician's salary for 40 hours} = \$12.40(40) = \$496.00$$
$$\text{technician's salary for 12 hours overtime} = \$18.60(12) = \$223.20$$
$$\text{technician's total salary} = \$496.00 + \$223.20 = \$719.20 \quad \blacksquare\blacksquare\blacksquare$$

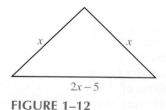

FIGURE 1–12

EXAMPLE 2 The third side of an isosceles triangle is 5 ft less than twice the length of one of the equal sides. Find the length of the sides of the triangle if the perimeter is 35 ft.

Solution Since the verbal statement asks for the length of the three sides, let x represent the two equal sides. Therefore, as shown in Figure 1–12,

$$x = \text{the length of each of the two equal sides of the triangle}$$
$$(2x - 5) = \text{the length of the third side of the triangle}$$

Since the perimeter of a triangle is the sum of the three sides, the equation is

$$x + x + (2x - 5) = 35$$

Solving for x gives

$$4x - 5 = 35$$
$$4x = 35 + 5$$
$$4x = 40$$
$$x = 10$$

Since x represents one of the two equal sides and $2x - 5$ represents the third side, the dimensions of the triangle are 10 ft, 10 ft, and 15 ft.

Checking $$15 = 2(10) - 5$$
$$\text{perimeter} = 10 + 10 + 15 = 35$$ ∎

Ratio and Proportion

A **ratio** is a comparison of any two (or more) quantities. For example, the ratio of Jeff's monthly income of $1,750 to Greg's monthly salary of $2,050 is $1,750/2,050 = 35/41$.

EXAMPLE 3 Mr. Jones invested $15,000 in two savings accounts paying 7.5% and 9% interest. He invested the money in the ratio of 2 to 3, respectively. How much money did he invest in each account?

Solution If $x =$ one share, then

$$2x = \text{the amount invested at 7.5\%}$$
$$3x = \text{the amount invested at 9\%}$$

Because the total investment is $15,000, the equation is

$$2x + 3x = 15,000$$
$$5x = 15,000$$
$$x = 3,000$$

Mr. Jones invested $2x$ or $6,000 at 7.5% and $3x$ or $9,000 at 9%. ∎

A **proportion** is a statement of equality between two ratios. A proportion can be written in fraction form as $a/b = c/d$ or in ratio form as $a:b = c:d$ (read "a is to b as c is to d"). In the proportion

$$\frac{a}{b} = \frac{c}{d}$$

multiplying each side of the equation by bd gives

$$bd\left(\frac{a}{b}\right) = \left(\frac{c}{d}\right)bd$$

$$ad = cb$$

This result is called the **cross-product rule** because multiplying diagonally across the fractions gives the same result, as shown below.

$$\frac{a}{b} = \frac{c}{d}$$

$$ad = cb$$

Cross-Product Rule

If

$$\frac{a}{b} = \frac{c}{d}$$

then

$$ad = bc$$

where $b \neq 0$ and $d \neq 0$.

Proportions are equal as long as the resulting cross-products are equal. The following proportions are all equivalent because the cross-products of each are equal.

$$\left.\begin{array}{c} \dfrac{a}{b} = \dfrac{c}{d} \\[2mm] \dfrac{b}{a} = \dfrac{d}{c} \\[2mm] \dfrac{a}{c} = \dfrac{b}{d} \\[2mm] \dfrac{c}{a} = \dfrac{d}{b} \end{array}\right\} \quad ad = bc$$

The application of proportion and the cross-product rule is illustrated in the next example.

EXAMPLE 4   A typist can complete 20 pages in 3.5 hours. How many hours would it take him to type 27 pages?

Solution   To solve the problem, set up a proportion letting x equal the number of hours it would take him to type 27 pages. The proportion can be set up in numerous ways, all of which result in the same cross-product. The following list gives several equivalent proportions for this problem:

$$\frac{\text{pages}}{\text{time}} = \frac{\text{pages}}{\text{time}} \qquad \frac{\text{time}}{\text{pages}} = \frac{\text{time}}{\text{pages}} \qquad \frac{\text{pages}}{\text{pages}} = \frac{\text{time}}{\text{time}}$$

Using $\dfrac{\text{pages}}{\text{time}} = \dfrac{\text{pages}}{\text{time}}$ gives the proportion

$$\frac{20}{3.5} = \frac{27}{x}$$

Solving for x gives

$$20x = 27(3.5) \quad \text{(cross-product rule)}$$
$$20x = 94.5$$
$$x \approx 4.7 \text{ hours}$$

It would take the typist approximately 4.7 hours to type 27 pages.

$$27 \;\boxed{\times}\; 3.5 \;\boxed{\div}\; 20 \;\boxed{=} \;\longrightarrow\; 4.725$$ ■ ■ ■

Motion

EXAMPLE 5 Two cars 600 miles apart travel toward each other. The rate of one car is 15 mi/h more than the rate of the other. Find the rate of each car if they meet in 5 hours.

Solution This problem is typical of "distance" problems. Figure 1–13 and the accompanying chart will help you organize the information. Set up the variables as follows:

$$x = \text{the rate of the slower car}$$
$$(x + 15) = \text{the rate of the faster car}$$

These values are placed in the rate column of the chart. Since the cars start at the same time, place 5 in the column for time. Since distance is the product of rate and time, fill in this column as the product of the rate and time columns.

FIGURE 1–13

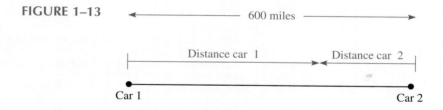

rate × time = distance			
	Rate	Time	Distance
Car 1	x	5	$5x$
Car 2	$(x + 15)$	5	$5(x + 15)$

We use the last column from the chart to write the equation. Therefore, we must read the verbal statement to find the relationship concerning distance. This relationship is

$$\text{total distance} = 600$$

Therefore, the equation is

$$\text{distance car 1} + \text{distance car 2} = \text{total distance}$$

$$5x \quad + \quad 5(x + 15) \quad = \quad 600$$

Solving for x gives

$$5x + 5x + 75 = 600$$
$$10x + 75 = 600$$
$$10x = 600 - 75$$
$$10x = 525$$
$$x = 52.5$$

Since x represents the speed of the slower car, its speed is 52.5 mi/h, and the speed of the faster car is $x + 15$ or 67.5 mi/h.

Checking

$$\text{distance car 1} = 5(52.5) = 262.5$$
$$\text{distance car 2} = 5(67.5) = 337.5$$
$$\text{total distance} = 262.5 + 337.5 = 600$$

■ ■ ■

Mixture

EXAMPLE 6   A chemist wants to make a 37% acid solution using 70 ml of a 48% acid solution and a 22% acid solution. How much of the 22% solution should she use?

Solution   This problem is typical of mixture problems. Again, a chart can help us organize the information. We let x represent the amount of the 22% acid solution. Since the chemist uses 70 ml of the 48% solution, the amount of the mixture is $70 + x$. The amount of each solution in the mixture is the product of the strength and amount columns.

strength $\times$ amount = amount of acid in solution		
Strength	Amount	Amount of acid in solution
48% acid	70	0.48(70)
22% acid	x	0.22 x
37% mixture	$70 + x$	$0.37(70 + x)$

As in the case of motion problems, the equation uses the last column of the chart. The equation is

$$\text{48\% in solution} + \text{22\% in solution} = \text{amount of 37\% in solution}$$

$$0.48(70) \quad + \quad 0.22x \quad = \quad 0.37(70 + x)$$
$$33.6 + 0.22x = 25.9 + 0.37x$$
$$33.6 - 25.9 = 0.37x - 0.22x$$
$$7.7 = 0.15x$$
$$51 \approx x$$

Checking   amount of 48% acid in the solution = 33.6

amount of 22% acid in the solution = 11.3

amount of 37% acid (mixture) = 44.9   ■■■

Finance

EXAMPLE 7   Jim Parker invested a certain amount of money at 10% and $1,000 more than three times this amount at 8%. The total annual interest from the two investments is $760. How much did Jim invest at each rate?

Solution   This problem is typical of finance problems. As in the previous two examples, a chart is helpful. The variables are

$$x = \text{the amount invested at 10\%}$$
$$(3x + 1,000) = \text{the amount invested at 8\%}$$

These values are placed in the amount column of the chart. Since interest is the product of rate and amount, we multiply the contents of the first two columns and place the result in the interest column.

rate $\times$	amount	=	interest
Rate	Amount		Interest
10%	x		$0.10(x)$
8%	$(3x + 1,000)$		$0.08(3x + 1,000)$

In calculating the interest, we must express the rate in decimal form. Since the total interest is $760, the equation is

interest at 10%   +   interest at 8%   = total interest

$$0.10(x) \quad + \quad 0.08(3x + 1,000) = 760$$
$$0.10x + 0.24x + 80 = 760$$
$$0.34x + 80 = 760$$
$$0.34x = 760 - 80$$
$$0.34x = 680$$
$$x = 2,000$$
$$3x + 1,000 = 3(2,000) + 1,000 = 7,000$$

Jim invested $2,000 at 10% interest and $7,000 at 8% interest.

Checking   interest on 10% investment = 0.10(2,000) = $200

interest on 8% investment = 0.08(7,000) = $560

total interest = $760   ■■■

Electrical

Kirchhoff's current law states that the sum of all currents at a junction point P in a circuit must equal zero. Figure 1-14 shows a schematic diagram of two lamps (R_1 and R_2) connected in parallel. Using Kirchhoff's current law, the sum of all currents at P equals zero.

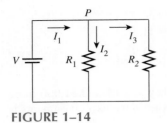

FIGURE 1–14

Currents flowing toward P are considered positive and currents flowing away from P are considered negative. This results in the equation

$$I_1 - I_2 - I_3 = 0$$

EXAMPLE 8   Given the circuit in Figure 1–14, determine the currents I_2 and I_3 if I_2 is 2.5 times I_3 and $I_1 = 0.18$ A.

Solution   For the given circuit

$$I_1 - I_2 - I_3 = 0$$

and $I_1 = 0.18$ A, $I_2 = 2.5\ I_3$.

$$0.18 - 2.5I_3 - I_3 = 0$$
$$-3.5I_3 = -0.18$$
$$I_3 = 0.051 \text{ A}$$
$$I_2 = 2.5I_3 = 0.13 \text{ A}$$ ■■■

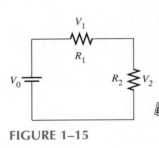

FIGURE 1–15

According to Kirchhoff's voltage law, the sum of the voltage drops equals the voltage supply. Figure 1–15 shows the schematic diagram of two lamps connected in series. Using Kirchhoff's voltage law gives

$$V_1 + V_2 = V_0$$

⊞ **EXAMPLE 9**   Given the circuit in Figure 1–15, determine the voltages V_1 and V_2 if $V_0 = 110$ V and V_2 is 48.3 V more than V_1.

Solution   From Kirchhoff's voltage law

$$V_0 = V_1 + V_2$$

Given that $V_0 = 110$ and $V_2 = 48.3 + V_1$,

$$110 = V_1 + 48.3 + V_1$$
$$110 - 48.3 = 2V_1$$
$$61.7 = 2V_1$$
$$30.9 \text{ V} = V_1$$
$$V_2 = V_1 + 48.3 = 30.9 + 48.3 = 79.2 \text{ V}$$

The two voltages are 30.9 V and 79.2 V.

$\boxed{(}$ 110 $\boxed{-}$ 48.3 $\boxed{)}$ $\boxed{\div}$ 2 $\boxed{=}$ ⟶ 30.85000   ■■■

NOTE ✦ In this section we have discussed solving applied problems that give a linear equation. This concept is difficult to master because each problem is different, but lots of practice does make it easier.

1–5 EXERCISES

1. A farmer wishes to fence in a triangular tract of land. One side is 60 rods, another is 40 rods, and the third side is 25 rods. How much fence is needed?

E 2. The resistance in one resistor is 3 Ω more than that in a second resistor. The total resistance of the two is 16 Ω. Find the resistance in each resistor.

3. The length of a rectangular tract of land is 2 ft more than the width. If it takes 480 ft of wire to enclose the tract, find the length of the land.

4. A 6-ft board is to be cut into two pieces such that one piece is 2 ft longer than the other piece. Find the length of each piece.

M 5. The transmission ratio is the ratio of the engine speed to the drive shaft speed. The engine speed of a car is 4,000 revolutions per minute (r/min), and the drive shaft speed is 1,000 r/min. What is the transmission ratio?

A 6. In constructing a building, the architect must know the bearing capacity of the soil, which is the number of pounds the soil can support. The total weight of a building divided by the bearing capacity of the soil determines the area of footing. What area of footing is required if the bearing capacity is 15,000 lb/ft^2 and the weight of the building is 75,000 tons?

7. The diagonal of a square is 1.414 times the length of a side of the square. Find the length of the diagonal for a square whose sides are 40 in.

M 8. A service station owner wants to be sure his hydraulic lift will support a truck. The pressure on the hydraulic fluid is the force in pounds divided by the area. What pressure is required to support 8,000 lb if the area is 400 in^2?

B 9. A boutique selling clothing on consignment finds that children's clothes have been selling more than usual and should be given more floor space. In the most recent month, children's clothes accounted for 20% of total sales. It is decided that children's clothes will be given the same percentage of the store's 875 sq. ft. How many square feet should be given to children's clothes?

10. The ratio of milk to cream in ice cream is 4 quarts to 3 quarts. How many quarts of milk and cream are used in 5 gallons of ice cream? (Be careful of the units of measure.)

11. It took Bob 2 hours to mow a lawn that is 17,500 ft^2. How long would it take him to mow an area of 30,000 ft^2?

12. Two cars start in Chicago and travel in opposite directions. At the end of 3 hours, they are 330 miles apart. If one car travels 10 mi/h faster than the other, find the speed of the slower car.

13. Sally saves dimes and quarters. She has three times as many quarters as dimes and a total of $12.75. How many dimes and quarters does she have?

B 14. Express the ratio of gross sales to profit if a profit of $750 was made on gross sales of $1,100.

CS 15. If a computer can print 900 lines in 180 seconds, express the ratio of lines per minute.

C 16. Cement, sand, and gravel are mixed in a 1:3:4 proportion to make concrete. How much sand would be used to make 64 yd^3 of concrete?

B 17. John is a buyer for a pharmaceutical company. He purchases 80 pounds of a chemical for $95. At this same rate, how much would 480 pounds of the chemical cost?

18. A 50-ft building casts a shadow 30 ft long. At the same time and place, how long a shadow would a 20-ft tree cast?

CH 19. How many liters of a 20% alcohol solution must be mixed with 40 liters of a 35% solution to get a 28% alcohol solution?

20. It takes 850 gallons of insecticide for a crop duster to spray 20 acres. Find the ratio of application in gallons per acre.

21. A video tape recorder moves 900 ft of tape past the recording head in 2 hours. Express the ratio of the amount of tape per minute.

22. It takes 7 people 12 hours to complete a job. If they worked at the same rate, how many people would it take to complete the job in 16 hours?

23. The sum of three consecutive integers is 75. Find the integers. (Represent the integers as x, $x + 1$, and $x + 2$.)

B 24. Fred's Appliance Store pays each salesperson a salary plus a commission of 5% on total sales for the week. Determine Donna's total pay if her salary is $330 and her sales are $3,880.

25. The Nielsen rating service found that for every 13 people who liked a new television show, 7 people did not like it. You conduct a random sample survey in your school and find that 156 people liked the television show. Based on the Nielsen rating, how many people in your school did not like the show?

26. A bar sells a 24-oz mug of beer for $1.00, while the grocery next door sells a six-pack of 12-oz cans for $3.29. Which is a better buy?

P 27. Two truckers drive the same route from Miami to Atlanta. One trucker travels at 60 mi/h, while the other trucker travels at 65 mi/h. The slower truck takes an hour longer. How long does it take each driver to complete the trip?

P 28. The power of one engine is 3 hp more than that of a second engine. A third engine is 5 hp less than twice the power of the second engine. If the total power of the three engines is 34 hp, find the horsepower of each engine.

A 29. A carpenter needs to cut an 8-ft board into two pieces so that one piece is 6 in. less than twice the length of the other piece. Find the length in feet of each piece.

CH 30. A cylinder contains 50 liters of a 60% chemical solution. How much of this solution should be drained off and replaced with a 40% solution to obtain a final strength of 46%?

E 31. One inductance has a value ⅔ that of a second inductance. If the smaller inductance is increased by 10, the sum is increased to twice the original value. Find the values of the inductances.

32. A TV repairperson charges $40 per hour to repair black-and-white sets and $60 per hour for color sets. If the repairperson earns $620 for working 12 hours, how many hours were spent repairing color sets?

B 33. An investment counselor advises you to invest in two stocks. If one stock returns 10% and the other stock returns 13% on your investment, and you invest $1,500 in each stock for a year, how much do you earn in interest?

34. Oil tank A has a capacity twice that of tank B. Oil tank C has a capacity 20 gallons less than that of tank A. The total capacity of the tanks is 1,080 gallons. Find the capacity of tank C.

35. The perimeter of an isosceles triangle is 30 in. If the third side is 3 in. more than the length of the other sides, find the length of each side.

E 36. A resistor costs one-tenth as much as a transistor. If three resistors and seven transistors cost $2.28, how much does one transistor cost?

P 37. On a trip Jerry drove a steady speed for 3 hours. An accident slowed his speed by 30 mi/h for the last part of the trip. If the 190-mile trip took 4 h, what was his speed during the first part of the trip?

E 38. Given the circuit in Figure 1–14, find I_1 and I_3 if $I_2 = 0.015$ A and I_3 is 0.10 times I_1.

E 39. Given the circuit in Figure 1–14, find I_1, I_2, and I_3 if I_1 is 2.5 A more than I_2 and I_2 is 1.5 A more than twice I_3.

E 40. Given the circuit in Figure 1–15, find V_0 and V_2 if V_1 is 12.3 V and V_0 is 6.30 V more than twice V_2.

E 41. Given the circuit in Figure 1–15, find V_0, V_1, and V_2 if V_0 is 15 V more than V_2, and V_1 is 3.0 V more than half V_2.

E 42. Find the currents in Figure 1–16 if I_1 is 6 A more than I_4, I_2 is twice I_3, and I_4 is the sum of I_2 and I_3.

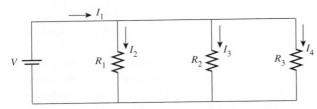

FIGURE 1–16

E 43. Find the voltages V_1, V_2, V_3, and V_0 in Figure 1–17 if V_0 is 5.0 V more than V_3, V_1, is half V_3, and V_2 is the sum of V_1 and V_3.

FIGURE 1–17

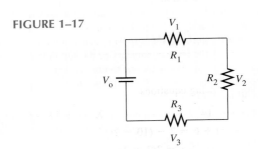

CS 44. A 56 k dial-up modem can download 10 megabytes of data in about 25 minutes. How many megabytes of data is downloaded per minute?

CS 45. A cellular phone company offers new customers 600 free minutes (anytime) for a monthly fee of $49.99. What is the cost per minute for the first 600 minutes?

CS 46. One model of laser printer can print 3.5 pages per minute. How many minutes would it take to print a term paper 22 pages long?

CS 47. Privacy is a key issue in computer technology. Much data is encrypted to ensure privacy. An important concept in encrypting is *keyspace,* which is the span of available keys used. A code system using 40 bits (or characters) provides 1,099,511,627,776 possible combinations. A similar code system using 56 bits gives 72,057,594,037,927,900 combinations. Prove or disprove that the number of bits is proportional to the number of possible combinations.

CHAPTER SUMMARY

Summary of Terms

absolute value (p. 5)

combined variation (p. 26)

compound inequality (p. 20)

constant of variation (p. 23)

cross-product rule (p. 32)

equation (p. 2)

inequality (p. 15)

inversely proportional (p. 24)

joint variation (p. 25)

linear equation (p. 2)

open statement (p. 2)

Principles of Equality (p. 2)

proportion (p. 31)

proportional (p. 23)

ratio (p. 31)

root (p. 2)

solution (p. 2, 15)

subscript (p. 9)

transposition (p. 2)

vary directly (p. 23)

vary inversely (p. 24)

Summary of Formulas

$y = kx$ direct variation

$y = \dfrac{k}{x}$ inverse variation

$y = kxz$ joint variation

$y = \dfrac{kx}{z}$ combined variation

CHAPTER REVIEW

Section 1–1

Solve the following equations.

1. $2x - 8 = 14$

2. $7x + 12 = 9$

3. $4(2x - 3) + 6 = 7 - (10 - 2x)$

4. $2x - 5 + 6x + x = 3(2 + 2x + 1)$

5. $2.3x - 4(x + 1.2) = 6.7 - 4.8x$

6. $7.3(x - 2) + 8.6 = 3 - 6.3x$

7. $\dfrac{2x - 6}{4} = \dfrac{1}{2} + \dfrac{1 - x}{4}$

8. $\dfrac{2x - 6}{5} = \dfrac{4 + 3x}{4}$

9. $3|2x - 1| + 7 = 14$

10. $|3x + 4| = 8$

Section 1–2

Solve for the specified variable.

11. $a = \dfrac{v - v_0}{t}$ for v

12. $V = \dfrac{1}{3}BH$ for B

13. $y - y_0 = m(x - x_0)$ for x

14. $y = mx + b$ for x

15. $I = \dfrac{E}{R + r}$ for E

16. $Q = wc(T_1 - T_2)$ for T_1

17. $Q = \dfrac{I_2Rt}{J}$ for R

18. $A = ab + \dfrac{d}{2}(a + c)$ for c

19. $L = a + (n - 1)d$ for n

20. $S = \dfrac{n}{2}(a + 1)$ for a

P 21. The acceleration of an object equals the ratio of the force applied and the mass of the object. Find the force in newtons when the acceleration is 16 m/s^2 and the mass is 4 kg. [*Note:* newton = (m · kg)/s^2.]

22. Calculate a golfer's handicap if his three-round scores are 70, 85, and 83. [$H = 0.85(A - 72)$ where A is the three-round average.]

A 23. Determine the area of the footing for a building weighing 120,000,000 lb, if the bearing weight of the soil is 15,000 lb/ft^2. (Footing area is the quotient of building weight and soil-bearing capacity.)

Section 1–3

Solve the following inequalities and graph the solution.

24. $4x - 8 + x - 14 > 8x - 1$

25. $-2x + 7 - 5x < 7(x + 2) - 13$

26. $3(2x - 4) \le -2(4 + x - 5)$

27. $2.5(3x + 1.8) \ge 7.9 - (1.6x - 5.6)$

28. $6.1(x - 3.7) - (8.5x - 5) > 7 + 3.2x$

29. $|8x + 6| > 34$

30. $|4x + 6| \le 7$

31. $|5x + 2| \ge 18$

32. $\dfrac{3x}{5} + \dfrac{1}{2} \ge \dfrac{x}{4} - \dfrac{2}{5}$

33. $\dfrac{7x}{4} + \dfrac{7}{8} \le \dfrac{x + 2}{2}$

Section 1–4

34. Write an equation to represent the following statement: a varies jointly with the square of b and the cube of c.

35. Write an equation for the following statement: x varies directly with the square root of y and inversely with the square of z.

36. Write an equation to represent the following statement: The volume of a pyramid varies jointly with the area of the base and the height.

37. a varies inversely with the square root of b. If $a = 3$ when $b = 4$, find a when $b = 16$.

38. m varies jointly with n and the square of p. If $m = 36$ when $n = 3$ and $p = 2$, find n when $m = 240$ and $p = 4$.

39. x varies directly as the cube root of y. If $x = 64$ when $y = 8$, find x when y is 27.

E 40. The voltage across a series circuit varies directly as the current. If the voltage is 30 V when the current is 5 A, find the current when the voltage is 66 V.

B 41. The cost of labor at a factory varies jointly with the number of workers and the number of hours worked. If 8 people work 320 hours to earn $11,648, how many people must work for 460 hours to earn $23,023?

P 42. The volume of a gas varies directly as the absolute temperature in rankine degrees (°R) and inversely with the pressure. If a gas at 639° R has a pressure of 12 lb/in^2 and a volume of 106 in^3, find the volume when the temperature is 600° R and the pressure is 144 lb/in^2.

Section 1–5

P 43. Distance is the product of rate and time. If a car travels 486 miles at a rate of 56 mi/h, how many hours does the trip require?

B 44. A technician earns $11.40 an hour for the first 40 hours and $17.10 for each hour over 40. How many overtime hours must the technician work to earn $609.90 a week?

P 45. Pressure is the quotient of force and area. If the pressure is 60 lb/in^2 and the area is 20 in^2, find the force exerted.

E 46. The sum of the resistance of two resistors is 147 Ω. Find each resistance if one is 3 Ω less than four times the other.

E 47. The total current in a two-component circuit is 54 A. Find the current in each component if one is 6 A less than three times the other.

48. The perimeter of a triangle is 48 cm. If the first side is three cm more than the second side and the second side is twice the third side, find the length of each of the sides.

B 49. The cost of producing one type of computer is $300 more than the cost of producing a second type of computer. The total cost of producing one of the first type and three of the second type is $8,300. Find the cost of producing each type of computer.

50. Peanuts and cashews are combined to make 20 lb of a mixture that will sell for $3.03/lb. If peanuts cost $2.40/lb and cashews cost $4.50/lb, how much of each should be used in the mixture?

B 51. A person pays social security tax, state income tax, and federal income tax on his salary in a 2 : 3 : 5 proportion. If the total deductions of these items is $400, how much is paid in social security tax?

B 52. Maria invests $800 in two stocks, one paying a return of 9.6% and the other paying 4.8%. If her total income from the two stocks is $66.00, how much did she invest in each stock?

53. A computer line printer can print 350 characters in 2 seconds. How many seconds are required to print 1,400 characters?

P 54. Two planes traveling in opposite directions leave the same airport at the same time. If they are 2,550 miles apart at the end of 3 hours and one flies 50 mi/h faster than the other, how fast is the slower plane traveling?

CHAPTER TEST

The number(s) in parentheses refers to the appropriate learning objective(s) given at the beginning of the chapter.

1. Solve $6x - 2(4x + 5) = 3(x - 6) + 2$. (1)

2. Solve the following literal equation for a: (3)

$$s = v + \frac{1}{2}at^2$$

B 3. Bob invests $6,000 in two accounts, one paying 7.3% and the other paying 9.5%. If his total interest is $521.60, how much did he invest in each account? (8)

4. Solve the following inequality and graph the solution: (4)

$$-6x + 12 \leq 2x + 20$$

5. The area of a triangle is ½ the product of the base and height. Find the base of a triangle whose area is 56 ft^2 and whose height is 8 ft. Solve and then substitute. (3, 8)

6. Write an equation to represent the following relationship: d varies directly with the cube of b and inversely with c. (6)

7. Solve $3|2x + 9| - 2 = 18$. (2)

8. Solve the following inequality and graph the solution: (4)

$$\frac{1}{2}x - 3 > \frac{5x}{3}$$

E 9. Resistors connected in series have voltage drops proportional to their resistances. If the voltage drop across a 30-Ω resistor is 75 V, what is the voltage drop across a 10-Ω resistor? (7)

10. The volume of a cone varies jointly with the square of the radius and the height. If the volume is 31.0 when the radius is 2.3 and the height is 5.6, find the volume when the radius is 6.8 and the height is 3.9. (7)

11. Solve the following inequality and graph the solution: (4)

$$|7x - 2| \geq 12$$

12. Express the ratio of $2.25 to $0.75 in lowest terms. (8)

P 13. Fred drives from Atlanta to Orlando in 8 hours. Part of the trip he drives 55 mi/h and the remainder at 60 mi/h. If the distance from Atlanta to Orlando is 450 miles, how long did he drive at each speed? (8)

14. Solve $V = \pi r^2 h$ for h. (3)

15. Solve the following: (1)

$$\frac{3x}{4} - \frac{1}{2} = \frac{2}{3} + \frac{5x}{6}$$

CH 16. Mr. Green mixes antifreeze and water in a ratio of 2:5. How many gallons of antifreeze are used in mixing 21 quarts? (8)

17. Solve the following literal equation for P: $a = V(k - PV)$. (3)

18. Solve $2.9x - (7.2 - 6.8x) = 4 + 3.1x$. (1)

19. Truck A can carry twice as much as truck B, and truck C carries 100 lb less than three times as much as truck A. The total capacity for the three trucks is 8,000 lb. Find the capacity of truck A. (8)

20. Solve the following and graph the solution: $3|2x - 5| < 11$. (5)

GROUP ACTIVITY

Smells So Good

The Smells So Good soap company has hired you to help with the problem of eroding profit margins. You have been asked to come up with a method of decreasing the amount of actual soap used in their premier product. The current product is sized as a simple rectangular solid. It measures 3 in. by 1.5 in. by 1 in. Therefore, the volume is $3 \times 1.5 \times 1 = 4.5$ in^3. They want you to decrease the volume by 10%.

- Discuss methods of decreasing the volume so that the customer does not feel that the bar of soap is getting smaller.
- Discuss decreasing each dimension by 10%. What effect does this have on the total volume of the bar of soap?
- Discuss decreasing only one dimension by 10%. What effect does this have on the total volume?
- Discuss packaging details. What are the alternatives for packaging the new size?
- Are there other approaches that might be used to decrease the volume (for example, changing the volume by placing a slight depression in one side of the bar)?
- Write a report outlining the pros and cons of each approach and your final recommendation. You may want to make a trade matrix, which would show the advantages of each approach. Then assign a value to each choice, which indicates how good or bad the choice would be. An example follows:

Choice	Customer Detects Change; Stops Buying Product	Repackaging Costs	Manufacturing Methods Costs	Etc.	Total
Change length					
Change width					
Change height					
Change all dimensions					
Etc.					

SOLUTION TO CHAPTER INTRODUCTION

Solving the equation $R = \sqrt{R_x^2 + R_y^2 + R_z^2}$ involves substituting known values and calculating the result. Substitution gives

$$R = \sqrt{(-2,100)^2 + (2,300)^2 + (210)^2}$$

The following keystrokes give the result.

$$\boxed{-}\ 2100\ \boxed{x^2}\ \boxed{\times}\ 2300\ \boxed{x^2}\ \boxed{+}\ 210\ \boxed{x^2}\ \boxed{=}\ \boxed{\sqrt{}}\ \longrightarrow\ 3121.5541$$

The cable must be able to withstand a force of 3,122 lb.

Y ou own a small data processing company that processes payroll for local firms. Currently your computer system takes 16 hours to process the monthly payroll. In anticipation of future growth and increased efficiency, you plan to add several new components to your system. Without the current system, the new system can process the payroll in 10 hours. How long would it take the new combined systems to process payroll?

Many technical problems, including the one given above, result in fractional expressions. For example, the measurement of parallax in surveying, the resistivity of metal conductors, and the reactance of a capacitor and inductor connected in series also involve fractional equations.

In this chapter, we will first discuss factoring because it is a tool used in simplifying fractional expressions. Then we will discuss equivalent fractions and operations with algebraic fractions. We will end the chapter with a discussion of solving equations containing algebraic fractions.

Learning Objectives

After completing this chapter, you should be able to

1. Factor a polynomial using the greatest common factor (Section 2–1).

2. Factor a binomial using the difference of two squares (Section 2–1).

3. Factor a trinomial using a perfect square trinomial (Section 2–2).

4. Factor a general trinomial using the *ac* method (Section 2–2).

5. Reduce fractions to lowest terms (Section 2–3).

6. Multiply and divide algebraic fractions (Section 2–4).

7. Add and subtract algebraic fractions (Section 2–5).

8. Simplify a complex fraction (Section 2–5).

9. Solve an equation containing fractions (Section 2–6).

Chapter 2

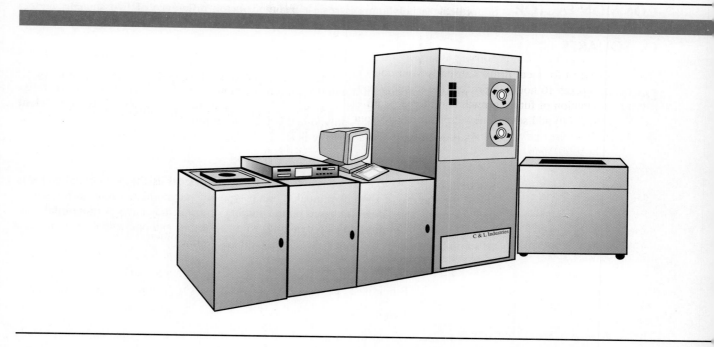

Factors and Fractions

2–1

FACTORING: GREATEST COMMON FACTOR AND DIFFERENCE OF SQUARES

In job situations you will rarely see a factoring problem alone. However, factoring an algebraic expression is used extensively in operations with algebraic fractions and in solving quadratic equations. The importance of factoring results from its usefulness as a tool to do other things. In the first two sections of this chapter, we will discuss four methods of factoring a polynomial: greatest common factor, difference of two squares, perfect square, and the *ac* method.

Definitions

A **term** is the product of a number and one or more variables raised to any powers. Examples of terms include $2x$, $5x^2y$, and $7y^4$. Any set of factors in a term is the **coefficient** of the remaining factors. A polynomial is the sum of terms.

A **polynomial** in one variable is an algebraic expression of the form

$$ax^n + bx^{n-1} + cx^{n-2} + \ldots dx + e$$

where a, b, c, $\ldots$ d, e are constants called the **coefficients** of the polynomial and n is called the **degree** of the polynomial. Certain types of polynomials occur so frequently they are given special names based on the number of terms they have. A **monomial** contains one term. A **binomial** is a polynomial with two terms, and one with three terms is a **trinomial.** Examples of these types of polynomials follow.

Monomials: $2x$, $5x^2y$, $7y^4$

Binomials: $2x + 8$, $5x^2y + 9x$, $7y^4 - 15$

Trinomials: $x^2 + 2x + 8$, $5x^2y + 9x - 3$, $7y^4 - 15 + 9$

The process of writing a polynomial as the product of other "simpler" polynomials is called **factoring.** A polynomial or a factor is called **prime** if its only factors are 1 and itself. A polynomial is said to be **factored completely** if each factor in its product is prime.

Greatest Common Factor

Factoring is the opposite of multiplication, and, specifically, factoring the greatest common factor is opposite to multiplying a monomial and a polynomial using the Distributive Property. A review of this concept is given in Appendix A–8. For example, we can multiply $2a^2(3ab + 5c)$ to obtain the product $6a^3b + 10a^2c$. In factoring a greatest common factor, we reverse this process because $6a^3b + 10a^2c$ is factored to obtain the expression $2a^2(3ab + 5c)$.

EXAMPLE 1   Use the Distributive Property to simplify the following expressions:

(a) $4m(1 + 3n)$ (b) $3x^2(3 - 9x^2 + 4x)$ (c) $9a^2c^3(2b^3c - 4a^2 - 3ab^2)$

Solution   (a) To apply the Distributive Property, distribute $4m$ over each term in the parentheses.

$$4m(1 + 3n)$$

This gives the multiplication and result as follows:

$$4m(1) + 4m(3n)$$
$$4m + 12mn$$

(b) Multiplying $3x^2$ and each term inside the parentheses gives:

$$3x^2(3 - 9x^2 + 4x)$$

$$3x^2(3) - 3x^2(9x^2) + 3x^2(4x)$$

$$9x^2 - 27x^4 + 12x^3$$

(c)

$$9a^2c^3(2b^3c - 4a^2 - 3ab^2)$$

$$9a^2c^3(2b^3c) - 9a^2c^3(4a^2) - 9a^2c^3(3ab^2)$$

$$18a^2b^3c^4 - 36a^4c^3 - 27a^3b^2c^3$$

As you work through Examples 2, 3, and 4, you will see that factoring the greatest common factor is the reverse of applying the Distributive Property. ∎ ∎ ∎

NOTE ✦ The most difficult part of factoring polynomials is recognizing which method of factoring to use. As we discuss each method of factoring, you will see a box outlining how to recognize that particular method of factoring and then how to factor. It may be helpful to write these methods on index cards for reference when doing your homework.

Greatest Common Factor

Recognition

There is no specific method of recognizing when this technique of factoring applies. You must look for a greatest common factor in every polynomial.

Factoring

1. Determine the greatest common factor by finding the largest number that divides into each numerical coefficient and by finding the smallest exponent of any variable or binomial factor that appears in each term.
2. Divide the common factor from Step 1 into the original polynomial, term by term.
3. Write the factored answer as the product of the common factor from Step 1 and the result of the division from Step 2.
4. Check your answer by multiplying the factored answer.

EXAMPLE 2 Factor $4m + 12mn$.

Solution First, we determine the common factor. Since the number 4 divides into both 4 and 12 and the variable m appears in each term, the common factor is $4m$. Note that the variable n is not part of the common factor because it does not appear in both terms. Second, we divide $4m$ into the polynomial $4m + 12mn$. (A review of polynomial division is given in Appendix A–9.) This gives

$$\frac{4m}{4m} + \frac{12mn}{4m} = 1 + 3n$$

Third, the factored answer is the product of the common factor and the result from division. The factored expression is

greatest common factor   result from division

$$4m(1 + 3n)$$

Multiplication should be used to check the factoring. It does not guarantee a prime factorization, but it does check the method of factoring used.

Checking $4m(1 + 3n)$

                 $4m + 12mn$ ■■■

CAUTION ✦ When dividing $4m$ into $4m$, remember that a nonzero quantity divided by itself equals 1. You may forget to place the 1 inside the parentheses in the factored answer.

EXAMPLE 3   Factor $9x^2 - 27x^4 + 12x^3$.

Solution   The numerical coefficients are all divisible by 3. Since x appears in every term, it is a common factor, and the smallest exponent is 2. Thus, the common factor is $3x^2$. Dividing $3x^2$ into each term gives the following result.

$$\frac{9x^2}{3x^2} - \frac{27x^4}{3x^2} + \frac{12x^3}{3x^2}$$

$$3 - 9x^2 + 4x$$

The factored polynomial is

$$3x^2(3 - 9x^2 + 4x)$$ ■■■

EXAMPLE 4   Factor $18a^2b^3c^4 + 36a^4c^3 - 27a^3b^2c^3$.

Solution   Nine divides into each numerical coefficient; the variables a and c appear in each term, and the smallest exponents on a and c are 2 and 3, respectively. Therefore, the greatest common factor is $9a^2c^3$. Dividing $9a^2c^3$ into the polynomial gives

$$\frac{18a^2b^3c^4}{9a^2c^3} + \frac{36a^4c^3}{9a^2c^3} - \frac{27a^3b^2c^3}{9a^2c^3}$$

$2 \cdot 1 \cdot b^3 \cdot c$ $4 \cdot a^2 \cdot 1$ $3 \cdot a \cdot b^2 \cdot 1$

$$2b^3c + 4a^2 - 3ab^2$$

The factored expression is

$$9a^2c^3(2b^3c + 4a^2 - 3ab^2)$$

Checking $9a^2c^3(2b^3c + 4a^2 - 3ab^2)$

                 $18a^2b^3c^4 + 36a^4c^3 - 27a^3b^2c^3$ ■■■

As illustrated in the next example, a binomial may also be a common factor.

EXAMPLE 5 An expression relating volume, pressure, and temperature of a gas is $P_1V_1T_2 - P_2V_2T_1 = 0$. If the pressure and temperature are fixed and temperature is given in kelvins, this expression becomes

$$PV_1(C + 273) - PV_2(C + 273) = 0$$

Factor the left side of this equation.

Solution   The greatest common factor is $P(C + 273)$. Note that P and the binomial $(C + 273)$ appear in each term, and the largest exponent on each is 1. Dividing the polynomial by the common factor gives

$$\frac{PV_1(C + 273)}{P(C + 273)} - \frac{PV_2(C + 273)}{P(C + 273)}$$

$$V_1 - V_2$$

The factored expression is

$$P(C + 273)(V_1 - V_2)$$ ■■■

Difference of Two Squares

An expression such as $x^2 - y^2$ in which one square is subtracted from another is called the **difference of two squares.** Factoring the difference of two squares is the reverse process of multiplying the sum and difference of a binomial: $(x + y)(x - y)$. The form and method of factoring the difference of two squares are given below.

Difference of Two Squares

Recognition

1. The difference of two squares must have exactly two terms (a binomial).

2. Each term must be a perfect square (variables with even-numbered exponents are perfect squares).

3. The terms must be separated by subtraction.

Factoring

The binomial $x^2 - y^2$ can be written and factored as follows:

$$(x)^2 - (y)^2$$
$$(x + y)(x - y)$$

sum   difference

EXAMPLE 6   Factor $4x^2 - y^2$.

Solution   Remember that we must always check for a common factor first. However, this binomial does not have a common factor. Next, we look at the difference of two squares. Since $4 = (2)^2$, $x^2 = (x)^2$, and $y^2 = (y)^2$, each factor is a square. The two terms are also separated by subtraction. This binomial is the difference of two squares.

$$4x^2 - y^2$$
$$(2x)^2 - (y)^2$$
$$(2x + y)\ (2x - y)$$

sum   difference

Checking
$$(2x + y)(2x - y)$$

Using FOIL (refer to Appendix A–8 for a review of FOIL) to multiply gives

$$4x^2 - 2xy + 2xy - y^2$$
$$4x^2 - y^2$$

■■■

EXAMPLE 7   Factor $16x^2 - 9y^2$.

Solution   Since there is no common factor and the two terms $16x^2$ and $9y^2$ are perfect squares and are separated by subtraction, the binomial is the difference of two squares. The binomial can be written and factored as follows:

$$16x^2 - 9y^2$$
$$(4x)^2 - (3y)^2$$
$$(4x + 3y)(4x - 3y)$$

sum   difference

Checking
$$(4x + 3y)(4x - 3y)$$
$$16x^2 - 12xy + 12xy - 9y^2 = 16x^2 - 9y^2$$

■■■

EXAMPLE 8   Factor the following:

(a) $25x^2 + 36y^2$ (b) $x^4 - 81$

Solution   There is no common factor, so we try the difference of squares.
(a) The two terms $25x^2$ and $36y^2$ are perfect squares, but the squared terms are not separated by subtraction. The binomial is prime because there is no common factor.
(b) The binomial $x^4 - 81$ is the difference of two squares. It can be written and factored as follows:

$$x^4 - 81$$
$$(x^2)^2 - (9)^2$$
$$(x^2 + 9)(x^2 - 9)$$

sum   difference

Checking each factor shows that $x^2 - 9$ will factor again as the difference of squares. The factor $x^2 + 9$ does not factor.

$$(x^2 + 9)[(x)^2 - (3)^2]$$
$$(x^2 + 9)(x + 3)(x - 3)$$

The factored expression is

$$(x^2 + 9)(x + 3)(x - 3)$$

■■■

NOTE ✦ To obtain the exponent to be placed inside the parentheses in factoring the difference of two squares, divide the original exponent by 2.

Combined Factoring

Frequently, polynomials must be factored more than once. The next two examples illustrate this process.

EXAMPLE 9   Factor $64x^5 - 16xy^6$.

Solution   The polynomial contains a greatest common factor of $16x$. Factoring the common factor gives

$$16x(4x^4 - y^6)$$

Next we must attempt to factor the binomial $4x^4 - y^6$. The two terms $4x^4$ and y^6 are perfect squares separated by subtraction. Thus, $4x^4 - y^6$ factors as the difference of two squares as follows:

$$(2x^2)^2 - (y^3)^2$$
$$(2x^2 - y^3)(2x^2 + y^3)$$

Since $(2x^2 - y^3)$ and $(2x^2 + y^3)$ will not factor, the factored expression is

$$16x(2x^2 - y^3)(2x^2 + y^3)$$

■■■

CAUTION ✦ When you factor out a common factor and then factor the result again, remember to write the common factor in your answer. For example,

$$8x^3 - 32xy^2 = 8x(x^2 - 4y^2) = 8x(x - 2y)(x + 2y)$$

Also, when some factors cannot be factored again, be sure to include them in the answer. For example,

$$x^4 - 81y^4 = (x^2 - 9y^2)(x^2 + 9y^2) = (x - 3y)(x + 3y)(x^2 + 9y^2)$$

EXAMPLE 10   The difference in the volumes of two cylinders with the same height but different radii is given by

$$\pi r_1^2 h - \pi r_2^2 h$$

Factor this expression.

Solution The polynomial contains a common factor of πh. Factoring the common factor gives

$$\pi h(r_1^2 - r_2^2)$$

The binomial will factor as the difference of two squares. The factored expression is

$$\pi h(r_1 - r_2)(r_1 + r_2)$$

■ ■ ■

2–1 EXERCISES

Factor each of the following polynomials completely. Remember to check for a common factor and then the difference of squares.

1. $3a^2b + 9a^3$

2. $32a^2b - 28a^3b^4$

3. $6a^3bc - 2a^3c^3 + 4a^2bc$

4. $17m^3n^2 - 34m^4n$

5. $49x^2 - 25y^2$

6. $15x^2y^4 + 45x^3y^3 - 20x^2y^3$

7. $30x^5 - 40x^2 + 100x^3$

8. $81s^2 - 49$

9. $4p^2 - 9$

10. $36x^7 - 72x^{10} + 18x^5$

11. $9y^3 + 7y^6 + 10y^4x^2$

12. $9p^2 - 25n^2$

13. $36m^2 - 121$

14. $64x^2 - 9$

15. $32s^2t - 40s^3t^3 + 8st$

16. $27a^3b^2 - 81a^4b - 63a^2b^3$

17. $18x^3y - 2xy^3$

18. $16a^2 - 81b^2$

19. $25x^5 + 30x^2 - 15x^6$

20. $17x^3y^2 + 29x^5y - 46x^4y^3$

21. $25s^2 - 49t^2$

22. $16h^2 - 36k^2$

23. $38m^4n^3 + 57m^3n^4 - 19m^3$

24. $56a^5b^2c^9 + 70b^5c^4 - 35a^6b^8$

25. $75x^3y - 3xy^2$

26. $256f^2 - 49g^2$

27. $64m^8 + 32m^4 + 48m^{15}$

28. $18x^4 - 24x^3 + 14x^2 - 36x^5$

29. $49c^2 - 36d^2$

30. $9m^2 - 36n^2$

31. $63a^7b^5c^{10} - 49a^5c^{10}$

32. $9ax + 9a^2x - 18a^3x^2$

33. $4x^2 - 49y^2$

34. $p^2 - 81q^2$

35. $54x^3y^2z - 63x^4y^5z^3 + 81x^3z$

36. $169s^4 - 64t^4$

37. $a^4 - 36b^4$

38. $9x^4y^4 - 81$

39. $46s^9t^6 + 23s^{12}t^4 + 69s^8t^9$

40. $40m^3n - 12m^{10}n^8 + 28m^8$

41. $2x(x - 3) + 3(x - 3)$

42. $x(2x + 1) - 3(2x + 1)$

43. $3(x - 5) - 5x(x - 5)$

44. $6x(2x - 3) + 1(2x - 3)$

45. The expression for the difference in the surface area of a sphere with radius r_1 and the surface area of a second sphere with radius r_2 is given by $4\pi r_2^2 - 4\pi r_1^2$. Factor this expression.

P 46. If an object is dropped from a building 64 ft high, the expression for the height at time t is $64 - 16t^2$. Factor this expression.

B 47. In economics the break-even point occurs when demand equals supply. If the demand is represented by the expression $1{,}125 - 4p$, and the supply is represented by $125p^2 - 4p$, then the equation for the break-even point becomes $125p^2 - 1{,}125 = 0$. Factor the left side of the equation.

E 48. The change in resistance in a circuit with a temperature change is given by $R_1 + R_1at - R_1at_1$. Factor this expression.

M 49. The momentum M of a mass m that has a change in velocity from v_1 to v_2 is given by the equation $M = mv_2 - mv_1$. Factor the right side of the equation.

P 50. The displacement s of an object is given by the equation

$$s = v_0t + \frac{1}{2}at^2$$

Factor the right side of the equation.

51. The difference in the volume of two cylinders with the same radius and heights h_1 and h_2 is given by $\pi r^2h_1 - \pi r^2h_2$. Factor this expression.

52. The surface area of a rectangular solid is given by $2hw + 2hl + 2wl$. Factor this expression.

B 53. The amount of money in a savings account at a given time is given by $P + Prt$ where P is principal, r is simple interest rate, and t is time. Factor this expression.

M 54. The coefficient of linear expansion, α, is the ratio of change in length per degree Celsius to the length at $0°C$. The new length is given by $L_0 + L_0\alpha\Delta t$. Factor this expression.

2–2

FACTORING TRINOMIALS

The number of terms in a polynomial is the key to determining which methods of factoring to consider. After removing any common factors, try to factor a trinomial (a polynomial with three terms) as a perfect square, and then try the *ac* method. Each of these methods of factoring trinomials is discussed in this section. Since factoring a perfect square is easier, it is discussed first.

Perfect Square Trinomial

The product of two identical binomials is called a **perfect square trinomial.**

$$(x + y)(x + y) = x^2 + 2xy + y^2$$
$$(x - y)(x - y) = x^2 - 2xy + y^2$$

Since recognition is the key to factoring a perfect square trinomial, the following information summarizes how to recognize and factor one.

Perfect Square Trinomial

Recognition

1. A perfect square trinomial must have exactly three terms.

2. The polynomial can be arranged so that the first and last terms are perfect squares (with positive signs for both).

3. The magnitude (neglect the sign) of the middle term must be the product of 2 and the square root of each squared term.

Factoring

To factor a perfect square trinomial, fill in the following formula:

$$\left(\sqrt{\text{first term}} \pm \sqrt{\text{third term}} \right)^2$$

Choose + or − depending on the middle sign of the polynomial.

EXAMPLE 1   Factor $x^2 + 4x + 4$.

Solution   Remember that we must always check for a common factor, but this trinomial does not have a common factor. Second, we check for a perfect square because the polynomial has three terms. Note that we do not even test for difference of two squares because the polynomial is not a binomial. Perfect square trinomial test:

1. Exactly three terms appear.

$$x^2 + 4x + 4$$

2. First and last terms are squares.

$$x^2 = (x)^2 \qquad 4 = (2)^2$$

3. The magnitude of the middle term is the product of 2 and the square root of the squared terms.

$$2 \cdot x \cdot 2$$
$$4x$$

Factoring gives

$$\left(\sqrt{\text{first term}} + \sqrt{\text{third term}}\right)^2$$
$$(x + 2)^2$$

■■■

EXAMPLE 2 Factor $9x^2 - 12x + 4$.

Solution First, we check for a common factor and then check the criteria for a perfect square trinomial:

1. The polynomial contains three terms: $9x^2$, $12x$, and 4.
2. $9x^2$ and 4 are perfect squares.
3. The magnitude of the middle term is

$$2\sqrt{9x^2}\sqrt{4} = 2(3x)(2) = 12x$$

Therefore, this polynomial meets the requirements for a perfect square trinomial. Next, to factor the polynomial, fill in the formula as follows:

$$\left(\sqrt{\text{first term}} - \sqrt{\text{third term}}\right)^2$$
$$\left(\sqrt{9x^2} - \sqrt{4}\right)^2$$
$$(3x - 2)^2$$

Checking

$$(3x - 2)(3x - 2)$$
$$9x^2 - 6x - 6x + 4$$
$$9x^2 - 12x + 4$$

The factored expression is $(3x - 2)^2$.

■■■

EXAMPLE 3 Factor $25x^2 + 30xy + 9y^2$.

Solution This trinomial does not contain a common factor. Next we check the criteria for a perfect square.

$$25x^2 + 30xy + 9y^2$$
$$(5x)^2 + 2(5x)(3y) + (3y)^2$$

The trinomial is a perfect square. Factoring gives the following result.

$$25x^2 + 30xy + 9y^2$$

$$(5x)^2 + 2(5x)(3y) + (3y)^2$$

$$(5x + 3y)^2$$

■■■

EXAMPLE 4   Factor $81x^2 - 72xy - 16y^2$.

Solution   We check to see if this polynomial fits the requirements of a perfect square trinomial:

1. The polynomial contains three terms: $81x^2$, $72xy$, and $16y^2$ (ignoring signs).
2. The first term, $81x^2$, is a square, but the third term, $-16y^2$, is not a square because a square must be positive.

Therefore, this trinomial is not a perfect square trinomial and does not factor using any of the other methods of factoring.   ■ ■ ■

The *ac* Method

If a trinomial does not have a greatest common factor and is not a perfect square trinomial, then we attempt to factor it using the *ac* method. If we have a trinomial of the form $ax^2 + bx + c$, from reversing the FOIL process we know that the sum of the product of factors a and c must combine to yield b. Figure 2–1 illustrates this point.

FIGURE 2–1

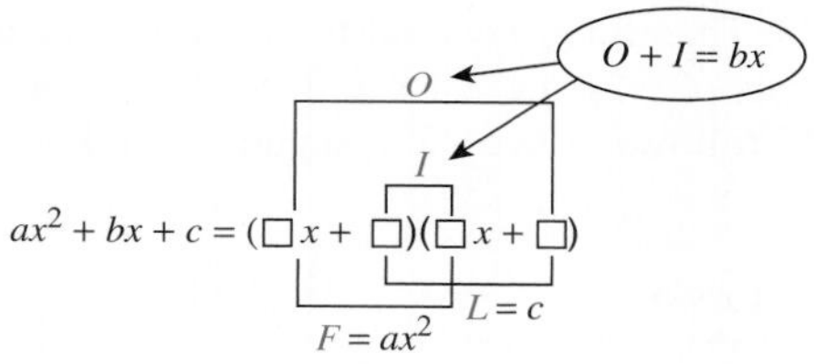

$$ax^2 + bx + c = (\square x + \square)(\square x + \square)$$

The goal of the *ac* method is to find the combination of factors of a and c so that the outer and inner products combine to give bx. The *ac* method is illustrated in the following examples. Remember, this method of factoring is trial and error.

EXAMPLE 5   Factor $x^2 + 6x + 8$.

Solution   Since this polynomial does not contain a common factor and does not fit the criteria for a perfect square trinomial, we try factoring by the *ac* method. By examining the polynomial, we see that $a = 1$, $c = 8$, and $ac = 8$, and factors of 8 must combine to yield b, which is 6. The factors of 8 are $\pm 8 \cdot \pm 1$ and $\pm 2 \cdot \pm 4$. A chart will help determine which of these factors work. Since c is positive, the factors must have like signs.

Factors of 8	+8, +1	−8, −1	+2, +4	−2, −4
Sum	+9	−9	+6	−6

Since b is $+6$, we see from the chart that the factors are $+2$ and $+4$. Then, we can write the polynomial as follows.

$$x^2 + 6x + 8$$
$$x^2 + 2x + 4x + 8$$

Then grouping the terms and factoring a common factor from each group gives the following result.

$$(x^2 + 2x) + (4x + 8)$$

$$\textcircled{x}(x + 2) \textcircled{+} \textcircled{4}(x + 2)$$

$$(x + 4)(x + 2)$$

Checking

$$(x + 4)(x + 2)$$
$$x^2 + 2x + 4x + 8$$
$$x^2 + 6x + 8$$

∎∎∎

EXAMPLE 6   Factor $x^2 + 3x - 28$.

Solution   Since the trinomial does not meet the requirements for a perfect square trinomial, we use the *ac* method of factoring. In this polynomial, $a = 1$, $c = -28$, and $ac = -28$. Therefore, we want to determine factors of -28 that combine to yield b, or 3. The factors of -28 are $\pm 1 \cdot \pm 28$, $\pm 4 \cdot \pm 7$, and $\pm 2 \cdot \pm 14$. The chart for possible factors of -28 follows. Since 28 is negative, we know the factors have unlike signs.

Factors of -28	$+1, -28$	$-1, +28$	$+4, -7$	$-4, +7$	$+2, -14$	$-2, +14$
Sum	-27	27	-3	3	-12	12

From the chart, we see that the factors that combine to give $b = +3$ are -4 and $+7$. We write the polynomial as

$$x^2 + 7x - 4x - 28$$

Then we group the terms and factor.

$$(x^2 + 7x) - (4x + 28)$$
$$x(x + 7) - 4(x + 7)$$
$$(x + 7)(x - 4)$$

Checking

$$(x - 4)(x + 7)$$
$$x^2 + 3x - 28$$

∎∎∎

EXAMPLE 7   Factor $3x^2 + 11x - 4$.

Solution   We use the *ac* method to factor this trinomial also. In this polynomial, $a = 3$, $c = -4$, and $ac = -12$. We must determine factors of -12 that combine to yield 11. The factors of -12 are $\pm 1 \cdot \pm 12$, $\pm 3 \cdot \pm 4$, and $\pm 2 \cdot \pm 6$. The possible factors of -12 are given in the chart below.

Factors of -12	$12, -1$	$-12, 1$	$3, -4$	$-3, 4$	$2, -6$	$-2, 6$
Sum	11	-11	-1	1	-4	4

The factors of -12 that combine to give 11 are 12 and -1. Thus, we can write the polynomial as

$$3x^2 + 11x - 4$$
$$3x^2 + 12x - x - 4$$

Factoring gives

$$3x(x + 4) - 1(x + 4)$$
$$(x + 4)(3x - 1)$$

Checking
$$(3x - 1)(x + 4)$$
$$3x^2 + 11x - 4$$

■■■

EXAMPLE 8   Factor $8x^2 - 14xy + 3y^2$.

Solution   Since $a = 8$, $c = 3$, and $ac = 24$, we must determine factors of 24 that combine to give -14. The factors that yield 24 are $\pm 1 \cdot \pm 24$, $\pm 2 \cdot \pm 12$, $\pm 3 \cdot \pm 8$, and $\pm 4 \cdot \pm 6$. The factors that combine to give -14 are -12 and -2. The polynomial can be written as

$$8x^2 - 12xy - 2xy + 3y^2$$

Factoring gives

$$4x(2x - 3y) - y(2x - 3y)$$
$$(4x - y)(2x - 3y)$$

Checking
$$(2x - 3y)(4x - y)$$
$$8x^2 - 14xy + 3y^2$$

■■■

Combined Factoring

Figure 2–2 will help you determine the correct method of factoring to use.

EXAMPLE 9   Factor $6x^2 + 33x - 18$.

Solution   This polynomial contains a common factor of 3.

$$3(2x^2 + 11x - 6)$$

Next, we try to factor $2x^2 + 11x - 6$ using the *ac* method. Since $a = 2$, $c = -6$, $ac = -12$, and $b = 11$, the factors of -12 that sum to 11 are $+12 \cdot -1$. Then we can write the polynomial and factor as follows:

$$2x^2 + 11x - 6$$
$$2x^2 + 12x - 1x - 6$$
$$\boxed{2x}\,(x + 6) \ominus \text{①} \,(x + 6)$$

$$(2x - 1)(x + 6)$$

The factored answer is $3(2x - 1)(x + 6)$.

■■■

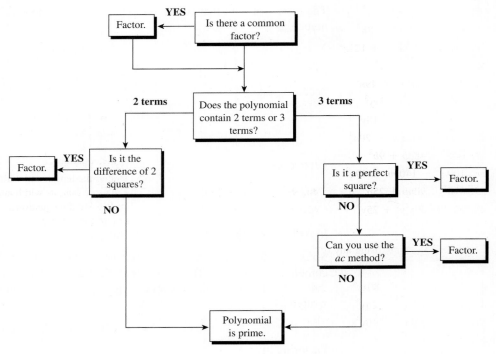

FIGURE 2–2

CAUTION ✦ Remember to include the common factor in the factored answer.

EXAMPLE 10   An object is thrown upward with an initial velocity of 74 ft/s. To find the time it takes to reach a height of 40 ft, we must solve the equation $16t^2 - 74t + 40 = 0$. Factor the left side of this equation.

Solution   First, we factor a greatest common factor of 2.

$$2(8t^2 - 37t + 20) = 0$$

Then factoring the trinomial inside the parentheses using the *ac* method gives the factored expression

$$2(8t - 5)(t - 4) = 0$$ ■■■

2–2 EXERCISES

Factor each of the following polynomials completely. Use the flow chart given in Figure 2–2 to help determine the correct method of factoring to use.

1. $x^2 - 5x + 6$

2. $a^2 + 5a - 14$

3. $y^2 - y - 20$

4. $x^2 + 11x + 18$

5. $x^2 + 4x - 32$

6. $y^2 + 9y + 18$

7. $a^2 + 8ab - 33b^2$

8. $b^2 + 8b + 16$

9. $25x^2 - 40x + 16$

10. $m^2 + 7mn + 12n^2$

11. $x^2 + xy - 30y^2$

12. $a^2 + 6ab - 16b^2$

13. $9x^2 - 12xy + 4y^2$

14. $81x^2 - 126xy + 49y^2$

15. $25x^2 - 30xy + 9y^2$

16. $c^2 + 8cd + 12d^2$

17. $m^2 - 2mn - 24n^2$

18. $m^2 + 21mn + 11n^2$

19. $4n^2 - 10n - 6$

20. $5p^2 - 9p - 2$

21. $6 + y - 12y^2$

22. $12x^2 + 5xy - 28y^2$

23. $144a^2 + 168a + 49$

24. $64p^2 + 80np + 25n^2$

25. $35k^2 + 9kg - 18g^2$

26. $40m^2 + 61mn + 18n^2$

27. $64n^2 - 112np + 49p^2$

28. $49x^2y^2 - 154xyz + 121z^2$

29. $5x^4 + x^2y^2 - 18y^4$

30. $25x^2 + 90xy - 36y^2$

31. $4m^4 + 28m^2n^2 + 49n^4$

32. $16s^4 - 40s^2t^2 + 25t^4$

33. $6x^3y + 3x^2y^2 - 3xy^3$

34. $2x^4 - 16x^2y^2 + 30y^4$

35. $21x^4 - 13x^2y^2 - 18y^4$

36. $27m^4 - 33m^2n^2 - 20n^4$

37. $64a^2 - 48ab + 9b^2$

38. $100m^2 - 140mn + 49n^2$

39. $81n^2 + 90np + 25p^2$

40. $9a^2 + 30ab + 25b^2$

41. $36x^4 - 60x^2y^2 + 25y^4$

42. $9s^4t^4 - 24s^2t^2 + 16$

Factor the following, using any of the methods discussed thus far.

43. $8x^3y^2 - 64x^4y - 80xy^4$

44. $36x^2 + 108xy + 81y^2$

45. $49x^2y + 70xy^2 + 25y^3$

46. $16x^3y - 4xy$

47. $17xyz + 34x^2z + 16y^3z$

48. $16x^2 + 64xy + 64y^2$

49. $9x^2 - 54xz + 81z^2$

50. $98x^3y - 112x^2y^2 + 32xy^3$

51. $16x^2z - 144xyz + 324y^2z$

52. $81x^4 - 6xy^4$

53. $5x^2 + 30xy + 45y^2$

54. $4x^2 - 12xy + 9y^2$

55. $12x^2 - 6xy - 90y^2$

56. $15x^2y^2 - 10x^2y + 20x^2$

57. $144x^2 - 49y^2$

58. $100m^4 - 180m^2 + 81$

59. To find the dimensions of a rectangle whose length is 2 more than twice the width and whose area is 60, you must factor the expression $2x^2 + 2x - 60$. Factor this expression.

P 60. If an object is projected upward from the ground a distance of 128 ft using an initial velocity of 144 ft/s, the expression for time of rise, t, becomes $128 - 144t + 16t^2 = 0$. Factor the left side of the equation.

P 61. The position of a projectile moving in a straight line is given by $s = 2t^2 - 5t$. If the displacement, s, is 12 m, the equation becomes $2t^2 - 5t - 12 = 0$. Factor the left side of the equation.

62. To find the length and width of a rectangle whose length is 7 m more than its width and whose area is 120 m^2, you must solve the equation $x^2 + 7x - 120 = 0$. Factor the left side of the equation.

63. A piece of cardboard 8 m square is to be made into a box by cutting a square from each corner and folding. If the area of the bottom of the box is to be 16 m^2, you must solve the equation $4x^2 - 32x + 48 = 0$. Factor the left side of the equation.

E 64. The resistance of a conductor varies with the temperature. Under certain conditions the resistance is given by $30 + 236t - 32t^2$. Factor this expression.

CS 65. A computer monitor is a rectangle with an area of 168 in^2. The length is 2 inches more than the width. An equation representing this information is $x^2 - 2x - 168 = 0$. Factor the left side of this equation.

CS 66. An equation that may be used to estimate the time needed to boot up a computer is $t^2 + 4t - 96 = 0$. Factor the left side of this equation.

2–3

EQUIVALENT FRACTIONS

In the remainder of this chapter, we will discuss operations with algebraic fractions as we continue to add to our basic algebra skills. Many technical formulas and problems require simplifying expressions containing algebraic fractions. The process of factoring is essential to operations with algebraic fractions, and you must understand it thoroughly.

Definitions

If a and b ($b \neq 0$) represent algebraic expressions, then a/b represents an **algebraic fraction.** A fraction is said to be in **lowest terms** or **reduced** if the numerator and denominator do not contain common factors. Two fractions are said to be **equivalent** if they can both be reduced to the same fraction. We can produce an equivalent fraction by multiplying or dividing the numerator and denominator of the fraction by the same quantity.

Equivalent Fractions

EXAMPLE 1 Change $\dfrac{4xy}{7z}$ to an equivalent fraction with a denominator of $21z^2$.

Solution To change the denominator of $7z$ to $21z^2$, we must multiply by $3z$. To produce an equivalent fraction, we must multiply the numerator and denominator by $3z$ as follows:

$$\frac{4xy}{7z} \cdot \frac{3z}{3z} = \frac{12xyz}{21z^2}$$

■ ■ ■

EXAMPLE 2 Change $\dfrac{5}{6x - 10}$ to an equivalent fraction with a denominator of $6x^2 - 4x - 10$.

Solution   We must determine what factor to multiply the denominator $6x - 10$ by to yield $6x^2 - 4x - 10$. This process is more difficult than in the previous example. Let us factor each denominator.

$$
\begin{array}{cc}
6x - 10 & 6x^2 - 4x - 10 \\
2(3x - 5) & 2(3x - 5)(x + 1)
\end{array}
$$

From the analysis of the denominators, we can see that we must multiply the numerator and denominator of the fraction by $x + 1$. The equivalent fraction is

$$
\frac{5}{6x - 10} \cdot \frac{x + 1}{x + 1} = \frac{5x + 5}{6x^2 - 4x - 10}
$$ ■■■

In producing an equivalent fraction in the two previous examples, we multiplied the numerator and denominator by the same quantity. In reducing a fraction to lowest terms, we divide the numerator and denominator by the same quantity.

Reducing Fractions

Reducing algebraic fractions to lowest terms uses the same procedure as that used in arithmetic. For example, if you were asked to reduce 27/36 to lowest terms, you would find the largest factor contained in both 27 and 36 and divide it out.

$$
\frac{27}{36} = \frac{9 \cdot 3}{9 \cdot 4} = \frac{3}{4}
$$

More commonly, we say that we divide away factors common to the numerator and denominator.

NOTE ✦ Refer to Appendix A–4 to review the laws of exponents used to divide out common factors in monomial expressions.

EXAMPLE 3   Reduce the following to lowest terms:

$$
\frac{16x^4 y^2 z^8}{32 y^5 z^6}
$$

Solution   We must divide out the greatest common factor (GCF) contained in both the numerator and denominator.

$$
\frac{\cancel{16}\, x^4\, y^2\, \cancel{z^8}^{z^2}}{\underset{2}{\cancel{32}}\, \underset{y^3}{\cancel{y^5}}\, \cancel{z^6}}
$$

The reduced fraction is

$$
\frac{x^4 z^2}{2 y^3}
$$ ■■■

The concept presented in Example 3 also applies to reducing polynomial fractions. We divide out the greatest common factor contained in both the numerator and denominator. However, you must factor polynomials before canceling. Study the next examples carefully.

EXAMPLE 4   Reduce the following fraction to lowest terms:

$$\frac{8x^2 - 10x + 3}{6x^2 - x - 1}$$

Solution   To reduce polynomial fractions, factor the numerator and denominator, cancel or divide out the greatest common factors, and write the product of the factors that remain. Factoring the numerator and denominator of this fraction gives

$$\frac{(4x - 3)(2x - 1)}{(2x - 1)(3x + 1)}$$

Canceling the greatest common factor contained in both the numerator and the denominator, $(2x - 1)$, gives

$$\frac{(4x - 3)(2x - 1)}{(2x - 1)(3x + 1)}$$

The reduced fraction is

$$\frac{4x - 3}{3x + 1}$$

■ ■ ■

EXAMPLE 5   Reduce the following fraction to lowest terms:

$$\frac{6x^2 + 16x + 10}{8x^2 - 12x - 20}$$

Solution   We begin by completely factoring the numerator and denominator of the fraction.

$$\frac{2(3x^2 + 8x + 5)}{4(2x^2 - 3x - 5)}$$

$$\frac{2(3x + 5)(x + 1)}{4(2x - 5)(x + 1)}$$

Canceling the greatest common factor gives

$$\frac{2(3x + 5)(x + 1)}{\underset{2}{4}(2x - 5)(x + 1)}$$

The reduced fraction is

$$\frac{3x + 5}{2(2x - 5)}$$

■ ■ ■

CAUTION ✦ Be sure to cancel only common *factors,* not terms, when reducing a fraction to lowest terms.

In reducing fractions to lowest terms, you may find some factors that differ only in their signs, such as $x - y$ and $y - x$. In this instance, it is helpful to remember that $-x = (-1)x$. Therefore, the expression $y - x$ can be written as $(-1)(x - y)$. Reducing the fraction $(x - y)/(y - x)$ using this technique gives

$$\frac{x - y}{y - x} = \frac{\cancel{(x - y)}}{-1\cancel{(x - y)}} = -1$$

Factors that differ only in their signs in all terms cancel and yield -1. These factors are considered opposites like a and $-a$. This concept is illustrated in the next two examples.

EXAMPLE 6 Reduce the following to lowest terms:

$$\frac{16 - x^2}{x - 4}$$

Solution First, we factor the numerator of the fraction completely.

$$\frac{(4 - x)(4 + x)}{(x - 4)}$$

Then we cancel the greatest common factor. Notice that $(4 - x)$ and $(x - 4)$ differ only in signs. From the previous discussion, we can write the factors as

$$\frac{-1(x - 4)(4 + x)}{(x - 4)}$$

Canceling common factors gives the reduced fraction as

$$-(4 + x) \qquad \qquad \blacksquare\blacksquare\blacksquare$$

EXAMPLE 7 Reduce the following fraction to lowest terms:

$$\frac{x^2 - 2x - 15}{25 - x^2}$$

Solution To reduce the fraction to lowest terms, factor the numerator and denominator, cancel the greatest common factor, and write the product of the factors that remain. Performing these steps gives

$$\frac{(x + 3)(x - 5)}{(5 - x)(5 + x)}$$

Notice that $(x - 5)$ and $(5 - x)$ differ only in signs. The cancellation of these factors gives -1. After the greatest common factor is canceled, the reduced fraction is

$$-\frac{x + 3}{5 + x} \qquad \qquad \blacksquare\blacksquare\blacksquare$$

CAUTION ✦ The factors $x - a$ and $x + a$ are not opposites and do not cancel, but $x - a$ and $a - x$ or $x + a$ and $-x - a$ are opposites and divide to produce -1.

2–3 EXERCISES

Change each fraction to an equivalent fraction with the indicated denominator.

1. $\dfrac{x}{5y}$ to $\dfrac{?}{15y^2z}$

2. $\dfrac{2x}{7}$ to $\dfrac{?}{14y}$

3. $\dfrac{xy}{3z}$ to $\dfrac{?}{12z^3}$

4. $\dfrac{5ab}{9c}$ to $\dfrac{?}{45c^3}$

5. $\dfrac{(a+b)}{2ab}$ to $\dfrac{?}{8a^2b}$

6. $\dfrac{(3x-y)}{mn}$ to $\dfrac{?}{3m^2n^3}$

7. $\dfrac{6m}{2(m-3n)}$ to $\dfrac{?}{6m(m-3n)}$

8. $\dfrac{9xy}{4(x-y)}$ to $\dfrac{?}{8x^2(x-y)}$

9. $\dfrac{x-4}{2x-5}$ to $\dfrac{?}{6xy-15y}$

10. $\dfrac{3x}{4x-7}$ to $\dfrac{?}{12x^3-21x^2}$

11. $\dfrac{3x}{x-5y}$ to $\dfrac{?}{2x^2-9xy-5y^2}$

12. $\dfrac{2x-1}{2x+1}$ to $\dfrac{?}{4x^2-1}$

Reduce each fraction to lowest terms. Some fractions may be written in lowest terms.

13. $\dfrac{9x^4y}{27y^4}$

14. $\dfrac{12m^4n^2}{20m^7}$

15. $\dfrac{54a^4bc^8}{18b^5}$

16. $\dfrac{40s^{10}t^9}{75s^2t^{11}}$

17. $\dfrac{8x^3y^2z^5}{72xy^2z^8}$

18. $\dfrac{62k^5p^8m^{10}}{3t^3m^{12}}$

19. $\dfrac{24mn^8p^3}{30m^6n^5p}$

20. $\dfrac{17a^{17}b^3}{51a^{10}b^3}$

21. $\dfrac{84a^3cd^5}{12ac^8d^2}$

22. $\dfrac{48s^{13}t^3w}{64s^{10}t^8}$

23. $\dfrac{6ab(2x-y)}{3a^2(2x-y)}$

24. $\dfrac{5s^3t(3x-2)}{10t(3x-2)}$

25. $\dfrac{14x^2(x-1)}{16x}$

26. $\dfrac{36m^4n^2(2m-n)}{8m^6n}$

27. $\dfrac{(2x-y)(3-y)}{(y-3)(x+2y)}$

28. $\dfrac{(3x-1)(x+7)}{(x-7)(1-3x)}$

29. $\dfrac{(x-5)(x+5)}{(x-5)(5-x)}$

30. $\dfrac{10x^3(2x-3y)}{-5x^4(3y-2x)}$

31. $\dfrac{14a^3b(a-2b)}{7b^3(a-2b)}$

32. $\dfrac{6x^2-7x+2}{8-2x-3x^2}$

33. $\dfrac{4x^2-8x}{8x-16}$

34. $\dfrac{25x^2-36y^2}{15x^2+43xy+30y^2}$

35. $\dfrac{2a^2b^2-6a^2b}{6ab^5-18ab^4}$

36. $\dfrac{x^2+4xy+4y^2}{4y^2-x^2}$

37. $\dfrac{7x^2+41xy-6y^2}{2x^2+15xy+18y^2}$

38. $\dfrac{7x^2+20xy-3y^2}{7x^2+6xy-y^2}$

2–4

MULTIPLICATION AND DIVISION OF FRACTIONS

The rules for multiplying algebraic fractions are the same as those used in arithmetic:

Multiplying Fractions

multiply numerators

$$\frac{a}{b} \cdot \frac{c}{d} = \frac{ac}{bd}$$

multiply denominators

However, the resulting fraction may need to be reduced to lowest terms. Thus, it is better to divide out any common factors prior to multiplying. In summary, when you multiply fractions in arithmetic, you divide out common factors and then multiply the remaining factors. For example, to multiply

$$\frac{16}{27} \cdot \frac{30}{48}$$

we remove common factors

$$\frac{\overset{1}{\cancel{16}}}{\underset{9}{\cancel{27}}} \cdot \frac{\overset{10}{\cancel{30}}}{\underset{3}{\cancel{48}}}$$

and multiply the remaining factors to obtain

$$\frac{10}{27}$$

We perform these same steps in multiplying algebraic fractions. A summary of this process follows.

Multiplying Algebraic Fractions

---

1. Factor each polynomial.
2. Cancel the greatest common factor contained in both the numerator and denominator.
3. Write the fraction as the product of the resulting factors.

EXAMPLE 1   Multiply the following:

$$\frac{8a^3b^4}{54a^2c^4} \cdot \frac{16c^2}{20ab^5}$$

Solution   To multiply these fractions, we remove common factors.

$$\frac{\overset{2}{\cancel{8}}a^3b^4}{\underset{27}{\cancel{54}}a^2c^4} \cdot \frac{\overset{8}{\cancel{16}}c^2}{\underset{5}{\cancel{20}}ab^5} \qquad \text{reducing numerical coefficients}$$

$$\frac{2\overset{a}{\cancel{a^3}}\,b^4}{27\cancel{a^2}c^4} \cdot \frac{8c^2}{5\cancel{a}b^5} \qquad \text{reducing the } a\text{'s}$$
$$\underset{1}{}$$

$$\frac{2b^4}{27c^4} \cdot \frac{8c^2}{5\underset{b}{\cancel{b^5}}} \qquad \text{reducing the } b\text{'s}$$

$$\frac{2}{27\underset{c^2}{\cancel{c^4}}} \cdot \frac{8\cancel{c^2}}{5b} \qquad \text{reducing the } c\text{'s}$$

$$\frac{2}{27c^2} \cdot \frac{8}{5b}$$

Multiplying the remaining numerator and denominator gives the product of the fraction as

$$\frac{16}{135c^2b} \qquad \blacksquare\,\blacksquare\,\blacksquare$$

CAUTION ✦ If you divide out the greatest common factor, the resulting fraction should be reduced to lowest terms; however, you should always check the answer just in case you have overlooked a common factor prior to multiplying.

EXAMPLE 2   Multiply the following fractions:

$$\frac{6x^2y}{4x^2 - 4x} \cdot \frac{5x - 5}{15x^3}$$

Solution   To multiply these fractions, we begin by factoring the binomials.

$$\frac{6x^2y}{4x(x - 1)} \cdot \frac{5(x - 1)}{15x^3}$$

Then cancel common factors.

$$\frac{\overset{2}{\cancel{6}}x^{\overset{x}{\cancel{2}}}y}{\underset{2}{\cancel{4}}x\cancel{(x - 1)}} \cdot \frac{5\cancel{(x - 1)}}{\underset{3}{\cancel{15}}x^{\underset{x^2}{\cancel{3}}}}$$

The resulting product is

$$\frac{y}{2x^2}$$ ■ ■ ■

EXAMPLE 3   Multiply the following:

$$\frac{4x^2 - 25y^2}{3x^2 + 11xy + 10y^2} \cdot \frac{3x^2 + 2xy - 5y^2}{25y - 10x}$$

Solution   First, we factor each polynomial.

$$\frac{(2x - 5y)(2x + 5y)}{(3x + 5y)(x + 2y)} \cdot \frac{(3x + 5y)(x - y)}{5(5y - 2x)}$$

Then we cancel the greatest common factor.

$$\frac{\overset{-1}{\cancel{(2x - 5y)}}(2x + 5y)}{\cancel{(3x + 5y)}(x + 2y)} \cdot \frac{\cancel{(3x + 5y)}(x - y)}{5\cancel{(5y - 2x)}}$$

The product is

$$\frac{-(2x + 5y)(x - y)}{5(x + 2y)}$$ ■ ■ ■

NOTE ✦ Although it is permissible to multiply the products in the answer, it is preferable to leave the fraction in factored form.

⌨ **EXAMPLE 4**   In converting the units on the rate of flow of a liquid from 150 in^3/s to ft^3/min, it is necessary to multiply in^3/s by the following conversion factors:

$$\frac{1\ \text{ft}^3}{1{,}728\ \text{in}^3} \qquad \text{and} \qquad \frac{60\ \text{s}}{1\ \text{min}}$$

Perform this conversion.

Solution   To convert 150 in³/s to ft³/min, we multiply the three fractions, canceling units just as we have canceled algebraic factors.

$$\frac{150 \ \cancel{in^3}}{s} \cdot \frac{1 \ ft^3}{1,728 \ \cancel{in^3}} \cdot \frac{60 \ \cancel{s}}{1 \ min}$$

Canceling units and performing the arithmetic with the coefficients gives

$$5.2 \ ft^3/min$$

$$150 \ \boxed{\times} \ 60 \ \boxed{\div} \ 1728 \ \boxed{=} \ \longrightarrow \ 5.2083 \qquad \blacksquare\blacksquare\blacksquare$$

Dividing Fractions

Recall from arithmetic that you divide fractions by inverting the divisor and multiplying the resulting fractions.

$$\frac{a}{b} \div \overset{\text{invert divisor}}{\frac{c}{d}} = \frac{a}{b} \cdot \frac{d}{c} = \frac{ad}{bc}$$

Applying this procedure to an arithmetic example gives

$$\frac{2}{3} \div \overset{\text{invert divisor}}{\frac{5}{9}} = \frac{2}{3} \cdot \frac{9}{5} = \frac{2}{\underset{1}{\cancel{3}}} \cdot \frac{\overset{3}{\cancel{9}}}{5} = \frac{6}{5}$$

This same process applies to dividing algebraic fractions and is summarized below.

Dividing Fractions

---

1. Invert the divisor.
2. Multiply the resulting fractions by following the procedure given for multiplying fractions.

EXAMPLE 5   Divide the following fractions:

$$\frac{9m^2 - 18m^3}{16m^2n^5} \div \frac{3 - 6m}{18m^3}$$

Solution   To divide the fractions, invert the divisor and multiply.

$$\frac{9m^2 - 18m^3}{16m^2n^5} \cdot \frac{18m^3}{3 - 6m}$$

To multiply the fractions, factor each polynomial, cancel the greatest common factors, and write the product of the factors that remain.

$$\frac{\overset{3}{\cancel{9}}m^2(1-2m)}{\underset{8}{\cancel{16}}m^2n^5} \cdot \frac{\overset{9}{\cancel{18}}m^3}{3(1-2m)}$$

The quotient of the fractions is

$$\frac{27m^3}{8n^5}$$

■ ■ ■

CAUTION ✦ To avoid errors in dividing fractions, always invert the divisor first and then factor and cancel as necessary. Do *not* factor or cancel first.

EXAMPLE 6 The length of an arc of a circle is given by $\dfrac{2n\pi r}{360}$, while the area of a sector is given by $\dfrac{n\pi r^2}{360}$, where n is the measure of the central angle in degrees and r is the radius. Express the ratio of the length of the arc to the area of the sector.

Solution The term *ratio* means division. Setting up the ratio gives

$$\begin{array}{ll} \dfrac{2n\pi r}{360} & \text{length of arc} \\[2ex] \dfrac{n\pi r^2}{360} & \text{area of sector} \end{array}$$

Converting to the use of a division symbol instead of the fraction bar gives

$$\frac{2n\pi r}{360} \div \frac{n\pi r^2}{360}$$

Dividing gives

$$\frac{2n\pi r}{360} \cdot \frac{360}{n\pi r^2}$$

Canceling common factors gives the ratio

$$\frac{2}{r}$$

■ ■ ■

EXAMPLE 7 The approximate annual interest rate r on a monthly installment loan is given by

$$r = \frac{\left[\dfrac{24(NM - P)}{N}\right]}{\left(P + \dfrac{NM}{12}\right)}$$

where N = total number of payments, M = monthly payment, and P = amount financed. Approximate the annual interest rate for a four-year car loan of $28,000 that has monthly payments of $640.

Solution   First, simplify the fraction by finding the common denominator for the fraction in the denominator.

$$r = \frac{\left[\dfrac{24(NM - P)}{N}\right]}{\left(\dfrac{12P + NM}{12}\right)}$$

Second, we divide the fractions by inverting the denominator and multiplying.

$$r = \frac{24(NM - P)}{N} \cdot \frac{12}{12P + NM}$$

Next, we substitute $N = 4$ years $= 48$ payments, $M = \$640$, and $P = \$28,000$.

$$r = \frac{24[48(640) - 28,000]}{48} \cdot \frac{12}{12(28,000) + (48)(640)}$$

$$r = \frac{24(2,720)}{48} \cdot \frac{12}{336,000 + 30,720}$$

$$r = 1,360(0.000033)$$

$$r = 0.044880$$

The interest rate is approximately 4.5%

■ ■ ■

2–4 EXERCISES

Perform the indicated operations.

1. $\dfrac{4x^2y^3}{2x^3} \cdot \dfrac{16x^4}{8y^3}$

2. $\dfrac{35a^3b}{6b^7} \cdot \dfrac{22b^4}{55a^5b}$

3. $\dfrac{5xy^3}{7x^3y} \cdot \dfrac{6x^2y}{30x^4y^3}$

4. $\dfrac{14m^4n^2}{42mn^5} \cdot \dfrac{m^2}{7m^5}$

5. $\dfrac{4x(2x + 3)}{(x - 1)} \cdot \dfrac{(x - 1)}{8x^2}$

6. $\dfrac{13ab^3}{2a^4(3 - 2x)} \cdot \dfrac{(2x - 3)}{26a^4}$

7. $\dfrac{(7x - 1)}{x(x + 2)} \cdot \dfrac{5x^2(x + 2)}{(7x + 3)}$

8. $\dfrac{(3x + 4)(x - 1)}{(1 - x)(2x + 3)} \cdot \dfrac{(2x + 3)}{3x}$

9. $\dfrac{8x^2y}{3y} \div \dfrac{24x}{27xy}$

10. $\dfrac{18m^4n^2}{4mn} \div \dfrac{6m^3n}{12mn}$

11. $\dfrac{2ab^2c}{9a} \div \dfrac{10b^3}{3a^4c^3}$

12. $\dfrac{48a(2 - b)}{27a^3} \div \dfrac{16(b - 2)}{9a^4}$

13. $\dfrac{8m^2(a + 2b)}{32m} \div \dfrac{16(a - 2b)}{64m^2}$

14. $\dfrac{(x - 3)(2x - 7)}{3x} \div \dfrac{(2x - 7)}{(x - 3)}$

15. $\dfrac{14x^2(x + y)}{3x^3(2x - y)} \div \dfrac{28(x + y)}{18(y - 2x)}$

16. $\dfrac{12a^3(a + 2b)}{(a + 2b)} \div \dfrac{6ab^3}{2b}$

17. $\dfrac{6x^2 + x - 2}{16x^3} \cdot \dfrac{4x^2 + 12x}{2x^2 + 5x - 3}$

18. $\dfrac{8m^2 - 4m}{20m^3} \cdot \dfrac{2m + 1}{4m^2 - 1}$

19. $\dfrac{24x^2 + 2x - 2}{16x^2 - 1} \cdot \dfrac{4x^2 - 19x - 5}{x - 5}$

20. $\dfrac{3a^2 + 5ab - 2b^2}{5a^2 + 6ab - 8b^2} \cdot \dfrac{5a^2 + ab - 4b^2}{2a^2 + 7ab + 5b^2}$

21. $\dfrac{a^2 - b^2}{a^2 + b^2} \cdot \dfrac{2a^2 + ab - b^2}{a^2 + 2ab + b^2}$

22. $\dfrac{7s^2 + 8st - 12t^2}{21s^2 - 25st + 6t^2} \cdot \dfrac{3s - t}{4s^2 + 17st + 18t^2}$

23. $\dfrac{7m^2 + 19m - 6}{2m^2 + m - 15} \cdot \dfrac{3m^2 + 5m - 8}{14m^2 + 17m - 6}$

24. $\dfrac{12x^2 + 23xy - 2y^2}{6x^2 + 13xy - 5y^2} \cdot \dfrac{4x^2 + 20xy + 25y^2}{x^2 + 3xy + 2y^2}$

25. $\dfrac{72s^2 - 42st + 5t^2}{18s^2 - 9st + t^2} \cdot \dfrac{9s^2 - t^2}{36s^2 - 3st - 5t^2}$

26. $\dfrac{10m^2 - 17mn + 3n^2}{m + 5n} \cdot \dfrac{7m^2 + 36mn + 5n^2}{14m^2 - 19mn - 3n^2}$

27. $\dfrac{14a^2 + 5ab - b^2}{2a + b} \div \dfrac{3a^2 - 11ab - 4b^2}{15a^2 + 2ab - b^2}$

28. $\dfrac{4s^2 - 11st - 3t^2}{8s^2 + 14st + 3t^2} \div \dfrac{5s^2 - 6st + t^2}{2s^2 + st - 3t^2}$

29. $\dfrac{9x^2 - 10xy + y^2}{4x^2 - 3xy - y^2} \div \dfrac{18x^2 + 7xy - y^2}{6x^2 + 13xy + 5y^2}$

30. $\dfrac{2a^2 + 11ab + 5b^2}{2a^2 - 3ab + b^2} \div \dfrac{3a^2 - 4ab - 15b^2}{2a^2 - 7ab + 3b^2}$

31. $\dfrac{x^2 - 9y^2}{x^2 + 4xy + 4y^2} \div \dfrac{x^2 + 2xy - 3y^2}{2x^2 - 5xy - 3y^2}$

32. $\dfrac{6x^2 - 7xy + y^2}{6x - 6y} \div \dfrac{36x^2 - y^2}{12x^2 + 8xy + y^2}$

33. $\dfrac{5s^2 + 11st + 6t^2}{3s^2 - 4st + t^2} \div \dfrac{s^2 - t^2}{6s^2 - 23st + 7t^2}$

34. $\dfrac{7m^2 - 21mn}{2m^2 - 5mn - 3n^2} \div \dfrac{28m + 14n}{6m^2 + 5mn + n^2}$

35. $\dfrac{9m^2 + 12mn + 4n^2}{3m^2 + 10mn - 8n^2} \div \dfrac{3m^2 - mn - 2n^2}{m^2 + 3mn - 4n^2}$

36. $\dfrac{6s^2t - 3st^2}{2s^2 + st - t^2} \div \dfrac{3s + 4t}{s^2 - t^2}$

37. $\dfrac{x^2 - 16}{x^2 + 6x + 9} \div \dfrac{4 - x}{x^2 - 9}$

38. $\dfrac{(5 - x)(x + 1)}{(x - 1)} \cdot \dfrac{10x^2 - 10}{x^2 - 25}$

P 39. The centripetal force F of an object traveling in a circular path is its mass m times its centripetal acceleration a. If $m = w/g$ and $a = v^2/r$, find the expression for F.

40. Find the ratio of the area of a circle (πr^2) and the circumference of a circle $(2\pi r)$.

M 41. Average power is defined to be the time rate of doing work. If the formula for power is solved for time, the units in the metric system become

$$\dfrac{N \cdot m}{N\left(\dfrac{m}{s}\right)}$$

Simplify this expression to obtain the correct units for time.

42. Convert 250 mi/h to ft/s by multiplying 5,280 ft/1 mi and 1 h/3,600 s.

43. Convert 30 yd³/s to ft³/min by multiplying by 27 ft³/1 yd³ and 60 s/1 min.

B 44. Using Example 7, find the annual interest rate for a five-year car loan of $18,000 with monthly payments of $315.

B 45. Using Example 7, find the annual interest rate for a four-year car loan of $12,000 with monthly payments of $291.

2–5

ADDITION AND SUBTRACTION OF FRACTIONS

So far, each operation with algebraic fractions has followed the same procedure as the respective operation with arithmetic fractions. The operations of addition and subtraction of algebraic fractions are no exception. To add and subtract both algebraic fractions and arithmetic fractions, we must find a common denominator. Before actually adding and subtracting fractions, however, we will examine how to find the **lowest common denominator (LCD),** which is the smallest quantity that is divisible by each denominator.

To find the lowest common denominator in arithmetic, first break each denominator into prime factors. Then form the common denominator by finding the product of the primes to the highest exponent in which the primes occur in any given denominator. For example, to find a common denominator for

$$\dfrac{7}{42} + \dfrac{5}{18}$$

we break the denominators into prime factors as $42 = 7 \cdot 3 \cdot 2$ and $18 = 3^2 \cdot 2$. The common denominator consists of each factor raised to its largest power. Thus, $7 \cdot 3^2 \cdot 2 = 126$ is the lowest common denominator for 42 and 18. As in the past, the process of finding a lowest common denominator for an algebraic fraction is the same as that used in arithmetic.

EXAMPLE 1 Find the lowest common denominator for the following fractions:

$$\frac{8}{15x^2y}; \frac{15}{40xy^4}; \frac{18}{x^3y^2}$$

Solution First, we factor each denominator.

$$15x^2y = \boxed{5} \cdot \boxed{3} \qquad \cdot x^2 \quad \cdot y$$
$$40xy^4 = 5 \cdot \quad \boxed{2^3} \cdot x \cdot \boxed{y^4}$$
$$x^3y^2 = \qquad \boxed{x^3} \cdot y^2$$

Choose the largest
exponent for each factor: $5 \cdot 3 \cdot 2^3 \cdot x^3 \cdot y^4$

The lowest common denominator is the product of these factors, or $120x^3y^4$.   ∎∎∎

EXAMPLE 2 Find the lowest common denominator for the following fractions:

$$\frac{6x}{2x - 4}; \frac{2x - 3}{x^2 - 4x + 4} \text{ and } \frac{7}{x^2 - 4}$$

Solution First, we factor each denominator.

$$2x - 4 = \boxed{2} \cdot (x - 2)$$
$$x^2 - 4x + 4 = \boxed{(x - 2)^2}$$
$$x^2 - 4 = (x - 2) \cdot \boxed{(x + 2)}$$

Choose the largest exponent
for each factor: $2 \cdot (x - 2)^2 \cdot (x + 2)$

The lowest common denominator is $2(x - 2)^2(x + 2)$.   ∎∎∎

To begin the discussion of adding and subtracting fractions, let us look once again at the arithmetic example. To add

$$\frac{7}{42} + \frac{5}{18}$$

we first find the common denominator of 126, as done previously. Then we change each fraction to an equivalent fraction with 126 as the denominator (refer to Section 2–3 on equivalent fractions). Thus,

$$\frac{7}{42} = \frac{7 \cdot 3}{42 \cdot 3} = \frac{21}{126}$$

$$\frac{5}{18} = \frac{5 \cdot 7}{18 \cdot 7} = \frac{35}{126}$$

Next, we combine the fractions with the common denominator.

$$\frac{21}{126} + \frac{35}{126} = \frac{21 + 35}{126} = \frac{56}{126}$$

The final step is to reduce the fraction to lowest terms, if necessary.

$$\frac{56}{126} = \frac{28}{63} = \frac{4}{9}$$

The sum of the two fractions is $\frac{4}{9}$. We follow this same procedure in adding and subtracting algebraic fractions.

Addition or Subtraction of Algebraic Fractions

1. Find the lowest common denominator.
2. Change each fraction to an equivalent fraction with the common denominator as the denominator.
3. Simplify the numerator by removing parentheses and combining like terms.
4. Reduce the fraction to lowest terms, if possible.

EXAMPLE 3 Perform the indicated operation:

$$\frac{2x}{3x - 1} - \frac{6}{2x + 5}$$

Solution Since each denominator is prime, the lowest common denominator is $(3x - 1)(2x + 5)$. Change each fraction to an equivalent fraction with $(3x - 1)(2x + 5)$ as the denominator.

$$\frac{2x}{3x - 1} \cdot \frac{2x + 5}{2x + 5} = \frac{2x(2x + 5)}{(3x - 1)(2x + 5)}$$

$$\frac{6}{2x + 5} \cdot \frac{3x - 1}{3x - 1} = \frac{6(3x - 1)}{(3x - 1)(2x + 5)}$$

Next, we combine the two fractions over the common denominator.

$$\frac{2x(2x + 5) - 6(3x - 1)}{(3x - 1)(2x + 5)}$$

Then we remove parentheses and combine like terms.

$$\frac{4x^2 + 10x - 18x + 6}{(3x - 1)(2x + 5)}$$

$$\frac{4x^2 - 8x + 6}{(3x - 1)(2x + 5)}$$

Last, we try to reduce the fraction to lowest terms.

$$\frac{2(2x^2 - 4x + 3)}{(3x - 1)(2x + 5)}$$

Since factoring the numerator reveals no common factors, the fraction is reduced and either form is acceptable. ■ ■ ■

EXAMPLE 4 Subtract the following fractions:

$$\frac{x - 3}{x^2 + 7x + 10} - \frac{2x + 1}{2x^2 + x - 6}$$

Solution First, we *determine the common denominator.*

$$x^2 + 7x + 10 = (x + 5)(x + 2)$$
$$2x^2 + x - 6 = \quad (x + 2)(2x - 3)$$

The lowest common denominator is $(x + 5)(x + 2)(2x - 3)$. Next, we *change each fraction to an equivalent fraction* with the lowest common denominator as the denominator.

$$\frac{x - 3}{(x + 5)(x + 2)} \cdot \frac{2x - 3}{2x - 3} = \frac{(x - 3)(2x - 3)}{(x + 2)(x + 5)(2x - 3)}$$

$$\frac{2x + 1}{(2x - 3)(x + 2)} \cdot \frac{x + 5}{x + 5} = \frac{(2x + 1)(x + 5)}{(x + 2)(x + 5)(2x - 3)}$$

Then we place the numerator of the fraction over the common denominator.

$$\frac{(x - 3)(2x - 3) - (2x + 1)(x + 5)}{(x + 2)(x + 5)(2x - 3)}$$

We *simplify the numerator* by removing parentheses and combining like terms.

$$\frac{2x^2 - 3x - 6x + 9 - 2x^2 - 10x - x - 5}{(x + 2)(x + 5)(2x - 3)}$$

$$\frac{-20x + 4}{(x + 2)(x + 5)(2x - 3)}$$

Last, we try to *reduce the fraction to lowest terms* by factoring the numerator.

$$\frac{-4(5x - 1)}{(x + 2)(x + 5)(2x - 3)}$$

Since factoring the numerator does not reveal a common factor, either of the last two fractions is an acceptable answer. ∎

EXAMPLE 5 The equivalent capacitance C_e of three capacitors in series is given by

$$\frac{1}{C_e} = \frac{1}{C_1} + \frac{1}{C_2} + \frac{1}{C_3}$$

Solve for C_e by combining the fractions on the right side of the equation and inverting both sides of the equation.

Solution   To add the fractions on the right side of the equation, we must obtain a common denominator, $C_1C_2C_3$, and write each fraction as an equivalent fraction with the common denominator.

$$\frac{1}{C_e} = \frac{C_2C_3 + C_1C_3 + C_1C_2}{C_1C_2C_3}$$

Inverting each side of the equation gives

$$C_e = \frac{C_1C_2C_3}{C_2C_3 + C_1C_3 + C_1C_2}$$ ∎

Complex Fractions

As you will see in the next chapter, removing negative exponents in an algebraic expression can lead to a **complex fraction,** which is a fraction whose numerator and/or denominator contain one or more fractions. Simplifying a complex fraction can involve addition, subtraction, multiplication, and division of fractions.

Simplifying a Complex Fraction

1. Simplify the numerator and denominator of the complex fraction by adding or subtracting the fractions, as necessary.
2. Divide the fractions by inverting the divisor and applying the rules for multiplication.

EXAMPLE 6   Simplify the following complex fraction:

$$\frac{\dfrac{x+2}{x} - \dfrac{3}{2}}{\dfrac{7}{x} + \dfrac{5}{4x^2}}$$

Solution   First, we must *obtain a single fraction* in the numerator and denominator by using addition or subtraction.

$$\frac{\dfrac{2(x+2) - 3x}{2x}}{\dfrac{7(4x) + 5(1)}{4x^2}} = \frac{\dfrac{-x+4}{2x}}{\dfrac{28x+5}{4x^2}}$$

Then we *divide the complex fraction* by inverting the divisor and multiplying.

$$\frac{-x+4}{\cancel{2x}} \cdot \frac{\overset{2x}{\cancel{4x^2}}}{28x+5}$$

$$\frac{2x(-x+4)}{28x+5}$$

∎

EXAMPLE 7   Simplify the following complex fraction:

$$\frac{\dfrac{2x+1}{x-3} + \dfrac{5}{x+3}}{\dfrac{6x-8}{x^2-9}}$$

Solution   First, we *add the fractions in the numerator.*

$$\frac{\dfrac{(2x+1)(x+3) + 5(x-3)}{(x-3)(x+3)}}{\dfrac{6x-8}{x^2-9}}$$

$$\frac{2x^2 + 7x + 3 + 5x - 15}{(x - 3)(x + 3)}$$

$$\frac{6x - 8}{x^2 - 9}$$

$$\frac{2x^2 + 12x - 12}{(x - 3)(x + 3)}$$

$$\frac{6x - 8}{x^2 - 9}$$

Last, we *divide the fraction* by inverting the divisor and multiplying.

$$\frac{2x^2 + 12x - 12}{(x - 3)(x + 3)} \cdot \frac{x^2 - 9}{6x - 8}$$

$$\frac{2(x^2 + 6x - 6)}{(x - 3)(x + 3)} \cdot \frac{(x + 3)(x - 3)}{2(3x - 4)}$$

$$\frac{x^2 + 6x - 6}{3x - 4}$$

■ ■ ■

2–5 EXERCISES

Add or subtract the following fractions.

1. $\dfrac{2x}{8x} + \dfrac{3}{6x^2} - \dfrac{2x^2}{4x^3}$

2. $\dfrac{4}{7y} + \dfrac{8}{2y^3} + \dfrac{6x}{14}$

3. $\dfrac{5x - 1}{2} + \dfrac{6x}{3x} - \dfrac{1}{4x^2}$

4. $\dfrac{3x}{4x^4} - \dfrac{7x + 4}{6x} - \dfrac{2}{3x^3}$

5. $\dfrac{2m + 3}{6m^2} - \dfrac{5m + 1}{5m^3}$

6. $\dfrac{4x - 9}{7x^4} - \dfrac{3x - 1}{2x}$

7. $\dfrac{4x - 1}{x + 1} - \dfrac{6x}{x^2}$

8. $\dfrac{8}{3n^3} + \dfrac{7n}{2n - 1}$

9. $\dfrac{4y - 3}{7y + 2} - \dfrac{6y}{3y^2}$

10. $\dfrac{x}{3x + 5} - \dfrac{5x - 6}{4x}$

11. $\dfrac{2x}{2x^2 - x} + \dfrac{x + 1}{3x}$

12. $\dfrac{3x - 4}{4x - 8} + \dfrac{6}{8x}$

13. $\dfrac{a + 1}{3a - 6} - \dfrac{5a}{a - 2}$

14. $\dfrac{7m}{7m - 14} + \dfrac{4}{21m^3}$

15. $\dfrac{4x + 1}{5x^2 - 15x} + \dfrac{8x}{10x}$

16. $\dfrac{5x - 2}{6x^2 - 18x} - \dfrac{3x}{3x^2}$

17. $\dfrac{3y}{2y - 1} + \dfrac{4y}{3y + 2}$

18. $\dfrac{8x}{4x + 5} - \dfrac{7x}{6x - 1}$

19. $\dfrac{9}{3x + 6} - \dfrac{4}{2x - 5}$

20. $\dfrac{3}{4c^2 - c} + \dfrac{c}{5c + 7}$

21. $\dfrac{3z + 1}{7z^2 - 14z} + \dfrac{z - 1}{2z + 3}$

22. $\dfrac{2x}{5x - 1} + \dfrac{3x - 1}{2x + 7}$

23. $\dfrac{2x}{8x + 7} + \dfrac{x - 1}{3x - 8}$

24. $\dfrac{d - 7}{4d + 1} - \dfrac{d^2 - 6}{7d + 3}$

25. $\dfrac{b - 3}{2b + 1} - \dfrac{2b + 9}{5b + 3}$

26. $\dfrac{x + 3}{4x + 5} + \dfrac{2x - 4}{5 - 3x}$

27. $\dfrac{9x - 2}{7x - 4} + \dfrac{3x + 6}{4x + 3}$

28. $\dfrac{x^2}{2x + 11} - \dfrac{3x + 1}{2x + 7}$

29. $\dfrac{4n}{n - 4} + \dfrac{2}{4 - n}$

30. $\dfrac{7x}{6x - 1} - \dfrac{x^2}{1 - 6x}$

31. $\dfrac{2k - 1}{3k^2 + 5k - 2} + \dfrac{5k}{6k^2 + k - 1}$

32. $\dfrac{y + 3}{y^2 - 16} - \dfrac{6y}{2y^2 - 11y + 12}$

33. $\dfrac{10x + 7}{12x^2 - 40x + 12} - \dfrac{5x + 2}{6x^2 + 16x - 6}$

34. $\dfrac{x^2 + 5x}{x^2 - x - 12} + \dfrac{4 + 6x}{3x^2 - 7x - 20}$

35. $\dfrac{2y - 4}{9 - y^2} + \dfrac{6y + 5}{y - 3}$

36. $\dfrac{7b + 3}{25 - b^2} - \dfrac{b - 9}{b - 5}$

37. $\dfrac{2x}{x^2 - 3x - 4} - \dfrac{x}{3x^2 + 5x + 2}$

38. $\dfrac{4c}{c^2 - 25} + \dfrac{3}{c^2 - 10c + 25}$

39. $\dfrac{2x + 1}{x^2 - x - 6} + \dfrac{3x}{x^2 + 3x - 4}$

40. $\dfrac{3x - 4}{x^2 + 5x + 6} + \dfrac{2x + 5}{x^2 + 4x + 3}$

Simplify the following complex fractions.

41. $\dfrac{\dfrac{5}{x} - \dfrac{3x}{y}}{\dfrac{6}{x^2} + 5}$

42. $\dfrac{\dfrac{7x + 1}{x^2} + \dfrac{4}{x}}{\dfrac{22x + 2}{5x}}$

43. $\dfrac{\dfrac{4x}{x + 1} - \dfrac{2}{x - 1}}{\dfrac{3}{x - 1} + \dfrac{6x}{x + 1}}$

44. $\dfrac{\dfrac{7}{x} - \dfrac{4x}{x - 1}}{\dfrac{3}{x + 4} + \dfrac{2x}{x}}$

45. $\dfrac{\dfrac{7}{2x + 1} - \dfrac{3x}{x - 1}}{\dfrac{4x + 5}{2x^2 - x - 1}}$

46. $\dfrac{\dfrac{8x}{5x^2} + \dfrac{4}{x^2 - x}}{\dfrac{4x + 3}{10x^3 - 10x^2}}$

47. $\dfrac{\dfrac{6}{x - 4} + \dfrac{2x}{2x + 3}}{\dfrac{2x - 5}{3} - \dfrac{4}{x}}$

48. $\dfrac{\dfrac{3x - 2}{x^2 - 1} + \dfrac{6x}{x + 1}}{\dfrac{3x - 9}{x^2 - 2x - 3}}$

49. $\dfrac{\dfrac{4a}{2a - b} + \dfrac{6b}{2a + b}}{\dfrac{4b}{6a + 3b} - \dfrac{3}{a}}$

50. $\dfrac{\dfrac{7x - 1}{2x - 4} + \dfrac{3x}{4x - 8}}{\dfrac{3x}{x^2 - 4} + \dfrac{8}{x + 2}}$

E 51. The reciprocal of the total resistance in a parallel circuit is the sum of the reciprocals of the individual resistances. If the individual resistances are represented by $8x$, $16x^3$, and $8x^2 - 32x$, find a simplified expression for the reciprocal of the total resistance.

M 52. A cord wrapped around the rim of a wheel of radius r has a mass m hanging on the end. (See Figure 2–3). The acceleration of the mass is given by

$$a = \dfrac{9.8m}{m + \dfrac{I}{r^2}}$$

where I is the moment of inertia for the wheel. Simplify this expression.

B 53. The approximate annual interest rate r of a monthly installment loan is given by

$$r = \dfrac{\left[\dfrac{24(NM - P)}{N}\right]}{\left(P + \dfrac{NM}{12}\right)}$$

where N = total number of payments, M = monthly payment, and P = the amount financed. Simplify the expression for the annual interest rate r and approximate the annual interest rate for a 5-year car loan of \$21,600 with monthly payments of \$450.

E 54. The reactance X of an inductor and capacitor in series is given by

$$X = \omega L - \dfrac{1}{\omega C}$$

Find a simplified expression for the reactance.

P 55. The distance s that a ball falls, neglecting friction, is given by

$$s = vt + \dfrac{at^2}{2}$$

Simplify the expression on the right side of the equation.

P 56. An expression that results from Einstein's theory of relativity relating the velocity of two objects approaching each other is

$$\dfrac{v_1 - v_2}{1 - \dfrac{v_1 v_2}{c^2}}$$

where c is the speed of light. Simplify this expression.

P 57. A car travels a distance d_1 at a rate v_1; then it travels another distance d_2 at another rate v_2. The average speed for the entire trip is given by

$$\dfrac{d_1 + d_2}{\dfrac{d_1}{v_1} + \dfrac{d_2}{v_2}}$$

Simplify this expression.

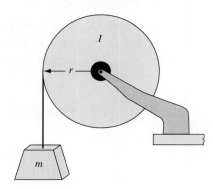

FIGURE 2–3

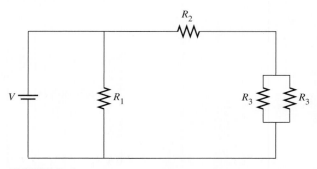

FIGURE 2–4

M 58. An equation in fluids is $P = Dhg$ where P is the pressure, D is the density of the liquid, h is the height of the column of liquid, and g is the gravitational acceleration. If this equation is solved for h, we have $h = P/Dg$. Substituting in the proper units gives

$$\frac{\dfrac{kg \cdot m/s^2}{m^2}}{\dfrac{kg}{m^3} \cdot \dfrac{m}{s^2}}$$

Simplify this expression to get the correct units for the height of the column of liquid.

E 59. The total resistance through the circuit shown in Figure 2–4 is given by:

$$R_t = \frac{R_1\left(R_2 + \dfrac{R_3}{2}\right)}{R_1 + \left(R_2 + \dfrac{R_3}{2}\right)}$$

Simplify this complex fraction.

2–6

FRACTIONAL EQUATIONS AND APPLICATIONS

Thus far in this chapter, we have discussed operations with algebraic fractions. The technique learned in this chapter as well as in Chapter 1 allows us to add another type of equation to the list that we can solve: equations with algebraic fractions. We follow the same procedure to solve these equations as we used in Chapter 1. These steps are repeated below.

Solving Equations with Fractions

1. Eliminate fractions by multiplying each term of the equation by the lowest common denominator (LCD). Find the LCD as outlined in Section 2–5.
2. Eliminate parentheses.
3. Isolate the variable using the Transposition Property.
4. Divide both sides of the equation by the coefficient of the variable.
5. Check your answer by substituting into the original equation.

EXAMPLE 1 Solve the following equation:

$$\frac{x}{2} + \frac{3x}{5} = 4$$

Solution First, we *eliminate the fractions* by multiplying by the lowest common denominator, 10.

$$⑩\left(\frac{x}{2}\right) + ⑩\left(\frac{3x}{5}\right) = ⑩\,4$$

$$5x + 6x = 40$$

$$11x = 40$$

Since the variable is isolated, we *divide both sides of the equation* by 11.

$$\frac{11x}{11} = \frac{40}{11}$$

$$x = \frac{40}{11}$$

■ ■ ■

EXAMPLE 2 Solve the following equation:

$$\frac{4}{x} - \frac{7}{x^2} = \frac{5}{x}$$

Solution First, we *eliminate the fractions* by multiplying each term by the LCD, x^2.

$$\left(x^2\right)\left(\frac{4}{x}\right) - \left(x^2\right)\left(\frac{7}{x^2}\right) = \left(x^2\right)\left(\frac{5}{x}\right)$$

$$4x - 7 = 5x$$

Then we *isolate the variable* by transposing $4x$.

$$-7 = 5x - 4x$$

$$-7 = x$$

Checking

$$\frac{4}{-7} - \frac{7}{(-7)^2} = \frac{5}{-7}$$

$$-\frac{4}{7} - \frac{1}{7} = \frac{-5}{7}$$

Since checking results in a true statement, the solution is $x = -7$. ■■■

EXAMPLE 3 Solve the following equation:

$$\frac{7.8}{x} + 3.6 = \frac{1.5}{x}$$

Solution We begin by removing fractions and multiply the equation by x.

$$x\left(\frac{7.8}{x}\right) + x(3.6) = x\left(\frac{1.5}{x}\right)$$

$$7.8 + 3.6x = 1.5$$

$$x \approx -1.8$$

$$1.5 \boxed{-} 7.8 \boxed{=} \boxed{\div} 3.6 \boxed{=} \longrightarrow -1.75$$ ■■■

Extraneous Roots

In solving equations that contain fractions with the variable in the denominator, always check for extraneous solutions that result in division by zero.

EXAMPLE 4 Solve the following:

$$\frac{1}{x - 3} - \frac{1}{x + 3} = \frac{2x}{x^2 - 9}$$

Solution First, we eliminate the fractions by multiplying each term by the lowest common denominator, $(x - 3)(x + 3)$.

$$\frac{(x - 3)(x + 3)(1)}{(x - 3)} - \frac{(x + 3)(x - 3)(1)}{(x + 3)} = \frac{2x(x + 3)(x - 3)}{(x + 3)(x - 3)}$$

$$1(x + 3) - 1(x - 3) = 2x$$

Then we remove parentheses and solve for x.

$$x + 3 - x + 3 = 2x \quad \text{remove parentheses}$$
$$6 = 2x \quad \text{combine like terms}$$
$$3 = x \quad \text{divide}$$

Checking
$$\frac{1}{3 - 3} - \frac{1}{3 + 3} = \frac{6}{9 - 9}$$

Since $x = 3$ leads to division by zero, $x = 3$ is an extraneous solution, and this equation has no solution. ■■■

Solving Formulas

As noted in Chapter 1, formulas must often be rearranged and solved for a different variable. The next two examples illustrate solving a formula containing fractions.

EXAMPLE 5 In business, straight-line depreciation of equipment is given by the equation

$$d = \frac{C - C_L}{L}$$

where C = cost of new equipment, d = annual amount of depreciation, and C_L = value of the equipment at L years. Solve the formula for C_L and find the value of a computer after 5 years if the depreciation is \$125/year and the computer cost \$1,995 originally.

Solution To solve the formula for C_L, we begin by removing fractions.

$$L(d) = L\left(\frac{C - C_L}{L}\right) \quad \text{multiply by LCD}$$

$$dL = C - C_L \quad \text{eliminate fractions}$$
$$dL - C = -C_L \quad \text{transpose}$$
$$C_L = C - dL \quad \text{divide by } -1$$

Substitute the numerical values $d = 125$, $L = 5$, and $C = 1{,}995$.

$$C_L = 1{,}995 - 125(5)$$
$$C_L = 1{,}995 - 625$$
$$C_L = 1{,}370$$

The computer is worth \$1,370 at the end of 5 years. ■■■

EXAMPLE 6 In surveying, parallax p must be considered in aerial photos. Parallax is an apparent change in the direction of an object caused by a change in the observer's position. The relationship is given by

$$\frac{H - h}{f} = \frac{B}{p}$$

Solve the formula for p and find the value for p if $H = 3{,}000$ ft, $h = 653$ ft, $f = 96$ ft, and $B = 300{,}000$ ft.

Solution
$$fp\left(\frac{H-h}{f}\right) = fp\left(\frac{B}{p}\right) \quad \text{multiply by LCD}$$

$$p(H-h) = Bf \quad \text{eliminate fraction}$$

$$p = \frac{Bf}{H-h} \quad \text{divide}$$

Substituting the numerical value gives

$$p = \frac{(300,000)(96)}{3,000 - 653}$$

$$p = 12,271 \text{ ft}$$

∎∎∎

EXAMPLE 7 Nearsightedness and farsightedness are common defects of the human eye. Both are corrected by eyeglass lenses. The power *P* of a lens needed to correct these conditions depends on the distance of the object d_1 and on the image distance d_2 given by

$$P = \frac{1}{d_1} - \frac{1}{d_2}$$

Joan is nearsighted and cannot focus clearly on objects farther than 0.6 m. If her glasses have a power of -1.2 diopters, how far can she see clearly with her glasses?

Solution From the power of a lens formula, we know $P = -1.2$ diopters and $d_2 = 0.6$ m. Substituting and solving for d_1 gives

$$-1.2 = \frac{1}{d_1} - \frac{1}{0.6}$$

$$-1.2\,(d_1)(0.6) = 0.6 - d_1$$

$$-0.72\,d_1 = 0.6 - d_1$$

$$0.28\,d_1 = 0.6$$

$$d_1 = 2 \text{ m}$$

Joan can see clearly up to 2 m away.

∎∎∎

Solving Rate Problems

One application that results in a fractional equation is called a *rate problem*. In these problems you are asked to find the time it takes for two people or machines to complete a task while working at constant rates. The next problem illustrates a rate problem.

EXAMPLE 8 One draftsperson can complete a drawing in 3 hours, while a second draftsperson takes 5 hours to complete the same job. How long would it take to complete the job if the two draftspeople worked together?

Solution The amount of work completed is the product of time and the rate of work. Therefore, the work done in one unit of time is 1 divided by the rate of work. If the first draftsperson can do the entire job in 3 hours, then he or she can complete 1/3 of the job in one hour. Similarly, the second draftsperson will have completed 1/5 of the job in one hour; and if *x* represents the time the job will require if they work together, then 1/x of the job will be done in one hour. Since their combined rate is equal to the sum of their individual rates, the equation is

$$\frac{1}{3} + \frac{1}{5} = \frac{1}{x}$$

First, multiply by the common denominator, $15x$.

$$15x\left(\frac{1}{3}\right) + 15x\left(\frac{1}{5}\right) = 15x\left(\frac{1}{x}\right)$$

Then solve the resulting equation for x.

$$5x + 3x = 15$$
$$8x = 15$$
$$x = \frac{15}{8}$$

Therefore, it takes 1.875 hours to complete the drawing if the draftspeople work together. ■■■

Solving Motion Problems

In this section we will extend our knowledge of motion problems to include fractional equations. The method of setting up and solving these equations is the same as discussed in Chapter 1.

EXAMPLE 9 Fred's boat can go 10 miles upstream in the same time that it takes to go 15 miles downstream. If the speed of the current is 5 mi/h, what is the speed of Fred's boat in still water?

Solution We will fill in a chart as we did in Chapter 1. The distance upstream is 10 miles, while the distance downstream is 15 miles. Let $x = $ the speed of the boat in still water. Then $x - 5$ will represent the speed of the boat upstream (against the current), and $x + 5$ will represent the speed of the boat downstream (with the current). Time is the quotient of distance and rate. These values are entered in the accompanying chart.

	distance ÷ rate = time		
	Distance	Rate	Time
Upstream	10	$x - 5$	$\dfrac{10}{x - 5}$
Downstream	15	$x + 5$	$\dfrac{15}{x + 5}$

Since the amount of time required to go upstream equals the time required to go downstream, the equation is

$$\frac{10}{x - 5} = \frac{15}{x + 5}$$

To solve the equation, multiply by the common denominator, $(x - 5)(x + 5)$.

$$(x + 5)(x - 5)\left(\frac{10}{x - 5}\right) = (x + 5)(x - 5)\left(\frac{15}{x + 5}\right)$$

$$10(x + 5) = 15(x - 5)$$

Eliminate parentheses and solve the equation.

$$10x + 50 = 15x - 75$$
$$10x - 15x = -75 - 50$$
$$-5x = -125$$
$$x = 25$$

Fred's boat goes 25 mi/h in still water.

2–6 EXERCISES

Solve the equation for the variable.

1. $\dfrac{x}{3} - 1 = \dfrac{2}{5} + x$

2. $4 - x + \dfrac{1}{7} = \dfrac{1}{3}$

3. $\dfrac{8}{9} + \dfrac{x - 2}{3} = 2$

4. $\dfrac{5}{6} + \dfrac{x}{4} = \dfrac{2x - 1}{2}$

5. $\dfrac{8}{x} + \dfrac{1}{3} = \dfrac{4}{x}$

6. $\dfrac{3}{x} + 8 = \dfrac{3x + 18}{x}$

7. $\dfrac{9}{x} = \dfrac{5}{x} - \dfrac{3}{4}$

8. $\dfrac{2x + 6}{x} = \dfrac{4}{3}$

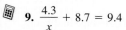 **9.** $\dfrac{4.3}{x} + 8.7 = 9.4$

10. $6.8x + \dfrac{14.7}{8.3} = 15.1$

11. $\dfrac{9.3}{x} + \dfrac{6.7}{x} = 5.2$

12. $11.3 + \dfrac{12.8}{x} = 16.4$

13. $\dfrac{7}{2x^2} = \dfrac{4}{x^2} - \dfrac{3}{4x}$

14. $\dfrac{8}{3x - 1} = \dfrac{7x}{3x - 1} - 2$

15. $\dfrac{1}{x} - \dfrac{4}{x^2} = \dfrac{1}{3x}$

16. $\dfrac{15}{x} - 7 = \dfrac{1}{x}$

17. $\dfrac{7}{3x^3} + \dfrac{1}{x^2} = \dfrac{5}{2x^3}$

18. $\dfrac{7}{x^4} + \dfrac{6}{x^3} = \dfrac{-11}{x^4}$

19. $\dfrac{2}{x^2} - \dfrac{9}{x^3} = \dfrac{5}{x^2}$

20. $\dfrac{2}{2x + 5} = \dfrac{6}{3x - 3}$

21. $\dfrac{3}{x - 4} + \dfrac{1}{x + 4} = \dfrac{5x + 4}{x^2 - 16}$

22. $\dfrac{8}{x - 2} + \dfrac{6}{x + 2} = \dfrac{32}{x^2 - 4}$

23. $\dfrac{2x + 3}{x - 1} = \dfrac{4x}{2x + 1}$

24. $\dfrac{9x}{x^2 - 9} = \dfrac{6}{x + 3}$

25. $\dfrac{7}{3x - 2} + \dfrac{5}{3x + 5} = 0$

26. $\dfrac{10}{4x - 3} + \dfrac{7}{2x + 5} = 0$

27. $\dfrac{5}{8x + 1} = \dfrac{2}{x - 4}$

28. $\dfrac{3}{x + 5} - \dfrac{4}{x - 5} = \dfrac{30}{x^2 - 25}$

29. $\dfrac{3}{2x - 1} + \dfrac{6}{2x + 1} = \dfrac{15}{4x^2 - 1}$

30. $\dfrac{7}{x + 3} - \dfrac{5}{x - 4} = \dfrac{9}{x^2 - x - 12}$

31. $\dfrac{2x}{x - 1} - \dfrac{3}{x + 1} = \dfrac{2(x^2 + 2)}{x^2 - 1}$

32. $\dfrac{5}{2x + 1} + \dfrac{2}{x - 3} = \dfrac{14}{2x^2 - 5x - 3}$

33. $\dfrac{1}{x + 2} + \dfrac{3x}{x - 5} = \dfrac{3x^2 + 30}{x^2 - 3x - 10}$

34. $\dfrac{7x}{3x + 5} - \dfrac{3}{x - 1} = \dfrac{7x^2 - 3x - 8}{3x^2 + 2x - 5}$

P 35. The harmonic mean m of two numbers x and y is given by

$$m = \frac{2xy}{x + y}$$

Solve for y.

P 36. The equation for displacement s of an object thrown upward is given by

$$s = v_0 t + \frac{1}{2}at^2$$

Solve for a.

M 37. An equation from hydrodynamics is

$$\frac{p}{d} = \frac{M}{d} - \frac{m^2}{2\pi^2 r^2}$$

Solve for p.

B 38. In straight-line depreciation, solve for C if $d = \$45$. $C_L = \$200$, and $L = 3$ yr given that

$$d = \frac{C - C_L}{L}$$

P 39. In physics, the density of a substance is the mass of the substance divided by its volume. Find the volume of a substance if its density is 9.6 kg/m^3 and its mass is 4 kg.

M 40. The mechanical advantage of a mechanism is the quotient of the output force and input force. If the mechanical advantage is 830 and the output force is 25 lb, find the input force.

E 41. The power required by an electrical appliance is the product of voltage and current (watts = volts · amperes). What current is required for an iron when the power is 660 watts and the voltage is 110 volts.

E 42. Five electric light bulbs, each with a rating of 60 W and 110 V, are being operated by a 110-V dc power source. What is the value of the total load resistance? Use $P_T = V^2/R_T$ where P_T is total power, V is voltage, and R_T is total resistance.

E 43. A 50-m coil of copper wire has a resistance of 1.05 Ω. If 15 m of wire is cut from the coil, what is the resistance of the remaining 35 m of wire? (*Note:* $R_1/R_2 = L_1/L_2$.)

44. The formula relating the temperature in degrees Celsius C to the temperature in degrees Fahrenheit F is

$$C = \frac{5F - 160}{9}$$

Find the temperature Fahrenheit that corresponds to 20°C.

E 45. Conductors of different metals are compared by their resistivity. The formula is $\sigma = RA/L$ where σ is resistivity in ohm · meters, R is resistance in ohms, A is cross-sectional area in square meters, and L is length in meters. Determine the resistance of a nichrome wire if $\sigma = 0.000001$ Ω · m, $A = 0.000000128 \text{ m}^2$, and $L = 2$ m.

C 46. In construction, the area of footing of a building is the ratio of the total weight of the building to the bearing capacity of the soil. If the area of footing is 24 ft^2 and the weight of the building is 60,000 lb, what is the bearing capacity of the soil?

E 47. The reactance of an inductor and capacitor in series is given by

$$X = \omega L - \frac{1}{\omega C}$$

If $L = 6$ henries (H), $\omega = 3$ rad/s, and $X = 3.5$ Ω, find C in microfarads (μF).

48. Find two consecutive integers such that four times the reciprocal of the larger integer is twice the reciprocal of the smaller integers (integers are x and $x + 1$).

49. If a number is added to the numerator and subtracted from the denominator of 7/9, the result is 3. Find the number.

50. Find three consecutive odd integers such that the quotient of the second and third integers is 3 (integers are x and $x + 2$).

51. Find three consecutive even integers such that the quotient of the third and first integers is 2.

52. If twice a number is added to the numerator and three times the number is subtracted from the denominator of $-9/5$, the result is 5. Find the number.

Bi 53. One biology lab assistant can clean the monkey cages in 3 hours, while the other lab assistant can clean the same cages in 2 hours. How long does it take if they work together?

A 54. A carpenter can complete a certain job in 5 days, while his apprentice takes 8 days to do the same job. How long will it take if they work together?

55. A swimming pool can be filled by an inlet pipe in 2 hours, while it takes 5 hours to drain the pool. Starting with an empty pool, how long will it take to fill the pool if both inlet and drain are open? (Set this problem up as a rate problem, except subtract the two times since they work against each other.)

56. A battery can be charged in 4 hours, while the load will discharge it in 9 hours. Assuming the battery is dead and being charged and discharged at the same time, how long will it take to recharge the battery?

57. Fred is farsighted and cannot focus on objects closer than 0.8 m. He is currently wearing lenses with a power of +2.5 diopters. How close can he see with these glasses?

58. Juan is nearsighted and can focus clearly on objects 18 m away when wearing glasses with a power of −1.3 diopters. How far can he see clearly without his glasses?

CS 59. One printer takes 12 minutes to print a report and a second printer takes 8 minutes to print the same report. How long would it take to print the report using both printers?

CHAPTER SUMMARY

Summary of Terms

algebraic fraction (p. 59)

binomial (p. 46)

coefficient (p. 46)

degree (p. 46)

difference of two squares (p. 49)

equivalent fraction (p. 59)

factored completely (p. 46)

factoring (p. 46)

lowest common denominator (p. 69)

lowest terms (p. 59)

monomial (p. 46)

perfect square trinomial (p. 53)

polynomial (p. 46)

prime (p. 46)

reduced fraction (p. 59)

term (p. 46)

trinomial (p. 46)

Summary of Formulas

$x^2 - y^2 = (x + y)(x - y)$ difference of two squares

$\left.\begin{array}{l} x^2 + 2xy + y^2 = (x + y)^2 \\ x^2 - 2xy + y^2 = (x - y)^2 \end{array}\right\}$ perfect square trinomial

CHAPTER REVIEW

Section 2–1

Factor using greatest common factor.

1. $7x^3 + 38x^8 - 28x^2$

2. $27x^5 - 18x^{10} + 3x^2$

3. $15a^2b + 40b^2c - 25ac^3$

4. $48yz^5 + 32y^4z^3$

5. $32m^8 + 8m - 24m^4$

6. $90a^5b^4 + 27a^3b^8 - 63a^6b^5$

7. $18x^3y^2 + 36x^2z - 72x^5y^3z^2$

8. $56x^3y^5 - 72x^8y^3$

9. $7x(2a + b) - 6y(2a + b)$

10. $4s^2(3x - 4y) - 3t(3x - 4y)$

11. $12x^2(m - 3n) + 6(m - 3n)$

12. $8a(2c - d) + 4(2c - d)$

13. $3r^2(2s - t) - 3r^2(7s + 5t)$

14. $r^3(5x - 4y) - r^3(2x + y)$

15. $8r^3(3x - y) + 4r^2(y + 4z)$

16. $6m^3(p - 4s) - 3m(m - p)$

Factor as the difference of two squares, if possible.

17. $16x^2 - 25y^2$

18. $64m^7 - 25n$

19. $9x^2 + 49y^2$

20. $121r^{16} - 81s^{14}$

21. $169s^2 - 121t^2$

22. $144x^2 - 49y^2$

23. $25c^4 - 36d^4$

24. $81z^4 - 16a^8$

25. The difference in the area of two cones with the same height but different radii r_1 and r_2 is

$\frac{1}{3}\pi r_1^2 - \frac{1}{3}\pi r_2^2$. Factor this expression.

26. The difference in the volume of two cylinders with the same radius but different heights is $\pi r^2h_1 - \pi r^2h_2$. Factor this expression.

B 27. The total weekly revenue of a company is represented by $400p - 8p^2$ where p is the price in dollars of the product. Factor this expression.

Section 2–2

Factor as a perfect square trinomial, if possible.

28. $64x^2 - 144xy + 81y^2$

29. $9p^2 - 6np + 4n^2$

30. $4x^2y^2 + 4xyz + z^2$

31. $169a^2 + 130a + 25$

32. $9a^2 - 6a + 1$

33. $49m^2 - 112mn + 64n^2$

34. $16m^4 - 40m^2n^2 - 25n^4$

35. $100s^4 - 260s^2t^2 + 169t^4$

36. The area of a square is given by $x^2 + 6x + 9$. Factor this expression.

Factor by the *ac* method, if possible.

37. $10x^2 + 23xy - 42y^2$

38. $72a^2 - ab - 56b^2$

39. $2a^2 - 7ab - 99b^2$

40. $28r^2 + rs - 45s^2$

41. $m^2 + 16m + 39$

42. $4x^2 - 7x + 3$

43. $32s^2 + 12st - 35t^2$

44. $b^2 + 9b + 8$

45. $72x^2 - 19xy - 40y^2$

46. $54s^4 + 69s^2 + 20$

47. $18s^4 - 73s^2y^2 + 35y^4$

48. $20a^2 - 58ac + 42c^2$

P 49. An object thrown upward with an initial velocity v_0 of 48 ft/s travels a distance s of 64 ft. The expression for the time t is $16t^2 - 48t + 64$. Factor the left side of this expression.

Section 2–3

Reduce each fraction to lowest terms.

50. $\dfrac{40x^2y^3}{45x^4y}$

51. $\dfrac{27m^4n^3}{54mn^5}$

52. $\dfrac{8x^3}{4x^5 - 16x^2}$

53. $\dfrac{6x^2y}{3x^3y^4 - 9xy^2}$

54. $\dfrac{16x^2 - 1}{12x^2 - 5x - 2}$

55. $\dfrac{8x^2 - 8}{5x^2 + x - 6}$

56. $\dfrac{6x^2 + 13x - 5}{1 - 9x^2}$

57. $\dfrac{4x^2 - x - 5}{25 - 16x^2}$

Section 2–4

Perform the indicated operation.

58. $\dfrac{7x^2y^3}{36x} \cdot \dfrac{3xy^4}{14y^9}$

59. $\dfrac{36x^5y^2}{18y^3} \cdot \dfrac{14y^3}{48x^8y^4}$

60. $\dfrac{4x - 4}{16x} \div \dfrac{9x^3 - 9x^2}{5x^5}$

61. $\dfrac{25a^5c^3}{36b^4c} \div \dfrac{30a^2b}{24a^4b^2c^5}$

62. $\dfrac{4x - 8}{3x - 5} \cdot \dfrac{25 - 9x^2}{3x^2 - x - 10}$

63. $\dfrac{14x^2 - 6x}{6x^2 + 7x - 20} \cdot \dfrac{2x^2 + 3x - 5}{6x^3 - 6x^2}$

64. $\dfrac{4x^2 - 4xy + y^2}{4x^2 - y^2} \div \dfrac{x^2 + 2xy + y^2}{x^2 - y^2}$

65. $\dfrac{15x^2 + 7xy - 2y^2}{3x^2 + 13xy + 4y^2} \cdot \dfrac{3x^2 - 5xy + 2y^2}{9x^2 - 4y^2}$

66. $\dfrac{15x^2y - 5xy^2}{14x^2 + 9xy + y^2} \div \dfrac{6x^2 + xy - y^2}{7x^3y + x^2y^2}$

67. $\dfrac{18x^2 - 24xy}{6x^2 - 10xy - 24y^2} \div \dfrac{9x^2 - 16y^2}{18x^3 - 54x^2y}$

Section 2–5

Perform the indicated operation.

68. $\dfrac{8x}{y^2} - \dfrac{3}{xy} + \dfrac{5}{x^3}$

69. $\dfrac{14}{5x} + \dfrac{9}{3x^2y} - \dfrac{5}{2y^3}$

70. $\dfrac{7x}{4x - 4} + \dfrac{5}{16x^2}$

71. $\dfrac{3x - 1}{2x - 8} + \dfrac{4}{x - 4}$

72. $\dfrac{3x}{2x + 1} - \dfrac{3x - 1}{4x^2 - 1}$

73. $\dfrac{6x}{x^2 - x - 2} + \dfrac{4x}{x^2 - 2x - 3}$

74. $\dfrac{x - 1}{x^2 - x - 12} + \dfrac{2x + 1}{x^2 - 3x - 4}$

75. $\dfrac{5x - 3}{9x^2 - 16} - \dfrac{2x + 7}{6x^2 + 5x - 4}$

76. $\dfrac{3x + 4}{4x^2 - 1} - \dfrac{x - 7}{2x^2 + 9x - 5}$

77. $\dfrac{9x + 2}{3x^2 + 11x - 20} + \dfrac{2x - 6}{x^2 + 4x - 5}$

78. $\dfrac{\dfrac{x + 2}{6}}{\dfrac{x^2 - 4}{3x^3}}$

79. $\dfrac{\dfrac{2x + 4}{3xy^2}}{\dfrac{8x + 16}{5x}}$

80. $\dfrac{\dfrac{7x - 14}{5x^2y^3}}{\dfrac{3x - 6}{x^2 - xy}}$

81. $\dfrac{\dfrac{9x^2 - 1}{6xy^3}}{\dfrac{3x + 1}{4y}}$

82. $\dfrac{\dfrac{x}{6} - \dfrac{2y^2}{3x}}{\dfrac{6x^2 - 12xy}{x + 2y}}$

83. $\dfrac{\dfrac{x}{8} - \dfrac{2}{4x}}{\dfrac{x - 2}{6x^3}}$

84. $\dfrac{\dfrac{7x}{x - 1} + \dfrac{5}{x + 1}}{\dfrac{3}{x} - \dfrac{2}{x - 1}}$

85. $\dfrac{\dfrac{9}{3x - 1} + \dfrac{6x}{x + 2}}{\dfrac{3}{x} - \dfrac{4}{21x^2}}$

86. $\dfrac{\dfrac{2x + 1}{3x - 2} + \dfrac{6x}{x - 1}}{\dfrac{4x + 3}{x - 3} + \dfrac{7}{x - 1}}$

87. $\dfrac{\dfrac{5x - 3}{x - 3} - \dfrac{2x + 3}{x + 2}}{\dfrac{2x}{5x + 4} + \dfrac{x + 1}{6x - 5}}$

Section 2–6

Solve the following equations.

88. $\dfrac{7}{x^2} + \dfrac{5}{x} = \dfrac{-3}{x^2}$

89. $\dfrac{9}{x} = \dfrac{-1}{2}$

90. $\dfrac{6}{x + 1} - \dfrac{3}{x - 1} = \dfrac{3}{x^2 - 1}$

91. $\dfrac{3}{x - 2} - \dfrac{2}{x + 2} = \dfrac{8}{x^2 - 4}$

92. $\dfrac{9}{x - 3} + \dfrac{7}{x - 4} = \dfrac{5x}{x^2 - 7x + 12}$

93. $\dfrac{5}{x + 1} + \dfrac{7}{x - 2} = \dfrac{10x}{x^2 - x - 2}$

94. $\dfrac{1}{3x} - \dfrac{4}{x - 1} = \dfrac{21}{3x^2 - 3x}$

95. $\dfrac{6x}{8x - 16} + \dfrac{3}{4} = \dfrac{3x}{x - 2}$

96. $\dfrac{-2}{x+1} + \dfrac{4}{x-3} = \dfrac{8x}{x^2 - 2x - 3}$

97. $\dfrac{7}{x+4} - \dfrac{3}{x-5} = \dfrac{-7}{x^2 - x - 20}$

CHAPTER TEST

The number(s) in parentheses refers to the appropriate learning objective(s) given at the beginning of the chapter.

1. Reduce the following to lowest terms: (5)

$$\dfrac{-18x^3 y^4 z}{48x^4 yz}$$

2. Factor completely $36x^3 y - 27x^2 y^3 + 18x^2 y$. (1)

3. Subtract the following: (7)

$$\dfrac{x-4}{x-3} - \dfrac{2x+1}{2x-3}$$

4. Factor completely $70x^3 - 115x^2 y + 15xy^2$. (1, 4)

5. If 7 is added to two-fifths of a number, the result is 9. Find the number. (9)

6. Multiply the following: (6)

$$\dfrac{6x^2 - 5x - 4}{-3x^2 + 19x - 20} \cdot \dfrac{5x^2 - 9x - 18}{10x^2 + 17x + 6}$$

7. Factor completely $8m^3 - 28m^2 n - 60mn^2$. (1, 4)

8. Simplify the following: (8)

$$\dfrac{\dfrac{4x+2}{x}}{\dfrac{6x+3}{x^2}}$$

9. Factor completely $28x^2 - 51xy - 27y^2$. (4)

10. Divide the following: (6)

$$\dfrac{7m^3 n^2 (n+1)}{8m^5 n^6} \div \dfrac{14n^4 (n-1)}{16mn^8}$$

P 11. The net rate of radiation is the difference between the rate of energy emission and the rate of energy absorption. This relationship is given by $e\pi T_1^4 - e\pi T_2^4$. Factor this expression. (1, 2)

12. One draftsperson can complete a drawing in 4 hours, while it takes another draftsperson 6 hours to do the same drawing. How long will it take them working together? (9)

13. Solve the following for x: (9)

$$\dfrac{3x-1}{x-1} - \dfrac{6x}{x+2} = \dfrac{-3x^2}{x^2 + x - 2}$$

14. Factor completely $25x^2 - 20xy + 4y^2$. (3)

15. The ratio of yield per acre is given as $(16x^2 + 24x - 40): (x - 1)$. Simplify this ratio and calculate the yield per acre when $x = 5$. (5)

16. Solve the following for x: (9)

$$\frac{8x - 1}{x^2} + \frac{3}{x} = \frac{2}{x}$$

P 17. Work is defined to be the product of force and displacement. If 50 ft-lb of work is done in lifting an 8-lb object, how far was the object moved? (9)

GROUP ACTIVITY

Automatic Water Works

The sprinkler company Automatic Water Works (AWW) has hired you to decide if there is a better method of covering a lawn with a sprinkler system. The diagrams show some of the alternatives that must be considered. The requirements are as follows:

1. Minimize the overlap of the sprinkler heads.

2. Minimize cost by using the minimum number of sprinkler heads.

In each of the diagrams find the area of overlap. The dimensions of the lawn are W by W feet and the area of a circle is given by:

$$Area = \pi r^2$$

where r is equal to the radius of the circle.

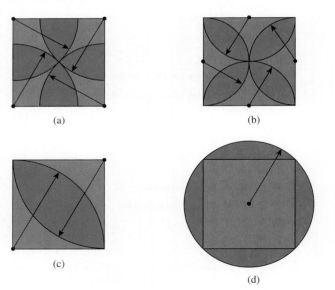

(a)　　　(b)

(c)　　　(d)

Try placing your data in a decision table to identify which method is best. An example follows. The last entry has been filled out as an example of the calculation needed.

Diagram	A.	B.	C.	D.
Number of heads	4	4	2	1
Placement	corners	middle of sides	corners	center
Area of overlap				$\pi r^2 - W^2$

Calculate each area of overlap. Is the area of overlap for each configuration the same? If not, then which arrangement gives you the smallest overlap? Are there other configurations that may give you a smaller area of overlap? What about using 5 sprinkler heads?

SOLUTION TO CHAPTER INTRODUCTION

Let x represent the number of hours the new computer system will require to process payroll. The amount of work done by each of the individual systems in one hour is given by 1/16 and 1/10. The amount of work done by the combined systems is $1/x$. The equation is

$$\frac{1}{16} + \frac{1}{10} = \frac{1}{x} \qquad \text{multiply by } 80x$$

$$5x + 8x = 80$$

$$13x = 80$$

$$x \approx 6 \text{ hours}$$

Y

ou are a financial advisor responsible for calculating the amount P that must be invested now at 9% interest (i) compounded annually to yield \$48,000 (S) in 10 years (n). ($P = S(1 + i)^{-n}$). (The solution to this problem is given at the end of the chapter.)

In this chapter we will discuss integral, fractional, and negative exponents. Since fractional exponents can be expressed in radical form, we will also discuss simplifying radicals and operations with radicals. We will conclude the chapter with a discussion of solving equations containing radicals, such as the one resulting from the problem above.

Learning Objectives

After completing this chapter, you should be able to

1. Simplify an algebraic expression with integral exponents (Section 3–1).

2. Simplify an algebraic expression with fractional exponents (Section 3–2).

3. Simplify radical expressions to simplest form (Section 3–3).

4. Add, subtract, multiply, and divide radical expressions (Section 3–4).

5. Solve equations involving radicals (Section 3–5).

Chapter 3

Great Investments

Exponents and Radicals

3–1

INTEGRAL EXPONENTS

In this section we will discuss some basic notation and terminology related to exponents. We often see the same quantity multiplied by itself repeatedly, such as $3 \cdot 3 \cdot 3 \cdot 3$. Repeated multiplication can be written in **exponential form.** For example,

$$a \cdot a \cdot a = a^3$$
$$b \cdot b = b^2$$
$$(-2)(-2)(-2)(-2) = (-2)^4$$

In general, if a is a real number and n is a positive integer, then

$$a^n = a \text{ multiplied } n \text{ times}$$

where n is the **exponent** and a is the **base.**

The expression a^n is read "the nth **power** of a."

Laws of Exponents

For any real numbers a and b and integers m and n, the following rules apply:

Rule 3–1. $a^m \cdot a^n = a^{m+n}$

If the bases are the same, add exponents to multiply.

Rule 3–2. $(ab)^m = a^m b^m$

The exponent applies to each factor.

Rule 3–3. $(a^n)^m = a^{nm}$

Multiply exponents to raise to a power.

Rule 3–4. $\dfrac{a^m}{a^n} = a^{m-n}$ if $m > n$, $a \neq 0$ ⎫

Rule 3–5. $\dfrac{a^m}{a^n} = \dfrac{1}{a^{n-m}}$ if $m < n$, $a \neq 0$ ⎬ To divide, subtract exponents

Rule 3–6. $\left(\dfrac{a}{b}\right)^m = \dfrac{a^m}{b^m}$ $(b \neq 0)$

The exponent applies to the numerator and denominator.

Rule 3–7. $a^0 = 1$ $(a \neq 0)$

Any base, other than zero, to the zero power equals one.

Rule 3–8. $a^{-n} = \dfrac{1}{a^n}$ $(a \neq 0)$

To remove a negative exponent, move the quantity across the fraction bar.

You should be familiar with operations concerning positive, integer exponents. Appendix A–4 reviews operations with exponents. Example 1 reviews the use of Rules 3–1 through 3–7. Then we will discuss in greater detail how to eliminate negative exponents using Rule 3–8.

EXAMPLE 1   Simplify each expression by applying the Laws of Exponents. Your answer may contain negative exponents.

a. $x^5 \cdot x^{-1}$ **b.** $(a^2b^{-3})^{-4}$ **c.** $\dfrac{x^8}{x^4}$ **d.** $\dfrac{y^2}{y^8}$ **e.** $\left(\dfrac{x^{-2}}{y^3}\right)^{-4}$

f. $4m^0$ **g.** $(4m)^0$

Solution

a. $x^{⑤} \cdot x^{⊖①} = x^{5+(-1)} = x^4$ Rule 3–1

b. $(a^{②}b^{⊖③})^{⊖④} = (a^2)^{-4}(b^{-3})^{-4} = a^{-8}b^{12}$ Rules 3–2 and 3–3

c. $\dfrac{x^{⑧}}{x^{④}} = x^{8-4} = x^4$ Rule 3–4

d. $\dfrac{y^{②}}{y^{⑧}} = \dfrac{1}{y^{8-2}} = \dfrac{1}{y^6}$ Rule 3–5

e. $\left(\dfrac{x^{⊖②}}{y^{③}}\right)^{⊖④} = \dfrac{(x^{-2})^{-4}}{(y^3)^{-4}} = \dfrac{x^8}{y^{-12}}$ Rules 3–6 and 3–3

f. $4m^{⓪} = 4(1) = 4$ Rule 3–7

g. $(4m)^{⓪} = 4^0m^0 = 1$ Rules 3–2 and 3–7   ■ ■ ■

CAUTION ✦ Be careful in simplifying expressions containing zero exponents. In an expression such as $4m^0$, the zero exponent applies only to m. On the other hand, in an expression such as $(4m)^0$, the zero exponent applies to both the 4 and the m.

    As a general rule, you should not leave zero or negative exponents in an algebraic expression. Rule 3–8 provides us with a method of eliminating negative exponents. To remove the negative exponents on a factor, move the factor across the fraction line. This technique is illustrated in the next example.

EXAMPLE 2   Simplify each expression using the Laws of Exponents. Express your answers with positive exponents only.

a. $y^4 \cdot y^{-9}$ **b.** $(2x^{-3}y^2)^3$ **c.** $\left(\dfrac{4}{x^3}\right)^{-2}$ **d.** $\left(\dfrac{m^{-4}}{n^3}\right)^{-2}$

Solution

a. $y^4 \cdot y^{-9} = y^{4+(-9)} = y^{-5} = \dfrac{1}{y^5}$ change the sign of the exponent

b. $(2x^{-3}y^2)^3 = (2)^3(x^{-3})^3(y^2)^3 = 8x^{-9}y^6 = \dfrac{8y^6}{x^9}$ change the sign of the exponent

$$\dfrac{1}{x^{-6}} = x^6$$

c. $\left(\dfrac{4}{x^3}\right)^{-2} = \dfrac{(4)^{-2}}{(x^3)^{-2}} = \dfrac{4^{-2}}{x^{-6}} = \dfrac{x^6}{4^2} = \dfrac{x^6}{16}$

$$4^{-2} = \dfrac{1}{4^2}$$

Remember that 4^{-2} is the same as $1/4^2$ or $1/16$. The negative exponent is not connected with the sign for the base (i.e., 4^{-2} is **not** -16).

d. $\left(\dfrac{m^{-4}}{n^3}\right)^{-2} = \dfrac{(m^{-4})^{-2}}{(n^3)^{-2}} = \dfrac{m^8}{n^{-6}} = m^8 n^6$　　　　　■■■

NOTE ✦ There may be more than one correct method of simplifying an expression using the Laws of Exponents, as shown in the next example.

EXAMPLE 3　Simplify the following:

$$\left(\frac{-2a^3b^{-4}}{5a^{-5}b^{-6}}\right)^{-2}$$

Express the answer with positive exponents.

Solution　We can simplify this expression by

1. applying the Laws of Exponents inside the parentheses and then simplifying the -2 power; or

2. simplifying the -2 power and then applying the Laws of Exponents to the result.

Both methods are shown below.
　　Applying the first method gives

$$\left(\frac{-2a^3b^{-4}}{5a^{-5}b^{-6}}\right)^{-2}$$

$$\left(\frac{-2a^3b^6a^5}{5b^4}\right)^{-2} \qquad \text{eliminate negative exponents inside parentheses (Rule 3–8)}$$

$$\left(\frac{-2a^8b^2}{5}\right)^{-2} \qquad \text{simplify inside parentheses (Rules 3–1 and 3–4)}$$

$$\frac{(-2)^{-2}(a^8)^{-2}(b^2)^{-2}}{(5)^{-2}} \qquad \text{eliminate parentheses (Rule 3–3)}$$

$$\frac{(5)^2}{(-2)^2(a^8)^2(b^2)^2} \qquad \text{eliminate negative exponents (Rule 3–8)}$$

$$\frac{25}{4a^{16}b^4} \qquad \text{simplify (Rule 3–3)}$$

Applying the second method of removing the negative power on the parentheses gives

$$\frac{(-2)^{-2}(a^3)^{-2}(b^{-4})^{-2}}{(5)^{-2}(a^{-5})^{-2}(b^{-6})^{-2}} \qquad \text{Rule 3–2}$$

$$\frac{(-2)^{-2}a^{-6}b^8}{(5)^{-2}a^{10}b^{12}} \qquad \text{Rule 3–3}$$

$$\frac{(5)^2b^8}{(-2)^2a^{10}b^{12}a^6} \qquad \text{Rule 3–8}$$

$$\frac{25}{4a^{16}b^4} \qquad \text{Rules 3–1 and 3–5}$$　　　■■■

NOTE ✦ When you have a negative exponent on a fraction, you can eliminate the negative exponent by inverting the fraction inside the parentheses. Then simplify the result as we did it in the previous example. For example,

$$\left(\frac{-2a^3b^{-4}}{5a^{-5}b^{-6}}\right)^{-2} = \left(\frac{5a^{-5}b^{-6}}{-2a^3b^{-4}}\right)^2 \text{ or } \left(\frac{a}{b}\right)^{-m} = \left(\frac{b}{a}\right)^m$$

Multinomial Expressions

The Laws of Exponents also apply to simplifying a multinomial expression. However, we must be careful to distinguish between terms and factors when we simplify negative exponents. You may also need to review operations with fractions in the previous chapter.

EXAMPLE 4 Simplify $a^{-2} + b - c^{-1}$.

Solution Each of the previous examples contained a single term. However, this example contains three terms: a^{-2}, b, and, ignoring the sign, c^{-1}. We must remove negative exponents within each *term* by moving *factors* with negative exponents across the fraction line. Removing the negative exponents within each term gives

$$a^{-2} + b - c^{-1}$$

$$\frac{1}{a^2} + b - \frac{1}{c}$$

After the negative exponents have been removed, no further simplification of exponents is necessary. However, we must simplify this fractional expression by adding the fractions using a common denominator. The simplified expression is

$$\frac{c + a^2bc - a^2}{a^2c}$$

■■■

EXAMPLE 5 Simplify $(3x + y^{-2})^{-3}$.

Solution To remove the negative exponent outside the parentheses, we take the factor across the fraction line.

$$\frac{1}{(3x + y^{-2})^3} \qquad \text{remove negative exponent outside the parentheses (Rule 3–8)}$$

$$\frac{1}{\left(3x + \dfrac{1}{y^2}\right)^3} \qquad \text{remove negative exponent within the parentheses (Rule 3–8)}$$

Notice that eliminating the negative exponents has led to a complex fraction. We simplify the complex fraction as we did in Chapter 2.

$$\frac{1}{\left(\dfrac{3xy^2 + 1}{y^2}\right)^3} \qquad \text{add the fractions within the parentheses}$$

$$\frac{1}{\dfrac{(3xy^2 + 1)^3}{y^6}} \qquad \text{eliminate parentheses (Rule 3–6) and Rule 3–3}$$

$$\frac{y^6}{(3xy^2 + 1)^3} \qquad \text{divide the fraction}$$

We do not need to expand the denominator of the fraction. ■■■

EXAMPLE 6 The current I in a circuit with an external resistance R, a cell of electromotive force E, and internal resistance r can be written $I = Er^{-1}(Rr^{-1} + 1)^{-1}$. Simplify this expression.

Solution   First, we remove the negative exponents outside the parentheses by taking the factors across the fraction line.

$$I = \frac{E}{r(Rr^{-1} + 1)}$$

Second, we remove the negative exponent in the denominator.

$$I = \frac{E}{r\left(\dfrac{R}{r} + 1\right)}$$

Last, we simplify the complex fraction.

$$I = \frac{E}{r\left(\dfrac{R + r}{r}\right)}$$

$$I = \frac{E}{R + r} \qquad\qquad ■■■$$

The next example illustrates a type of expression that you will need to simplify in calculus.

EXAMPLE 7 The amount borrowed A in a credit purchase is given by

$$A = R\left[\frac{1 - \left(1 + \dfrac{i}{n}\right)^{-nt}}{\dfrac{i}{n}}\right]$$

where R = monthly payment, i = annual percentage rate (APR), n = number of times per year that interest on the loan is compounded, and t = number of years for the loan.

Suppose that you want to purchase office furniture. The APR is 9.5%, the interest is compounded monthly, you plan to repay the loan in 3 years, and you want monthly payments of $750. How much can you borrow?

Solution   From the information given, we must calculate A when $R = 750$, $i = 0.095$ (interest rate as a decimal), $n = 12$, and $t = 3$. Substituting these values gives

$$A = 750\left[\frac{1 - \left(1 + \dfrac{0.095}{12}\right)^{-12(3)}}{\dfrac{0.095}{12}}\right]$$

$$A = 750\left(\frac{1 - 0.7529}{0.0079}\right)$$

$$A = 750(31.28)$$

$$A = \$23,459$$

∎∎∎

3–1 EXERCISES

Simplify and express all answers with positive exponents.

1. $b^9 \cdot b^{-4}$

2. $y^{-6} \cdot y^2$

3. $m^{-10} \cdot m^6$

4. $x^{12} \cdot x^3$

5. $x^4 \cdot x^{-3}$

6. $y^{-5} \cdot y^{-2}$

7. $\dfrac{p^5}{p^9}$

8. $(m^2n)^3$

9. $\left(\dfrac{a^2}{b}\right)^3$

10. $\left(\dfrac{x}{y^4}\right)^2$

11. $a^{\frac{1}{3}} \cdot a^{\frac{2}{3}}$

12. $x^{\frac{3}{4}} \cdot x^{\frac{1}{4}}$

13. $(x^2)^{\frac{1}{3}}$

14. $\left(y^{\frac{1}{2}}\right)^2$

15. $(3x^{-2}y^4)^{-3}$

16. $(2ab^{-3})^{-2}$

17. $\left(\dfrac{2x^{-1}}{y}\right)^2$

18. $\left(\dfrac{4m}{n^{-3}}\right)^2$

19. $\left(\dfrac{6a^{-4}}{c^{-3}}\right)^{-2}$

20. $\left(\dfrac{5x}{y^{-4}}\right)^{-3}$

21. $(2a + b)^0$

22. $(14a + 8b)^0$

23. $m^0 + n^0$

24. $m + 3n^0$

25. $3x^0 + y$

26. $7x^0 + 8y^0$

27. $\left(\dfrac{3a}{2b^2}\right)^0$

28. $\left(\dfrac{17x^3}{54y^2}\right)^0$

29. 18^0

30. $(46^2)^0$

31. $\dfrac{x^{-4}y^3z^3}{x^6y^{-4}z^{-5}}$

32. $\dfrac{(a^0b^{-1})^{-3}}{(-a)^{-2}b^5}$

33. $\dfrac{(-ab^{-2})^0c^3}{a^{-3}}$

34. $\dfrac{x^4y^{-3}z^0}{-2x^{-3}z^{-4}}$

35. $\dfrac{m^{-3}n^{-2}p^3}{m^0p^{-3}}$

36. $\dfrac{m^{-2}n^{-3}}{m^4n^{-6}}$

37. $\left(\dfrac{a^{-3}b^4}{c^{-3}a}\right)^{-2}$

38. $\left(\dfrac{3x^{-2}y^4}{9x^3y^2}\right)^{-2}$

39. $\left(\dfrac{x^{-4}y^3z^0}{z^4y^{-5}}\right)^3$

40. $\left(\dfrac{a^{-4}b^2c}{ab^{-6}c^3}\right)^4$

41. $\left(\dfrac{2x^2y^{-3}}{4x^{-5}y}\right)^{-3}$

42. $\left(\dfrac{9m^{-6}n^3}{81m^{-7}n^{-4}}\right)^{-1}$

43. $3(x + 4)^2(2x + 5)^{-2} + (x + 4)(2x + 5)^{-1}$

44. $2(x + 1)^2(3x - 1)^{-2} - (x + 1)(3x - 1)^{-1}$

45. $a^{-1} + b^2 - c^0$

46. $x^{-2} + y^{-2}$

47. $x^2 + y^{-3}$

48. $(4x - 5)(x + 2)^{-2} + 3(4x - 5)(x + 2)^{-1}$

P 49. The following equation gives the magnetic field H at a distance r cm from the center of the magnet whose length is $2L$ and whose pole strengths are at $+m$ and $-m$:

$$H = \frac{Lr(r^2 - L^2)^{-2}}{(4m)^{-1}}$$

Simplify this expression.

E 50. The total resistance R_T of two conductors in parallel is given by

$$R_T = \frac{1}{r_2^{-1} + r_1^{-1}}$$

Simplify this expression.

B 51. Using Example 8, calculate the amount you can borrow for a car loan if $i = 9\%$, $n = 4$, $t = 15$ years, and $R = \$425$.

P 52. The focal length F of two lenses with focal lengths f_1 and f_2 separated by a distance d is given by

$$F = \frac{1}{f_2^{-1} + f_1^{-1} - d(f_1f_2)^{-1}}$$

Simplify this expression.

P 53. The units of measure for the height of a liquid under pressure are given by

$$\frac{(\text{kg} \cdot \text{m})(\text{s} \cdot \text{m})^{-2}}{\text{kg} \cdot \text{m}^{-2} \cdot \text{s}^{-2}}$$

Remove negative exponents and then simplify.

B 54. The following equation gives the amount A of money you would need to invest to yield y dollars at the end of x years at n rate of interest compounded annually:

$$A = y(1 + n)^{-x}$$

Write this expression without negative exponents.

CS 55. An n-bit video adapter can display up to 2^n colors. How many colors would an 8-bit video adapter be capable of displaying?

CS 56. Some computers work with 2^6 bits of information at one time. If 8 bits equals 1 byte, how many bytes can this computer work with at one time?

3–2

RATIONAL EXPONENTS

Before we begin the discussion of rational exponents, we need to discuss the terminology of radicals because rational exponent expressions can be converted to radicals. A **radical** is an expression having the form shown below.

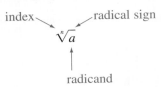

The **index** is a positive number representing the nth root of the **radicand.** In other words, if $b^n = a$, then b is the nth root of a. If no index is written, it is understood to be 2, or a square root. Use of radicals is another method of writing expressions with fractional or rational exponents.

All Laws of Exponents discussed in the previous section apply to rational (fractional) exponents as well as integral exponents. However, an additional rule is used to convert a fractional exponent into a radical expression. You can convert fractional exponents into a radical expression by using the following rule.

Fractional Exponents

If $\sqrt[n]{a}$ exists as a real number, then

$$\overset{\text{power}}{a^{m/n}} = \sqrt[n]{a^m} = \left(\sqrt[n]{a}\right)^m$$

root

where m and n are integers and n is positive.

Given the conditions stated above, the expressions $\sqrt[n]{a^m}$ and $\left(\sqrt[n]{a}\right)^m$ are equivalent. But the second expression is easier to evaluate because we are taking the root of a smaller number.

So far we have discussed positive radicands. If the radicand is negative and the index is an odd number, the root is real and negative. *On the other hand, if the radicand is negative and the index is an even number, the root is an imaginary number.* We will discuss imaginary and complex numbers in more detail in Chapter 14. For now, just state that the solution is an imaginary number.

EXAMPLE 1 Evaluate the following expressions:

a. $27^{2/3}$ **b.** $(-32)^{1/5}$ **c.** $(-16)^{1/2}$

Solution

a. From the rule of fractional exponents,

$$27^{2/3} = \left(\sqrt[3]{27}\right)^2$$

Since $(3)^3 = 27$, $\sqrt[3]{27} = 3$ and the expression becomes $(3)^2 = 9$.

$$27^{2/3} = 9$$

b. From the rule of fractional exponents,

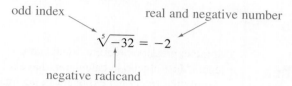

c. We convert from a fractional exponent to a radical.

even index

$$\sqrt[2]{-16} = \sqrt{-16} = \text{imaginary number}$$

negative radicand

EXAMPLE 2   Evaluate $64^{-5/6}$.

Solution   Before converting the fractional exponent to a radical, remove the negative exponent.

$$\frac{1}{64^{5/6}}$$ eliminate negative exponent

$$\frac{1}{\left(\sqrt[6]{64}\right)^5}$$ convert fractional exponent to radical

$$\frac{1}{(2)^5}$$ evaluate the radical ($\sqrt[6]{64} = 2$)

$$\frac{1}{32}$$ evaluate the exponent ($2^5 = 32$)   ■■■

EXAMPLE 3   The instantaneous current in a capacitor is given by

$$I = \frac{V}{R}e^{-t/(RC)}$$

where C = capacitance, V = terminal voltage of the battery, t = time, R = resistance, and e is the irrational number approximately equal to 2.71828. Simplify this expression by removing negative exponents and converting the fractional exponent to a radical.

Solution   First, we remove the negative exponent by moving the factor across the fraction line.

$$I = \frac{V}{Re^{t/(RC)}}$$

Next, we convert the fractional exponent into a radical expression.

$$I = \frac{V}{R\sqrt[RC]{e^t}}$$

■■■

Fractional Exponents with a Calculator

Some expressions require a calculator to evaluate roots because the answer is not an integer. Most algebraic calculators have a $\boxed{y^x}$ button for this purpose. However, you must express the fractional exponent in decimal form to use this button.

⌨ **EXAMPLE 4**   Evaluate the following:

a. $54^{3/5}$ **b.** $9^{3/4}$

Solution

a. 54 $\boxed{y^x}$ $\boxed{(}$ 3 $\boxed{\div}$ 5 $\boxed{)}$ $\boxed{=}$ ⟶ 10.950558

b. 9 $\boxed{y^x}$ $\boxed{(}$ 3 $\boxed{\div}$ 4 $\boxed{)}$ $\boxed{=}$ ⟶ 5.196152

■■■

EXAMPLE 5   The total yield y from an investment is given by

$$y = a(1 + n)^x$$

where a = initial investment, n = interest rate, and x = time in years. Find the yield on an initial investment of \$5,000 at a rate of 9% compounded annually for $1\frac{1}{2}$ years.

Solution   Substituting into the formula $a = 5,000$, $n = 0.09$, and $x = \frac{3}{2}$ gives

$$y = 5,000(1 + 0.09)^{3/2}$$
$$y = 5,000(1.09)^{3/2}$$

The yield is \$5,689.97.

5000 $\boxed{\times}$ 1.09 $\boxed{y^x}$ $\boxed{(}$ 3 $\boxed{\div}$ 2 $\boxed{)}$ $\boxed{=}$ ⟶ 5689.97

■■■

Operations with Fractional Exponents

The Laws of Exponents discussed in Section 3–1 also apply to fractional exponents. The next two examples illustrate their use.

EXAMPLE 6   Simplify each expression and give all answers with positive exponents.

a. $4^{3/5} \cdot 4^{2/5}$ **b.** $(27^{2/3})^{1/2}$ **c.** $(16a^3)^{3/4}$ **d.** $\left(\dfrac{m^{2/3}}{m}\right)^{3/4}$

Solution

a. $4^{3/5} \cdot 4^{2/5} = 4^{(3/5)+(2/5)} = 4^{5/5} = 4^1 = 4$

b. $(27^{2/3})^{1/2} = 27^{(2/3) \cdot (1/2)} = 27^{1/3} = 3$

c. $(16a^3)^{3/4} = 16^{3/4}(a^3)^{3/4} = 8a^{9/4}$

d. $\left(\dfrac{m^{2/3}}{m}\right)^{3/4} = \dfrac{(m^{2/3})^{3/4}}{m^{3/4}} = \dfrac{m^{1/2}}{m^{3/4}} = \dfrac{1}{m^{3/4-1/2}} = \dfrac{1}{m^{3/4-2/4}} = \dfrac{1}{m^{1/4}}$ ■■■

EXAMPLE 7 Simplify each expression and give all answers with positive exponents.

a. $\left(\dfrac{9x^{-2}y^{1/3}}{4x^{1/2}y^{-3/4}}\right)^{3/2}$ **b.** $(2x-1)(x+6)^{-1/3} + (x+6)^{2/3}$

Solution

a.

$$\left(\frac{9x^{-2}y^{1/3}}{4x^{1/2}y^{-3/4}}\right)^{3/2}$$

$$\left(\frac{9y^{13/12}}{4x^{5/2}}\right)^{3/2} \qquad \text{Rules 3–1 and 3–8}$$

$$\frac{9^{3/2}(y^{13/12})^{3/2}}{4^{3/2}(x^{5/2})^{3/2}} \qquad \text{Rule 3–2}$$

$$\frac{27y^{13/8}}{8x^{15/4}} \qquad \text{Rule 3–3}$$

b. $(2x-1)(x+6)^{-1/3} + (x+6)^{2/3}$

$$\frac{2x-1}{(x+6)^{1/3}} + (x+6)^{2/3} \qquad \text{Rule 3–8}$$

$$\frac{2x-1 + (x+6)^{1/3}(x+6)^{2/3}}{(x+6)^{1/3}} \qquad \text{add fractions}$$

$$\boxed{(x+6)^{(1/3)+(2/3)} = (x+6)}$$

$$\frac{2x-1+(x+6)}{(x+6)^{1/3}} \qquad \text{multiply}$$

$$\frac{3x+5}{(x+6)^{1/3}} \qquad \text{combine like terms} \qquad ■■■$$

EXAMPLE 8 The annual depreciation rate r using the declining balance method is given by the formula

$$r = 1 - \left(\frac{s}{c}\right)^{1/n}$$

where n = useful life in years, s = salvage value in dollars, and c = original cost in dollars. Find the annual depreciation rate at 3 years if the original cost of a computer system is \$4,595, the salvage value is \$1,250, and the useful life is 5 years.

Solution From the information given, $s = 1{,}250$, $c = 4{,}595$, $n = 5$, and we must calculate r. Substituting these values gives the following result.

$$r = 1 - \left(\frac{s}{c}\right)^{1/n}$$

$$r = 1 - \left(\frac{1{,}250}{4{,}595}\right)^{1/5}$$

$$r = 1 - (0.2720)^{.2}$$

$$r = 1 - 0.7708$$

$$r \approx 0.2292$$

The annual rate of depreciation is approximately 0.23, or 23%.

■■■

3–2 EXERCISES

Evaluate the given expression.

1. $27^{2/3}$ **2.** $32^{2/5}$ **3.** $64^{1/3}$ **4.** $256^{1/4}$

5. $216^{1/3}$ **6.** $243^{3/5}$

7. $125^{2/3} - 81^{3/4}$ **8.** $9^{1/2} + 125^{2/3}$

9. $8^{2/3} + 16^{1/4}$ **10.** $16^{3/4} - 27^{2/3}$

11. $\dfrac{25^{1/2} \cdot 8^{-2/3}}{81^{-1/3}}$ **12.** $49^{1/2} \cdot 36^{-1/2}$

13. $\dfrac{25^{1/2} \cdot 216^{2/3}}{16^{-3/2}}$ **14.** $\dfrac{64^{-2/3} \cdot 81^{1/4}}{27^{-1/3}}$

15. $4^{-1/2} + 27^{1/3}$ **16.** $16^{-3/4} + 8^{-2/3}$

 Calculate and round to hundredths.

17. $143^{2/3}$ **18.** $67^{3/5}$

19. $\dfrac{8^{-1/5}}{8^{3/4}}$ **20.** $6^{1.7} \cdot 14^{1/3}$

21. $2^{1/4} \cdot 3^{-3/4}$ **22.** $\dfrac{11^{2/3}}{11^{1/2}}$

23. $(7^{3/4})^{-1/6}$ **24.** $(18^{2/3})^{-2/5}$

Simplify each expression and give all answers with positive, rational exponents.

25. $x^{-\frac{1}{3}} x^{-\frac{1}{2}}$ **26.** $\dfrac{y^{3/5}}{y^{-1/3}}$

27. $a^{4/7} \cdot a^{3/5}$ **28.** $m^{3/7} \cdot m^{1/3}$

29. $(m^{-2/3} n^2)^{-3/4}$ **30.** $(x^{1/2} y^{-3/4})^{-1}$

31. $y^{1/8} y^{-3/4}$ **32.** $b^{2/9} b^{1/3}$

33. $\dfrac{b^3}{b^{-3/4}}$ **34.** $\dfrac{x^{-3/5}}{x^{-3/4}}$

35. $\dfrac{x^{1/2} x^{3/4}}{x^{1/8}}$ **36.** $\dfrac{a^{2/7} a^{-1/3}}{a^{-1/2}}$

37. $\dfrac{a^{-3} a^{3/7}}{a^{-1/3}}$ **38.** $\dfrac{x^{1/2} y^{-1/4}}{(xy)^{3/8}}$

39. $\dfrac{x^{-2/3} y^2}{x^{-1/4} y^{1/4}}$ **40.** $\dfrac{(m^{-2} n^{1/3})^2}{m^{1/3} n^{-1}}$

41. $9^{1/3} \cdot 9^{5/3}$ **42.** $8^{3/4} \cdot 8^{1/4}$

43. $\dfrac{4^{3/2}}{4^{-1/2}}$ **44.** $\dfrac{10^{1/2}}{10^{-3/2}}$

45. $\dfrac{6^{1/6}}{6^{-5/6}}$ **46.** $\dfrac{8^{-3/4}}{8^{1/4}}$

47. $(2x + 5)(x - 1)^{-1/4} + (x - 1)^{3/4}$

48. $(x + 7)(3x + 5)^{-1/2} - (3x + 5)^{1/2}$

49. $(3y + 4)(y + 2)^{-3/5} - 8(y + 2)^{2/5}$

50. $(4m - 1)(m + 4)^{-2/3} + 6(m + 4)^{1/3}$

P 51. The magnetic field is given by

$$H = \frac{2mL}{(r^2 + L^2)^{3/2}}$$

Convert the fractional exponent into a radical expression.

P 52. The emissive power of a blackbody at wavelength λ (the Greek letter lambda) can be written

$$E = \frac{c\lambda^{-5}}{e^{c/\lambda t} - 1}$$

Remove negative exponents and convert the fractional exponent into a radical.

B 53. Find the yield on an initial investment of $a = \$8{,}000$ at $n = 8\%$ for $x = 3\frac{1}{2}$ years. (Refer to Example 5.)

M 54. An expression used to estimate the shear stress in pounds per square inch that must be transferred between the immediate stiffeners and web during shear transfer is

$$f = h\left(\frac{F}{340}\right)^{3/2}$$

Write this expression in radical form.

P 55. The number N of radioactive nuclei left after undergoing a decay process is given by $N = N_0 e^{-Dt}$, where N_0 is the initial number of unstable nuclei and D is the disintegration constant for the particular nuclei. The half-life of the nuclei is defined as the length of time required for half the initial number of nuclei to decay. This leads to $N = N_0 e^{-0.69t/h}$, where h is the half-life of the nuclei. The fraction of nuclei remaining after time t can be written as $N/N_0 = (0.5)^{t/h}$ where h and t have the same units. Find the fractional amount of strontium 90 left after 8 years if its half-life is 28 years.

M 56. An approximation of engine efficiency E is given by $E = 100(1 - R^{-2/5})$ where R is the compression ratio. Find the efficiency of an engine whose compression ratio is 240/35.

B 57. Using Example 8, find the annual rate of depreciation in the second year on a copier whose original cost is \$6,995 if the salvage value is \$5,000 and the useful life is 7 years.

CS 58. A software program evaluates a particular formula in $10^{-1/12}$ seconds. How many seconds would it take to evaluate the formula 5 times?

3–3

SIMPLEST RADICAL FORM

So far in this chapter, we have discussed expressions written in exponential form. However, in the previous section we discussed the relationship between fractional exponents and radical form. In the remainder of this chapter we will discuss algebraic expressions written in radical form. Namely, in this section we will discuss converting a radical to its simplest form.

Laws of Exponents in Radical Form

Several of the Laws of Exponents can be written in radical form. These rules, which will form the basis for simplifying radicals, are summarized below.

Radical Form of Laws of Exponents

For positive integers m and n, the following rules apply:

Rule 3–9. $\sqrt[n]{a^n} = a$ for $a \geq 0$

Rule 3–10. $\sqrt[n]{ab} = \sqrt[n]{a}\sqrt[n]{b} =$ for $a, b \geq 0$

Rule 3–11. $\sqrt[m]{\sqrt[n]{a}} = \sqrt[mn]{a}$

Rule 3–12. $\sqrt[n]{\dfrac{a}{b}} = \dfrac{\sqrt[n]{a}}{\sqrt[n]{b}}$ where $b \neq 0$

Criteria for Simplest Radical Form

A simplified radical implies certain requirements of the answer. These requirements are as follows:

- An nth-root radical does not contain any nth factors.
- There are no radicals contained inside other radicals.
- The denominator of any fraction does not contain a radical.
- The order of the radical is reduced.

The following examples show how to simplify a radical to meet each of these requirements.

Removing nth-Root Factors

EXAMPLE 1 Simplify the radical expression $\sqrt{72}$.

Solution In simplifying $\sqrt{72}$, we are trying to remove any squares contained in 72. If you know that $72 = 36 \cdot 2$ and $36 = 6^2$, then you can immediately simplify

$\sqrt{72} = \sqrt{36}\sqrt{2} = 6\sqrt{2}$. However, if you have difficulty determining the squares contained in 72, break 72 into prime factors. Then write these prime factors in terms of exponents of 2 because the index of the radical is 2. Breaking 72 into prime factors gives

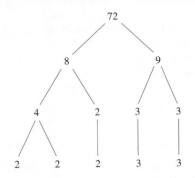

NOTE ✦ When the index of a radical is omitted, it is understood to be 2.

Writing the factors in terms of an exponent of 2 gives

$$72 = 2 \cdot 2 \cdot 2 \cdot 3 \cdot 3 = 2^2 \cdot 3^2 \cdot 2$$

Therefore,

$$\sqrt{72} = \sqrt{2^2 \cdot 3^2 \cdot 2}$$

Using Rule 3–9 for simplifying radicals gives

$$2 \cdot 3\sqrt{2} = 6\sqrt{2} \qquad \blacksquare\blacksquare\blacksquare$$

CAUTION ✦ Only the base is removed from the radical when the index and the exponent are equal ($\sqrt[n]{a^n} = a$).

The same process of removing nth-root factors applies to variable expressions also.

EXAMPLE 2 Simplify the radical expression $\sqrt[4]{64a^7b^8}$.

Solution To simplify this radical, we must remove from the radical any 4th-root factors. Therefore, *each factor under the radical should be written in terms of exponents of 4.*

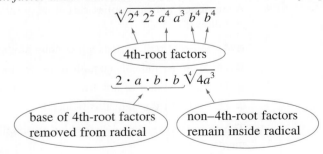

The simplified radical is $2ab^2 \sqrt[4]{4a^3}$. $\qquad \blacksquare\blacksquare\blacksquare$

Radical Within Radicals

According to Rule 3–11, you can simplify a radical inside another radical to a single radical by multiplying indices, as shown in the next example.

EXAMPLE 3   Simplify $\sqrt[3]{\sqrt[5]{4x}}$.

Solution

$$\sqrt[3]{\sqrt[5]{4x}} = \sqrt[(3 \cdot 5)]{4x} = \sqrt[15]{4x}$$

■ ■ ■

Rationalizing the Denominator

Simplifying a radical also involves eliminating radicals from the denominator. This process is called **rationalizing the denominator** because we multiply the numerator and denominator by a quantity that results in a rational denominator.

EXAMPLE 4   Simplify the radical expression $\sqrt{\dfrac{5a}{7}}$.

Solution   First, let us apply Rule 3–12 and write the expression as two radicals.

$$\frac{\sqrt{5a}}{\sqrt{7}}$$

To remove the radical in the denominator, we must have a square under the radical ($\sqrt[2]{a^2} = a$, from the more general statement $\sqrt[n]{a^n} = a$). Therefore, we must multiply the numerator and denominator by $\sqrt{7}$.

$$\frac{\sqrt{5a}}{\sqrt{7}} \cdot \frac{\sqrt{7}}{\sqrt{7}} \qquad \text{rationalize denominator}$$

$$\frac{\sqrt{35a}}{\sqrt{49}} \qquad \text{multiply radicals}$$

$$\frac{\sqrt{35a}}{7} \qquad \text{simplify denominator}$$

■ ■ ■

EXAMPLE 5   Simplify $\sqrt[3]{\dfrac{5a^4}{4b^3}}$.

Solution   First, simplify the radical by removing any cubes contained under the radical.

$$\frac{\sqrt[3]{5a^3a}}{\sqrt[3]{4b^3}}$$

$$\frac{a\sqrt[3]{5a}}{b\sqrt[3]{4}}$$

Before we can eliminate the radical in the denominator, each factor under the radical must have an exponent of 3. To determine the factor for rationalizing, write each factor in terms of exponents.

$$\frac{a\sqrt[3]{5a}}{b\sqrt[3]{2^2}}$$

From this information, we can see that we rationalize by multiplying the numerator and denominator by $\sqrt[3]{2}$.

$$\frac{a\sqrt[3]{5a}}{b\sqrt[3]{2^2}} \cdot \frac{\sqrt[3]{2}}{\sqrt[3]{2}} \qquad \text{rationalize denominator}$$

$$\frac{a\sqrt[3]{10a}}{b\sqrt[3]{2^3}} \qquad \text{multiply radicals}$$

The simplified radical is

$$\frac{a\sqrt[3]{10a}}{2b}$$

■ ■ ■

NOTE ✦ Always simplify a radical by removing nth-root factors before rationalizing the denominator. In some instances, rationalizing may not be necessary after this operation has been performed.

Reducing the Index of a Radical

The final requirement for simplifying a radical is that the order of the radical be reduced. Fractional exponents make this operation easier.

NOTE ✦ Before you can convert several fractional exponents to a radical, the denominator of the fractions must be the same.

$$a^{m/n}b^{p/n} = \sqrt[n]{a^m b^p}$$

In some instances, you may have to obtain a common denominator, as shown in the following example.

EXAMPLE 6   Simplify the following:

a. $\sqrt[6]{4x^2}$ **b.** $\sqrt[6]{81y^2}$

Solution

a. First, *write the radicand in exponential form.*

$$\sqrt[6]{2^2 x^2}$$

Then *express the radical expression with fractional exponents.*

$$(2^2 x^2)^{1/6}$$
$$(2^2)^{1/6}(x^2)^{1/6} \qquad \text{Rule 3–2}$$
$$2^{2/6} x^{2/6} \qquad \text{Rule 3–3}$$

Last, *reduce the fractional exponent,* if possible, and convert back to radical form.

$$2^{1/3} x^{1/3}$$
$$\sqrt[3]{2x}$$

b.

$\sqrt[6]{81y^2}$	
$\sqrt[6]{3^4 y^2}$	convert to exponential form
$(3^4 y^2)^{1/6}$	convert to fractional exponent
$3^{4/6} y^{2/6}$	simplify exponents
$3^{2/3} y^{1/3}$	reduce fractional exponent
$\sqrt[3]{9y}$	convert to radical form

■ ■ ■

Polynomial Radicands

Polynomial radicands must also meet the requirements for simplest form.

EXAMPLE 7 Simplify the following radical expression:

$$\sqrt{3x^2 - 2x + \frac{1}{3}}$$

Solution

$$\sqrt{\frac{9x^2 - 6x + 1}{3}} \qquad \text{combine the fractions}$$

$$\sqrt{\frac{(3x - 1)^2}{3}} \qquad \text{factor the numerator}$$

$$(3x - 1)\sqrt{\frac{1}{3}} \qquad \text{remove the square from the radical}$$

$$(3x - 1)\frac{\sqrt{1}}{\sqrt{3}} \cdot \frac{\sqrt{3}}{\sqrt{3}} \qquad \text{rationalize the denominator}$$

$$\frac{(3x - 1)\sqrt{3}}{3}$$

■ ■ ■

CAUTION ✦ Only *n*th-root *factors* can be removed from the radical. It is necessary to combine terms before simplifying a radical.

3–3 EXERCISES

Express each of the following radicals in simplest radical form.

1. $\sqrt{180}$　　　　　　　**2.** $\sqrt{405}$

3. $\sqrt[3]{24}$　　　　　　　**4.** $\sqrt[3]{108}$

5. $\sqrt[4]{90}$　　　　　　　**6.** $\sqrt[4]{162}$

7. $\sqrt{80a^4b^3}$　　　　　**8.** $\sqrt{98m^4}$

9. $\sqrt{150x^2y^5}$　　　　**10.** $\sqrt{104a^5b}$

11. $\sqrt[3]{32x^2y^4}$　　　　**12.** $\sqrt[4]{64x^6}$

13. $\sqrt{162m^3n^5y^8}$　　**14.** $\sqrt{250p^5m^8}$

15. $\sqrt{363ab^5c^{12}}$　　　**16.** $\sqrt{375k^3n^8}$

17. $\sqrt{288xy^5z^8}$　　　**18.** $\sqrt[4]{81a^6b^{12}}$

19. $\sqrt{\sqrt[3]{y}}$　　　　　　**20.** $\sqrt[3]{4\sqrt{a}}$

21. $\sqrt{\sqrt[3]{4x}}$　　　　　**22.** $\sqrt[4]{\sqrt{6m}}$

23. $\sqrt{\dfrac{72a^7}{9b^6}}$　　　　**24.** $\sqrt[3]{\dfrac{128m^5}{27p^6}}$

25. $\sqrt[3]{192x^5y^2z^8}$　　**26.** $\sqrt[3]{324a^6b^{10}c^3}$

27. $\sqrt[4]{80m^4n^7p^{15}}$　　**28.** $\sqrt{300m^7n^{13}p^4}$

29. $\sqrt[5]{64k^5m^{12}}$　　　**30.** $\sqrt[4]{256x^8y^5z^{15}}$

Simplify by rationalizing the denominator.

31. $\sqrt{\dfrac{15}{8}}$　　　　　**32.** $\sqrt{\dfrac{18}{5}}$

33. $\sqrt{\dfrac{21}{5}}$　　　　　**34.** $\sqrt{\dfrac{19}{11}}$

35. $\sqrt{\dfrac{46a^2b}{3a}}$　　　　**36.** $\sqrt{\dfrac{40}{13a}}$

37. $\sqrt{\dfrac{7a^3}{8}}$　　　　**38.** $\sqrt{\dfrac{13n^2}{54m^3}}$

39. $\sqrt{\dfrac{81m^4}{7n}}$　　　**40.** $\sqrt{\dfrac{36y^3}{27x}}$

41. $\sqrt{\dfrac{5a}{20}}$　　　　**42.** $\sqrt{\dfrac{4b^2}{18a^3}}$

43. $\dfrac{4b}{\sqrt{32b}}$　　　　**44.** $\dfrac{\sqrt{60m^3}}{\sqrt{20p^2}}$

45. $\sqrt{\dfrac{40m^2}{7n^3}}$　　　**46.** $\sqrt{\dfrac{108a^5b^3}{27c^5}}$

47. $\sqrt[3]{\dfrac{6m^3}{4}}$　　　**48.** $\sqrt[3]{\dfrac{16y^6}{81x^4}}$

49. $\sqrt[3]{\dfrac{32m^5}{9n^4}}$　　**50.** $\sqrt[3]{\dfrac{x^3}{4y}}$

51. $\sqrt[4]{\dfrac{81a^2}{4b^2}}$　　　**52.** $\sqrt{\dfrac{64a^6}{27b^3}}$

53. $\sqrt{\dfrac{72x^2y^3}{5z^3}}$　　**54.** $\sqrt[4]{\dfrac{32y^8}{9x^2}}$

55. $\dfrac{6m^3}{\sqrt[3]{n^5}}$　　　**56.** $\dfrac{14m}{\sqrt[4]{8n^2}}$

57. $\dfrac{\sqrt{x^2 - 4}}{x}$

58. $\dfrac{\sqrt{y^2 - 8}}{y}$

59. $\sqrt{x^2 + 8x + 16}$

60. $\sqrt{m^2 + 2m + 1}$

61. $\sqrt{y^2 + 5y + \dfrac{25}{4}}$

62. $\sqrt{3x^2 + 8x + \dfrac{16}{3}}$

63. $\sqrt{8m^2 - 4m + \dfrac{1}{2}}$

64. $\sqrt{5a^2 - 4a + \dfrac{4}{5}}$

P 65. The angular velocity ω after movement through the arc θ is

$$\omega = \sqrt{\omega_0^2 + 2\alpha\theta}$$

where ω_0 is the initial angular velocity and α (alpha) is the angular acceleration. Find the angular velocity if $\omega_0 = 30$ rad/s, $\theta = 1.5$ rad, and $\alpha = 60$ rad/s^2. Leave your answer in simplest radical form.

P 66. The fundamental frequency n of a vibrating string is given by

$$n = \dfrac{1}{2L}\sqrt{\dfrac{T}{m}}$$

where L = length, T = tension, and m = mass/unit length. A string 20 cm long with a mass per unit length

of 0.04 kg/m is under a tension of 450 N. Find the fundamental frequency. Round to hundredths.

C 67. When the map of a region is available, you can find its area by dividing the region into geometric shapes. When the three sides of a triangle have been scaled from the map, the area of a triangle with sides a, b, and c is found by

$$\sqrt{s(s - a)(s - b)(s - c)}$$

where $s = (a + b + c)/2$. Find the area of a triangle with sides 40 m, 25 m, and 35 m. Round to hundredths.

P 68. The period T of a pendulum is given by

$$T = 2\pi\sqrt{\dfrac{L}{g}}$$

where T = time in seconds, L = length in centimeters, and g = acceleration due to gravity. Find the period of a pendulum 40 cm long if $g = 980$ cm/s^2. Leave your answer in simplified radical form.

CS 69. A cell in a spreadsheet program contained the radical expression $\sqrt{2x^2 + 3x + \dfrac{11}{4}}$. Simplify this radical expression.

3–4

OPERATIONS WITH RADICALS

In this section we will discuss adding, subtracting, multiplying, and dividing expressions containing a radical. To perform these operations with radicals, we use many of the same techniques that we used with the corresponding operations with polynomials.

Adding and Subtracting Radicals

To add and subtract expressions containing a radical, combine like terms, just as with polynomials. However, we must define like radical terms. **Like radicals** have the same index and radicand.

Adding and Subtracting Radicals

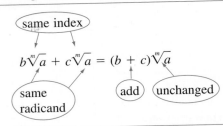

We cannot always identify like radicals at a glance. Usually the radicals must be simplified first.

EXAMPLE 1   Add the following radical expressions:

$$4\sqrt{8} + 5\sqrt{18}$$

Solution   First, we simplify each radical by removing any squares.

$$4\sqrt{8} + 5\sqrt{18}$$
$$4\sqrt{2^2 \cdot 2} + 5\sqrt{3^2 \cdot 2}$$
$$8\sqrt{2} + 15\sqrt{2}$$

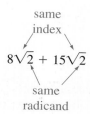

Second, we combine the two like terms.

$$(8 + 15)\sqrt{2}$$
$$23\sqrt{2}$$

■ ■ ■

EXAMPLE 2   Simplify $2a\sqrt[3]{16} + \sqrt[3]{81a^3} - 7a\sqrt[3]{2}$.

Solution   First, we simplify each radical by removing any cubes contained under the radical.

$$2a\sqrt[3]{2^3 \cdot 2} + \sqrt[3]{3^3 \cdot 3 \cdot a^3} - 7a\sqrt[3]{2}$$
$$4a\sqrt[3]{2} + 3a\sqrt[3]{3} - 7a\sqrt[3]{2}$$

Next, we combine like terms. The terms $4a\sqrt[3]{2}$ and $-7a\sqrt[3]{2}$ contain like radicals, but $3a\sqrt[3]{3}$ is not like the other two terms because it has a different radicand. Since only the first and last terms are like, we can combine only these two terms. The result is

$$-3a\sqrt[3]{2} + 3a\sqrt[3]{3}$$

■ ■ ■

EXAMPLE 3   Simplify the following:

$$\sqrt{\frac{a^3}{12}} + \sqrt{27a^3} - \sqrt{\frac{27}{a^3}}$$

Solution   First, we simplify each radical by removing any squares.

$$\frac{a}{2}\sqrt{\frac{a}{3}} + 3a\sqrt{3a} + \frac{3}{a}\sqrt{\frac{3}{a}}$$

Second, we rationalize the denominator, where necessary.

$$\frac{a\sqrt{a}\sqrt{3}}{2\sqrt{3}\sqrt{3}} + 3a\sqrt{3a} + \frac{3\sqrt{3}\sqrt{a}}{a\sqrt{a}\sqrt{a}}$$

$$\frac{a}{2}\sqrt{\frac{3a}{3^2}} + 3a\sqrt{3a} + \frac{3}{a}\sqrt{\frac{3a}{a^2}}$$

$$\frac{a}{6}\sqrt{3a} + 3a\sqrt{3a} + \frac{3}{a^2}\sqrt{3a}$$

Finally, we combine like terms. Since all three terms contain like radicals, we add as follows:

$$\left(\frac{a}{6} + 3a + \frac{3}{a^2}\right)\sqrt{3a}$$

Adding the coefficients using a common denominator of $6a^2$ gives

$$\left(\frac{a^3 + 18a^3 + 18}{6a^2}\right)\sqrt{3a}$$

$$\left(\frac{19a^3 + 18}{6a^2}\right)\sqrt{3a}$$
∎∎∎

Multiplying Radicals

We multiply radicals using the same techniques used to multiply polynomials: the Distributive Property or FOIL. By restating Rule 3–10,

$$\sqrt[n]{a} \cdot \sqrt[n]{b} = \sqrt[n]{ab}$$

we can see that to multiply radicals of the same order, we multiply their radicands. Therefore,

$$\underbrace{\sqrt[3]{6} \cdot \sqrt[3]{2}}_{} = \sqrt[3]{12}$$

(same index) (multiply radicands)

In addition to multiplying the radicals, we must also multiply all factors outside the radical separately.

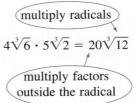

$$4\sqrt[3]{6} \cdot 5\sqrt[3]{2} = 20\sqrt[3]{12}$$

(multiply radicals)

(multiply factors outside the radical)

NOTE ✦ Once the multiplication has been performed, each radical must be simplified and like radicals combined.

EXAMPLE 4 Multiply the following radical expressions:

$$2\sqrt{3}(\sqrt{5} - 6\sqrt{7})$$

Solution To multiply this monomial and binomial, we use the Distributive Property.

$$2\sqrt{3}(\sqrt{5} - 6\sqrt{7})$$

$$2\sqrt{3}(\sqrt{5}) - 2\sqrt{3}(6\sqrt{7}) \qquad \text{Distributive Property}$$

$$2 \cdot 1\sqrt{3 \cdot 5} - 2 \cdot 6\sqrt{3 \cdot 7} \qquad \text{multiply}$$

$$2\sqrt{15} - 12\sqrt{21} \qquad \text{simplify}$$

Since each radical is simplified and there are no like radicals, the product is $2\sqrt{15} - 12\sqrt{21}$.

■■■

EXAMPLE 5 Multiply $(4\sqrt{3} - \sqrt{6})(3\sqrt{3} + 5\sqrt{6})$.

Solution We multiply these two binomials by using FOIL.

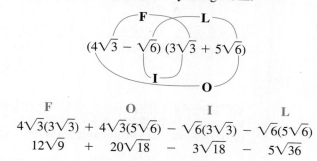

$$4\sqrt{3}(3\sqrt{3}) + 4\sqrt{3}(5\sqrt{6}) - \sqrt{6}(3\sqrt{3}) - \sqrt{6}(5\sqrt{6})$$
$$12\sqrt{9} \quad + \quad 20\sqrt{18} \quad - \quad 3\sqrt{18} \quad - \quad 5\sqrt{36}$$

Next, we simplify the radicals and combine like radicals.

$$36 + 60\sqrt{2} - 9\sqrt{2} - 30$$
$$6 + 51\sqrt{2}$$

■■■

Multiplying Radicals of Different Order

In the previous discussion on multiplication, only radicals of the same order could be multiplied. However, if we convert to fractional exponents, we can multiply radicals of different order. The next example illustrates this process.

EXAMPLE 6 Multiply $\sqrt[3]{6} \cdot \sqrt{4x}$.

Solution First, *write each radical in fractional exponent form.*

$$(6)^{1/3}(4x)^{1/2}$$

Next, *change the fractional exponents to equivalent fractions* with a common denominator.

$$(6)^{2/6}(4x)^{3/6}$$
$$[(6)^2(4x)^3]^{1/6}$$

Then *change back to radical form and simplify.*

$$\sqrt[6]{6^2 \cdot (4x)^3}$$
$$\sqrt[6]{6^2 \cdot 4^3 \cdot x^3}$$
$$\sqrt[6]{(2 \cdot 3)^2 \cdot (2^2)^3 \cdot x^3}$$
$$\sqrt[6]{2^2 \cdot 3^2 \cdot 2^6 \cdot x^3}$$
$$2\sqrt[6]{36x^3}$$

■■■

Dividing Radicals

To divide radicals, we use Rule 3–12, repeated here for convenience.

$$\frac{\sqrt[n]{a}}{\sqrt[n]{b}} = \sqrt[n]{\frac{a}{b}} \quad \text{where } b \neq 0$$

Then we simplify the result as necessary.

EXAMPLE 7 Divide the following:

a. $\dfrac{\sqrt{42}}{\sqrt{7}}$ **b.** $\dfrac{4\sqrt{10x}}{3\sqrt{2x^2}}$

Solution

a. $\dfrac{\sqrt{42}}{\sqrt{7}} = \sqrt{\dfrac{42}{7}} = \sqrt{6}$

b. $\dfrac{4\sqrt{10x}}{3\sqrt{2x^2}} = \dfrac{4}{3}\sqrt{\dfrac{10x}{2x^2}} = \dfrac{4}{3}\sqrt{\dfrac{5}{x}} = \dfrac{4\sqrt{5}\sqrt{x}}{3\sqrt{x}\sqrt{x}} = \dfrac{4\sqrt{5x}}{3x}$ ∎

Binomial Denominators

Rationalizing a binomial denominator requires multiplying by the conjugate of the denominator. Two binomials, for example, $(a + b)(a - b)$, whose product is the difference of two squares are called **conjugates.** For example, $(\sqrt{2} - \sqrt{3})$ is the conjugate of $(\sqrt{2} + \sqrt{3})$.

EXAMPLE 8 Perform the following division:

$$(\sqrt{5} + 3) \div (\sqrt{18} - \sqrt{2})$$

Solution To perform the division, multiply the numerator and denominator by the conjugate of $\sqrt{18} - \sqrt{2}$, which is $\sqrt{18} + \sqrt{2}$.

$$\frac{\sqrt{5} + 3}{\sqrt{18} - \sqrt{2}} \cdot \frac{\sqrt{18} + \sqrt{2}}{\sqrt{18} + \sqrt{2}}$$

$$\frac{\sqrt{90} + 3\sqrt{18} + 3\sqrt{2} + \sqrt{10}}{(\sqrt{18})^2 - (\sqrt{2})^2}$$

$$\frac{\sqrt{9 \cdot 10} + 3\sqrt{9 \cdot 2} + 3\sqrt{2} + \sqrt{10}}{18 - 2}$$

$$\frac{3\sqrt{10} + 9\sqrt{2} + 3\sqrt{2} + \sqrt{10}}{16}$$

$$\frac{4\sqrt{10} + 12\sqrt{2}}{16}$$

$$\frac{4(\sqrt{10} + 3\sqrt{2})}{16}$$

$$\frac{\sqrt{10} + 3\sqrt{2}}{4}$$

EXAMPLE 9 Simplify the following:

$$\frac{(\sqrt{x} - 3\sqrt{y})}{(\sqrt{x} - \sqrt{y})}$$

Solution The conjugate of $\sqrt{x} - \sqrt{y}$ is $\sqrt{x} + \sqrt{y}$. We will multiply the numerator and denominator of the fraction by the conjugate of the denominator and simplify the result.

$$\frac{(\sqrt{x} - 3\sqrt{y})}{(\sqrt{x} - \sqrt{y})} \cdot \frac{(\sqrt{x} + \sqrt{y})}{(\sqrt{x} + \sqrt{y})}$$

multiply using FOIL

$$\frac{\overbrace{\sqrt{x^2} + \sqrt{xy} - 3\sqrt{xy} - 3\sqrt{y^2}}}{\underbrace{(\sqrt{x})^2 - (\sqrt{y})^2}}$$

difference of squares

$$\frac{x - 2\sqrt{xy} - 3y}{x - y}$$

■■■

3–4 EXERCISES

Perform the indicated operation and leave the answer in simplest form.

1. $\sqrt{27} + \sqrt{12} + \sqrt{75}$

2. $\sqrt{27} - \sqrt{75} + \sqrt{48}$

3. $2\sqrt{80} + \sqrt{45} - 2\sqrt{125}$

4. $\sqrt{28} + \sqrt{63} + \sqrt{175}$

5. $3\sqrt{75} - 4\sqrt{20} - \sqrt{80}$

6. $-\sqrt{24} + 2\sqrt{216} - 5\sqrt{54}$

7. $\sqrt{24} - \sqrt{54} - \sqrt{150}$

8. $3\sqrt{40a^3} - 2\sqrt{90a^3} - 5a\sqrt{250a}$

9. $a\sqrt{40a^3} + 3\sqrt{10a^5} + 6a\sqrt{90}$

10. $3x\sqrt{18x^3} + \sqrt{2x^5} + x^2\sqrt{32x}$

11. $\sqrt{27a^3b^4} + \sqrt{48a^5b^2}$

12. $\sqrt{48m^4n^7} + \sqrt{192m^4n^7}$

13. $\sqrt[3]{16} - \sqrt[3]{54}$

14. $\sqrt[3]{81} + \sqrt[3]{648}$

15. $7\sqrt[3]{81} - 2\sqrt[3]{24}$

16. $5\sqrt[3]{16x^4} - 9\sqrt[3]{128x^4}$

17. $6\sqrt{32x^2} + 4\sqrt{162x^2}$

18. $8m\sqrt[3]{24m^3} + 4m\sqrt[3]{54m^3} - m\sqrt[3]{375m^3}$

19. $\sqrt{\dfrac{8}{a^2}} + \dfrac{3}{a}\sqrt{18}$

20. $\sqrt{9a} + \sqrt{\dfrac{4}{a}}$

21. $3\sqrt{\dfrac{2}{a^3}} - \sqrt{8a} + 4\sqrt{\dfrac{9}{a}}$

22. $6x\sqrt{\dfrac{27}{x}} - x\sqrt{\dfrac{12}{x}} + \dfrac{1}{x}\sqrt{3x}$

23. $\sqrt{5}(\sqrt{3} + 6)$

24. $\sqrt{6}(\sqrt{7} - 2\sqrt{10})$

25. $2\sqrt{6}(\sqrt{8} - 4\sqrt{5})$

26. $7\sqrt{3}(\sqrt{3} + 2\sqrt{5})$

27. $(\sqrt{3} + \sqrt{6})(\sqrt{2} - \sqrt{5})$

28. $(\sqrt{7} - \sqrt{2})(\sqrt{4} + \sqrt{3})$

29. $(\sqrt[3]{2x} + 7)(\sqrt[3]{10x} + 6)$

30. $(\sqrt[5]{8x} - 4)(\sqrt[5]{2x} + 7)$

31. $(2\sqrt{3} - 6\sqrt{7})(4\sqrt{5} + \sqrt{7})$

32. $(8\sqrt{3a} + \sqrt{6})(5\sqrt{3a} + \sqrt{2})$

33. $(7\sqrt{3} + 8)(6\sqrt{3} - 5)$

34. $(7\sqrt{12} - \sqrt{5})(2\sqrt{2} + \sqrt{5})$

35. $(3\sqrt[4]{8} + 7\sqrt[4]{6})(2\sqrt[4]{5} - 4\sqrt[4]{12})$

36. $(9\sqrt[3]{40} + \sqrt[3]{48})(6\sqrt[3]{4} + \sqrt[3]{32})$

37. $(8\sqrt{7m} + 6\sqrt{10n})(\sqrt{7m} + 2\sqrt{10n})$

38. $(\sqrt{6} + \sqrt{10})^2$

39. $(5\sqrt{8} - 2\sqrt{10})^2$

40. $(3\sqrt{5} - 7\sqrt{6})^2$

41. $(6\sqrt{3} - 7)^2$

42. $(\sqrt{7} + 4\sqrt{12})(\sqrt{7} - 4\sqrt{12})$

43. $(4\sqrt{5} + \sqrt{6})(4\sqrt{5} - \sqrt{6})$

44. $\dfrac{\sqrt{18x^3}}{\sqrt{x^2}}$

45. $\dfrac{\sqrt{3m^2}}{\sqrt{21m^2}}$

46. $\dfrac{4\sqrt{3}}{2\sqrt{5}}$

47. $\dfrac{12\sqrt{6}}{6\sqrt{5}}$

48. $\dfrac{\sqrt{6}}{3\sqrt{5} - \sqrt{3}}$

49. $\dfrac{\sqrt{7}}{6 - 9\sqrt{2}}$

50. $\dfrac{2\sqrt{x} - \sqrt{y}}{\sqrt{x} + 2\sqrt{y}}$

51. $\dfrac{3\sqrt{a} + 2\sqrt{b}}{4\sqrt{a} + b}$

P 52. The effective mass of the electron has been determined experimentally to be

$$m = \frac{m_0}{\sqrt{1 - (v/c)^2}}$$

where m = effective mass at velocity v, m_0 = rest mass, v = velocity of the electron, and c = velocity of light. Simplify this expression.

CS 53. The formula $f = \dfrac{20}{d + \sqrt{3d + 400}}$ was part of a software program used to analyze viruses. Rationalize the denominator.

P 54. The time required to drain a tank through a hole in the bottom from level l_1 to level l_2 is given by

$$\frac{l_1 - l_2}{\sqrt{2g}\,(\sqrt{l_1} + \sqrt{l_2})}$$

where g is the acceleration due to gravity. Simplify this expression by rationalizing the denominator.

P 55. The damped angular velocity of a suspended weight is given by

$$\sqrt{w^2 - \frac{c^2 g^2}{w^2}}$$

Simplify this expression.

3–5

EQUATIONS WITH RADICALS

Many technical formulas contain radical expressions. In rearranging these formulas, you must often solve an equation containing radicals. Equations containing radicals may be solved using the **Power Rule,** which states that *an equivalent equation results from raising both sides of an equation to the same power.* We can use the Power Rule to remove the radicals, thus making it easier to solve the equation. However, raising each side of an equation to a power can lead to an extraneous solution. An **extraneous solution** appears to be a solution but does not satisfy the original equation. For this reason, all solutions to equations containing a radical must be checked and extraneous roots discarded.

EXAMPLE 1 Solve $\sqrt{3x - 4} - 5 = 0$.

Solution First, we *isolate the radical on one side of the equation.*

$$\sqrt{3x - 4} = 5$$

Next, we *square both sides of the equation.*

$$(\sqrt{3x - 4})^2 = (5)^2$$
$$3x - 4 = 25$$

Then we *solve the resulting equation for x.*

$$3x = 29$$
$$x = 29/3$$

Checking
$$\sqrt{3(29/3) - 4} - 5 = 0$$
$$\sqrt{25} - 5 = 0$$
$$5 - 5 = 0$$
$$0 = 0$$

Remember that we use the principal or positive square root in evaluating these radicals. The solution is 29/3.

■ ■ ■

Solving an Equation with Radicals

1. Isolate the radical term on one side of the equation. If there are two expressions containing a radical, place one radical expression on each side of the equation.
2. Raise each side of the equation to a power equal to the index of the radical.
3. Combine like terms.
4. If the equation still contains a radical, repeat Steps 1, 2, and 3.
5. Solve the equation for the variable.
6. Check the solution in the original equation. Discard any extraneous roots.

EXAMPLE 2 Solve the following equation for x:

$$\sqrt[3]{2x - 1} = 3$$

Solution

$$\sqrt[3]{2x - 1} = 3$$
$$\left(\sqrt[3]{2x - 1}\right)^3 = (3)^3 \qquad \text{cube both sides}$$
$$2x - 1 = 27 \qquad \text{simplify}$$
$$2x = 28 \qquad \text{transpose and add}$$
$$x = 14 \qquad \text{divide}$$

Checking

$$\sqrt[3]{2(14) - 1} = 3$$
$$\sqrt[3]{27} = 3$$
$$3 = 3$$

Since checking results in a true statement, the solution is $x = 14$. ∎

EXAMPLE 3 Solve $\sqrt{y - 1} + 2 = \sqrt{y - 3}$.

Solution Since the two radicals are on opposite sides of the equation, we begin by squaring both sides of the equation.

$$\left(\sqrt{y - 1} + 2\right)^2 = (\sqrt{y - 3})^2 \qquad \text{square both sides}$$
$$y - 1 + 4\sqrt{y - 1} + 4 = y - 3 \qquad \text{simplify}$$
$$4\sqrt{y - 1} = -6 \qquad \text{combine like terms}$$
$$\left(4\sqrt{y - 1}\right)^2 = (-6)^2 \qquad \text{square both sides}$$
$$16(y - 1) = 36 \qquad \text{simplify}$$
$$16y - 16 = 36 \qquad \text{eliminate parentheses}$$
$$16y = 52 \qquad \text{transpose and add}$$
$$y = \frac{13}{4} \qquad \text{divide}$$

Checking

$$\sqrt{\frac{13}{4} - 1} + 2 = \sqrt{\frac{13}{4} - 3}$$
$$\frac{3}{2} + 2 \neq \frac{1}{2}$$

Since $\frac{13}{4}$ is an extraneous solution, this equation has no solution. ∎

CAUTION ✦ In squaring a binomial such as ($\sqrt{y-1}+2$), remember the product of the outside and inside terms:

$$(a + b)^2 \neq a^2 + b^2$$

EXAMPLE 4 Solve $\sqrt{x^2 - 4x} - 5 = x - 2$.

Solution   Isolating the radical on one side of the equation gives

$$\sqrt{x^2 - 4x} = x + 3 \qquad \text{transpose and add}$$
$$\left(\sqrt{x^2 - 4x}\right)^2 = (x + 3)^2 \qquad \text{square both sides}$$
$$x^2 - 4x = x^2 + 6x + 9 \qquad \text{simplify}$$
$$x^2 - 4x - x^2 - 6x = 9 \qquad \text{transpose}$$
$$-10x = 9 \qquad \text{combine like terms}$$
$$x = \frac{-9}{10} \qquad \text{divide}$$

Checking

$$\sqrt{\left(-\frac{9}{10}\right)^2 - 4\left(-\frac{9}{10}\right)} - 5 = -\frac{9}{10} - 2$$

$$\sqrt{\frac{81}{100} + \frac{36}{10}} - 5 = -\frac{29}{10}$$

$$\sqrt{\frac{441}{100}} - 5 = -\frac{29}{10}$$

$$\frac{21}{10} - 5 = -\frac{29}{10}$$

$$-\frac{29}{10} = -\frac{29}{10}$$

Since checking results in a true statement, the solution is

$$x = \frac{-9}{10}$$

■ ■ ■

Applications

🔲 **EXAMPLE 5**   The velocity of an object moving in a horizontal circle is

$$V = \sqrt{\frac{Fr}{m}}$$

where F = centripetal force, r = radius of the circle, and m = mass of the object. Find the radius of the circle if an object has a centripetal force of 6.8 N, a velocity of 4.1 m/s, and a mass of 1.7 kg. Solve for r and then substitute.

Solution   First, square both sides of the equation.

$$(V)^2 = \left(\sqrt{\frac{Fr}{m}}\right)^2 \qquad \text{square both sides}$$

$$V^2 = \frac{Fr}{m} \qquad \text{simplify}$$

Then solve for r.

$$V^2 m = Fr \qquad \text{eliminate fractions}$$

$$\frac{V^2 m}{F} = r \qquad \text{divide by } F$$

Last, substitute the values $F = 6.8$ N, $V = 4.1$ m/s, and $m = 1.7$ kg $\left(\text{N} = \dfrac{\text{kg} \cdot \text{m}}{\text{s}^2}\right)$.

$$\frac{(4.1 \text{ m/s})^2 (1.7 \text{ kg})}{6.8 \text{ N}} = r$$

$$4.2 \text{ m} = r$$

The radius of the circle is 4.2 m.

4.1 $\boxed{x^2}$ $\boxed{\times}$ 1.7 $\boxed{\div}$ 6.8 $\boxed{=}$ $\longrightarrow$ 4.2025

$\blacksquare\blacksquare\blacksquare$

EXAMPLE 6 The formula to determine the range d of a ship-to-shore radio message is given by

$$d = \sqrt{2rh + h^2}$$

Solve this literal equation for r.

Solution

$$
\begin{array}{ll}
d = \sqrt{2rh + h^2} & \\
(d)^2 = \left(\sqrt{2rh + h^2}\right)^2 & \text{square both sides} \\
d^2 = 2rh + h^2 & \text{simplify} \\
d^2 - h^2 = 2rh & \text{transpose} \\
\dfrac{d^2 - h^2}{2h} = r & \text{divide by } 2h
\end{array}
$$

$\blacksquare\blacksquare\blacksquare$

EXAMPLE 7 The marketing department of a computer manufacturer determines that the demand for its computer depends on the price it charges. The relationship of price to demand is given by the formula

$$p = 1{,}595 - \sqrt{0.10x + 150} \qquad x \geq 0$$

where p is the price and x is the number of computers sold at that price. Find the number of computers sold if the price is $1,295.

Solution We substitute $p = 1{,}295$ and solve the equation for x.

$$
\begin{array}{ll}
1{,}295 = 1{,}595 - \sqrt{0.10x + 150} & \\
\sqrt{0.10x + 150} = 1{,}595 - 1{,}295 & \text{isolate the radical} \\
\sqrt{0.10x + 150} = 300 & \text{simplify} \\
0.10x + 150 = 90{,}000 & \text{square both sides} \\
0.10x = 90{,}000 - 150 & \text{isolate } x \\
0.10x = 89{,}850 & \text{simplify} \\
x = 898{,}500 & \text{divide}
\end{array}
$$

The company can sell 898,500 computers at a price of $1,295 each.

$\blacksquare\blacksquare\blacksquare$

3–5 EXERCISES

Solve the following equations for the variable.

1. $\sqrt{3y + 1} - 5 = 0$

2. $2\sqrt{y} + 8 = 0$

3. $\sqrt{4y - 7} = 3$

4. $\sqrt{5x - 1} = 3$

5. $\sqrt{5a - 8} = 3$

6. $\sqrt{7m - 6} = 6$

7. $\sqrt{2m + 10} = -4$

8. $\sqrt{4a - 2} = 6$

9. $\sqrt{6x - 9} = 9$

10. $\sqrt[3]{5a - 1} = 3$

11. $\sqrt[3]{7a - 1} = 3$

12. $\sqrt[3]{6b - 4} = -4$

13. $\sqrt[3]{4y + 12} = -2$

14. $\sqrt[3]{2y - 5} = 3$

15. $\sqrt[3]{7m - 8} = 3$

16. $2\sqrt{x - 1} = 5$

17. $\sqrt{3y + 7} + 5 = 8$

18. $3\sqrt{y - 2} = 9$

19. $\sqrt{6a - 1} - 9 = 5$

20. $\sqrt{3m + 12} - 5 = 4$

21. $\sqrt{8b + 4} - 2 = 8$

22. $6 - \sqrt{8m - 1} = 10$

23. $\sqrt{2y + 1} + 2 = \sqrt{2y - 5}$

24. $\sqrt{5x - 2} + 3 = \sqrt{7 + 5x}$

25. $\sqrt{3x - 1} + 10 = \sqrt{3x + 8}$

26. $\sqrt[3]{9a - 6} = \sqrt[3]{3a + 18}$

27. $\sqrt[4]{5m - 1} = \sqrt[4]{2m + 8}$

28. $\sqrt[4]{5x - 4} - \sqrt[4]{2x + 5} = 0$

29. $\sqrt{7p - 8} = \sqrt{2p + 5}$

30. $\sqrt{10m + 6} = \sqrt{2m + 22}$

31. $\sqrt[3]{3x + 6} - \sqrt[3]{9x + 22} = 0$

32. $\sqrt[3]{11x - 5} = \sqrt[3]{8x + 4}$

33. $\sqrt{3a + 4} - \sqrt{2a + 4} = 0$

34. $\sqrt{a^2 - 3a + 5} = a - 7$

35. $\sqrt{m^2 + 8m - 17} = m$

36. $\sqrt{4b^2 - 6b - 2} = 2b - 10$

37. $\sqrt{1 - 2p + p^2} = p$

38. $\sqrt{16m^2 - 9} = 4m - 1$

39. $\sqrt{y^2 + 5y + 9} = 3$

40. $3x + 2 = \sqrt{9x^2 + 5x}$

41. $\sqrt{9a^2 + 4a} - 3a = 2$

42. $\sqrt{5 - 2a + 16a^2} = 4a - 1$

43. $\sqrt{a^2 + 2a + 4} = a + 4$

44. $\sqrt{a^2 - 16} = 4 + a$

45. $\sqrt{4a^2 + 8a - 5} = 2a + 3$

46. $a + 6 = \sqrt{a^2 + 7a - 4}$

P 47. The average velocity of molecules is given by

$$v = \sqrt{\frac{3kT}{m}}$$

where T is temperature, m is mass, and k is a constant involving the universal gas constant. Solve this literal equation for T.

E 48. The resonance frequency of a circuit containing inductance L and capacitance C is

$$f = \frac{1}{2\pi\sqrt{LC}}$$

Solve for C.

49. The radius of a sphere in terms of its volume is given by

$$r = \sqrt[3]{\frac{3V}{4\pi}}$$

Solve for V.

P 50. The velocity V of an object in simple harmonic motion is given by

$$V = \frac{2\pi}{T}\sqrt{r^2 - x^2}$$

where r = amplitude, T = period, and x = displacement of the vibrating body from its position of rest. Solve for x.

P 51. The velocity v of an object with initial velocity v_0, displacement s, and acceleration a is given by

$$v = \sqrt{v_0^2 + 2as}$$

Solve this literal equation for s.

52. The lateral surface area S of a cone as a function of the radius of the base r and height h is given by

$$S = \pi r\sqrt{r^2 + h^2}$$

Solve this formula for h.

B 53. Using Example 7, determine how many computers the company would sell if it lowered the price to $995.

CHAPTER SUMMARY

Summary of Terms

base (p. 90)

conjugates (p. 110)

exponent (p. 90)

exponential form (p. 90)

extraneous solution (p. 112)

index (p. 96)

like radicals (p. 106)

nth power (p. 90)

Power Rule (p. 112)

radical (p. 96)

radicand (p. 96)

rationalizing the denominator (p. 103)

CHAPTER REVIEW

Section 3–1

Simplify by removing zero and negative exponents and applying the rules for exponents.

1. $\dfrac{2x^{-3}y^2}{4^{-1}y^{-6}}$

2. $\dfrac{6a^{-4}b^3c^0}{9a^3b^{-5}c^3}$

3. $\dfrac{16m^{-6}n^3}{24m^{-4}n^{-8}}$

4. $\dfrac{15x^{-8}y^0z^4}{20x^{10}y^{-4}z^{-7}}$

5. $\dfrac{3m^{-1}p^3}{mnp^{-6}}$

6. $\dfrac{3xy^{-2}z^4}{9x^{-4}y^{-3}}$

7. $\left(\dfrac{a^{-4}b^3c^{-1}}{a^0c^{-6}}\right)^{-3}$

8. $\left(\dfrac{3x^3y^{-4}}{x^{-6}y^{-3}}\right)^2$

9. $x^{-1} + 3y^{-2}$

10. $4m^{-1} - 6n^{-2}$

Section 3–2

Evaluate.

11. $81^{3/4}$

12. $343^{-2/3}$

13. $16^{-1/4}$

14. $64^{5/6}$

15. $8^{-2/3} - 125^{1/3}$

16. $16^{3/4} + 216^{-2/3}$

Evaluate and round to hundredths.

17. $6^{1/3} \cdot 6^{1/2}$

18. $8^{3/4} \cdot 9^{2/3}$

19. $(16^{3/4})^{1/2}$

20. $(7^{3/5})^{-2}$

Simplify

21. $\dfrac{m^{3/2}n^{2/5}}{m^{-1/3}n^2}$

22. $\dfrac{a^3b^{-1/2}}{a^{-2/3}b^0}$

23. $\dfrac{(x^{-2}y^{1/3})^2}{x^{-3/4}z^{-2}}$

24. $\dfrac{(m^{-3/5}n^2)^{-5}}{n^{-8}}$

Section 3–3

Express each answer in simplest radical form.

25. $\sqrt{180x^2y^5}$

26. $\sqrt{108m^5n^8}$

27. $\sqrt{54x^5y^8}$

28. $\sqrt{64x^9y^3}$

29. $\sqrt{\dfrac{8x^2y^3}{z^6}}$

30. $\sqrt{\dfrac{64b^7c^3}{a^8}}$

31. $\sqrt{\dfrac{27x^3y^5}{16z^5}}$

32. $\sqrt{\dfrac{81ab^4}{c^7}}$

33. $\sqrt{\dfrac{a^5b^9}{4c}}$

34. $\sqrt{\dfrac{9m^8n^3}{8p}}$

35. $\sqrt{2x^2 + 10x + \dfrac{25}{2}}$

36. $\sqrt{5y^2 - 2y + \dfrac{1}{5}}$

Section 3–4

Perform the indicated operation. Leave your answer in simplest radical form.

37. $3\sqrt{50} - 5\sqrt{32} + \sqrt{8}$

38. $3\sqrt{8} + 5\sqrt{18} - 3\sqrt{72}$

39. $\sqrt{27} + 2\sqrt{72} - 3\sqrt{48}$

40. $\sqrt{125} - \sqrt{20} - \sqrt{45}$

41. $2\sqrt{16} - \sqrt{54} + 3\sqrt{128}$

42. $\sqrt{81} - 2\sqrt{24} + 5\sqrt{3}$

43. $3\sqrt{2}(\sqrt{5} + 4\sqrt{3})$

44. $\sqrt{7}(2\sqrt{3} - \sqrt{5})$

45. $(\sqrt{3} + \sqrt{8})(\sqrt{3} - 3\sqrt{5})$

46. $(3\sqrt{6} + \sqrt{5})(\sqrt{6} - 2\sqrt{5})$

47. $(2\sqrt{3} + 6\sqrt{5})(\sqrt{3} - 3\sqrt{5})$

48. $(\sqrt{8} - \sqrt{6})(2\sqrt{8} + 3\sqrt{6})$

49. $\dfrac{\sqrt{20x^3}}{\sqrt{5x^5}}$

50. $\dfrac{\sqrt{60y^5}}{\sqrt{4y}}$

51. $\dfrac{6\sqrt{3}}{2\sqrt{7}}$

52. $\dfrac{4\sqrt{5}}{16\sqrt{3}}$

53. $\dfrac{3\sqrt{x} - 4}{\sqrt{x} + 2}$

54. $\dfrac{6 - \sqrt{y}}{5 + \sqrt{y}}$

55. $\dfrac{2\sqrt{m} - \sqrt{n}}{\sqrt{m} + \sqrt{n}}$

56. $\dfrac{\sqrt{a} + 3\sqrt{b}}{\sqrt{a} - 3\sqrt{b}}$

Section 3–5

Solve the following equations.

57. $3\sqrt{y} = 9$

58. $6\sqrt{x} = 12$

59. $\sqrt{2y + 3} = 5$

60. $\sqrt{3y + 4} = 4$

61. $\sqrt{3z + 1} + 5 = 8$

62. $\sqrt{4x - 1} + 5 = 12$

63. $6 - \sqrt{x + 3} = 7$

64. $7 + \sqrt{2y - 1} = -1$

65. $3\sqrt{a - 4} = \sqrt{2a + 9}$

66. $4\sqrt{m - 5} = 20$

67. $\sqrt{x + 5} = \sqrt{x - 1} + 6$

68. $\sqrt{y + 3} - 6 = \sqrt{y - 7}$

69. $\sqrt{x - 3} + 6 = \sqrt{x - 8}$

CHAPTER TEST

The number in parentheses refers to the appropriate learning objective at the beginning of the chapter.

1. Combine $2\sqrt{27} - 7\sqrt{24} + 5\sqrt{3}$.　**(4)**

2. Simplify $\sqrt[3]{108x^5y^8z}$.　**(3)**

3. Evaluate the following expression: $8^{-2/3} + 16^{3/4}$.　**(2)**

4. Simplify the following and rationalize the denominator.　**(4)**

$$\sqrt[3]{\frac{16x^4y^3}{z^4}}$$

P 5. Pressure has units of force per unit area. This gives　**(1)**

$$\frac{(M)(L)(T)^{-2}}{(L)^2}$$

Remove negative exponents and simplify.

6. Solve the following for x:　**(5)**

$$\sqrt{3x - 1} + 8 = 10$$

7. Apply the Laws of Exponents to simplify the following:　**(1)**

$$\frac{9x^{-1}y^{-3}z^0}{15x^4y^{-8}z^4}$$

8. Multiply the following binomials:　**(4)**

$$(3\sqrt{2} + \sqrt{5})(\sqrt{2} - 4\sqrt{5})$$

9. Simplify the radical $\sqrt{216x^5y^3z}$.　**(3)**

B 10. A sinking fund is an annuity established to produce an amount of money at some future date. If amount A is deposited at the end of each period and interest is compounded at rate i per period, the total P will be produced at the end of n periods.

$$P = A\left[\frac{(1 + i)^n - 1}{i}\right]$$

Evaluate this expression if $A =$ \$5,000, $i = 8.5\%$, and $n = 2.5$ yr.

11. Simplify　**(4)**

$$\sqrt{\frac{x^9z^2}{18y^5}}$$

12. Multiply the following radicals:　**(4)**

$$6\sqrt{3}(\sqrt{8} - 3\sqrt{6})$$

13. Simplify　**(1)**

$$\left(\frac{4x^{-1}y^3}{18x^3y^{-4}}\right)^{-2}$$

14. Evaluate $81^{-3/4}$.　**(2)**

P 15. The frequency n of a stretched string of length L, tension T, density d, and radius r is given by

$$n = (2rL)^{-1} \sqrt{\frac{T}{\pi d}}$$

Simplify this expression when $T = 21$ N (2.1×10^6 g·cm/s^2), $d = 8$ g/cm^3, $r = 0.05$ cm, and $L = 20$ cm.

16. Simplify the following: (2)

$$\frac{9^{3/4} \cdot 9^{-1/4}}{9}$$

17. Solve the equation $\sqrt{4x + 1} + 6 = \sqrt{4x + 2}$. (5)

GROUP ACTIVITY

Most long-term loans, like home mortgages and some automobile loans, are amortized by paying the same amount at each payment period (e.g., monthly payment) over a specified period of time (e.g., 15 years). Part of the payment is interest and part is the principal (i.e., amount borrowed). The amount of money that goes toward the principal and toward the interest in each payment varies over the life of the loan. This means that the first payment is made up of mostly interest while the last payment is made up of mostly principal. This is not unexpected because the bank generally wants to get its fee (interest) as soon as possible. The payment that you must make is calculated by the following equation.

$$\text{Payment} = \frac{PV}{\left[\dfrac{1 - (1 + i)^{-n}}{i} \right]}$$

where PV is the amount of the loan, n is the number of payments that are to be made, and i is equal to the amount of interest at each payment period. As an example, a 13.5% yearly interest rate with monthly payments would result in:

$$i = \frac{0.135}{12} \qquad \text{Remember that you have to convert percentage into a decimal}$$

Assume that you have $600 available for a down payment. Also assume that after all expenses are taken out, you have $250 available for monthly payments. Which of the following deals would fit your finances?

Dealer	Price of Car	Loan Amount	Loan Duration	Interest
Bob's Used Cars	$8,290.00	$7,690.00	3.0 years	12.5%
Mary's Used Cars	$9,000.00	$8,400.00	3.0 years	10.5%
Carlos's Wheels	$10,300.00	$9,700.00	4.0 years	10.5%
Linda's Deals	$6,200.00	$5,600.00	2.0 years	8.5%
Cars by Exie	$7,200.00	$6,600.00	2.5 years	9.0%

1. For each dealer in the table, calculate the monthly payment.
2. What is the total interest that you will pay over the life of each of the loans?
3. Which one(s) are you qualified to purchase?
4. Generate an amortization table for the loan with the lowest monthly payment. An amortization table shows how much of each payment goes toward the interest and how much goes toward the principal. You need to calculate only the first 6 payments and the last 6 payments. Is the ratio of principal to interest the same for each payment?
5. What other cars could you purchase if you could afford another $100 as down payment?
6. Which car should you purchase? What is the best value or lowest cost over the life of the loan?
7. How could you help to reduce the interest paid to the dealer? Would accelerated payment of the loan help?

SOLUTION TO CHAPTER INTRODUCTION

The formula for determining the amount to be invested now is $P = S(1 + i)^{-n}$ where $S =$ the yield in dollars after n years of interest at rate i. From the information given $S = \$48,000$, $i = 9\%$, and $n = 10$ years. Before substituting, remove the negative exponent.

$$P = S(1 + i)^{-n}$$

$$P = \frac{S}{(1 + i)^n}$$

$$P = \frac{48,000}{(1 + 0.09)^{10}}$$

$$P \approx \$20,276$$

1.09 $\boxed{y^x}$ 10 $\boxed{=}$ $\boxed{\div}$ 48000 $\boxed{x \leftrightarrows y}$ $\boxed{=}$ $\longrightarrow$ 20275.719

G

raphs provide a visual representation of the relationship between two variables. They can be used to determine the break-even point for a company. For example, imagine that you own a company that makes and sells widgets for $2.50. The fixed cost of renting the building, paying utilities, and monthly salaries is $1,500 per month. The variable cost of raw materials, hourly salaries, and transportation is $1.25 per unit. Determine the quantity of widgets to manufacture and the resulting revenue to break even. (The answer to this problem is given at the end of the chapter.)

We deal with functions every day without realizing it. For example, the distance you can drive on a full tank of gas is a function of the type of car, the speed you drive, and road conditions. The velocity with which a ball strikes the ground is a function of its initial velocity, acceleration, and elapsed time.

Learning Objectives

After completing this chapter, you should be able to

1. Express one variable as a function of a second variable (Section 4–1).

2. Identify the domain of a function (Section 4–1).

3. Evaluate $f(x)$ for a given value of x (Section 4–1).

4. Plot points on the rectangular coordinate system (Section 4–2).

5. Graph a given function on the rectangular coordinate system (Section 4–3).

6. Determine the slope of a line given two points (Section 4–4).

7. Determine the distance between two points and the midpoint of two points (Section 4–4).

8. Find the solution to a given inequality or system of two inequalities on the rectangular coordinate system (Section 4–5).

9. Apply these concepts to solve an equation, find the break-even point of a business, and graph empirical data (Section 4–6).

Chapter 4

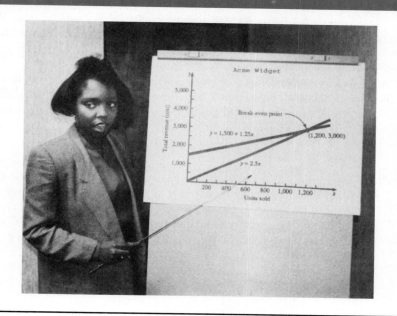

Functions and Graphs

4–1

FUNCTIONS

Definitions

A **function** describes the relationship between variables. In other words, a function is a rule, usually in the form of an equation, that generates an ordered pair representing a unique number for the dependent variable for each chosen value of the independent variable. The variable whose numerical value is arbitrarily chosen is called the **independent variable,** and the variable whose value is determined from this choice is called the **dependent variable.** In an equation involving x and y, x is usually the independent variable, and y is usually the dependent variable. With variables other than x and y, the physical constraints determine the independent variable. For example, the distance s that an object falls is given by $s = 16t^2$, where t represents elapsed time. Since the distance that the object has fallen depends on the elapsed time, time is the independent variable, and distance is the dependent variable. Although we will discuss functions of two variables, it is also possible for a function to have several independent variables and one dependent variable, such as the volume of a pyramid, given by $V = (\pi r^2 h)/3$. A function can also consist of only a dependent variable, such as $y = 3$.

Functional Notation

Returning to the example of the distance that an object falls, given by $s = 16t^2$, we can say that the distance an object falls is a function of the elapsed time. To denote this functional relationship, the equation $s = 16t^2$ could be written $f(t) = 16t^2$ or $s(t) = 16t^2$. The notation $f(t)$ is called **functional notation** and is read "f of t" to denote that the relationship is a function of t, the independent variable. In the case of the variables x and y, the equation $y = 2x - 7$ could also be represented as $f(x) = 2x - 7$ or $y(x) = 2x - 7$ because y is a function of the independent variable x.

EXAMPLE 1   Write the indicated function using functional notation.

(a) Express $m = 6p^2 - p + 5$ as a function of p.

(b) Express the area A of a triangle whose height is 10 ft as a function of the base b.

(c) Express the perimeter P of a rectangle of width 8 m as a function of length L.

Solution

(a) Since m is a function of p, one functional notation is $f(p) = 6p^2 - p + 5$. To denote m as the dependent variable, you could also write $m(p) = 6p^2 - p + 5$.

(b) The area of a triangle whose height is 10 ft is given by the formula

$$A = \frac{10b}{2}$$

$$A = 5b$$

The functional notation could be $f(b) = 5b$ or $A(b) = 5b$.

(c) The perimeter of a rectangle of width 8 m is given by

$$P = 2(8) + 2L$$
$$P = 16 + 2L$$

The functional notation could be written as $f(L) = 16 + 2L$ or $P(L) = 16 + 2L$.

■ ■ ■

Functional notation is useful in naming a specific value of the independent variable. For example, for the function $f(x) = 2x - 7$, the notation $f(3)$ represents the value of the dependent variable when the independent variable equals 3. To determine $f(3)$, we substitute 3 for x as follows:

$$f(x) = 2x - 7$$
$$f(3) = 2(3) - 7 \qquad \text{substitute } x = 3$$
$$f(3) = -1$$

EXAMPLE 2 Given the function $f(x) = 3x^2 - 2x + 1$, find the following:

(a) $f(0)$ **(b)** $f(-2)$ **(c)** $f(4)$

Solution

(a) To find $f(0)$, we substitute 0 for x in the function.

$$f(x) = 3x^2 - 2x + 1$$
$$f(0) = 3(0)^2 - 2(0) + 1 \qquad \text{substitute } x = 0$$
$$f(0) = 0 - 0 + 1$$
$$f(0) = 1$$

(b)
$$f(-2) = 3(-2)^2 - 2(-2) + 1 \qquad \text{substitute } x = -2$$
$$f(-2) = 12 + 4 + 1$$
$$f(-2) = 17$$

(c)
$$f(4) = 3(4)^2 - 2(4) + 1 \qquad \text{substitute } x = 4$$
$$f(4) = 48 - 8 + 1$$
$$f(4) = 41$$

■■■

A method similar to that used in Examples 1 and 2 is used to evaluate a function when a literal expression is substituted for the independent variable, as shown in the next example.

EXAMPLE 3 If $f(x) = x^2 - 4x$, find $f(x - 3)$.

Solution To find $f(x - 3)$, we substitute $x - 3$ for x in the function and simplify the result.

$$f(x) = x^2 - 4x$$
$$f(x - 3) = (x - 3)^2 - 4(x - 3) \qquad \text{substitute}$$
$$f(x - 3) = x^2 - 6x + 9 - 4x + 12 \qquad \text{multiply}$$
$$f(x - 3) = x^2 - 10x + 21 \qquad \text{simplify}$$

■■■

EXAMPLE 4 Evaluate the following functions.

(a) Find $m(2)$ for $m(x) = \sqrt{4x - 5}$.

(b) Find $p(1.8)$ for $p(n) = \dfrac{n^3 - 3n}{9 - 2n}$.

(c) The displacement of a free-falling object with an initial velocity of 9 ft/s as a function of time is given by $s(t) = 9t + 16t^2$. Find $s(4) + s(t)$.

Solution

(a) To find $m(2)$, substitute $x = 2$ into the function and simplify.

$$m(x) = \sqrt{4x - 5}$$
$$m(2) = \sqrt{4(2) - 5}$$
$$m(2) = \sqrt{3}$$

(b) To determine $p(1.8)$, substitute $n = 1.8$ into the function.

$$p(n) = \frac{n^3 - 3n}{9 - 2n}$$

$$p(1.8) = \frac{(1.8)^3 - 3(1.8)}{9 - 2(1.8)}$$

$$p(1.8) \approx 0.08$$

1.8 $\boxed{y^x}$ 3 $\boxed{-}$ 3 $\boxed{\times}$ 1.8 $\boxed{=}$ $\boxed{\div}$ $\boxed{(}$ 9 $\boxed{-}$ 2 $\boxed{\times}$ 1.8 $\boxed{)}$ $\boxed{=}$ $\longrightarrow$ 0.08

(c)
$$s(t) = 9t + 16t^2$$

$$s(4) + s(t) = \overbrace{[9(4) + 16(4)^2]}^{s(4)} + \overbrace{[9t + 16t^2]}^{s(t)}$$
$$= (36 + 256) \quad\quad + (9t + 16t^2)$$
$$= 292 + 9t + 16t^2$$

∎

Restriction on the Domain

The **domain** of a function is the set of permissible values for the independent variable. The **range** of a function is the set of possible values for the dependent variable resulting from the domain. The domain can be called the *input values,* and the range can be called the *output values.* Unless otherwise noted, the domain and range of a function are assumed to be the set of all real numbers.

HINT ✦ Since the range of a function is limited to real numbers, certain functions may require a restriction of the domain. The primary restrictions result from avoiding any attempt to divide by zero or take the square root of a negative number.

EXAMPLE 5 Determine the domain for the following function:

$$f(x) = \frac{1}{x} + 3x$$

Solution The domain must exclude zero because it leads to division by zero. The domain is all real numbers except zero. ∎

Restrictions on the domain of a function can be seen in the graph of the function. For example, the graph of $f(x) = \frac{1}{x} + 3x$ is shown in Figure 4–1. From Example 5, we determine that x cannot be zero. If you examine the graph, you will see that $x = 0$ divides the graph into two sections, one on each side of zero.

FIGURE 4–1

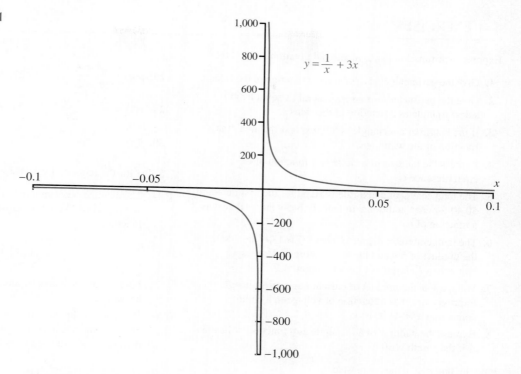

$$y = \frac{1}{x} + 3x$$

EXAMPLE 6 Determine the domain for the following functions:

(a) $f(x) = \sqrt{x - 2}$

(b) $f(p) = \dfrac{1}{p} + \dfrac{3}{p - 4}$

(c) $y(x) = \dfrac{\sqrt{9 - x}}{2x - 12}$

Solution

(a) The square root of a negative number does not result in a real number. Therefore, the quantity under the radical must be positive or zero: $x - 2 \geq 0$, $x \geq 2$. The domain is all real numbers greater than or equal to 2.

(b) Since division by zero is not allowed, we determine the values of p in the denominator that would result in division by zero.

$$p \neq 0 \qquad \text{and} \qquad p - 4 \neq 0$$
$$p \neq 4$$

The domain is all real numbers except 0 and 4.

(c) Since the quantity under the radical cannot be negative and the quantity in the denominator cannot equal zero, the domain is

$$9 - x \geq 0 \qquad\qquad 2x - 12 \neq 0$$
$$-x \geq -9 \qquad\qquad 2x \neq 12$$
$$x \leq 9 \qquad\qquad x \neq 6$$

all real numbers less than or equal to 9 except 6. ∎ ∎ ∎

4–1 EXERCISES

Express each function using functional notation.

1. Give the perimeter P of a square as a function of the sides s.

2. Give the perimeter P of an equilateral (3 equal length sides) triangle as a function of the sides s.

3. If the length of a rectangle is 5 ft, express the area A as a function of the width w.

4. Express the radius r of a circle as a function of its circumference C.

5. The total cost C of a taxi ride is $0.60 for the first mile and $0.40 for each additional mile m. Express the total cost as a function of m.

6. The temperature in degrees Celsius C is 17.8° less than the product of 5/9 and the temperature in degrees Fahrenheit F. Express C as a function of F.

E 7. Voltage V is the product of current I and resistance R. Express current as a function of voltage in a circuit containing a 10-Ω resistor.

8. Express the width w of a rectangle as a function of its area A if the length is 30 ft.

Find the function values requested.

9. $f(x) = 5x^2 + 6$ (a) $f(2)$ (b) $f(-3)$ (c) $f(-1)$

10. $y(x) = 4x - 6$ (a) $y(-2)$ (b) $y(0)$ (c) $y(4)$

11. $f(x) = -x^2 + 3x - 1$ (a) $f(6)$ (b) $f(1)$ (c) $f(-3)$

12. $m(p) = 2\sqrt{x} - 4$ (a) $m(6)$ (b) $m(9.7)$ (c) $m(2)$

13. $g(a) = \dfrac{3}{a} + 2a$ (a) $g(2.8)$ (b) $g(-3)$ (c) $g(-2.1)$

14. $b(c) = \sqrt{4c} - 15$ (a) $b(0)$ (b) $b(1.4)$ (c) $b(4)$

15. $g(x) = \sqrt{x} - \dfrac{3}{x + 1}$ (a) $g(0)$ (b) $g(7)$ (c) $g(3)$

16. $f(h) = \dfrac{3h}{h^2 - 2}$ (a) $f(3.7)$ (b) $f(-4)$ (c) $f(0)$

17. $s(t) = 3t^2 - 4t + 9$ (a) $s(-3)$ (b) $s(1)$ (c) $s(6)$

18. If $f(x) = 4x^2 - 7$, find $f(m)$ and $f(x - 3)$.

19. If $m(s) = 3s - 14$, find $m(x - 8)$ and $m(2s + 5)$.

20. If $g(x) = x^2 - 7x + 9$, find $g(t^3)$ and $g(x - 4)$.

21. If $p(n) = \dfrac{3n + 8}{n - 1}$, find $p(2n)$ and $p(n + 1) - p(n)$.

22. If $f(y) = 2y + 9$, find $f(3y - 4)$ and $f(y^2)$.

23. If $g(y) = 3y^2 - 7y$, find $g(y - 1) + g(3y)$.

If the range is restricted to real numbers, determine the domain of the following functions.

24. $f(x) = \dfrac{2}{x} + 3$

25. $g(y) = \sqrt{y + 7}$

26. $p(a) = \dfrac{a^2 + 6}{a - 2}$

27. $m(t) = \dfrac{3 + t}{2t - 1} + \sqrt{3t}$

28. $y(x) = \dfrac{\sqrt{x^2 - 9}}{4x - 20}$

29. $h(x) = \dfrac{3}{x} + \dfrac{4x}{x + 6}$

30. $f(p) = \dfrac{\sqrt{p + 4} + \sqrt{p - 5}}{3p + 9}$

Solve for the requested variable.

P 31. The average acceleration a of an object starting from rest and reaching 30 mi/h in elapsed time t is given by the function

$$a(t) = \frac{30}{t}$$

Find the average acceleration if the time required is 6 hours.

M 32. Horsepower (hp) is a function of force, distance, and elapsed time t. If a force of 50 lb moves an object 6 ft in time t seconds, the equivalent horsepower is

$$hp = \frac{300}{550t}$$

Find the horsepower if $t = 8$ s.

B 33. Tax rate R is a function of the levied tax L and the assessed value V of the property and is given by

$$R = \frac{L}{V}$$

If the levied tax is $800, determine the tax rate when the property is assessed at $1,300.

34. The surface area s of a cube as a function of the length of an edge e is given by $s = 6e^2$. Find the surface area of a cube whose edge is 8 in.

35. The volume V of a pyramid with a square base of 6 m as a function of height h is given by $V = 12h$. Find the volume of a pyramid 7 m high.

CS 36. Some Internet access providers charge an hourly rate for their services. The total cost c of Internet access with these companies is a function of time t (in hours) online. If the hourly rate is $0.99, express the total cost as a function of time.

4–2

THE RECTANGULAR COORDINATE SYSTEM

In discussing functions such as $y = 3x - 2$, it is helpful to have a visual representation. The solution to a function of two variables contains a value for each variable. Since a pair of numbers cannot be graphed on the number line, in this section we will discuss a system on which to graph a function of two variables.

Definitions

A function in two variables can be graphed on the **rectangular coordinate system,** also called the **Cartesian coordinate system,** named after the French philosopher René Descartes. The rectangular coordinate system consists of two perpendicular number lines, called **axes.** The horizontal number line, usually called the **x axis,** represents the independent variable and is positive to the right of zero. The vertical number line, usually called the **y axis,** represents the dependent variable and is positive above zero. The intersection of the axes is called the **origin** because it represents zero for each variable. The axes divide the plane into four equal regions, called **quadrants.** The quadrants are numbered I, II, III, and IV in a counterclockwise direction, with Quadrant I in the upper right. In Quadrant I, both the x and y coordinates are positive; in Quadrant II, the x coordinate is negative and the y coordinate is positive; in Quadrant III, both the x and y coordinates are negative; and in Quadrant IV, x is positive and y is negative. Refer to Figure 4–2 for an illustration of this terminology.

FIGURE 4–2

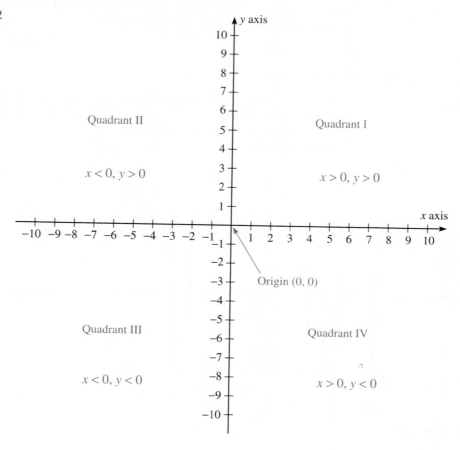

The solution to a function of two variables consists of ordered pairs (x, y) where the first element in the ordered pair is called the **x coordinate,** or abscissa, and the second element is called the **y coordinate,** or ordinate. Furthermore, there is a **one-to-one correspondence** between an ordered pair and a point on the rectangular coordinate system; that is, each ordered pair results in exactly one point on the rectangular coordinate system, and each point results in exactly one ordered pair. The process of locating points on the rectangular coordinate system is called *plotting the points.*

Plotting Points

To Plot an Ordered Pair

---

1. Start at the origin and move x units left or right, depending on the sign.

2. From the x location, move y units up or down, depending on the sign. Place a dot at the point.

EXAMPLE 1   Plot the point corresponding to $(4, -3)$.

Solution   We start at the origin and, because the x coordinate is positive, move to the right four units. From this location move down three units because the y coordinate is negative. Place a dot. This point is shown in Figure 4–3(a). From our earlier discussion of the signs of the coordinates in the various quadrants, we know that the x coordinate is positive and y coordinate is negative in Quadrant IV, which is where our point is located. This serves as a check for plotting the point.   ■ ■ ■

EXAMPLE 2   Plot the point corresponding to $(-1, 4)$.

Solution   We start at the origin and move left one unit. Then from this location move up four units and place a dot. This point is shown in Figure 4–3(b). Checking our point location against the coordinate signs in the four quadrants shows that $(-1, 4)$ should be located in Quadrant II, which is where our point is located.   ■ ■ ■

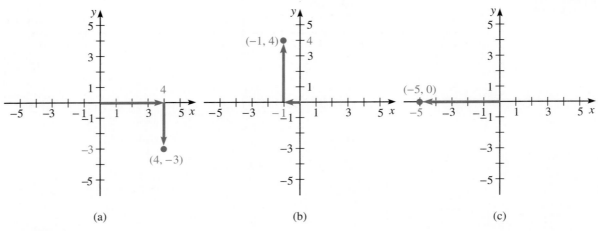

(a)   (b)   (c)

FIGURE 4–3

EXAMPLE 3   Plot the point corresponding to $(-5, 0)$.

Solution   Start at the origin and move left five units because the x coordinate is negative. Place a dot at this location because the y coordinate is zero. This point is shown in Figure 4–3(c).

■■■

CAUTION ✦ Be very careful when plotting points. The x coordinate always appears first in the ordered pair. You may have a tendency to confuse the order of the coordinates in plotting points. When plotting a point, always move left or right first, then up or down.

EXAMPLE 4   Identify the coordinates of points A, B, and C in Figure 4–4(a), (b), and (c), respectively.

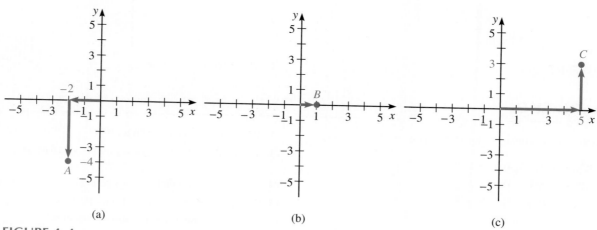

(a)   (b)   (c)

FIGURE 4–4

Solution

(a) Point A is two units to the left of the origin and four units below the x axis. The coordinates of point A are $(-2, -4)$. These coordinates confirm that the x and y coordinates are both negative in Quadrant III, where the point is located.

(b) Point B is one unit to the right of the y axis and zero units above or below the x axis. The coordinates of point B are $(1, 0)$.

(c) Point C is five units to the right of the y axis and three units above the x axis; therefore, the coordinates of point C are $(5, 3)$.

■■■

EXAMPLE 5   Three vertices of a rectangle are given by $(-3, 4)$, $(2, 4)$, and $(-3, 2)$. Determine the coordinates of the fourth vertex.

Solution   To find the fourth vertex, it may be helpful to plot the known vertices and join them with line segments, as shown in Figure 4–5. Since opposite sides of a rectangle are equal, we can count down vertically from the upper right vertex a length equal to the opposite side, two. The coordinates of the fourth vertex are $(2, 2)$.

■■■

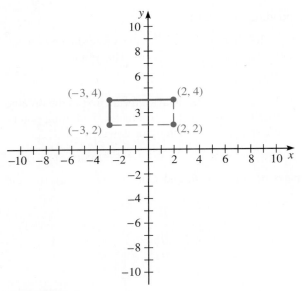

FIGURE 4–5

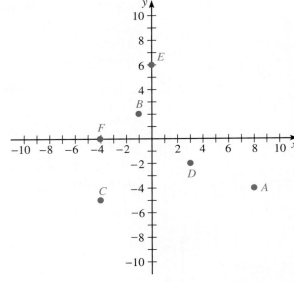

FIGURE 4–6

4–2 EXERCISES

Plot the following points on the rectangular coordinate system.

1. $A(3, -2)$, $B(-6, 8)$, $C(-3, -4)$
2. $A(0, 4)$, $B(-3, -3)$, $C(-2, -5)$
3. $A(0, -2)$, $B(-1, 5)$, $C(5, -9)$
4. $A(4, -7)$, $B(4, 8)$, $C(8, 0)$

Find the coordinates of the points given on the graph shown in Figure 4–6.

5. A, B, C
6. D, E, F
7. In what quadrant are the points located where x is negative and y is positive?
8. In what quadrant are the points located where x is positive and y is negative?
9. In what quadrant are the points located where both x and y are negative?
10. In what quadrant are the points located where both x and y are positive?
11. Where are the points $(0, y)$ located?

12. Where are the points $(x, 0)$ located?
13. What is the y coordinate of any point located on the x axis?
14. What is the x coordinate of any point located on the y axis?
15. In what quadrant(s) is the ratio y/x positive?
16. In what quadrant(s) is the ratio y/x negative?
17. A square has three vertices at $(0, 0)$, $(5, 0)$, and $(5, 5)$. Find the coordinates of the fourth vertex.
18. Three vertices of a rectangle are $(4, 3)$, $(10, 6)$, and $(10, 3)$. What are the coordinates of the fourth vertex?
19. Two vertices of an equilateral triangle are $(-5, 4)$ and $(4, 4)$. What is the x coordinate of the third vertex?
20. The vertices of the base of an isosceles triangle are $(-7, 6)$ and $(6, 6)$. Find the x coordinate of the third vertex.
CS 21. A CAD operator found the endpoint of a line to be at the point $(3, -2)$. In which quadrant is the endpoint of the line?

4–3

GRAPHING A FUNCTION

In this section we will integrate the concept of a function with graphing in the rectangular coordinate system to obtain a visual representation of a function that will be used later in this chapter.

The **graph** of a function consists of all ordered pairs that satisfy the functional rule. However, an equation in two variables may have an infinite number of ordered pairs that satisfy the functional rule. Since it is impossible to plot all these ordered pairs, we plot a representative sample and draw a smooth curve through the points. (In this chapter, we will discuss graphs of linear equations only. Chapter 5 will discuss graphing quadratic equations, which result in a curved line.)

Equations of the form $ax + by = c$, where a and b are constants and not both zero, are called **linear equations in two variables.** Since linear equations have graphs that are lines, it is necessary to plot only two ordered pairs that satisfy the functional rule. However, you will have fewer careless errors if you graph at least three ordered pairs.

Graphing a Function

EXAMPLE 1   Graph the function $y = f(x) = 3x - 4$.

Solution   To graph the line that represents this function, we need three ordered pairs that satisfy the equation. To determine these ordered pairs, we arbitrarily choose three values for x and substitute them into the functional rule. Then we plot the points and draw a line through them. Calculating the ordered pairs gives

$$
\begin{aligned}
x &= -2 & y &= f(x) = 3(-2) - 4 = -10 \\
x &= 0 & y &= f(x) = 3(0) - 4 = -4 \\
x &= 1 & y &= f(x) = 3(1) - 4 = -1
\end{aligned}
$$

The ordered pairs that we plot are $(-2, -10)$, $(0, -4)$, and $(1, -1)$. The graph of $y = 3x - 4$ is shown in Figure 4–7.

FIGURE 4–7

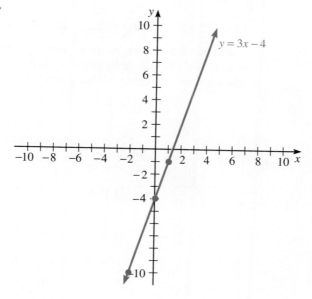

NOTE ✦ When selecting x values for your table of values, choose a negative value, zero, and a positive value.

To obtain the ordered pairs, it is often helpful to solve the equation for the dependent variable. Then, to organize the ordered pairs, fill in a table of values with columns for the independent and dependent variables. If the equation is nonlinear, you will need to obtain enough ordered pairs to draw a smooth curve. Depending on the size of the numbers required on the graph, it may also be necessary to scale one or both axes. To determine the scale for each axis, subtract the smallest value in the table of values from the largest to find the range of values required. Then count the total number of units available on the graph. Divide the number of values required by the total number of units available. You may want to round off this number since the scale should be convenient to use. The following box summarizes the steps required to graph a function.

To Graph a Function

1. Determine the dependent variable and solve the equation for it. In the case of x and y, solve the equation for y.

2. Determine any restrictions on the domain.

3. Construct a table of values reflecting the domain restrictions, if they exist.

4. Determine an appropriate scale for each axis. The numbers for one or both of the variables may be too large or too small to fit on the graph.

5. Plot the ordered pairs and draw a line or smooth curve through and extending beyond the points.

EXAMPLE 2   Graph $y = 2(x - 1) + 3 - x$.

Solution   It is not necessary to simplify the right side of the equation to graph the function. Calculating the ordered pairs when $x = -2$, $x = 0$, and $x = 3$ is given below. The resulting table of values is also given. The resulting graph is shown in Figure 4–8.

$$x = -2 \qquad y = 2(-2 - 1) + 3 - (-2)$$
$$y = 2(-3) + 3 + 2$$
$$y = -6 + 3 + 2$$
$$y = -1$$

$$x = 0 \qquad y = 2(0 - 1) + 3 - 0$$
$$y = 2(-1) + 3$$
$$y = 1$$

$$x = 3 \qquad y = 2(3 - 1) + 3 - 3$$
$$y = 2(2) + 3 - 3$$
$$y = 4 + 3 - 3$$
$$y = 4$$

$-x$	$-y$
-2	-1
0	1
3	4

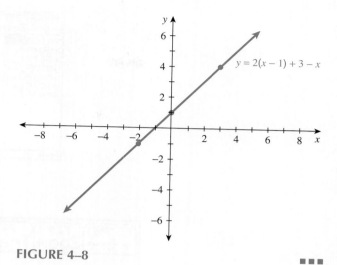

$y = 2(x - 1) + 3 - x$

FIGURE 4–8

■■■

The Graphing Calculator

A graphing calculator can be used to construct graphs quickly. Most graphing calculators perform the functions of a scientific calculator in addition to graphing equations and data. Because the procedures vary from one graphing calculator to another, this text gives general procedures. For keystrokes specific to your graphing calculator, refer to the owner's guide.

In general, to graph an equation, you first solve the equation for y or $f(x)$ in terms of x. Then you must determine the range and scale of numbers to use for the x and y axes. This determines the **viewing rectangle** where the graph is displayed. Next, you must input the equation to be graphed. Then you must execute the actual graphing. Depending on what you are trying to learn from the graph, you may use the zoom or trace capabilities.

EXAMPLE 3   Graph $y = 2x + 7 - 3(x + 2) + \dfrac{x}{2}$ using a graphing calculator.

Solution   The equation is solved for y in terms of x. For the graphing calculator, there is no need to simplify the right side of the equation. Next, we set our viewing rectangle. The values for x min, x max, x scale, y min, y max, and y scale are shown on page 136. After entering the equation and executing the graph, you should have the result shown in Figure 4–9.

The steps for graphing using a graphing calculator are summarized below.

■■■

Creating Graphs Using a Graphing Calculator

1. Solve the equation for y or $f(x)$ in terms of x.
2. Set the viewing rectangle.
3. Input the equation.
4. Execute the graph.

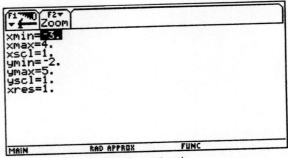

Viewing rectangle

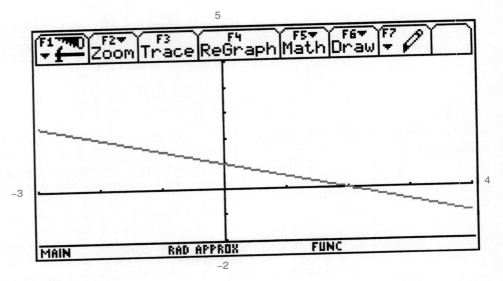

FIGURE 4–9

EXAMPLE 4   Use a graphing calculator to graph $5x + 2y = 6$.

Solution   First, we solve the equation for y.

$$5x + 2y = 6$$
$$2y = 6 - 5x$$
$$y = 3 - \frac{5}{2}x$$

Then we set the viewing rectangle. These values are shown in the viewing rectangle as part of Figure 4–10. We input the equation and execute the graph, which is shown in Figure 4–10.

EXAMPLE 5   Use a graphing calculator to graph the function $3x - 7y + 4(2x - 1) = 6x + 4(y - 2) + 15$.

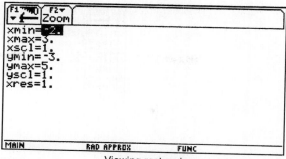

Viewing rectangle

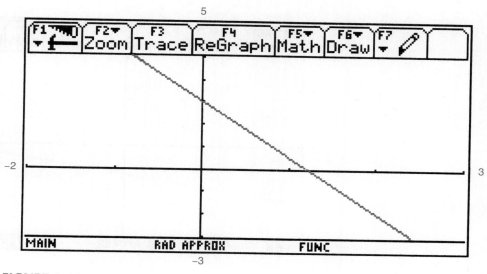

FIGURE 4–10

∎∎∎

Solution   First, we must rearrange the equation and solve for y. We begin by removing the parentheses and combining like terms.

$$3x - 7y + 4(2x - 1) = 6x + 4(y - 2) + 15$$
$$3x - 7y + 8x - 4 = 6x + 4y - 8 + 15$$
$$11x - 7y - 4 = 6x + 4y + 7$$

Next, we collect all the y terms on the left and all other terms on the right side of the equation.

$$-7y - 4y = 6x + 7 - 11x + 4$$
$$-11y = -5x + 11$$

Last, we divide by the coefficient of y and simplify.

$$y = \frac{5}{11}x - 1$$

The graph of this function is shown in Figure 4–11.

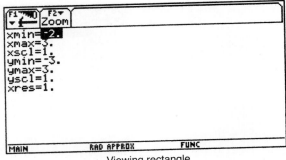

Viewing rectangle

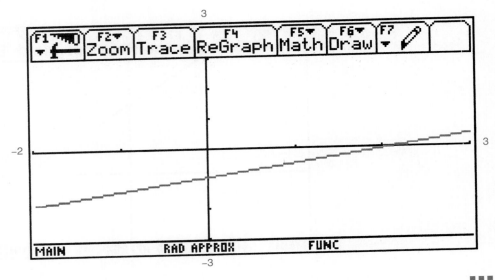

FIGURE 4–11

■■■

EXAMPLE 6 The manager of a chocolate candy store has found that x boxes can be sold if the price y is

$$y = 30 - \frac{x}{15} \text{ dollars}$$

What price should be charged to sell 75 boxes of chocolate? Check your answer algebraically.

Solution To graph the function, we fill in a table of values starting with zero for x. The table of values and the resulting graph are shown in Figure 4–12. From the graph, select 75 along the x axis and go up to the line and then horizontally to the y axis. From the graph the price should be $25. To solve algebraically, substitute $x = 75$ and solve for y.

$$y = 30 - \frac{x}{15}$$

$$y = 30 - \frac{75}{15}$$

$$y = 30 - 5$$
$$y = 25$$

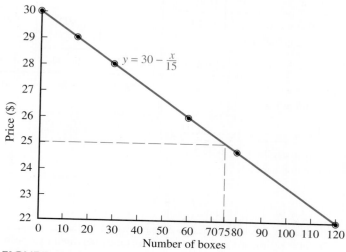

x	y
0	30
15	29
30	28
60	26
80	24.67
100	23.33
120	22.00

FIGURE 4–12

■ ■ ■

4–3 EXERCISES

Graph the following functions on the rectangular coordinate plane.

1. $4x - y = 8$

2. $3x - 1 = 2y$

3. $3x - y = 5$

4. $2y - 8 = 3x$

5. $5x - 3y = 9$

6. $x - 4y = 12$

7. $y = 7x - 1$

8. $y = 3x + 5$

9. $y = 9 + 2x$

10. $5x - 7y = 10$

11. $2x = 8y - 3$

12. $2x = 3y$

13. $5 - 6x + 9y = 0$

14. $6x - 10y = 2$

If possible, graph the following functions on a graphing calculator. If a graphing calculator is not available, use the rectangular coordinate plane and a table of values.

15. $y = 2(3x - 1) + 7 - 3(x + 4)$

16. $y = 5x + 2(x - 1) + 4(x - 6)$

17. $y = 6(x - 3) + 9x - 2(4x - 6)$

18. $y = 3(5x + 2) - (4x - 7) + 2x$

19. $2x - 7y = 24$

20. $3y - 4x = 15$

21. $2y - 9x = 36$

22. $x + 4y = 20$

23. $5x + 2(4x - 1) + 5y + 6 = 0$

24. $12x - 17 + 4(x - 8) + 5(y - 2) = 0$

25. $2y + 8(x - 1) = y + 2(3x + 5)$

26. $3x + 2(y - 4) = 15 + 2(3x + 1)$

27. $5(2x + 1) - 7 + 4y = 3(x + 2) - 10$

28. $6y - (3x + 7) = 18 - 4y + 6$

P 29. The velocity v (ft/s) of an object thrown upward under the influence of gravity is given by the equation $v = v_0 + at$. If an object is thrown upward with an initial velocity v_0 of 20 ft/s with the acceleration due to gravity a of -32 ft/s^2, the equation becomes $v = 20 - 32t$. Graph velocity as a function of time.

M 30. The normal stress σ (lowercase Greek letter sigma) on a bar is the load P divided by the cross-sectional area a ($\sigma = P/a$). Graph the relationship between stress and load if the cross-sectional area is 10 m^2.

B 31. The normal range of crude oil stockpiles is between 300 million barrels and 320 million barrels. The table on p. 140 gives the actual stockpile of crude oil for July 99–July 00. Graph this data and analyze it relative to the normal range.

C 32. In construction, the area of footing A of a building is the ratio of the total weight W of the building to the bearing capacity B of the soil. If a building weighs 20 tons, graph area as a function of bearing capacity.

M 33. An equation in fluid mechanics is $P = Dhg$ where P is the pressure, D is the density of the liquid, h is the height of the column of liquid, and g is the gravitational acceleration. If $g = 32$ ft/s^2 and $D = 0.8$ slugs/ft^3, graph pressure as a function of height.

Date	July 99	Aug 99	Sept 99	Oct 99	Nov 99	Dec 99
Oil stockpile (million barrels)	335	315	302	302	298	287
Date	Jan 00	Feb 00	Mar 00	Apr 00	May 00	Jun 00
Oil stockpile (million barrels)	288	290	298	305	299	295
Date	Jul 00					
Oil stockpile (million barrels)	285					

4–4

DISTANCE AND SLOPE

In this section we will discuss the distance between two points, the midpoint of a line segment joining two points, and the slope of a line. Our discussion of the straight line is limited in this section. However, in Chapter 15, we will discuss the different forms of a straight line (point-slope, general, and slope-intercept) in greater detail. Here, we will discuss distance first.

Distance

Let us develop the formula for the distance between two points. We take two points P and Q represented by the ordered pairs (x_1, y_1) and (x_2, y_2), respectively. The distance between P and Q is represented by line segment d shown in Figure 4–13. If we draw a line segment through P parallel to the x axis and another line segment through Q parallel to the y axis, a right triangle is formed. The lengths of the sides of the triangle are shown in the figure. Since a right triangle is formed, we can use the Pythagorean theorem to solve for d, obtaining

$$d^2 = (x_2 - x_1)^2 + (y_2 - y_1)^2$$
$$d = \sqrt{(x_2 - x_1)^2 + (y_2 - y_1)^2}$$

FIGURE 4–13

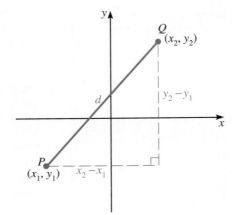

Distance Formula

Given two points represented by the ordered pairs (x_1, y_1) and (x_2, y_2), the distance d between the points is given by

$$d = \sqrt{(x_2 - x_1)^2 + (y_2 - y_1)^2}$$

EXAMPLE 1   Find the distance between $(-6, 2)$ and $(8, -5)$.

Solution   First, choose one ordered pair to serve as (x_1, y_1), and the other to serve as (x_2, y_2). This choice is completely arbitrary and does not affect the final result.

$$\underset{\underset{(x_1, y_1)}{\uparrow}}{(-6, 2)} \qquad \underset{\underset{(x_2, y_2)}{\uparrow}}{(8, -5)}$$

Then we substitute the appropriate members into the distance formula and simplify the result.

$$d = \sqrt{(x_2 - x_1)^2 + (y_2 - y_1)^2}$$
$$d = \sqrt{(8 - (-6))^2 + (-5 - 2)^2}$$
$$d = \sqrt{196 + 49}$$
$$d = 7\sqrt{5} \approx 15.7$$

The line segment joining these two points is shown in Figure 4–14.

FIGURE 4–14

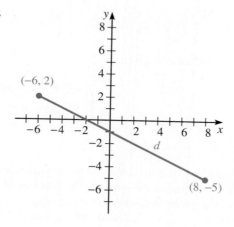

CAUTION ✦ Be consistent when choosing x_1 and y_1. Both must come from the same ordered pair. In this example, if x_1 is -6, then y_1 must be 2.

Midpoint

The **midpoint** of a line segment is the point halfway between the endpoints. The midpoint is given by the following formula.

Midpoint of a Line Segment

If (x_1, y_1) and (x_2, y_2) represent the endpoints of a line segment, the midpoint is represented by

$$\left(\frac{x_1 + x_2}{2}, \frac{y_1 + y_2}{2} \right)$$

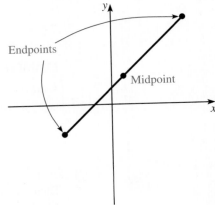

Endpoints

Midpoint

EXAMPLE 2   Find the midpoint of the line segment joining $\left(-5, \dfrac{3}{4} \right)$ and $\left(\dfrac{1}{2}, 3 \right)$.

Solution   First, choose the ordered pairs to serve as (x_1, y_1) and (x_2, y_2).

$$(x_1, y_1) \qquad (x_2, y_2)$$

$$\uparrow \qquad\qquad \uparrow$$

$$\left(-5, \frac{3}{4} \right) \qquad \left(\frac{1}{2}, 3 \right)$$

The midpoint is given by

$$\left(\frac{-5 + \dfrac{1}{2}}{2}, \frac{\dfrac{3}{4} + 3}{2} \right) = \left(-\frac{9}{4}, \frac{15}{8} \right) \qquad\qquad \blacksquare\blacksquare\blacksquare$$

Slope

The **slope** of a line refers to the direction and steepness of the line's slant. Slope can also be expressed as the ratio of rise over run. If we choose any two points (x_1, y_1) and (x_2, y_2) on the given line, and if we construct a right triangle using these points, the rise is the length of the vertical side and the run is the length of the horizontal side of the triangle. The slope formula is given below.

Slope of a Line

Given line L, choose any two points (x_1, y_1) and (x_2, y_2) on this line. Then the slope m of line L is given by

$$m = \frac{\text{rise}}{\text{run}} = \frac{y_2 - y_1}{x_2 - x_1}$$

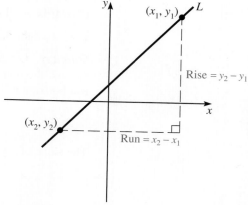

EXAMPLE 3   Find the slope of the line passing through the points $(6, -5)$ and $(8, 7)$.

FIGURE 4–15

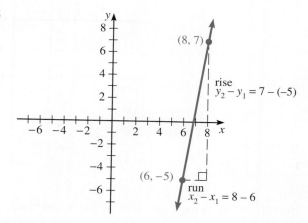

Solution   Figure 4–15 shows the line passing through these points, the right triangle that results, and the rise and run. To find the slope of the line, let

$$(x_1, y_1) \qquad (x_2, y_2)$$
$$\uparrow \qquad\qquad \uparrow$$
$$(6, -5) \qquad (8, 7)$$

Then substitute into the slope formula.

$$m = \frac{y_2 - y_1}{x_2 - x_1}$$

$$m = \frac{7 - (-5)}{8 - 6}$$

$$m = 6$$

■ ■ ■

NOTE ✦ The sign of the slope denotes the direction of slant. A line that slants from lower left to upper right has a positive slope; a line that slants from upper left to lower right has a negative slope. The slope of -6 in Example 3 illustrates this fact.

EXAMPLE 4 Find the slope of the line passing through the following points:

a. $(3, 6)$ and $(3, -4)$ **b.** $(-2, 5)$ and $(3, 5)$

Solution Let L_1 represent the line through the points in (a) and L_2 represent the line through the points in (b). From Figure 4–16, you can see that L_1 is a vertical line and L_2 is a horizontal line. The slope of L_1 is

$$\text{(a)} \quad m = \frac{6 - (-4)}{3 - 3} = \frac{10}{0} = \text{undefined}$$

The slope of L_2 is

$$\text{(b)} \quad m = \frac{5 - 5}{3 - (-2)} = \frac{0}{5} = 0$$

FIGURE 4–16

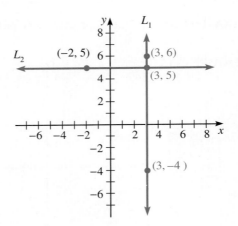

NOTE ✦ The slope of any vertical line is undefined, and the slope of any horizontal line is 0.

EXAMPLE 5 Find the slope of the line passing through the following points:

a. $(1, 6)$ and $(3, 2)$ **b.** $(2, -9)$ and $(-3, 1)$

Solution Let L_1 represent the line through the points in (a) and L_2 represent the line through the points in (b). Figure 4–17 shows that L_1 and L_2 appear to be parallel. Using the slope formula to find the slope of L_1 gives

$$\text{(a)} \quad m = \frac{6 - 2}{1 - 3} = \frac{4}{-2} = -2$$

The slope of L_2 is

$$\text{(b)} \quad m = \frac{-9 - 1}{2 - (-3)} = \frac{-10}{5} = -2$$

FIGURE 4–17

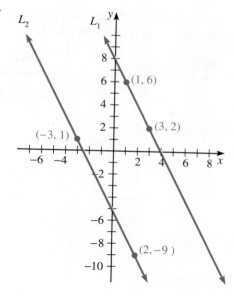

NOTE ✦ Parallel lines have the same slope.

EXAMPLE 6   Find the slope of the line passing through the following points:

a. (0, −8) and (3, 1)   **b.** (−3, 3) and (6, 0)

Solution   Again, let L_1 represent the line passing through the points in (a) and L_2 represent the line through the points in (b). Figure 4–18 suggests that lines L_1 and L_2 appear to be perpendicular. The slope of L_1 is

$$(a) \quad m = \frac{-8 - 1}{0 - 3} = 3$$

The slope of L_2 is

$$(b) \quad m = \frac{3 - 0}{-3 - 6} = -\frac{1}{3}$$

FIGURE 4–18

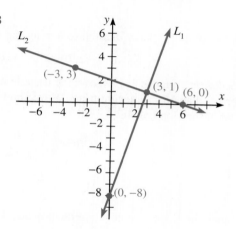

NOTE ✦ The slopes of perpendicular lines are negative reciprocals of each other. Thus, the product of the slopes of perpendicular lines is -1 (unless they are a vertical line and a horizontal line).

Summary of Slopes

Negative slope

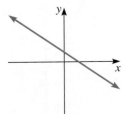

Positive slope

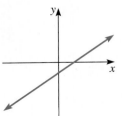

Horizontal line, slope = 0

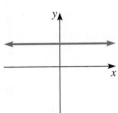

Vertical line, slope = undefined

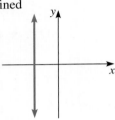

Parallel lines, same slope

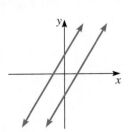

Perpendicular lines, negative reciprocal slopes

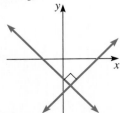

EXAMPLE 7   Show that the following represent vertices of a right triangle:

$$C: (-2, 0) \qquad A: (-4, 2) \qquad B: (1, 3)$$

Solution   Figure 4–19 shows the triangle ABC formed by joining points A, B, and C. To prove that ABC is a right triangle, we show that the slope of CB is the negative reciprocal of the slope of AC.

$$CB = \frac{3 - 0}{1 - (-2)} = \frac{3 - 0}{1 + 2} = 1$$

$$AC = \frac{2 - 0}{-4 - (-2)} = \frac{2 - 0}{-4 + 2} = -1$$

Since the slopes are negative reciprocals, ABC is a right triangle.

FIGURE 4–19

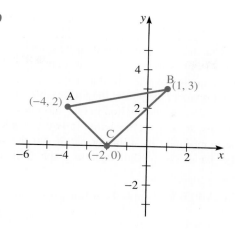

4–4 EXERCISES

Find the distance between the following ordered pairs.

1. $(6, -3)$ and $(4, 2)$

2. $(-6, -4)$ and $(3, -7)$

3. $(6, -5)$ and $(3, -8)$

4. $(5, 12)$ and $(-7, -8)$

5. $(-1, -7)$ and $(6, -13)$

6. $\left(-\frac{1}{2}, \frac{3}{4}\right)$ and $(3, -5)$

7. $(2, 6)$ and $(-1, 6)$

8. $(-7, 10)$ and $(-7, -4)$

9. $(-5, -2)$ and $(-5, 4)$

10. $(3, -9)$ and $(-2, -9)$

Find the midpoint of the line segment joining the given ordered pairs.

11. $(9, -3)$ and $(2, 5)$

12. $(0, 5)$ and $(-3, 10)$

13. $(3, 7)$ and $(-8, 5)$

14. $(-2, -7)$ and $(-8, -3)$

15. $(1.7, -5)$ and $(3, -2.3)$

16. $(3.2, -1.8)$ and $(6.7, 8.1)$

17. $(-5, -8)$ and $(-1, -12)$

18. $\left(\frac{3}{7}, 2\right)$ and $\left(-3, \frac{1}{3}\right)$

19. $\left(\frac{1}{6}, \frac{7}{8}\right)$ and $\left(\frac{15}{2}, \frac{11}{4}\right)$

20. $\left(\frac{3}{4}, -\frac{8}{3}\right)$ and $\left(\frac{15}{2}, -\frac{3}{2}\right)$

Find the slope of the line passing through the given points.

21. $(-5, -8)$ and $(6, -3)$

22. $(-3, 11)$ and $(7, -8)$

23. $(4, -2)$ and $(4, 10)$

24. $(9, 6)$ and $(12, 5)$

25. $(-3, -10)$ and $(-1, 5)$

26. $(7, -0.5)$ and $(-8, -0.5)$

27. $(3.6, -2.8)$ and $(-1.7, 4.3)$

28. $(-14, 3)$ and $(6, -9)$

29. $(14, -8)$ and $(-7, 4)$

30. $(-1, 9)$ and $(-1, 0)$

Determine if the lines passing through the given points are parallel, perpendicular, or neither.

31. L_1: $(-1, 8)$ and $(3, -1)$
L_2: $(-2, -3)$ and $(1, -9)$

32. L_1: $(-3, 6)$ and $(4, -7)$
L_2: $(2, -2)$ and $(-1, 8)$

33. L_1: $(4, 4)$ and $(-8, -5)$
L_2: $(-3, -1)$ and $(0, -5)$

34. L_1: $(-2, 13)$ and $(3, -2)$
L_2: $(0, -1)$ and $(-2, 5)$

35. L_1: $(-6, 0)$ and $(4, 3)$
L_2: $(5, 4)$ and $(-3, -2)$

36. Show that $(0.5, 5)$, $(-2, -2)$, and $(3, -2)$ represent the vertices of an isosceles triangle.

37. Show that $(0, 2)$, $(4, -2)$, and $(3, -1)$ represent the vertices of a right triangle.

38. Show that $(-2, 3)$, $(2, -2)$, $(4, 3)$, and $(-4, -2)$ represent the vertices of a parallelogram.

39. Show that $(-1, 5)$, $(2, 2)$, $(-1, 2)$, and $(2, 5)$ represent the vertices of a square.

CS 40. Using a computer drawing program, a student found the endpoints of the diameter of a circular image to be at $(5, -4)$ and $(3, 2)$. What are the coordinates of the center of the circle?

4–5

GRAPHING AN INEQUALITY

In Chapter 1, we discussed solving linear inequalities in one variable. In this section we will discuss solving inequalities in two variables. Recall from Chapter 1 that the solution to an inequality in one variable may be represented on a number line. On the other hand, the solution to an inequality in two variables must be represented on the rectangular coordinate plane. Just as an equation in two variables may have an infinite number of ordered-pair solutions, so may an inequality in two variables. The graphical solution to an equation may be represented by a line or curve, but the solution to an inequality could be a region of the plane. The method used to graph an inequality is developed in the next example.

Inequalities in Two Variables

EXAMPLE 1   Graph the following inequality:

$$2x - 5y \geq 10$$

Solution   We begin by solving the inequality for the dependent variable.

$$2x - 5y \geq 10$$
$$-5y \geq 10 - 2x$$
$$y \leq \frac{-10 + 2x}{5}$$

Next, we obtain the boundary of the solution set by graphing the equation

$$y = \frac{-10 + 2x}{5}$$

The table of values for this equation is given in Figure 4–20. We plot the points and draw a line through them. The solution to this linear inequality is a half-plane bounded by this line. To determine which half-plane to shade, we choose a test point (which may be any point not on the boundary). Let us use the point $(7, -4)$. Substituting $(7, -4)$ into the inequality gives

$$2(7) - 5(-4) \geq 10$$
$$14 + 20 \geq 10$$
$$34 \geq 10 \quad \text{(true)}$$

Since this is a true statement, shade the half of the plane containing the test point—the lower half, in this case. Since the inequality includes the "equal to" symbol, the line is part of the solution set. The graph of $2x - 5y \geq 10$ is shown in Figure 4–20.

FIGURE 4–20

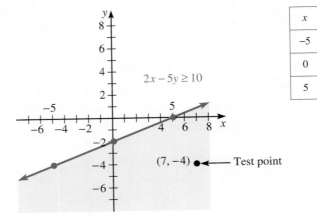

x	y
-5	-4
0	-2
5	0

■ ■ ■

Graphing an Inequality

1. Solve the inequality for the dependent variable. Transpose the dependent variable to the left side of the inequality.

2. Replace the inequality symbol with the equals symbol, and graph the resulting equation.

3. Join the points with a dotted curve if the inequality is less than or greater than, or with a solid curve if the inequality contains the equals symbol.

4. Shade the graph by choosing a test point and substituting it into the original inequality. If the inequality is true, shade the portion of the plane containing the test point. Otherwise, shade the other portion of the plane.

EXAMPLE 2 Use a graphing calculator to graph

$$y \geq 6(x + 2) - 17 - 3x + 4(2x - 1).$$

Solution We graph the equation $y = 6(x + 2) - 17 - 3x + 4(2x - 1)$ using the methods described in Section 4–3. The graphing calculator does not shade automatically, so you must tell it whether to shade above or below the line. This instruction requires us to choose a test point, as in Example 1. Choosing $(0, 0)$ gives a true statement, so we shade above the line. The graph is shown in Figure 4–21. ■■■

CAUTION ✦ Always check your solution by substituting a point from the shaded region. One of the most frequent errors in graphing inequalities is shading the wrong area.

Systems of Inequalities

In Chapter 6, we will discuss solving systems of equations using both graphical and algebraic techniques. However, we will discuss solving systems of inequalities in this section because it uses the techniques we have just developed.

A **system of linear inequalities** contains more than one linear inequality in two variables. The solution to a system of linear inequalities is the region of the plane contained in the solution set of both inequalities. The remainder of this section is devoted to solving systems of linear inequalities.

EXAMPLE 3 Find the solution to the following system of inequalities:

$$x - 5y > 12$$
$$3x + 4y \leq 6$$

Solution We graph each inequality using the same methods used in the previous examples. The table of values is given on page 150, and the graph is given in Figure 4–22. Part (a) of the figure shows the solution to the inequality $x - 5y > 12$; part (b) shows the solution to $3x + 4y \leq 6$; and part (c) shows the area common to both graphs. We can check the solution by choosing a point in the shaded region and substituting into both inequalities. Substituting $(3, -5)$ into each inequality gives

$$x - 5y > 12 \qquad\qquad 3x + 4y \leq 6$$
$$3 - 5(-5) > 12 \qquad\qquad 3(3) + 4(-5) \leq 6$$
$$28 > 12 \quad \text{(true)} \qquad\qquad -11 \leq 6 \quad \text{(true)}$$

FIGURE 4–21

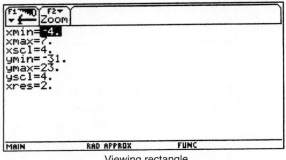

Viewing rectangle

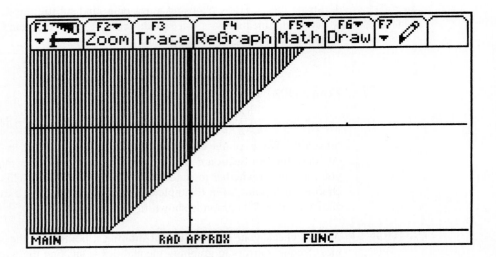

$x - 5y = 12$	
x	y
-2	$-14/5$
0	$-12/5$
1	$-11/5$

$3x + 4y = 6$	
x	y
-2	3
0	$3/2$
2	0

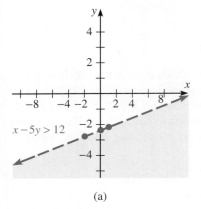

(a)

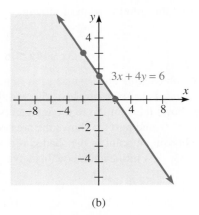

(b)

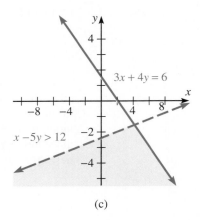

(c)

FIGURE 4–22

EXAMPLE 4   The manager of a donut shop pays $1.10 a dozen for cinnamon donuts and $1.40 a dozen for chocolate donuts. The manager wants to stock at least 52 dozen donuts but cannot spend more than $120. Find the solution set.

Solution   We must set up a system of inequalities to solve this problem. Let x represent the number of dozen of cinnamon donuts and y represent the number of dozen of chocolate donuts. If the manager wants to have at least 52 dozen donuts, then

$$x + y \geq 52$$

If the manager can spend no more than $120 for the donuts, then

$$1.10x + 1.40y \leq 120$$

The system of inequalities to be solved graphically is

$$x + y \geq 52$$
$$1.10x + 1.40y \leq 120$$

The table of values for each equation is given below. Notice that negative values for x and y are not reasonable in this case. The solution set is shown in Figure 4–23. A scale of 5 was used for each axis. From the graph, if the manager chooses any ordered pair in the shaded region, he will satisfy the constraints of both stock and money. For example, the ordered pair (30, 50) lies in the shaded region.

$$30 + 50 > 52 \qquad \text{and} \qquad 33 + 70 \leq 120$$

Therefore, the manager could buy 30 dozen cinnamon donuts and 50 dozen chocolate donuts. He could also choose any other combination of dozens of donuts from the shaded region.

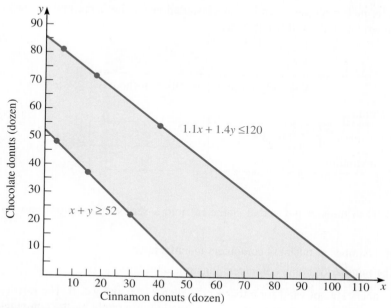

$x + y = 52$	
x	y
4	48
15	37
30	22

$1.1x + 1.4y = 120$	
x	y
6	81
20	70
48	48

FIGURE 4–23

■■■

Linear Programming   One of the most important applications of systems of inequalities is linear programming. Many problems in business and industry require the most cost-effective solution. **Linear programming** uses systems of inequalities to maximize or minimize such quantities as cost, profit, and allocation of raw materials subject to certain conditions. The system of inequalities is often large and complicated and is usually solved using the simplex algorithm implemented on a computer. However, in this section we will limit the discussion to the graphical solution of simple systems.

From previous experience a business derives an equation representing profit and a set of inequalities representing the allocation of raw materials. After deriving the system of inequalities, we plot each inequality on the same rectangular coordinate system and find the solution set, which should be a polygon. Then we find the coordinates of the vertices of this polygon from the graph because the solution to a linear programming problem is always found at one of these vertices. Last, we find the solution to the linear programming problem by substituting the coordinates of the polygon vertices into the profit or cost equation to find the maximum or minimum, respectively.

EXAMPLE 5   A small computer company manufactures two models of a computer, one for home use and one for business use. In the manufacturing process, a certain number of labor-hours are allocated to assembling the computer components, and a certain number of labor-hours are allocated to installing the computer into its case. The home computer requires 1 labor-hour to be assembled and 2 labor-hours to be put into its case. The business computer requires 3 labor-hours to be assembled and 1 labor-hour to be put into its case. Workers have a maximum of 9 labor-hours per day for assembling the computer and a maximum of 8 labor-hours per day for installing the computer into its case. If the company makes a profit of $100 on each home computer and $150 on each business computer, how many of each computer should the company manufacture to maximize its profit?

Solution   The information on the assembly and installation for the home and business computers is summarized in the following table:

	Assembly (hours)	Installation (hours)
Home	1	2
Business	3	1
Maximum	9	8

The first task is to transfer the verbal statement into a series of equations and inequalities. Let

x represent the number of business computers manufactured

y represent the number of home computers manufactured

Since negative numbers are not valid for either x or y, then $x \geq 0$ and $y \geq 0$. The restriction on production relates to the number of hours available for assembling the computer components and then installing the computer into its case. These restrictions translate into the following inequalities:

$$3x + y \leq 9 \quad \text{(assembling components)}$$
$$x + 2y \leq 8 \quad \text{(installing computer in its case)}$$

We are attempting to maximize the profit represented by

$$P = 150x + 100y$$

To solve this problem, we find the point of intersection of the two lines. The region of solution is shown in Figure 4–24. From the graph we can determine that the coordinates for the vertices of the polygon are (0, 0), (0, 4), (2, 3), and (3, 0). Substituting each point into the profit equation gives

$$P = 150x + 100y$$
$$P = 150(0) + 100(0) = 0$$
$$P = 150(0) + 100(4) = 400$$
$$P = 150(2) + 100(3) = 600$$
$$P = 150(3) + 100(0) = 450$$

FIGURE 4–24

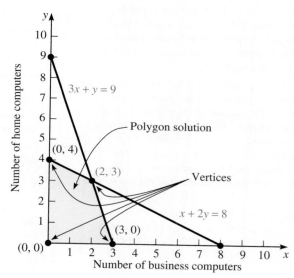

Since the point (2, 3) results in the largest profit, it is the solution to the linear programming problem. The computer company should produce 2 business computers and 3 home computers per day for a profit of $600.

■ ■ ■

4–5 EXERCISES

Graph each of the following inequalities on the rectangular coordinate system or graphing calculator.

1. $y > 5x - 6$

2. $y < 2x + 1$

3. $2x - y \leq 3$

4. $y \geq 3(x - 1) + 6$

5. $x \geq -5$

6. $4x - 5y \leq 8$

7. $2x < 5y + 6$

8. $y \leq -3$

9. $3y \geq 4x - 9$

10. $7x \leq 2y - 3$

11. $y < 1$

12. $x > -6$

13. $y < 3(2x + 4) - 6 - (4x + 1)$

14. $y \geq 6(2x - 4) + 18 - 2x$

15. $7x - 3y \geq 5$

16. $4x - 5y < 15$

17. $9y + 15 < 2x$

18. $2y - 18 > 4x$

 19. $y \geq 2.6x + 14.8$

 20. $9.4x \leq 4.1y - 18.4$

21. $6.7x + 4.3y < 21.8$

22. $2.8x - 3.2y > 16.1$

Solve the following systems of inequalities.

23. $x \geq y, 2x + y > 3$

24. $5x - y > 10, x - 3y > 4$

25. $2x + y < 6, x - y > 1$

26. $4x \leq 3y, x \geq 2y + 1$

27. $y \geq 3x - 5, x < -3$

28. $y \leq 4, x > 5y - 4$

29. Given the inequalities $2x - 3y \leq 4, x + 4y \leq 20, x \geq 0$, $y \geq 0$, graph them, find the corner points, and find the point that gives the maximum profit if the profit is given by $P = 25x + 16y$.

B 30. If the profit for a company is given by $P = 30x + 85y$, find the values for x and y that give the maximum profit subject to the following constraints: $x \geq 0, y \geq 0, x + 2y \leq 11$, and $7x + y \leq 12$.

B 31. A company produces two kinds of sewing machines, an economy model and a deluxe model. The economy model requires 1 labor-hour to be assembled and 3 labor-hours to be put into its case. The deluxe machine requires 2 labor-hours to be assembled and 4 labor-hours to be put into its case. Workers have a maximum of 15 hours per day to assemble the machines and 24 hours per day to put the machines into their cases. If the company makes a profit of $50 on the economy machine and $110 on the deluxe machine, how many of each sewing machine should the company make to maximize profit?

B 32. A watchmaker makes two types of watches. One watch requires 1 labor-hour to prepare the case and 4 labor-hours to assemble the parts. The second watch requires 1 labor-hour to prepare the case and 2 labor-hours to assemble the parts. She has a maximum of 5 hours to prepare the cases and 12 hours to assemble the parts. If she makes a profit of $60 on the first watch and $40 on the second watch, how many of each watch should she make to maximize profit?

B 33. A company manufactures two types of integrated circuits. The first integrated circuit requires 2 hours to manufacture and 1.25 hours to test per thousand. The second type requires 1.5 hours to manufacture and 1.75 hours to test per thousand. The company has a maximum of 7 hours to manufacture the integrated circuits and a maximum of 7.875 hours to test them. It makes a profit of $2 on the first integrated circuit and $2.50 on the second one. How many (in thousands) should the company manufacture to maximize its profit?

B 34. KLM Corporation manufactures two circuit analyzers. The first analyzer requires 28 labor-hours to design and 18 labor-hours to produce. The second analyzer requires 16 labor-hours to design and 12 labor-hours to produce. The company has a maximum of 8,436 hours to design the analyzers and 5,622 hours to produce them. It makes a profit of $75 on the first analyzer and $125 on the second analyzer. How many of each should the company manufacture?

B 35. ABC Company manufactures two portable computers. The first computer requires 265 labor-hours to design and 170 labor-hours to manufacture. The second computer requires 190 labor-hours to design and 150 labor-hours to manufacture. The company has 2,845 hours available for design and 2,050 hours available for manufacture. It makes a profit of $200 on the first computer and $350 on the second computer. How many of each computer should the company manufacture?

CS 36. When writing a computer program, the programmer employed the following inequalities:

$$x + y \leq 6 \qquad x + y \geq 4 \qquad x \geq 0 \qquad y \geq 0$$

Without graphing, explain why the solution to the system must be in the first quadrant.

4–6

APPLICATIONS

In this section we will discuss several applications of graphing. Although graphs are not as accurate as algebraic solutions, we can use them to solve equations. We can also use them to determine the break-even point, that is, the point at which a business neither makes a profit nor suffers a loss on its product. Another application is in graphing data gained from an experiment or observations. For example, if you stand on a street corner and count the number of cars passing every hour, you can graph these data and use them to develop a traffic plan. Each of these applications will be discussed in this section.

Solving Equations Graphically

The **solution** or **root** of an equation is the x coordinate of the point where the graph intersects the x axis. This point is also called the **x intercept** of the graph. Since the equation is represented as a function, this same point is also called a **zero** of the function. To determine the solution to an equation or the zero of a function, we must find the x intercept. If the graph does not touch the x axis, the equation does not have a real number solution.

To Solve an Equation Graphically

---

1. Collect all terms on one side of the equation.

2. Write the equation as a function, equal to y or $f(x)$.

3. Graph the equation.

4. The point(s) where the graph intersects the x axis is the solution to the equation.

EXAMPLE 1 Solve $4x = 3(x - 1) + 6$ graphically.

Solution First, collect all terms on one side of the equation. Transposing all terms to the left side of the equation gives

$$4x - 3(x - 1) - 6 = 0$$

Setting this equation equal to y or $f(x)$ and simplifying gives

$$y = f(x) = 4x - 3(x - 1) - 6$$
$$= x - 3$$

Next, we graph the equation using a table of values. The table of values is given below, and the resulting graph is shown in Figure 4–25. Note that the graph intersects the x axis at 3. It may be necessary to estimate the solution from the graph if the solution is not an integer. Thus, the root or solution to the equation is $x = 3$. We can verify this solution by solving the linear equation algebraically.

$$
\begin{array}{lll}
4x = 3(x - 1) + 6 & \\
4x = 3x - 3 + 6 & \text{eliminate parentheses} \\
4x = 3x + 3 & \text{combine like terms} \\
4x - 3x = 3 & \text{transpose} \\
x = 3 & \text{combine like terms}
\end{array}
$$

FIGURE 4–25

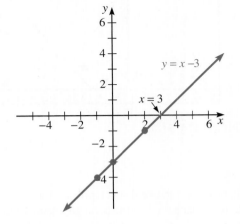

x	y
-1	-4
0	-3
2	-1

■■■

EXAMPLE 2 Solve $y = 3(2x + 1) - 9 + (x - 7)$ using a graphing calculator.

Solution Recall from Section 4–3 that we must solve the equation for y. Since this equation is given in terms of y, we proceed to the next step and set the viewing rectangle.

The max, min, and scale values are shown below. Then we execute the graph. The solution is where the graph crosses the *x* axis. Most graphing calculators have a **TRACE** function that follows the graph while displaying the *x* and *y* coordinates. We TRACE the graph until we reach the point where $y \approx 0$, or where the graph crosses the *x* axis, which is shown in Figure 4–26. The solution is $x \approx 1.9$.

NOTE ✦ Solutions by graphing are only approximations.

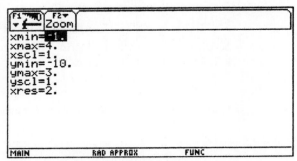

Viewing rectangle

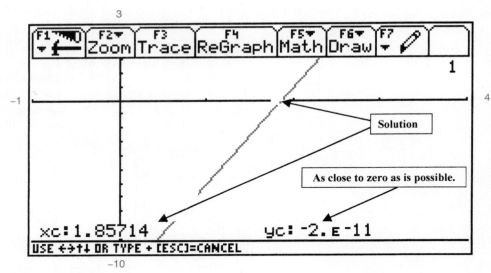

FIGURE 4–26   ■ ■ ■

Break-Even Point

In business the **break-even point** is the point in production where total cost equals total revenue. It occurs when the company neither makes a profit nor suffers a loss. Using principles of business, we can determine one equation to represent the total cost and a second equation to represent total revenue. Graphically, the point of intersection of the cost equation and the revenue equation represents the break-even quantity and the break-even revenue. Thus, in linear equations, any production above the break-even quantity usually results in a profit, while any production less than the break-even quantity usually results in a loss.

To Determine the Break-Even Point

1. Determine the equation for total cost and the equation for total revenue using x to represent the number of units to be manufactured and y to represent the total cost.

2. Graph each equation and find the point of intersection.

3. The x coordinate of the break-even point denotes the number of units of the product to manufacture, and the y coordinate denotes the resulting revenue.

EXAMPLE 3   A manufacturer sells her product for $5 per unit and sells all that she produces. The fixed cost of production is $3,500 per month, and the variable cost per unit is $1.50. Find the break-even quantity and revenue per month.

Solution   We begin by determining equations to represent total cost and total revenue. Let y represent total revenue and total cost (in dollars), and let x represent the number of units of the product sold. The principles of business state that

$$\text{total cost} = \text{fixed cost} + \text{variable cost}$$
$$y = 3{,}500 + 1.50x$$

and

$$\text{total revenue} = (\text{price/unit})(\text{number of units sold})$$
$$y = 5x$$

We graph the equations $y = 5x$ and $y = 3{,}500 + 1.50x$ as shown in Figure 4–27. Locating the break-even point from the graph, we find that the coordinates are (1,000, 5,000). Therefore, the company should produce 1,000 units to break even and would receive and spend a revenue of $5,000.

FIGURE 4–27

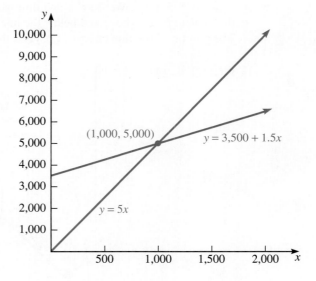

Empirical Data

Graphing is also used in analyzing empirical data, or data that are obtained by experiment or observation. Since measurements in a laboratory are subject to human and experimental errors, the data may not follow a functional equation. Therefore, in graphing empirical data, we graph the points and try to connect them with a smooth curve. However, all of the points may not fit on the curve. Therefore, we average the curve through the data points to obtain the **line of best fit.**

Graphing Empirical Data

1. Make a table of values from the data, if necessary.
2. Determine a scale for each axis by scanning the range of values.
3. Construct and label the axes with the scale and the units of measure.
4. Plot the points. Draw a circle around each point.
5. Draw a smooth curve by averaging the plotted points. To average the plotted points, draw the curve so that it passes midway between two points such that the vertical distances from the curve to points above and below the curve are equal. (Figure 4–28 in Example 4 illustrates.)

EXAMPLE 4   The following empirical data were gathered concerning the pressure on the bottom of a cylindrical tank exerted by a certain force:

Force (lb)	0	5	12	25	42	65
Pressure (lb/in^2)	0	0.4	1.5	2.7	4	6.5

Solution   The values for force range from 0 to 65, while the values for pressure range from 0 to 6.5. The independent variable, force, is placed on the horizontal axis in increments of 5 lb. The dependent variable, pressure, is placed on the vertical axis in increments of 0.5 lb/in^2. We draw a line through the data points so that the distance from the line to data points above and below the line is equal. The graph is shown in Figure 4–28. The portion of the line of best fit beyond the data set given is denoted by a dashed line.

FIGURE 4–28

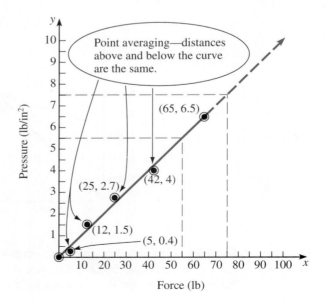

You can use a graph to determine additional information about a function. The process of estimating values between data points on a graph is called **interpolation.** (We will discuss this in more detail in Section 16–4 on emperical curve fitting.) In Example 4, we can estimate the pressure exerted by a particular force by locating the force on the horizontal axis, moving vertically to the graph and horizontally across to the pressure axis, and then reading the value. For example, a 55-lb force results in a 5.5 lb/in² pressure as read from the graph (see the dotted line in Figure 4–28). **Extrapolation** is the process of estimating the value of a function beyond the known data. We read data from the graph in exactly the same manner as for interpolation. For example, a 75-lb force results in a pressure of 7.5 lb/in² as read from the graph (see the dotted line in Figure 4–28). Both interpolation and extrapolation should be used cautiously because they both depend on the accuracy of the data and the graph.

4–6 EXERCISES

Find the solution to the following equations graphically.

1. $6(x - 2) + 3 = 2x - 1$

2. $3x - 5 + 2x = 8x + 10$

3. $7x - 3 + 4(x - 1) = 4$ **4.** $4x - (x - 5) = 2x + 7$

5. $2x + 6 - 5(4x - 3) = x - 2(7x - 1)$

6. $(9 - x) + 3x + 5 = 2(x - 3) + x - 10$

7. $16 - 7(2x + 3) = 5x - 17 + 6(x + 4)$

8. $5x - 13 = 6(2x + 1) - 7x + 15$

9. $4x + 1 - 3(5x + 2) - 9 + 2(x + 6) = 0$

10. $0 = 3x + 4(x + 5) - 10 + 9x$

11. $-3x + 10 = 6$ **12.** $2x + 14 = 6$

13. $4x + (8 - 2x) = 3(x + 1)$

14. $5x - (x + 9) = 3x + 19$

15. $2x - 1 + 3(x + 3) = 6x - 1$

16. $8x + 6 = 5(3x - 4) + 16 - (2x + 5)$

B 17. A manufacturer sells his product for $3 per unit and sells all that he produces. The fixed cost of production is $8,000, and the variable cost is $1.00 per unit. Find the break-even point.

B 18. A manufacturer sells his product for $8 per unit and sells all that he produces. The fixed cost of production is $12,000, and the variable cost is $2 per unit. Find the break-even quantity and revenue.

B 19. A manufacturer sells his product for $14 per unit, selling all that he produces. If the fixed cost of production is $18,000 and the variable cost is $8 per unit, find the break-even quantity.

B 20. A manufacturer sells his product for $3 per unit and has fixed costs of $2,000 and a variable cost of $1 per unit. Under normal conditions, would the manufacturer make a profit or loss if he produces 1,200 units?

B 21. A manufacturer sells her product for $12 per unit and has fixed costs of $15,000 and variable costs of $7 per unit. Does the manufacturer make a profit if she produces 1,000 units? Why or why not?

E 22. Graph the following empirical data relating the voltage and current of a small light bulb. Use the graph to approximate the current at 0.9 V.

Voltage (V)	0.1	0.4	0.7	1	2	3
Current (mA)	30	50	60	70	100	135

23. Graph the relationship between the time of day and the corresponding temperature (in degrees Fahrenheit).

Time	6:00 A.M.	9:00	10:00	11:00	1:00 P.M.	3:00	4:00	7:00	10:00
Temp (°F)	35	40	43	50	54	67	65	58	50

E 24. Graph the following empirical data relating the voltage and current of a light-emitting diode (LED).

Voltage (V)	1.4	1.43	1.48	1.5	1.52	1.55	1.6	1.64
Current (mA)	0.01	0.08	0.4	0.6	1.2	3	10	30

E 25. Graph the following empirical data relating the voltage and current of a fast switching diode. Approximate the voltage of a current of 4 mA.

Voltage (V)	0.4	0.5	0.53	0.64	0.69	0.75	0.79
Current (mA)	0.01	0.05	0.08	0.5	1.2	3	6

P 26. Graph the relationship between mass and weight.

Mass (slugs)	2	3	5	6	9	10	13	14
Weight (lb)	64	96	160	192	288	320	416	448

27. Using the following data, graph the relationship between the radius and area of a circle.

Radius (ft)	1	2	3	4	5	6
Area (ft^2)	3.14	12.56	28.3	50.3	78.5	113.1

B 28. In business, demand (amount purchased) for a product depends on the price of the product. Graph the following data and determine the price at a demand of 27.

Price	70	60	40	30	20	8
Demand	5	10	20	24	30	36

CS 29. A byte almost always consists of 8 bits of data. Graph the following data between the bits in an operating system and the bytes used.

System bits	8	16	32	64
Bytes	1	2	4	8

CHAPTER SUMMARY

Summary of Terms

axes (p. 129)

break-even point (p. 156)

Cartesian coordinate system (p. 129)

dependent variable (p. 124)

domain (p. 126)

extrapolation (p. 159)

function (p. 124)

functional notation (p. 124)

graph (p. 133)

independent variable (p. 124)

interpolation (p. 159)

linear equation in two variables (p. 133)

linear programming (p. 151)

line of best fit (p. 158)

midpoint (p. 141)

one-to-one correspondence (p. 130)

origin (p. 129)

quadrants (p. 129)

range (p. 126)

rectangular coordinate system (p. 129)

root (p. 154)

slope (p. 142)

solution (p. 154)

system of linear inequalities (p. 149)

TRACE (p. 156)

viewing rectangle (p. 135)

x axis (p. 129)

x coordinate (p. 130)

x intercept (p. 154)

y axis (p. 129)

y coordinate (p. 130)

zero (p. 154)

Summary of Formulas

$$d = \sqrt{(x_2 - x_1)^2 + (y_2 - y_1)^2} \quad \text{distance}$$

$$\left(\frac{(x_1 + x_2)}{2}, \frac{(y_1 + y_2)}{2}\right) \quad \text{midpoint}$$

$$m = \frac{\text{rise}}{\text{run}} = \frac{y_2 - y_1}{x_2 - x_1} \quad \text{slope}$$

CHAPTER REVIEW

Section 4–1

Express each function using functional notation.

1. Give the perimeter P of an equilateral triangle as a function of the sides s.

2. Give the perimeter P of an isosceles triangle as a function of the equal sides s if the base is three more than the other sides.

3. Express the area of a triangle as a function of the base if the height is three more than the base.

4. Express the radius r of a sphere as a function of the volume V.

Determine the function values.

5. $f(x) = 3x + 6$. Find $f(-4)$ and $f(2)$.

6. $g(y) = 4y^2 - 5$. Find $g(-2)$ and $g(m)$.

7. $s(t) = t^2 - 5t + 7$. Find $s(-3)$ and $s(4)$.

8. $m(p) = 6p^2 - 1$. Find $m(p - 3)$.

9. $f(y) = 3y - 4$. Find $f(y - 6) + f(7y)$.

10. $g(x) = x^2 + 8x$. Find $g(x - 2) + g(3)$.

If the range is restricted to real numbers, determine the domain of the following functions.

11. $f(x) = \sqrt{3x + 8}$

12. $g(y) = \dfrac{2y + 7}{y - 1}$

13. $s(t) = \sqrt{t - 4}$

14. $f(h) = \dfrac{7}{h} + \dfrac{h + 1}{h - 3} - \dfrac{6}{2h}$

15. $p(y) = \dfrac{3}{y} - \dfrac{2y - 9}{3 - y}$

16. $h(m) = \sqrt{5m + 15}$

Section 4–2

Plot the following points on the rectangular coordinate system.

17. $A(6, 2)$, $B(-1, 0)$, and $C(3, -5)$

18. $D(8, -1)$, $E(-5, -2)$, and $F(0, -4)$

19. $G(7, 2)$, $H(6, 0)$, and $I(-9, 8)$

20. $J(-3, -5)$, $K(4, -8)$, and $L(0, 7)$

21. $M(-6, 4)$, $N(-3, -3)$, and $O(8, 4)$

22. Three vertices of a rectangle are given by $(-2, 4)$, $(3, 4)$, and $(-2, -2)$. Find the coordinate of the fourth vertex.

Section 4–3

Graph the following equations.

23. $y = 3(x - 1) + 4$

24. $y = 5x - 3$

25. $y = 2(x - 1) + 6 - 9(2x + 3)$

26. $y = 16 - 3(x + 4) + 7x$

27. $3x - 2y = 10$

28. $x + 4y = 5$

29. $4x - 2(3x + 1) - 4y = 12$

30. $7y - 3x + 6(x - 1) = 4y - 9$

Section 4–4

Determine the distance between the points given, the slope of the line passing through them, and the midpoint of the line segment joining them.

31. $(-2, 6)$ and $(8, 1)$

32. $(-3, 5)$ and $(-1, -7)$

33. $(-6, -5)$ and $(1, -10)$

34. $(2, 8)$ and $(-3, -9)$

35. $(7, -3)$ and $(-8, -3)$

36. $(4, 3)$ and $(3, 10)$

37. $\left(\dfrac{3}{5}, -4\right)$ and $\left(-\dfrac{1}{3}, 3\right)$

38. $(-2, 4)$ and $(-2, 7)$

39. $(1.7, -3.6)$ and $(2.9, 8.1)$

40. $\left(-\dfrac{1}{3}, \dfrac{3}{4}\right)$ and $\left(\dfrac{1}{2}, -\dfrac{1}{5}\right)$

Section 4–5

Graph the following inequalities.

41. $3x + y > 6$

42. $x - 2y \geq 10$

43. $y > 5x - 1$

44. $2y \geq -5x + 1$

45. $y > 3(x + 6) - 20 + 4x - 4$

46. $y \leq 3(2x + 5) - 7$

47. Find the solution for the following system of inequalities: $x > y + 1$ and $y \geq x$.

48. Find the solution for the following system: $4x - 3y \leq 12$ and $y > 2x - 1$.

Section 4–6

Solve the following equations graphically.

49. $4(x - 1) + 3 = 2x - (5x + 6)$

50. $2x + 3(x - 1) = -(x - 5) + 8$

51. $8x - 2(3x + 5) = 3x - 9$

52. $5x - 3 + 7x = 4(2x + 1)$

53. $6x - 7 + 4(x + 3) = 0$

54. $9(x - 1) + 6 = 3(x + 2) - 5x$

55. $3(x - 1) - 5 = 0$ **56.** $6 - 4(x - 1) = 0$

B 57. A company sells its product for $8 per unit and sells all that it produces. If the fixed cost of production is $12,000 and the variable cost is $2 per unit, find the break-even point.

B 58. A company sells its product for $12 per unit and sells all that it produces. If the fixed cost is $20,000 and the variable cost is $8 per unit, find the break-even cost.

59. Graph the following empirical data as a function of the distance traveled versus speed. Is this relationship linear?

Speed (mi/h)	15	30	45	50	60	65	70
Distance (mi)	30	60	90	100	120	130	140

CHAPTER TEST

The number in parentheses refers to the appropriate learning objective given at the beginning of the chapter.

1. If $f(x) = -(3x - 2) + 6$, find $f(-2)$. (3)

2. Graph $y = 4(2x - 3) + 15 - 2(x + 8)$. (5)

3. Solve $4x + 3y = 18$ graphically. (9)

4. Graph $5x - 6 \geq y$. (8)

5. Plot the following points on the rectangular coordinate system: (4)

$$(-1, 6), (-1, -1), (7, 6), (7, -1)$$

Join the points with lines and describe the resulting figure.

6. A company sells its product for $3 per unit, selling all that it manufactures. If the fixed cost is $6,000 and variable cost is $1 per unit, find the break-even point. (9)

7. Find the solution for $3x + y \geq 6$ and $y \geq x$. (8)

8. Graph $f(x) = 2x - 15$. (5)

9. Solve $5x - 8 = 2(3x + 6) - 2x$ graphically. (9)

10. Determine the slope of the line passing through $(-3, -8)$ and $(6, -4)$. (6)

11. If the range is restricted to real numbers, find the domain for (2)

$$f(x) = \sqrt{4x - 20}$$

12. If $f(x) = 3x - 5$ and $g(x) = 5x^2 + 1$, find $2f(x) - 3g(x)$. (3)

13. Determine the distance between the points $(-6, 7)$ and $(4, -10)$ and the midpoint of the line segment joining them. (7)

14. Graph the relationship between displacement in meters and time in seconds from the following data. Is this relationship linear? (9)

Displacement (m)	4	36	100	144	400
Time (s)	1	3	5	6	10

15. If $f(x) = 2x^2 + 1$, find $f(3m) - f(m)$. (3)

16. Plot the points $A(-6, 2)$, $B(0, 3)$, and $C(-8, 0)$. (4)

17. Express interest rate as a function of principal and time. (1)

GROUP ACTIVITY

Bob's Motor Works manufactures two models of conversion vans—Windstar and Caravan. Bob's Motor Works can convert only 20,000 vans a year, so they cannot meet the demand for both vans. To compete with Fred's Conversions down the street, Bob's Motor Works must convert at least 8,000 Windstars, but Bob hopes to sell at least 10,000 Caravans this year. Bob makes a profit of $4,500 for each Windstar conversion and $6,000 for each Caravan conversion. Determine the number of each type of conversion that Bob should manufacture to maximize his profit, and prepare a presentation that Bob could use when he meets with his investors.

1. Write out the inequalities and graph them.

2. Determine the vertices of the polygon and substitute these points into the profit equation, and determine the values that produce the maximum profit.

3. Prepare a presentation for the investors. Remember that they will not be as mathematically astute as you are. Show how you arrived at your conclusions. Show your graph and explain your calculations for arriving at the maximum profit. Remember that Bob needs a lot of help. He may be presenting your data.

SOLUTION TO CHAPTER INTRODUCTION

To determine the break-even point, we graph the following equations:

$$\text{total cost} = \text{fixed cost} + \text{variable cost}$$
$$y = 1,500 + 1.25x$$
$$\text{total revenue} = (\text{price per unit})(\text{number of units})$$
$$y = 2.5x$$

The graph of these equations is given in Figure 4–29. From the graph, we see that the intersection of the lines occurs at the point (1,200, 3,000). The company should produce 1,200 widgets and receive a total revenue (and cost) of $3,000.

FIGURE 4–29

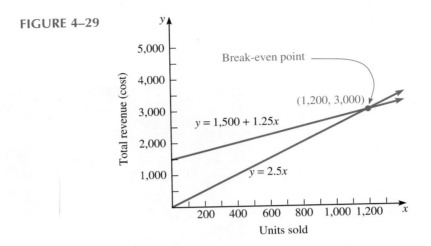

The Scribner log rule, proposed by J. M. Scribner in 1846, is used to estimate the volume in board feet of standing timber. The rule can be expressed as

$$V = (0.79D^2 - 2D - 4)\frac{L}{16}$$

where V is volume in board feet, D is the small end diameter of the log in inches, and L is the log length in feet. Your company needs 625 board feet of fir lumber. What does the diameter of a 16-ft-long log need to be in order to produce 625 board feet? (The solution to this problem is given at the end of the chapter.)

Solving problems of this type results in a quadratic equation. In this chapter we will discuss three methods for solving quadratic equations: factoring, completing the square, and the quadratic formula. Also, we will discuss solving nonlinear inequalities and the application of these concepts to technical problems.

Learning Objectives

After completing this chapter, you should be able to

1. Solve a quadratic equation by factoring (Section 5–1).

2. Solve a quadratic equation by completing the square (Section 5–1).

3. Solve a quadratic equation by using the quadratic formula (Section 5–2).

4. Using the discriminant, describe the number and type of roots in a quadratic equation (Section 5–2).

5. Solve equations in quadratic form (Section 5–3).

6. Solve a quadratic inequality (Section 5–4).

7. Apply the techniques for solving quadratic equations to technical problems (Section 5–5).

Chapter 5

Quadratic Equations

5–1

SOLUTION BY FACTORING AND COMPLETING THE SQUARE

An equation of the form $ax^2 + bx + c = 0$ where a, b, and c are constants and $a \neq 0$ is called a **quadratic equation in standard form.** A **solution** or **root** of a quadratic equation is any value of the variable that results in a true statement. The solution may be two real numbers or a single real number or a pair of complex numbers. We begin the discussion of solving quadratic equations with the solution by factoring method.

Solution by Factoring

In real word problems, quadratic equations are usually solved using the quadratic formula, which is discussed in the next section. However, solution by factoring is included because it is the quickest way to solve a quadratic equation if you see the factors. We begin the discussion with the Zero Factor Property.

Zero Factor Property

The **Zero Factor Property** serves as the basis for solving quadratic equations by factoring. It states that *if the product of two or more factors equals zero, then at least one of the factors must equal zero.* For example, if $(x + 3)(x - 4) = 0$, then by the Zero Factor Property, $x + 3 = 0$ or $x - 4 = 0$. As a result of applying the Zero Factor Property, we can say that the solutions to the equation are $x = -3$ and $x = 4$ because both numbers result in a true statement. The next example illustrates the use of the Zero Factor Property in solving quadratic equations by factoring.

EXAMPLE 1 Solve $6x^2 + 7x - 20 = 0$.

Solution First, we factor the left side of the equation.

$$(3x - 4)(2x + 5) = 0$$

Next, we apply the Zero Factor Property.

$$3x - 4 = 0 \qquad 2x + 5 = 0$$

Then we solve each equation for x.

$$
\begin{aligned}
3x - 4 &= 0 & 2x + 5 &= 0 \\
3x &= 4 & 2x &= -5 \\
x &= 4/3 & x &= -5/2
\end{aligned}
$$

Checking:

$$
\begin{aligned}
x = 4/3: \quad & 6(4/3)^2 + 7(4/3) - 20 = 0 \\
& 32/3 + 28/3 - 20 = 0 \\
& 0 = 0
\end{aligned}
$$

$$
\begin{aligned}
x = -5/2: \quad & 6(-5/2)^2 + 7(-5/2) - 20 = 0 \\
& 75/2 - 35/2 - 20 = 0 \\
& 0 = 0
\end{aligned}
$$

Since both answers result in a true statement, the solutions are $x = 4/3$ and $x = -5/2$.

■ ■ ■

Solution by Factoring

1. Arrange the quadratic equation in standard form, $ax^2 + bx + c = 0$.
2. Factor $ax^2 + bx + c$.
3. Apply the Zero Factor Property and solve for the variable.
4. Check the answer.

NOTE ✦ Since this method of solving quadratic equations is based on factoring, you may want to review the methods of factoring discussed in Chapter 2.

EXAMPLE 2   Solve $9x^2 = 10x$.

Solution   Arranging the quadratic equation in standard form gives

$$9x^2 - 10x = 0$$

Notice that the equation does not contain a constant term; however, this does not affect the way we solve the equation. Next, we factor the left side of the equation using the greatest common factor technique of factoring.

$$x(9x - 10) = 0$$

Last, we apply the Zero Factor Property and solve for x.

$$x = 0 \qquad 9x - 10 = 0$$
$$9x = 10$$
$$x = 10/9$$

The solutions are $x = 0$ and $x = 10/9$. ■■■

CAUTION ✦ Be sure to arrange the quadratic equation in standard form before you begin to factor. The Zero Factor Property applies only to factors whose product equals zero. You may have a tendency to apply the Zero Factor Property to equations like $3(x + 4)(x - 8) = 10$, but the Zero Factor Property cannot be applied to this equation in its current form because it does not equal zero.

EXAMPLE 3   Solve $9y^2 = 4$.

Solution   First, we must arrange the equation in standard form.

$$9y^2 - 4 = 0$$

Next, we factor the left side of the equation using the difference of two squares method.

$$(3y - 2)(3y + 2) = 0$$

Finally, we apply the Zero Factor Property and solve for y.

$$3y - 2 = 0 \qquad 3y + 2 = 0$$
$$3y = 2 \qquad 3y = -2$$
$$y = 2/3 \qquad y = -2/3$$

The solutions are $y = 2/3$ and $y = -2/3$. ■■■

Fractional Equations   Like linear equations, quadratic equations are easier to solve if fractions are eliminated. Some equations do not appear to be quadratic until the fractions are actually eliminated, as shown in the next example.

EXAMPLE 4   Solve the following equation:

$$\frac{x + 2}{x} = \frac{6x}{x + 1}$$

Solution   First, arrange the equation in standard form.

$$x(x + 1)\left(\frac{x + 2}{x}\right) = x(x + 1)\left(\frac{6x}{x + 1}\right) \qquad \text{multiply by LCD}$$

$$(x + 1)(x + 2) = x(6x) \qquad \text{eliminate fractions}$$

$$x^2 + 2x + x + 2 = 6x^2 \qquad \text{eliminate parentheses}$$

$$-5x^2 + 3x + 2 = 0 \qquad \text{transpose and simplify}$$

$$5x^2 - 3x - 2 = 0 \qquad \text{multiply by } -1$$

Next, factor the left side of the equation.

$$(5x + 2)(x - 1) = 0$$

Finally, apply the Zero Factor Property and solve for x.

$$5x + 2 = 0 \qquad\qquad x - 1 = 0$$
$$5x = -2$$
$$x = -2/5 \qquad\qquad x = 1$$

After checking, the solutions are $x = -2/5$ and $x = 1$.   ■■■

CAUTION ✦ Remember to check for division by zero when the original equation contains a fraction.

Solution by Completing the Square

The factoring method of solving quadratic equations is limited in its usefulness. A second method for solving quadratic equations, called *completing the square,* can be used to solve any quadratic equation. However, the method of completing the square is too cumbersome to be used routinely for solving quadratic equations. We discuss it here because it leads to the quadratic formula discussed in the next section.

Before actually discussing completing the square, we must solve a particular type of quadratic equation: equations of the form $(ax + b)^2 = c$. Solving this type of equation is based on the **Square Root Property,** which states that *if m and n are real numbers and $m^2 = n^2$, then $m = n$ or $m = -n$.* In other words, taking the square root of both sides of an equation produces an equivalent equation.

EXAMPLE 5   Solve $(x + 3)^2 = 25$.

Solution   First, we *apply the Square Root Property.*

$$\sqrt{(x + 3)^2} = \pm\sqrt{25}$$
$$x + 3 = \pm 5$$

The notation ± 5 is a shorthand method of writing both $+5$ and -5. Then we *solve the resulting equation for x.*

$$x = -3 \pm 5$$

The following solutions result from solving the two equations:

$$x = -3 + 5 \qquad x = -3 - 5$$
$$x = 2 \qquad x = -8$$

The solutions are $x = 2$ and $x = -8$.

■ ■ ■

The equation in Example 5 was easy to solve because the expression containing the variable was a perfect square. The process of completing the square transforms an equation to this form.

Solution by Completing the Square

1. Arrange the equation in $ax^2 + bx = c$ form.
2. If $a \ne 1$, divide each term of the equation by a, giving

$$x^2 + \frac{b}{a}x = \frac{c}{a}$$

3. Complete the square by *adding* the magnitude (omit the sign) of $\left(\dfrac{b}{2a}\right)^2$ to both the sides of the equation.

$$x^2 + \frac{b}{a}x + \left(\frac{b}{2a}\right)^2 = \frac{c}{a} + \left(\frac{b}{2a}\right)^2$$

4. Write the left-hand side of the equation as a perfect square:

$$\left(\sqrt{\text{first term}} \pm \sqrt{\text{third term}}\right)^2 = \left(x + \frac{b}{2a}\right)^2 = \frac{c}{a} + \left(\frac{b}{2a}\right)^2$$

5. Apply the Square Root Property.

$$\sqrt{\left(x + \frac{b}{2a}\right)^2} = \pm \sqrt{\frac{c}{a} + \left(\frac{b}{2a}\right)^2}$$

6. Solve the resulting linear equations for the variable.

EXAMPLE 6 Determine what to add to each equation to complete the square. (Do not solve the equation.)

a. $x^2 + 12x = 40$ **b.** $x^2 + 18x - 20 = 0$
c. $2x^2 + 10x = 15$ **d.** $3x^2 + 7x - 12 = 0$

Solution (a) The equation is arranged in $ax^2 + bx = c$ form and $a = 1$. Therefore, we complete the square by adding $\left(\dfrac{b}{2a}\right)^2$.

$$b = 12$$

$$\frac{b}{2a} = \frac{12}{2(1)} = 6$$

$$\left(\frac{b}{2a}\right)^2 = (6)^2 = 36$$

We would add 36 to both sides of the equation.

(b) This equation must be rearranged to the required $ax^2 + bx = c$ form.

$$x^2 + 18x - 20 = 0$$
$$x^2 + 18x = 20$$

Then, we complete the square by adding $(b/2a)^2$ to both sides.

$$b = 18$$

$$\frac{b}{2a} = \frac{18}{2(1)} = 9$$

$$\left(\frac{b}{2a}\right)^2 = (9)^2 = 81$$

We could add 81 to both sides of the equation.

(c) This equation is arranged in $ax^2 + bx = c$ form. Since $a \neq 1$, we must divide both sides of the equation by $a = 2$.

$$\frac{2x^2}{2} + \frac{10x}{2} = \frac{15}{2}$$

$$x^2 + 5x = \frac{15}{2}$$

Then, we complete the square.

$$b = 5$$

$$\frac{b}{2a} = \frac{5}{2}$$

$$\left(\frac{b}{2a}\right)^2 = \left(\frac{5}{2}\right)^2 = \frac{25}{4}$$

We add 25/4 to both sides of the equation.

(d) We begin by arranging this equation in $ax^2 + bx = c$ form.

$$3x^2 + 7x - 12 = 0$$
$$3x^2 + 7x = 12$$

Next, we divide the equation by $a = 3$.

$$\frac{3x^2}{3} + \frac{7x}{3} = \frac{12}{3}$$

$$x^2 + \frac{7}{3}x = 4$$

Last, we complete the square

$$b = \frac{7}{3}$$

$$\frac{b}{2a} = \frac{7}{6}$$

$$\left(\frac{b}{2a}\right)^2 = \left(\frac{7}{6}\right)^2 = \frac{49}{36}$$

We add 49/36 to both sides of the equation.

EXAMPLE 7   Solve $x^2 + 6x = 24$ by completing the square.

Solution   Since the equation is arranged in the required form and $a = 1$, we begin with Step 3 by *completing the square*. We must add the magnitude of

$$\left(\frac{b}{2a}\right)^2 = \left(\frac{6}{2}\right)^2 = 9$$

to both sides of the equation.

$$x^2 + 6x \boxed{+ 9} = 24 \boxed{+ 9}$$

Next, we *write the left side of the equation as a perfect square* and *simplify* the right side.

$$(x + 3)^2 = 33$$

Then we *apply the Square Root Property.*

$$\sqrt{(x + 3)^2} = \pm\sqrt{33}$$
$$x + 3 = \pm\sqrt{33}$$

Finally, we solve the linear equation for x.

$$x = -3 \pm \sqrt{33}$$ ∎

CAUTION ✦ Do not forget to add $(b/2a)^2$ to *both* sides of the equation when completing the square.

EXAMPLE 8   Solve $2x^2 + 3x - 8 = 0$. Express the solution in decimal form rounded to thousandths.

Solution   First, we arrange the equation in the required form.

$$2x^2 + 3x = 8$$

Since $a \neq 1$, we divide both sides of the equation by the value of a, which is 2.

$$\frac{2x^2}{2} + \frac{3x}{2} = \frac{8}{2}$$

$$x^2 + \frac{3}{2}x = 4$$

Next, we complete the square by adding

$$\left(\frac{b}{2a}\right)^2 = \left(\frac{3}{4}\right)^2 = \frac{9}{16}$$

to both sides of the equation.

$$x^2 + \frac{3}{2}x + \frac{9}{16} = 4 + \frac{9}{16}$$

Then we write the left side of the equation as a perfect square and simplify the right-hand side.

$$\left(x + \frac{3}{4}\right)^2 = \frac{73}{16}$$

Take the square root of both sides of the equation.

$$x + \frac{3}{4} = \pm\frac{\sqrt{73}}{4}$$

Solve for x.

$$x = \frac{-3 \pm \sqrt{73}}{4}$$

The roots are approximately 1.386 and -2.886.

First root: 3 +/− + 73 √ = ÷ 4 = ⟶ 1.386

Second root: 3 +/− − 73 √ = ÷ 4 = ⟶ −2.886   ∎∎∎

Applications

EXAMPLE 9   The power P delivered to the load by a generator with a voltage V and internal resistance R is given by $P = IV - I^2R$, where I is the current in the circuit, as shown in Figure 5–1. Find the current in the circuit when $P = 9$ W, $V = 6$ V, and $R = 1$ Ω.

Solution   Substituting the given values into the formula yields the following quadratic equation.

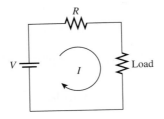

$$P = IV - I^2R$$
$$9 = 6I - 1I^2$$

We arrange the equation in the required order.

$$I^2 - 6I = -9$$

FIGURE 5–1

Next, we complete the square by adding $(6/2)^2 = 9$ to both sides of the equation.

$$I^2 - 6I + 9 = -9 + 9$$

Solving the equation gives

$$(I - 3)^2 = 0$$
$$I = 3$$

The current in the circuit is 3 A.   ∎∎∎

EXAMPLE 10   The amount of propellant P (in pounds) remaining in a rocket at time t seconds is given by $P = 25 - 2t - t^2$. How many seconds after ignition will it take for 10 lb of propellant to remain?

Solution   Since $P = 10$ lb in this case, the equation to be solved is

$$10 = 25 - 2t - t^2$$

Arranging the equation in the standard form to factor gives

$$t^2 + 2t + 10 - 25 = 0$$
$$t^2 + 2t - 15 = 0$$

Factoring and solving for t gives

$$(t + 5)(t - 3) = 0$$
$$t = -5 \qquad \text{or} \qquad t = 3$$

Since time is not measured in negative numbers, it takes 3 s for the propellant remaining to be 10 lb.

■ ■ ■

5–1 EXERCISES

Solve the following quadratic equations by factoring.

1. $x^2 - 9x + 20 = 0$

2. $x^2 - 8x + 15 = 0$

3. $x^2 + 2x - 63 = 0$

4. $18x - 14x^2 = 0$

5. $6x^2 + 7x = 5$

6. $6y^2 + 13y - 5 = 0$

7. $4x^2 + 12x + 9 = 0$

8. $9x^2 = 1$

9. $15x^2 + 13x = -2$

10. $15d^2 = 24 - 2d$

11. $25x^2 - 15 = 23x - 3x^2$

12. $2s^2 = 70 - 4s$

13. $2f^2 = 8f$

14. $x^2 = 64$

15. $16x^2 - 25 = 0$

16. $3m(m - 2) = 2(m + 5) - 9m$

17. $(4x - 1)(x + 2) = x(2x - 3) - 10$

18. $(3x + 4)(x - 2) = 2x(x + 1) + 4$

19. $\dfrac{3c}{c - 2} + \dfrac{6}{c + 2} = \dfrac{24}{c^2 - 4}$

20. $\dfrac{5x}{x + 1} + \dfrac{7x}{x - 1} = \dfrac{3x + 1}{x^2 - 1}$

21. $\dfrac{1}{y} + \dfrac{1}{y + 6} = \dfrac{1}{4}$

22. $\dfrac{60}{b} + 5 = \dfrac{60}{b - 1}$

Solve the following quadratic equations by completing the square.

23. $x^2 - 4x = 20$

24. $a^2 + 6a = 12$

25. $m^2 + 6m = 14$

26. $x^2 - 8x = 15$

27. $x^2 - 2x = 3$

28. $x^2 + 5x = 10$

29. $k^2 - 9k - 8 = 0$

30. $3x^2 + 6x - 17 = 0$

31. $2x^2 + 8x = 13$

32. $5x^2 - 15x = 23$

33. $3x^2 - 9x = 14$

34. $4x^2 + 10x = 31$

35. $20m = 24 - 5m^2$

36. $7x = 24 - 3x^2$

37. $7x^2 - 21x + 17 = 0$

38. $6z^2 = 25 - 3z$

39. $-9c = -2 - 2c^2$

40. $3a^2 + 4a = -8$

41. $17y - 4y^2 = 20$

42. $2y^2 = 8 - 7y$

Solve using either method.

E 43. The power dissipated in a circuit with two currents flowing through a resistor is given by

$$P = (I_1 + I_2)^2 R$$

Find I_1 if $I_2 = 0.3$ A, $R = 100$ Ω, and $P = 11$ W. Round your answer to hundredths.

44. The sum of two numbers is 10 and their product is 24. Find the numbers.

B 45. The weekly profit P of Lou's Surfboards is given by $P = x^4 - 27x^3$, $x \geq 1$, where x is the week in the year. During what week is the profit equal to zero?

46. Find two consecutive integers such that the sum of their squares is 613.

B 47. Total revenue R is the product of price and quantity. If the price is a function of the quantity sold q, the equation is $R = (1,000 - q)q$. Determine the production quantity to earn $187,500 in revenue.

48. If three times the square of a number is added to ten times the number, the result is 25. Find the number.

B 49. Working together, Fred and Larry can paint the exterior of a house in 8 hours, while it takes Larry 12 hours longer than Fred if each works alone. How long does it take Larry to paint the house alone.

P 50. When an object is dropped from a 64-ft-tall building, the resulting equation for its height is $h = 64 - 16t^2$. Find the time, t, in seconds, required for the object to strike the ground ($h = 0$).

E 51. The power dissipated in a 25-Ω resistor is 49 W. The equation to find the current passing through the resistor is $49 = 25I^2$. Solve for the current, I.

C 52. A tract of land is triangular. If the area of the tract is 1,200 m^2 and the height is 40 m more than the base, find the dimensions of the tract.

E 53. Find the current in the circuit in Example 8 when $P = 25$ W, $V = 10$ V, and $R = 1.0$ Ω.

B 54. Doyle's log rule is one of the most widely used as well as one of the oldest rules used to estimate the volume V from a log. Doyle's log rule is given by $V = L\left(\dfrac{D - 4}{4}\right)^2$ where D = diameter of the small end in inches and L = log length in feet. Find the diameter of a log 16 ft long if it produces a volume of 144 board feet.

A 55. An architect must design a fixed window to fit into a 5.0-ft by 8.0-ft opening. The plans call for 32 ft^2 of glass to be encased in an oak frame of uniform width. What size glass and frame should be used to meet these specifications?

E 56. The power, in watts, consumed by an electrical circuit as a function of elapsed time t, in seconds, is given by $p = 15t + 3t^2$. How long will it take power consumption to be 42 W?

M 57. The bending moment m of a beam as a function of the distance x, in feet, from one end is given by $m = -15x + x^2$. Find the distance x when the bending moment is 76.

5–2

SOLUTION BY THE QUADRATIC FORMULA

A third method of solving quadratic equations is the quadratic formula. You can use the quadratic formula, as well as completing the square, to solve any quadratic equation. However, you will find that the quadratic formula is easier to use. The derivation of the quadratic formula follows.

Deriving the Quadratic Formula

We derive the quadratic formula by taking the standard form of a quadratic equation and completing the square.

$$ax^2 + bx + c = 0$$

$$ax^2 + bx = -c \qquad \text{transpose } c$$

$$x^2 + \frac{b}{a}x = -\frac{c}{a} \qquad \text{divide by } a$$

$$x^2 + \frac{b}{a}x + \left(\frac{b}{2a}\right)^2 = -\frac{c}{a} + \left(\frac{b}{2a}\right)^2 \qquad \text{add } \left(\frac{b}{2a}\right)^2 \text{ to both sides}$$

$$x^2 + \frac{b}{a}x + \frac{b^2}{4a^2} = -\frac{c}{a} + \frac{b^2}{4a^2} \qquad \text{simplify}$$

$$\left(x + \frac{b}{2a}\right)^2 = \frac{-4ac + b^2}{4a^2} \qquad \begin{array}{l}\text{factor left side and} \\ \text{simplify right side}\end{array}$$

$$x + \frac{b}{2a} = \pm\sqrt{\frac{b^2 - 4ac}{4a^2}} \qquad \begin{array}{l}\text{take square root of} \\ \text{both sides}\end{array}$$

$$x = \frac{-b}{2a} \pm \sqrt{\frac{b^2 - 4ac}{4a^2}} \qquad \text{solve for } x$$

$$x = \frac{-b \pm \sqrt{b^2 - 4ac}}{2a} \qquad \begin{array}{l}\text{simplify and add} \\ \text{fractions}\end{array}$$

The quadratic formula is summarized below.

Quadratic Formula

If $ax^2 + bx + c = 0$, where $a \neq 0$, then

$$x = \frac{-b \pm \sqrt{b^2 - 4ac}}{2a}$$

Notice that the quadratic formula results in the following two solutions:

$$\frac{-b + \sqrt{b^2 - 4ac}}{2a} \qquad \text{and} \qquad \frac{-b - \sqrt{b^2 - 4ac}}{2a}$$

Solution by the Quadratic Formula

To solve a quadratic equation using the formula, use the following steps:

1. Arrange the equation in standard form,

$$ax^2 + bx + c = 0$$

2. Identify a, b, and c. The coefficient of the squared term is a; the coefficient of the linear term is b; and the constant term is c.

3. Substitute the values of a, b, and c into the formula.

4. Simplify the expression.

Note: The coefficients a, b, and c include their signs.

The process of solving several types of quadratic equations is illustrated in the next three examples.

EXAMPLE 1 Solve $5x^2 - 6x = 3$.

Solution First, we arrange the equation in standard form and identify the values of a, b, and c.

$$5x^2 \quad - 6x \quad -3 = 0$$

$$a = 5 \quad b = -6 \quad c = -3$$

Then we substitute values for a, b, and c into the quadratic formula.

$$x = \frac{-b \pm \sqrt{b^2 - 4ac}}{2a}$$

$$x = \frac{-(-6) \pm \sqrt{(-6)^2 - 4(5)(-3)}}{2(5)}$$

$$x = \frac{6 \pm \sqrt{36 + 60}}{10}$$

$$x = \frac{6 \pm \sqrt{96}}{10}$$

Next, we simplify the radical as follows: $\sqrt{96} = \sqrt{16 \cdot 6} = 4\sqrt{6}$ (remove any squares just as we did in Chapter 3).

$$x = \frac{6 \pm 4\sqrt{6}}{10}$$

Last, we reduce the fractional expression to lowest terms.

$$x = \frac{2(3 \pm 2\sqrt{6})}{10}$$

$$x = \frac{3 \pm 2\sqrt{6}}{5}$$

■■■

EXAMPLE 2   Solve the following:

$$\frac{y+1}{y-2} - \frac{6}{y^2 - 5y + 6} = \frac{2y}{y-3}$$

Solution   First, we must *arrange the equation in standard form*. We *eliminate fractions* by multiplying by the common denominator, $(y-3)(y-2)$. Then we identify a, b, and c.

$$(y+1)(y-3) - 6 = 2y(y-2) \qquad \text{eliminate fractions}$$
$$y^2 - 2y - 3 - 6 = 2y^2 - 4y \qquad \text{eliminate parentheses}$$
$$-y^2 + 2y - 9 = 0 \qquad \text{transpose and combine like terms}$$
$$y^2 \qquad -2y \qquad +9 = 0 \qquad \text{multiply by } -1$$

$$\uparrow \qquad\qquad \uparrow \qquad\qquad \uparrow$$

$$a = 1 \qquad b = -2 \qquad c = 9$$

Substituting into the quadratic formula and *simplifying* the result gives

$$y = \frac{-(-2) \pm \sqrt{(-2)^2 - 4(1)(9)}}{2(1)}$$

$$y = \frac{2 \pm \sqrt{4 - 36}}{2}$$

$$y = \frac{2 \pm \sqrt{-32}}{2}$$

In Chapter 3, we briefly discussed the square root of a negative number. Recall that the square root of a negative number is an imaginary number. In most real-world problems, an imaginary number as a solution is not workable. Since this example does not relate to an application, we will continue its solution even though that solution is an imaginary or complex number. By definition, $\sqrt{-1} = j$ (the letter i is also used but j is commonly used in electronics). Therefore, we can write $\sqrt{-32}$ as $j\sqrt{32}$ and proceed to simplify the radical portion of $j\sqrt{32}$ as we did in Chapter 3 to become $4j\sqrt{2}$.

$$y = \frac{2 \pm 4j\sqrt{2}}{2} \qquad \text{simplify radical}$$

$$y = \frac{2(1 \pm 2j\sqrt{2})}{2} \qquad \text{factor numerator}$$

$$y = 1 \pm 2j\sqrt{2} \qquad \text{reduce fraction} \qquad\qquad ■■■$$

EXAMPLE 3   Solve $(3x + 2)(5x - 1) - 1 = 3x(x - 1)$.

Solution   First, we eliminate parentheses and arrange the equation in standard form.

$$(3x + 2)(5x - 1) - 1 = 3x(x - 1)$$
$$15x^2 - 3x + 10x - 2 - 1 = 3x^2 - 3x \qquad \text{eliminate parentheses}$$
$$15x^2 + 7x - 3 - 3x^2 + 3x = 0 \qquad \text{add and transpose}$$
$$12x^2 + 10x - 3 = 0 \qquad \text{combine like terms}$$

Next, we substitute $a = 12$, $b = 10$, and $c = -3$ into the quadratic formula and simplify.

$$x = \frac{-10 \pm \sqrt{100 - 4(12)(-3)}}{2(12)}$$

$$x = \frac{-10 \pm \sqrt{100 + 144}}{24}$$

$$x = \frac{-10 \pm \sqrt{244}}{24}$$

$$x = \frac{-10 \pm 2\sqrt{61}}{24}$$

$$x = \frac{2(-5 \pm \sqrt{61})}{24}$$

Reducing the fraction gives the solutions as

$$x = \frac{-5 \pm \sqrt{61}}{12} \approx 0.2 \text{ and } -1.1$$

■ ■ ■

Equations with Radicals

The following example illustrates solving an equation that contains radicals and that must be squared twice to eliminate the radical. The resulting equation is quadratic and can be solved using the quadratic formula.

EXAMPLE 4   Solve $\sqrt{y + 1} - 8 = \sqrt{3y - 4}$. Express the answer to the nearest thousandths.

Solution   First, we must arrange the equation in standard form.

$$\sqrt{y + 1} - 8 = \sqrt{3y - 4}$$

$y + 1 - 16\sqrt{y + 1} + 64 = 3y - 4$	square both sides
$y + 65 - 16\sqrt{y + 1} = 3y - 4$	combine like terms
$-16\sqrt{y + 1} = 3y - 4 - y - 65$	transpose
$-16\sqrt{y + 1} = 2y - 69$	combine like terms
$256(y + 1) = 4y^2 - 276y + 4{,}761$	square both sides
$256y + 256 = 4y^2 - 276y + 4{,}761$	eliminate parentheses
$-4y^2 + 532y - 4{,}505 = 0$	transpose and simplify

Next, we substitute $a = -4$, $b = 532$, and $c = -4{,}505$ into the quadratic formula and simplify the result.

$$y = \frac{-532 \pm \sqrt{(532)^2 - 4(-4)(-4{,}505)}}{2(-4)}$$

$$y = \frac{-532 \pm \sqrt{210{,}994}}{-8}$$

$$y \approx 9.089 \qquad \text{and} \qquad y \approx 123.911$$

To calculate $\sqrt{}$:   532 $\boxed{x^2}$ $\boxed{-}$ 4 $\boxed{\times}$ 4 $\boxed{+/-}$ $\boxed{\times}$ 4505

$\boxed{+/-}$ $\boxed{=}$ $\boxed{\sqrt{}}$ $\boxed{\text{STO}}$ $\longrightarrow$ 459.2864030

To determine y: 532 $\boxed{+/-}$ $\boxed{+}$ $\boxed{\text{RCL}}$ $\boxed{=}$ $\boxed{\div}$ 8 $\boxed{+/-}$ $\boxed{=}$ $\longrightarrow$ 9.089

To determine y: 532 $\boxed{+/-}$ $\boxed{-}$ $\boxed{\text{RCL}}$ $\boxed{=}$ $\boxed{\div}$ 8 $\boxed{+/-}$ $\boxed{=}$ $\longrightarrow$ 123.911

Checking $y = 9.089$: $\sqrt{9.089 + 1} - 8 = \sqrt{3(9.089)} - 4$

$$-4.824 \neq 4.824$$

$y = 123.911$: $\sqrt{123.911 + 1} - 8 = \sqrt{3(123.911)} - 4$

$$3.176 \neq 19.176$$

Since neither of the answers checks, this equation has no solution. ■■■

CAUTION ✦ When you square both sides of an equation, remember to check the solution for extraneous roots.

Quadratic Literal Equation

You can also use the quadratic formula to solve an equation in which the coefficients are literal, rather than numerical.

EXAMPLE 5 The distance fallen, s, of a free-falling object is given by

$$s = vt + \frac{gt^2}{2}$$

where v = initial velocity, t = time, and g = acceleration due to gravity. Solve this formula for t.

Solution This formula is quadratic in the variable t, so we must use one of the three methods for solving a quadratic equation. Using the quadratic formula, we must arrange the equation in standard form.

$$2(s) = 2(vt) + 2\left(\frac{gt^2}{2}\right) \qquad \text{eliminate fractions}$$

$$2s = 2vt + gt^2 \qquad \text{simplify}$$

$$-gt^2 - 2vt + 2s = 0 \qquad \text{transpose}$$

Remember that a represents the coefficient of the squared variable, b represents the coefficient of the linear variable, and c represents the constant term. Next, we substitute $a = -g$, $b = -2v$, and $c = 2s$ into the quadratic formula.

$$t = \frac{-(-2v) \pm \sqrt{(-2v)^2 - 4(-g)(2s)}}{2(-g)}$$

$$t = \frac{2v \pm \sqrt{4v^2 - 4(-g)(2s)}}{-2g}$$

$$t = \frac{2v \pm \sqrt{4v^2 + 8gs}}{-2g}$$

Then we simplify the radical expression.

$$t = \frac{2v \pm \sqrt{4(v^2 + 2gs)}}{-2g}$$

$$t = \frac{2v \pm 2\sqrt{v^2 + 2gs}}{-2g}$$

$$t = \frac{2(v \pm \sqrt{v^2 + 2gs})}{-2g}$$

Reducing the fraction to lowest terms gives the following solutions:

$$t = \frac{v \pm \sqrt{v^2 + 2gs}}{-g}$$

$$t = \frac{-v \pm \sqrt{v^2 + 2gs}}{g}$$

■ ■ ■

When a load is applied to a steel or wooden beam, the force created by the load produces a bending moment. Under the bending moment, the beam will curve slightly. To ensure that beams do not break, structural engineers determine the bending moment at various points along the beam to design and select a beam of the proper size and material. For a simply supported beam of length L carrying an evenly distributed load of w pounds per foot, the bending moment m at any distance x from one end of the beam is given by $m = \frac{1}{2}Lwx - \frac{1}{2}wx^2$.

EXAMPLE 6   A simply supported Douglas fir beam, 12 ft long, carries an evenly distributed load of 800 lb/ft (including the weight of the beam). At what distance from one end is the bending moment 8,000 ft-lb?

Solution

$$m = \frac{1}{2}Lwx - \frac{1}{2}wx^2$$

$$8{,}000 = \frac{1}{2}(12)(800)x - \frac{1}{2}(800)x^2 \qquad \text{substitute}$$

$$8{,}000 = 4{,}800x - 400x^2 \qquad \text{simplify}$$

$$20 = 12x - x^2 \qquad \text{divide by 400}$$

$$x^2 - 12x + 20 = 0 \qquad \text{arrange in standard form}$$

$$x = \frac{12 \pm \sqrt{(12)^2 - 4(1)(20)}}{2}$$

$$x = \frac{12 \pm 8}{2}$$

$$x = 10 \text{ ft} \qquad \text{or} \qquad 2 \text{ ft}$$

■ ■ ■

Graphing Quadratic Equations

In Chapter 4, we discussed graphing linear equations using a table of values and the graphing calculator. Graphing a quadratic equation uses exactly the same techniques. Since the graph of a quadratic equation is a curved line, we must choose more x-values for the table of values. The next example illustrates the technique.

EXAMPLE 7   Graph $y = 2x^2 + 8x + 1$ using a table of values.

Solution   The table of values and calculations are given below. The resulting graph is shown in Figure 5–2.

x	y
-4	1
-3	-5
-2	-7
-1	-5
0	1
1	11
2	25

$$2(-4)^2 + 8(-4) + 1 = 1$$
$$2(-3)^2 + 8(-3) + 1 = -5$$
$$2(-2)^2 + 8(-2) + 1 = -7$$
$$2(-1)^2 + 8(-1) + 1 = -5$$
$$2(0)^2 + 8(0) + 1 = 1$$
$$2(1)^2 + 8(1) + 1 = 11$$
$$2(2)^2 + 8(2) + 1 = 25$$

The U-shaped graph shown in Figure 5–2 is called a **parabola.** Its shape is characteristic of quadratic equations. The turning point of the curve is called the **vertex** of the parabola. The coordinates of the vertex are given by

$$\left(\frac{b}{-2a}, f\left(\frac{b}{-2a} \right) \right)$$

FIGURE 5–2

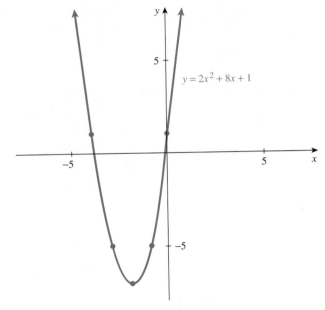

$$y = 2x^2 + 8x + 1$$

In Example 7, the vertex is given by

$$x = \frac{b}{-2a} = \frac{8}{-2(2)} = -2$$

$$y = f\left(\frac{b}{-2a} \right) = 2(-2)^2 + 8(-2) + 1 = -7$$

The vertex is $(-2, -7)$, which is confirmed by the graph in Figure 5–2. We discuss the parabola in greater detail in Chapter 15. ∎∎∎

EXAMPLE 8   Use the graphing calculator to solve the equation $x^2 - 3x - 4 = 0$.

Solution   We input the equation $y = x^2 - 3x - 4$. Set the viewing rectangle (shown in Figure 5–3) and execute the graph. Using the TRACE function, we follow the curve until it crosses the x axis. Note in this case the graph crosses the x axis twice and has two solutions. From Figure 5–3, we see the solutions are $x = -1$ and $x = 4$. We can use the algebraic techniques developed in this chapter to check our graph. Using the quadratic formula gives the following result:

$$x = \frac{3 \pm \sqrt{(3)^2 - 4(1)(-4)}}{2(1)}$$

$$x = \frac{3 \pm \sqrt{25}}{2}$$

$$x = \frac{3 \pm 5}{2}$$

$$x = 4 \qquad \text{or} \qquad x = -1$$

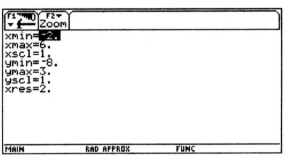

Viewing rectangle

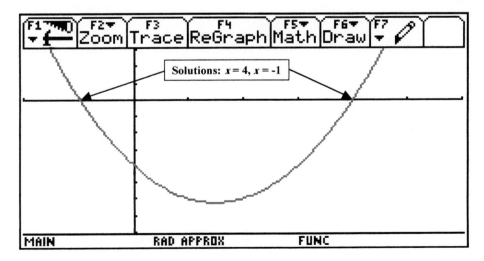

FIGURE 5–3

Discriminant

The **discriminant** is the expression that appears under the radical in the quadratic formula.

$$b^2 - 4ac$$

You can use the sign of the discriminant to determine the number and type of solutions to a quadratic equation. This information is summarized below.

Number and Type of Solutions	
Value of the Discriminant	Number and Type of Solutions
Zero (0)	One real solution
Negative (−)	Two complex solutions
Positive (+)	Two real solutions

CAUTION ✦ You must arrange the quadratic equation in standard form before determining the value of the discriminant.

EXAMPLE 9 Using the discriminant, describe the number and type of solutions for the quadratic equation $2x^2 - 6x + 10 = 0$.

Solution Since the quadratic equation is arranged in standard form, $a = 2$, $b = -6$, and $c = 10$. Substituting into the discriminant, we have

$$b^2 - 4ac = (-6)^2 - 4(2)(10)$$
$$= 36 - 80$$
$$= -44$$

Since the discriminant is negative, the quadratic equation has two complex solutions.

■ ■ ■

NOTE ✦ The discriminant does not give the actual solution to a quadratic equation; rather, it gives information about the number and type of solutions. When complex numbers are not a feasible solution, it would not be necessary to solve a quadratic equation whose discriminant is a negative number.

EXAMPLE 10 Using the discriminant, determine the number and type of solutions for $3x^2 - 6 = -5x$.

Solution First, we arrange the equation in standard form and identify a, b, and c.

$$3x^2 + 5x - 6 = 0$$

Then we substitute $a = 3$, $b = 5$, and $c = -6$ into the discriminant.

$$b^2 - 4ac = (5)^2 - 4(3)(-6)$$
$$= 25 + 72$$
$$= 97$$

Since the discriminant is positive, this quadratic equation has two real number solutions.

■ ■ ■

5–2 EXERCISES

Solve the following equations using the quadratic formula. Graph problems 1–12.

1. $4x^2 + 5x - 6 = 0$

2. $3x^2 + 8x - 1 = 0$

3. $8z - 2z^2 = 10$

4. $7 - 5p = 4p^2$

5. $12 = 9y^2 - 5y$

6. $15x - 4 = x^2$

7. $3x^2 - 16 = 2x$

8. $6n - 9 + 7n^2 = 0$

9. $10x^2 - x + 11 = 0$

10. $13k + 2k^2 - 8 = 0$

11. $5x - 7x^2 + 4 = 3x + x^2 - 10$

12. $x^2 + 3x = 6 + 9x - 4x^2$

13. $11z - 18 + z^2 = 12 + 23z$

14. $3y^2 + 10 = 14y - 2$

15. $8m^2 - 9 = 0$

16. $5x^2 - 7x = 0$

17. $8x^2 = 15x$

18. $3z^2 - 14 = 0$

19. $-3x^2 - 8 = 2x(x - 3)$

20. $4(3x + 2) + 5x(x - 1) = 8x$

21. $3x(2x + 5) = (x + 1)(4x - 7)$

22. $(2x + 1)(x - 5) = 7(3x + 2)$

23. $(2m - 3)m + 6(m^2 - 4m) = 8$

24. $9t(t - 2) = 17 - 3(5t + 6)$

25. $(x - 4)(x + 3) - 5(2x - 3) = 0$

26. $8(3x + 3) = 9x(2x + 5)$

27. $3(t - 5) + 4t(2t - 1) = 38$

28. $(y + 3)(4y - 1) = 6y(y - 2) + 13$

29. $\dfrac{6}{x} + 9 = \dfrac{5}{x^2}$

30. $7 = \dfrac{2}{x} - \dfrac{11}{x^2}$

31. $\dfrac{2s}{s - 1} + \dfrac{6}{s + 3} = \dfrac{4s}{s^2 + 2s - 3}$

32. $\dfrac{5x}{2x + 3} - \dfrac{16}{2x^2 - x - 6} = \dfrac{x}{x - 2}$

33. $\dfrac{1}{z} + \dfrac{1}{z - 2} = \dfrac{1}{7}$

34. $\dfrac{1}{y} + \dfrac{1}{y + 2} = \dfrac{1}{12}$

35. $\dfrac{1}{x} + \dfrac{1}{x - 1} = \dfrac{1}{4}$

36. $\dfrac{1}{x} + \dfrac{1}{x - 3} = \dfrac{1}{12}$

37. $y = \sqrt{\dfrac{3y - 4}{7}}$

38. $x = \sqrt{\dfrac{6x - 9}{5}}$

39. $\sqrt{x^2 - 9x} = \sqrt{7 - 3x^2}$

40. $\sqrt{2x - 3} = \sqrt{4x - 7x^2}$

Using the discriminant, describe the number and type of solutions for the following equations.

41. $3x^2 - 2x = 7$

42. $9 - y^2 = 6y$

43. $6y^2 = 5y$

44. $10x^2 - 7x = 0$

45. $9x^2 = 6x$

46. $8z^2 - 11 = 0$

47. $10z^2 = 25 - 4z^2$

48. $5x^2 + 13x + 25 = 0$

49. $18x = -25 - 4x^2$

50. $14n - 17 = 3n^2$

51. $\dfrac{d^2}{6} + \dfrac{5d}{3} = 2$

52. $\dfrac{7x^2}{2} = 10 - \dfrac{5x}{3}$

53. $\dfrac{t^2}{4} - \dfrac{3}{5} = 2t$

54. $\dfrac{4s}{5} - 3 = \dfrac{s^2}{2}$

55. $\dfrac{2x^2}{7} - \dfrac{x}{3} = 2$

56. The number of diagonals N of a polygon given the number of sides n is

$$N = \frac{n(n - 3)}{2}$$

How many sides does a polygon with 35 diagonals ($N = 35$) have?

57. The formula for the surface area S of a cylinder is $S = 2\pi r^2 + 2\pi rh$ where r is the radius and h is the height. Solve the formula for r.

58. The formula for the area A of a sector of a circle is

$$A = \frac{r^2 \theta}{2}$$

where r is the radius and θ is the measure of the central angle in radians. Solve the formula for r.

B 59. Total revenue is the product of price and quantity. If q represents quantity and price is a function of quantity given by $1{,}500 - 3q$, find the quantity that would produce a total revenue of \$5,000.

60. The formula to determine the range d of a ship's radio communication is given by

$$d = \sqrt{2rh + h^2}$$

Solve the formula for h.

M 61. A simply supported Douglas fir beam, 9 ft long, carries a uniformly distributed load of 700 lb/ft (including the weight of the beam). At what distance from the end of the beam is the bending moment 900 ft-lb? (Refer to Example 6.)

A 62. Based on the limitations of the building site and customer requirements, an architect must design a rectangular storage warehouse with a flat roof that contains 1,750 m³ of space. The length of the building is to be 6.50 m more than the width and the height is to be 5.00 m. What should be the dimensions of the warehouse?

P 63. A dive bomber projects a bomb downward with an initial velocity of 350 ft/s from a height of 4,500 ft. How long will it take the bomb to reach the ground [$d = vt + 16t^2$ where d = distance (ft), v = initial velocity (ft/s), and t = time (sec)]?

CS 64. An equation that may be used to estimate the time needed to boot up a computer is $t^2 + 4t - 96 = 0$ where t is the time in seconds. How long will it take to boot up the computer?

CS 65. A computer monitor is a rectangle with an area of 168 in^2. The length is 2 inches more than the width. What are the dimensions of the monitor?

5–3

EQUATIONS IN QUADRATIC FORM

Recognition

The methods of solving quadratic equations discussed in this chapter can actually be used to solve certain nonquadratic equations. These equations are said to be in **quadratic form** if one exponent on the variable is twice the other. Recall that the standard form of a quadratic equation is $ax^2 + bx + c = 0$, and notice that the exponent on the variable x for ax^2 is twice the exponent on x for bx. Using this relationship between the exponents, we can transform many equations into quadratic form and use quadratic methods to solve them. For example, the equation $8x^6 + 7x^3 + 5 = 0$ is in quadratic form because the exponent 6 is twice the exponent 3. To transform this equation to a quadratic, we rewrite the equation as

$$8(x^3)^2 + 7(x^3)^1 + 5 = 0$$

and then let $y = x^3$ to obtain the following quadratic equation

$$8y^2 + 7y + 5 = 0$$

EXAMPLE 1 Transform the following equations into quadratic form.

a. $6x^4 - 5x^2 + 1 = 0$ **b.** $\sqrt{x} + 3\sqrt[4]{x} - 4 = 0$

c. $2x^{-6} + x^{-3} - 9 = 0$ **d.** $3(x^2 + 1)^2 - 5(x^2 + 1) + 8 = 0$

Solution
(a) Since the exponent 4 is twice the exponent 2, we rewrite the equation as

$$6(x^2)^2 - 5(x^2) + 1$$

If we let $y = x^2$, then the equation becomes

$$6y^2 - 5y + 1 = 0$$

(b) First, write this equation with fractional exponents instead of radicals.

$$x^{1/2} + 3x^{1/4} - 4 = 0$$

Since 1/2 is twice 1/4, let $y = x^{1/4}$. The equation written in quadratic form is

$$y^2 + 3y - 4 = 0$$

(c) Since -6 is twice -3, let $y = x^{-3}$ and write the equation as $2y^2 + y - 9 = 0$.

(d) The exponents are 2 and 1. Let $y = x^2 + 1$ and write the equation in quadratic form as

$$3y^2 - 5y + 8 = 0$$

■ ■ ■

Solving Equations in Quadratic Form

The method for solving equations in quadratic form is developed in the following example and summarized in the succeeding box.

EXAMPLE 2 Solve $x^4 - 13x^2 = -36$.

Solution First, we need to rewrite the equation in quadratic form. Since 4 is twice 2, we let $y = x^2$ and write the equation as

$$y^2 - 13y + 36 = 0$$

Second, we solve this equation for y using the quadratic methods discussed in this chapter. We use factoring for this particular equation, which gives

$$(y - 9)(y - 4) = 0$$
$$y = 9 \qquad \text{or} \qquad y = 4$$

Third, we replace y with x^2 and solve the equation for x.

$$x^2 = 9 \qquad \text{or} \qquad x^2 = 4$$
$$x = \pm 3 \qquad\qquad\qquad x = \pm 2$$

Fourth, we check each solution in the original equation.

$x = 3$: $\qquad (3)^4 - 13(3)^2 = -36 \qquad$ $x = 2$: $\qquad (2)^4 - 13(2)^2 = -36$
$\qquad\qquad\qquad 81 - 117 = -36 \qquad\qquad\qquad\qquad 16 - 52 = -36$
$\qquad\qquad\qquad\quad -36 = -36 \qquad\qquad\qquad\qquad\quad -36 = -36$

$x = -3$: $\qquad (-3)^4 - 13(-3)^2 = -36 \qquad$ $x = -2$: $\qquad (-2)^4 - 13(-2)^2 = -36$
$\qquad\qquad\qquad\quad 81 - 117 = -36 \qquad\qquad\qquad\qquad\qquad 16 - 52 = -36$
$\qquad\qquad\qquad\qquad -36 = -36 \qquad\qquad\qquad\qquad\qquad\quad -36 = -36$

The solution is $x = 3$, $x = -3$, $x = 2$, or $x = -2$. ▪▪▪

Solving Equations in Quadratic Form

1. Rewrite the equation in quadratic form using a substitute variable.
2. Solve the resulting quadratic equation for the substitute variable.
3. Replace the substitute variable with the original variable and solve.
4. Check your answers in the original equation.

EXAMPLE 3 Solve $2\sqrt{x} - 3\sqrt[4]{x} = 2$.

Solution First, we rewrite the equation in quadratic form. Since the exponent for $x^{1/2}$ is twice that of $x^{1/4}$, let $y = x^{1/4}$. The equation can be rewritten as

$$2y^2 - 3y - 2 = 0$$

Second, we solve this quadratic equation for y by factoring.

$$2y^2 - 3y - 2 = 0$$
$$(2y + 1)(y - 2) = 0$$
$$y = -\frac{1}{2} \qquad \text{or} \qquad y = 2$$

Third, we substitute $y = x^{1/4}$ and solve for x.

$$x^{1/4} = -\frac{1}{2} \qquad\qquad x^{1/4} = 2$$

$$(x^{1/4})^4 = \left(-\frac{1}{2}\right)^4 \qquad (x^{1/4})^4 = (2)^4$$

$$x = \frac{1}{16} \qquad\qquad x = 16$$

Fourth, we check the answers in the original equation.

$$x = \frac{1}{16}: \qquad 2\sqrt{\frac{1}{16}} - 3\sqrt[4]{\frac{1}{16}} = 2$$

$$2\left(\frac{1}{4}\right) - 3\left(\frac{1}{2}\right) = 2$$

$$\frac{1}{2} - \frac{3}{2} \neq 2$$

$$x = 16: \qquad 2\sqrt{16} - 3\sqrt[4]{16} = 2$$

$$8 - 6 = 2$$

The only solution is $x = 16$. ∎

CAUTION ✦ One of the most frequent errors in solving equations in quadratic form is forgetting to check the answers. The previous example shows the need to check for extraneous solutions.

EXAMPLE 4   Solve $7(x^2 + 3)^2 - 6(x^2 + 3) - 55 = 0$. Round your answer to the nearest hundredth.

Solution   We rewrite the equation in quadratic form. Let $y = x^2 + 3$.

$$7y^2 - 6y - 55 = 0$$

We solve the resulting equation for y using the quadratic formula.

$$y = \frac{6 \pm \sqrt{36 - 4(7)(-55)}}{14}$$

$(\!|$ 6 $\boxed{+}$ $\boxed{\sqrt{}}$ $(\!|$ 36 $\boxed{-}$ 4 $\boxed{\times}$ 7 $\boxed{\times}$ $\boxed{-}$ 55 $\boxed{)}$ $\boxed{)}$ $\boxed{\div}$ 14 $\boxed{\text{ENTER}}$ $\to$ 3.26420

$(\!|$ 6 $\boxed{-}$ $\boxed{\sqrt{}}$ $(\!|$ 36 $\boxed{-}$ 4 $\boxed{\times}$ 7 $\boxed{\times}$ $\boxed{-}$ 55 $\boxed{)}$ $\boxed{)}$ $\boxed{\div}$ 14 $\boxed{\text{ENTER}}$ $\to$ -2.40706

We replace y with $x^2 + 3$ and solve for x.

$$x^2 + 3 \approx 3.26420 \qquad\qquad x^2 + 3 \approx -2.40700$$

$$x^2 \approx 0.26420 \qquad\qquad x^2 \approx -5.40706 \quad\text{not a real number}$$

$$x \approx \pm 0.51$$

After checking, the real solution is $x \approx 0.51, -0.51$. ∎

EXAMPLE 5   Solve the equation $3x^{-2} + x^{-1} - 7 = 0$. Round your answer to the nearest hundredth.

Solution   Let $y = x^{-1}$; then the equation in quadratic form is

$$3y^2 + y - 7 = 0$$

We solve this equation by the quadratic formula.

$$y = \frac{-1 \pm \sqrt{1 - 4(3)(-7)}}{6}$$

$$y = \frac{-1 \pm \sqrt{85}}{6}$$

Substituting $y = x^{-1}$ or $x = 1/y$ gives the following solution.

$$x = \frac{6}{-1 \pm \sqrt{85}}$$

$$x \approx 0.73 \qquad \text{and} \qquad x \approx -0.59$$

∎∎∎

CAUTION ✦ Do not forget to substitute the original variable. When we use y as a substitution to solve a quadratic equation, we do not have the solution to the original equation for x.

EXAMPLE 6   If a resistance R and reactance X are connected in parallel, the impedance Z is given by

$$Z = \frac{RX}{\sqrt{R^2 + X^2}}$$

Solve for R.

Solution   First, we remove the radical by squaring both sides of the equation.

$$(Z)^2 = \left(\frac{RX}{\sqrt{R^2 + X^2}}\right)^2$$

$$Z^2 = \frac{R^2 X^2}{R^2 + X^2}$$

Next, we remove the fraction by multiplying both sides of the equation by the LCD, $R^2 + X^2$.

$$Z^2(R^2 + X^2) = \frac{R^2 X^2}{(R^2 + X^2)} \cdot (R^2 + X^2)$$

$$R^2 Z^2 + X^2 Z^2 = R^2 X^2$$

Then we combine the R^2 terms, and factor and solve for R by taking the square root of both sides of the equation.

$$X^2 Z^2 = R^2 X^2 - R^2 Z^2$$

$$X^2 Z^2 = R^2(X^2 - Z^2)$$

$$\frac{X^2 Z^2}{X^2 - Z^2} = R^2$$

$$\pm \sqrt{\frac{X^2 Z^2}{X^2 - Z^2}} = R$$

$$\pm \frac{XZ}{\sqrt{X^2 - Z^2}} = R$$

∎∎∎

5–3 EXERCISES

Solve the following equations for real number solutions. Round all irrational answers to the nearest hundredth.

1. $x^4 - x^2 - 12 = 0$

2. $x^4 - x^2 - 72 = 0$

3. $x^4 + 5x^2 - 24 = 0$

4. $x^4 + 14x^2 - 32 = 0$

5. $x^5 - 7x^3 - 18x = 0$

6. $x^5 + x^3 - 20x = 0$

7. $3x^4 + 2x^2 - 21 = 0$

8. $2x^4 + 5x^2 - 42 = 0$

9. $1 = 15x^{-2} - 2x^{-1}$

10. $3x^{-2} + 17x^{-1} = 6$

11. $x + 7\sqrt{x} = 144$

12. $2\sqrt[6]{x} + 5\sqrt[3]{x} = 12$

13. $(x - 2)^{-2/3} - (x - 2)^{-1/3} - 6 = 0$

14. $(3x + 1)^2 - 22(3x + 1) + 112 = 0$

15. $4(x^2 + 1)^2 + 4(x^2 + 1) = 6$

16. $6(x^2 + 3)^2 + 13(x^2 + 3) = 5$

17. $\sqrt{x} - 13\sqrt[4]{x} = -36$

18. $\sqrt{x} + \sqrt[4]{x} = 30$

19. $x^{2/3} + 19x^{1/3} = 216$

20. $2x^{2/3} + x^{1/3} = 15$

21. $x^6 + 10x^3 + 16 = 0$

22. $x^6 - 5x^3 = 36$

23. $x^6 + 26x^3 - 27 = 0$

24. $x^6 - 56x^3 - 512 = 0$

E 25. The reactance X in ohms of a capacitance C in microfarads and an inductance L in henries connected in series is given by

$$X = wL - \frac{1}{wC}$$

where w is the angular frequency in rad/s. Solve for the angular frequency when $X = 1,000\ \Omega$, $L = 0.60$ H, and $C = 0.10\ \mu$F.

P 26. The focal length f of a lens q distance from an object and p distance from the image is given by

$$\frac{1}{f} = \frac{1}{p} + \frac{1}{q}$$

Find q if $f = 5$ in. and p is 7 in. longer than q.

M 27. The pressure P in a waste disposal bin varies with time t according to the function

$$P = 6 + \sqrt{t + 4}$$

At what time (in hours) will the pressure be 13 lb/ft²?

5–4

NONLINEAR INEQUALITIES

In Chapter 1, we solved linear equations and inequalities by using much the same technique for both. In this section, we will discuss solving nonlinear inequalities. The technique is based on factoring, just as with quadratic equations, but it differs from the techniques used to solve quadratic equations in the previous sections of this chapter. Solving nonlinear inequalities is based on the concept of critical values.

Critical Values

The zeroes of a function (roots of an equation), called the **critical values,** *divide the number line into intervals over which the function is always positive or always negative.* For example, the critical values of the equation $x^2 - 2x - 3 = 0$ are $x = 3$ and $x = -1$. These two critical values divide the number line into three regions where the function can be positive or negative, as shown in the following figure:

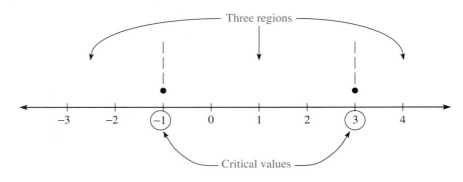

Quadratic Inequalities

In the next example, we illustrate how to use the critical values to solve a quadratic inequality.

EXAMPLE 1 Solve $x^2 + 3x - 4 > 0$.

Solution First, we must *find the critical values.* Since the critical values occur where the function $x^2 + 3x - 4$ equals zero, we solve the following equation:

$$x^2 + 3x - 4 = 0$$
$$(x + 4)(x - 1) = 0$$
$$x + 4 = 0 \qquad x - 1 = 0$$
$$x = -4 \qquad x = 1$$

The critical values are -4 and 1. Since the original inequality is strictly greater than, place circles at -4 and 1 on the number line. Notice that these critical values divide the number line into three regions, as shown below:

Next, we must *determine which of these regions satisfy the original inequality,* $x^2 + 3x - 4 > 0$. To do so, we will determine the sign of each factor $(x + 4)$ and $(x - 1)$ in each region. Since the sign of the inequality, and thus each factor, does not change throughout a region, we can choose any number in a given region to substitute in the factors. This process is shown below:

	Region 1		Region 2		Region 3	
Factor	Substitute $x = -6$	Sign of factor	Substitute $x = 0$	Sign of factor	Substitute $x = 3$	Sign of factor
$x + 4$	$-6 + 4 = -2$	$-$	$0 + 4 = 4$	$+$	$3 + 4 = 7$	$+$
$x - 1$	$-6 - 1 = -7$	$-$	$0 - 1 = -1$	$-$	$3 - 1 = 2$	$+$

Since $x^2 + 3x - 4 = (x + 4)(x - 1)$, the sign of the function in each region is as given in the following figure:

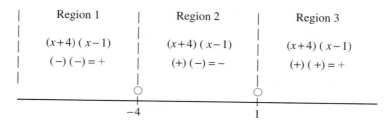

Since the inequality is greater than zero, the solution consists of those regions where the sign of the result is positive, or regions 1 and 3. Since the original inequality does not contain the "equals to" symbol ($>$ rather than $\geq$), the critical values are not included in the solution; that is, $x < -4$ or $x > 1$. ■■■

Solving Quadratic Inequalities

1. Arrange the inequality in $ax^2 + bx + c \leq 0$ form. Change to an equation.
2. Factor and determine the critical values.
3. Draw a number line and mark each critical value. Determine the sign of each factor in each region.
4. Determine the sign of the inequality in each region and the solution.

EXAMPLE 2 Solve the following inequality:

$$2x^2 \leq 15 - x$$

Solution First, arrange the inequality in $ax^2 + bx + c \leq 0$ form, and then change it to an equation.

$$2x^2 + x - 15 \leq 0$$
$$2x^2 + x - 15 = 0$$

Second, factor and determine the critical values.

$$(2x - 5)(x + 3) = 0$$

$$2x - 5 = 0 \qquad\qquad x + 3 = 0$$
$$x = 5/2 \qquad\qquad x = -3$$

Third, place dots at 5/2 and −3 on the number line, and mark the resulting regions. To determine the sign of each factor in each region, substitute a number from the region into the factor. Fourth, determine the sign of the factored inequality in each region as shown below:

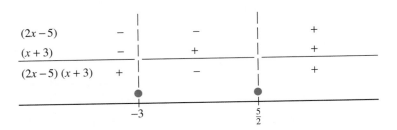

The solution is $-3 \leq x \leq 5/2$. The critical values are included in the solution because the original inequality contains the "equals to" symbol. ■■■

Rational Inequalities

In a rational expression, the critical values occur where the numerator or denominator equals zero. A rational expression changes signs only at its critical values. In solving a rational inequality, remember domain restrictions necessary to avoid division by zero.

EXAMPLE 3 Solve the following rational inequality:

$$\frac{(x-5)^2(x+3)}{(x-1)} \geq 0$$

Solution Since the critical values occur where the numerator or denominator equals zero, the critical values are $x = 5$, -3, and 1. On the number line we place a dot at 5 and -3, but we must place a circle at 1 because $x = 1$ leads to division by zero. The critical values, the resulting regions, the sign of each factor in each region, and the sign of the original fraction in each region are as follows:

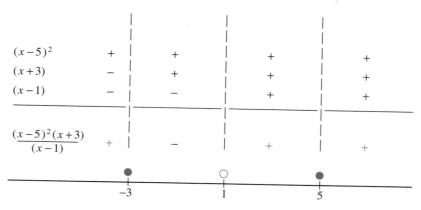

Since the original inequality is greater than or equal to zero, the solution includes those regions where the fraction is positive. The solution is $x \leq -3$ or $x > 1$. ∎ ∎ ∎

Application

EXAMPLE 4 The formula for displacement (the number of feet fallen), s, of a free-falling object is given by

$$s = v_0 t + \frac{gt^2}{2}$$

where v_0 = initial velocity = -32 ft/s, t = time, and g = acceleration due to gravity = -32 ft/s². An object thrown from a 250-ft building has an initial velocity of 32 ft/s. How many seconds after release will the object be more than 122 ft above the ground?

Solution Substituting the given information into the formula yields the inequality (250 $-$ 122 = 128):

$$-128 < -32t - 16t^2$$

The signs are negative due to the downward direction of movement. Finding the critical values gives

$$
\begin{aligned}
16t^2 + 32t - 128 &< 0 \qquad \text{transpose} \\
16t^2 + 32t - 128 &= 0 \qquad \text{change to equation} \\
t^2 + 2t - 8 &= 0 \qquad \text{divide by 16} \\
(t + 4)(t - 2) &= 0 \qquad \text{factor}
\end{aligned}
$$

Since the critical values are $t = -4$ and 2, we place circles at -4 and 2 on the number line. The sign of each factor and the product in each region are given in the following figure:

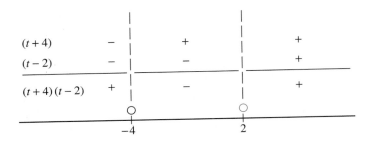

Since the inequality is less than zero, the solution includes the region where the product is negative. The solution is $-4 < t < 2$. However, from the physical constraints of this problem, time is not negative. Therefore, the solution is $0 < t < 2$. The object is more than 122 ft above the ground up to the first 2 seconds after release. ∎∎∎

EXAMPLE 5   The tensile strength S (in lb/in^2) of a plastic varies with temperature T according to the equation $S = 800 + 500T - 10T^2$. At what temperature is the tensile strength greater than 6,050 lb/in^2?

Solution   Given that $S > 6,050$, we must solve the following quadratic inequality:

$$6,050 < 800 + 500T - 10T^2$$

Solving this inequality gives

$$
\begin{aligned}
10T^2 - 500T + 5,250 &< 0 \\
10T^2 - 500T + 5,250 &= 0 \qquad \text{change to equation} \\
T^2 - 50T + 525 &= 0 \qquad \text{divide by 10} \\
(T - 35)(T - 15) &= 0 \qquad \text{factor} \\
T = 35 \qquad \text{or} \qquad T &= 15 \qquad \text{critical values}
\end{aligned}
$$

The regions resulting from these critical values and the sign of each factor are shown below.

$(T-35)$	$-$	$-$	$+$
$(T-15)$	$-$	$+$	$+$
	0	15	35

Since the quadratic inequality is less than zero, the solution is the region where the product is negative. Therefore, the tensile strength of the plastic is greater than 6,050 lb/in^2 when the temperature is between 15 and 35 degrees. ∎∎∎

5–4 EXERCISES

Solve the following nonlinear inequalities.

1. $2x^2 > 12 - 2x$
2. $x^2 + 2x \le 35$
3. $3x^2 + 19x + 6 \le 0$
4. $18x^2 + 34x > 4$
5. $4x^2 - 12x \ge 0$
6. $x^2 - x - 30 \le 0$
7. $21x^2 - 10 > -29x$
8. $16x^2 > 9$
9. $49x^2 \ge 81$
10. $36x^2 - 25 \le 0$
11. $6x^2 + 43x - 40 < 0$
12. $12x^2 + 28x \ge 5$
13. $x^2 + 11x \le -28$
14. $4x^2 - 20x + 24 > 0$
15. $x(2x + 1)(x - 5) \ge 0$
16. $6x(x - 3)(5x + 6) < 0$

17. $2x^3 - 8x \leq 0$

18. $3x^3 - 27x \geq 0$

19. $\dfrac{2x - 1}{x + 4} > 0$

20. $\dfrac{3x - 9}{x - 5} < 0$

21. $\dfrac{x^2 + 8x - 9}{x + 3} \geq 0$

22. $\dfrac{x^2 - 5x - 36}{x + 1} > 0$

23. $\dfrac{x(x + 5)^2(x + 2)}{x - 3} < 0$

24. $\dfrac{2x(x - 1)^2(x + 4)}{x + 7} \leq 0$

25. $\dfrac{(3x - 4)(x + 6)}{x + 1} \leq 0$

26. $\dfrac{(x - 5)(2x + 6)}{x - 3} > 0$

P 27. The velocity v of an object can be found by $v^2 = v_0^2 + 2as$ where v_0 is initial velocity, a is acceleration, and s is displacement. Find the value to the nearest tenth for the initial velocity if the velocity must be greater than 150 ft/s and $a = 2.5$ ft/s^2 when $s = 5$ ft.

28. Find the values of the radius r that make the surface area s of a cylinder greater than 32π m^2 if the height h is 6 m. (*Note:* $s = 2\pi r^2 + 2\pi rh$.)

29. For what values of the length of a rectangle is the area less than 40 m^2 if the length of the rectangle is 3 m more than the width?

E 30. The transient current i (A) in a circuit varies with time t (ms) according to the equation $i = 8 + 4t^2$. How long will it take for the transient current to exceed 12 A?

P 31. The weight w, in pounds, of propellant in a rocket after launch is given by $w = 4{,}000 - t^2 - 175t$, where t is measured in seconds. When is the weight of fuel greater than 1,500 lb?

M 32. The deflection of a beam is given by $d = x^2 - 0.20x + 0.02$ where x is the distance from the end of the beam. For what values of x is $d > 0.01$?

B 33. The weekly profit P of a widget manufacturing company is given by $P = x^4 - 15x^3$ where x is the week in the year. During what period of the year is the company operating at a loss?

B 34. A technician receives $400 per week plus $24.50 per hour for overtime. How many hours must she work for her pay to exceed $596.00?

E 35. The resistance R (Ω) of a certain wire as a function of temperature T ($\degree$C) is given by $R = 5.0 + 0.50T + 0.00070T^2$. Find $T > 0$ such that $R > 15$ Ω.

5–5

APPLICATIONS

In the previous sections of this chapter, the method of solving a quadratic equation was specified. However, in a job situation where solving a quadratic equation is necessary, your supervisor will not come to you and say, "Here is a problem, use a quadratic equation to model the problem, and solve the quadratic equation by factoring."

Since we have three methods for solving a quadratic equation, we need a strategy for selecting an appropriate method for a given situation. You can use both completing the square and the quadratic formula to solve any quadratic equation. However, factoring can be used only for equations that have rational roots. One strategy is first to attempt to solve the equation by factoring. If you do not easily see how to factor the equation, then use the quadratic formula.

EXAMPLE 1 The formula for the total displacement, s, of a free-falling object is given by

$$s = v_0 t + \frac{gt^2}{2}$$

where $v_0 =$ initial velocity, $t =$ time, and $g =$ acceleration due to gravity. In the English system of measurement, where $g = -32$ ft/s^2, the displacement is given by $s = v_0 t - 16t^2$. If a baseball thrown straight down from the top of a 100-ft building has an initial velocity of 20 ft/s, how long does it take the ball to strike the ground? Substitute and then solve for t.

Solution We substitute $s = -100$ and $v_0 = -20$ into the formula.

$$-100 = -20t - 16t^2$$

Arranging the equation in standard form gives

$$16t^2 + 20t - 100 = 0 \qquad \text{transpose}$$
$$4t^2 + 5t - 25 = 0 \qquad \text{divide by 4}$$

This equation does not appear to factor, so we will solve for t using the quadratic formula.

$$a = 4 \qquad b = 5 \qquad c = -25$$

$$t = \frac{-5 \pm \sqrt{25 - 4(4)(-25)}}{2(4)}$$

$$t = \frac{-5 \pm \sqrt{25 + 400}}{8}$$

$$t = \frac{-5 \pm \sqrt{425}}{8}$$

$$t = \frac{-5 \pm 20.62}{8}$$

The solutions are $t = -3.20$ s and $t = 1.95$ s. Since time is not measured in negative numbers, it will take 1.95 s for the baseball to strike the ground.

To calculate $\sqrt{\ }$: $\quad$ 25 $\boxed{-}$ 4 $\boxed{\times}$ 4 $\boxed{\times}$ 25 $\boxed{+/-}$ $\boxed{=}$ $\boxed{\sqrt{\ }}$ $\boxed{\text{STO}}$

Solutions: $\quad$ 5 $\boxed{+/-}$ $\boxed{-}$ $\boxed{\text{RCL}}$ $\boxed{=}$ $\boxed{\div}$ 8 $\boxed{=}$ $\longrightarrow$ -3.20

$\qquad\qquad$ 5 $\boxed{+/-}$ $\boxed{+}$ $\boxed{\text{RCL}}$ $\boxed{=}$ $\boxed{\div}$ 8 $\boxed{=}$ $\longrightarrow$ 1.95 $\qquad$ ■■■

EXAMPLE 2 In business, the **equilibrium point** is the point where the demand for a product equals the supply. If a company knows that the demand for handmade surfboards is given by the expression $(500 - x^2)$/day and the supply is given by the expression $(10x + 300)$/day, find the equilibrium point (x represents the quantity produced and sold).

Solution Because the equilibrium point is the point where the demand equals the supply, the equation is

$$500 - x^2 = 10x + 300$$

We arrange the equation in standard form and solve by factoring.

$$-x^2 - 10x + 200 = 0 \qquad \text{transpose and add}$$
$$x^2 + 10x - 200 = 0 \qquad \text{multiply by } -1$$
$$(x + 20)(x - 10) = 0 \qquad \text{factor}$$

$$x + 20 = 0 \qquad\qquad x - 10 = 0 \qquad \text{Zero Factor Property}$$
$$x = -20 \qquad\qquad\quad x = 10$$

Since the manufacturer cannot produce -20 surfboards, the supply equals the demand when 10 surfboards are produced per day. Thus, the manufacturer is underproducing if he produces fewer than 10 surfboards, and he is overproducing if he produces more than 10 surfboards. ■■■

EXAMPLE 3 Diane drives from Atlanta to Orlando, a distance of 360 miles, and back along the same route to Atlanta. Her average speed from Atlanta to Orlando is 15 mi/h less than her speed on the return trip. If the total driving time is 14 hours, what is her average speed from Atlanta to Orlando?

Solution If we let

$$x = \text{average speed from Atlanta to Orlando}$$

then

$$x + 15 = \text{average speed from Orlando to Atlanta}$$

Since Diane drives the same route both to and from Orlando the distance for each leg of the trip is 360 miles. This information is included in the chart below. Since we have expressions for both distance and rate, we can use the relationship time = distance/rate to fill in an expression for time in the chart.

Distance ÷ rate = time			
	Distance	Rate	Time
Atlanta to Orlando	360	x	$\dfrac{360}{x}$
Orlando to Atlanta	360	$(x + 15)$	$\dfrac{360}{(x + 15)}$

Since the equation is set up using the last column, we will need to read the problem again to find a relationship for time. The problem states that the total time for the trip is 14 hours. The equation is as follows:

time (Atlanta to Orlando) + time (Orlando to Atlanta) = total time

$$\frac{360}{x} \quad + \quad \frac{360}{(x + 15)} \quad = \quad 14$$

First, eliminate the fractions by multiplying by the common denominator, $x(x + 15)$.

$$360(x + 15) + 360x = 14x(x + 15) \qquad \text{eliminate fractions}$$
$$360x + 5{,}400 + 360x = 14x^2 + 210x \qquad \text{eliminate parentheses}$$
$$-14x^2 + 510x + 5{,}400 = 0 \qquad \text{transpose}$$
$$7x^2 - 255x - 2{,}700 = 0 \qquad \text{divide by } -2$$

Using the quadratic formula to solve, we have

$$x = \frac{255 \pm \sqrt{(-255)^2 - 4(7)(-2{,}700)}}{2(7)}$$

$$x = 45 \qquad \text{or} \qquad x \approx -8.57$$

Since speed is not measured in negative numbers, the solution is $x = 45$. Diane's average speed from Atlanta to Orlando is 45 mi/h.

To calculate $\sqrt{\ }$: 255 $\boxed{\pm}$ $\boxed{x^2}$ $\boxed{-}$ 4 $\boxed{\times}$ 7 $\boxed{\times}$ 2700 $\boxed{\pm}$ $\boxed{=}$ $\boxed{\sqrt{\ }}$ $\boxed{\text{STO}}$

Solution: $255\ \boxed{+}\ \boxed{RCL}\ \boxed{=}\ \boxed{\div}\ 14\ \boxed{=}\ \longrightarrow\ 45$

$255\ \boxed{-}\ \boxed{RCL}\ \boxed{=}\ \boxed{\div}\ 14\ \boxed{=}\ \longrightarrow\ -8.57$ ■ ■ ■

EXAMPLE 4   A landscape architect wants to surround a rectangular swimming pool with a brick walk of uniform width. If the area of the brick walk is 405 ft^2 and the swimming pool is 15 ft by 20 ft, how wide should the walk be?

Solution   If we let

$$x = \text{width of the walkway}$$

then the dimensions of both the pool and the walkway are $(15 + 2x)$ and $(20 + 2x)$ (the dimensions of the pool are 15 by 20). As shown in Figure 5–4, the area of the walkway is the difference in the area of the larger rectangle and the area of the small rectangle. Since the area of a rectangle is the product of length and width, the area of the large rectangle is $(20 + 2x)(15 + 2x)$, and the area of the small rectangle is $20(15)$. The equation is

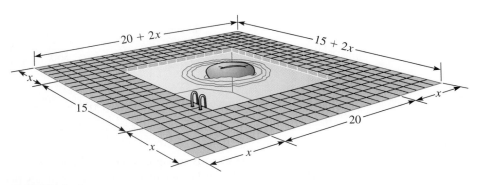

FIGURE 5–4

$$\text{area of large rectangle} - \text{area of small rectangle} = 405$$

$$\underset{\uparrow}{(20 + 2x)(15 + 2x)} - \underset{\uparrow}{20(15)} \quad \underset{\uparrow}{= 405}$$

First, we arrange the equation in standard form.

$$300 + 40x + 30x + 4x^2 - 300 = 405$$
$$4x^2 + 70x - 405 = 0$$

Then we solve the equation by the quadratic formula.

$$a = 4 \qquad b = 70 \qquad c = -405$$

$$x = \frac{-70 \pm \sqrt{4{,}900 - 4(4)(-405)}}{2(4)}$$

$$x = \frac{-70 \pm \sqrt{4{,}900 + 6{,}480}}{8}$$

$$x = \frac{-70 \pm \sqrt{11{,}380}}{8}$$

$$x \approx 4.58 \qquad \text{or} \qquad x \approx -22.08$$

The walkway should be 4.6 ft wide.

To calculate $\sqrt{}$: 70 $\boxed{x^2}$ $\boxed{-}$ 4 $\boxed{\times}$ 4 $\boxed{\times}$ 405 $\boxed{+/-}$ $\boxed{=}$ $\boxed{\sqrt{}}$ $\boxed{\text{STO}}$

Solution: 70 $\boxed{+/-}$ $\boxed{+}$ $\boxed{\text{RCL}}$ $\boxed{=}$ $\boxed{\div}$ 8 $\boxed{=}$ ——→ 4.58

  70 $\boxed{+/-}$ $\boxed{-}$ $\boxed{\text{RCL}}$ $\boxed{=}$ $\boxed{\div}$ 8 $\boxed{=}$ ——→ −22.08 ∎∎∎

EXAMPLE 5 The total surface area of a cylinder with a top and a bottom is given by the formula $A = 2\pi r^2 + 2\pi rh$. If a cylinder has a surface area of 64π cm^2 and a height of 6.0 cm, find the radius. Substitute and solve.

Solution First, substitute $A = 64\pi$ and $h = 6.0$ into the formula.

$$64\pi = 2\pi r^2 + 12\pi r$$

$$2\pi r^2 + 12\pi r - 64\pi = 0 \qquad \text{transpose and multiply all terms by } -1$$

$$r^2 + 6r - 32 = 0 \qquad \text{divide by } 2\pi$$

$$a = 1 \qquad b = 6 \qquad c = -32$$

$$r = \frac{-6 \pm \sqrt{36 - 4(1)(-32)}}{2}$$

$$r \approx 3.4 \qquad \text{or} \qquad r \approx -9.4$$

The radius of the cylinder is 3.4 cm. ∎∎∎

Maximum and Minimum

The graph of a quadratic function is a parabola. The tip of the parabola is called the vertex. For a quadratic equation written in $y = ax^2 + bx + c$ form, the *parabola opens upward and the vertex is a minimum if $a > 0$. If $a < 0$, the parabola opens downward and the vertex is a maximum.* The x coordinate of the vertex, given by $-b/(2a)$, defines where the maximum or minimum value occurs. (We will discuss the parabola in greater detail in Chapter 15.) Figure 5–5 illustrates the maximum and minimum.

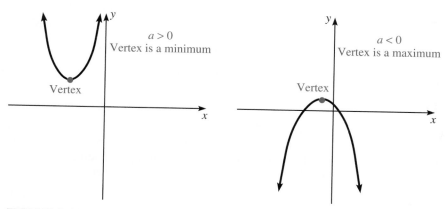

FIGURE 5–5

EXAMPLE 6 A rectangular dog run is to be fenced with 40 yd of barbed wire. Find the maximum area that can be fenced if one width is not enclosed. See Figure 5–6.

FIGURE 5–6

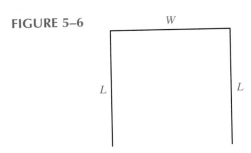

Solution   Since we are fencing three sides of the tract, the amount of barbed wire is given by

$$2L + W = 40$$

If we solve for W, we obtain

$$W = 40 - 2L$$

However, we want to fence the tract so that the area is a maximum. The area of the rectangular tract is given by

$$A = LW$$

If we substitute for W in the area formula, we have

$$A = L(40 - 2L)$$
$$A = 40L - 2L^2$$

The equation in standard form is $y = -2L^2 + 40L - A$. Since the coefficient of L^2 is negative, the function does have a maximum, and that maximum occurs at the x coordinate of the vertex given by $-b/(2a)$.

$$\frac{-b}{2a} = \frac{-40}{2(-2)} = 10$$

The maximum area occurs when $L = 10$. The maximum area is

$$A = 40L - 2L^2$$
$$A = 40(10) - 2(10)^2$$
$$A = 200$$

The maximum rectangular area that can be fenced with 40 yd of barbed wire is 200 yd^2.

■ ■ ■

EXAMPLE 7   A company sells 1,000 widgets each week at the price of $2.00 each. However, the company needs to raise the price. Market research shows that for each $0.10 increase in price, the company will sell 20 fewer widgets. If the company wants to receive $2,450 in revenue each week, what price should it charge?

Solution   Let n = the number of $0.10 price increases. Each increase in price is represented by $0.10n$, and this results in $20n$ fewer units sold. From business principles, we know that

$$\text{total revenue} = (\text{price/unit}) \cdot (\text{number of units sold})$$

$$2{,}450 \quad = (2 + 0.10n) \quad \cdot \quad (1{,}000 - 20n)$$

Solving the equation gives

$$2{,}450 = 2{,}000 - 40n + 100n - 2n^2 \qquad \text{eliminate parentheses}$$
$$2n^2 - 60n + 450 = 0 \qquad \text{transpose and add}$$
$$n^2 - 30n + 225 = 0 \qquad \text{divide by 2}$$
$$(n - 15)(n - 15) = 0 \qquad \text{factor}$$
$$n = 15$$

Since the price/unit is given by $2 + 0.10n$, the company should charge

$$2 + 0.10(15)$$
$$2 + 1.50$$
$$\$3.50$$

for each widget.

■ ■ ■

5–5 EXERCISES

P 1. If a ball thrown from the top of a 120-ft building has an initial velocity of 8 ft/s, how long will it take the ball to strike the ground? (Solve $120 = 8t + 16t^2$.)

B 2. If the demand equation for an integrated circuit is given by $p = -600 - x^2$ and the supply equation is given by $p = 400 - 3x$, find the equilibrium quantity.

3. The city decides to put a brick walkway of uniform width around a rectangular flower garden that is 40 ft by 20 ft. If the area of the walkway is half the area of the garden, find the width of the walkway.

4. $N = n(n - 3)/2$ is the formula for the number of diagonals N of a polygon with n sides. Find the number of sides for a polygon with five diagonals.

5. The sum of six times the number and the square of the number is -8. Find the number.

6. The length of the base of a triangle is 3 cm less than the length of the height. Find the length of the base if the area of the triangle is 40 cm^2. (The area of a triangle $= 1/2 \cdot$ base $\cdot$ height.)

C 7. Jeff can survey a tract of land in three days less than Marvin. If it takes eight days for them to survey the land together, how long does it take Marvin alone? Compute the answer to two decimal places.

E 8. When two dc currents flow in the same direction in a common resistor, the power dissipated in the resistor is given by $P = R(I_1 + I_2)^2$. If $P = 147$ W, $R = 7$ Ω, and $I_1 = 4$ A, find I_2 by substituting and solving. (Compute the answer to two decimal places.)

B 9. A company sells its product for $3 and sells 1,500 per week at this price. To increase its revenue, the company needs to increase its price; however, for each $0.40 increase in price, the company will sell 100 fewer items. If the company wants to earn a total revenue of $5,060 per week, what price should it charge? (See Example 7.)

10. Find two consecutive integers whose product is 132. Let $x + 1$ represent the consecutive integers.

11. Lucy wants to mat a picture that is 10 cm by 14 cm. If the area of the matting is 28 cm^2, find the width of the matting.

12. Jose drove from Los Angeles to the Mexican border and back, a total distance of 240 miles. His average speed from Los Angeles to the border was 15 mi/h more than his speed on the return trip. If the total driving time was 6 hours, what was Jose's average speed from the border back to Los Angeles? (Compute the answer to two decimal places.)

M 13. The area of a washer needs to be 3π cm^2. If the radius of the outside circle is 2 cm, find the radius of the inner circle. (*Note:* $A = \pi r^2$, area of a circle.)

P 14. A workman drops a hammer from the top of a 90-ft building. How long does it take the hammer to strike the ground if the initial velocity is zero? (Solve $90 = 16t^2$. Compute your answer to two decimal places.)

15. If the product of two consecutive odd integers is increased by six, the result is nine. Find the integers. (Let x and $x + 2$ represent the consecutive odd integers.)

16. Two inlet pipes can fill a swimming pool in $23\frac{1}{2}$ hours. If the smaller pipe alone takes 12 h more to fill the pool, how much time is required for the larger pipe to fill the pool alone?

17. To conserve space, an oil dealer decides to tear down two oil storage tanks and replace them with one tank equal in capacity to the two tanks. If the radii of the two tanks are 18 ft and 20 ft and the height of each is 15 ft, what should be the radius of the new tank if its height is 25 ft? (The volume of a cylinder is $\pi r^2 h$.) Give your answer to two decimal places.

E 18. An electrical circuit contains two resistors. The sum of their resistances is 46 Ω, and their product is 465 Ω. Find the two resistances.

B 19. If the demand equation is given by $p = 8x - 5x^2$, and the supply equation is $p = 18x - 40$, find the equilibrium quantity, where supply equals demand. (Refer to Example 2.)

E 20. The power in the load P_L of a series circuit is given by the equation $P_L = -I_L^2 R_L + V_S I_L$. Find I_L when $P_L = 5$ W, $R_L = 5$ Ω, and $V_S = 10$ V.

P 21. Use the fact that the vertex results in a maximum or minimum to find the maximum height s an object will reach if the equation of flight is given by $s = 20t - 16t^2$ (thrown upward with initial velocity of 20 ft/s).

22. Find the maximum area of a rectangle whose perimeter is 36 ft. (Refer to Example 6.)

23. Divide 20 into two parts such that the product will be a maximum.

B 24. A company sells a product for $8 and sells 1,200 per week at this price. The company finds that for each $0.50 increase in price, 40 fewer of the product will be sold. If the company wants to make $10,260 per week, what price should it charge? (Refer to Example 7.)

B 25. Mark's Mobile Phones charges $300 per box of phones for orders of 175 boxes or less. Mark offers a discount for large orders. If a customer orders x boxes in excess of 178, the cost for each box is reduced by x dollars. If a customer's total bill was $57,121, how many boxes were ordered?

CS 26. Total cost C for manufacturing x computer components is $1,680. If the equation for cost is $C = 800 + 0.04x + 0.002x^2$, how many components were manufactured?

M 27. A solar-cell collector on the roof of a house is to have the shape shown in Figure 5–7. If the area of the collector must be 35 m^2 to collect enough energy to meet the needs of the house, what should be the dimension x?

FIGURE 5–7

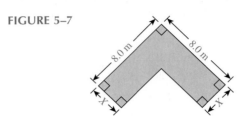

E 28. In a dc circuit containing a resistance R, inductance L, and capacitance C connected in series, one must solve the equation $I^2 + (R/L)I + 1/LC = 0$ to find the instantaneous value of the current. Find I when $R = 25.0$ Ω, $L = 1.00$ H, and $C = 0.100$ F.

CS 29. The rate, in kilobits per second (kbs), at which information is transferred on a certain computer CD-ROM drive is given by the equation $x^2 - 570x = 18,000$. Determine the speed of information transfer.

M 30. The bending moment M of a certain beam at a distance x (ft) from a support is given by $M = 25x - 16(x - 2) - 1.5x^2$. Find the distance x when $M = 0$.

CS 31. The time for a particular printer to prepare to print one page of information depends on what is being printed (text only, text and graphics, or graphics only). This time may be represented by the equation $-t^2 + 4t = 0$. What is the maximum amount of time to prepare to print one page?

CHAPTER SUMMARY

Summary of Terms

critical values (p. 190)

discriminant (p. 183)

equilibrium point (p. 196)

parabola (p. 182)

quadratic equation in standard form (p. 168)

quadratic form (p. 186)

root (p. 168)

solution (p. 168)

Square Root Property (p. 170)

vertex (p. 182)

Zero Factor Property (p. 168)

Summary of Formulas

$$x = \frac{-b \pm \sqrt{b^2 - 4ac}}{2a} \quad \text{quadratic formula}$$

$b^2 - 4ac$ discriminant

CHAPTER REVIEW

Section 5–1

Solve by factoring.

1. $3x^2 + 16x = 12$

2. $8x = 20 - x^2$

3. $x^2 = 16$

4. $5x^2 - 21 = 8x$

5. $x(2x + 4) = -(7x + 5)$

6. $9x^2 = 81$

7. $10x = 14x^2$

8. $7x^2 + 18x - 12 = 2x^2 + 7x$

9. $4x(x + 1) - 6(2x + 3) = 3x + 2$

10. $5x(x + 4) - 10 = -2(x - 19)$

11. $\dfrac{x}{x - 5} + \dfrac{4}{x + 5} = \dfrac{x}{x^2 - 25}$

12. $\dfrac{6}{x^2} + 8 = \dfrac{14}{x}$

13. $\dfrac{x}{x + 3} + \dfrac{x}{x + 1} = \dfrac{3 - x}{x^2 + 4x + 3}$

14. $\dfrac{3}{x + 2} - \dfrac{4}{x - 2} = \dfrac{-2 - x^2}{x^2 - 4}$

15. $24x^2 = 31x + 15$

16. $x^2 = \dfrac{11x + 3}{4}$

Solve by completing the square.

17. $x^2 + 6x = 12$

18. $x^2 + 10x - 15 = 0$

19. $x^2 = 5x$

20. $7x(x + 2) = 25$

21. $2x^2 - 6x = 14$

22. $9x^2 + 18x = 30$

23. $3x^2 + 7x + 17 = 0$

24. $4x^2 + 11x - 20 = 0$

25. $5x(x - 1) = 11 - 10x$

26. $8x^2 + 7x = -21$

27. $\dfrac{3x^2}{2} + x = \dfrac{1}{3}$

28. $\dfrac{4}{x} - \dfrac{1}{x^2} = 5$

29. $\dfrac{x - 1}{x + 2} + \dfrac{x}{x - 3} = \dfrac{2}{x^2 - x - 6}$

30. $\dfrac{6}{2x - 1} + \dfrac{2x}{x + 3} = \dfrac{3x^2}{2x^2 + 5x - 3}$

Section 5–2

Solve by the quadratic formula.

31. $14x^2 - 8 = 0$

32. $7x^2 - 3x + 10 = 0$

33. $5x^2 - 12x + 13 = 0$

34. $5x^2 - x = 7$

35. $9x^2 + 8x = 0$

36. $36x^2 - 1 = 0$

37. $x^2 + 3(x - 4) = 7 - 2x(x - 1)$

38. $7x^2 - 14x = 0$

39. $2x(x - 3) + 6 = 4(x^2 + 5)$

40. $6x(x + 2) - 5 = (x + 1)(x - 3)$

41. $x^2 = 18$

42. $11x + 15x^2 - 5 = 8x^2 + 5x$

43. $\dfrac{2x - 3}{x + 1} + \dfrac{3x}{x - 6} = \dfrac{10}{x^2 - 5x - 6}$

44. $\dfrac{13x}{2x + 3} + \dfrac{5}{x - 5} = \dfrac{6}{2x^2 - 7x - 15}$

45. $\dfrac{4x^2}{9} - \dfrac{1}{3} = \dfrac{2x - 1}{6}$

46. $\dfrac{9}{x^2} = 16 + \dfrac{8}{x}$

Using the discriminant, describe the number and type of solutions.

47. $4x - x(8 - x) = 20$

48. $14x^2 - 8x = 15 + 6x + 2x^2$

49. $\dfrac{7}{x^2} + 6 = \dfrac{1}{x}$

50. $\dfrac{9x}{x - 1} + \dfrac{7}{x + 1} = \dfrac{6}{x^2 - 1}$

51. $5x(x + 6) = (x - 3)(x - 7)$

52. $15x^2 - 9x = 6(x - 3) + 12x(x - 1)$

53. $17x^2 + 6x - 3(x + 1) = 15 + 4x(x + 2)$

54. $13x^2 - 10x + 6 = 2x + 12 + 6x^2$

Section 5–3

Solve the following equations.

55. $x^4 - 29x^2 = -100$

56. $x^4 + x^2 = 20$

57. $3x - 5\sqrt{x} + 2 = 0$

58. $2x + 7\sqrt{x} = 15$

59. $2x^{2/3} + 15x^{1/3} = 8$

60. $2\sqrt{x} + \sqrt[4]{x} = 1$

61. $6x^{-2} - 10x^{-1} = 4$

62. $6x^{-2} + 5x^{-1} = 4$

63. $(x^2 - 2)^2 - 9(x^2 - 2) + 14 = 0$

64. $6(x^2 - 1) + 11(x^2 - 1) - 35 = 0$

Section 5–4

Solve the following inequalities.

65. $4x^2 \geq 3 - x$

66. $3x^2 + 12x < -(x + 4)$

67. $2x^3 + 7x^2 - 15x < 0$

68. $2x^3 + 8x^2 \geq 5x^2 + 20x$

69. $\dfrac{x - 3}{x + 1} < 0$

70. $\dfrac{2x - 1}{x + 5} \geq 0$

71. $\dfrac{(x - 1)(x + 3)}{(x - 2)} \geq 0$

72. $\dfrac{(x + 4)(x - 2)}{x^2} < 0$

Section 5–5

73. Find two consecutive integers whose product is 342.

74. The length of a rectangle is five more than the width. Find the dimensions of the rectangle if its area is 104 in².

E 75. The equivalent resistance, R_T, of two resistors connected in parallel is given by the relationship

$$\frac{1}{R_T} = \frac{1}{R_1} + \frac{1}{R_2}$$

while two resistors connected in series have an equivalent resistance given by $R_T = R_1 + R_2$. Find two resistances that will give an equivalent resistance of 25 Ω in series and 4 Ω in parallel.

E 76. The power P_L is given by $P_L = EI - RI^2$ where $E =$ voltage, $I =$ current, and $R =$ resistance. Find the current in a circuit where $P_L = 9$ W, $E = 12$ V, and $R = 4\ \Omega$.

CS 77. One computer can run a program in 3 hours less than a second computer. If the computers are networked together, they can run the program in 8 hours. How long does it take to run the program on each computer alone? (Round your answer to hundredths.)

CHAPTER TEST

The number(s) in parentheses refers to the appropriate learning objective(s) at the beginning of the chapter.

1. Solve the following by factoring: (1)

$$7x(x - 2) - 10 = 2 + 3x$$

2. Using the discriminant, describe the number and kind of solutions for $3x^2 + 9(x - 3) = 4(2x + 5)$. (4)

3. Solve the following: (1, 2, 3)

$$\frac{6x}{(2x + 5)} - \frac{3}{(2x^2 + 3x - 5)} = \frac{4x}{(x - 1)}$$

4. Find two consecutive even integers whose product is 120. (6)

5. Solve the following by factoring: (1)

$$10x^2 - 29x + 21 = 0$$

6. Solve the following by completing the square: (2)

$$2x^2 + 6x - 13 = 0$$

7. Solve the following by using the quadratic formula: (3)

$$(8x - 3)(2x + 5) + 8(x - 1) = 5(3 - 2x) + 9$$

8. The length of a rectangle is 7 ft more than the width. Find the width of the rectangle if its area is 120 ft^2. (6)

9. Solve the following quadratic equation: (1, 2, 3)

$$9x^2 - 14 = 0$$

10. Using the discriminant, determine the number and type of solutions for $7x^2 - 8x + 6 = 4x(x - 3)$. (4)

11. Solve the following quadratic equation: (1, 2, 3)

$$3x^2 - 8x + 5 = 0$$

CS 12. One computer can execute a program in 4 minutes less than a second computer. If it takes 12 minutes for the two computers together to complete the job, how long does it take for the faster computer to complete the job alone? (Round your answer to hundredths.) (7)

13. Solve the following by the quadratic formula: (3)

$$\frac{7x^2}{2} + \frac{5x}{3} = 4$$

14. Solve the following quadratic equation: (1, 2, 3)

$$\frac{4x}{x-1} + \frac{6}{x+3} = \frac{-24}{x^2 + 2x - 3}$$

15. Solve the following equation: (5)

$$5x^4 - 8x^2 + 3 = 0$$

P16. A stone weighing 2 pounds thrown down from the top of a 20-ft house has an initial (7) velocity of 8 ft/s. How long will it take the stone to strike the ground? (Solve $20 = 8t + 16t^2$.) Round your answer to tenths.

17. Solve the following quadratic equation: (1, 2, 3)

$$8x^2 + 28x = 6x + 21$$

18. Use the discriminant to describe the number and type of solutions for the following: (4)

$$4x(x-1) + 6x = (x-2)(3x+6)$$

19. Solve by completing the square: (2)

$$\frac{1}{2}x^2 + 8x - 9 = 0$$

20. Solve the following quadratic inequality: (6)

$$4x^2 + 6x \le 18$$

GROUP ACTIVITY

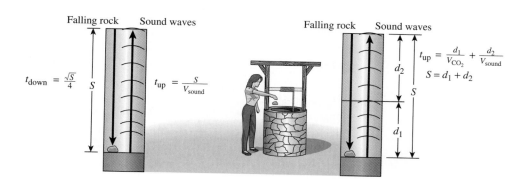

An engineer wants to determine the depth of a well that is located on the factory's property. She decides that the best way to do this is to drop a rock down the shaft and measure the time it takes to hear the splash. She takes a sound detector that is capable of triggering a timer to the well and performs the measurement. The following data is gathered:

$$\text{Time to hear the splash} = 5.3543 \text{ s}$$

$$\text{Speed of sound in air} = 1{,}129 \text{ ft/s}$$

$$\text{Acceleration due to gravity} = 32 \text{ ft/s}^2$$

Other important data includes (1) the time for the sound to travel up the well to the sound detector as given by:

$$t_{up} = \frac{S}{V_{\text{sound in air}}}$$

where S = depth of the well; (2) the time for the rock to reach the water as given by:

$$S = V_0 t_{down} + \frac{1}{2}at^2_{down}$$

where t is the time to fall a distance S, a is equal to the acceleration due to gravity, and V_0 is equal to the initial speed of the rock.

The engineer is not throwing the rock, therefore, the initial velocity is equal to 0, which results in:

$$t_{down} = \sqrt{\frac{2S}{32 \text{ ft/s}^2}} = \sqrt{\frac{S}{16}} = \sqrt{\frac{S}{4}}$$

The measured time is the sum of t_{up} and t_{down}. (*Note:* Air resistance is ignored.)

1. Write the equation for the measured time and solve for the depth of the well.
2. After the rock hits the water, how long does it take to hear the splash?
3. The engineer repeats the measurement a few months later with different results. She rationalizes that the time difference is due to the introduction of CO_2 gas into the well. Describe how you would find the depth of CO_2 gas if the original depth of water is unchanged, the speed of sound in CO_2 is 846 ft/s^2, and the new time measurement is 5.4431. (*Hint:* 0.4431 seconds is the time for the sound waves to travel d_1 plus d_2.)

SOLUTION TO CHAPTER INTRODUCTION

From the equation

$$V = (0.79D^2 - 2D - 4)\frac{L}{16}$$

we know $L = 16$ ft and $V = 625$ board feet. Substituting these values and solving by the quadratic formula gives

$$625 = (0.79D^2 - 2D - 4)\frac{16}{16}$$

$$0.79D^2 - 2D - 4 - 625 = 0$$

$$0.79D^2 - 2D - 629 = 0$$

$$D = \frac{-b \pm \sqrt{b^2 - 4ac}}{2a}$$

$$D = \frac{2 \pm \sqrt{4 - 4(0.79)(-629)}}{2(0.79)}$$

$$D = \frac{2 \pm \sqrt{1{,}991.64}}{1.58}$$

$$D \approx \frac{2 \pm 44.628}{1.58}$$

$$D \approx 29.5$$

The fir log needs to be approximately 29.5 inches in diameter.

R

honda's Ostrich Ranch has hired you to fence
two square ostrich pens. Rhonda would like to fence a
total of 2,425 square feet. She wants the two pens next to
each other so they can share one side of each pen. Her budget limits her to 220 running feet of fence. What are the
sizes of the two pens? (The solution to this problem is given
at the end of the chapter.)

So far, we have discussed solving linear and quadratic
equations of various types, but all the equations discussed
have been similar because they contained only one variable.
However, in this chapter we will discuss solving systems of
equations that contain two or more variables. We will discuss solving systems graphically and algebraically.

Learning Objectives

After completing this chapter, you should be able to

1. Determine whether an ordered pair is the solution to
   a system of equations (Section 6–1).

2. Solve a system of linear equations graphically (Section 6–1).

3. Solve a system of two or three linear equations using
   substitution or addition (Sections 6–2 and 6–3).

4. Evaluate a determinant (Sections 6–2 and 6–3).

5. Solve a system of linear equations by using Cramer's
   rule (Sections 6–2 and 6–3).

6. Apply the concepts of solving a system of equations
   to technical situations (Section 6–4).

7. Perform operations with matrices (Section 6–5).

Chapter 6

Systems of Equations

6–1

GRAPHICAL SOLUTION OF SYSTEMS OF LINEAR EQUATIONS

In Chapter 4, we discussed linear equations in two variables, such as $4x - 7y = 10$. Two equations of this form are called a **system of linear equations.** The **solution** to the system is the ordered pair (x, y) that satisfies both equations simultaneously.

Checking Solutions

Since the solution to a system of equations must satisfy both equations simultaneously, we can *check* a possible solution *by substituting the ordered pair into both equations.* The next example illustrates this process.

EXAMPLE 1 For each of the following systems, determine whether the given ordered pair is a solution:

(a) $\left.\begin{array}{l} 3x - y = 8 \\ x + 5y = 24 \end{array}\right\}$ $(4, 4)$ (b) $\left.\begin{array}{l} 4x - 3y = 10 \\ x + 2y = 0 \end{array}\right\}$ $(2, -1)$

Solution

(a) We substitute the ordered pair $(4, 4)$ into each equation.

$$3x - y = 8 \qquad\qquad x + 5y = 24$$
$$3(4) - (4) = 8 \qquad\qquad (4) + 5(4) = 24$$
$$8 = 8 \quad \text{(true)} \qquad\qquad 24 = 24 \quad \text{(true)}$$

Since both substitutions result in a true statement, the ordered pair is the solution to this system.

(b) $\qquad\qquad 4x - 3y = 10 \qquad\qquad\qquad x + 2y = 0$
$$4(2) - 3(-1) = 10 \qquad\qquad (2) + 2(-1) = 0$$
$$8 + 3 = 10 \quad \text{(false)} \qquad\qquad 0 = 0 \quad \text{(true)}$$

Since $(2, -1)$ does not satisfy the equation $4x - 3y = 10$, it cannot be the solution to this system of equations. ■ ■ ■

Solving Graphically

Recall from Chapter 4 that a linear equation in two variables is represented graphically by a line. Therefore, the graph of a system of two linear equations contains two lines. The **graphical solution** to such a system is *the point of intersection of the lines* because that point satisfies both equations simultaneously.

EXAMPLE 2 Solve the following system of equations graphically:

$$2x - y = 6$$
$$3x + y = 4$$

Solution   To graph each equation, we can fill in a table of values just as we did in Chapter 4.

$2x - y = 6$	
x	y
-1	-8
0	-6
2	-2

$3x + y = 4$	
x	y
-2	10
0	4
1	1

The graph of these two equations is given in Figure 6–1. As shown in the graph, the point of intersection (solution) is the ordered pair $(2, -2)$. The solution is $x = 2$ and $y = -2$.

FIGURE 6–1

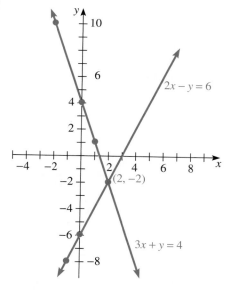

Graphing Using the Intercepts

An alternate method of graphing linear equations is to use the x and y intercepts. The **x intercept** is the x coordinate of the point where the graph crosses the x axis, and the **y intercept** is the y coordinate of the point where the graph crosses the y axis. To calculate the x intercept, substitute $y = 0$ into the equation; to find the y intercept, substitute $x = 0$.

EXAMPLE 3   Solve the following system graphically using intercepts. Estimate the solution to the nearest tenth.

$$3x + 2y = -6$$
$$4x - y = 8$$

Solution   To determine the *x* intercept for each equation, substitute $y = 0$ into the equation.

x intercept ($y = 0$)	
$3x + 2y = -6$	$4x - y = 8$
$3x + 2(0) = -6$	$4x - (0) = 8$
$x = -2$	$x = 2$

To determine the *y* intercept for each equation, substitute $x = 0$ into the equation.

y intercept ($x = 0$)	
$3x + 2y = -6$	$4x - y = 8$
$3(0) + 2y = -6$	$4(0) - y = 8$
$y = -3$	$y = -8$

The *x* intercept for the graph of $3x + 2y = -6$ is -2, and the *y* intercept is -3. The *x* intercept for the graph of $4x - y = 8$ is 2, and the *y* intercept is -8. Graphing both lines as shown in Figure 6–2, we can see that the intersection does not occur at integer values. Therefore, it is necessary to estimate the solution to be $x \approx 0.9$ and $y \approx -4.3$.

FIGURE 6–2

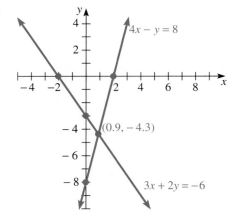

Checking

$$3x + 2y = -6 \qquad\qquad 4x - y = 8$$
$$3(0.9) + 2(-4.3) = -6 \qquad 4(0.9) - (-4.3) = 8$$
$$-5.9 \approx -6 \qquad\qquad 7.9 \approx 8$$

■ ■ ■

Graphing Calculator

In Chapters 4 and 5, we discussed the solution of equations using a graphing calculator. We will extend the discussion here to include a system of linear equations. As we know from our discussion in this section, we graph the two equations on the same coordinate plane and find the coordinates of the point of intersection. The next example illustrates the process on the graphing calculator.

EXAMPLE 4   Using the graphing calculator, solve the following system of equations.

$$3x + 2y = 8$$
$$x - 3y = -1$$

Solution   The graphing calculator requires the equation to be solved for y. Solving each equation for y gives:

$$y = -\frac{3}{2}x + 4$$

$$y = \frac{1}{3}x + \frac{1}{3}$$

Graphing these equations results in Figure 6–3. Using the TRACE function to locate the point of intersection gives $x \approx 2$ and $y \approx 1$. (In the next section we will discuss the algebraic solution and discover $x = 2$ and $y = 1$ is the actual solution.)

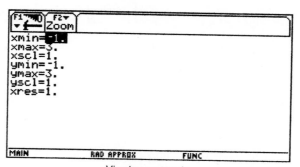

Viewing rectangle

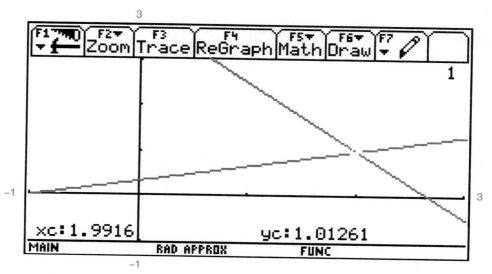

FIGURE 6–3

Application

EXAMPLE 5   A 12.0-ft beam weighing 50.0 lb is supported on each end. The uniform beam supports a 75.0-lb pile of bricks 4.00 ft from the right end and a 160-lb man 3.00 ft from the left end (see Figure 6–4). A system of equations that results from this situation is

$$A + B = 285$$
$$6A - 6B = 330$$

where A represents the force supported by the left end of the beam and B represents the force supported by the right end. Use a graph to find how much weight each end supports.

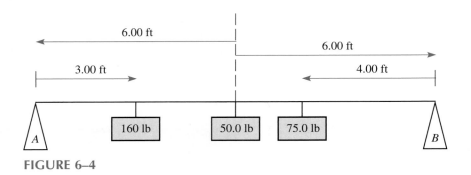

FIGURE 6–4

Solution   Placing A on the horizontal axis and B on the vertical axis, we find that the x and y intercepts for the graph of $A + B = 285$ are 285 and 285, respectively. The x and y intercepts of $6A - 6B = 330$ are 55 and -55, respectively. From Figure 6–5, we can see that the *point of intersection* is estimated to be $A = 170$ and $B = 115$. Therefore, the left end supports 170 lb and the right end supports 115 lb. Large numbers also make graphing difficult.

FIGURE 6–5

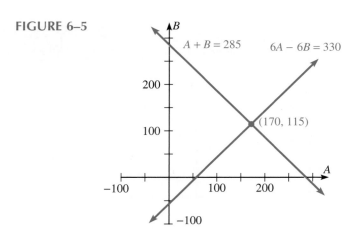

■ ■ ■

EXAMPLE 6   Find currents I_1 and I_2 shown in Figure 6–6 given the following equations (resulting from Kirchhoff's law):

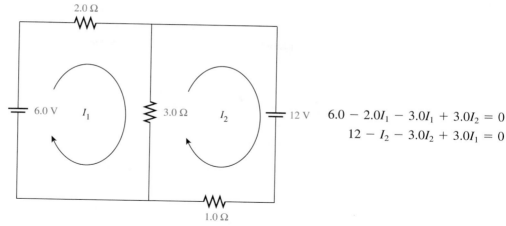

$$6.0 - 2.0I_1 - 3.0I_1 + 3.0I_2 = 0$$
$$12 - I_2 - 3.0I_2 + 3.0I_1 = 0$$

FIGURE 6–6

Solution   First, we must write the equations in the same order.

$$-5.0I_1 + 3.0I_2 = -6.0$$
$$3.0I_1 - 4.0I_2 = -12$$

We graph the equations using intercepts by placing I_1 on the horizontal axis and I_2 on the vertical axis.

I_1 intercept	
$-5.0I_1 + 3.0I_2 = -6.0$	$3.0I_1 - 4.0I_2 = -12$
$-5.0I_1 = -6.0$	$3.0I_1 = -12$
$I_1 = \dfrac{6.0}{5.0}$	$I_1 = -4.0$

I_2 intercept	
$-5.0I_1 + 3.0I_2 = -6.0$	$3.0I_1 - 4.0I_2 = -12$
$3.0I_2 = -6.0$	$-4.0I_2 = -12$
$I_2 = -2.0$	$I_2 = 3.0$

From Figure 6–7, we see that the point of intersection is $I_1 \approx 5.5$ A and $I_2 \approx 7.1$ A.

FIGURE 6–7

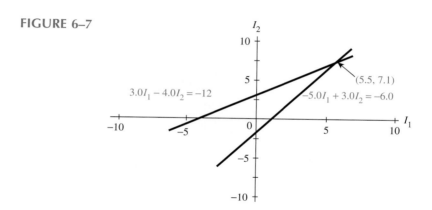

The total cost C of producing x units of a product typically consists of an initial cost and a cost per unit. At the break-even point enough units have been sold so that the revenue generated equals the total cost. Graphically the break-even point corresponds to the point of intersection of the total cost and revenue curves.

EXAMPLE 7 Mark's Frame Shop invests $5,000 in equipment to do custom framing. Each custom frame costs $12.50 to produce and is sold for $45. How many custom frames must be sold before Mark breaks even?

Solution The equation representing the total cost C of producing x custom frames is given by

$$C = 5,000 + 12.5x$$

The equation representing revenue R resulting from selling x custom frames is given by

$$R = 45x$$

The break-even point is given by the point of intersection of the two lines. Figure 6–8 represents the two lines, the break-even point, and the areas where profit and loss occur. Algebraically, the break-even point occurs when $R = C$. Solving gives

$$R = C$$

$$45x = 5,000 + 12.5x$$

$$32.5x = 5,000$$

$$x = 153.846$$

Mark must sell about 154 custom frames.

FIGURE 6–8

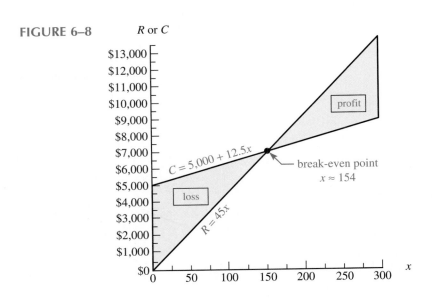

Inconsistent and Dependent Systems

The systems of equations solved thus far are called *consistent* and *independent* because the graphs intersect in exactly one point and the system has exactly one solution. However, not every system of equations has one solution. Since parallel lines never intersect, a *system of equations whose graph consists of parallel lines* has no solution and is called **inconsistent.** If the *graph of a system of two linear equations produces one line,* every point on the line is a solution. A system of this type is called **dependent.** Inconsistent and dependent systems are illustrated in the next example.

EXAMPLE 8   Solve the following systems of equations graphically:

(a) $5x - 3y = 11$ (b) $4x - y = 3$
   $10x - 6y = 7$ $8x - 2y = 6$

Solution

(a) The x and y intercepts of the graph for $5x - 3y = 11$ are 11/5 and −11/3, respectively, while the intercepts of the graph for $10x - 6y = 7$ are 7/10 and −7/6, respectively. The graph of these equations is shown in Figure 6–9(a). Since the lines have no point of intersection, this system of equations is inconsistent and has no solution.

(b) The x and y intercepts of the graph for both $4x - y = 3$ and $8x - 2y = 6$ are 3/4 and −3, respectively. This system of equations is dependent, as shown in Figure 6–9(b), and the solutions are all the points on the line.

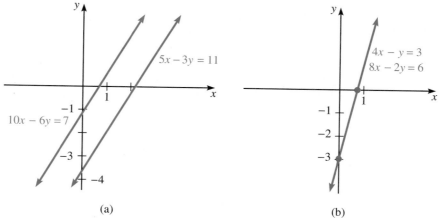

(a)   (b)

FIGURE 6–9

■ ■ ■

6–1 EXERCISES

Determine whether the given ordered pair is the solution to the system of equations.

1. $\left.\begin{array}{l} 7x - y = 8 \\ 3x + 2y = 6 \end{array}\right\}$ (0, 1)

2. $\left.\begin{array}{l} x + 4y = -5 \\ y = 2x + 10 \end{array}\right\}$ (−5, 0)

3. $\left.\begin{array}{l} 4x - y = 5 \\ x = 3y - 7 \end{array}\right\}$ (2, 3)

4. $\left.\begin{array}{l} 5x - y = 7 \\ 2x + 9y = 22 \end{array}\right\}$ (2, 2)

5. $\left.\begin{array}{l} 6x + 7y = 13 \\ 5x + y = -1 \end{array}\right\}$ (−1, 4)

6. $\left.\begin{array}{l} x - y = 15 \\ 2x - 3y = -3 \end{array}\right\}$ (7, 8)

7. $2x - 4y = 14$
$2y = -13 - x$ $\quad (-3, -5)$

8. $3x - 5y = 7$
$x - 3y = 1$ $\quad (4, 1)$

9. $9x - y = 14$
$6x + 3y = 21$ $\quad (3, 13)$

10. $x + 3y = -3$
$2x - y = 15$ $\quad (6, -3)$

Solve the following systems of equations graphically. Estimate the answer to the nearest tenth, if necessary. Check your solution with the graphing calculator.

11. $2x + 3y = -6$
$x - 4y = -3$

12. $2x - 3y = 1$
$5x - y = -17$

13. $x + y = 6$
$3x - y = 10$

14. $5x - 4y = 12$
$9x + 2y = -6$

15. $7x - 2y = 11$
$3y = x - 7$

16. $3x + 2y = 13$
$x - 5y = 10$

17. $x + 3y = 10$
$2x - y = 8$

18. $8x - 3y = 12$
$2x + 3y = 8$

19. $2x + 3y = 8$
$4x + 6y = 10$

20. $10x = 3y + 4$
$2x - y = 6$

21. $5x + 7y = 13$
$y = 2x - 9$

22. $3x - 7y = 14$
$9x - 21y = 42$

23. $3x - y = 7$
$12x - 4y = 28$

24. $7x + y = 10$
$x - 5y = 8$

25. $4x - y = 10$
$3x + 5y = 8$

26. $8x - y = 15$
$16x - 2y = 10$

P 27. A uniform 100-ft bridge weighing 200 tons is supported on each end. If a truck weighing 14 tons is 10 ft from the west end and a car weighing 3 tons is 30 ft from the east end, the system of equations could be written as

$$A + B = 217$$
$$50A + 500 = 50B$$

where A and B represent the weight supported by the west and east ends, respectively. Solve this system for A and B.

E 28. Applying Ohm's law to a particular circuit results in the following system of equations:

$$E - 6I = 0$$
$$E + 10I = 8$$

Solve this system graphically.

C 29. The perimeter of a rectangular plot of land is 28 rods. If the length L of the tract is 4 rods more than the width W, the system of equations is

$$2L + 2W = 28$$
$$L = W + 4$$

Solve this system graphically to find the dimensions of the rectangle.

B 30. Using Example 7, determine the break-even point if Mark invests $7,800 in framing equipment and each frame has a cost of $16.00 and sells for $55.

E 31. To find the currents I_1 and I_2 shown in Figure 6–10, solve the following equations graphically.

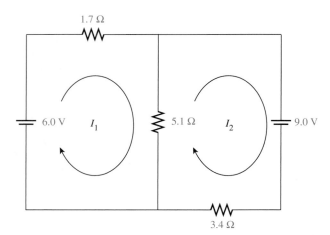

$$6.0 - 1.7I_1 - 5.1I_1 + 5.1I_2 = 0$$
$$9.0 - 3.4I_2 - 5.1I_2 + 5.1I_1 = 0$$

FIGURE 6–10

M 32. Two different size pipes are used to pump crude oil from a tanker to storage tanks. One day, Pipe A pumped oil for 5.0 hours, while Pipe B pumped oil for 3.0 hours, for a total volume of 125,000 gallons. The next day, Pipe A pumped oil for 3.5 hours and Pipe B pumped oil for 4.5 hours, for a total volume of 135,500 gallons. How much oil per hour does each pipe pump?

M 33. A machinist must mill a metal plate to meet the following specifications: The length is 14 cm less than twice the width and the perimeter is 152 cm. Find the dimensions of the plate.

E 34. Two resistors connected in series have a combined resistance of 70 Ω. If the resistance of one resistor is 4 Ω less than the resistance of the other, find the value of each resistor.

M 35. Design the height dimensions for two supports such that S_1 is 3.70 ft longer than S_2 and the combined length is 20.9 ft.

6–2

ALGEBRAIC SOLUTION OF TWO LINEAR EQUATIONS

As shown in Section 6–1, graphing may lead to an approximate solution for a system of linear equations. For this reason, we will now discuss three methods of algebraic solution: substitution, addition, and Cramer's Rule.

Solving by Substitution

To solve a system of two linear equations, we must eliminate one of the variables to obtain one linear equation with one variable and solve as in Chapter 1. One such method of eliminating a variable is called **substitution.** The process used in substitution is developed in the next example and summarized in the box following it.

EXAMPLE 1   Solve the following system of equations by substitution:

$$2x - y = 8$$
$$3x + 4y = 10$$

Solution   First, we solve one of the two equations for a variable. Solving the first equation for y gives

$$2x - y = 8$$
$$-y = 8 - 2x$$
$$y = \boxed{-8 + 2x}$$

Then, we substitute this quantity for y into the second equation.

$$3x + 4y = 10$$
$$3x + 4(-8 + 2x) = 10$$

The resulting equation is a linear equation in one variable that we now solve.

$$3x - 32 + 8x = 10 \qquad \text{eliminate parentheses}$$
$$11x = 10 + 32 \qquad \text{transpose}$$
$$x = 42/11 \qquad \text{divide}$$

To solve for y, we substitute the value of x into the expression previously solved for y.

$$y = \boxed{-8 + 2x}$$
$$y = -8 + 2(42/11)$$
$$y = -4/11$$

The solution is $x = 42/11$ and $y = -4/11$.   ∎∎∎

Solution by Substitution

1. Solve one of the equations for a variable. (Choose a variable with a numerical coefficient of positive or negative one [$+1$ or -1], if possible, to make the problem easier to solve.)
2. Substitute for the variable from Step 1 into the remaining equation.
3. Solve the resulting equation for the variable.
4. Solve for the other variable by substituting the value of the variable from Step 3 into the equation from Step 1.
5. Check your answer by substituting the solution into both original equations.

Note: The solution to a system of equations is not dependent on which variable you solve for in Step 1.

In some instances, the substitution method is simple and straightforward, as in Example 2. However, the substitution method can be long and tedious.

EXAMPLE 2 Solve the following system of equations using substitution:

$$4x + y = 10$$
$$6x - 3y = 6$$

Solution First, we solve the first equation for y because it has a coefficient of $+1$.

$$4x + y = 10$$
$$y = \boxed{10 - 4x}$$

Second, we substitute this expression for y in the second equation.

$$6x - 3y = 6$$
$$6x - 3(10 - 4x) = 6$$

Third, we solve the resulting equation for x.

$$6x - 30 + 12x = 6$$
$$18x = 6 + 30$$
$$18x = 36$$
$$x = 2$$

Fourth, we solve for y by substituting $x = 2$ into the equation from Step 1.

$$y = \boxed{10 - 4x}$$
$$y = 10 - 4(2)$$
$$y = 2$$

Checking $\qquad\qquad 4(2) + (2) = 10 \qquad\qquad 6(2) - 3(2) = 6$
$$10 = 10 \qquad\qquad\qquad 6 = 6$$

Since both equations result in a true statement, the solution is $x = 2$ and $y = 2$. ■■■

CAUTION ✦ In Step 2, be sure to substitute into the equation that you did not use in Step 1.

Solution by Addition

Some systems of equations are difficult to solve by substitution. Such systems are generally solved more easily by the **addition method.** Like the substitution method, the addition method of algebraic solution eliminates one of the variables. However, unlike substitution, the elimination is accomplished by addition of the equations. The process used in the addition method is illustrated in the next example.

EXAMPLE 3 Solve the following system of equations by addition:

$$3x - y = 10$$
$$2x + y = 15$$

Solution Notice that in these two equations, y has opposite coefficients. Therefore, *we can eliminate the variable y by adding the two equations.*

$$3x - y = 10$$
$$\underline{2x + y = 15}$$
$$5x \quad\;\; = 25$$

Solving the resulting linear equation in one variable gives

$$5x = 25$$
$$x = 5$$

To solve for y, substitute $x = 5$ into either of the original equations in the system. Try to pick the simplest of the two equations for this step.

$$3x - y = 10$$
$$3(5) - y = 10$$
$$15 - y = 10$$
$$-y = 10 - 15$$
$$y = 5$$

Checking

$$3x - y = 10 \qquad\qquad 2x + y = 15$$
$$3(5) - (5) = 10 \qquad\qquad 2(5) + (5) = 15$$
$$10 = 10 \qquad\qquad\qquad 15 = 15$$

Since both equations result in a true statement, the solution is $x = 5$ and $y = 5$. ∎∎∎

Solution by Addition

1. Arrange the equations in $ax + by = c$ form, if necessary.
2. Arrange opposite coefficients on one of the variables. To do so, you may have to multiply one or both equations by a constant.
3. Add the resulting equations and solve for the variable.
4. Solve for the remaining variable by substituting the value from Step 3 into either equation in the original system.
5. Check your solution in each equation.

EXAMPLE 4   Solve the following system of equations by addition:

$$4x - 2y = -4$$
$$2x + y = 10$$

Solution   Since both equations are in $ax + by = c$ form, we obtain opposite coefficients on the variable y by multiplying the second equation by 2, as follows:

$$4x - 2y = -4 \longrightarrow 4x - 2y = -4$$
$$\underline{2(2x) + 2(y) = 2(10)} \longrightarrow \underline{4x + 2y = 20}$$

Next, add the equations and solve for x.

$$4x - 2y = -4$$
$$\underline{4x + 2y = 20}$$
$$8x \qquad = 16$$
$$x = 2$$

Then substitute $x = 2$ into one of the original equations.

$$2x + y = 10$$
$$2(2) + y = 10$$
$$4 + y = 10$$
$$y = 6$$

After checking, the solution is $x = 2$ and $y = 6$. ■■■

CAUTION ✦ Be sure that the variable you wish to eliminate has opposite coefficients; that is, you must have opposite signs.

Solution by Cramer's Rule

A third algebraic method of solving systems of equations is called **Cramer's rule.** This method is based on matrices and determinants. Before discussing Cramer's rule, we need to discuss matrices and determinants briefly.

A **matrix** is a rectangular array of numbers consisting of **elements** arranged in rows and columns. A matrix is enclosed in brackets, and its size is given as the number of rows and the number of columns. The following matrix is a two-by-three (also written 2×3) matrix, that is, two rows by three columns:

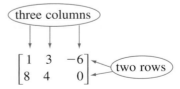

Each square matrix (that is, one having the same number of rows and columns) has a unique number associated with that matrix. That number, called its **determinant,** is written in the following form:

$$\begin{vmatrix} a & b \\ c & d \end{vmatrix}$$

To find the value of a 2 × 2 determinant, take the difference in the diagonal products as follows:

$$\begin{array}{c} + \quad - \\ \left| \overset{a}{\underset{c}{\times}} \overset{b}{d} \right| = ad - bc \end{array}$$

EXAMPLE 5   Evaluate the following determinant:

$$\begin{vmatrix} -6 & 2 \\ -1 & 5 \end{vmatrix}$$

Solution

$$\begin{array}{c} + \quad - \\ \left| \overset{-6}{\underset{-1}{\times}} \overset{2}{5} \right| = (-6)(5) - (-1)(2) = -28 \end{array}$$

The value of the determinant is -28.

∎∎∎

Cramer's rule can be derived by taking the system of equations below and solving it using addition.

$$ax + by = c$$
$$\underline{dx + ey = f}$$

$$\begin{array}{l} -d(ax + by = c) \\ \underline{a(dx + ey = f)} \end{array} \quad \longrightarrow \quad \begin{array}{l} -adx - bdy = -cd \\ \underline{adx + aey = af} \\ (ae - bd)y = af - cd \end{array}$$

$$y = \frac{af - cd}{ae - bd}$$

Converting the numerator and denominator to determinants gives

$$y = \frac{\begin{vmatrix} a & c \\ d & f \end{vmatrix}}{\begin{vmatrix} a & b \\ d & e \end{vmatrix}}$$

Similarly, solving for x by eliminating y gives the following result.

$$\begin{array}{l} e(ax + by = c) \\ \underline{-b(dx + ey = f)} \end{array} \quad \longrightarrow \quad \begin{array}{l} aex + bey = ce \\ \underline{-bdx - bey = -bf} \\ (ae - bd)x = ce - bf \end{array}$$

$$x = \frac{ce - bf}{ae - bd}$$

$$x = \frac{\begin{vmatrix} c & f \\ b & e \end{vmatrix}}{\begin{vmatrix} a & b \\ d & e \end{vmatrix}}$$

Cramer's rule is summarized below.

Cramer's Rule

$$x = \frac{\begin{vmatrix} c & b \\ f & e \end{vmatrix}}{\begin{vmatrix} a & b \\ d & e \end{vmatrix}} = \frac{D_x}{D} \qquad\qquad y = \frac{\begin{vmatrix} a & c \\ d & f \end{vmatrix}}{\begin{vmatrix} a & b \\ d & e \end{vmatrix}} = \frac{D_y}{D}$$

where $D \neq 0$.

Rather than memorize the determinants used in Cramer's rule, notice the following:

■ The determinant D is the same for both variables and consists of the coefficients of x and y from the original system.

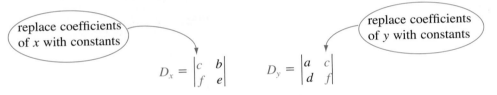

$$ax + by = c$$
$$dx + ey = f \longrightarrow \begin{vmatrix} a & b \\ d & e \end{vmatrix} = D$$

■ To find the determinant in the numerator, replace the column of coefficients of the subscripted variable with the constants.

replace coefficients of x with constants

replace coefficients of y with constants

$$D_x = \begin{vmatrix} c & b \\ f & e \end{vmatrix} \qquad D_y = \begin{vmatrix} a & c \\ d & f \end{vmatrix}$$

EXAMPLE 6   Solve the following system of equations using Cramer's rule:

$$4x - 5y = 10$$
$$3x + 7y = 14$$

Solution   To solve this system for x and y, we must calculate the determinants D, D_x, and D_y.

$$D = \begin{vmatrix} 4 & -5 \\ 3 & 7 \end{vmatrix} = 28 + 15 \qquad \text{coefficients of } x \text{ and } y$$

$$D = 43$$

$$D_x = \begin{vmatrix} 10 & -5 \\ 14 & 7 \end{vmatrix} = 70 + 70 \qquad \text{replace coefficients of } x \text{ with constants}$$

$$D_x = 140$$

$$D_y = \begin{vmatrix} 4 & 10 \\ 3 & 14 \end{vmatrix} = 56 - 30 \qquad \text{replace coefficients of } y \text{ with constants}$$

$$D_y = 26$$

The solution is

$$x = \frac{D_x}{D} = \frac{140}{43}$$

$$y = \frac{D_y}{D} = \frac{26}{43}$$

The solution is $x = {}^{140}\!/_{43}$ and $y = {}^{26}\!/_{43}$.

■ ■ ■

CAUTION ✦ When applying Cramer's rule, be sure that the equations are arranged in standard form: $ax + by = c$.

Inconsistent and Dependent Systems

In the previous section, we discussed the graphical solution to inconsistent and dependent systems of equations. These systems also occur in algebraic solutions, and we must be able to identify them. Identifying inconsistent and dependent systems using substitution and addition is different from Cramer's rule. The following box summarizes the identification process.

	Inconsistent, no solution	Dependent, line is solution
Substitution and addition	Both variables eliminated → false statement.	Both variables eliminated → true statement.
Cramer's rule	Determinant in denominator equals 0.	Both determinants equal 0.

Both inconsistent and dependent systems of equations are illustrated in Example 7.

EXAMPLE 7   Solve the following systems of equations:

(a) $3x - 2y = 8$
$-9x + 6y = 20$

(b) $4x - 5y = -10$
$-8x + 10y = 20$

(c) $2x - 3y = 7$
$-4x + 6y = 10$

Solution

(a) Solving by addition gives

$$
\begin{array}{l}
3(3x) - 3(2y) = 3(8) \longrightarrow \quad 9x - 6y = 24 \\
\underline{-9x + 6y \qquad = 20} \longrightarrow \underline{-9x + 6y = 20} \\
\hphantom{-9x + 6y \qquad = 20 \longrightarrow} \quad 0 = 44 \quad \text{(false)}
\end{array}
$$

Since *both variables are eliminated* and a *false* statement, $0 = 44$, results, this system of equations is *inconsistent* and has no solution.

(b) Solving by addition gives

$$
\begin{array}{l}
2(4x) - 2(5y) = 2(-10) \longrightarrow \quad 8x - 10y = -20 \\
\underline{-8x + 10y \quad = 20} \longrightarrow \underline{-8x + 10y = 20} \\
\hphantom{-8x + 10y \quad = 20 \longrightarrow} \quad 0 = 0 \quad \text{(true)}
\end{array}
$$

Since *both variables are eliminated* and a *true* statement, $0 = 0$, results, this system of equations is *dependent*.

(c)
$$x = \frac{\begin{vmatrix} 7 & -3 \\ 10 & 6 \end{vmatrix}}{\begin{vmatrix} 2 & -3 \\ -4 & 6 \end{vmatrix}} = \frac{42 + 30}{12 - 12} = \frac{72}{0} = \text{undefined}$$

$$y = \frac{\begin{vmatrix} 2 & 7 \\ -4 & 10 \end{vmatrix}}{\begin{vmatrix} 2 & -3 \\ -4 & 6 \end{vmatrix}} = \frac{20 + 28}{12 - 12} = \frac{48}{0} = \text{undefined}$$

Since only the determinant in the denominator equals zero, the system of equations is *inconsistent* and has *no solution.* ■ ■ ■

Applications

▣ **EXAMPLE 8** A chemist wants to make 10 gallons of a 25% solution of sulfuric acid using a 15% solution and a 30% solution. How much of each solution should the chemist use?

Solution We have discussed similar mixture problems in previous chapters. However, in this section we can use two variables to set up the problem. Let

$$x = \text{amount of the 30\% solution used}$$
$$y = \text{amount of the 15\% solution used}$$

With this information, we fill in the following table, just as we did in Chapter 1.

% × amount = amount in mixture		
%	Amount	Amount in Mixture
30%	x	$0.30x$
15%	y	$0.15y$
25%	10	$0.25(10)$

Unlike the previous treatment of mixture problems, we have used two variables. Therefore, we must write two equations. First, the total amount of the mixture is 10 gallons and is represented by the equation

$$x + y = 10$$

Second, the total amount of each solution in the mixture is given by the equation

amount of 30% + amount of 15% = amount of 25%

$$0.30x \quad + \quad 0.15y \quad = \quad 0.25(10)$$

The system of equations is

$$x + y = 10$$
$$0.30x + 0.15y = 0.25(10)$$

Using substitution, we solve the first equation for y and substitute into the second equation.

$$y = 10 - x$$
$$0.30x + 0.15y = 2.50$$

$0.30x + 0.15(10 - x) = 2.50$ substitute

$0.30x + 1.50 - 0.15x = 2.50$ eliminate parentheses

$0.15x = 1$ transpose and add

$x = 6.7$ divide

2.5 ⊟ 1.5 🟰 ÷ 🄰 0.30 ⊟ 0.15 🄳 🟰 ⟶ 6.7

Then substitute $x = 6.7$ to find y.

$$y = 10 - x$$
$$y = 10 - 6.7$$
$$y = 3.3$$

The chemist should use 6.7 gal of the 30% solution and 3.3 gal of the 15% solution.

∎∎∎

LEARNING HINT ✦ Notice that one equation resulted from the Amount column and the second equation resulted from the Amount in Mixture column. This is generally true for mixture problems.

Graphing Calculator

The graphing calculator can be used to solve systems of equations algebraically, as well as graphically. There are several ways to solve a system of linear equations on the TI-92. You may use **Solve, Simult,** or **rref**. To use **rref,** press [2nd], [MATH], [4], and [4]. Then, you enter the coefficients as an extended matrix. Using Example 8, you would enter [1, 1, 10; .3, .15, 2.5]. The calculator screen is shown in Figure 6–11. The solution is given as a matrix $x = 6.6\overline{6}$ and $y = 3.3\overline{3}$. You may want to use the graphing calculator to check your answers.

FIGURE 6–11

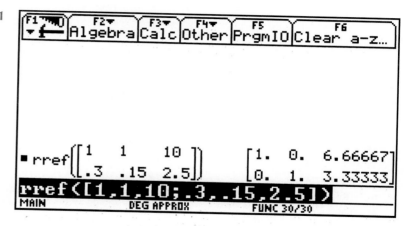

EXAMPLE 9   To find the current I_1 and I_2 in Figure 6–12, we must solve the following equations:

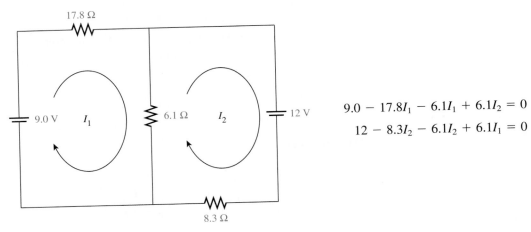

$$9.0 - 17.8I_1 - 6.1I_1 + 6.1I_2 = 0$$
$$12 - 8.3I_2 - 6.1I_2 + 6.1I_1 = 0$$

FIGURE 6–12

Solution   Simplifying these equations and arranging them in the $ax + by = c$ form gives

$$-23.9I_1 + 6.1I_2 = -9.0$$
$$6.1I_1 - 14.4I_2 = -12$$

Using Cramer's rule to solve this system of equations gives

$$I_1 = \frac{\begin{vmatrix} -9.0 & 6.1 \\ -12 & -14.4 \end{vmatrix}}{\begin{vmatrix} -23.9 & 6.1 \\ 6.1 & -14.4 \end{vmatrix}} = \frac{+129.6 - (-73.2)}{344.16 - 37.21} = \frac{202.8}{306.95}$$

$$I_1 \approx 0.66 \text{ A}$$

$$I_2 = \frac{\begin{vmatrix} -23.9 & -9 \\ 6.1 & -12 \end{vmatrix}}{306.95} = \frac{286.8 - (-54.9)}{306.95} = \frac{341.7}{306.95}$$

$$I_2 \approx 1.1 \text{ A}$$

■ ■ ■

6–2 EXERCISES

Solve the following systems of equations using any method.

1. $3x - y = 10$
    $2x + y = 5$

2. $x + y = 8$
    $3x - y = 8$

3. $4x + 3y = -4$
    $8x - 2y = -24$

4. $x - 4y = 8$
    $2x + 6y = 9$

5. $6x - 8y = -14$
    $2y = 5x + 14$

6. $3x - y = 8$
    $4y = 12x + 10$

7. $3x - 2y = 9$
    $x + 5y = 13$

9. $y = 2x + 6$
    $3x + 4y = 13$

8. $\frac{1}{2}x + \frac{2}{3}y = \frac{5}{9}$
    $4x - \frac{1}{2}y = \frac{25}{3}$

10. $x = 2y + 3$
    $7y + 3x = 9$

11. $2x + y = 17$
$5x - 3y = 8$

12. $3x + y = -2$
$y = 4x + 5$

13. $6x - y = 14$
$3x + 4y = -2$

14. $4x - 5y = 13$
$y - 2x = 1$

15. $\frac{1}{2}x - \frac{2}{3}y = 1$
$\frac{1}{4}x + \frac{1}{6}y = 1$

16. $\frac{2}{5}x - \frac{1}{3}y = \frac{7}{3}$
$\frac{1}{2}x + \frac{3}{4}y = \frac{7}{4}$

17. $3x + 4y = 10$
$7x - 2y = 14$

18. $6x - 3y = 8$
$4x = 18 - 7y$

19. $x + 3y = 20$
$x - 4y = -22$

20. $-2x + 7y = 10$
$x - 9y = -5$

21. $2x - 3y = 15$
$4y = 3x - 19$

22. $4x - 6y = -3$
$x + 4y = 2$

23. $4x - 3y = -1$
$7x + y = 17$

24. $x + 3y = -2$
$5x + 7y = 6$

25. $x - \frac{1}{3}y = \frac{5}{6}$
$\frac{1}{3}x - \frac{1}{4}y = \frac{5}{12}$

26. $4x - 9y = 20$
$2x - 7y = 8$

27. $x - 2y = 7$
$3x + 4y = 1$

28. $3x + 2y = 11$
$5x + y = 9$

29. $6x - 5y = -19$
$9x + 11y = -47$

30. $3x - y = 6$
$4x + 5y = 20$

Evaluate the following determinants.

31. $\begin{vmatrix} 2 & -3 \\ 6 & -4 \end{vmatrix}$

32. $\begin{vmatrix} 0 & -5 \\ -6 & -8 \end{vmatrix}$

33. $\begin{vmatrix} 9 & 2 \\ 5 & 8 \end{vmatrix}$

34. $\begin{vmatrix} 1 & 3 \\ -6 & 7 \end{vmatrix}$

35. $\begin{vmatrix} 0 & 1 \\ 1 & 0 \end{vmatrix}$

36. $\begin{vmatrix} 4.7 & 3.8 \\ 5.9 & 2.4 \end{vmatrix}$

37. $\begin{vmatrix} -7.4 & 0 \\ 18.5 & 2.3 \end{vmatrix}$

38. $\begin{vmatrix} -6 & -4 \\ 2 & -5 \end{vmatrix}$

39. $\begin{vmatrix} 0 & -3 \\ -7 & -10 \end{vmatrix}$

40. Find the equation of the line of the form $ax + by = 10$ that passes through the points $(1, 6)$ and $(2, 4)$. (*Hint:* Each point must satisfy the equation. Substitute the ordered pairs for x and y into this equation to obtain a system of two linear equations that you then solve for a and b.)

B 41. Your administrative assistant went to the office supply store and bought 1 ream of laser printer paper and 3 boxes of ballpoint pens and paid a total of $39.70. You also went to the office supply store and bought 3 reams of laser printer paper and 2 boxes of ballpoint pens and paid $35.10. If you want to return 1 ream of paper and 2 boxes of pens, what is your refund?

42. Mrs. Smith buys apples and nectarines at the market. She buys one more pound of apples at $0.69 a pound than nectarines at $0.89 a pound. If the total cost is $3.85, how many pounds of each did she buy?

P 43. Two cars leave the same town and travel in opposite directions. At the end of 8 hours, the cars are 896 miles apart. Using a system of equations, find the rate of each car if the rate of one car is 4 mi/h more than the rate of the other.

B 44. Betsy invests $13,000 in two stocks, one yielding 7% and the other yielding 8.3%. If her total yield for a year is $1,000, how much did she invest in each stock?

45. Find two numbers whose sum is 32 and whose difference is 14.

C 46. One alloy contains 25% copper and another 32% copper. How many pounds of each must be used to form 40 lb of an alloy containing 27% copper?

E 47. Find currents I_1 and I_2 in Figure 6–13 given that the system of equations is

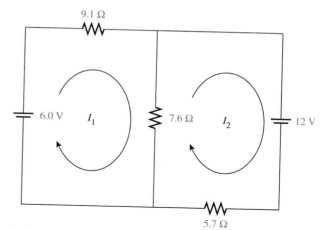

FIGURE 6–13

$$6.0 - 9.1I_1 - 7.6I_1 + 7.6I_2 = 0$$
$$12 - 5.7I_2 - 7.6I_2 + 7.6I_1 = 0$$

B 48. A recently retired couple needs a yearly income of $24,000 to supplement their social security income. If they have $250,000 to invest and want to invest in two mutual funds currently yielding 15% per year and 9% per year, how should they divide their investment between the two funds?

CS 49. One office supply company rents computers at a base rate of $15 plus $0.30 per hour. A second company rents the same computer for $10 plus $0.35 per hour. How many hours would it take for the total cost of renting to be the same?

E 50. Two resistors connected in series have a combined resistance of 4.31 Ω. Find each resistance if the resistance of one resistor is 3.70 Ω more than that of the other.

M 51. The forces acting on part of a robot are shown in Figure 6–14. An analysis of the forces leads to the equations

$$0.5F_1 + 0.8F_2 = 30$$
$$0.9F_1 = 0.6F_2 + 10$$

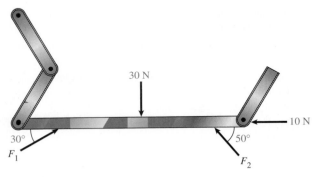

30 N

10 N

30°

F_1

50°

F_2

FIGURE 6–14

Solve for F_1 and F_2.

CS 52. A computer assembler needed two different components, a and b, to complete the assembly of computers. The total number of components was 82 and the total cost was $273. How many of each component did the assembler need if a costs $3 and b costs $4?

CS 53. A computer component manufacturer produces two types of motherboards for a PC. If he ships 4,500 motherboards that bring a revenue of $358,500 and the motherboards cost $67 and $105, how many of each are shipped?

P 54. Cables support a crate as shown in Figure 6–15. Find the tension in the two cables if the equations are

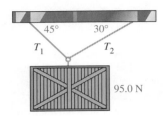

45°   30°

T_1 T_2

95.0 N

FIGURE 6–15

$$0.707T_1 = 0.900T_2$$
$$0.707T_1 + 0.500T_2 = 95.0$$

B 55. Two conveyer belts fill a bin in 4 h working together. If one conveyer works 2.3 times as fast as the other, how long would it take each conveyer to fill the bin working alone?

CH 56. Two kinds of milk containing 1% and 6% butterfat by volume are to be mixed to obtain 65 gallons of 2% milk. How much of each kind of milk should be used?

CH 57. Two grades of gasoline are mixed to obtain 1,500 L of a mixture containing 1.7% of a special additive. If one gasoline contains 2.1% of the additive and the other gasoline contains 1.4% of the additive, how much of each gasoline should be used?

B 58. During a marketing presentation, Mr. Campbell made the statement that sales this month were $53,000 more than last month, which means total sales for the last two months were $127,000 more than twice the sales last month. What is Mr. Campbell really saying about sales?

E 59. Find currents I_1 and I_2 for Figure 6–16 given the following equations:

$$I_1 + I_2 + I_3 = 0$$
$$7.1I_1 - 4.3I_2 = 3.0$$
$$-4.3I_2 + 2.9I_3 = 12$$

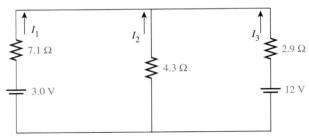

I_1

7.1 Ω

3.0 V

I_2

4.3 Ω

I_3

2.9 Ω

12 V

FIGURE 6–16

where $I_3 = 1.9$ A.

A 60. Find the dimensions of a rectangular solar panel if the perimeter is 15 ft and the length is 10 in. longer than the width.

6–3

ALGEBRAIC SOLUTION OF THREE LINEAR EQUATIONS

Addition, substitution, and Cramer's rule can be extended to solve a system of three linear equations using those techniques presented in the previous section. For substitution and addition, we eliminate one variable to obtain a system of two equations in two variables and solve this system just as we did in the last section. Then we obtain the solution to the system with three variables. For Cramer's rule, we extend the concept of determinants to 3×3 s. The next example illustrates the first method.

EXAMPLE 1   Solve the following system of equations:

$$x + y - z = 8 \qquad \qquad \textbf{(1)}$$
$$2x - y + z = 10 \qquad \qquad \textbf{(2)}$$
$$x + 3y + z = 6 \qquad \qquad \textbf{(3)}$$

Solution   First, we reduce this system of three equations to a system of two equations containing the same two variables. Thus, we must eliminate one of the variables from two equations. We can eliminate z by adding equations (1) and (2), obtaining

$$
\begin{array}{ll}
x + y - z = 8 & \textbf{(1)} \\
\underline{2x - y + z = 10} & \textbf{(2)} \\
3x \qquad\quad = 18 &
\end{array}
$$

Then we can also eliminate z by adding equations (1) and (3).

$$
\begin{array}{ll}
x + y - z = 8 & \textbf{(1)} \\
\underline{x + 3y + z = 6} & \textbf{(3)} \\
2x + 4y \quad = 14 &
\end{array}
$$

Now, we solve the resulting system of two equations.

$$3x = 18$$
$$\underline{2x + 4y = 14}$$
$$3x = 18$$
$$x = 6$$
$$2(6) + 4y = 14 \qquad \text{solve for } y$$
$$12 + 4y = 14$$
$$4y = 2$$
$$y = 1/2$$

At this point, we know that $x = 6$ and $y = 1/2$. To complete the solution of the original system of three equations, we must solve for z. Substituting the numerical values for x and y into equation (1) gives

$$x + y - z = 8 \qquad \qquad \textbf{(1)}$$
$$(6) + (1/2) - z = 8$$
$$13/2 - z = 8$$
$$-z = 8 - 13/2$$
$$-z = 3/2$$
$$z = -3/2$$

To check, we must substitute the values for x, y, and z into all three equations. The solution is $x = 6$, $y = 1/2$, and $z = -3/2$. ∎

Solving a System of Three Equations

1. Eliminate one variable from two equations using substitution or addition.
2. Solve the resulting system of two equations by using substitution or addition.
3. Solve for the remaining variable by substituting into one of the original equations.
4. Check the solution.

EXAMPLE 2   Solve the following system of equations:

$$x - y + z = -4 \qquad \text{(1)}$$
$$2x + y + 2z = -5 \qquad \text{(2)}$$
$$3x - y - z = -6 \qquad \text{(3)}$$

Solution   The variable we choose to eliminate is completely arbitrary, but one variable may be easier to eliminate than another.

Step 1: Eliminate One Variable. In this system, let us eliminate y from two equations by using addition. Adding equations (1) and (2) gives

$$\begin{array}{ll} x - y + z = -4 & \text{(1)} \\ \underline{2x + y + 2z = -5} & \text{(2)} \\ 3x + 3z = -9 & \end{array}$$

Adding equations (2) and (3) gives

$$\begin{array}{ll} 2x + y + 2z = -5 & \text{(2)} \\ \underline{3x - y - z = -6} & \text{(3)} \\ 5x + z = -11 & \end{array}$$

The resulting system of two equations is

$$3x + 3z = -9$$
$$5x + z = -11$$

Step 2: Solve the System of Two Variables. We solve this system using addition, and we eliminate z by multiplying the second equation by -3.

$$\begin{array}{lcl} 3x + 3z = -9 & \longrightarrow & 3x + 3z = -9 \\ \underline{-3(5x) + (-3)(z) = -3(-11)} & \longrightarrow & \underline{-15x - 3z = 33} \\ & & -12x = 24 \\ & & x = -2 \end{array}$$

Solving for z, we substitute $x = -2$ into $5x + z = -11$.

$$5(-2) + z = -11$$
$$z = -11 + 10$$
$$z = -1$$

Step 3: Solve for y. Next, we solve for y by substituting $x = -2$ and $z = -1$ into equation (1).

$$x - y + z = -4 \qquad \textbf{(1)}$$
$$(-2) - y + (-1) = -4$$
$$-3 - y = -4$$
$$-3 + 4 = y$$
$$1 = y$$

Step 4: Check.

$$x - y + z = -4 \qquad 2x + y + 2z = -5 \qquad 3x - y - z = -6$$
$$-2 - 1 - 1 = -4 \qquad -4 + 1 - 2 = -5 \qquad -6 - 1 + 1 = -6$$
$$-4 = -4 \qquad\qquad -5 = -5 \qquad\qquad -6 = -6$$

Since the answer checks, the solution is $x = -2$, $y = 1$, and $z = -1$. ∎

EXAMPLE 3 Solve the following system of equations:

$$x - 3y + 2z = 13 \qquad \textbf{(1)}$$
$$2x + y - 3z = -9 \qquad \textbf{(2)}$$
$$4x - 2y + z = 8 \qquad \textbf{(3)}$$

Solution *Step 1: Eliminate One Variable.* We can eliminate x by solving equation (1) for x and substituting that quantity into equations (2) and (3).

$$x = 13 + 3y - 2z \qquad \textbf{(1)}$$

$$2x + y - 3z = -9 \qquad \textbf{(2)}$$
$$2(13 + 3y - 2z) + y - 3z = -9$$
$$26 + 6y - 4z + y - 3z = -9$$
$$7y - 7z = -35$$

$$4x - 2y + z = 8 \qquad \textbf{(3)}$$
$$4(13 + 3y - 2z) - 2y + z = 8$$
$$52 + 12y - 8z - 2y + z = 8$$
$$10y - 7z = -44$$

The resulting system of equations is

$$7y - 7z = -35$$
$$10y - 7z = -44$$

Step 2: Solve the System of Two Variables. To solve this system of equations by addition, we can eliminate z by multiplying the second equation by -1.

$$7y - 7z = -35 \longrightarrow 7y - 7z = -35$$
$$\underline{-1(10y - 7z = -44)} \longrightarrow \underline{-10y + 7z = 44}$$
$$-3y \quad\;\; = 9$$
$$y = -3$$

Substituting $y = -3$ into the equation $7y - 7z = -35$ and solving for z gives

$$7(-3) - 7z = -35$$
$$-21 - 7z = -35$$
$$-7z = -14$$
$$z = 2$$

Step 3: Solve for x. We solve for x by substituting $y = -3$ and $z = 2$ into the expression for x in Step 1.

$$x = 13 + 3y - 2z$$
$$x = 13 + 3(-3) - 2(2)$$
$$x = 13 - 9 - 4$$
$$x = 0$$

The solution is $x = 0$, $y = -3$, and $z = 2$.　　　　　　　　　■ ■ ■

Three-by-Three Determinant

Evaluating a 3×3 determinant is a little more difficult than the 2×2 in the previous section. First, we will discuss a method of evaluation that **works only for a 3×3 determinant.** Then we will discuss expansion by cofactors, a method that can be used for all higher-order determinants. Probably the easier method of evaluating a 3×3 determinant is as follows:

■ Rewrite the first two columns of the determinant to the right of the existing determinant.

■ Multiply the six diagonals, three in the downward (left to right) direction and three in the upward direction (left to right). Diagonals in the downward direction are added, and diagonals in the upward direction are subtracted. The following determinant illustrates:

$$\begin{vmatrix} a & b & c \\ d & e & f \\ g & h & i \end{vmatrix} \begin{matrix} a & b \\ d & e \\ g & h \end{matrix} = aei + bfg + cdh - gec - hfa - idb$$

EXAMPLE 4　Evaluate the following determinant:

$$\begin{vmatrix} 1 & 3 & -2 \\ 4 & -1 & -3 \\ 2 & 5 & 2 \end{vmatrix}$$

Solution　First, we rewrite the first two columns of the determinant to the right of the existing determinant.

$$\begin{vmatrix} 1 & 3 & -2 \\ 4 & -1 & -3 \\ 2 & 5 & 2 \end{vmatrix} \begin{matrix} 1 & 3 \\ 4 & -1 \\ 2 & 5 \end{matrix}$$

Then we multiply the six diagonals.

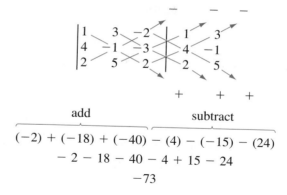

$$\underbrace{(-2) + (-18) + (-40)}_{\text{add}} \underbrace{- (4) - (-15) - (24)}_{\text{subtract}}$$

$$- 2 - 18 - 40 - 4 + 15 - 24$$

$$-73$$

The value of the determinant is -73.

■■■

LEARNING HINT ✦ In multiplying the diagonals of a 3×3 determinant, draw the diagonals one at a time as you multiply.

EXAMPLE 5 Evaluate the following determinant:

$$\begin{vmatrix} 3 & 0 & -2 \\ -1 & 6 & 4 \\ 5 & -3 & 1 \end{vmatrix}$$

Solution We rewrite the first two columns and multiply the diagonals.

$$\begin{vmatrix} 3 & 0 & -2 \\ -1 & 6 & 4 \\ 5 & -3 & 1 \end{vmatrix} \begin{matrix} 3 & 0 \\ -1 & 6 \\ 5 & -3 \end{matrix}$$

$$D = (18) + (0) + (-6) - (-60) - (-36) - (0)$$

$$= 18 - 6 + 60 + 36$$

$$= 108$$

■■■

Minors

Before evaluating a determinant using expansion by cofactors, we must define *minor*. The **minor** of a particular element is *the determinant that remains after the row and the column that contain the element have been deleted.* For example, in the following determinant, the minor of e is the determinant remaining after we have deleted the row and the column that contain the element e:

delete this column
↓
$$\begin{vmatrix} a & b & c \\ d & e & f \\ g & h & i \end{vmatrix}$$ ← delete this row

The minor for the element e is the determinant

$$\begin{vmatrix} a & c \\ g & i \end{vmatrix}$$

EXAMPLE 6 Determine the minor for the element k in the following determinant:

$$\begin{vmatrix} a & f & m \\ t & d & x \\ c & k & g \end{vmatrix}$$

Solution If we delete the row and the column containing k, we have

$$\begin{vmatrix} a & f & m \\ t & d & x \\ c - k - g \end{vmatrix}$$

The minor of the element k is

$$\begin{vmatrix} a & m \\ t & x \end{vmatrix}$$

∎∎∎

Cofactors

The **cofactor** of an element is *its minor with a sign attached.* To obtain the sign diagram, start in the upper left corner of the determinant with a plus sign, and alternate signs in both the vertical and the horizontal direction (*not* diagonal). The following sign diagram is for a 4 × 4 determinant:

$$\begin{vmatrix} + & - & + & - \\ - & + & - & + \\ + & - & + & - \\ - & + & - & + \end{vmatrix}$$

You can expand the sign diagram to any size by continuing to alternate signs.

LEARNING HINT ✦ If the sum of the row number and the column number is even, the sign of the minor is positive. If the sum is odd, the sign of the minor is negative. In Example 6, element k is in column 2 and row 3. Therefore, the sign of the minor is negative ($2 + 3 = 5$).

Expansion by Cofactors

To **expand a determinant by cofactors,** *choose any column or row* about which to expand. The determinant is the sum of the product of each element in the chosen row or column and its cofactor.

EXAMPLE 7 Evaluate the following determinant by using expansion by cofactors:

$$\begin{vmatrix} 2 & -4 & 3 \\ -1 & 5 & -2 \\ 7 & -8 & 1 \end{vmatrix}$$

Solution   The value of the determinant will be the same no matter which row or column we choose. Thus, if a row or column has smaller numbers or zero elements, choose it because the calculations are easier. Let us expand this determinant using the third column. Expanding about the third column gives

column elements

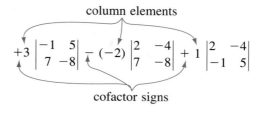

cofactor signs

$$D = +3(8 - 35) + 2[-16 - (-28)] + 1(10 - 4)$$
$$= 3(-27) + 2(12) + 1(6)$$
$$= -81 + 24 + 6$$
$$= -51$$

■ ■ ■

EXAMPLE 8   Use Cramer's rule to solve the following system of equations:

$$x + y - z = 5$$
$$2x - 3y + 6z = -6$$
$$x + 4y - 3z = 9$$

Solution   Since these equations are in standard form, we need to calculate D, D_x, D_y, and D_z,

$$D = \begin{vmatrix} 1 & 1 & -1 \\ 2 & -3 & 6 \\ 1 & 4 & -3 \end{vmatrix} \begin{matrix} 1 & 1 \\ 2 & -3 \\ 1 & 4 \end{matrix}$$

$$D = 9 + 6 - 8 - 3 - 24 + 6$$
$$D = -14$$

$$D_x = \begin{vmatrix} 5 & 1 & -1 \\ -6 & -3 & 6 \\ 9 & 4 & -3 \end{vmatrix} \begin{matrix} 5 & 1 \\ -6 & -3 \\ 9 & 4 \end{matrix}$$

$$D_x = 45 + 54 + 24 - 27 - 120 - 18$$
$$D_x = -42$$

$$D_y = \begin{vmatrix} 1 & 5 & -1 \\ 2 & -6 & 6 \\ 1 & 9 & -3 \end{vmatrix} \begin{matrix} 1 & 5 \\ 2 & -6 \\ 1 & 9 \end{matrix}$$

$$D_y = 18 + 30 - 18 - 6 - 54 + 30$$
$$D_y = 0$$

$$D_z = \begin{vmatrix} 1 & 1 & 5 \\ 2 & -3 & -6 \\ 1 & 4 & 9 \end{vmatrix} \begin{matrix} 1 & 1 \\ 2 & -3 \\ 1 & 4 \end{matrix}$$

$$D_z = -27 - 6 + 40 + 15 + 24 - 18$$
$$D_z = 28$$

The solution is given by

$$x = \frac{D_x}{D} = \frac{-42}{-14} = 3$$

$$y = \frac{D_y}{D} = \frac{0}{-14} = 0$$

$$z = \frac{D_z}{D} = \frac{28}{-14} = -2$$

The solution is $x = 3$, $y = 0$, and $z = -2$. ▪▪▪

Applications

EXAMPLE 9 The following system of equations results from applying Kirchhoff's laws to the circuit shown in Figure 6–17. Solve for the currents.

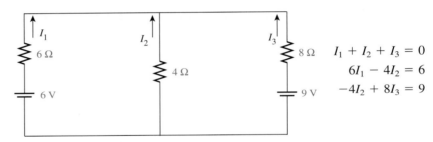

$$I_1 + I_2 + I_3 = 0$$
$$6I_1 - 4I_2 = 6$$
$$-4I_2 + 8I_3 = 9$$

FIGURE 6–17

Solution Let us solve the first equation for I_1 and substitute the result into the second equation. This results in a system of two equations containing I_2 and I_3.

$$I_1 = -I_2 - I_3$$
$$6(-I_2 - I_3) - 4I_2 = 6$$
$$-10I_2 - 6I_3 = 6$$

Solving the resulting system gives

$$\begin{aligned}
2(-10I_2 - 6I_3 = 6) &\longrightarrow & -20I_2 - 12I_3 &= 12 \\
-5(-4I_2 + 8I_3 = 9) &\longrightarrow & \underline{20I_2 - 40I_3} &= -45 \\
& & -52I_3 &= -33 \\
& & I_3 &\approx 0.6 \\
& & I_2 &\approx -1 \\
& & I_1 &\approx 0.4
\end{aligned}$$

The current is 0.4 A for I_1, 1 A for I_2, and 0.6 A for I_3. The negative sign for I_2 indicates that the direction for I_2 in Figure 6–17 should be reversed. ▪▪▪

EXAMPLE 10 The foot brake on a certain truck has the shape shown in Figure 6–18. An analysis of the system of forces yields the following set of equations:

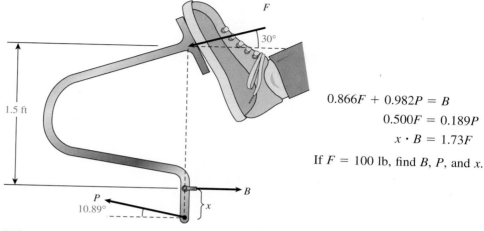

$$0.866F + 0.982P = B$$
$$0.500F = 0.189P$$
$$x \cdot B = 1.73F$$

If $F = 100$ lb, find B, P, and x.

FIGURE 6–18

Solution   Substituting the value for F into the equations produces:

$$86.6 + 0.982P = B$$
$$50.0 = 0.189P$$
$$x \cdot B = 173$$

Solving the second equation for P produces:

$$P = \frac{50.0}{0.189} = 264.55$$

Substituting this value for P into the first equation and solving for B produces:

$$B = 86.6 + (0.982)(264.55) = 346.39$$

Finally, x may be found from the third equation.

$$x = \frac{173}{346.39} = 0.499$$

The solution is $B = 346$ lb, $P = 265$ lb, and $x = 0.499$ ft.   ∎∎∎

6–3 EXERCISES

Solve the following systems of equations.

1. $x - y + 2z = -5$
   $2x + y + z = -1$
   $x + y - z = 3$

2. $x + y - z = 0$
   $x - y - z = 4$
   $x + 2y + z = -4$

3. $2x - y + 2z = 8$
   $x + 2y - z = -1$
   $x + y + z = 1$

4. $x - 2y + 2z = -3$
   $x + y - 2z = 4$
   $3x - y + z = 1$

5. $3x + y - z = 9$
   $x + 3y + 2z = 1$
   $2x - y + z = 1$

6. $4x + 2y - z = -7$
   $2x + y + z = -2$
   $x - y - 2z = -5$

7. $2x - y + 3z = 9$
   $x + 2y + 2z = 5$
   $3x - 2y + z = 10$

8. $x + 3y - z = -3$
   $2x - 3y + 2z = 6$
   $x + y + z = 3$

9. $x - 2y + 3z = -13$
$4x + y - 3z = 5$
$x + 5y - z = 16$

10. $2x - y + 3z = 13$
$x + 3y - 2z = -4$
$4x + 2y - z = 12$

11. $4x - 3y + z = -6$
$2x + 2y - 4z = 6$
$3x + y + 3z = -9$

12. $2x + y + 4z = 10$
$x - 3y + 2z = 12$
$3x - 2y + z = 17$

13. $x - 2z = 6$
$2y - 3z = 12$
$2x + 3y = 13$

14. $3x + 4y = 9$
$-2y + z = -7$
$2x - 3z = 1$

15. $2y - 3z = -7$
$x + y = 1$
$3x + 2z = 6$

16. $2x + 3y = 8$
$x + 4z = -6$
$2y - z = 9$

17. $y + 4z = 9$
$2x + 3z = 2$
$x - 3y = -5$

18. $3x - y = -17$
$x + 3z = -5$
$2y - z = 4$

19. $2x + y = -7$
$3z - x = -5$
$2y + z = -1$

20. $y - 4z = -6$
$2x - 3y = 18$
$x + 2y = 2$

21. $2x + z = -6$
$x + 2y = 9$
$3y + z = 18$

22. $4x + 2y = -6$
$x - 5z = -5$
$6z - 2y = 0$

23. $6x - 4y = -8$
$z - 3x = 6$
$2y + z = -2$

24. $z + x = 4$
$3y + 4z = 1$
$2y - x = -5$

25. $2x - 3y = 2$
$4z + 6y = -2$
$5z + 4x = 2$

26. $4x - 3z = -4$
$2y + 5z = 5$
$y - 8x = 2$

27. $5y + 2z = -3$
$4x - z = 2$
$2x - 10y = -2$

28. $3x + 2y = 2$
$4x - 3z = 2$
$4y - 6z = -6$

29. $\begin{vmatrix} 0 & 3 & -1 \\ 2 & 1 & 0 \\ 4 & 0 & 1 \end{vmatrix}$

30. $\begin{vmatrix} 1 & 4 & 2 \\ 0 & -1 & 3 \\ -3 & -2 & 0 \end{vmatrix}$

🖩 **31.** $\begin{vmatrix} 6.1 & -5.6 & 0 \\ 4.6 & 1.4 & 0 \\ -1 & 2.3 & 0 \end{vmatrix}$

32. $\begin{vmatrix} -1 & 3 & 0 \\ 0 & 4 & 2 \\ -3 & 0 & -2 \end{vmatrix}$

33. $\begin{vmatrix} 0 & -5 & 2 \\ -1 & 6 & 0 \\ 0 & 1 & -3 \end{vmatrix}$

34. $\begin{vmatrix} -4 & 1 & -3 \\ 1 & 2 & -1 \\ 0 & 6 & 0 \end{vmatrix}$

35. $\begin{vmatrix} 1 & 0 & -5 \\ -6 & 2 & 0 \\ 4 & -1 & 3 \end{vmatrix}$

36. $\begin{vmatrix} -5 & 2 & -1 \\ 1 & 0 & 0 \\ 4 & -2 & -3 \end{vmatrix}$

37. $\begin{vmatrix} 0 & 3 & 2 \\ -4 & -1 & 5 \\ 1 & -6 & 1 \end{vmatrix}$

38. $\begin{vmatrix} 7 & 2 & -9 \\ 0 & -1 & -3 \\ 3 & 8 & 1 \end{vmatrix}$

39. $\begin{vmatrix} 1 & 0 & -2 & 0 \\ 2 & 1 & 3 & -4 \\ 0 & 5 & -1 & 4 \\ -1 & 0 & -2 & 1 \end{vmatrix}$

40. $\begin{vmatrix} 1 & 2 & 2 & 1 \\ 0 & -1 & 0 & -3 \\ 3 & 1 & 1 & -4 \\ 4 & 3 & -3 & 0 \end{vmatrix}$

41. $\begin{vmatrix} 5 & 1 & -1 & 2 \\ 3 & -2 & 0 & 1 \\ 3 & 0 & -3 & 0 \\ 0 & 2 & -1 & 3 \end{vmatrix}$

42. $\begin{vmatrix} 0 & 4 & -3 & 2 \\ 1 & 2 & 2 & -3 \\ 1 & -4 & 2 & -1 \\ 0 & 1 & -3 & 2 \end{vmatrix}$

43. $\begin{vmatrix} -1 & 3 & 2 & -4 \\ 2 & 1 & -5 & -2 \\ 6 & 1 & 0 & -2 \\ -2 & 4 & -1 & 1 \end{vmatrix}$

E 44. When Kirchhoff's laws are applied to the circuit shown in Figure 6–19, the following equations result:

$$9 - 6I_1 - 5I_2 = 0$$
$$9 - 6I_1 - 10I_3 = 0$$
$$I_1 - I_2 - I_3 = 0$$

Solve this system of equations for the currents I_1, I_2, and I_3.

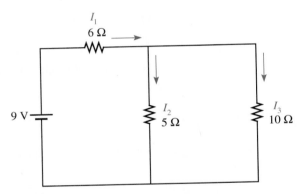

FIGURE 6–19

45. The perimeter of a triangle is 72 yd. The longest side is 10 yd more than twice the shortest side, and twice the middle side is 2 yd more than the longest side. Find the length of the sides of the triangle.

46. Find three numbers whose sum is 77 if the largest number is 2 more than twice the smallest number, and is 12 less than twice the middle number.

47. Jane has a collection of 48 coins consisting of nickels, dimes, and quarters valued at $6.20. How many of each coin does she have if the number of quarters is eight less than twice the number of dimes?

48. Find the equation of the parabola, $y = ax^2 + bx + c$, that passes through the points $(0, 5)$, $(-2, 21)$, and $(1, 6)$.

49. Find the equation of the parabola passing through the points $(0, 6)$, $(1, 5)$, and $(-3, -15)$.

E 50. Solve for the currents shown in Figure 6–20 by solving the following system of equations:

$$I_1 - I_2 + I_3 = 0$$
$$3I_1 + 4I_2 = 6$$
$$-4I_2 - 7I_3 = -3$$

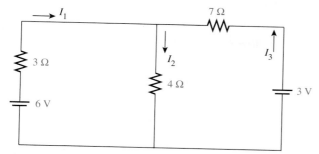

FIGURE 6–20

B 51. A portion of $4,325 is invested at 8% and another portion is invested at 6.5%. If the total interest is $311.35, how much is invested at each rate?

B 52. Lucia invests a portion of her $2,500 at 5.5% and another portion at 7%. If the total interest is $161.50, how much was invested at each rate?

E 53. Applying Kirchhoff's laws to the circuit shown in Figure 6–21 gives the following system of equations.

$$I_1 - I_2 + I_3 = 0$$
$$3I_1 + 9I_2 = 2$$
$$9I_2 + 5I_3 = 6$$

Solve for the currents.

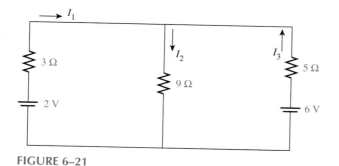

FIGURE 6–21

M 54. Solve the following system of equations to find F_1 and F_2 in Figure 6–22.

$$F_1 + 0.4F_2 = 583.0$$
$$0.9F_2 = 321.4$$

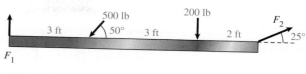

FIGURE 6–22

B 55. Pepe invests $1,500 in two investments paying 9.5% and 7.0%. If the total interest is $116.75, how much did he invest at 7.0%?

P 56. Find the tension in the cables shown in Figure 6–23 given the following system of equations:

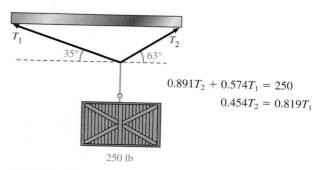

$$0.891T_2 + 0.574T_1 = 250$$
$$0.454T_2 = 0.819T_1$$

FIGURE 6–23

CS 57. The number of pages printed in one minute by three printers, x, y, and z, working together can be modeled by the following system of equations:

$$x + y + z = 18$$
$$2x - 3y + z = 0$$
$$x + y - z = 10$$

Find the speed of each printer.

C 58. Use Cramer's rule to find the width of a rectangular building site if the perimeter is 120 ft and the length is 9 ft less than twice the width.

C 59. Use Cramer's rule to find out the smaller acute angle of a right triangle if the angles of a triangle sum to 180° and twice the smallest angle is 21° more than the other acute angle.

C 60. Use Cramer's rule to find the largest of three numbers if the sum of the three numbers is 87 and the largest number is three times the smallest number. The second largest number is nine less than twice the smallest number.

B 61. Ray has invested $20,000 in investments paying 8%, 9.5%, and 11%. If the total annual interest is $2,005 and the sum of the amounts at 9.5% and 11% is four times the amount at 8%, how much does Ray have invested at 9.5%?

B 62. A theater owner sold 245 tickets, and receipts totaled $895.00. How many children's tickets did he sell if a child's ticket is $2 and an adult's ticket is $4.25?

6–4

APPLICATIONS

In this section, we will apply the methods of solving a system of equations to technical problems. The first application involves **torque,** which is defined as the tendency of a force to produce a change in rotational motion. Mathematically, torque is the product of a force and its lever arm, defined to be the perpendicular distance from the pivot point to the line of action of the force. For example, the torque produced by the 100-lb force in Figure 6–24 is

$$\text{torque} = (100 \text{ lb})(8 \text{ ft}) = 800 \text{ ft-lb}$$

FIGURE 6–24

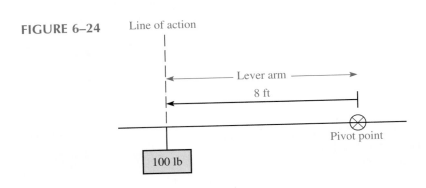

EXAMPLE 1 A 100-ft uniform bridge weighing 100 tons is supported at each end by a pillar. A 15-ton truck is 25 ft from one end, and a car weighing 1.5 tons is 40 ft from the opposite end of the bridge. Under these conditions, how much weight does each pillar support?

Solution First, we draw a figure to represent this problem, shown in Figure 6–25. Let

$$A = \text{weight supported by the left pillar}$$
$$B = \text{weight supported by the right pillar}$$

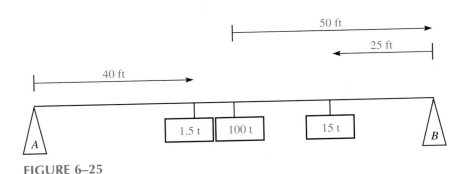

FIGURE 6–25

Next, we draw a force diagram showing the direction in which each force acts on the bridge. This force diagram is shown in Figure 6–26. Notice that the weight of the bridge, the truck, and the car act downward on the bridge, and the supports act upward on the bridge.

FIGURE 6–26

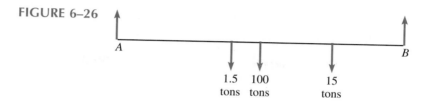

The First Condition of Equilibrium states that the sum of the forces acting upward or left on an object equals the sum of the forces acting downward or right on the object. Using the First Condition of Equilibrium and the force diagram, we can write the following equation:

$$\overbrace{A + B}^{\text{upward}} = \overbrace{1.5 + 100 + 15}^{\text{downward}}$$
$$A + B = 116.5$$

The Second Condition of Equilibrium states that the sum of the torques that tend to cause clockwise (CW) rotation equals the sum of the torques that tend to cause counterclockwise (CCW) rotation. To arrive at an equation, we must choose a pivot point, and we must determine the torque produced by each force and the direction of the force. Let us choose the center of the bridge as the pivot point (the choice of a pivot point is arbitrary). Figure 6–27 shows the pivot point, the lever arm for each force, and the direction of the resulting torque.

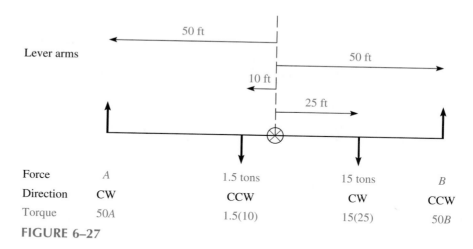

Force	A	1.5 tons	15 tons	B
Direction	CW	CCW	CW	CCW
Torque	$50A$	1.5(10)	15(25)	$50B$

FIGURE 6–27

Note that the 100-ton force produces no torque since its lever arm is zero. Since torque is the product of a force and its lever arm, applying the Second Condition of Equilibrium gives the following equation:

$$\overbrace{50(A) + 15(25)}^{\text{clockwise}} = \overbrace{1.5(10) + 50(B)}^{\text{counterclockwise}}$$
$$50A - 50B = -360$$

We must solve the following system to determine A and B:

$$A + B = 116.5 \tag{1}$$
$$50A - 50B = -360 \tag{2}$$

We use Cramer's rule to solve for A.

replace column for
A with constants

$$A = \frac{\begin{vmatrix} 116.5 & 1 \\ -360 & -50 \end{vmatrix}}{\begin{vmatrix} 1 & 1 \\ 50 & -50 \end{vmatrix}} = \frac{-5{,}825 - (-360)}{-50 - 50} = \frac{-5{,}465}{-100} = 54.65$$

coefficients of variables

$$A = 54.65$$

$116.5 \boxed{\times} 50 \boxed{\pm} \boxed{=} \boxed{-} 360 \boxed{\pm} \boxed{=} \boxed{\div} 100 \boxed{\pm} \boxed{=} \longrightarrow 54.65$

Solve for B by substituting $A = 54.65$ into equation (1).

$$A + B = 116.5 \qquad \textbf{(1)}$$
$$54.65 + B = 116.5$$
$$B = 61.85$$

One pillar supports 62 tons, and the other pillar supports 55 tons. ∎

LEARNING HINT ✦ To determine the direction of rotation of torque, hold one end of your pencil fixed at the pivot point and rotate the other end of the pencil in the direction of the arrow for the force.

EXAMPLE 2 From the plat of a triangular tract of land, a surveyor knows that the perimeter is 70 m. The lengths on the plat have been blurred, but the surveyor knows that the length of the longest side is 10 m less than the sum of the other two sides, and the longest side is 2 m less than twice the shortest side. Find the length of the three sides of the triangular tract.

Solution Let

$$x = \text{length of the longest side}$$
$$y = \text{length of the middle side}$$
$$z = \text{length of the shortest side}$$

The system of equations from the verbal statement is

The perimeter is 70.

$$x + y + z = 70 \qquad \textbf{(1)}$$

The longest side is 10 less than the sum of the other two sides.

$$x = y + z - 10 \qquad \textbf{(2)}$$

The longest side is 2 less than twice the shortest side.

$$x = 2z - 2 \qquad \textbf{(3)}$$

We use the substitution method to reduce this system of three equations to two. Substituting the quantity for x from equation (3) into equations (1) and (2) gives

$$(2z - 2) + y + z = 70 \qquad \textbf{(1)}$$
$$3z + y = 72$$

$$(2z - 2) = y + z - 10$$
$$z - y = -8 \qquad \textbf{(2)}$$

Then we solve the resulting system.

$$3z + y = 72$$
$$\underline{z - y = -8}$$
$$4z \quad\;\; = 64$$
$$z \quad\;\; = 16$$

We substitute $z = 16$ into the equation and solve for y.

$$z - y = -8$$
$$16 - y = -8$$
$$y = 24$$

We substitute $z = 16$ into equation (3) to solve for x.

$$x = 2z - 2$$
$$x = 2(16) - 2$$
$$x = 30$$

The triangular tract is 30 m by 24 m by 16 m.

■ ■ ■

EXAMPLE 3 A small textile company uses two different spinning machines, A and B, in the manufacture of cotton thread and wool yarn. Each unit of cotton thread requires 2 hours on machine A and 1 hour on machine B, while wool yarn requires 1 hour on machine A and 1 hour on machine B. Machine A can run a maximum of 22 hours per day, while machine B can run a maximum of 16 hours per day. How many units of cotton thread and wool yarn should the company make to keep the machines running at maximum capacity?

Solution Let

$$x = \text{amount of cotton thread produced}$$
$$y = \text{amount of wool yarn produced}$$

The system of equations is

$$\text{Machine } A\!: \quad 2x + y = 22 \qquad \textbf{(1)}$$
$$\text{Machine } B\!: \quad\;\; x + y = 16 \qquad \textbf{(2)}$$

Using the addition method to solve the system, we multiply equation (2) by -1.

$$\begin{array}{lcl} 2x + y = 22 & \longrightarrow & 2x + y = 22 \\ \underline{-1(x + y = 16)} & \longrightarrow & \underline{-x - y = -16} \\ & & x \quad\;\; = 6 \end{array}$$

Solve for y by substituting $x = 6$ into equation (2).

$$x + y = 16 \qquad \textbf{(2)}$$
$$6 + y = 16$$
$$y = 10$$

Therefore, the textile company should produce 6 units of cotton thread and 10 units of wool yarn per day to keep the machines running at maximum capacity. ∎∎∎

Kirchhoff's Laws

Kirchhoff's laws state the following:

1. The sum of the voltage rises and drops around a closed loop is zero.
2. The sum of the currents entering a node equals the sum of currents leaving the node.

A **closed loop** is any path that touches a point exactly once and returns to the starting point. A **node** is a point where three or more conductors (lines) are joined.

EXAMPLE 4 Find currents I_1, I_2, I_3 in the circuit given in Figure 6–28.

Solution We begin by applying Kirchhoff's current law. We must apply this law at enough nodes so that each current appears in at least one equation. There are nodes at b and e in Figure 6–28. From Kirchhoff's current law, the equation at node b is

$$I_1 = I_2 + I_3$$
$$I_1 - I_2 - I_3 = 0$$

Since each current appears in the previous equation, we do not need to write the equation for node e.

FIGURE 6–28

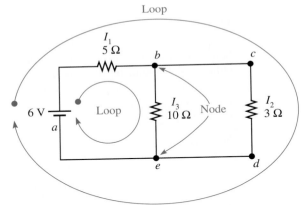

Next, we apply Kirchhoff's voltage law to closed loops until every current is included in at least one equation. Figure 6–28 has two closed loops: *abea* and *abcdea*. Applying the voltage law ($V = IR$) for both loops gives the following equations, respectively:

$$6 - 5I_1 - 10I_3 = 0$$
$$6 - 5I_1 - 3I_2 = 0$$

We must solve the following system of equations:

$$I_1 - I_2 - I_3 = 0$$
$$-5I_1 - 10I_3 = -6$$
$$-5I_1 - 3I_2 = -6$$

We solve this system by using Cramer's rule.

$$D = \begin{vmatrix} 1 & -1 & -1 \\ -5 & 0 & -10 \\ -5 & -3 & 0 \end{vmatrix} \begin{matrix} 1 & -1 \\ -5 & 0 \\ -5 & -3 \end{matrix}$$

$$D = 0 - 50 - 15 - 0 - 30 - 0$$

$$D = -95$$

$$I_1 = \frac{\begin{vmatrix} 0 & -1 & -1 \\ -6 & 0 & -10 \\ -6 & -3 & 0 \end{vmatrix} \begin{matrix} 0 & -1 \\ -6 & 0 \\ -6 & -3 \end{matrix}}{-95}$$

$$I_1 = -78/-95 \approx 0.8 \text{ A}$$

$$I_2 = \frac{\begin{vmatrix} 1 & 0 & -1 \\ -5 & -6 & -10 \\ -5 & -6 & 0 \end{vmatrix} \begin{matrix} 1 & 0 \\ -5 & -6 \\ -5 & -6 \end{matrix}}{-95}$$

$$I_2 = -60/-95 \approx 0.6 \text{ A}$$

$$I_3 = \frac{\begin{vmatrix} 1 & -1 & 0 \\ -5 & 0 & -6 \\ -5 & -3 & -6 \end{vmatrix} \begin{matrix} 1 & -1 \\ -5 & 0 \\ -5 & -3 \end{matrix}}{-95}$$

$$I_3 = -18/-95 \approx 0.2 \text{ A}$$

■ ■ ■

Partial Fractions

In calculus, it is sometimes necessary to break an algebraic fraction into partial fractions to integrate it. In Chapter 2, we combined several algebraic fractions into a single fraction. For example, if we add two fractions such as $\dfrac{4}{x + 1}$ and $\dfrac{3}{x + 3}$, we get

$$\frac{4}{x + 1} + \frac{3}{x + 3} = \frac{4(x + 3) + 3(x + 1)}{(x + 1)(x + 3)} = \frac{7x + 15}{(x + 1)(x + 3)}$$

Reversing this process to *transform a single fraction into several separate fractions* is called **decomposing into partial fractions.**

EXAMPLE 5 Decompose the following into partial fractions:

$$\frac{-4x + 23}{(2x - 1)(x + 3)}$$

Solution To decompose this fraction, we must write the fraction as a partial fraction of the form

$$\frac{\text{constant}}{\text{linear factor}}$$

Since there are two linear factors in the denominator of this fraction, we have

$$\frac{-4x + 23}{(2x - 1)(x + 3)} = \frac{A}{2x - 1} + \frac{B}{x + 3}$$

To simplify this equation, we eliminate fractions by multiplying the equation by the lowest common denominator, $(2x - 1)(x + 3)$.

$$-4x + 23 = A(x + 3) + B(2x - 1)$$
$$-4x + 23 = Ax + 3A + 2Bx - B \qquad \text{eliminate parentheses}$$
$$-4x + 23 = (A + 2B)x + (3A - B) \qquad \text{combine like terms}$$

For this equation to be satisfied, the coefficient of the same variable must be equal. Therefore,

$$-4x + 23 = (A + 2B)x + (3A - B)$$
$$A + 2B = -4$$
$$3A - B = 23$$

To decompose the original fraction into partial fractions, we solve the resulting system of equations. We solve this system by the addition method.

$$\begin{array}{ll} A + 2B = -4 \longrightarrow & A + 2B = -4 \\ 2(3A - B = 23) \longrightarrow & 6A - 2B = 46 \\ \hline & 7A \qquad = 42 \\ & A \qquad = 6 \end{array}$$

$$A + 2B = -4$$
$$6 + 2B = -4$$
$$2B = -10$$
$$B = -5$$

Decomposing the fraction into partial fractions gives

$$\frac{-4x + 23}{(2x - 1)(x + 3)} = \frac{6}{2x - 1} - \frac{5}{x + 3}$$

∎∎∎

6–4 EXERCISES

B 1. On an assembly line it takes 2 hours to complete two tasks. One task takes 2/3 hour more than the other. Use a system of equations to find how long each task requires.

2. A pile of quarters and nickels has a value of $9.00. Twice the number of quarters is four more than three times the number of nickels. How many of each coin are there?

CH 3. A chemist wants to make 400 ml of a 20% sulfuric acid solution using a 5% solution and a 30% solution. How much of each solution should she use?

P 4. A 12-ft scaffold of negligible weight is supported at each end. Two brick masons weighing 180 lb and 155 lb stand 3 ft and 5 ft, respectively, from opposite ends. If a pile of bricks weighing 100 lb is positioned at the center of the scaffold, how much weight does each end support?

5. Decompose the following into partial fractions:

$$\frac{16x + 3}{(2x - 2)(x + 3)}$$

CH 6. How much solder containing 60% tin should be mixed with a solder containing 40% tin to make 65 lb of solder containing 52% tin?

C 7. A surveyor is trying to find the dimensions of a rectangular tract. The surveyor knows that the perimeter of the tract is 122 yd and the length is 7 yd more than the width. Find the dimensions of the tract.

B 8. Helen invested $25,000, part at 10% and part at 12%, yielding an annual income of $2,640. How much did Helen invest at each rate?

M 9. Weights of 100 N, 150 N, and 75 N are placed on a board resting on two supports as shown in Figure 6–29. Neglecting the weight of the board, what are the forces exerted by the supports? Round your answer to tenths.

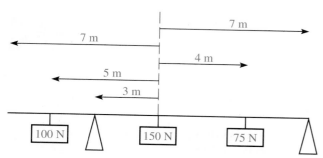

FIGURE 6–29

10. Decompose the following into partial fractions:

$$\frac{-22x + 47}{(3x - 5)(2x + 7)}$$

C 11. A builder wants to combine a concrete mixture that is 25% cement with a concrete mixture that is 40% cement to make a 100-yd^3 concrete mixture that is 35% cement. How much of each concrete mixture should he use?

E 12. Find currents I_1 and I_2 shown in Figure 6–30.

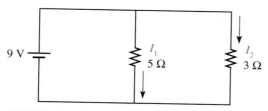

FIGURE 6–30

P 13. A plane traveled 1,200 miles with the wind in 3 hours and made the return trip against the wind in 4 hours. Determine the speed of the plane and the speed of the wind.

14. Decompose the following into partial fractions:

$$\frac{37x - 35}{(x + 1)(5x - 7)}$$

B 15. The cost of manufacturing a product consists of a fixed charge plus a variable charge for each item. If it costs $1,370 to produce 650 units and $1,640 to produce 800 units, what are the fixed and variable charges?

C 16. The perimeter of a rectangular tract of land is 130 yd. Find the length and width of the rectangle if the length is five more than twice the width.

17. Decompose the following into partial fractions:

$$\frac{-2x - 42}{(x + 3)(x - 1)}$$

M 18. A 50-ft uniform bridge weighing 100 tons is supported on each end. Two trucks weighing 7 tons and 5 tons are 10 ft and 18 ft, respectively, from opposite ends. How much weight does each end support?

E 19. Find currents I_1, I_2, and I_3 shown in Figure 6–31.

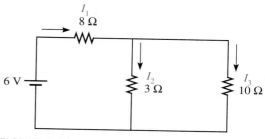

FIGURE 6–31

M 20. A system of equations for forces A and B shown in Figure 6–32 is

$$0.5A = 0.71B$$
$$0.87A = 100 + 0.71B$$

Find forces A and B.

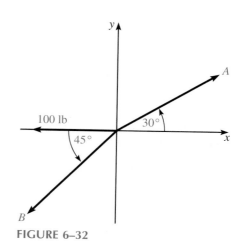

FIGURE 6–32

E 21. Find currents I_1, I_2, and I_3, in amperes, in Figure 6–33. Round your answer to tenths.

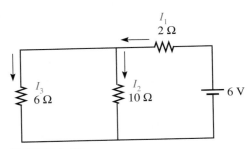

FIGURE 6–33

22. Find three numbers whose sum is 42. The largest number is two less than the sum of the other two numbers. Twice the smallest number is four less than the largest number.

23. Bill has a collection of dimes and quarters. If he has 30 coins worth $5.70, how many dimes and quarters does he have?

E 24. The following system of equations results from the circuit shown in Figure 6–34.

$$I_1 + I_2 + I_3 = 0$$
$$9I_1 - 5I_2 = 12$$
$$-5I_2 + 7I_3 = 6$$

Solve for the currents.

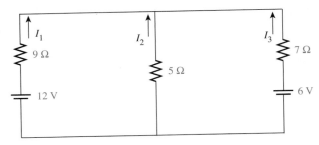

FIGURE 6–34

M 25. A nonuniform strut is used with a cable to support a 50 N crate, as shown in Figure 6–35. Solve the resulting system of equations to find the unknown forces.

$$0.6F_2 + 0.6F_1 = F_3 + 50$$
$$0.8F_1 = 0.8F_2$$
$$27.6F_1 + 2{,}850 = 34.3F_2$$

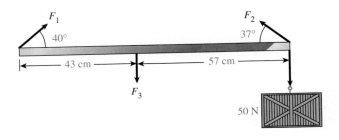

FIGURE 6–35

B 26. Susan wants to invest $22,500 in three investments. She invests the largest amount in an annuity at 10%, a second amount in a savings account at 5.5%, and a third amount in a certificate of deposit at 8.0%. If the total interest is $1,960.50 and the amount in the annuity is $1,300 more than the other two amounts combined, how much is invested at each rate?

E 27. In applying Kirchhoff's laws to the circuit shown in Figure 6–36, the following equations result.

$$I_1 + I_2 + I_3 = 0$$
$$15I_1 - 10I_2 = 0$$
$$6.0I_3 - 10I_2 = 9.0$$

Find the values for I_1, I_2, and I_3.

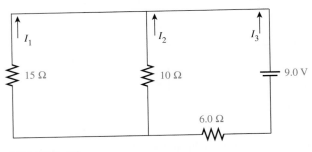

FIGURE 6–36

M 28. The equations resulting from an analysis of the forces acting on a steel girder used in constructing a skyscraper (see Figure 6–37) are as follows:

$$25.0 + F_3 = 0.620F_2 + 0.640F_1$$
$$0.770F_1 = 0.790F_2$$
$$3.54F_1 = 100 + 0.926F_2$$

Solve the equations for F_1, F_2, and F_3.

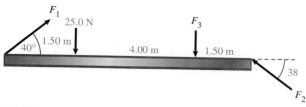

FIGURE 6–37

CS 29. One computer processes information at x bits per second, while a second computer processes y bits per second. If the first processes for 5 s and the second for 4 s, the total number of bits of information processed is 7,168. If the times are reversed, the total number of bits of information processed is 6,656. How fast is each computer processing information?

6–5

MATRIX ALGEBRA

In Section 6–3 we discussed matrices very briefly as they related to determinants and Cramer's Rule. In this section we will discuss matrix operations. The field of matrix algebra was invented by Arthur Cayley in England. In 1838, he began his studies at Cambridge where he was encouraged to pursue mathematics rather than enter the family business as merchants. He won a fellowship and taught at Cambridge for four years; however, its tenure was limited. Therefore, Cayley became a lawyer in 1849. It was not until 1863 that Cayley obtained another professorship in mathematics. It was during this time that he developed matrices, which remained a curiosity until Werner Heisenburg used them in the theory of quantum mechanics. Today, matrices are used in a wide variety of applications.

Matrix Equality

Recall from Section 6–2 that a matrix is a rectangular array of elements arranged in rows and columns. Generally, matrices are denoted by uppercase letters, such as **A** and **B**. For two matrices, **A** and **B**, to be equal, they must have the same dimension and each element in **A** must be equal to the corresponding element in **B**.

EXAMPLE 1 Solve for x, y, and z given that the following matrices are equal.

$$\mathbf{A} = \begin{bmatrix} x & 3 & 6 \\ y & 2 & z \end{bmatrix} \qquad \mathbf{B} = \begin{bmatrix} 5 & 3 & 6 \\ 7 & 2 & 10 \end{bmatrix}$$

Solution Since two matrices are equal only if their corresponding elements are equal, we can conclude that

$$x = 5$$
$$y = 7$$
$$z = 10$$

■■■

Matrix Addition

You can add (or subtract) two matrices of the same dimension by adding (or subtracting) corresponding elements.

EXAMPLE 2   Add the following matrices.

$$\begin{bmatrix} 2 & -5 \\ 6 & 3 \end{bmatrix} + \begin{bmatrix} -3 & 2 \\ 4 & -1 \end{bmatrix}$$

Solution   To add these matrices, we add the corresponding elements.
This gives

$$\begin{bmatrix} 2-3 & -5+2 \\ 6+4 & 3-1 \end{bmatrix}$$

$$\begin{bmatrix} -1 & -3 \\ 10 & 2 \end{bmatrix}$$

■■■

NOTE ✦ The addition (or subtraction) of matrices of different dimensions is undefined. You **cannot** add the following matrices.

$$\begin{bmatrix} 1 & 2 \\ 3 & 4 \end{bmatrix} \quad \begin{bmatrix} 3 & 4 & 5 \\ 6 & 7 & 8 \end{bmatrix}$$

Scalar Multiplication

We can multiply a matrix by a real number called a **scalar**. To multiply a scalar and a matrix, multiply each element of the matrix by the scalar.

EXAMPLE 3   For the following matrices, **A** and **B**, find (a) **2A**, (b) **3B**, and (c) **2A+3B**.

$$\mathbf{A} = \begin{bmatrix} -6 & 7 & 8 \\ 2 & -3 & 4 \end{bmatrix} \quad \mathbf{B} = \begin{bmatrix} 0 & 2 & -5 \\ 8 & -1 & 6 \end{bmatrix}$$

Solution   (a) To calculate **2A**, we multiply 2 and each element of matrix **A**.

$$2\mathbf{A} = \begin{bmatrix} 2(-6) & 2(7) & 2(8) \\ 2(2) & 2(-3) & 2(4) \end{bmatrix} = \begin{bmatrix} -12 & 14 & 16 \\ 4 & -6 & 8 \end{bmatrix}$$

(b) To calculate **3B**, we multiply 3 and each element of matrix **B**.

$$3\mathbf{B} = \begin{bmatrix} 3(0) & 3(2) & 3(-5) \\ 3(8) & 3(-1) & 3(6) \end{bmatrix} = \begin{bmatrix} 0 & 6 & -15 \\ 24 & -3 & 18 \end{bmatrix}$$

(c) To calculate **2A+3B**, we add corresponding elements of the matrices found in parts (a) and (b).

$$\begin{bmatrix} -12 & 14 & 16 \\ 4 & -6 & 8 \end{bmatrix} + \begin{bmatrix} 0 & 6 & -15 \\ 24 & -3 & 18 \end{bmatrix} = \begin{bmatrix} -12+0 & 14+6 & 16-15 \\ 4+24 & -6-3 & 8+18 \end{bmatrix}$$

$$\begin{bmatrix} -12 & 14 & 16 \\ 4 & -6 & 8 \end{bmatrix} + \begin{bmatrix} 0 & 6 & -15 \\ 24 & -3 & 18 \end{bmatrix} = \begin{bmatrix} -12 & 20 & 1 \\ 28 & -9 & 26 \end{bmatrix}$$

■■■

Matrix Multiplication

The matrix operations we have discussed so far have involved element-by-element manipulation. However, matrix multiplication involves a row-by-column multiplication. For the matrix multiplication of **A** and **B** to be defined, the number of columns of **A** must equal the number of rows of **B.** In other words,

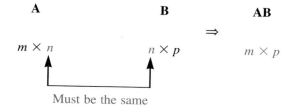

The following graphic shows matrix multiplication of two 2×2 matrices.

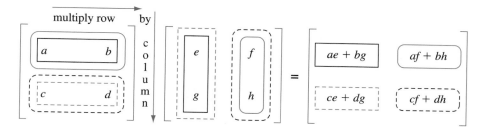

EXAMPLE 4   Multiply the following matrices to obtain **AB.**

$$\mathbf{A} = \begin{bmatrix} 3 & -1 \\ 5 & 2 \end{bmatrix} \qquad \mathbf{B} = \begin{bmatrix} -1 & 3 \\ 0 & -2 \end{bmatrix}$$

Solution   First, we must determine whether we can multiply the matrices.

We can multiply the matrices. Each step is shown as follows.

$$\begin{bmatrix} 3 & -1 \\ 5 & 2 \end{bmatrix}\begin{bmatrix} -1 & 3 \\ 0 & -2 \end{bmatrix} = \begin{bmatrix} 3(-1) + (-1)(0) & \\ & \end{bmatrix}$$

$$\begin{bmatrix} 3 & -1 \\ 5 & 2 \end{bmatrix}\begin{bmatrix} -1 & 3 \\ 0 & -2 \end{bmatrix} = \begin{bmatrix} -3 & 3(3) + (-1)(-2) \\ & \end{bmatrix}$$

$$\begin{bmatrix} 3 & -1 \\ 5 & 2 \end{bmatrix}\begin{bmatrix} -1 & 3 \\ 0 & -2 \end{bmatrix} = \begin{bmatrix} -3 & 11 \\ 5(-1) + 2(0) & \end{bmatrix}$$

$$\begin{bmatrix} 3 & -1 \\ 5 & 2 \end{bmatrix}\begin{bmatrix} -1 & 3 \\ 0 & -2 \end{bmatrix} = \begin{bmatrix} -3 & 11 \\ -5 & 5(3) + 2(-2) \end{bmatrix}$$

$$\begin{bmatrix} 3 & -1 \\ 5 & 2 \end{bmatrix}\begin{bmatrix} -1 & 3 \\ 0 & -2 \end{bmatrix} = \begin{bmatrix} -3 & 11 \\ -5 & 11 \end{bmatrix}$$

■ ■ ■

CAUTION ✦ Be sure to check that the matrix multiplication is defined by verifying that the column dimension of the first matrix equals the row dimension of the second matrix.

EXAMPLE 5 Find **AB** given

$$\mathbf{A} = \begin{bmatrix} -1 & 0 & 3 \\ 2 & -4 & -2 \end{bmatrix} \quad \mathbf{B} = \begin{bmatrix} -1 & 6 \\ 2 & -3 \\ 0 & 4 \end{bmatrix}$$

Solution Since **A** is 2×3 and **B** is 3×2, we can multiply the matrices, and the product will be 2×2.

$$\begin{bmatrix} -1 & 0 & 3 \\ 2 & -4 & -2 \end{bmatrix} \begin{bmatrix} -1 & 6 \\ 2 & -3 \\ 0 & 4 \end{bmatrix} = \begin{bmatrix} -1(-1) + 0(2) + (3)(0) & \end{bmatrix}$$

$$\begin{bmatrix} -1 & 0 & 3 \\ 2 & -4 & -2 \end{bmatrix} \begin{bmatrix} -1 & 6 \\ 2 & -3 \\ 0 & 4 \end{bmatrix} = \begin{bmatrix} 1 & -1(6) + 0(-3) + (3)(4) \end{bmatrix}$$

$$\begin{bmatrix} -1 & 0 & 3 \\ 2 & -4 & -2 \end{bmatrix} \begin{bmatrix} -1 & 6 \\ 2 & -3 \\ 0 & 4 \end{bmatrix} = \begin{bmatrix} 1 & 6 \\ 2(-1) + (-4)(2) + (-2)(0) & \end{bmatrix}$$

$$\begin{bmatrix} -1 & 0 & 3 \\ 2 & -4 & -2 \end{bmatrix} \begin{bmatrix} -1 & 6 \\ 2 & -3 \\ 0 & 4 \end{bmatrix} = \begin{bmatrix} 1 & 6 \\ -10 & 2(6) + (-4)(-3) + (-2)(4) \end{bmatrix}$$

$$\begin{bmatrix} -1 & 0 & 3 \\ 2 & -4 & -2 \end{bmatrix} \begin{bmatrix} -1 & 6 \\ 2 & -3 \\ 0 & 4 \end{bmatrix} = \begin{bmatrix} 1 & 6 \\ -10 & 16 \end{bmatrix}$$

■ ■ ■

NOTE ✦ Matrix multiplication is not commutative. This means that **AB** ≠ **BA.** For example, in Example 5, **BA** would be a 3×3 matrix.

Inverse Matrix

For an $n \times n$ square matrix, the **identity matrix I** has 1's on the diagonal and 0's for all other elements. Several examples of an identity matrix follow.

$$\begin{bmatrix} 1 & 0 \\ 0 & 1 \end{bmatrix} \quad \begin{bmatrix} 1 & 0 & 0 \\ 0 & 1 & 0 \\ 0 & 0 & 1 \end{bmatrix}$$

The **inverse** of an $n \times n$ matrix, denoted $\mathbf{A}^{-1}$, is an $n \times n$ matrix such that

$$\mathbf{AA}^{-1} = \mathbf{I}$$

This means that a matrix times its inverse yields the identity matrix. If a matrix **A** has an inverse, it is called **invertible** or **nonsingular.**

EXAMPLE 6 Variety that the following matrices are inverses by multiplying $\mathbf{AA}^{-1}$ and determine if it results in the identity matrix.

$$\mathbf{A} = \begin{bmatrix} 2 & -1 \\ 3 & 5 \end{bmatrix} \qquad \mathbf{A}^{-1} = \begin{bmatrix} \dfrac{5}{13} & \dfrac{1}{13} \\ \dfrac{-3}{13} & \dfrac{2}{13} \end{bmatrix}$$

Solution   To verify that $\mathbf{A}^{-1}$ is the inverse of $\mathbf{A}$, we multiply $\mathbf{AA}^{-1}$ and determine whether it results in the identity matrix.

$$\mathbf{AA}^{-1} = \begin{bmatrix} 2 & -1 \\ 3 & 5 \end{bmatrix} \begin{bmatrix} \dfrac{5}{13} & \dfrac{1}{13} \\ \dfrac{-3}{13} & \dfrac{2}{13} \end{bmatrix}$$

$$\mathbf{AA}^{-1} = \begin{bmatrix} 2\left(\dfrac{5}{13}\right) - 1\left(\dfrac{-3}{13}\right) & 2\left(\dfrac{1}{13}\right) - 1\left(\dfrac{2}{13}\right) \\ 3\left(\dfrac{5}{13}\right) + 5\left(\dfrac{-3}{13}\right) & 3\left(\dfrac{1}{13}\right) + 5\left(\dfrac{2}{13}\right) \end{bmatrix}$$

$$\mathbf{AA}^{-1} = \begin{bmatrix} 1 & 0 \\ 0 & 1 \end{bmatrix}$$

Since $\mathbf{AA}^{-1}$ results in the identity matrix, $\mathbf{A}^{-1}$ is the inverse matrix of $\mathbf{A}$.   ■ ■ ■

EXAMPLE 7   Find the inverse of

$$\mathbf{A} = \begin{bmatrix} 1 & 3 \\ -2 & 5 \end{bmatrix}$$

Solution   To find the inverse of $\mathbf{A}$, we must find matrix $\mathbf{X}$ that when multiplied by $\mathbf{A}$ gives the identity matrix $\mathbf{I}$ ($\mathbf{AX} = \mathbf{I}$).

$$\overset{\mathbf{A}}{\begin{bmatrix} 1 & 3 \\ -2 & 5 \end{bmatrix}} \overset{\mathbf{X}}{\begin{bmatrix} x_{11} & x_{12} \\ x_{21} & x_{22} \end{bmatrix}} = \overset{\mathbf{I}}{\begin{bmatrix} 1 & 0 \\ 0 & 1 \end{bmatrix}}$$

The notation x_{11} means row 1, column 1, while x_{12} means row 1, column 2. If we multiply $\mathbf{AX} = \mathbf{I}$, we have

$$\begin{bmatrix} x_{11} + 3x_{21} & x_{12} + 3x_{22} \\ -2x_{11} + 5x_{21} & -2x_{12} + 5x_{22} \end{bmatrix} = \begin{bmatrix} 1 & 0 \\ 0 & 1 \end{bmatrix}$$

In order for these two matrices to be equal, corresponding elements must be equal. This gives

$$\begin{array}{cc} x_{11} + 3x_{21} = 1 & x_{12} + 3x_{22} = 0 \\ -2x_{11} + 5x_{21} = 0 & -2x_{12} + 5x_{22} = 1 \end{array}$$

Then, we solve these two systems of equations for x_{11}, x_{12}, x_{21}, and x_{22}.

$$
\begin{array}{c}
2(x_{11}) + 2(3x_{21}) = 2(1) \\
\underline{-2x_{11} + 5x_{21} = 0} \\
2x_{11} + 6x_{21} = 2 \\
\underline{-2x_{11} + 5x_{21} = 0} \\
11x_{21} = 2
\end{array}
\qquad
\begin{array}{c}
2(x_{12}) + 2(3x_{22}) = 0 \\
\underline{-2x_{12} + 5x_{22} = 1} \\
2x_{12} + 6x_{22} = 0 \\
\underline{-2x_{12} + 5x_{22} = 0} \\
11x_{22} = 1
\end{array}
$$

$$x_{21} = \frac{2}{11} \qquad\qquad x_{22} = \frac{1}{11}$$

$$-2x_{11} + 5\left(\frac{2}{11}\right) = 0 \qquad\qquad x_{12} + 3\left(\frac{1}{11}\right) = 0$$

$$-2x_{11} = \frac{-10}{11} \qquad\qquad x_{12} = \frac{-3}{11}$$

$$x_{11} = \frac{10}{22} = \frac{5}{11}$$

The inverse matrix is

$$\mathbf{A}^{-1} = \begin{bmatrix} \dfrac{5}{11} & \dfrac{-3}{11} \\ \dfrac{2}{11} & \dfrac{1}{11} \end{bmatrix} = \frac{1}{11}\begin{bmatrix} 5 & -3 \\ 2 & 1 \end{bmatrix} \qquad\qquad \blacksquare\blacksquare\blacksquare$$

The method for finding the inverse matrix shown in Example 7 applies to any size matrix. However, there is a shorter method for 2 × 2 matrices.

Inverse of a 2 × 2 Matrix

To find the inverse of a 2 × 2 matrix

$$\mathbf{A} = \begin{bmatrix} a & b \\ c & d \end{bmatrix}$$

use the following formula

$$\mathbf{A}^{-1} = \frac{1}{\mathbf{D}}\begin{bmatrix} d & -c \\ -b & a \end{bmatrix}$$

where $\mathbf{D}$ = determinant. If $\mathbf{D} = 0$, then $\mathbf{A}$ does not have an inverse.

Graphing Calculator

Most graphing calculators will perform the matrix operations discussed in this section.

EXAMPLE 8   Use a graphing calculator to perform the following operations.

(a) $\begin{bmatrix} -4 & 2 \\ 3 & 1 \end{bmatrix} + \begin{bmatrix} -2 & 5 \\ -1 & 2 \end{bmatrix}$

(b) $\begin{bmatrix} 6 & -8 \\ 2 & 4 \end{bmatrix}\begin{bmatrix} -3 & 2 \\ -1 & 5 \end{bmatrix}$

(c) $\begin{bmatrix} -3 & 2 \\ 4 & -6 \end{bmatrix}$ find its inverse.

Solution   (a) On the TI-92, you enter the matrices separated by ⊞ and press ⎣ENTER⎦. The matrices are entered as follows:

$$[-4, 2; 3, 1] \boxed{+} [-2, 5; -1, 2] \boxed{\text{ENTER}}$$

The TI screen is shown in Figure 6–38.

FIGURE 6–38

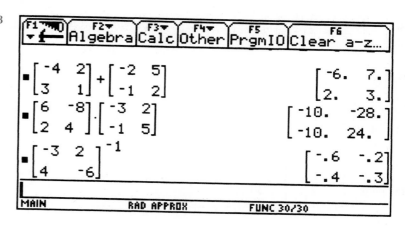

(b) The two matrices are entered as follows:

$$[6, -8; 2, 4] \boxed{\times} [-3, 2; -1, 5] \boxed{\text{ENTER}}$$

The TI screen is shown in Figure 6–38.

(c) Enter the matrix and press the $\boxed{x^{-1}}$ key and ⎣ENTER⎦. The solution is shown in Figure 6–38.

Application

EXAMPLE 9   Bill's Electronics sells three brands of televisions for $569, $450, and $389, respectively. If Bill sells 20, 17, and 30 of each brand of television, respectively, find the revenue for the sale of each television.

Solution   The revenue for each type of television can be represented by

$$\mathbf{A} = [569 \quad 450 \quad 389][20 \quad 17 \quad 30] = [11{,}380 \quad 7{,}650 \quad 11{,}670]$$

The revenue from the televisions is $11,380, $7,650, and $11,670, respectively.

6–5 EXERCISES

Add the following matrices. You may want to use a graphing calculator to check your answers.

1. $\begin{bmatrix} 6 & -4 \\ 3 & 1 \end{bmatrix} + \begin{bmatrix} -2 & 0 \\ 5 & 7 \end{bmatrix}$

2. $\begin{bmatrix} -1 & 5 \\ 3 & -2 \end{bmatrix} + \begin{bmatrix} 0 & -2 \\ 8 & -4 \end{bmatrix}$

3. $\begin{bmatrix} 9 & 8 \\ -1 & 6 \end{bmatrix} + \begin{bmatrix} -3 & 6 \\ 1 & 2 \end{bmatrix}$

4. $\begin{bmatrix} 10 & 4 \\ -6 & 2 \end{bmatrix} + \begin{bmatrix} 3 & -5 \\ 1 & 7 \end{bmatrix}$

5. $\begin{bmatrix} 3 & 6 \\ 0 & -1 \end{bmatrix} + \begin{bmatrix} 2 & -7 \\ 4 & 5 \end{bmatrix}$

6. $\begin{bmatrix} 1 & 5 \\ -4 & 2 \end{bmatrix} + \begin{bmatrix} 7 & 4 \\ 3 & -1 \end{bmatrix}$

7. $\begin{bmatrix} -8 & 5 & 3 \\ 6 & 0 & -1 \end{bmatrix} + \begin{bmatrix} 2 & 9 & -3 \\ 4 & -5 & 1 \end{bmatrix}$

8. $\begin{bmatrix} 7 & -3 \\ 6 & -4 \end{bmatrix} + \begin{bmatrix} 2 & 5 \\ 1 & 8 \end{bmatrix}$

9. $\begin{bmatrix} -3 & 1 & 10 \\ 7 & -2 & -1 \end{bmatrix} + \begin{bmatrix} 4 & -5 & 9 \\ 2 & 3 & 6 \end{bmatrix}$

10. $\begin{bmatrix} 3 & 0 & 10 \\ -1 & 6 & -4 \end{bmatrix} + \begin{bmatrix} 2 & 1 & -2 \\ 4 & -3 & 9 \end{bmatrix}$

11. $\begin{bmatrix} 9 & -4 & 5 \\ 0 & -7 & -2 \end{bmatrix} + \begin{bmatrix} -1 & 8 & 2 \\ 6 & 3 & 4 \end{bmatrix}$

12. $\begin{bmatrix} 4 & -5 & 9 \\ -1 & 8 & -4 \end{bmatrix} + \begin{bmatrix} 7 & -2 & 6 \\ 5 & 10 & 3 \end{bmatrix}$

13. $\begin{bmatrix} 2 & 1 & 5 \\ 8 & 3 & -2 \\ 4 & 7 & -6 \end{bmatrix} + \begin{bmatrix} 17 & -2 & 0 \\ -7 & 6 & 11 \\ 9 & -5 & 4 \end{bmatrix}$

14. $\begin{bmatrix} 1 & -2 & 4 \\ 10 & 6 & 3 \\ 5 & -8 & 7 \end{bmatrix} + \begin{bmatrix} 8 & -1 & 3 \\ -5 & 9 & 12 \\ 2 & -4 & 10 \end{bmatrix}$

15. $\begin{bmatrix} 5 & 3 & -6 \\ -4 & 7 & 2 \\ -2 & 4 & 1 \end{bmatrix} + \begin{bmatrix} -11 & 8 & 4 \\ -3 & 6 & 2 \\ 9 & -1 & -5 \end{bmatrix}$

16. $\begin{bmatrix} 7 & -3 & -10 \\ -8 & -1 & 8 \\ 12 & -9 & 0 \end{bmatrix} + \begin{bmatrix} 11 & -5 & 9 \\ -4 & 0 & -6 \\ 10 & -7 & 4 \end{bmatrix}$

Multiply the following matrices. You may want to use a graphing calculator to check your work.

17. $\begin{bmatrix} -1 & 4 \\ 6 & -3 \end{bmatrix} \begin{bmatrix} 2 & -7 \\ 5 & 1 \end{bmatrix}$

18. $\begin{bmatrix} 4 & 10 \\ -5 & 12 \end{bmatrix} \begin{bmatrix} 7 & -9 \\ -4 & 3 \end{bmatrix}$

19. $\begin{bmatrix} 0 & 6 \\ -8 & 5 \end{bmatrix} \begin{bmatrix} 4 & 2 \\ 1 & -3 \end{bmatrix}$

20. $\begin{bmatrix} 9 & -7 \\ -3 & 6 \end{bmatrix} \begin{bmatrix} 2 & -6 \\ -10 & 8 \end{bmatrix}$

21. $\begin{bmatrix} 1 & 5 \\ 9 & 2 \end{bmatrix} \begin{bmatrix} -3 & 7 \\ 4 & 8 \end{bmatrix}$

22. $\begin{bmatrix} -3 & 10 \\ 1 & -8 \end{bmatrix} \begin{bmatrix} 11 & -1 \\ -2 & 5 \end{bmatrix}$

23. $\begin{bmatrix} 1 & 6 & 9 \\ -7 & 1 & 2 \end{bmatrix} \begin{bmatrix} -3 & 6 \\ 2 & 4 \\ 5 & -1 \end{bmatrix}$

24. $\begin{bmatrix} 1 & 5 & -1 \\ -7 & 2 & 11 \end{bmatrix} \begin{bmatrix} -12 & 9 \\ 8 & 2 \\ -6 & -8 \end{bmatrix}$

25. $\begin{bmatrix} -2 & 5 & 1 \\ 3 & 0 & 6 \end{bmatrix} \begin{bmatrix} -3 & 2 \\ 8 & -7 \\ 1 & -4 \end{bmatrix}$

26. $\begin{bmatrix} 13 & -1 & -8 \\ 3 & -6 & 3 \end{bmatrix} \begin{bmatrix} 7 & 7 \\ 2 & -3 \\ -5 & 10 \end{bmatrix}$

27. $\begin{bmatrix} 8 & 1 & -7 \\ 5 & -6 & 2 \end{bmatrix} \begin{bmatrix} -6 & 3 \\ 9 & -2 \\ 10 & 4 \end{bmatrix}$

28. $\begin{bmatrix} 5 & -8 & 6 \\ -9 & 4 & 3 \end{bmatrix} \begin{bmatrix} -7 & 2 \\ 1 & -3 \\ -6 & 8 \end{bmatrix}$

29. $\begin{bmatrix} -3 & -8 & 2 \\ -4 & 1 & 5 \end{bmatrix} \begin{bmatrix} -2 & 0 \\ 3 & 9 \\ 4 & -1 \end{bmatrix}$

30. $\begin{bmatrix} 5 & 9 & 7 \\ 10 & -4 & -3 \end{bmatrix} \begin{bmatrix} 4 & -5 \\ 11 & 8 \\ 6 & 10 \end{bmatrix}$

Find the inverse of the following matrices.

31. $\begin{bmatrix} -1 & 2 \\ 3 & -4 \end{bmatrix}$

32. $\begin{bmatrix} -6 & 3 \\ 1 & 2 \end{bmatrix}$

33. $\begin{bmatrix} 4 & -5 \\ 6 & 1 \end{bmatrix}$

34. $\begin{bmatrix} 3 & -7 \\ -2 & 10 \end{bmatrix}$

35. $\begin{bmatrix} 2 & -3 \\ 7 & 1 \end{bmatrix}$

36. $\begin{bmatrix} 1 & -6 \\ -5 & 9 \end{bmatrix}$

37. $\begin{bmatrix} 3 & -1 & 4 \\ 2 & 0 & 8 \\ 6 & 2 & -3 \end{bmatrix}$

38. $\begin{bmatrix} -1 & 5 & 4 \\ 6 & -2 & 0 \\ 7 & -5 & 3 \end{bmatrix}$

39. $\begin{bmatrix} 1 & 5 & 4 \\ 6 & 2 & 3 \\ -3 & 0 & -1 \end{bmatrix}$

40. $\begin{bmatrix} 5 & 7 & -3 \\ 2 & 0 & 6 \\ 8 & -4 & 1 \end{bmatrix}$

CHAPTER SUMMARY

Summary of Terms

addition method (p. 221)

closed loop (p. 246)

cofactor (p. 236)

Cramer's rule (pp. 222, 224)

decomposing into partial fractions (p. 247)

dependent system (p. 217)

determinant (p. 222)

elements (p. 222)

expanding a determinant by cofactors (p. 236)

graphical solution (p. 210)

inconsistent system (p. 217)

inverse matrix (p. 254)

invertible (p. 254)

matrix (p. 222)

minor (p. 235)

node (p. 246)

nonsingular (p. 254)

scalar (p. 252)

solution (p. 210)

substitution method (p. 219)

system of linear equations (p. 210)

torque (p. 242)

x intercept (p. 211)

y intercept (p. 211)

CHAPTER REVIEW

Section 6–1

1. Is $(-2, 3)$ the solution to the following system of equations?

$$2x + y = -1$$
$$x - 4y = 8$$

Why or why not?

2. Is $(-4, 0)$ the solution to the following system of equations?

$$2x + 3y = -8$$
$$y = 3x + 12$$

Why or why not?

Solve the following systems of equations graphically.

3. $x - 6y = 6$
$4x + y = -1$

4. $x - 3y = -7$
$2x + y = 7$

5. $3x - 5y = -7$
$y = 2x + 7$

6. $x = 2y - 11$
$3x + y = 2$

7. $4x - y = 6$
$3x + 4y = 8$

8. $x - 2y = 6$
$5x + y = 8$

Section 6–2

Solve the following systems of equations by using the addition method.

9. $2x + y = 0$
$3x + 2y = 3$

10. $3x + y = 11$
$4x - 3y = -7$

11. $3x - 2y = -2$
$3y = -5x - 1$

12. $3x + 7y = 8$
$-9x = -21y + 10$

Solve the following systems of equations.

13. $y = 3x - 4$
$3x - 2y = 2$

14. $x = 4y - 1$
$2x - 5y = 4$

15. $x - 8y = 5$
$16y = 2x - 10$

16. $3x + 4y = 8$
$2x + 3y = 6$

Evaluate the following determinants.

17. $\begin{vmatrix} -6 & 2 \\ 4 & -1 \end{vmatrix}$

18. $\begin{vmatrix} a & c \\ 3 & -2 \end{vmatrix}$

B 19. Mr. Roberts invests \$15,000 in two savings accounts, one at 8.5% and the other at 9.75%. If he wants to earn \$1,387.50 in interest per year, how much should he invest at each rate?

20. Find two numbers such that the first number increased by 25 is 3 more than the second number, and twice the first number is 14 more than the second number.

Section 6–3

Solve each of the following systems of equations.

21. $x - 3y + z = -2$
$2x + y - z = -6$
$3x - y + 2z = -1$

22. $2x + 3y - z = 0$
$x - 3y + 4z = -9$
$-x + y + z = 5$

23. $4x - 3y + 2z = -12$
$x + 5y - 3z = 20$
$2x + 2y - z = 8$

24. $x + 2y + 3z = 3$
$3x - y - 2z = -6$
$2x + y + z = -1$

25. $x - 2y = -5$
$3z = 9$
$2x + 4z = 10$

26. $z - 3x = -14$
$4y + z = -4$
$2x - y = 4$

27. $4x - z = -1$
$6y + 7z = -11$
$2x - 3y = 9$

Evaluate the following determinants.

28. $\begin{vmatrix} 2 & 0 & 5 \\ -1 & 3 & 1 \\ 4 & -2 & 0 \end{vmatrix}$

29. $\begin{vmatrix} 8 & -6 & 2 \\ 0 & 3 & 4 \\ -1 & -2 & 1 \end{vmatrix}$

30. $\begin{vmatrix} 0 & -1 & 3 \\ 4 & -2 & 2 \\ 1 & 1 & 5 \end{vmatrix}$

31. $\begin{vmatrix} -3 & 2 & 1 \\ 0 & 4 & -2 \\ -1 & 2 & -1 \end{vmatrix}$

Using expansion by cofactors, evaluate the following determinants.

32. $\begin{vmatrix} -1 & 2 & 0 & 3 \\ 0 & 4 & -2 & 6 \\ 1 & 1 & 0 & 2 \\ 3 & -2 & 1 & 4 \end{vmatrix}$

33. $\begin{vmatrix} 3 & 2 & 0 & -4 \\ 1 & 0 & 1 & 6 \\ -1 & 6 & -2 & -3 \\ -2 & 3 & 0 & 2 \end{vmatrix}$

B 34. Fred has a collection of five-, ten-, and twenty-dollar bills. If the total value of the bills is \$175, and there are twice as many tens as fives and two more twenties than fives, how many of each bill does Fred have?

35. The sum of the angles of any triangles is 180°. In a certain triangle, the largest angle is 9° more than three times the smallest angle, and the middle-sized angle is 11° more than the smallest angle. Find the measure of each angle.

Section 6–4

M 36. Find the forces shown in Figure 6–39 if the resulting system of equations is

$$0.64A - 0.5B = 0$$
$$0.77A + 0.87B = 200$$

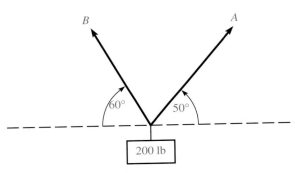

FIGURE 6–39

37. Find two numbers whose sum is 55 if the larger number is one more than twice the smaller number.

M 38. Find the weight supported by each end, A and B, shown in Figure 6–40.

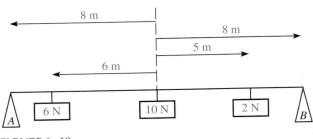

FIGURE 6–40

E 39. Find currents I_1, I_2, and I_3 shown in Figure 6–41.

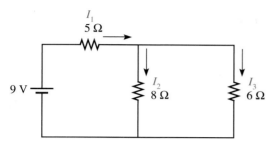

FIGURE 6–41

B 40. Jim has a collection of nickels, dimes, and quarters. He has a total of 24 coins worth $3.90. Find how many of each coin Jim has if he has the same number of nickels and dimes.

41. Decompose the following into partial fractions:

$$\frac{34x - 18}{8x^2 - 2x - 3}$$

Factor the denominator first.

Section 6–5

Perform the indicated matrix operations.

42. $\begin{bmatrix} -5 & 4 \\ -2 & 2 \end{bmatrix} + \begin{bmatrix} -1 & 3 \\ -7 & 6 \end{bmatrix}$

43. $\begin{bmatrix} 9 & 0 \\ 1 & -8 \end{bmatrix} + \begin{bmatrix} 6 & -7 \\ 2 & 3 \end{bmatrix}$

44. $\begin{bmatrix} 6 & -1 \\ -8 & 13 \end{bmatrix} + \begin{bmatrix} -11 & 8 \\ 10 & -20 \end{bmatrix}$

45. $\begin{bmatrix} 12 & -6 \\ 7 & 3 \end{bmatrix} + \begin{bmatrix} 9 & -10 \\ -2 & 14 \end{bmatrix}$

46. $\begin{bmatrix} -3 & 6 & 18 \\ 10 & -1 & -5 \end{bmatrix} + \begin{bmatrix} -3 & 10 & 12 \\ 17 & -4 & 6 \end{bmatrix}$

47. $\begin{bmatrix} 16 & 12 & 11 \\ -4 & -3 & -2 \end{bmatrix} + \begin{bmatrix} -7 & 8 & -9 \\ 4 & 2 & 15 \end{bmatrix}$

48. $\begin{bmatrix} 4 & 0 \\ 2 & -1 \end{bmatrix} \cdot \begin{bmatrix} 1 & -5 \\ -2 & 7 \end{bmatrix}$

49. $\begin{bmatrix} -1 & 6 \\ -4 & 3 \end{bmatrix} \cdot \begin{bmatrix} -5 & -2 \\ 1 & 3 \end{bmatrix}$

50. $\begin{bmatrix} 4 & -3 \\ -2 & 6 \end{bmatrix} \cdot \begin{bmatrix} -7 & -1 \\ -5 & 0 \end{bmatrix}$

51. $\begin{bmatrix} 7 & -2 \\ -4 & 1 \end{bmatrix} \cdot \begin{bmatrix} 0 & -8 \\ 5 & 3 \end{bmatrix}$

52. $\begin{bmatrix} 5 & -7 & 9 \\ 12 & 1 & -2 \end{bmatrix} \cdot \begin{bmatrix} -1 & 3 \\ -6 & 0 \\ -4 & 8 \end{bmatrix}$

53. $\begin{bmatrix} -3 & 4 & -1 \\ 6 & 0 & 10 \end{bmatrix} \cdot \begin{bmatrix} 7 & -2 \\ -3 & 2 \\ 11 & -5 \end{bmatrix}$

Find the inverse of the following matrices.

54. $\begin{bmatrix} 4 & -6 \\ -1 & 7 \end{bmatrix}$

55. $\begin{bmatrix} -7 & 3 \\ 0 & 2 \end{bmatrix}$

56. $\begin{bmatrix} 4 & 8 \\ 1 & -2 \end{bmatrix}$

57. $\begin{bmatrix} 5 & -3 \\ -8 & 9 \end{bmatrix}$

58. $\begin{bmatrix} 10 & 2 \\ -4 & -9 \end{bmatrix}$

59. $\begin{bmatrix} -10 & 6 \\ 11 & -5 \end{bmatrix}$

CHAPTER TEST

The number in parentheses refers to the appropriate learning objective given at the beginning of the chapter.

1. Use the addition method to solve the following system: (3)

$$3x + 2y = 4$$
$$4x + 5y = -11$$

2. Evaluate the following determinant: (5)

$$\begin{vmatrix} -1 & 3 & -2 \\ 2 & 0 & 4 \\ -3 & 1 & -1 \end{vmatrix}$$

3. Is the ordered pair $(-1, 0)$ the solution to the following system of equations: (1)

$$-6x + y = 6$$
$$3x + 7y = -3$$

Why or why not?

4. The sum of the angles of any triangle is $180°$. In a certain triangle, the two smallest angles are of equal measure and the largest angle is $20°$ more than the sum of the other two angles. Find the angles of the triangle.   (7)

5. Solve the following system of equations graphically: (2)

$$2x + 3y = -6$$
$$x - 4y = 13$$

6. Solve the following system of equations algebraically: (3)

$$2x + y - 3z = -5$$
$$x - 3y + 2z = -6$$
$$3x + 2y - z = 1$$

7. Add the matrices: (7)

$$\begin{bmatrix} 0 & 3 & -2 \\ 4 & 8 & 6 \end{bmatrix} + \begin{bmatrix} 7 & -4 & 2 \\ 1 & 5 & -6 \end{bmatrix}$$

8. Solve the following system of equations by using the substitution method: (3)

$$6x + y = 2$$
$$2y = 10 - 6x$$

9. A grocer wants to mix coffee costing \$1.99/lb with coffee costing \$3.49/lb to make 50 lb of a coffee blend selling for \$2.80/lb. How many pounds of each coffee should he use?   (7)

10. Multiply the following matrices: (7)

$$\begin{bmatrix} 1 & -12 \\ -10 & 2 \end{bmatrix} \begin{bmatrix} 4 & -5 \\ -3 & 8 \end{bmatrix}$$

11. Evaluate the following determinant by using expansion by cofactors: (5)

$$\begin{vmatrix} -1 & 3 & 0 & 5 \\ 0 & 2 & -1 & 0 \\ 3 & 0 & -2 & 6 \\ 0 & 1 & 3 & 2 \end{vmatrix}$$

12. Use Cramer's rule to solve the following for x: (6)

$$3x - 5y = 13$$
$$4x + 7y = 10$$

13. Find forces A and B in Figure 6–42. (7)

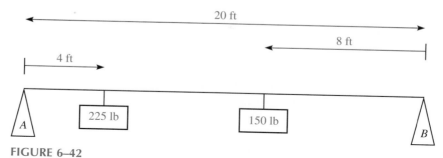

FIGURE 6–42

14. Find the inverse of the matrix. (7)

$$\begin{bmatrix} -10 & -4 \\ 6 & 5 \end{bmatrix}$$

15. Solve the following system of equations for x and y: (3)

$$2x + y = 7$$
$$6x + 3y = 14$$

16. Solve the following system of equations graphically: (2)

$$3x - 5y = -11$$
$$4y = 5x + 14$$

GROUP ACTIVITY

Rick's Sneaks wants to determine the relationship between price and demand for its Glow'n the Dark Sneakers. Rick has gathered the following data on the price and demand for these sneakers.

Price (x)	$20	$36	$40
Demand (y)	35	23	20

1. Determine the equation that relates the price and demand. (*Hint*: The equation is of the form $y = ax + b$. You know several values of x and y. Substitute several values and solve the resulting system of equations for a and b.)

2. Graph the data given and determine the demand when the price is $56.00.

3. Use the linear regression function on a graphing calculator to confirm your results in Steps 1 and 2.

4. Determine the price where the demand is equal to zero.

5. Explain why or why not this model is realistic.

SOLUTION TO CHAPTER INTRODUCTION

Figure 6–43 shows a representation of the two pens with y representing the length of the sides for one pen and x representing the length of the sides for the second pen. The equation representing the perimeter of the pens is

$$4x + 3y = 220$$

FIGURE 6–43

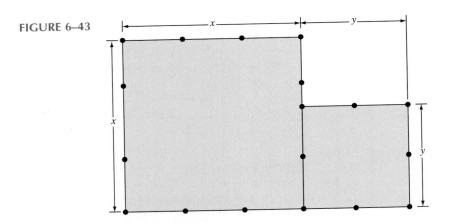

The equation representing the area enclosed by the two pens is

$$x^2 + y^2 = 2{,}425$$

Solving this system of equations using substitution gives

$$4x + 3y = 220$$

$$3y = 220 - 4x$$

$$y = \boxed{\dfrac{220 - 4x}{3}}$$

$$x^2 + \left(\dfrac{220 - 4x}{3}\right)^2 = 2{,}425$$

$$x^2 + \frac{48,400}{9} + \frac{16x^2}{9} - \frac{1,760x}{9} = 2,425$$

$$9x^2 + 48,400 + 16x^2 - 1,760x = 21,825$$

$$25x^2 - 1,760x + 26,575 = 0$$

Solving by the quadratic formula gives

$$x = \frac{1,760 \pm \sqrt{(1,760)^2 - 4(25)(26,575)}}{2(25)}$$

$$x = \frac{1,760 \pm \sqrt{440,100}}{50}$$

$$x \approx 48.4 \text{ or } x \approx 21.9$$

$$y = \frac{220 - 4(48.4)}{3} \approx 8.8$$

or

$$y = \frac{220 - 4(21.9)}{3} \approx 44.1$$

Rhonda can fence an area 48.4 ft and 8.8 ft or 21.9 ft and 44.1 ft.

Y

ou have been hired by Fred's Phones that produces cellular phones to estimate the profit for selling a particular model phone. Fred has estimated that the profit is given by the function $P = -96x^3 + 5,386x^2 - 250,000$ where the profit P is given in dollars and x is the advertising expense in tens of thousands of dollars. Using this model, you must find the smallest amount of advertising dollars that yields a profit of $1,800,000. (The solution to this problem is given at the end of the chapter.)

To solve a problem of this type, you must know how to solve a higher-degree equation. In this chapter we will develop a technique for solving problems that result in equations of degree three or more.

Learning Objectives

After you complete this chapter, you should be able to

1. Apply the Remainder Theorem (Section 7–1).

2. Use the Factor Theorem to determine whether a quantity is a factor of a given polynomial, a zero of a given function, or the root of a given equation (Section 7–1).

3. Perform synthetic division and use it to determine whether a quantity is a root, a zero, or a factor of a second quantity (Section 7–2).

4. Find the roots of a higher-degree equation (Sections 7–3 and 7–4).

5. Use Descartes's Rule of Signs to determine the maximum number of positive and negative roots of an equation (Section 7–4).

6. List all possible rational roots of a polynomial equation (Section 7–4).

7. Find irrational roots of a polynomial equation by linear approximation (Section 7–5).

Chapter 7

Solving Higher-Degree Equations

7–1

THE REMAINDER THEOREM AND THE FACTOR THEOREM

In previous chapters we discussed both linear and quadratic equations. Both types of equations are special cases of polynomial functions. A **polynomial function** is of the form

$$f(x) = a_0 x^n + a_1 x^{n-1} + \cdots + a_n$$

where $a_0, a_1, \cdots, a_n$ are constant, real numbers that serve as coefficients of the equation, and n is a nonnegative integer that determines the degree of the equation. In this chapter we will extend the techniques of solving equations to include equations of degree greater than two. Division will play a major role in solving these polynomial equations.

In polynomial division, the polynomial we are dividing by is called the **divisor;** the polynomial we divide into is called the **dividend;** and the answer is called the **quotient.** However, some polynomial division results in a **remainder.** If $f(x)$ is divided by $x - r$, we obtain a polynomial quotient $Q(x)$ and a remainder R. This means that

$$f(x) = Q(x)(x - r) + R$$

An important consequence of this formula is the Remainder Theorem. If $x = r$ in the above formula, then

$$f(r) = Q(r)(r - r) + R$$
$$\text{or}$$
$$f(r) = R$$

In other words, R, the remainder, is equal to the function value of $f(x)$ evaluated at $x = r$.

Remainder Theorem   The **Remainder Theorem** states that *if a polynomial $f(x)$ is divided by the linear factor $x - r$, then $f(r) = R$.* For example, if we divide $3x^2 - 2x + 5$ by $x - 3$, the remainder is 26. Similarly, if we evaluate $f(3)$, the answer is 26. Thus, $f(3)$ equals the remainder resulting from division by $x - 3$.

NOTE ✦ Recall from Chapter 4 that $f(r)$ is functional notation denoting the value of the function when r is substituted for the variable.

EXAMPLE 1   Verify the Remainder Theorem by dividing $x^2 - 6x - 8$ by $x + 1$ and then evaluating $f(-1)$.

Solution   First, dividing $x^2 - 6x - 8$ by $x + 1$ gives

$$
\begin{array}{r}
x - 7 \\
x + 1 \overline{) x^2 - 6x - 8} \\
\ominus \quad \ominus \\
x^2 + x \\
\hline
-7x - 8 \\
\oplus \quad \oplus \\
-7x - 7 \\
\hline
-1
\end{array}
$$

The remainder resulting from this division is -1. Evaluating $f(-1)$ (as we did in Chapter 4) gives

$$f(x) = x^2 - 6x - 8$$
$$f(-1) = (-1)^2 - 6(-1) - 8$$
$$f(-1) = 1 + 6 - 8$$
$$f(-1) = -1$$

Since the remainder from polynomial division and $f(r)$ are equal, the Remainder Theorem is verified for this example.

■ ■ ■

CAUTION ✦ For the Remainder Theorem to apply, the divisor must be in the form $x - r$.

EXAMPLE 2 Using the Remainder Theorem, determine the remainder when $2x^3 + 5x^2 - 6x + 8$ is divided by $x - 3$.

Solution We will find the remainder by evaluating $f(r)$. First, we must determine the value of r. Since $x - r = x - 3$, $r = 3$, and we evaluate $f(3)$.

$$f(x) = 2x^3 + 5x^2 - 6x + 8$$
$$f(3) = 2(3)^3 + 5(3)^2 - 6(3) + 8$$
$$f(3) = 54 + 45 - 18 + 8$$
$$f(3) = 89$$

Since $f(3) = 89$, then by the Remainder Theorem, the remainder is also 89.

■ ■ ■

Factor Theorem

An important special case of the Remainder Theorem is the **Factor Theorem,** which states that if $f(r) = R = 0$, then $(x - r)$ is a factor of the polynomial. This theorem is very important in solving higher-degree equations because it states that if the remainder is zero,

■ $(x - r)$ is a factor of the polynomial;

■ $x = r$ is a root (solution) of the polynomial equation; and

■ r is a zero (the x intercept of the graph) of the polynomial function.

The Factor Theorem can be shown as follows:

$$f(r) = 0$$

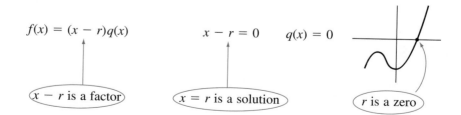

$$f(x) = (x - r)q(x) \qquad x - r = 0 \qquad q(x) = 0$$

$x - r$ is a factor $x = r$ is a solution r is a zero

EXAMPLE 3 Determine if $x + 4$ is a factor of the following polynomial:

$$f(x) = 2x^3 - 6x + 8$$

Solution   According to the Factor Theorem, $x + 4$ is a factor of $2x^3 - 6x + 8$ if and only if the remainder is zero. To determine the remainder, we evaluate $f(-4)$.

$$f(x) = 2x^3 - 6x + 8$$
$$f(-4) = 2(-4)^3 - 6(-4) + 8$$
$$f(-4) = -128 + 24 + 8$$
$$f(-4) = -96$$

Since the remainder is *not* zero, $x + 4$ is *not* a factor of $2x^3 - 6x + 8$. Furthermore, -4 is not a zero of the polynomial function, and -4 is not a root of the equation $2x^3 - 6x + 8 = 0$. ■■■

CAUTION ✦ Be sure to distinguish between the r in the expression $x - r$ and the expression $f(r)$. When determining zeros, roots, remainders, and factors by evaluating $f(r)$, always be sure that you have solved $x - r = 0$ for r.

EXAMPLE 4   Determine if -1 is a root of the following polynomial equation:

$$6x^4 + 5x^3 - x^2 + 3 = 0$$

Solution   To determine if -1 is a root of the polynomial equation, we evaluate $f(-1)$ and determine if the result is zero.

$$f(-1) = 6(-1)^4 + 5(-1)^3 - (-1)^2 + 3$$
$$f(-1) = 6 - 5 - 1 + 3$$
$$f(-1) = 3$$

Since $f(-1) \neq 0$, -1 is not a root of the polynomial equation. ■■■

EXAMPLE 5   Determine if 2 is a zero of the following polynomial function:

$$3x^3 - 8x^2 + 5x - 2$$

Solution   We must find $f(2)$ and determine if the result is zero.

$$f(2) = 3(2)^3 - 8(2)^2 + 5(2) - 2$$
$$f(2) = 24 - 32 + 10 - 2$$
$$f(2) = 0$$

Since $f(2) = 0$, 2 is a zero of the polynomial function. In addition, we also know that $x - 2$ is a factor of the polynomial $3x^3 - 8x^2 + 5x - 2$ and 2 is a root of the polynomial equation $3x^3 - 8x^2 + 5x - 2 = 0$. ■■■

7–1 EXERCISES

Verify the Remainder Theorem by using long division and evaluating $f(r)$.

1. $(x^3 - 5x^2 + x + 16) \div (x - 2)$

2. $(5x^3 + 6x^2 - 3x + 1) \div (x - 3)$

3. $(2x^3 - 7x^2 + x - 18) \div (x + 1)$

4. $(2x^3 - 7x + 3) \div (x + 4)$

5. $(12x^3 - 7x^2 + 16x + 12) \div (3x - 1)$

6. $(3x^3 + 4x^2 - x + 5) \div (x - 1)$

7. $(5x^3 - x + 8) \div (x + 2)$

8. $(10x^3 + 28x^2 - 41x + 21) \div (5x - 1)$

9 $(x^4 - x^3 + 7x - 8) \div (x + 3)$

10. $(3x^4 + 8x^2 - 7x + 5) \div (x + 1)$

11. $(2x^4 + 8x^2 - 3x + 1) \div (x + 2)$

13. $(x^4 + 6x^3 - 4x^2 + 8) \div (x - 7)$

13. $(15x^4 + 10x^3 + 21x^2 + 5x - 8) \div (3x + 2)$

14. $(6x^4 - 5x^3 - 12x^2 + 28x + 7) \div (6x - 5)$

Using the Remainder Theorem, find the remainder R.

15. $(7x^3 + 8x^2 - 3x + 5) \div (x - 3)$

16. $(7x^3 - 4x^2 + 6x - 2) \div (x + 1)$

17. $(x^3 + x^2 + x - 6) \div (x - 2)$

18. $(4x^4 + 5x^2 - 9x + 8) \div (x - 4)$

19. $(2x^3 - 7x^2 + 5x + 1) \div (x - 4)$

20. $(x^3 - 3x^2 + 9) \div (x - 3)$

21. $(12x^3 - 7x^2 + 9) \div (x - 3)$

22. $(2x^3 - 6x + 14) \div (x + 7)$

23. $(2x^4 - 5x^2 + 7x + 9) \div (x + 2)$

24. $(3x^4 + 2x^3 - x + 6) \div (x + 3)$

25. $(x^4 - 5x^3 + 4x^2 - 7) \div (x + 2)$

26. $(5x^4 - 4x^2 + 8) \div (x + 4)$

27. $(x^3 + 3x^2 - 8) \div (2x - 1)$

28. $(x^3 + 5x - 7) \div (2x + 3)$

29. $(3x^3 + 2x - 6) \div (3x - 2)$

30. $(2x^3 + 7x^2 + 9) \div (4x - 1)$

Use the Factor Theorem to determine if the second expression is a factor of the first.

31. $7x^3 - 6x^2 + 9x - 1; x + 1$

32. $x^3 - 3x^2 - 4x + 12; x + 2$

33. $3x^3 + 18x^2 - 47x + 8; x + 8$

34. $x^3 + 8x - 5; x - 1$

35. $9x^3 + 18x^2 - 13x + 2; 3x - 1$ (*Hint:* Let $r = 1/3$.)

36. $2x^3 - 6x^2 + 3; x + 3$

37. $x^4 - 2x^3 - 5x^2 + 17x + 14; x - 2$

38. $x^4 + 5x^3 + 3x^2 - 6x - 8; x + 4$

39. $2x^4 - 6x^3 + 8x^2 + 5x - 15; x - 3$

40. $3x^4 + x^3 - 5x + 8; x - 2$

41. $6x^3 - 3x^2 + 14x - 10; 2x - 1$

42. $2x^3 - 7x^2 - 13x + 3; 2x + 3$

43. $3x^3 - 13x^2 - 7x + 2; 3x + 2$

44. $4x^3 - 3x^2 - 6x + 18; 4x - 3$

Determine if the given number is a zero of the function $f(x)$.

45. $2x^3 - 7x + 8; -2$

46. $2x^3 + x^2 - 14x + 3; -3$

47. $x^3 + 3x^2 - 19x + 3; 3$

48. $6x^3 - 7; -2$

49. $3x^3 - 2x^2 + 2x + 7; -1$

50. $x^3 + 6x^2 + 5x - 12; -4$

51. $x^4 + x^3 - 30x^2 - 4x + 20; 5$

52. $x^4 - 1; -1$

53. $x^4 + 8x^2 - 7x + 6; 2$

54. $3x^4 - 7x^2 + 8x - 1; 1$

55. $6x^3 + x^2 - 22x + 7; 1/3$

56. $2x^4 - 5x^3 - 8x^2 + 24x - 10; 5/2$

57. $3x^4 - 5x^3 + 9x - 6; 1/2$

58. $x^3 + 2x^2 - 3x + 6; -1/3$

Determine if the given number is a root of the equation.

59. $5x^3 + 15x^2 - 6x - 18 = 0; -3$

60. $3x^3 + 5x^2 - 7 = 0; -1$

61. $2x^3 + 13x^2 + x - 30 = 0; -6$

62. $3x^3 + 5x^2 - 10x + 2 = 0; 1$

63. $x^3 - 4x^2 - 22x + 7 = 0; -7$

64. $2x^3 - 4x^2 + 7x - 2 = 0; 2$

65. $x^3 - 6x^2 + 8 = 0; -4$

66. $5x^3 + 9x - 11 = 0; -3$

67. $x^4 + 3x^3 - 4x^2 - 32 = 0; 2$

68. $3x^3 + 22x^2 + 18x - 8 = 0; -4/3$

69. $2x^3 - 5x^2 + 6x - 9 = 0; -1/2$

70. $2x^3 + 3x^2 - 11x + 3 = 0; 3/2$

7–2

SYNTHETIC DIVISION

In this section we will discuss a shortcut to polynomial long division called **synthetic division.** With a higher-degree polynomial, the computation to determine $f(r)$ is difficult. Generally, synthetic division is much easier than evaluating $f(r)$ or performing long division. We develop the method of synthetic division in the next example.

EXAMPLE 1   Divide $3x^3 + 5x^2 - x + 4$ by $x + 2$.

Solution

$$
\begin{array}{r}
3x^2 - x + 1 \\
x + 2 \overline{)\,3x^3 + 5x^2 - x + 4} \\
\ominus \quad\ \ominus \\
\underline{3x^3 + 6x^2} \\
-x^2 - x \\
\oplus \quad\ \oplus \\
\underline{-x^2 - 2x} \\
x + 4 \\
\ominus \quad\ \ominus \\
\underline{x + 2} \\
2
\end{array}
$$

In performing this division, we need to consider only the coefficients, and many of them are repeated. Let us rewrite the long division from above, omitting the variables and any repeated coefficients.

$$
\begin{array}{r}
3 \quad\ -1 \quad\ 1 \\
2 \overline{)\,3 \quad 5 \quad -1 \quad 4} \\
\underline{6} \\
-1 \\
\underline{-2} \\
1 \\
\underline{2} \\
2
\end{array}
$$

Then let us compress this problem by moving all the numbers below the dividend into two rows.

$$
\begin{array}{r}
③ \ \ ⊖1 \ \ ①\\
2 \overline{)\,3 \qquad 5 \quad -1 \quad 4} \\
\underline{6 \quad -2 \quad\ 2} \\
-1 \qquad 1 \quad ②
\end{array}
$$

All the coefficients of the quotient and the remainder appear in the bottom line except for the first one. Therefore, we repeat the first coefficient, 3, in the bottom line and omit the quotient above the dividend. Also, to avoid subtraction, we change the divisor of 2 to -2, the actual value of r, and change each sign in the second row.

$$
\begin{array}{r}
\underline{-2|} \ \ 3 \qquad 5 \quad -1 \quad 4 \\
\underline{-6 \quad\ 2 \ -2} \\
③ \ \ ⊖1 \ \ ① \ ②
\end{array}
$$

coefficients of quotient ⟵ remainder

The first three numbers in the bottom row represent the coefficients of the quotient, and the last number represents the remainder. To write the quotient in polynomial form, we write the variable in descending powers, starting with an exponent one less than the degree of the dividend. In this polynomial, the degree of the dividend is three; therefore, the degree of the quotient is two. Filling in descending powers of x starting with 2 gives the quotient as $3x^2 - x + 1$ with a remainder of 2. ∎

Example 2 shows the step-by-step procedure for synthetic division.

EXAMPLE 2 Use synthetic division to divide $2x^4 - 6x^3 + x^2 + 7x - 5$ by $x - 1$.

Solution First, write the dividend coefficients in descending powers and find the value of r.

$$\underline{1|} \quad 2 \quad -6 \quad 1 \quad 7 \quad -5$$

Bring down the first dividend coefficient.

$$\underline{1|} \quad 2 \quad -6 \quad 1 \quad 7 \quad -5$$
$$\overline{ 2}$$

Then multiply this coefficient by r, placing the product under the next dividend coefficient.

$$\underline{1|} \quad 2 \quad -6 \quad 1 \quad 7 \quad -5$$
$$ 2$$
$$\overline{ 2}$$
multiply

Add the result.

$$\underline{1|} \quad 2 \quad -6 \quad 1 \quad 7 \quad -5$$
$$ 2$$
$$\overline{ 2 \quad -4 \quad \text{add}}$$

Then continue this process until all the dividend coefficients have been used.

$$\underline{1|} \quad 2 \quad -6 \quad 1 \quad 7 \quad -5$$
$$ 2 \quad -4 \quad -3 \quad 4$$
$$\overline{ 2 \quad -4 \quad -3 \quad 4 \quad -1}$$

Write the quotient using the coefficients in the bottom row. Insert powers of x, starting with a power one less than the degree of the original polynomial.

$$2 \quad -4 \quad -3 \quad 4 \quad -1$$
$$\downarrow \quad \downarrow \quad \downarrow \quad \downarrow \quad \downarrow$$
$$x^3 \quad x^2 \quad x \quad x^0 \quad \text{remainder}$$

The quotient is $2x^3 - 4x^2 - 3x + 4$ with a remainder of -1. ∎ ■■■

Synthetic Division

1. Arrange the coefficients of the dividend in descending powers, placing a zero for any missing terms. Determine r and place it to the left of the dividend coefficients.

2. Bring down the first coefficient of the dividend.

3. Multiply r by this coefficient and place the product under the next dividend coefficient. Add the result.

4. Repeat Step 3 until all dividend coefficients have been used.

Function Values

Since the Remainder Theorem equates $f(r)$ and R, we can determine a function value by using synthetic division to determine the remainder.

EXAMPLE 3 Find $f(-3)$ if $f(x) = 7x^4 + 3x^3 - 5x + 6$.

Solution We arrange the dividend coefficients in descending powers and insert a 0 for the missing x^2 term.

$$7 \quad 3 \quad 0 \quad -5 \quad 6$$

Dividing gives

$$
\begin{array}{r|rrrrr}
-3 & 7 & 3 & 0 & -5 & 6 \\
   &   & -21 & 54 & -162 & 501 \\
\hline
   & 7 & -18 & 54 & -167 & 507 \\
\end{array}
$$

Because the remainder is 507, $f(-3) = 507$. You can check this result by evaluating $f(-3) = 7(-3)^4 + 3(-3)^3 - 5(-3) + 6$. ■■■

NOTE ✦ Graphically, $f(-3) = 507$ translates into the ordered pair $(-3, 507)$. Therefore, we could also use synthetic division to fill in a table of values for graphing.

Recall from Chapter 5 on quadratic equations that we could solve these equations by factoring. For example, $x^2 + x - 12 = 0$ factored into $(x + 4)(x - 3) = 0$, which means that $x + 4$ and $x - 3$ are factors of the quadratic equation and that $x^2 + x - 12 = (x + 4)(x - 3)$. Also recall that once the equation was factored, we set each factor equal to zero using the Zero Factor Property and then solved each factor for x. Graphically, these solutions, also called roots, are x intercepts where the curve crosses the x axis. The quadratic equation is a special type of polynomial function where the degree is two. The relationship among factors, roots, and zeros holds true for all polynomial functions.

Testing Factors

The Factor Theorem states that $x - r$ is a factor of a polynomial if the remainder is zero. We can also use synthetic division to test factors.

EXAMPLE 4 Use synthetic division to determine if $4x + 1$ is a factor of the following polynomial function:

$$12x^4 + 11x^3 - 2x^2 + 11x + 3$$

Solution The Factor Theorem is valid only for factors of the form $x - r$. To test a factor such as $ax - b$, write the factor as $a\left(x - \dfrac{b}{a}\right)$ and use synthetic division to test $\left(x - \dfrac{b}{a}\right)$. If the remainder is zero and the quotient is divisible by a, then $ax - b$ is a factor of the polynomial. We write $4x + 1$ as $4\left(x + \dfrac{1}{4}\right)$ and divide by $-1/4$.

$$
\begin{array}{r|rrrrr}
-\dfrac{1}{4} & 12 & 11 & -2 & 11 & 3 \\
   &   & -3 & -2 & 1 & -3 \\
\hline
   & 12 & 8 & -4 & 12 & 0 \\
\end{array}
$$

Since the remainder is zero and the quotient is divisible by 4, $4x + 1$ is a factor of $12x^4 + 11x^3 - 2x^2 + 11x + 3$. ■■■

Testing Zeros

From the previous section, we know that r is a zero of a polynomial function if the remainder is zero. We can determine the remainder by synthetic division.

EXAMPLE 5   Determine if $\frac{1}{2}$ is a zero of the function $6x^3 + 13x^2 - 26x + 9$.

Solution   To determine if $\frac{1}{2}$ is a zero, we perform synthetic division using $r = \frac{1}{2}$ and look at the remainder.

$$
\begin{array}{r|rrrr}
\frac{1}{2} & 6 & 13 & -26 & 9 \\
 & & 3 & 8 & -9 \\
\hline
 & 6 & 16 & -18 & 0
\end{array}
$$

Since the remainder is zero, $\frac{1}{2}$ is a zero of the function $6x^3 + 13x^2 - 26x + 9$.   ■ ■ ■

7–2 EXERCISES

Divide by using synthetic division.

1. $(x^3 - 2x^2 - 23x + 60) \div (x - 3)$

2. $(x^3 + 5x^2 - 3x + 8) \div (x + 2)$

3. $(2x^3 - 5x^2 + 3x + 8) \div (x + 1)$

4. $(6x^3 - x^2 + 2x - 7) \div (x - 4)$

5. $(9x^3 - 8x^2 + 5x + 10) \div (x - 2)$

6. $(x^3 + 9x^2 + 7x - 11) \div (x - 2)$

7. $(4x^3 - x^2 + 6x - 7) \div (x - 1)$

8. $(7x^3 + x^2 - 8x - 12) \div (x + 1)$

9. $(x^3 + 3x^2 - 10) \div (x - 2)$

10. $(2x^3 + 7x^2 - 9) \div (x + 3)$

11. $(6x^3 - 5x^2 + 16) \div (x + 4)$

12. $(9x^3 + 11x - 5) \div (x - 4)$

13. $(3x^4 - 2x^3 + 7x^2 - x + 9) \div (x - 5)$

14. $(x^4 - 6x^3 + 2x^2 + 11x - 8) \div (x - 1)$

15. $(x^4 - 7x^2 + 8) \div (x + 2)$

16. $(6x^4 - 7x + 10) \div (x - 6)$

17. $(x^4 + x^3 - 3x^2 + 2x - 6) \div (x + 3)$

18. $(2x^4 + 9x^3 + 5x - 3) \div (x + 5)$

19. $(7x^5 + 4x^2 - 16) \div (x - 2)$

20. $(3x^5 - 8x^2 + 4) \div (x + 2)$

21. $(x^5 + x^4 - 8x^2 + 6) \div (x + 4)$

22. $(x^6 + 4x^4 - x^3 + 8) \div (x + 4)$

23. $(3x^6 - 3x^3 + 8) \div (x - 2)$

24. $(2x^6 - 2x^2 + 9) \div (x + 1)$

Using synthetic division, determine if the second expression is a factor of the first.

25. $x^3 + 3x^2 - 16x + 12; x + 6$

26. $2x^3 - 9x^2 - 8x + 15; x - 5$

27. $3x^3 - 10x^2 - 9x + 4; x - 4$

28. $2x^3 + 7x^2 - 6x - 3; x + 3$

29. $x^3 + 6x^2 + 8x - 7; x + 7$

30. $x^3 - x^2 - 4x + 4; x - 1$

31. $6x^3 - 29x^2 - 17x + 60; x - 4$

32. $3x^3 + 7x^2 - 8x + 5; x - 3$

33. $2x^3 - 9x + 15; x - 5$

34. $x^3 + 8x - 9; x + 1$

35. $x^4 + 8x^3 - x^2 + 3x + 10; x + 2$

36. $x^4 + 2x^2 - 24; x - 2$

37. $x^4 + 3x^3 - 39x^2 - 187x - 210; x - 7$

38. $x^4 + 6x^2 - 12; x - 4$

39. $2x^4 + 3x^3 - 2x^2 - x - 6; 2x + 3$

40. $6x^4 - 2x^3 - 7x + 7; 3x - 1$

41. $3x^3 + 2x^2 - 7x + 2; 3x - 1$

42. $2x^4 - x^3 - 2x + 1; 2x - 1$

Using synthetic division, determine if the given number is a zero of the given function $f(x)$.

43. $3x^3 + 8x - 9; 3$

44. $5x^3 + 9x^2 + 7x + 18; -2$

45. $x^3 - 4x^2 - x + 4; 4$

46. $x^3 + 8x^2 - 4x + 11; -1$

47. $3x^4 - 2x^3 - 3x^2 + 8x - 4; 2/3$

48. $2x^4 - 7x^3 + 9x - 5; -1/2$

49. $2x^4 + 5x^3 - 2x - 3; -3/2$

50. $5x^4 - 2x^3 - 5x^2 + 7x - 2; 2/5$

7–3

ROOTS OF AN EQUATION

In this section we will discuss solving equations of degree three or more. In solving higher-degree equations, you will find it advantageous to know as much as possible about the number and type of roots. We will discuss several rules that will prove helpful in this area.

Number of Roots

The first rule, called the **Fundamental Theorem of Algebra,** states that *every polynomial equation has at least one root,* which may be a real or a complex number. By this theorem, the equation $f(x) = 0$ of degree n has at least one root r_1, and $(x - r_1)$ is a factor of $f(x)$. Thus, we can write

$$f(x) = (x - r_1)q_1(x) = 0$$

where $q_1(x)$ is a polynomial of degree $n - 1$. However, by the Fundamental Theorem, $q_1(x)$ has at least one root r_2. Thus, $f(x)$ can be written

$$f(x) = (x - r_1)(x - r_2)q_2(x) = 0$$

This process continues until $f(x)$ can be written in terms of only linear factors. As a result of the Fundamental Theorem of Algebra, we know the following:

■ If $f(x)$ is of degree n, there will be n linear factors.
■ Every polynomial equation of degree n has exactly n roots.

For example, this theorem states that the equation $6x^4 - 3x^2 + 5x - 1 = 0$ has exactly 4 linear factors and 4 roots. However, this theorem does not state that all n roots are different. When a root is repeated m times, the root is said to have a **multiplicity** of m. For example, if $f(x)$ can be written as

$$f(x) = (x - 1)(x + 5)^2(x - 8)^3 = 0$$

then -5 is a root of multiplicity two and 8 is a root of multiplicity three.

Reducing Equations to Quadratic Form

If we know all the roots but two in solving a higher-degree equation, we can use the methods of solving quadratic equations to find the remaining roots.

EXAMPLE 1   Solve the equation $x^3 - 2x^2 - 5x + 6 = 0$ if 3 is one of the roots.

Solution   Since the degree of this equation is three, there are three roots, one of which is the number 3. Performing synthetic division by 3 gives

$$\begin{array}{r|rrrr} 3 & 1 & -2 & -5 & 6 \\ & & 3 & 3 & -6 \\ \hline & 1 & 1 & -2 & 0 \end{array}$$

Then we can write the original equation as the product of the following factors:

$$x^3 - 2x^2 - 5x + 6 = (x - 3)(x^2 + x - 2)$$

The remaining two solutions are roots of the quadratic equation $x^2 + x - 2 = 0$. We can solve this quadratic equation by factoring.

$$x^2 + x - 2 = 0$$
$$(x + 2)(x - 1) = 0$$

The original equation can be written in terms of the following linear factors:

$$(x - 3)(x + 2)(x - 1) = 0$$

The roots of the equation are 3, -2, and 1.

■ ■ ■

EXAMPLE 2   Solve the equation $x^4 + 4x^3 - 7x^2 - 4x + 6 = 0$ given that -1 and 1 are roots. Round the solution to hundredths.

Solution   Since -1 and 1 are both roots of the equation, we use synthetic division twice to yield a quadratic equation.

$$
\begin{array}{r|rrrrr}
-1 & 1 & 4 & -7 & -4 & 6 \\
   &   & -1 & -3 & 10 & -6 \\
\hline
1 & 1 & 3 & -10 & 6 & 0 \\
  &   & 1 & 4 & -6 & \\
\hline
  & 1 & 4 & -6 & 0 &
\end{array}
$$

The resulting quadratic equation is $x^2 + 4x - 6 = 0$, which we can solve by using the quadratic formula.

$$x = \frac{-4 \pm \sqrt{16 - 4(1)(-6)}}{2}$$

$$x = \frac{-4 \pm 2\sqrt{10}}{2}$$

$$x = -2 \pm \sqrt{10}$$

$$2 \;\boxed{+/-}\; \boxed{+}\; 10 \;\boxed{\sqrt{\;}}\; \boxed{=} \longrightarrow 1.162$$

$$2 \;\boxed{+/-}\; \boxed{-}\; 10 \;\boxed{\sqrt{\;}}\; \boxed{=} \longrightarrow -5.162$$

The four roots of the equation are 1, -1, $-2 + \sqrt{10} \approx 1.16$, and $-2 - \sqrt{10} \approx -5.16$.

■ ■ ■

Complex Roots

We have briefly discussed complex numbers in Chapters 3 and 5. Recall that a complex number results from taking the square root of a negative number, for example, $\sqrt{-16}$. Previously we did not go any further than just recognizing this result as a complex number. However, in this chapter we want to obtain a solution. By definition

$$j = \sqrt{-1}$$

(The letter j is used in engineering and the letter i is used in mathematics. They mean the same thing.) To simplify $\sqrt{-16}$, we can write

$$\sqrt{-16} = \sqrt{16 \cdot -1}$$
$$= \sqrt{16}\sqrt{-1}$$
$$= 4j$$

The Fundamental Theorem of Algebra states that the roots of an equation may be complex numbers. *Complex roots will always occur in conjugate pairs.* For example, if $a + bj$ is a root, then $a - bj$ is also a root of the equation.

EXAMPLE 3   Solve the equation $2x^4 + 7x^3 + 6x^2 + 4x + 8 = 0$ if -2 is a root of multiplicity two.

Solution Using synthetic division, we divide by -2 twice.

$$
\begin{array}{r|rrrrr}
-2 & 2 & 7 & 6 & 4 & 8 \\
   &   & -4 & -6 & 0 & -8 \\
\hline
-2 & 2 & 3 & 0 & 4 & 0 \\
   &   & -4 & 2 & -4 & \\
\hline
   & 2 & -1 & 2 & 0 & \\
\end{array}
$$

The resulting quadratic equation is $2x^2 - x + 2 = 0$, which we solve by the quadratic formula.

$$x = \frac{1 \pm \sqrt{1 - 4(2)(2)}}{4}$$

$$x = \frac{1 \pm \sqrt{-15}}{4}$$

$$x = \frac{1 \pm j\sqrt{15}}{4}$$

The roots of the equation are $-2, -2, \dfrac{1 + j\sqrt{15}}{4}$, and $\dfrac{1 - j\sqrt{15}}{4}$.   ∎∎∎

EXAMPLE 4 Solve the equation $x^4 - 5x^3 + 12x^2 - 20x + 32 = 0$ if $2j$ is a root of the equation.

Solution Since $2j$ is a root, $-2j$ is also a root of the equation. We now use synthetic division twice, dividing by $2j$ and then $-2j$, to reduce to a quadratic equation. (Since $j = \sqrt{-1}$, squaring both sides of this equation gives $j^2 = -1$.)

$$
\begin{array}{r|rrrrr}
2j & 1 & -5 & +12 & -20 & +32 \\
   &   & 0 + 2j & -4 - 10j & 20 + 16j & -32 \\
\hline
-2j & 1 & -5 + 2j & 8 - 10j & 0 + 16j & 0 \\
   &   & 0 - 2j & 0 + 10j & 0 - 16j & \\
\hline
   & 1 & -5 & 8 & 0 & \\
\end{array}
$$

The resulting quadratic equation is $x^2 - 5x + 8 = 0$, which we solve by the quadratic formula.

$$x = \frac{5 \pm \sqrt{25 - 4(1)(8)}}{2}$$

$$x = \frac{5 \pm \sqrt{-7}}{2}$$

$$x = \frac{5 \pm j\sqrt{7}}{2}$$

The roots of the equation are $2j, -2j, \dfrac{5 + j\sqrt{7}}{2}$, and $\dfrac{5 - j\sqrt{7}}{2}$.   ∎∎∎

The Graphing Calculator

As in Chapters 4 and 5, we can extend the use of the graphing calculator to solving these higher-degree equations. Graphically, the solution to an equation is the x intercept or the x coordinate of the point where the line crosses the x axis.

EXAMPLE 5   Solve $x^3 - 2x^2 - 5x + 6 = 0$ (from Example 1) using the graphing calculator.

Solution   Set the viewing rectangle, enter the equation into the graphing calculator, and graph the equation. Figure 7–1 shows the graph of this equation. The roots are shown and agree with the solutions obtained algebraically in Example 1. ■■■

FIGURE 7–1

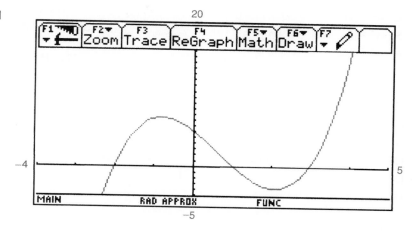

CAUTION ✦ You must remember that graphing yields only real number solutions.

EXAMPLE 6   Solve the equation $2x^4 + 7x^3 + 6x^2 + 4x + 8 = 0$ (Example 3) using the graphing calculator.

Solution   Follow the same procedure used in the previous example. The resulting graph is shown in Figure 7–2. ■■■

FIGURE 7–2

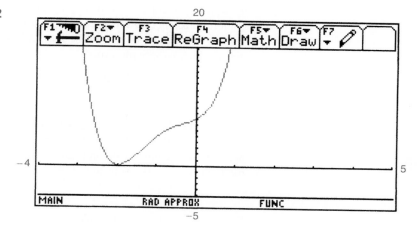

Note from the graph in Figure 7–2 that the only solution appears to be -2. This is one of the drawbacks to solving higher-degree equations graphically. First, the multiplicity of roots is not shown. Second, complex roots are not shown because they cannot be graphed using the rectangular coordinate system. We will discuss complex numbers in more detail in Chapter 14.

7–3 EXERCISES

Solve each equation given the roots indicated.

1. $x^3 + 2x^2 - 10x - 56 = 0$; 4
2. $x^3 + 6x^2 + 9x + 200 = 0$; -8
3. $6x^3 + 17x^2 - 104x + 60 = 0$; -6
4. $2x^3 + 11x^2 - x + 30 = 0$; -6
5. $4x^3 - x^2 - 11x + 8 = 0$; 1
6. $3x^3 + 16x^2 + 6x + 5 = 0$; -5
7. $2x^3 - 7x^2 + 19x - 14 = 0$; 1
8. $2x^3 - 2x^2 - 13x + 3 = 0$; 3
9. $x^3 - 12x^2 + 44x - 45 = 0$; 5
10. $4x^3 - 7x^2 + 8x - 5 = 0$; 1
11. $8x^3 + 82x^2 + 137x - 315 = 0$; -7
12. $x^3 + 9x^2 + 17x - 12 = 0$; -4
13. $x^4 + 4x^3 - 7x^2 - 34x - 24 = 0$; $-2, -1$
14. $x^4 - 2x^3 - 8x^2 + 19x - 6 = 0$; $-3, 2$
15. $x^4 + 2x^3 - 12x^2 - 40x - 32 = 0$; -2 is a root of multiplicity two.
16. $x^4 - 7x^3 + 17x^2 - 17x + 6 = 0$; 1 is a root of multiplicity two.
17. $x^4 + 2x^3 - 32x^2 + 30x + 63 = 0$; 3 is a root of multiplicity two.
18. $x^4 + x^3 - 46x^2 - 4x + 168 = 0$; 6, -7
19. $x^4 - 4x^3 - 13x^2 + 28x + 60 = 0$; 5, 3
20. $x^4 - 6x^3 + 5x^2 - 8x + 80 = 0$; 4 is a root of multiplicity two.
21. $15x^4 - 53x^3 - 125x^2 + 77x^{-10} = 0$; 5, -2
22. $x^4 + 3x^3 + 6x^2 + 3x + 5 = 0$; $-j$
23. $x^4 - x^3 - 9x^2 - 11x - 4 = 0$; 4, -1
24. $x^4 + 2x^3 - 14x^2 + 7x + 10 = 0$; 2, -5
25. $x^4 + 2x^3 - 10x^2 - 17x + 6 = 0$; $-2, 3$
26. $x^4 + 4x^3 - 53x^2 - 60x + 108 = 0$; 6, -2
27. $7x^3 - 36x^2 - 45x + 54 = 0$; 6
28. $x^4 - 3x^3 + 6x^2 - 3x + 5 = 0$; j
29. $x^4 - 6x^3 - 16x^2 + 150x - 225 = 0$; 3, -5
30. $x^4 + 7x^3 + 19x^2 + 63x + 90 = 0$; $3j$

7–4

RATIONAL ROOTS

In the previous section we solved higher-degree equations when we knew enough roots to obtain a quadratic equation. However, in reality we seldom know initially any of the roots of a higher-degree equation. Finding the rational roots of a higher-degree equation by trial and error can be a lengthy process, but in this section we will develop some guidelines to shorten this process. (Appendix A–1 explains that a rational number is any number that can be written in the form a/b where a and b are integers and $b \neq 0$.) We begin the discussion with a rule.

RULE ✦ If the coefficients of the equation

$$f(x) = a_0 x^n + a_1 x^{n-1} + a_2 x^{n-2} + \cdots + a_n = 0$$

are integers, then any rational roots are factors of a_n divided by factors of a_0 (a_n/a_0).

This rule does not guarantee the existence of rational roots, but it does limit the number of trials we must make in finding rational roots. Moreover, if $a_0 = 1$, we need try only the factors of a_n, the constant term.

EXAMPLE 1   List all possible rational roots of the following equation:

$$x^3 - x^2 - 14x + 24 = 0$$

Solution   Since the coefficient of the highest-degree term is 1, any rational roots must be factors of the constant term, 24. The possible rational roots are ± 1, ± 2, ± 3, ± 4, ± 6, ± 8, ± 12, and ± 24. We must include the plus-and-minus sign before each factor because, for example, both $+2$ and -2 could be roots of the equation.   ■■■

EXAMPLE 2 List all possible rational roots of the following equation:

$$4x^4 - 7x^3 + 9x^2 - x + 10 = 0$$

Solution The possible rational roots are all combinations of *factors of 10 divided by factors of 4*. The factors of 10 are ± 1, ± 2, ± 5, and ± 10, and the factors of 4 are ± 1, ± 2, and ± 4. The possible rational roots are

$$\pm 1, \quad \pm 2, \quad \pm 5, \quad \pm 10, \quad \pm 1/2, \quad \pm 5/2, \quad \pm 1/4, \quad \pm 5/4 \qquad \blacksquare\blacksquare\blacksquare$$

Descartes's Rule of Signs

An additional aid in limiting the number of possible rational roots is **Descartes's Rule of Signs,** which states that the number of positive real roots of $f(x) = 0$ cannot exceed the number of sign changes in the terms of $f(x)$, and the number of negative roots cannot exceed the number of sign changes in the terms of $f(-x)$.

EXAMPLE 3 Determine the maximum number of positive and negative roots for the following equation:

$$3x^3 + 4x^2 - 6x + 8 = 0$$

Solution To determine the maximum number of positive, real roots, we write the sign associated with each term of $f(x)$ as follows:

$$f(x) = 3x^3 + 4x^2 - 6x + 8$$
$$+ \quad + \quad - \quad +$$
$$① \quad ②$$

Since the sign changes twice, the number of positive, rational roots cannot exceed two. To determine the maximum number of negative roots, we determine the number of sign changes in $f(-x)$. To do so, we replace x with $-x$ in the equation.

$$f(x) = 3x^3 + 4x^2 - 6x + 8$$
$$f(-x) = 3(-x)^3 + 4(-x)^2 - 6(-x) + 8$$
$$f(-x) = -3x^3 + 4x^2 + 6x + 8$$

The signs are

$$- \quad + \quad + \quad +$$
$$①$$

Since there is one change, there is at most one negative, rational root. $\qquad \blacksquare\blacksquare\blacksquare$

Maximum and Minimum Values

One additional rule will help restrict the number of rational numbers that we must consider in finding the roots of $f(x) = 0$. If we divide $f(x)$ by a number and the quotient has only positive signs, that number is a maximum of the roots. Similarly, if we divide $f(x)$ by a number and the signs on the quotient alternate, that number is a minimum of the roots.

⌨ **EXAMPLE 4** Determine the roots of the following equation:

$$2x^4 + x^3 - 35x^2 - 28x + 96 = 0$$

Solution   First, we determine the maximum number of positive and negative roots.

$$f(x) \quad + \quad + \quad - \quad - \quad +$$

a maximum number of two positive roots

a maximum number of two negative roots

$$f(-x) \quad + \quad - \quad - \quad + \quad +$$

Second, we list the possible factors, the quotient of factors of 96 and factors of 2. The possible factors are

$$\pm 1, \quad \pm 2, \quad \pm 3, \quad \pm 4, \quad \pm 6, \quad \pm 8, \quad \pm 12, \quad \pm 16,$$
$$\pm 24, \quad \pm 32, \quad \pm 48, \quad \pm 96, \quad \pm 1/2, \quad \pm 3/2$$

Third, we test the possibilities until we find all but two roots. We begin by testing positive roots. With an extensive list of possibilities, let us choose a number in the middle of the list, 6.

$$\begin{array}{r|rrrrr}
6 & 2 & 1 & -35 & -28 & 96 \\
 & & 12 & 78 & 258 & 1{,}380 \\
\hline
 & +2 & +13 & +43 & +230 & +1{,}476
\end{array}$$

Although 6 is not a root, the quotient contains only positive signs; therefore, 6 is a maximum value for the roots. This means we do not need to test any numbers larger than 6. Next, let us test 4.

$$\begin{array}{r|rrrrr}
4 & 2 & 1 & -35 & -28 & 96 \\
 & & 8 & 36 & 4 & -96 \\
\hline
 & 2 & 9 & 1 & -24 & 0
\end{array}$$

We could continue to test another positive root, but let us test some negative values because there could be two negative roots. Again, let us test a number in the middle of the list, -6.

$$\begin{array}{r|rrrr}
-6 & 2 & 9 & 1 & -24 \\
 & & -12 & 18 & -114 \\
\hline
 & +2 & -3 & +19 & -138
\end{array}$$

Although -6 is not a root, it is a minimum value for the roots because the signs on the quotient alternate. This means we do not need to test any numbers smaller than -6. Next, let us test -3.

$$\begin{array}{r|rrrr}
-3 & 2 & 9 & 1 & -24 \\
 & & -6 & -9 & 24 \\
\hline
 & 2 & 3 & -8 & 0
\end{array}$$

There are two roots remaining represented by the quadratic equation $2x^2 + 3x - 8 = 0$, which we solve by using the quadratic formula.

$$x = \frac{-3 \pm \sqrt{9 - 4(2)(-8)}}{4} = \frac{-3 \pm \sqrt{73}}{4}$$

$$x \approx 1.39 \qquad \text{and} \qquad x \approx -2.89$$

The roots of the equation are 4, -3, $\dfrac{-3 + \sqrt{73}}{4} \approx 1.39$, and $\dfrac{-3 - \sqrt{73}}{4}$ ≈ -2.89. The graphing calculator can also be used to find the zeros of an equation algebraically. On the TI-92, select $\boxed{\text{F2}}$ and $\boxed{4}$. When **zeros** (appears at the bottom of the screen enter the polynomial, comma, *x*, a closing parenthesis, and press $\boxed{\text{ENTER}}$. The zeros (roots) will appear on the screen. Figure 7–3 shows the screen for this example. ∎∎∎

FIGURE 7–3

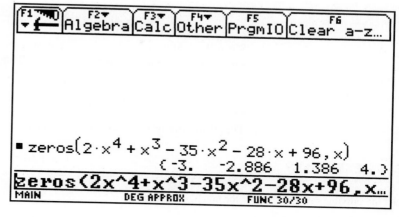

EXAMPLE 5 A rectangular piece of plastic 12 cm by 13 cm is to be molded into an open box, the same thickness as the original piece of plastic, with no material wasted. The box is to have a square base and a volume of 180 cm³. What are the possible dimensions of the box?

Solution Let

$$L = \text{length and width of the square base}$$
$$H = \text{height of the box}$$

Then the *volume* of the box is represented by

$$L^2H = 180$$

and the *surface area* of the box is represented by

$$L^2 + 4LH = 156$$

The surface area of the box equals the area of the original piece of plastic: (12 cm) (13 cm). We solve this nonlinear system of equations by substituting $H = 180/L^2$ into the second equation for *H*.

$$L^2 + 4L\left(\frac{180}{L^2}\right) = 156$$

This equation simplifies to yield

$$L^2 + \frac{720}{L} = 156$$
$$L^3 + 720 = 156L$$
$$L^3 + 720 - 156L = 0$$
$$L^3 - 156L + 720 = 0$$

There are a large number of factors of 720, but let us test 6.

$$
\begin{array}{r|rrrr}
6 & 1 & 0 & -156 & 720 \\
  &   & 6 & 36   & -720 \\
\hline
  & 1 & 6 & -120 & 0 \\
\end{array}
$$

One possibility for the length is 6. The resulting quotient is $L^2 + 6L - 120 = 0$. Solving by the quadratic formula gives

$$L = \frac{-6 \pm \sqrt{36 - 4(1)(-120)}}{2} = \frac{-6 \pm \sqrt{516}}{2}$$

$$L \approx 8.36 \qquad \text{and} \qquad L \approx -14.36$$

Since the length of the box is not measured in negative numbers, the only reasonable answer from the quadratic formula is 8.36. Thus,

$$L = 6 \qquad \text{or} \qquad L \approx 8.36$$
$$H = (180/L^2) = 5 \qquad \text{or} \qquad H \approx 2.58$$

Therefore, the box is 6 cm by 6 cm by 5 cm, or 8.36 cm by 8.36 cm by 2.58 cm. ■■■

7–4   EXERCISES

List the possible rational roots of the equations given.

1. $x^3 - 8x^2 + 11x + 20 = 0$

2. $x^3 + 8x^2 - 16 = 0$

3. $x^4 + 8x^3 + 7x - 24 = 0$

4. $x^4 - 7x^3 + 8x + 48 = 0$

5. $x^3 - 5x + 36 = 0$

6. $x^5 + 3x^4 - 5x^2 + 6x - 50 = 0$

7. $3x^3 + 8x^2 + 5x - 18 = 0$

8. $6x^3 + 12x - 35 = 0$

9. $2x^4 - 6x^2 + 9x + 14 = 0$

10. $9x^3 - 7x^2 + 6x + 36 = 0$

11. $2x^3 + 13x^2 + 17x - 12 = 0$

12. $4x^4 + 11x^2 + 12 = 0$

13. $6x^4 - 7x^3 + 9x + 18 = 0$

14. $7x^4 - 9x^3 + 6x - 24 = 0$

15. $14x^4 + 79x^3 + 25x^2 - 103x - 15 = 0$

16. $3x^5 - 10x^4 + 7x^2 + 18 = 0$

Determine the maximum number of positive and negative roots for the equations given.

17. $5x^4 - 6x^3 - 8x^2 + x - 5 = 0$

18. $8x^4 + 7x^3 - x^2 - x + 6 = 0$

19. $4x^3 - 6x^2 + x - 5 = 0$

20. $9x^3 - 6x^2 - 8x - 5 = 0$

21. $4x^5 + x^3 - 8x^2 + 6x - 1 = 0$

22. $6x^7 + x^5 - 4x^2 + 9 = 0$

23. $x^7 - 6x^4 + 3x^3 + x^2 - 4x + 8 = 0$

24. $3x^8 + 7x^5 - 3x^3 + 6x + 10 = 0$

25. $7x^8 - 3x^7 - x^5 + 8x^2 - 12 = 0$

26. $5x^6 + 4x^3 + 7x^2 - 3 = 0$

27. $-x^9 + 6x^7 + x^6 - 3x^2 + x + 5 = 0$

28. $x^{10} - 4x^6 + 7x^3 + 8x - 5 = 0$

Solve the following equations. You may use a graphing calculator to check your answers.

29. $2x^3 + 3x^2 - 8x + 3 = 0$

30. $3x^3 + x^2 - 12x - 4 = 0$

31. $3x^3 - 4x^2 - 28x - 16 = 0$

32. $4x^3 - 19x^2 - 8x + 15 = 0$

33. $x^3 + 10x^2 + 29x + 24 = 0$

34. $6x^3 - 5x^2 - 8x + 3 = 0$

35. $x^3 + 2x^2 - 21x + 18 = 0$

36. $2x^3 + 7x^2 - 10x - 50 = 0$

37. $x^3 + 8x^2 + 5x - 14 = 0$

38. $6x^3 + 7x^2 - 44x + 32 = 0$

39. $2x^4 + x^3 - x^2 + x - 3 = 0$

40. $2x^4 + 9x^3 - 40x^2 - 54x - 7 = 0$

41. $x^4 + 3x^3 - 21x^2 - 43x + 60 = 0$

42. $x^4 + 8x^3 - x^2 - 68x + 60 = 0$

43. $2x^5 + 15x^4 + 2x^3 - 36x^2 - 4x + 21 = 0$

44. $2x^4 + 5x^3 - 19x^2 - 45x + 9 = 0$

45. $x^5 - 6x^4 + 7x^3 - 9x^2 - 8x + 15 = 0$

46. $x^5 + 7x^4 - 18x^3 - 72x^2 + 80x + 128 = 0$

B 47. A company that manufactures computers estimates that the profit for a particular model is given by $P = -35x^3 + 4{,}650x^2 - 250{,}000$, where P is the profit in dollars and x is the advertising cost in tens of thousands of dollars. Find the minimum advertising cost that yields a profit of $4,544,080.

M 48. A rectangular piece of plastic 5 in. by 37 in. is to be molded into an open box with no material wasted. If the box is to have a square base, the same thickness as the original piece of plastic, and a volume of 200 in³, what should the dimensions of the box be? (See Example 5.)

M 49. A piece of plastic 16 cm by 10 cm is to be molded into an open box with no material wasted. The box is to have the same thickness as the original piece of plastic, a square base, and a volume of 144 cm³. Find the dimensions of the box.

P 50. Under certain conditions the horizontal displacement of an object as a function of time is given by $s(t) = t^3 - 3t^2 + 4$. Find the time t when the displacement equals zero, i.e., when $s(t) = 0$.

B 51. Bob's Digital Phones Inc. estimates that the demand for its super duper model is given by $D = -x^3 + 45x^2 - 130x - 3,500$, where D is demand and x is the price of the phone. Determine the minimum price for the phone that would produce a demand of 556 phones.

M 52. The radius of a cylindrical tank is 5 m less than its height. If the volume of the cylinder is $3,042\pi$ m³, find the dimensions of the tank.

M 53. You can make a rectangular box from a piece of cardboard 10 cm by 7 cm by cutting a square from each corner and bending up the sides. What are the resulting dimensions of the box if its volume is 36 cm³?

7–5

IRRATIONAL ROOTS BY LINEAR APPROXIMATION

In this section we will develop a method of solving for irrational roots. When we cannot reduce a polynomial equation to a quadratic equation by finding all but two rational roots, we need another method to find irrational roots.

Locating Roots Graphically

In Chapter 4, we solved equations graphically by locating the x intercept of the graph. This procedure is illustrated in Example 1.

EXAMPLE 1   Locate the real roots of the equation $x^3 + 2x^2 - x + 1 = 0$ graphically.

Solution   First, we fill in the accompanying table of values. As shown in the graph in Figure 7–4, the intercept and root occur between -3 and -2. The table of values also indicates that the sign of $f(x)$ changes between $x = -3$ and $x = -2$.

FIGURE 7–4

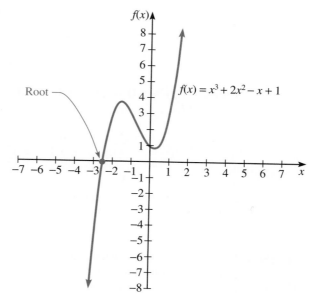

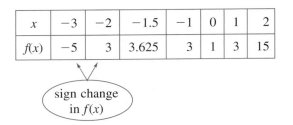

x	-3	-2	-1.5	-1	0	1	2
$f(x)$	-5	3	3.625	3	1	3	15

sign change
in $f(x)$

Linear Interpolation

One method of *approximating irrational roots* is called **linear interpolation.** This method is based on the assumption that if two points are sufficiently close to each other, then the straight line joining the points is a good approximation of the actual curve between the points. To find irrational roots, we begin by finding an interval on which a root exists, and then we successively narrow the interval until we find the root to the desired degree of accuracy. We know that when a curve crosses the x axis, the x value is a root of the equation. Therefore, finding an interval on which a root exists is based on the fact that the value of y or $f(x)$ changes sign whenever the curve crosses the x axis. This translates to a sign change on the remainder from synthetic division.

EXAMPLE 2   Using linear interpolation, find the root, to the nearest hundredth, between -2 and -1 of the equation $x^3 - 6x^2 + 10 = 0$.

Solution   We know that since $f(-2) = -22$ and $f(-1) = 3$, the graph crosses the x axis and a root does exist between -2 and -1. Since $f(-1)$ is closer to zero, the root is closer to $x = -1$. Let us test $x = -1.2$ by using synthetic division.

$$
\begin{array}{r|rrrr}
-1.2 & 1 & -6 & 0 & 10 \\
     &   & -1.2 & 8.64 & -10.37 \\
\hline
     & 1 & -7.2 & 8.64 & -0.37
\end{array}
$$

Since $f(-1.2) \approx -0.37$ and $f(-1) = 3$, the root is between -1 and -1.2. Next, we will test -1.1.

$$
\begin{array}{r|rrrr}
-1.1 & 1 & -6 & 0 & 10 \\
     &   & -1.1 & 7.81 & -8.59 \\
\hline
     & 1 & -7.1 & 7.81 & 1.41
\end{array}
$$

Since $f(-1.2) \approx -0.37$ and $f(-1.1) \approx 1.41$, the root is closer to $x = -1.2$. Let us test $x = -1.18$.

$$
\begin{array}{r|rrrr}
-1.18 & 1 & -6 & 0 & 10 \\
      &   & -1.18 & 8.47 & -10 \quad \text{(rounded)} \\
\hline
      & 1 & -7.18 & 8.47 & 0 \quad \text{(approximately)}
\end{array}
$$

Since the remainder is approximately zero, the root rounded to hundredths is -1.18.

CAUTION ✦ When $f(x)$ changes signs between $x = a$ and $x = b$, we know only that an odd number of roots occur in the interval. If $f(x)$ does not change signs on an interval, there may be no real roots or an even number of real roots in the interval.

EXAMPLE 3 By linear interpolation, find the root, to the nearest hundredth, between 1 and 2 of the equation $x^4 - 2x^3 + 6x - 6 = 0$.

Solution Since $f(1) = -1$ and $f(2) = 6$, a root does exist in the interval. Next, let us approximate the root by constructing a graph of the straight line joining the points $(1, -1)$ and $(2, 6)$, as shown in Figure 7–5. The graph illustrates that the root is closer to 1; therefore, let us test $x = 1.2$ and $x = 1.3$ by synthetic division.

$$
\begin{array}{r|rrrrr}
1.2 & 1 & -2 & 0 & 6 & -6 \\
 & & 1.2 & -0.96 & -1.15 & 5.82 \\
\hline
 & 1 & -0.8 & -0.96 & 4.85 & -0.18
\end{array}
$$

$$
\begin{array}{r|rrrrr}
1.3 & 1 & -2 & 0 & 6 & -6 \\
 & & 1.3 & -0.91 & -1.18 & 6.27 \\
\hline
 & 1 & -0.7 & -0.91 & 4.82 & 0.27
\end{array}
$$

FIGURE 7–5

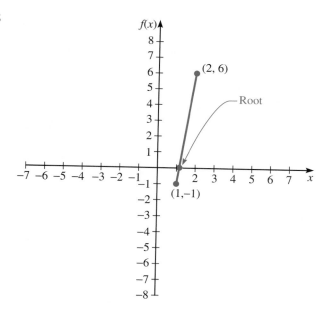

Since the sign on the remainder changes, a root exists in the interval between 1.2 and 1.3. Since the remainder is closer to zero for 1.2, let us test 1.24 and 1.25.

$$
\begin{array}{r|rrrrr}
1.25 & 1 & -2 & 0 & 6 & -6 \\
 & & 1.25 & -0.94 & -1.17 & 6.04 \\
\hline
 & 1 & -0.75 & -0.94 & 4.83 & 0.04
\end{array}
$$

$$
\begin{array}{r|rrrrr}
1.24 & 1 & -2 & 0 & 6 & -6 \\
 & & 1.24 & -0.94 & -1.17 & 5.99 \\
\hline
 & 1 & -0.76 & -0.94 & 4.83 & -0.01
\end{array}
$$

The approximate root in the given interval is 1.24. ■■■

EXAMPLE 4 Find all the real roots, to the nearest tenth, of the equation $6x^4 - 3x^2 + 5x - 4 = 0$.

Solution First, we complete the accompanying table of values and graph the equation so that we can approximately locate roots. The graph is shown in Figure 7–6.

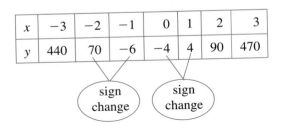

x	-3	-2	-1	0	1	2	3
y	440	70	-6	-4	4	90	470

sign change   sign change

FIGURE 7–6

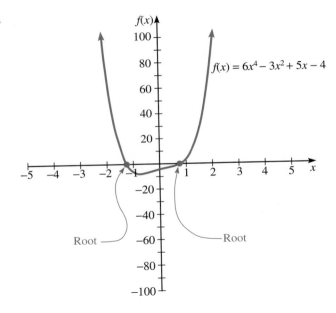

$f(x) = 6x^4 - 3x^2 + 5x - 4$

Root   Root

The curve crosses the x axis between -2 and -1 and again between 0 and 1. Since the curve crosses the x axis closer to -1 than -2, let us test -1.1 and -1.2.

$$
\begin{array}{r|rrrrr}
-1.1 & 6 & 0 & -3 & 5 & -4 \\
 & & -6.6 & 7.3 & -4.7 & -0.3 \\
\hline
 & 6 & -6.6 & 4.3 & 0.3 & -4.3 \\
\end{array}
$$

$$
\begin{array}{r|rrrrr}
-1.2 & 6 & 0 & -3 & 5 & -4 \\
 & & -7.2 & 8.6 & -6.8 & 2.1 \\
\hline
 & 6 & -7.2 & 5.6 & -1.8 & -1.9 \\
\end{array}
$$

Since the remainder is still negative, let us test $x = -1.3$.

$$
\begin{array}{r|rrrrr}
-1.3 & 6 & 0 & -3 & 5 & -4 \\
 & & -7.8 & 10.1 & -9.3 & 5.6 \\
\hline
 & 6 & -7.8 & 7.1 & -4.3 & 1.6 \\
\end{array}
$$

Since the sign changes and the remainder of 1.6 is the smaller, the approximate root is $x = -1.3$. Similarly, we approximate the root between 0 and 1, testing 0.7 and 0.8.

$$\begin{array}{r|rrrrr}
0.7 & 6 & 0 & -3 & 5 & -4 \\
& & 4.2 & 2.9 & -0.1 & 3.4 \\
\hline
& 6 & 4.2 & -0.1 & 4.9 & -0.6 \\
\end{array}$$

$$\begin{array}{r|rrrrr}
0.8 & 6 & 0 & -3 & 5 & -4 \\
& & 4.8 & 3.8 & 0.6 & 4.5 \\
\hline
& 6 & 4.8 & 0.8 & 5.6 & 0.5 \\
\end{array}$$

Since $f(0.8) \approx 0.5$, the approximate root is 0.8. The approximate real roots of this equation rounded to tenths are 0.8 and -1.3. Because there are only two real roots, we also know there are two complex number roots. The graphing calculator solution using the zero function is shown in Figure 7–7.

■ ■ ■

FIGURE 7–7

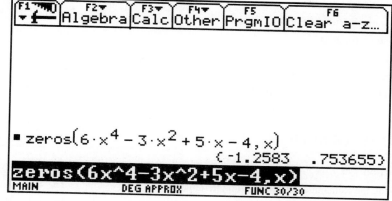

7–5 EXERCISES

Find the real roots, to the nearest tenth, of the following equations.

1. $2x^3 - 8x^2 + x - 12 = 0$

2. $x^3 + 6x^2 - 5x + 8 = 0$

3. $5x^3 - 8x - 18 = 0$

4. $x^3 + 2x^2 - 3x + 12 = 0$

5. $3x^3 + 7x - 9 = 0$

6. $4x^3 - 3x^2 + 2x - 5 = 0$

7. $x^3 - 2x^2 + x - 3 = 0$

8. $2x^4 - 3x^2 - 6 = 0$

9. $x^4 + 6x^3 - 5x + 9 = 0$

10. $x^4 - x^3 + 2x^2 + x - 8 = 0$

11. $x^4 + 6x^3 + x^2 - 3x + 10 = 0$

12. $5x^4 + 6x - 8 = 0$

13. $6x^4 - 2x^3 + x^2 + 3x - 6 = 0$

14. $3x^4 + 8x^2 - 14 = 0$

15. $2x^4 - 3x^3 + x^2 + 5x - 4 = 0$

16. $x^5 - x^3 + 6x^2 + x - 10 = 0$

17. $2x^5 + x^4 - 3x^3 - 4x^2 + x + 8 = 0$

18. $x^5 + x^4 - x^3 + 2x^2 + x - 8 = 0$

19. $2x^5 - 4x^4 + 6x^2 - x + 9 = 0$

20. $x^5 + 2x^3 - 2 = 0$

M 21. The dimensions of a rectangular box are 5, 8, and 6 cm. If each dimension of the box is increased by a fixed amount, say, x, the volume is 792 cm^3. Find the amount by which to increase each dimension.

M 22. A cylindrical storage tank 10 ft high holds 1,131 ft^3. Determine the thickness (to hundredths) of the tank if the outside radius is 6.2 ft.

P 23. The position x of a particle that moves in a line as a function of time t is given by $x = t^3 + 4t^2 - 10t + 20$. Find the time when the particle has moved 195 ft.

B 24. A company manufactures CD-ROM drives for OEM's (Original Equipment Manufacturers) that install them in their computer systems. The transportation costs C are given by

$$C = 150\left(\frac{100}{x^2} + \frac{x}{x + 80}\right)$$

where C is cost in thousands of dollars and x is the order size in thousands. In calculus, it can be shown that the minimum transportation cost occurs when $12x^3 - 30x^2 - 4,800x - 192,000 = 0$. Solve this equation to determine the shipment size (to the nearest hundred) that produces a minimal cost.

B 25. A company that manufactures ATV's estimates that its profit P for selling a particular model is given by

$$P = -56x^3 + 2,800x^2 - 250,000$$

where P is in dollars and x is the advertising expense in tens of thousands of dollars. Find the smallest amount that yields a profit of $525,000.

Bi 26. A biologist is growing an organism to be used in an experiment. The number of bacteria N (in millions) as a function of time, t, in hours is given by $N = -0.003t^3 + t^2 + 6.3$. Determine to the nearest hour the smallest value of t when $N = 17.8$.

CHAPTER SUMMARY

Summary of Terms

Descartes's Rule of Signs (p. 281)

dividend (p. 268)

divisor (p. 268)

Factor Theorem (p. 269)

Fundamental Theorem of Algebra (p. 276)

linear interpolation (p. 286)

multiplicity (p. 276)

polynomial function (p. 268)

quotient (p. 268)

remainder (p. 268)

Remainder Theorem (p. 268)

synthetic division (p. 271)

CHAPTER REVIEW

Section 7–1

Using the Remainder Theorem, find the remainder.

1. $(4x^3 - 7x^2 + 6x - 9) \div (x + 3)$

2. $(2x^4 + 3x^2 + 7) \div (x - 1)$

3. $(5x^4 + 7x^3 + 3x - 8) \div (x + 2)$

4. $(9x^3 + 8x^2 + x - 10) \div (x + 5)$

Use the Factor Theorem to determine if the second expression is a factor of the first.

5. $2x^3 - 6x^2 + 9; x - 3$

6. $x^3 - 5x^2 + 7x - 3; x - 1$

7. $3x^3 - 15x^2 - 11x + 5; 3x - 1$

8. $4x^3 - x^2 + 8; 2x + 1$

9. $x^4 + 6x^3 - 8x + 8; x + 5$

10. $2x^4 - x^2 + 3x - 144; x + 3$

Section 7–2

Using synthetic division, find the quotient and remainder.

11. $6x^3 - 5x^2 + 3x - 1; x + 1$

12. $9x^3 + 7x^2 - x + 6; x - 2$

13. $x^4 - 3x^3 + 7x - 8; x + 2$

14. $2x^4 - 7x^2 + 6x - 10; x + 5$

15. $4x^4 - x^3 + 2x^2 - 9; x - 3$

16. $4x^3 - 9x^2 + 3x - 5; x - 1$

Use synthetic division to determine if the given number is a zero of the function.

17. $6x^3 - 11x^2 + 9x - 2; 1/3$

18. $2x^3 + x^2 + 5x - 6; -3$

19. $5x^4 + 6x^3 - 8x + 7; -1$

20. $7x^5 + 6x^3 + x^2 - 6x + 14; 2$

21. $x^4 + 6x^3 + 8x^2 + 43x - 30; -6$

22. $4x^4 + 27x^3 - 7x^2 - 20x + 5; 1/4$

Section 7–3

Find the remaining roots of the equation given the root(s) indicated.

23. $2x^3 + 3x^2 - 8x + 3 = 0; -3$

24. $x^3 + 5x^2 + 3x + 54 = 0; -6$

25. $2x^3 - 3x^2 + 4x + 9 = 0; -1$

26. $6x^3 - 37x^2 + 47x + 20 = 0; 5/2$

27. $2x^4 - 5x^3 + 3x^2 - 15x - 9 = 0; -1/2$ and 3

28. $2x^4 + x^3 - 24x^2 - 20x + 21 = 0; 7/2$ and -3

29. $x^4 - 8x^3 + 25x^2 - 58x + 40 = 0$; 5 and 1

30. $x^4 - 4x^3 - x^2 + 16x - 12 = 0$; 2 and 3

Section 7–4

List all possible rational roots for the equations given.

31. $4x^3 + 7x^2 - 3x + 2 = 0$

32. $2x^3 + 6x^2 - 9x + 10 = 0$

33. $3x^4 + 7x^3 - 5x^2 + 12 = 0$

34. $8x^3 - 7x^2 + x + 6 = 0$

35. $5x^3 + 8x^2 - 10x + 15 = 0$

36. $3x^4 - 5x^2 + 7x - 8 = 0$

Using Descartes's Rule of Signs, list the maximum number of positive and negative roots for the given equation.

37. $8x^3 + 7x^2 - 6x + 10 = 0$

38. $2x^3 - 7x^2 - 5x - 3 = 0$

39. $4x^4 - 5x^3 - 6x^2 + 7x - 3 = 0$

40. $3x^4 + x^3 - 2x^2 - x + 6 = 0$

41. $x^5 - 4x^3 - 2x^2 - x + 9 = 0$

42. $2x^5 - x^4 + 2x^3 + 4x^2 - 3x - 11 = 0$

Solve the following equations.

43. $2x^3 - 9x^2 - 2x + 24 = 0$

44. $3x^3 - 4x^2 - 16x + 3 = 0$

45. $4x^3 - 28x^2 - 69x - 27 = 0$

46. $2x^3 - 11x^2 - 41x + 140 = 0$

47. $x^3 + x^2 - 14x - 24 = 0$

48. $x^4 + x^3 - 14x^2 + 21x - 45 = 0$

49. $2x^4 + 13x^3 - 14x^2 - 43x + 42 = 0$

50. $2x^4 - 4x^3 - 25x^2 - 3x + 30 = 0$

Section 7–5

Use linear approximation to find the irrational roots, to the nearest tenth, of the following equations.

51. $3x^3 + 6x - 10 = 0$ **52.** $x^3 + x^2 - 6x + 7 = 0$

53. $x^3 + 2x^2 + 3x + 4 = 0$

54. $x^3 - 5x^2 + x - 8 = 0$

CHAPTER TEST

The number in parentheses refers to the appropriate learning objective given at the beginning of the chapter.

1. Use synthetic division to determine if -2 is a zero of the function $x^4 - 3x^2 - x - 6$. Explain why or why not. (3)

2. Determine the rational roots of the following equation: (7)
$$x^3 + x^2 - 2x - 2 = 0$$

3. Find the remaining roots of the equation $6x^3 - 35x^2 + 21x + 20 = 0$ if 5 is one of the roots. (4)

4. Solve the following equation: (4)
$$x^3 - 14x^2 + 58x - 80 = 0$$

5. Use the Factor Theorem to determine if $2x + 5$ is a factor of $4x^3 + 8x^2 - x - 15$. (2)

6. List all possible rational roots for $2x^4 + 6x^3 - x^2 + 5x - 6 = 0$. (6)

7. Solve the equation $2x^3 - 7x^2 - 28x - 12 = 0$. (4)

8. Apply the Remainder Theorem to find the remainder when $8x^4 + 7x^3 + 5x - 10$ is divided by $x + 2$. (1)

9. Using Descartes's Rule of Signs, determine the maximum number of positive and negative roots for $2x^4 - 6x^3 - x^2 + 7x - 10 = 0$. (5)

10. You can make a rectangular box from a piece of cardboard by cutting a square from each corner and bending up the sides. What are the resulting dimensions of a box made from a 16-cm-by-17-cm piece of cardboard if the volume of the box is to be 312 cm³? (4)

11. Use synthetic division to determine the quotient and remainder when $7x^4 + 8x^3 - 6x^2 + 10$ is divided by $x + 2$. (3)

12. Use linear approximation to find the irrational root(s) of the following equation: (7)

$$x^4 + 6x^3 - 7x^2 + x + 8 = 0$$

13. Use Descartes's Rule of Signs to list the maximum number of positive and negative   (5)
roots for $3x^5 - 4x^4 + 7x^2 + 8x + 5 = 0$.

14. Solve the equation $6x^3 + 43x^2 + 60x - 25 = 0$.   (4)

15. Using synthetic division, find the quotient and remainder resulting from the division   (3)
of $x^5 + 6x^4 - 7x^2 + 8$ by $x - 2$.

16. List all possible rational roots of $3x^3 + 5x^2 - 10x + 16 = 0$.   (6)

GROUP ACTIVITY

You have been hired by the Lucky Ladybugs Company to determine the best method for packaging the ladybugs that they ship all around the country. Your team has been given the task of determining the best dimension for the shipping container to achieve the maximum volume. The constraints are such that the company that ships the load of ladybugs requires that the package be restricted to a girth (distance around the package) plus length measurement of 120 in.

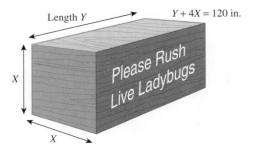

The volume of the box is given by:

$$V = X^2Y$$

1. Derive an equation for the volume of the box that contains only the variable X.
2. Determine the value for X that will give the maximum volume. You may want to graph the equation that you developed from Step 1 to find the maximum value.
3. Find the values of X that will yield a volume of 8,000 in^3. Explain why some of your answers are impossible.

SOLUTION TO CHAPTER INTRODUCTION

Since you have a graphing calculator, you decide to use the zero function to solve Fred's problem. You must solve the equation

$$P = -96x^3 + 5,386x^2 - 250,000$$

when $P = \$1,800,000$.

$$1,800,000 = -96x^3 + 5,386x^2 - 250,000$$
$$0 = -96x^3 + 5,386x^2 - 2,050,000$$

Figure 7–8 shows the graphing calculator screen. The smallest amount Fred should spend on advertising is approximately \$27,000.

FIGURE 7–8

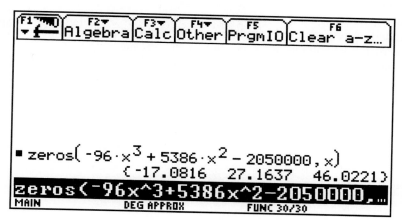

You are responsible for population control in a Florida scrub jay preserve. These endangered birds are very territorial and the reserve can only support 50 birds. A scientist has modeled the growth of the population by the equation

$$P = \frac{50}{1 + 42e^{-0.18t}} \text{ where } t = \text{months}$$

What is the population of scrub jays 3 years after the preserve's creation? (The solution to this problem is given at the end of the chapter.)

Population growth is one example of an exponential function. The exponential function and its inverse, the logarithmic function, are very important to many areas of technology. For example, bacteria in a petri dish and the world's population both grow at an exponential rate. Energy loss and absorption, radioactive decay, and compound interest are also examples of exponential functions. The logarithmic and exponential functions and their applications are discussed in this chapter.

Learning Objectives

After you complete this chapter, you should be able to

1. Graph exponential and logarithmic functions (Sections 8–1 and 8–2).

2. Evaluate exponential expressions using a calculator (Section 8–1).

3. Convert a logarithmic expression to an exponential expression, and vice versa (Section 8–2).

4. Solve simple logarithmic equations by converting them to exponential form (Section 8–2).

5. Apply the properties of logarithms to write logarithmic expressions in different forms, and evaluate the resulting expressions (Section 8–3).

6. Solve exponential and logarithmic equations (Section 8–4).

Chapter 8

Exponential and Logarithmic Functions

8–1

THE EXPONENTIAL FUNCTION

An **exponential function** is a function y or $f(x)$ in which *the base is a constant b and the exponent contains the independent variable x.* The exponential function is defined by the following expression.

Definitions

The Exponential Function

$$y = b^x$$

where

b = any positive real number other than 1

x = any real number

y = any positive real number

The exponential function differs from the exponential expressions discussed in Chapter 3, such as $3x^3y^4$, because the independent variable appears in the exponent. Some examples of exponential functions are

$$y = 3^x \qquad y = 6^{2x-1} \qquad y = 5e^{3x}$$

NOTE ✦ The base of e represents the irrational number $e \approx 2.71828...$ This number is called the natural base.

Graphs of Exponential Functions

To graph an exponential function, we follow the procedure outlined in Chapter 4. We obtain ordered pairs by substituting a chosen value for x into the equation and calculating the resulting value of y. Since the graph of an exponential function is a curved line, we must choose a sufficient number of values for x to sketch a smooth curve through the plotted points.

EXAMPLE 1 Sketch the graph of $y = 3^x$.

Solution First, we prepare a table of values by substituting integral values of x from -4 to 4. Then we plot the points and connect them with a smooth curve as shown in Figure 8–1.

x	-4	-3	-2	-1	0	1	2	3	4
y	1/81	1/27	1/9	1/3	1	3	9	27	81

■■■

NOTE ✦ Notice that in Figure 8–1, the value of y increases without bound as x increases, and y approaches zero as x decreases. This relation is characteristic of an **exponential growth function.**

FIGURE 8–1

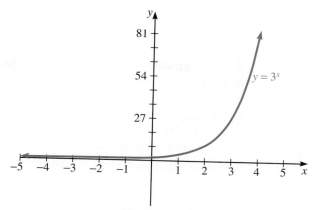

EXAMPLE 2 Graph $y = 2^{-x}$.

Solution To graph this function, we fill in a table of values just as we did in the previous example. Remember from Chapter 3 that 2^{-x} is equivalent to $\left(\dfrac{1}{2}\right)^{x}$. The graph of $y = 2^{-x}$ is shown in Figure 8–2.

x	-4	-3	-2	-1	0	1	2	3	4
y	16	8	4	2	1	1/2	1/4	1/8	1/16

FIGURE 8–2

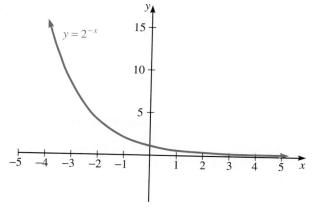

NOTE ◆ Notice the difference between the graph of $y = 2^{-x}$ (Figure 8–2) and the graph of $y = 3^{x}$ (Figure 8–1). Figure 8–2 represents an **exponential decay function** where y approaches zero as x increases and infinity as x decreases.

From the graphs shown in Figures 8–1 and 8–2, we can summarize the following properties of exponential functions.

Properties of Exponential Functions

■ For $b > 1$, an exponential growth function results as shown in Figure 8–1.

■ For $0 < b < 1$ and $x > 0$, an exponential decay function results as shown in Figure 8–2.

■ The graph of an exponential function of the form $y = b^x$ is above the x axis but approaches the x axis as an asymptote. A line is an asymptote of the curve if the curve keeps getting closer to the line as the variable approaches a certain value.

■ Exponential functions of the form $y = b^x$ intersect the y axis at $(0, 1)$.

Graphing Calculator

The graphing calculator can be used to graph exponential equations. We use the same procedures discussed in earlier chapters.

EXAMPLE 3   Use the graphing calculator to graph $y = 2^{x-1}$.

Solution   We enter the exponential equation, set the viewing rectangle, and execute the graph. The viewing rectangle and graph are shown in Figure 8–3.   ■ ■ ■

FIGURE 8–3

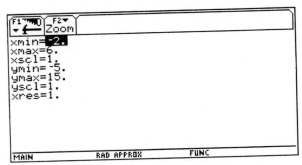

Viewing rectangle

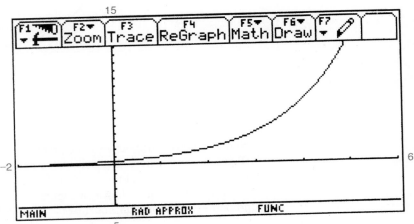

Application

EXAMPLE 4   Compound interest is given by the formula

$$y = P\left(1 + \frac{r}{m}\right)^{mt}$$

where y = accumulated value, P = principal, r = rate of interest, m = frequency of compounding, and t = time. Graph the relationship between the time and accumulated value if P = \$500, r = 10%, and m = 12 (interest is compounded monthly). Then use the graph to approximate how long it would take to have an accumulated value of \$1,000.

Solution   Substituting the given values into the formula, we have

$$y = 500\left(1 + \frac{0.10}{12}\right)^{12t}$$

$$y = 500(1.0083)^{12t}$$

Substituting positive integral values of t gives the following table of values:

t	0	1	2	3	4	5	6
y	500	552.36	610.20	674.09	744.68	822.65	908.80

The graph of this function is given in Figure 8–4. To determine how long it takes for the accumulated value to reach \$1,000, we find \$1,000 along the y axis; then we move horizontally to the graph and vertically down to the t axis. As shown by the graph, it takes about 7 years for \$500 to have an accumulated value of \$1,000.   ∎∎∎

CAUTION ✦ When interest rate is substituted into a formula, it must be expressed as a decimal (10% = 0.10).

FIGURE 8–4

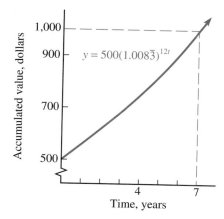

Base e

The base of an exponential function can be any positive real number other than 1. In the previous discussion, the base has been a rational number, but irrational numbers are also possible. The number e = 2.71828... is irrational, just as π and $\sqrt{2}$. The expression $(1 + (1/n))^n$ approaches the value of e as n approaches infinity. The number e, named after Swiss mathematician Leonhard Euler, is used extensively in applications of exponential growth and decay. You can find the value of e^x on the calculator using an $\boxed{e^x}$ key or [inv] [ln x] keys, depending on your calculator. In this text, calculator keystrokes use the [e^x] key.

Graphs of Exponential Functions

⊞ **EXAMPLE 5** Graph the function $y = e^x$.

Solution We fill in a table of values using the calculator. The graph of $y = e^x$ is given in Figure 8–5. As you can see by comparing the graphs in Figures 8–1 and 8–4, this graph also represents exponential growth. ∎∎∎

NOTE ✦ The equation $y = ae^{nx}$ represents exponential growth with a base of e, if n is positive.

FIGURE 8–5

x	y
−4	0.02
−3	0.05
−2	0.14
−1	0.37
0	1
1	2.7
2	7.4
3	20.1
4	54.6

4 [+/−] [e^x] → 0.018
3 [+/−] [e^x] → 0.050
2 [+/−] [e^x] → 0.14
1 [+/−] [e^x] → 0.37
0 [e^x] → 1
1 [e^x] → 2.7
2 [e^x] → 7.4
3 [e^x] → 20.1
4 [e^x] → 54.6

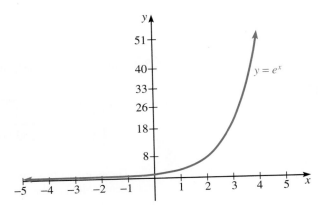

EXAMPLE 6 The circuit shown in Figure 8–6 consists of a resistance R, a capacitance C, a constant voltage V, and a switch. The instantaneous current I at any time t is given by

$$I = \frac{V}{R}e^{-t/(RC)}$$

Graph I in amps as a function of t time when $V = 120$ V, $R = 500\ \Omega$, and $C = 0.0002$ F. (RC is the capacitive time constant of the circuit and has units of time.)

Solution Substituting in the given values, we obtain the following equation:

$$I = \frac{120}{500}e^{-(t/[500(0.0002)])}$$

$$I = 0.24e^{-t/0.10}$$

$$I = 0.24e^{-10t}$$

The table of values is given below. The graph of this function is shown in Figure 8–7 and is characteristic of the exponential decay curve.

t	0	0.05	0.10	0.15	0.20	0.25	0.30
I	0.24	0.15	0.09	0.05	0.03	0.02	0.012

NOTE ✦ The general equation $y = ae^{-nx}$ represents exponential decay with a base of e if n is positive.

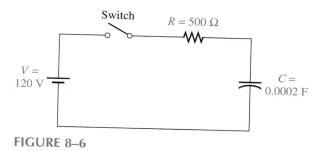

FIGURE 8–6

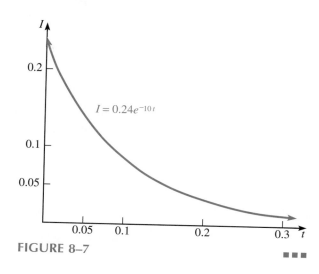

$I = 0.24e^{-10t}$

FIGURE 8–7

Application

▦ **EXAMPLE 7**   A biologist finds that bacteria in a petri dish grow at a rate of 1.8% each hour. If the biologist starts the culture with 750 cells, how many cells of the culture will she have at the end of 8 hours? (Use a base of e.)

Solution   The general equation for exponential growth with a base of e is

$$y = ae^{nx}$$

where a = initial amount, n = growth rate, and x = time. Filling in the values $a = 750$, $n = 0.018$, and $x = 8$, we obtain the following equation:

$$y = 750e^{(0.018)(8)}$$

Performing the calculation yields

$$y = 866 \text{ cells}$$

The biologist will have 866 cells of the culture at the end of 8 hours.

$$750 \; \boxed{\times} \; \boxed{(} \; 0.018 \; \boxed{\times} \; 8 \; \boxed{)} \; \boxed{e^x} \; \boxed{=} \; \rightarrow 866.16$$

▦ **EXAMPLE 8**   From Example 4, we know that compound interest is given by the formula $y = P\left(1 + \dfrac{r}{m}\right)^{mt}$, where y = accumulated value, P = principal, r = rate of interest, m = frequency of compounding, and t = time. The formula for continuous compounding is given by $y = Pe^{rt}$ where y = accumulated value, P = principal, r = rate of interest, and t = time. If \$10,000 is invested at 12% interest, find the accumulated value after 3 years if it is compounded quarterly and continuously.

Solution   From the information given, $P = \$10,000$, $r = 12\% = 0.12$, $t = 3$ years, and $m = 4$ for quarterly compounding.

$$y = P\left(1 + \frac{r}{m}\right)^{mt}$$

$$y = 10{,}000\left(1 + \frac{0.12}{4}\right)^{4(3)}$$

$$y = \$14{,}257.61$$

$$\boxed{(}\ 1\ \boxed{+}\ 0.12\ \boxed{\div}\ 4\ \boxed{)}\ \boxed{y^x}\ 12\ \boxed{=}\ \boxed{\times}\ 10000\ \boxed{=}\ \rightarrow 14257.609$$

For continuous compounding, the values of P, r, and t are the same. Substituting these values into the formula for continuous compounding gives

$$y = Pe^{rt}$$

$$y = 10{,}000e^{(0.12)(3)}$$

$$y = \$14{,}333.29$$

$$0.12\ \boxed{\times}\ 3\ \boxed{=}\ \boxed{EX}\ \boxed{\times}\ 10000\ \boxed{=}\ \rightarrow 14333.294$$

Note that continuous compounding gives

$$14{,}333.29 - 14{,}257.61 = \$75.68$$

more than quarterly compounding.

■ ■ ■

8–1 EXERCISES

▦ Graph the following functions.

1. $y = 4^x$

2. $y = 5^{-x}$

3. $y = 3^{-x}$

4. $y = 2^{3x}$

5. $y = \left(\frac{1}{2}\right)^{3x}$

6. $y = \left(\frac{1}{4}\right)^{-x}$

7. $y = \left(\frac{1}{3}\right)^{-x}$

8. $y = \left(\frac{1}{3}\right)^{2x}$

9. $y = e^x$

10. $y = e^{-x}$

11. $y = e^{2x}$

12. $y = e^{4x}$

13. $y = 4e^{-x}$

14. $y = 2e^{-x}$

15. $y = 3e^{-2x}$

16. $y = 4e^{-3x}$

Bi 17. The growth of a bacteria is given by the equation $N = N_0 3^t$ where $N_0 = $ initial number of bacteria and $t = $ elapsed time. Graph the relationship between N and t if $N_0 = 100$. Using the graph, find the number of bacteria at the end of 4 hours. Verify your answer using a calculator.

C 18. Population growth P is given by the equation $P = P_0 e^{kt}$ where $P_0 = $ initial population, $t = $ time, and $k = $ rate of growth. Graph the yearly population growth of the city of Glenwood from 1976 to 1983 if its population in 1976 was 250 and $k = 0.05$. Use the graph to predict the population of Glenwood in 1989.

B 19. On his daughter's sixteenth birthday, a father promises to give her $30,000 on her twenty-fifth birthday. If interest is compounded continuously at an annual rate of 9%, how

much money should the father put into an account initially? (*Note:* $P = Se^{-rt}$ where $P = $ initial investment, $S = $ yield at t years, and $r = $ rate of interest.)

P 20. A radioisotope decays at a rate equal to $N = N_0(1/2)^x$ where $N_0 = $ initial number of nuclei present after x half-lives. How many nuclei remain after 3 half-lives if the initial number is 1,000,000?

B 21. A *continuous* annuity is an annuity in which R dollars are dispersed annually by uniform payments that are payable continuously. The present value y of a continuous annuity for x years is

$$y = R\left(\frac{1 - e^{-rx}}{r}\right)$$

where $r = $ annual interest rate compounded continuously. Find the present value of an annuity in which $3,000 is invested annually for 10 years at 12% annual interest.

Bi 22. A yeast manufacturer finds that yeast grows exponentially at a rate of 20% per hour. How many pounds of yeast will the manufacturer have at the end of 12 hours if 100 pounds is used initially? (Use $y = ae^{nx}$. See Example 7.)

B 23. Using Example 8, determine the difference in monthly versus continuous compounding if $15,000 is invested at 8% for 6 years.

E 24. The circuit shown in Figure 8–8 consists of a voltage source V, a resistance R, a capacitance C, and a switch. The current is given by $I = (V/R)e^{-t/(RC)}$. Find the current after 2 seconds if $V = 100$ V, $R = 350$ Ω, and $C = 0.1$ mF.

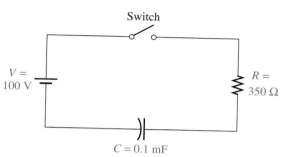

Switch

$V = 100$ V $R = 350$ Ω

$C = 0.1$ mF

FIGURE 8–8

E 25. In the circuit shown in Figure 8–9, determine the instantaneous voltage across the capacitor 2 seconds after the switch is closed. Use the relationship $V_c = V(1 - e^{-t/(RC)})$ when $V = 110$ V, $R = 50$ Ω, and $C = 0.05$ F.

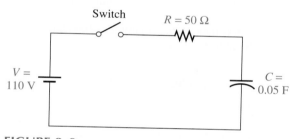

Switch $R = 50$ Ω

$V = 110$ V $C = 0.05$ F

FIGURE 8–9

E 26. Plot the relationship between I and t in the formula $I = (V/R)(1 - e^{-Rt/L})$ where $V = 9$ V, $R = 1.5$ Ω, and $L = 10$ H.

B 27. An amount of money R invested every year for x years at i rate of interest will accumulate to y dollars. The formula relating these factors is given by

$$y = R\left(\frac{(1 + i)^x - 1}{i}\right)$$

If Jose pays $2,000 each year into a retirement plan that earns interest at 12% compounded annually, what will be the value of his annuity after 30 years?

B 28. The demand for green widgets is given by $P = 750 - 0.3e^{0.006x}$ where $P = $ price and $x = $ demand for a number of units. Find the price for a demand of 1,500 units.

E 29. Determine the instantaneous current 3 seconds after the switch is closed in the circuit shown in Figure 8–10. Use

the relationship $I = (V/R)(1 - e^{-Rt/L})$ when $V = 12$ V, $R = 10$ Ω, and $L = 10$ H.

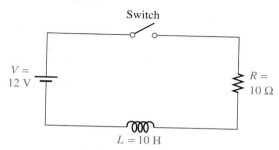

Switch

$V = 12$ V $R = 10$ Ω

$L = 10$ H

FIGURE 8–10

E 30. When the switch in Figure 8–11 is closed, the current I grows exponentially according to the formula $I = (V/R)(1 - e^{-(R/L)t})$ where L is inductance in henries and R is resistance in ohms. Find the current at 0.05 s if $R = 500$ Ω, $L = 100$ H, and $V = 120$ V.

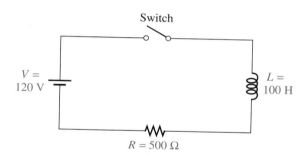

Switch

$V = 120$ V $L = 100$ H

$R = 500$ Ω

FIGURE 8–11

B 31. On the day his grandchild is born, Richard deposits $5,000 in a trust fund that pays 9.5% interest, compounded continuously. How much will be in the fund when the grandchild is 18?

B 32. The capital recovery factor is the annual payment y required to pay off completely some present amount P over x years at i rate of interest. The capital recovery factor is given by

$$y = P\left(\frac{i(1 + i)^x}{(1 + i)^x - 1}\right)$$

If a businessperson wants to pay off a $100,000 debt over 10 years at an interest rate of 8% compounded annually, what annual payment is required?

8–2

THE LOGARITHMIC FUNCTION

The equation $y = b^x$ is an exponential function, which means that it has an inverse function. The inverse of a function is found by interchanging the variables x and y. This means the inverse of the exponential function $y = b^x$ is given by $x = b^y$. This inverse is so important that it is given a name, the logarithmic function. The fact that these two functions are inverses allows us to express an equation in either logarithmic or exponential form, whichever is most convenient.

Definition

Definition of Logarithm

$$y = b^x$$

if and only if

$$x = \log_b y$$

where

$$x = \text{any real number}$$
$$b = \text{any positive real number, } b \neq 1$$
$$y = \text{any positive real number}$$

NOTE ✦ Since y must be a positive number, the logarithm of zero and negative numbers is undefined.

The definition of a logarithm provides a mechanism of solving an exponential function for the exponent. Therefore, we must be able to translate from exponential form to logarithmic form. This relationship is as follows:

$$y = b^x \qquad \text{exponential form}$$
$$x = \log_b y \qquad \text{logarithmic form}$$

EXAMPLE 1 Express each of the following equations in exponential form:

(a) $\log_4 16 = 2$ (b) $\log_2 8 = 3$ (c) $\log_e x = 3$
(d) $\log_{25} 5 = 1/2$ (e) $\log_b 1 = 0$

Solution

(a) $4^2 = 16$ (b) $2^3 = 8$ (c) $e^3 = x$
(d) $25^{1/2} = 5$ (e) $b^0 = 1$

In part (e), since any nonzero base raised to the zero power equals one, the logarithm of one is always zero. ■ ■ ■

EXAMPLE 2 Express each of the following equations in logarithmic form.

(a) $3^4 = 81$ (b) $5^3 = 125$
(c) $b^8 = 246$ (d) $e^4 = d$

Solution

(a) $\log_3 81 = 4$ (b) $\log_5 125 = 3$
(c) $\log_b 246 = 8$ (d) $\log_e d = 4$ ■ ■ ■

The logarithmic bases that are used most frequently are 10 and e. Logarithms with a base of 10 are called **common logarithms** and are written as log x. If the base of a logarithm is not written, it is understood to be ten. Likewise, logarithms with a base of e are called **natural logarithms** and are written as $\ln x$. Calculators calculate logarithms only to base 10 and base e. Base 10 on your calculator will use the $\boxed{\log}$ key, and base e will use the $\boxed{\ln}$ (natural logarithm) key. In other words, $\log_e x = \ln x$.

Solving Logarithmic Equations

To solve a logarithmic equation, convert the equation to exponential form. This procedure is developed in the next example.

EXAMPLE 3 Solve the following equations for the variable:

(a) $\log_x 27 = 3$ (b) $\log_2 16 = x$ (c) $\log_4 x = 3$

(d) $\log_9(1/81) = x$ (e) $\log_3 81^{-1} = x$

Solution

(a) $\log_x 27 = 3$ is equivalent to $x^3 = 27$, so $x = 3$.

(b) $\log_2 16 = x$ is equivalent to $2^x = 16$, so $x = 4$.

(c) $\log_4 x = 3$ is equivalent to $4^3 = x$, so $x = 64$.

(d) $\log_9(1/81) = x$ is equivalent to $9^x = 1/81$. Since $81 = 9^2$, it will be easier if we express both sides of the equation in terms of the same base.

$$9^x = \frac{1}{81}$$

$$9^x = \frac{1}{9^2}$$

$$9^x = 9^{-2}$$

$$x = -2$$

(e) $\log_3 81^{-1} = x$ is equivalent to $3^x = 81^{-1}$ or $3^x = (3^4)^{-1}$, so $x = -4$. ■■■

Solving Exponential Equations

To solve an exponential equation, convert the equation to a logarithmic expression.

 EXAMPLE 4 The formula to evaluate the initial investment P needed to accumulate an amount S at a rate r compounded continuously in a certain time t is

$$P = Se^{-rt}$$

Solve the equation for t and determine the time required for a $2,000 investment to accumulate to $3,500 at an interest rate of 8% per year compounded continuously.

Solution To solve the equation for t, we divide both sides by S.

$$\frac{P}{S} = e^{-rt}$$

Using the definition of logarithms, we convert to the following equation:

$$\log_e\left(\frac{P}{S}\right) = -rt$$

But log to base e is natural log, ln.

$$\ln\left(\frac{P}{S}\right) = -rt$$

Dividing by $-r$ gives the solution for t.

$$\frac{\ln\left(\dfrac{P}{S}\right)}{-r} = t$$

Substituting the values $P = 2{,}000$, $S = 3{,}500$, and $r = 0.08$ gives

$$\frac{\ln(0.5714)}{-0.08} = t$$

$$7 \text{ years} \approx t$$

2000 $\boxed{\div}$ 3500 $\boxed{=}$ $\boxed{\ln x}$ $\boxed{\div}$ 0.08 $\boxed{+/-}$ $\boxed{=}$ $\rightarrow$ 6.995 ■■■

Logarithmic Functions

The logarithmic relationship is also a function. However, in writing a function, we normally label x the independent variable and y the dependent variable. By interchanging x and y in the previous definition, we can write the logarithmic function in the traditional manner, as follows.

Logarithmic Function

$$y = \log_b x$$

where $x > 0$, $b > 0$, and $b \neq 1$.

Graphing Logarithmic Functions

The graphs in the next two examples illustrate some of the properties of logarithms.

EXAMPLE 5   Graph $y = 3 \log x$.

Solution   The domain of the logarithmic function is restricted. Remember that the function is undefined for $x \leq 0$. The table of values is given below, and the graph is shown in Figure 8–12.

x	1/2	1	2	4	5	8	10	15	20
y	-0.9	0	0.9	1.8	2.1	2.7	3	3.5	3.9

FIGURE 8–12

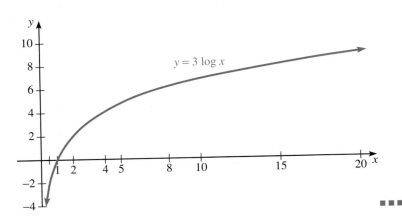

$y = 3 \log x$

■■■

EXAMPLE 6 Graph $y = 3 \log_{0.4} x$.

Solution Since the calculator performs only logarithmic operations with base 10 or e, we must convert this equation to exponential form as follows:

$$y = 3 \log_{0.4} x$$

$$\frac{y}{3} = \log_{0.4} x$$

$$0.4^{y/3} = x$$

Then we fill in a table of values by choosing values for y and calculating x. The graph of $y = 3 \log_{0.4} x$ is given in Figure 8–13.

x	y
3.39	−4	0.4 $\boxed{y^x}$ $\boxed{(}$ 4 $\boxed{+/-}$ $\boxed{\div}$ 3 $\boxed{)}$ $\boxed{=}$ → 3.39
1.84	−2	0.4 $\boxed{y^x}$ $\boxed{(}$ 2 $\boxed{+/-}$ $\boxed{\div}$ 3 $\boxed{)}$ $\boxed{=}$ → 1.84
1.36	−1	0.4 $\boxed{y^x}$ $\boxed{(}$ 1 $\boxed{+/-}$ $\boxed{\div}$ 3 $\boxed{)}$ $\boxed{=}$ → 1.36
1	0	0.4 $\boxed{y^x}$ $\boxed{(}$ 0 $\boxed{\div}$ 3 $\boxed{)}$ $\boxed{=}$ → 1
0.74	1	0.4 $\boxed{y^x}$ $\boxed{(}$ 1 $\boxed{\div}$ 3 $\boxed{)}$ $\boxed{=}$ → 0.74
0.40	3	0.4 $\boxed{y^x}$ $\boxed{(}$ 3 $\boxed{\div}$ 3 $\boxed{)}$ $\boxed{=}$ → 0.40
0.29	4	0.4 $\boxed{y^x}$ $\boxed{(}$ 4 $\boxed{\div}$ 3 $\boxed{)}$ $\boxed{=}$ → 0.29

FIGURE 8–13

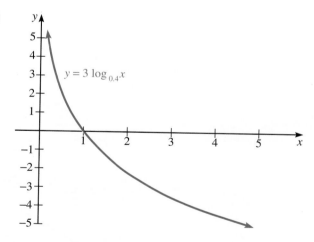

$y = 3 \log_{0.4} x$

Examples 5 and 6 demonstrate the following properties of the logarithmic function $y = \log_b x$:

- If $0 < b < 1$, the function is decreasing; that is, y decreases as x increases (Example 6).
- If $b > 1$, the function is increasing (Example 5).
- The y axis is an asymptote of the curve.
- The x intercept is the point $(1, 0)$.
- $\log_b x$ is undefined for $x \le 0$.

Graphs on Logarithmic and Semilogarithmic Paper

With logarithmic functions, generally one of the variables changes much more rapidly than the other variable. This discrepancy can sometimes make reading the graph more difficult. To compensate for a large change in the values of one variable, we can use **semilogarithmic,** or semilog, graph paper. If both variables have a large change in value, we use logarithmic, or **log-log,** graph paper. The next two examples illustrate the use of semilog and log-log graph paper.

FIGURE 8–14

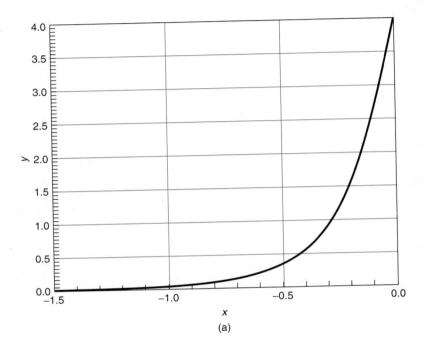

(a)

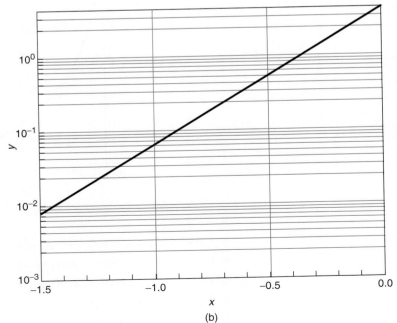

(b)

EXAMPLE 7   Construct the graph of $y = 4^{3x+1}$ on semilog graph paper.

Solution   Figure 8–14a shows the graph of $y = 4^{3x+1}$ on regular graph paper. Notice that it is very difficult to determine values for y when x is between -1.5 and -1. Figure 8–14b shows the semilog graph of this equation.   ■ ■ ■

EXAMPLE 8   Construct the graph of $x^3y^2 = 15$ on log-log paper.

Solution   Figure 8–15a shows the graph of $x^3y^2 = 15$. It is difficult to determine values of y when x is close to one. However, when the x and y axes are logarithmic in scale, the graph is a straight line.

FIGURE 8–15

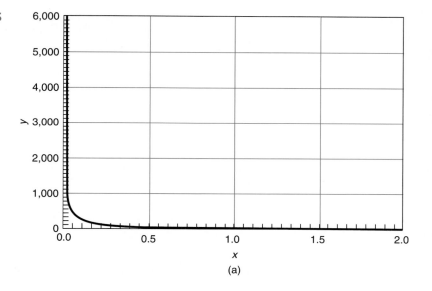

(a)

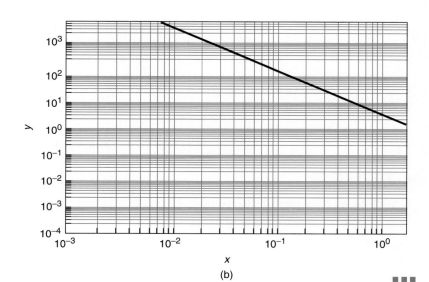

(b)   ■ ■ ■

Applications

The ear hears in a logarithmic fashion. This means that the multiplication of the acoustic power to a sound is heard as an increment of loudness of one decibel (dB). Over the years, the decibel has been used to indicate the performance of many electrical devices, including amplifiers.

EXAMPLE 9 Determine the power gain of an amplifier when an input power of 0.75 W produces an output power of 50 W.

Solution The formula comparing input and output power levels is

$$N = 10 \log\left(\frac{P_{out}}{P_{in}}\right)$$

where N = gain or loss in decibels, P_{out} = output power (W), and P_{in} = input power (W). Substituting the values given

$$N = 10 \log\left(\frac{50 \text{ W}}{0.75 \text{ W}}\right)$$

$$N = 10 \log(66.67)$$

$$N \approx 18.24 \text{ dB}$$

The power gain or loss in decibels may be computed from a voltage ratio if the measurements are made across equal resistances ($R_{in} = R_{out}$). This leads to the following formula:

$$N = 20 \log\left(\frac{V_{out}}{V_{in}}\right)$$

EXAMPLE 10 Determine the decibel loss in a 72-Ω coaxial cable that has an input of 9 V. The output is 3.0 V across a 72-Ω termination.

Solution Since the input and output resistances are equal, we can use the formula

$$N = 20 \log\left(\frac{V_{out}}{V_{in}}\right)$$

Substituting and calculating gives

$$N = 20 \log\left(\frac{3.0}{9}\right)$$

$$N = -0.48 \text{ dB}$$

The negative sign on the answer indicates a loss of 0.48 dB. ∎

8–2 EXERCISES

Convert the following logarithmic expressions into exponential form.

1. $\log_3 27 = 3$

2. $\log_4 64 = 3$

3. $\log_2 16 = 4$

4. $\log_6 216 = 3$

5. $\log_5 125 = 3$

6. $\log_2 128 = 7$

7. $\ln 20.1 = 3$

8. $\log_{343} 49 = 2/3$

9. $\log_6 1/36 = -2$

10. $\ln 2.72 = 1$

11. $\ln 7.39 = 2$

12. $\ln 148.41 = 5$

13. $\log_{8/27} 9/4 = -2/3$

14. $\log_{16/81}(27/8) = -3/4$

Convert the following exponential expressions into logarithmic form.

15. $6^3 = 216$

16. $3^4 = 81$

17. $2^5 = 32$

18. $4^3 = 64$

19. $3^{-4} = 1/81$

20. $2^{-4} = 1/16$

21. $(1/3)^3 = 1/27$

22. $e^4 = 54.6$

23. $e^2 = 7.39$

24. $27^{2/3} = 9$

25. $16^{-3/4} = 1/8$

26. $81^{-3/4} = 1/27$

27. $e^{-1} = 0.37$

28. $e^{-3} = 0.05$

29. $8^{-2/3} = 1/4$

30. $125^{-1/3} = 1/5$

Without using a calculator, solve the following equations for the variable.

31. $\log_2 64 = x$

32. $\log_4 y = 4$

33. $\log_8 y = 3$

34. $\log_{1/2} 4 = x$

35. $\log_b 16/9 = -2$

36. $\log_b 8 = -3$

37. $\log 10^8 = x$

38. $\log_b 12^{-3} = -3$

39. $\ln e^x = 6$

40. $\ln e^4 = x$

41. $\ln e = x$

42. $\ln(1/e^2) = x$

43. $\ln(1/e) = x$

44. $\ln e^x = 8$

Graph the following functions.

45. $y = \log_2 x$

46. $y = \log_4 x$

47. $y = \log_{0.05} x$

48. $y = 2 \log x$

49. $y = \log x$

50. $y = \log_{0.7}(x - 1)$

51. $y = 4 \log(x + 1)$

52. $y = \log(-x)$

53. $y = -2 \log_{0.4} x$

54. $y = 3 \log_{0.8} 2x$

55. $y = \ln 2x$

56. $y = \ln(x + 3)$

57. $y = -3 \ln x$

58. $y = 4 \ln x$

59. $y = 4 \ln(x - 2)$

60. $y = 5 \ln(2x + 1)$

B 61. Using Example 4, determine how long it takes an initial investment of $11,000 to accumulate to $25,000 at an interest rate of 10% compounded continuously.

C 62. The equation for population growth P is given by $P = P_0 e^{kt}$ where $P_0 =$ initial population, $t =$ time, and $k =$ rate of growth. Solve for t.

P 63. A certain radioisotope decays at a rate equal to $N = N_0(1/2)^x$ where N_0 is the initial number of nuclei present after x half-lives. Solve for x.

E 64. A circuit consists of a voltage source V, a resistance R, a capacitance C, and a switch. The current is given by $I = (V/R)e^{-t/(RC)}$. Solve for t.

P 65. According to Lambert's law, the intensity I of light with original intensity I_0 after passing through a thickness x of a material whose absorption coefficient is k is given by $I = I_0 e^{-kx}$. Solve for x.

Bi 66. Yeast grows at a rate given by $y = 100e^{0.2x}$. Solve for x in terms of y.

B 67. A principal invested at 12% and compounded continuously increases to k times the original principal after t years. The relationship is given by

$$t = \frac{\ln k}{0.12}$$

How long will it take to double the principal?

CS 68. The time in picoseconds, represented by t, required for a number of calculations, represented by N, by a certain computer is $t = N + \log_2 N$. Sketch the graph of this function.

E 69. The input to a preamp is 8 V and the output is 10 V. Assuming that each voltage is across equal resistances, determine the power gain of the preamp.

E 70. An RF linear amplifier requires 3.5-W input to develop a 56-W output. Determine the power gain of the amplifier in decibels.

8–3

PROPERTIES OF LOGARITHMS

Since a logarithm represents an exponential expression, the properties of logarithms are based on the Laws of Exponents. The following Laws of Exponents from Chapter 3 are of greatest importance.

Laws of Exponents

$$b^x \cdot b^y = b^{x+y} \qquad \text{multiplication}$$

$$\frac{b^x}{b^y} = b^{x-y} \qquad \text{division, } b \neq 0$$

$$(b^x)^p = b^{xp} \qquad \text{power}$$

Next, we will relate each of these rules to properties of logarithms. If we let $m = \log_b x$ and $n = \log_b y$, then converting these expressions to exponential form gives $b^m = x$ and $b^n = y$. Applying the multiplication rule for exponents gives

$$xy = b^m b^n$$
$$xy = b^{m+n}$$

Converting this expression to logarithmic form gives

$$\log_b(xy) = m + n$$

Substituting the assigned values $m = \log_b x$ and $n = \log_b y$ gives the following property of logarithms. Figure 8–16 illustrates graphically the validity of this property. Note that the graphs of log $(3x)$ and log $3 + \log x$ are the same.

FIGURE 8–16

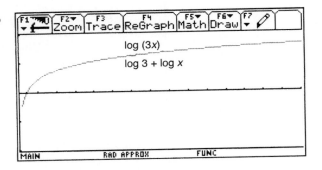

Logarithm of a Product

$$\log_b(xy) = \log_b x + \log_b y \qquad \text{Rule 8–1}$$

This property of logarithms is based on the multiplication law of exponents.

$$b^x \cdot b^y = b^{x+y}$$

Verbally, this rule states that the logarithm of a product is the sum of the logarithms of each factor.

Similarly, using the same definition for m and n, we have

$$\frac{x}{y} = \frac{b^m}{b^n} = b^{(m-n)}$$

Writing this equation in logarithmic form gives

$$\log_b\left(\frac{x}{y}\right) = m - n$$

Substituting $m = \log_b x$ and $n = \log_b y$ gives the following property of logarithms. Figure 8–17 illustrates graphically the validity of this property. Note that log $(x/3)$ results in the same graph as log $x - \log 3$.

FIGURE 8–17

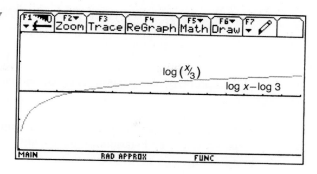

Logarithm of a Quotient

$$\log_b\left(\frac{x}{y}\right) = \log_b x - \log_b y \qquad \text{Rule 8–2}$$

This property of logarithms is based on the division law of exponents:

$$\frac{b^x}{b^y} = b^{x-y}$$

Verbally, this rule states that the logarithm of a quotient is the logarithm of the numerator minus the logarithm of the denominator.

Let $m = \log_b x$. Then converting to an exponential expression gives

$$b^m = x$$

Raising both sides of the equation to the p power gives

$$(b^m)^p = x^p$$
$$b^{mp} = x^p$$

Converting this exponential expression to a logarithmic form gives

$$\log_b(x^p) = mp$$

Substituting for m gives the following logarithmic expression. Note from Figure 8–18 that the graphs of $\log(3^x)$ and $x \log 3$ are the same.

FIGURE 8–18

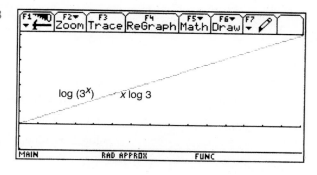

Logarithm of a Power

$$\log_b(x^p) = p \log_b x \quad \text{Rule 8–3}$$

This property of logarithms is based on the power law of exponents.

$$(b^x)^p = b^{xp}$$

Verbally, this rule states that the logarithm of a quantity raised to a power equals the power times the logarithm of the quantity.

Other properties of logarithms that result from the three properties are listed below.

Other Properties of Logarithms

$$\log_b(b^n) = n \quad \text{Rule 8–4}$$
$$\log_b b = 1 \quad \text{Rule 8–5}$$
$$\log_b 1 = 0 \quad \text{Rule 8–6}$$

CAUTION ✦ Be very careful in using the properties of logarithms. There is no rule for simplifying such expressions as $\log_b(x + y)$ or $\log_b(x - y)$ because there are no laws of exponents for $b^x + b^y$ or $b^x - b^y$. Keep relating the properties of logarithms back to the rules for exponents.

The following examples illustrate the properties of logarithms.

EXAMPLE 1   Express each of the following as a sum, difference, or multiple of logarithms:

(a) $\log_2(5x)$ (b) $\log_7\left(\dfrac{6}{x}\right)$ (c) $\log_3(2x^4)$

Solution

(a) $\log_2(5x) = \log_2 5 + \log_2 x$ Rule 8–1

(b) $\log_7\left(\dfrac{6}{x}\right) = \log_7 6 - \log_7 x$ Rule 8–2

(c) $\log_3(2x^4) = \log_3 2 + 4 \log_3 x$ Rules 8–1 and 8–3   ∎

EXAMPLE 2   Express the following as a sum, difference, or multiple of logarithms:

$$\log_b\left(\frac{64c^4}{a^2}\right)$$

Solution

$$\log_b(64c^4) - \log_b a^2 \qquad \text{Rule 8–2}$$
$$\log_b 2^6 + \log_b c^4 - \log_b a^2 \qquad \text{Rule 8–1}$$
$$6 \log_b 2 + 4 \log_b c - 2 \log_b a \qquad \text{Rule 8–3}$$
∎

EXAMPLE 3 Find the exact value of each of the following expressions by applying the properties of logarithms:

(a) $\log_3(9 \cdot 81)$ (b) $\log_3(3^4 \cdot 243)^5$

Solution

(a)

$$\log_3 9 + \log_3 81 \qquad \text{Rule 8–1}$$
$$\log_3 3^2 + \log_3 3^4 \qquad (3^2 = 9 \text{ and } 3^4 = 81)$$
$$2 + 4 \qquad \text{Rule 8–4}$$
$$6$$

(b)

$$5 \log_3(3^4 \cdot 243) \qquad \text{Rule 8–3}$$
$$5[\log_3 3^4 + \log_3 243] \qquad \text{Rule 8–1}$$
$$5(4 + 5) \qquad \text{Rule 8–4}, 3^5 = 243$$
$$45$$

■■■

EXAMPLE 4 Express each of the following as a single logarithm and simplify:

(a) $2 \ln 6 - 3 \ln 3 + \ln 9$ (b) $\log_b(x^2 - 1) + \dfrac{1}{2} \log_b x - \log_b(x + 1)$

Solution

(a)

$$\ln 6^2 - \ln 3^3 + \ln 9 \qquad \text{Rule 8–3}$$
$$\ln \left(\frac{6^2}{3^3} \right) + \ln 9 \qquad \text{Rule 8–2}$$
$$\ln \left(\frac{6^2 \cdot 9}{3^3} \right) \qquad \text{Rule 8–1}$$
$$\ln 12$$

(b)

$$\log_b(x^2 - 1) + \log_b \sqrt{x} - \log_b(x + 1) \qquad \text{Rule 8–3}$$
$$(\text{Remember, } 1/2 \log x = \log x^{1/2} = \log \sqrt{x}.)$$
$$\log_b(x^2 - 1)(\sqrt{x}) - \log_b(x + 1) \qquad \text{Rule 8–1}$$
$$\log_b \left(\frac{(x^2 - 1)(\sqrt{x})}{(x + 1)} \right) \qquad \text{Rule 8–2}$$
$$\log_b [\sqrt{x}(x - 1)] \qquad \text{simplify}$$

■■■

EXAMPLE 5 An expression for the current I with voltage source V, resistance R, capacitance C, and a switch is

$$\ln I + \ln R - \ln V = \frac{-t}{RC}$$

Solve for I.

Solution Let us begin by simplifying the left side of the equation into a single logarithmic expression.

$$\ln(I \cdot R) - \ln V = \frac{-t}{RC} \qquad \text{Rule 8–1}$$

$$\ln\left(\frac{IR}{V}\right) = \frac{-t}{RC} \qquad \text{Rule 8–2}$$

Then we convert from logarithmic form to exponential form.

$$e^{-t/(RC)} = \frac{IR}{V}$$

Solving the equation for *I* gives

$$Ve^{-t/(RC)} = \left(\frac{IR}{V}\right)V \qquad \text{remove fractions}$$

$$Ve^{-t/(RC)} = IR \qquad \text{simplify}$$

$$\frac{V}{R}e^{-t/(RC)} = I \qquad \text{divide by } R \qquad\blacksquare\blacksquare\blacksquare$$

Change of Base

Sometimes it is necessary to change from one logarithmic base to another. The formula for changing bases is as follows.

Change of Base

$$\log_b x = \frac{\log_a x}{\log_a b}$$

where *a* can be any base but is usually 10 or *e*.

 EXAMPLE 6 Find $\log_4 280$ by changing to base 10.

Solution By comparing $\log_4 280$ with the formula for change of base, we find that $b = 4$, $x = 280$, and $a = 10$.

$$\log_4 280 = \frac{\log_{10} 280}{\log_{10} 4}$$

$$\log_4 280 \approx 4.0646$$

$$280 \boxed{\log} \boxed{\div} 4 \boxed{\log} \boxed{=} \rightarrow 4.0646 \qquad\blacksquare\blacksquare\blacksquare$$

8–3 EXERCISES

Evaluate the logarithms by using the properties of logarithms.

1. $\log_2(16 \cdot 8)$

2. $\log_3(81 \cdot 27)$

3. $\log_3 81^2$

4. $\log_4\left(\frac{256}{16}\right)$

5. $\log_2\left(\frac{4 \cdot 64}{16}\right)$

6. $\log_2 16^3$

7. $\log_2(8 \cdot 32)^3$

8. $\log_3(27 \cdot 81)^{1/2}$

9. $\log_3 \sqrt{81 \cdot 3 \cdot 27}$

10. $\log_4 \sqrt{\dfrac{64}{16}}$

11. $\log_2 \sqrt[3]{\dfrac{64}{8 \cdot 2}}$

12. $\log_2(16\sqrt{8})$

Express as a single logarithmic expression and simplify, if possible.

13. $\log x - 2 \log(x - 3)$

14. $3 \log x - 2 \log(y + 1)$

15. $3 \log(x + 1) + \log(x - 3) - \log(x + 1)$

16. $\log(3x - 1) + 2 \log x$

17. $\log x + 3 \log y - \dfrac{1}{2} \log(x - 5)$

18. $\log x - 2 \log y + \log(x + 1) - \log(y + 3)$

19. $\log(5y) + \dfrac{1}{3} \log(2y + 5) - 2 \log y$

20. $\dfrac{1}{2} \log(x + 6) + 4 \log x - \log(2x)$

21. $2 \log x + \log(x + 1) - \log x - \dfrac{1}{3} \log x$

22. $\log(y + 3) - 2 \log(y + 3) - \log y$

23. $3 \log(x - 1) - (\log x + 8 \log x)$

24. $\log(2x - 3) - 2 \log(x + 1) - \log x$

25. $\log\left(\dfrac{x}{y}\right) + 3 \log\left(\dfrac{x}{z}\right)$

26. $2 \log\left(\dfrac{x}{a}\right) + 3 \log\left(\dfrac{y}{b}\right)$

Given that $\log 2 = 0.3010$, $\log 3 = 0.4771$, and $\log 5 = 0.6990$, find the following logarithms by using the properties of logarithms.

27. $\log 6$

28. $\log 12$

29. $\log\left(\dfrac{5}{2}\right)$

30. $\log 45$

31. $\log \sqrt{15}$

32. $\log \sqrt{8}$

33. $\log 24$

34. $\log\left(\dfrac{16}{3}\right)$

35. $\log 75$

36. $\log 200$

37. $\log 90$

38. $\log \sqrt{36}$

39. $\log\left(\dfrac{75}{4}\right)$

40. $\log\left(\dfrac{81}{25}\right)$

Use the formula for changing bases to evaluate the following expressions.

41. $\log_6 203$

42. $\log_2 0.0068$

43. $\log_4(16 \cdot 84)$

44. $\log_3(287 \cdot 5)$

45. $\log_8 576$

46. Graph $y = 2 \log x$ and $y = \log x^2$ on the same set of axes. How do the two graphs compare?

47. Graph $y = \log(x + 3)$ and $y = \log x + \log 3$ on the same set of axes, and compare the results.

P 48. The human ear hears in a logarithmic manner given by the equation

$$\frac{N}{10} = \log\left(\frac{P(\text{out})}{P(\text{in})}\right)$$

where N = gain or loss, $P(\text{in})$ = input power, and $P(\text{out})$ = output power. Express the right side of this equation as a sum, difference, or power of logarithms.

P 49. The atmospheric pressure P at h miles above sea level leads to the equation

$$\ln P - \ln 14.7 \approx -0.21h$$

Solve this equation for P.

CS 50. An equation used to calculate capacity C, in bits per second, of a telephone channel is $C = B \log_2(1 + R)$. Solve for R.

8–4

EXPONENTIAL AND LOGARITHMIC EQUATIONS

Exponential Equations

As defined earlier, an exponential function is a function in which the independent variable appears as an exponent. As we discussed earlier, one technique for solving an exponential equation is to transform each side into an expression with the same base, set the exponents equal, and solve for the variable. However, this technique is not always easy to apply. Therefore, the most applicable method of solving an exponential equation is to take the logarithm of both sides of the equation and then solve. This technique is illustrated in the next four examples.

CAUTION ✦ Always check your answer to make sure it does not result in the logarithm of a negative number.

▦ **EXAMPLE 1**   Solve the equation $3^x = 27$.

Solution   To solve this problem, we could try to make the bases the same and equate the exponents to solve for x. However, it is not always easy to see 27 as 3^3. Therefore, we will use a method of solving exponential equations that always works. Taking the logarithm of both sides gives

$$\log 3^x = \log 27$$

Applying the properties of logarithms, we obtain

$$x \log 3 = \log 27$$

Then, we solve for x using algebraic techniques.

$$x = \frac{\log 27}{\log 3}$$

$$x = 3$$

The graphing calculator solution to this equation is shown in Figure 8–19.

$$27 \;\boxed{\log}\; \boxed{\div}\; 3 \;\boxed{\log}\; \boxed{=} \longrightarrow 3$$

FIGURE 8–19

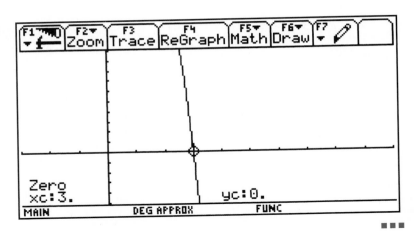

▦ **EXAMPLE 2**   Solve the following equation:

$$3(2^{(3x-4)}) = 12^{(x+1)}$$

Solution   Taking the logarithm of each side of the equation gives

$$\log[3(2^{(3x-4)})] = \log 12^{(x+1)}$$

Next, apply the properties of logarithms.

$$\log 3 + (3x - 4)\log 2 = (x + 1)\log 12$$

Then solve for x.

$$\log 3 + 3x \log 2 - 4 \log 2 = x \log 12 + \log 12$$

$$3x \log 2 - x \log 12 = \log 12 - \log 3 + 4 \log 2$$

$$x(3 \log 2 - \log 12) = \log 12 - \log 3 + 4 \log 2$$

$$x = \frac{\log 12 - \log 3 + 4 \log 2}{(3 \log 2 - \log 12)}$$

$$x \approx -10.3$$

12 $\boxed{\log}$ $\boxed{-}$ 3 $\boxed{\log}$ $\boxed{+}$ 2 $\boxed{\log}$ $\boxed{\times}$ 4 $\boxed{=}$ $\boxed{\div}$

$\boxed{(}$ 2 $\boxed{\log}$ $\boxed{\times}$ 3 $\boxed{-}$ 12 $\boxed{\log}$ $\boxed{)}$ $\boxed{=}$ ⟶ -10.2571

Although x is a negative number, this answer does not result in taking the logarithm of a negative number; therefore, $x \approx -10.3$. The graphing calculator solution to this equation is shown in Figure 8–20.

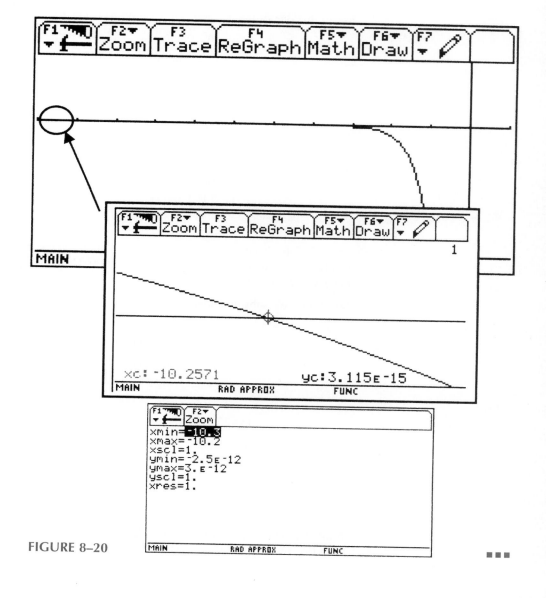

FIGURE 8–20

Applications

One application of exponential equations is exponential growth or decay, discussed earlier in this chapter. The next two examples illustrate exponential growth.

▦ **EXAMPLE 5** The equation for population growth is given by $P = P_0 e^{kt}$. How long will it take Smalltown to achieve a population of 775 if the current population is 490, and its growth rate is 3% per year?

Solution Substituting the given figures into the formula gives

$$775 = 490 e^{0.03t}$$

Then we employ the method used in the previous examples to solve for t. Since the base is e, we take the natural logarithm of both sides of the equation.

$$\ln 775 = \ln(490 e^{0.03t})$$
$$\ln 775 = \ln 490 + \ln e^{0.03t}$$

Recall that one of the properties of logarithms is $\log_b b^n = n$. Therefore,

$$\ln e^{0.03t} = 0.03t$$
$$\ln 775 = \ln 490 + 0.03t$$

Solving for t gives

$$\ln 775 - \ln 490 = 0.03t$$

$$\frac{\ln 775 - \ln 490}{0.03} = t$$

$$15.3 \approx t$$

It will require 15.3 years for Smalltown to grow from a population of 490 to one of 775 if it continues its current rate of growth.

$$775 \boxed{\ln x} \boxed{-} 490 \boxed{\ln x} \boxed{=} \boxed{\div} 0.03 \boxed{=} \longrightarrow 15.28 \qquad \blacksquare\,\blacksquare\,\blacksquare$$

▦ **EXAMPLE 6** Tom wants to invest $15,000 at an interest rate of 10% compounded annually and accumulate this principal to $25,000. How long will this take given the formula $S = P(1 + i)^n$?

Solution We substitute and solve the equation for n.

$$S = P(1 + i)^n$$
$$25{,}000 = 15{,}000(1 + 0.10)^n$$
$$\log 25{,}000 = \log[15{,}000(1 + 0.10)^n]$$
$$\log 25{,}000 = \log[15{,}000(1.10)^n]$$
$$\log 25{,}000 = \log 15{,}000 + n \log 1.10$$
$$\log 25{,}000 - \log 15{,}000 = n \log 1.10$$

$$\frac{\log 25{,}000 - \log 15{,}000}{\log 1.10} = n$$

$$5.4 \approx n$$

$$25000 \boxed{\log} \boxed{-} 15000 \boxed{\log} \boxed{=} \boxed{\div} 1.10 \boxed{\log} \boxed{=} \longrightarrow 5.4$$

Tom's $15,000 will accumulate to $25,000 in 5.4 years if it is invested at 10% compounded annually and if he does not make any deposits or withdrawals. $\blacksquare\,\blacksquare\,\blacksquare$

Logarithmic Equations

A logarithmic equation is an equation that contains the logarithm of the variable. To solve a logarithmic equation, apply the properties of logarithms to *obtain a single logarithmic expression:* then *convert to exponential form.* This technique is illustrated in the next four examples.

EXAMPLE 7   Solve the following equation:

$$\log(2x - 3) - 2 = \log(4x - 1)$$

Solution   Isolating the logarithmic terms on one side of the equation gives

$$\log(2x - 3) - \log(4x - 1) = 2$$

Converting to a single logarithmic expression gives

$$\log\left(\frac{2x - 3}{4x - 1}\right) = 2$$

Converting from a logarithmic to an exponential expression gives

$$\frac{2x - 3}{4x - 1} = 10^2 \qquad \text{base of logarithm is 10}$$

$$\frac{2x - 3}{4x - 1} = 100 \qquad \text{square 10}$$

$$400x - 100 = 2x - 3 \qquad \text{eliminate fractions}$$

$$398x = 97 \qquad \text{transpose and add}$$

$$x \approx 0.244 \qquad \text{divide}$$

Since $\log(2x - 3)$ and $\log(4x - 1)$ are logarithms of negative numbers when $x = 0.244$, this equation has **no solution.**   ∎∎∎

 EXAMPLE 8   Solve the following equation:

$$2 \log x - 1 = \log(x - 1)$$

Solution   Isolating the logarithmic terms and applying the properties of logarithms gives

$$2 \log x - \log(x - 1) = 1 \qquad \text{isolate logarithms}$$

$$\log\left(\frac{x^2}{x - 1}\right) = 1 \qquad \text{convert to single logarithmic expression}$$

$$\frac{x^2}{x - 1} = 10 \qquad \text{convert to exponential form}$$

$$x^2 = 10x - 10 \qquad \text{eliminate fraction}$$

$$x^2 - 10x + 10 = 0 \qquad \text{transpose}$$

Use the quadratic formula to solve for x.

$$x = \frac{10 \pm \sqrt{100 - 4(1)(10)}}{2}$$

$$x = \frac{10 \pm \sqrt{60}}{2}$$

$$x \approx 8.87 \qquad \text{and} \qquad x \approx 1.13$$

Since neither answer results in taking the logarithm of a negative number, the solutions are $x \approx 8.87$ and $x \approx 1.13$.   ∎∎∎

▦ **EXAMPLE 9**   Solve the equation $\ln(2x + 3) - 3 = \ln(x - 5)$.

Solution   We use the same process to solve an equation with natural logarithms. Isolating the logarithmic terms gives

$$\ln(2x + 3) - \ln(x - 5) = 3$$

Then we convert to a single logarithmic expression.

$$\ln\!\left(\frac{2x + 3}{x - 5}\right) = 3$$

Last, we convert to exponential form and solve for x. Note that the base of the logarithm is e instead of 10.

$$e^3 = \frac{2x + 3}{x - 5}$$

$$20 = \frac{2x + 3}{x - 5}$$

$$20x - 100 = 2x + 3$$

$$18x = 103$$

$$x \approx 5.72$$

The graphing calculator solution to this equation is given in Figure 8–21.

FIGURE 8–21

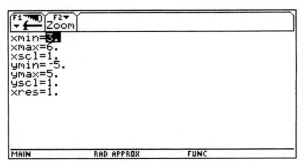

Viewing rectangle

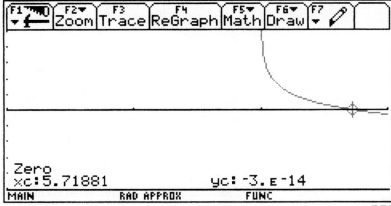

■■■

Applications

⊞ **EXAMPLE 10** In a circuit with a voltage source V, a resistance R, a capacitance C, and a switch, the current I at any time t is given by

$$\ln R + \ln I - \ln V = \frac{-t}{RC}$$

Find the current in a circuit where $R = 500 \ \Omega$, $V = 120$ V, $C = 0.0002$ F, and $t = 0.025$ s.

Solution To provide extra practice, let us solve the equation for I and then substitute the given values. Converting to a single logarithmic expression and solving for I gives

$$\ln\left(\frac{RI}{V}\right) = \frac{-t}{RC}$$

$$e^{-t/(RC)} = \frac{RI}{V} \qquad \text{convert to exponential form}$$

$$\frac{V}{R}e^{-t/(RC)} = I \qquad \text{divide}$$

Substituting the given values yields

$$\frac{120}{500}e^{-0.025/[500(0.0002)]} = I$$

$$0.19 \ \text{A} \approx I$$

The current is 0.19 A.

■■■

When a capacitance is connected in series with a resistance and a dc source of voltage, the instantaneous current is an exponential function. The instantaneous voltage across the capacitance is given by $V_c = V\left(1 - e^{-\frac{t}{RC}}\right)$.

⊞ **EXAMPLE 11** Determine how long it will take for the voltage across the capacitor of Figure 8–22 to reach 80% of the source voltage once the switch is closed.

FIGURE 8–22

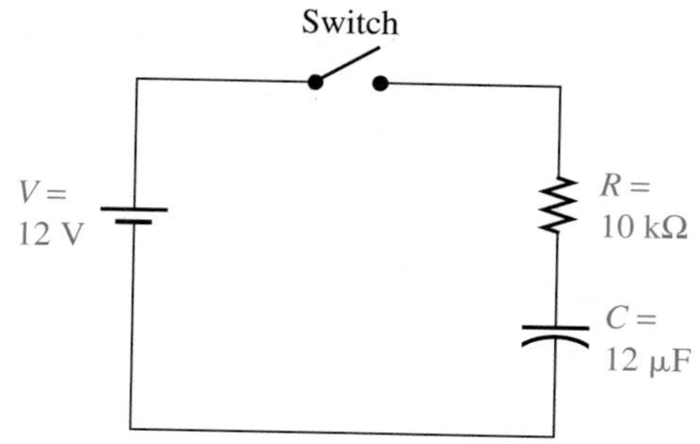

Solution Let $x = \frac{-t}{RC}$ in the equation for V_c.

$$V_c = V(1 - e^x)$$

But we want $V_c = 80\%$ of V. Therefore,

$$0.80V = V(1 - e^x)$$

Solving for x gives

$$\frac{0.80V}{V} = \frac{V(1 - e^x)}{V}$$

$$0.80 = 1 - e^x$$

$$0.2 = e^x$$

Then, take the natural log of each side.

$$\ln 0.2 = x$$

$$-1.61 = x$$

Then, substitute for x and solve for t.

$$x = \frac{-t}{RC}$$

$$-1.61 = \frac{-t}{RC}$$

$$1.61RC = t$$

$$1.61(10 \times 10^3 \ \Omega)(12 \times 10^{-6} \ \text{F}) = t$$

$$0.19 \ \text{s} = t$$

The units of s result from the product of ohms (Ω) and farads (F). This can be shown by writing these physical quantities in terms of their primary *mksq* (rationalized) system units.

$$\Omega F = \left(\frac{(\text{kilograms})(\text{meters})^2}{(\text{seconds})(\text{coulombs})^2} \right) \left(\frac{(\text{seconds})^2(\text{coulombs})^2}{(\text{kilograms})(\text{meters})^2} \right) = \text{seconds}$$

■■■

▦ 8–4 EXERCISES

Solve the following exponential equations.

1. $3^{2x} = 5^{x+1}$
2. $4^{5x} = 6^{x+1}$
3. $2^{3x+1} = 6^x$
4. $7^{3x} = 21$
5. $4^{3x} = 7^{x+1}$
6. $5^{3x+1} = 8^{2x}$
7. $6 \cdot 3^{x+5} = 8$
8. $2 \cdot 4^{3x} = 10^x$
9. $3 \cdot 2^{4x} = 6^{x-3}$
10. $6 \cdot 3^x = 7^{x-2}$
11. $e^{3x} = 5$
12. $e^{5x} = 2$
13. $e^{x+1} = 8$
14. $2e^{x+1} = 9$
15. $3e^{2x-1} = 7$
16. $7e^x = 15$

Solve the following logarithmic equations.

17. $\log(3x + 1) - 1 = \log 2x$
18. $\log 2x + \log 3 = \log 12$
19. $2 + \log(5x - 3) = \log x$
20. $\log(6x + 2) = 1 + \log 2$
21. $\log(x^2 - 4) - 3 = \log(x + 2)$
22. $\log(x^2 + 2x - 8) = 2 + \log(x + 4)$
23. $\log(x^2 - 9) - 1 = \log(x - 3)$
24. $\log(x^2 - 25) - \log(x + 5) = 1$
25. $\log x + \log(x + 1) = \log(2x - 1)$
26. $\log(x + 1) = 2 - \log x$
27. $2 \log x + 1 = \log x$
28. $\log(x^2 + 3) - 1 = 2 \log x$
29. $\log(x - 1) = 2 + 2 \log x$
30. $\ln(2x + 1) - \ln x = 3$
31. $\ln x - \ln(x - 1) = 1$
32. $\ln 3x + \ln x = \ln 9$
33. $\ln(3x + 2) - 2 = \ln(x + 1)$
34. $\ln(3x - 2) = \ln(x + 1)$

P 35. The logarithmic mean temperature difference ΔT_m is given by

$$\Delta T_m = \frac{\Delta T_a - \Delta T_b}{\ln(\Delta T_a / \Delta T_b)}$$

where ΔT_a = change in temperature of the hot liquid and ΔT_b = change in temperature of the cooling liquid. A heat exchanger is operating with a change in the temperature of the hot liquid of 100°, and the change in the temperature of the cooling liquid of 210°. Find the logarithmic mean temperature difference.

Bi 36. The growth A of a certain bacteria is given by

$$A = me^{nx}$$

where m = initial amount of bacteria, n = rate of growth, and x = time. If the biologist starts with 500 cells of the bacteria and they grow at a rate of 3% per hour, how long will it take to produce 750 cells of the bacteria?

B 37. An amount y that must be invested now to produce an amount S in x years at rate i compounded annually is given by

$$y = S\frac{1}{(1 + i)^x} \qquad \text{or} \qquad S(1 + i)^{-x}$$

If Jenny invests \$2,000 at 12% compounded annually, how long will it take her to reach \$7,000?

P 38. The amount of time it takes for a radioisotope to decay to half the initial amount of radioactivity is called the *half-life* and is expressed by the equation $A = A_0e^{-kt}$. Determine the half-life of a radioisotope by letting $A = A_0/2$ and solving for t.

P 39. Determine the power gain N of an amplifier in decibels when an input power of 2 W produces an output power of 150 W if $N = 10 \log(P_{out}/P_{in})$.

Ch 40. In chemistry the pH of a solution is given by

$$pH = -\log(H^+)$$

where H^+ is hydrogen ion concentration. If the pH of a solution is 8.763, find the hydrogen ion concentration.

B 41. The demand for a particular graphing calculator is given by $P = 750 - 0.2e^{0.006x}$. Find the demand x for a price of \$200.

B 42. Using the formula for continuous compounding, $A = Pe^{rt}$ (A = value at t years, P = principal, r = rate), how long will it take to double a \$5,000 investment if the rate is 8%.

CS 43. A particular computer design has 32 different sequences of 10 binary digits so that the total number of possible states is $(2^{10})^{32} = 1,024^{32}$. Evaluate $1,024^{32}$ using logarithms.

CS 44. When designing computers, the number of bits of memory N is usually expressed as a power of 2, as in $N = 2^x$. Find x if $N = 6.5536 \times 10^4$ bits.

E 45. Given the circuit shown in Figure 8–22, determine how long after the switch is closed the instantaneous voltage across the capacitor (Vc) becomes 90% of the source voltage.

E 46. Repeat #45 using 75% of the source voltage.

CHAPTER SUMMARY

Summary of Terms

common logarithm (p. 304)

exponential decay function (p.297)

exponential function (p. 296)

exponential growth function (p. 296)

natural logarithms (p. 305)

Summary of Formulas

$y = b^x$ exponential function

$x = \log_b y$ if $y = b^x$ logarithm

$y = \log_b x$ logarithmic function

$\log_b(xy) = \log_b x + \log_b y$ logarithm of a product

$\log_b\left(\dfrac{x}{y}\right) = \log_b x - \log_b y$ logarithm of a quotient

$\log_b(x^p) = p \log_b x$ logarithm of a power

$\left.\begin{array}{l}\log_b(b^n) = n \\ \log_b b = 1 \\ \log_b 1 = 0\end{array}\right\}$ other properties of logarithms

$\log_b x = \dfrac{\log_a x}{\log_a b}$ change of base

CHAPTER REVIEW

Section 8–1

Graph the following exponential functions.

1. $y = 3^x$

2. $y = 6^{-x}$

3. $y = (1/5)^x$

4. $y = 4^{2x}$

5. $y = 2e^x$

6. $y = e^{2x}$

7. $y = e^{-3x}$

8. $y = 3e^x$

B 9. Compound interest is given by $A = P(1 + i)^t$ where P = principal, i = interest rate, t = time in years, and A = accumulated amount. If $P = \$1,000$ and $i = 10\%$

compounded annually, plot the relationship between t and A.

E 10. In the circuit shown in Figure 8–23, the instantaneous voltage across the capacitor, V_c, is given by

$$V_c = V(1 - e^{-t/(RC)})$$

Find V_c when $V = 120$ V, $t = 2$ s, $R = 50\ \Omega$, and $C = 0.003$ F.

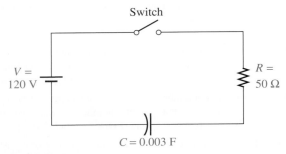

FIGURE 8–23

B 11. Depreciation of business equipment is given by the formula

$$S = C(1 - r)^n$$

where $S =$ scrap value, $C =$ original cost, $r =$ rate of depreciation, and $n =$ useful life in years. Find the scrap value of a machine costing \$8,000 and having a useful life of 10 years and a depreciation rate of 25%.

Section 8–2

Convert the exponential expressions to logarithmic form.

12. $8^3 = 512$

13. $4^{-2} = 1/16$

14. $(1/2)^{-2} = 4$

15. $(81)^{3/4} = 27$

16. $e^0 = 1$

17. $e^3 \approx 20.1$

18. $e^{-4} \approx 0.018$

19. $e^{-2} \approx 0.135$

Convert the logarithmic expressions to exponential form.

20. $\log_{10} 100 = 2$

21. $\log_3 9 = 2$

22. $\log_5(1/125) = -3$

23. $\log_2 32 = 5$

24. $\ln 1.65 \approx 1/2$

25. $\ln 0.25 \approx -1.4$

26. $\ln 0.607 \approx -1/2$

27. $\ln 20.1 \approx 3$

Solve for x (without using a calculator) by converting to exponential form.

28. $\log_x 49 = 2$

29. $\log_8 2 = x$

30. $\log_3 x = -3$

31. $\log_x 6^5 = 5$

32. $\ln e^{-4} = x$

33. $\ln(1/e^4) = x$

Section 8–3

Use the properties of logarithms to evaluate the following.

34. $\log_3 81^4$

35. $\log_5(125 \cdot 25)$

36. $\log_4(256 \div 16)$

37. $\log_2(16 \cdot 32 \div 8)$

38. $\log_2(8^3 \cdot 4)$

39. $\log_3(27^3 \cdot 243)$

Write as a single logarithmic expression and simplify, if possible.

40. $\log(x^2 + 1) - \log x + 3 \log x$

41. $3 \log(x + 1) - \log(x + 1)$

42. $\log(x^2 - 16) + \log(x + 4) - \log(x - 8)$

43. $4 \log x + 6 \log(x + 1) - \log x$

44. $\log(2x + 1) - (\log x + 3 \log y)$

45. $\log\left(\dfrac{x}{y}\right) + 2 \log\left(\dfrac{y}{z}\right)$

Section 8–4

Solve the following equations.

46. $2^x = 3^{2x-1}$

47. $7^{3x} = 6^{2x+1}$

48. $4(3^{5x-1}) = 18$

49. $7(4^{x+1}) = 6^{2x}$

50. $e^{5x-1} = 28$

51. $e^{2x} = 90$

52. $\log x - \log 3 = \log 8$

53. $\log(x + 1) + 1 = \log(3x + 7)$

54. $\log x + \log(x + 2) = -1$

55. $2 \log x = \log(x + 2)$

56. $\ln x - \ln(3x + 2) = 2$

57. $\ln(x + 3) - \ln x = 4$

B 58. Given the equation $A = P(1 + i)^t$, determine how long it would take \$750 to grow into \$1,200 at an interest rate of 10.75% compounded annually.

C 59. Given the equation $P = P_0 e^{kt}$, determine how long it would take a town of 1,200 people (P_0) to grow to 2,000 people (P) if the annual rate of growth k is 8%.

CHAPTER TEST

The number(s) in parentheses refers to the appropriate learning objective(s) at the beginning of the chapter.

1. Graph the logarithmic function $y = 3 \log(x + 1)$. (1)

2. Evaluate the following by using the properties of logarithms: (5)

$$\log_3(81)^8$$

B 3. Depreciation is given by the formula $S = C(1 - r)^n$. What is the useful life n of a computer whose original cost C is \$10,000, depreciation rate r is 15%, and scrap value S is \$2,500? (6)

4. Graph the exponential function $y = 5e^{-x}$. (1)

5. The formula for business depreciation can be written as (3, 5)

$$\log S - \log C = \log(1 - r)^n$$

Solve for S.

6. Solve the following for x: (6)

$$e^{x+1} = 40$$

7. Convert $\left(\dfrac{1}{8}\right)^{1/3} = \dfrac{1}{2}$ to logarithmic form. (3)

8. Graph $x = e^y$ and $\ln x = y$ on the same set of axes. How do the graphs compare? (1)

P 9. The intensity (in lumens) of a light, after passing through a thickness x (in cm) of a medium having an absorption coefficient of 0.3, is given by (2)

$$I = 1{,}500e^{-0.3x}$$

What is the intensity of a light beam that passes through 8 cm of this medium?

10. Write the following as a single logarithmic expression: (5)

$$\log(x + 1) - 2 \log(x - 1) + 3 \log x - \log(3x + 4)$$

11. Solve the following for x: (6)

$$\log(x + 3) - \log 2 = \log 6x$$

12. Convert $\ln 0.37 \approx -1$ into an exponential expression. (3)

13. Given that $\log 5 \approx 0.6990$ and $\log 11 \approx 1.0414$, find $\log (50 \cdot 11^2)$ using the properties of logarithms. (Do not use a calculator.) (5)

14. Solve the following for x: (6)

$$3^{x+1} = 7^{2x-1}$$

15. Solve the following for x by converting to exponential form: (4)

$$\log_6 216 = x$$

P 16. Newton's law of cooling states that when a heated object is immersed in a cooling solution, the temperature T of the object at any time t after immersion is given by (2)

$$T = T_0 e^{-0.4t}$$

where T_0 = initial temperature of the cooling solution. Find T when $T_0 = 230°$ and $t = 5$ s.

Ch 17. Given the equation pH $= -\log[H^+]$, find the hydrogen ion concentration H^+ of a solution whose pH is 5.73. (6)

GROUP ACTIVITY

You have been hired by Freddy's Fish Farm to help determine two important factors in the growth and production of the fish. They want to know what the population should be for the maximum sustained yield (MSY) and the period of time it will take to get to this population number. You use the Internet to research population growth and find the following information. Unconstrained population growth follows an exponential equation:

$$N_t = N_0 e^{rt}$$

where $r = b - d$ [(birth rate − death rate)/1000] and is known as the intrinsic rate of natural increase or the Malthusian parameter, from Thomas Malthus (who described the exponential growth consequences for the human population approximately 200 years ago). The variable N_0 is the initial size of the population, and N_t is the population size at time t.

You also learned that this is an unrealistic model for predicting population growth. Instead, you should use a model that takes into consideration the environmental factors that limit growth (e.g., food). This model is the logistic model of population growth (developed by Belgian mathematician Pierre Verhulst (in 1838). Verhulst suggested that the population increase depended on population density. This means that the exponential curve would rise to some upper limit (carrying capacity), represented by K. The curve is given by:

$$N_t = \frac{N_0 K}{N_0 + (K - N_0)e^{(-r_0 t)}}$$

After reviewing data on the Internet, you determine that a good value for r is 0.41, and the maximum number of fish K that the pond used by the fish farm can support is equal to 1,000.

1. Plot a graph of the population growth over a period of 20 years if the initial number of fish is equal to 100.

2. What is the general shape of the curve when N_t is plotted versus t?

3. The area close to $t = 0$ follows the curve of an unconstrained exponential growth. Plot this graph on the same axis and find what the difference is at 5 years.

4. Find the time when the pond has reached the point of maximum growth.

5. What is the population of the pond at the point of maximum growth? Theoretically, this point on the growth curve is the population level for maximum sustained yield (MSY).

6. How would you determine the time to sell a batch of fish? What would be the size of the shipment (i.e., how many fish would you sell)? How long would it take to make another shipment? Justify your answer.

SOLUTION TO CHAPTER INTRODUCTION

$$P = \frac{50}{1 + 42e^{-0.18(36)}}$$

$$P \approx 47$$

The preserve will have approximately 47 scrub jays.

You are a surveyor responsible for finding the height of a lighthouse. You take two sightings 500 ft apart. The first sighting results in an angle of elevation of 47.7° and the second results in an angle of elevation of 14.2°. How tall is the lighthouse? (The answer to this problem is given at the end of the chapter.)

Ancient Greek engineers used the field of trigonometry to measure objects that could not be measured directly. Trigonometry has evolved to its current status as an invaluable tool in almost every area of science and technology. In this chapter we discuss right angle trigonometry.

Learning Objectives

After you complete this chapter, you should be able to

1. Convert an angle measured in degrees, minutes, and seconds into decimal parts of a degree, and vice versa (Section 9–1).

2. Draw an angle and determine angles coterminal with it (Section 9–1).

3. Given a point on the terminal side of an angle in standard position, determine any of the six trigonometric functions of that angle (Section 9–2).

4. Given the value of one trigonometric function, determine the value of any of the remaining functions of that angle (Section 9–2).

5. Using a calculator, find the value of a given trigonometric function (Section 9–3).

6. Using a calculator, find the first quadrant angle for a given trigonometric value (Section 9–3).

7. Solve a right triangle from given information (Section 9–4).

8. Apply the methods of solving a right triangle to technical problems (Section 9–4).

Chapter 9

Right Angle
Trigonometry

9–1

ANGLES

This chapter begins the study of the branch of mathematics called **trigonometry,** which literally means *triangle measurement.* Trigonometry was developed by Greek astronomers who viewed the sky as the inside of a sphere. Trigonometry was initially used in the development of astronomy, navigation, and surveying. However, trigonometry has evolved to present applications involving sound waves, light rays, vibrating strings, pendulums, and orbits of atomic particles. Trigonometry deals primarily with six ratios called **trigonometric functions,** which are used extensively in physics and engineering. However, before discussing the trigonometric functions, we must define the terminology used.

Definitions

An **angle** is defined as the amount of rotation required to move a ray (a halfline with an endpoint) from one position to another. *The original position of the ray* is called the **initial side** of the angle, and *the final position of the ray* is called the **terminal side** of the angle. *The point about which the rotation occurs and at which the initial and terminal sides of the angle intersect* is called the **vertex.** The measure of the angle itself is the amount of rotation between the initial and terminal sides. The sign of an angle is determined by the direction of rotation of the initial side. *If the rotation from the initial to terminal side is in a counterclockwise direction,* the angle measure is said to be *positive,* and *if the rotation is in a clockwise direction,* the angle measure is said to be *negative.* To specify an angle, we need the initial side, the terminal side, and a curved arrow extending from the initial to the terminal side to show the direction of rotation. Figure 9–1 illustrates this terminology, with angle 1 a positive angle and angle 2 a negative angle.

FIGURE 9–1

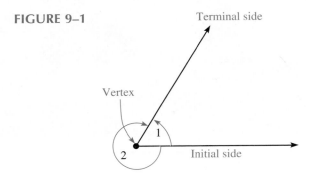

Angular Measure

Angles can be measured in degrees, radians, or grads. One rotation measures 360 degrees, 2π radians, or 400 grads. The discussion in this chapter will concentrate on degree measure, and radians will be discussed in Chapter 10. *The degree can be further divided into 60 equal parts* called **minutes,** and *each minute can be divided into 60 equal parts* called **seconds.** The symbols °, ′, and ″ are used to denote degrees, minutes, and seconds, respectively. Calculators usually use decimal parts of a degree for angular measure rather than degrees, minutes, and seconds. Therefore, to use a calculator, you must convert angles in degrees, minutes, and seconds to degrees. The next two examples illustrate this conversion process.

▣ **EXAMPLE 1**   Convert an angle whose measure is 47°39′53″ to a decimal part of a degree. Round to hundredths.

Solution   First, we convert 53 seconds into a decimal part of a minute by dividing by 60, obtaining 0.88. At this point we have converted 47°39′53″ to 47°39.88′. Then we must change 39.88 minutes into a decimal portion of a degree by dividing by 60. Thus, 47°39′53″ is equivalent to 47.66°.

$$53 \boxed{\div} 60 \boxed{=} 0.88\overline{3} \boxed{+} 39 \boxed{=} \boxed{\div} 60 \boxed{=} \boxed{+} 47 \boxed{=} \longrightarrow 47.66 \quad \blacksquare\blacksquare\blacksquare$$

EXAMPLE 2 Convert an angle whose measure is 158.48° to an angle measured to the nearest minute.

Solution We must convert 0.48° to minutes by multiplying 0.48° by 60′. The angle 158.48° is equivalent to 158°29′. ∎ ∎∎∎

NOTE ✦ Some calculators have a button to convert degrees, minutes, and seconds to a decimal part of a degree, and vice versa. Look in your owner's manual for details on the procedure.

Coterminal Angles

Two angles are **coterminal** if they have the *same initial and terminal sides*. In Figure 9–2, angles *A*, *B*, and *C* are all coterminal angles. To find angles coterminal to a given angle, add or subtract multiples of 360° to it.

FIGURE 9–2

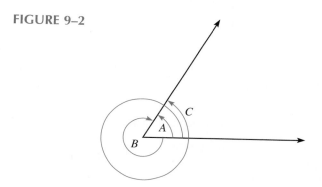

EXAMPLE 3 Find two angles that are coterminal with an angle of 112°.

Solution Since there are 360° in one complete rotation, you can find one angle coterminal with a 112° angle by adding 360° to 112°, which gives 472°. To find a second angle coterminal with 112°, find the negative angle with the same initial and terminal sides. To find the measure of this angle, subtract 360° from 112° to obtain −248°. These angles are illustrated in Figure 9–3. ∎∎∎

FIGURE 9–3

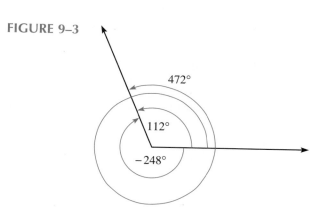

Angles and Rectangular Coordinates

Often we will draw angles on the rectangular coordinate plane. If an angle has *its initial side on the positive x axis and its vertex at the origin*, the angle is said to be in **standard position.** Therefore, the quadrant of an angle in standard position, disregarding the direction of rotation, is determined by the location of the terminal side. An angle whose *terminal side is located on one of the axes* is called a **quadrantal angle.** Figure 9–4 shows angle θ in standard position and angle β a quadrantal angle of 270°. The figure also gives the measure of the other quadrantal angles.

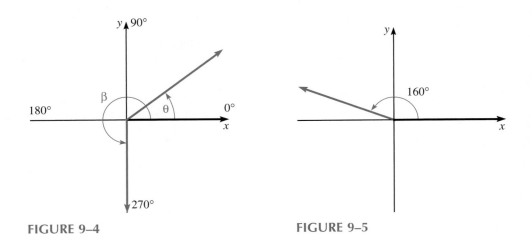

FIGURE 9–4   **FIGURE 9–5**

EXAMPLE 4   Draw an angle in standard position whose measure is 160°.

Solution   We start by positioning the initial side of the angle on the positive *x* axis. Then we rotate that ray in a counterclockwise direction 160°, which places the terminal side of the angle in the second quadrant. This angle is shown in Figure 9–5. This is a second quadrant angle.   ■ ■ ■

An angle in standard position whose terminal side is in the *first, second, third,* or *fourth quadrant* is called a *first, second, third,* or *fourth quadrant angle,* respectively. An angle can be specified by angular measure, as in Example 4, or by a point on the terminal side, as in the next example.

FIGURE 9–6

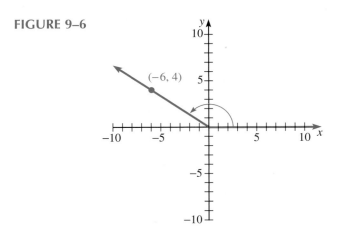

EXAMPLE 5   Draw an angle in standard position whose terminal side passes through the point $(-6, 4)$.

Solution   To draw this angle, we place the initial side of the angle on the positive x axis. Then we plot the point $(-6, 4)$ in the coordinate plane as we did in Chapter 4 and draw a line from the origin through the point. This angle is a second quadrant angle and is shown in Figure 9–6.   ■■■

9–1 EXERCISES

Convert the following angles measured in degrees, minutes, and seconds to angles measured to the nearest hundredth of a degree.

1. $63°16'37''$
2. $158°37'48''$
3. $120°18'24''$
4. $98°26'38''$
5. $148°17'36''$
6. $229°26'15''$
7. $81°43'54''$
8. $346°12'56''$
9. $256°25'37''$

Convert the following angles measured in degrees to angles measured to the nearest minute.

10. $35.68°$
11. $221.76°$
12. $57.20°$
13. $58.39°$
14. $63.56°$
15. $173.18°$
16. $263.8°$
17. $330.28°$
18. $208.43°$

Draw the following angles and find two angles coterminal with each.

19. $35.8°$
20. $167.46°$
21. $318.6°$
22. $193.27°$
23. $56.5°$
24. $75.8°$

25. $18.9°$
26. $347°$
27. $253.6°$

Draw the following angles in standard position and determine whether the angle is a first, second, third, or fourth quadrant angle.

28. $156.7°$
29. $93.4°$
30. $395°$
31. $43.9°$
32. $126.67°$
33. $169.75°$
34. $216.6°$
35. $184.6°$
36. $283.75°$

Draw an angle in standard position such that the terminal side passes through the point given.

37. $(3, 8)$
38. $(-4, 6)$
39. $(3, -6)$
40. $(6, 9)$
41. $(-5, -2)$
42. $(-5, 8)$
43. $(6, -1)$
44. $(-8, -5)$
45. $(7, -4)$

9–2

DEFINING THE TRIGONOMETRIC FUNCTIONS

Trigonometry is based on six ratios that we will now develop. If we place a first quadrant angle θ in standard position and drop a perpendicular line from a point (x, y) on the terminal side of the angle to the x axis, we form a right triangle. As illustrated in Figure 9–7, *the side opposite the right angle* is called the **hypotenuse** or the **radius,** r. *The side opposite angle* θ is called the **opposite side,** and its length can be denoted by the y coordi-

FIGURE 9–7

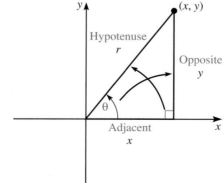

nate of the point (x, y). *The side next to angle θ is called the* **adjacent side,** and its length can be denoted by the x coordinate of the point (x, y). Given this right triangle, the trigonometric functions are defined as the ratio of the sides of the triangle. The **trigonometric functions** are sine (sin), cosine (cos), tangent (tan), secant (sec), cosecant (csc), and cotangent (cot). These functions are defined in the following box.

The Trigonometric Functions

Given the right triangle defined in Figure 9–7.

$$\sin \theta = \frac{\text{opposite}}{\text{hypotenuse}} = \frac{y}{r}$$

$$\cos \theta = \frac{\text{adjacent}}{\text{hypotenuse}} = \frac{x}{r}$$

$$\tan \theta = \frac{\text{opposite}}{\text{adjacent}} = \frac{y}{x}$$

$$\sec \theta = \frac{\text{hypotenuse}}{\text{adjacent}} = \frac{r}{x}$$

$$\csc \theta = \frac{\text{hypotenuse}}{\text{opposite}} = \frac{r}{y}$$

$$\cot \theta = \frac{\text{adjacent}}{\text{opposite}} = \frac{x}{y}$$

NOTE ✦ In determining the trigonometric functions for a given angle, it does not matter which point on the terminal side you choose because the resulting triangles are similar. Since the ratios of the sides of similar triangles are equal, an equivalent trigonometric ratio would result from any point chosen on the terminal side.

Reciprocal Trigonometric Ratios

From the definitions of the six trigonometric functions given earlier, we see that

$$\sin \theta = \frac{y}{r} \bigtimes \frac{r}{y} = \csc \theta$$

Notice that the sine and cosecant functions are reciprocals. Also note that

$$\cos \theta = \frac{x}{r} \bigtimes \frac{r}{x} = \sec \theta$$

and

$$\tan \theta = \frac{y}{x} \bigtimes \frac{x}{y} = \cot \theta$$

Since calculators contain only sin, cos, and tan keys, these reciprocal relationships are extremely important to remember and are summarized in the following box.

Reciprocal Functions

$$\sec \theta = \frac{1}{\cos \theta}$$

$$\csc \theta = \frac{1}{\sin \theta}$$

$$\cot \theta = \frac{1}{\tan \theta}$$

Finding Trigonometric Values

Given the sides of a right triangle, we can determine the value of all six trigonometric functions.

EXAMPLE 1   For the triangle given in Figure 9–8, find the six trigonometric functions. Round to hundredths.

FIGURE 9–8

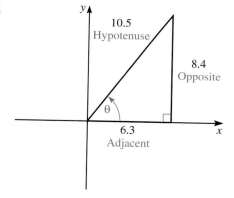

Solution   For angle θ, 8.4 is the length of the opposite side, or y; 6.3 is the length of the adjacent side, or x; and 10.5 is the length of the hypotenuse, or r. Therefore, the trigonometric functions are as follows:

$$\sin \theta = \frac{y}{r} = \frac{\text{opposite}}{\text{hypotenuse}} = \frac{8.4}{10.5} = 0.80$$

$$\cos \theta = \frac{x}{r} = \frac{\text{adjacent}}{\text{hypotenuse}} = \frac{6.3}{10.5} = 0.60$$

$$\tan \theta = \frac{y}{x} = \frac{\text{opposite}}{\text{adjacent}} = \frac{8.4}{6.3} = 1.33$$

$$\csc \theta = \frac{r}{y} = \frac{\text{hypotenuse}}{\text{opposite}} = \frac{10.5}{8.4} = 1.25$$

$$\sec \theta = \frac{r}{x} = \frac{\text{hypotenuse}}{\text{adjacent}} = \frac{10.5}{6.3} = 1.67$$

$$\cot \theta = \frac{x}{y} = \frac{\text{adjacent}}{\text{opposite}} = \frac{6.3}{8.4} = 0.75$$

■ ■ ■

CAUTION ✦ An angle must always be given with a trigonometric function. For example, $\sin = 0.7893$ is a meaningless expression because an angle is not specified. You must write $\sin 52.12° = 0.7893$.

EXAMPLE 2　The terminal side of angle β passes through the point (5, 8). Assuming that β is in standard position, determine the trigonometric functions of this angle to four decimal places.

Solution　To determine the value of the trigonometric functions, we must know the lengths of the three sides of the right triangle shown in Figure 9–9. Therefore, we calculate r using the Pythagorean theorem.

$$\text{hypotenuse}^2 = \text{leg}_1^2 + \text{leg}_2^2$$
$$r^2 = 8^2 + 5^2$$
$$r = \sqrt{8^2 + 5^2}$$
$$r = 9.4340$$

$$8 \boxed{x^2} \boxed{+} 5 \boxed{x^2} \boxed{=} \boxed{\sqrt{}} \longrightarrow 9.4340$$

$$\sin \beta = \frac{\text{opposite}}{\text{hypotenuse}} = \frac{8}{9.4340} = 0.8480$$

$$\cos \beta = \frac{\text{adjacent}}{\text{hypotenuse}} = \frac{5}{9.4340} = 0.5300$$

$$\tan \beta = \frac{\text{opposite}}{\text{adjacent}} = \frac{8}{5} = 1.6000$$

$$\csc \beta = \frac{\text{hypotenuse}}{\text{opposite}} = \frac{9.4340}{8} = 1.1792$$

$$\sec \beta = \frac{\text{hypotenuse}}{\text{adjacent}} = \frac{9.4340}{5} = 1.8868$$

$$\cos \beta = \frac{\text{adjacent}}{\text{opposite}} = \frac{5}{8} = 0.6250$$

FIGURE 9–9

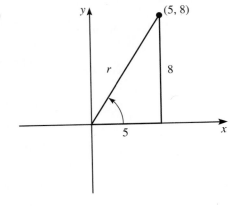

Determining Values from a Given Value

Given the value of one of the trigonometric functions, we can find the value of the remaining functions. The next example illustrates this procedure.

■ EXAMPLE 3   Given that $\cos \alpha = 6/11$, find the remaining trigonometric functions of α to four decimal places.

Solution   Since the cosine is defined to be the ratio of the adjacent side and the hypotenuse, we can label the appropriate sides of the right triangle as shown in Figure 9–10. To find the missing side y, we apply the Pythagorean theorem.

$$11^2 = 6^2 + y^2$$
$$y = 9.2195$$

$$11 \; \boxed{x^2} \; \boxed{-} \; 6 \; \boxed{x^2} \; \boxed{=} \; \boxed{\sqrt{}} \; \longrightarrow \; 9.2195$$

FIGURE 9–10

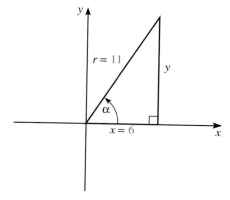

The remaining trigonometric values are calculated as follows:

$$\sin \alpha = \frac{y}{r} = \frac{9.2195}{11} = 0.8381 \qquad \csc \alpha = \frac{r}{y} = \frac{11}{9.2195} = 1.1931$$

$$\tan \alpha = \frac{y}{x} = \frac{9.2195}{6} = 1.5366 \qquad \sec \alpha = \frac{r}{x} = \frac{11}{6} = 1.8333$$

$$\cot \alpha = \frac{x}{y} = \frac{6}{9.2195} = 0.6508$$

■ ■ ■

We have discussed right triangles positioned in the rectangular coordinate plane. However, the trigonometric ratios are true for all right triangles, as shown in Figure 9–11.

FIGURE 9–11

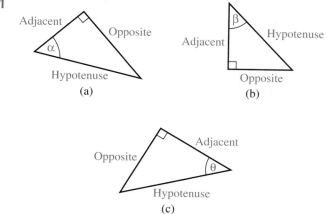

EXAMPLE 4   Given the triangle in Figure 9–11c determine cos θ and csc θ (to 3 decimal places) if the opposite side = 6 and the hypotenuse = 10.

Solution   Since $\cos \theta = \dfrac{\text{adjacent}}{\text{hypotenuse}}$ we must determine the length of the adjacent side using the Pythagorean theorem.

$$(\text{hypotenuse})^2 = (\text{leg}_1)^2 + (\text{leg}_2)^2$$
$$10^2 = 6^2 + (\text{leg}_2)^2$$
$$100 = 36 + (\text{leg}_2)^2$$
$$100 - 36 = (\text{leg}_2)^2$$
$$64 = (\text{leg}_2)^2$$
$$8 = \text{leg}_2$$

$$\cos \theta = \frac{\text{adjacent}}{\text{hypotenuse}} = \frac{8}{10} = 0.800$$

and

$$\csc \theta = \frac{\text{hypotenuse}}{\text{opposite}} = \frac{10}{6} = 1.667 \qquad \blacksquare\blacksquare\blacksquare$$

NOTE ✦ There are certain dimensions for the sides of a triangle that result in a right triangle. They are

$$3, 4, 5 \qquad 5, 12, 13 \qquad 8, 15, 17$$

The longest side is always the hypotenuse, and the other sides are the legs. Not only is 3, 4, 5 a right triangle, multiples of these dimensions, such as 9, 12, 15, also result in a right triangle.

EXAMPLE 5   Given that cos θ = 5/8, find the remaining trigonometric functions of θ. Leave your answer as exact numbers.

Solution   In Figure 9–12 you will find angle θ labeled. Since $\cos \theta = \dfrac{\text{adjacent}}{\text{hypotenuse}} = \dfrac{x}{r}$, these sides have been labeled accordingly. To find the remaining trigonometric functions, we must solve for *y*. Applying the Pythagorean theorem gives

$$r^2 = x^2 + y^2$$
$$(8)^2 = (5)^2 + y^2$$
$$64 - 25 = y^2$$
$$\sqrt{39} = y$$

Applying the definitions of the trigonometric functions gives the following results.

$$\sin \theta = \frac{y}{r} = \frac{\sqrt{39}}{8} \qquad\qquad \csc \theta = \frac{r}{y} = \frac{8}{\sqrt{39}}$$

$$\tan \theta = \frac{y}{x} = \frac{\sqrt{39}}{5} \qquad\qquad \cot \theta = \frac{x}{y} = \frac{5}{\sqrt{39}}$$

$$\sec \theta = \frac{r}{x} = \frac{8}{5} \qquad\qquad\qquad\qquad \blacksquare\blacksquare\blacksquare$$

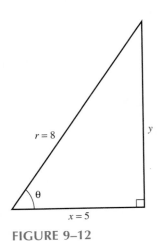

$r = 8$
y
θ
$x = 5$

FIGURE 9–12

9–2 EXERCISES

Each of the following is a point on the terminal side of an angle in standard position. Determine the value to four decimal places of all six trigonometric functions of the angle.

1. (7, 12)

2. $(4, \sqrt{5})$

3. (6, 13)

4. (15, 3)

5. (9, 8)

6. (10, 7)

7. $(\sqrt{3}, 1)$

8. (1, 6)

9. (8, 10)

10. (7.4, 11.8)

11. (16, 12)

12. (9.1, 6.5)

Use the given trigonometric function to find the value of the indicated trigonometric function. Give your answer to four decimal places.

13. $\sin \theta = \dfrac{8}{13}$; find $\cos \theta$ and $\csc \theta$.

14. $\cot \theta = \dfrac{14}{5}$; find $\sin \theta$ and $\cos \theta$.

15. $\sec \theta = \dfrac{11}{7}$; find $\sin \theta$ and $\cot \theta$.

16. $\cos \theta = \dfrac{5}{8}$; find $\tan \theta$ and $\csc \theta$.

17. $\tan \theta = 3$; find $\cos \theta$ and $\csc \theta$.

18. From Figure 9–11a, find $\tan \alpha$ and $\sec \alpha$ when the adjacent side = 4 and the hypotenuse = 5.

19. From Figure 9–11b, find $\sec \beta$ and $\cot \beta$ when the opposite side = 12 and the adjacent side = 5.

20. From Figure 9–11c, find $\tan \theta$ and $\cos \theta$ when the hypotenuse = 17 and the adjacent side = 8.

21. From Figure 9–11a, find $\sin \alpha$ and $\cot \alpha$ when the adjacent side = 16 and the opposite side = 30.

22. From Figure 9–11b, find $\sec \beta$ and $\csc \beta$ when the hypotenuse = 26 and the opposite side = 10.

23. From Figure 9–11c, find $\sin \theta$ and $\csc \theta$ when the adjacent side = 24 and the opposite side = 10.

9–3

VALUES OF THE TRIGONOMETRIC FUNCTIONS

In Section 9–2, we discussed finding the value of the trigonometric functions given the length of the sides of the right triangle. However, in reality we are usually trying to find the value of a trigonometric function given the measure of the angle. Later in this section we will discuss finding the value of trigonometric functions using a calculator. We will begin our discussion with some special angles.

Functions of 45°

If we construct an isosceles (two sides of equal length) right triangle with legs each measuring one unit, the resulting angles of the triangle are 45°, 45°, and 90°. Remember that the sum of the angles of a triangle is 180°. Using the Pythagorean theorem, we find that the hypotenuse of this right triangle is

$$\text{leg}_1^2 + \text{leg}_2^2 = \text{hypotenuse}^2$$
$$\sqrt{\text{leg}_1^2 + \text{leg}_2^2} = \text{hypotenuse}$$
$$\sqrt{1^2 + 1^2} = \sqrt{2}$$

The sides of a 45°, 45°, 90° triangle are 1, 1, and $\sqrt{2}$. The basic trigonometric functions for 45° are given in the table on the following page.

REMEMBER ✦ The longest side of a triangle is always opposite the largest angle, and the shortest side is opposite the smallest angle.

Functions of 45°

$$\sin 45° = \frac{1}{\sqrt{2}} = \frac{\sqrt{2}}{2}$$

$$\cos 45° = \frac{1}{\sqrt{2}} = \frac{\sqrt{2}}{2}$$

$$\tan 45° = \frac{1}{1} = 1$$

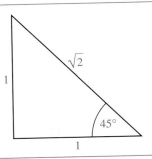

NOTE ✦ Remember that we obtain $\frac{\sqrt{2}}{2}$ from $\frac{1}{\sqrt{2}}$ by rationalizing the denominator.

Functions of 30° and 60°

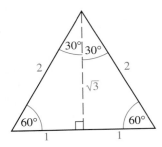

FIGURE 9–13

If we construct an equilateral (three sides of equal length) triangle with sides each two units long, each of the resulting angles is 60° (180°/3 = 60°). If we construct an altitude to the base of the triangle, it bisects both the angle and the base as shown in Figure 9–13. Using the Pythagorean theorem to determine the length of the altitude gives

$$\text{leg}_1^2 + \text{leg}_2^2 = \text{hypotenuse}^2$$
$$\sqrt{\text{hypotenuse}^2 - \text{leg}_1^2} = \text{leg}_2$$
$$\sqrt{2^2 - 1^2} = \sqrt{3}$$

The sides of a 30°, 60°, 90° triangle are 1, $\sqrt{3}$, and 2, respectively. Using this triangle, we can calculate the following trigonometric functions.

Functions of 30°

$$\sin 30° = \frac{1}{2}$$

$$\cos 30° = \frac{\sqrt{3}}{2}$$

$$\tan 30° = \frac{\sqrt{3}}{3}$$

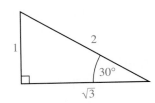

Functions of 60°

$$\sin 60° = \frac{\sqrt{3}}{2}$$

$$\cos 60° = \frac{1}{2}$$

$$\tan 60° = \frac{\sqrt{3}}{1} = \sqrt{3}$$

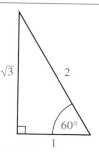

**Functions of 0°
and 90°**

For the quadrantal angle 0° with $r = 1$, we can find the following trigonometric functions.

Functions of 0°

$$\sin 0° = \frac{y}{r} = \frac{0}{1} = 0$$

$$\cos 0° = \frac{x}{r} = \frac{1}{1} = 1$$

$$\tan 0° = \frac{y}{x} = \frac{0}{1} = 0$$

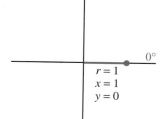

Similarly, for the quadrantal angle 90° with $r = 1$, we can find the following trigonometric functions.

Functions of 90°

$$\sin 90° = \frac{y}{r} = \frac{1}{1} = 1$$

$$\cos 90° = \frac{x}{r} = \frac{0}{1} = 0$$

$$\tan 90° = \frac{y}{x} = \frac{1}{0} = \text{undefined}$$

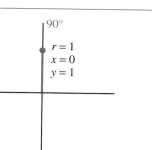

Table 9–1 summarizes the trigonometric functions of the angles discussed in this section. The values of these trigonometric functions are used often enough that you should

memorize them. Also remember that the secant, cosecant, and cotangent functions are reciprocals of the cosine, sine, and tangent functions, respectively, and each can be found from the value of its reciprocal.

TABLE 9–1

Function	Special Angles				
	0°	30°	45°	60°	90°
sin	0	$\dfrac{1}{2}$	$\dfrac{\sqrt{2}}{2}$	$\dfrac{\sqrt{3}}{2}$	1
cos	1	$\dfrac{\sqrt{3}}{2}$	$\dfrac{\sqrt{2}}{2}$	$\dfrac{1}{2}$	0
tan	0	$\dfrac{\sqrt{3}}{3}$	1	$\sqrt{3}$	undef.
cot	undef.	$\sqrt{3}$	1	$\dfrac{\sqrt{3}}{3}$	0
sec	1	$\dfrac{2\sqrt{3}}{3}$	$\sqrt{2}$	2	undef.
csc	undef.	2	$\sqrt{2}$	$\dfrac{2\sqrt{3}}{3}$	1

Cofunctions

Recall from geometry that two angles whose sum is 90° are called **complementary angles.** For example, 43° and 47° are complementary because 43° + 47° = 90°. If α and β are two complementary angles, then $\sin \alpha = \cos \beta$. Trigonometric functions like these are called **cofunctions.** The tangent and cotangent are also cofunctions, as are the secant and cosecant.

Cofunctions

If α and β are complementary angles,

$$\sin \alpha = \cos \beta \qquad \cos \alpha = \sin \beta$$
$$\tan \alpha = \cot \beta \qquad \cot \alpha = \tan \beta$$
$$\sec \alpha = \csc \beta \qquad \csc \alpha = \sec \beta$$

Evaluating Trigonometric Functions

Since the geometric methods used to find the trigonometric value of angles such as 30° or 45° are difficult for other angles, we use a calculator to find the trigonometric functions for most acute angles. To find the value of a trigonometric function with an algebraic calculator, enter the angle and then press the desired trigonometric function key. For example, to find sin 48.6°, make sure your calculator is in the degree mode and your angle is in decimal degrees. Press 48.6 and then the $\boxed{\sin}$ key.

▦ EXAMPLE 1   Using your calculator, find the following trigonometric functions rounded to three decimal places:

(a) sin 63°   (b) cos 38.7°   (c) tan 57.49°   (d) sin 36.7°

Solution   Be sure your calculator is in the degree mode when performing this operation.

(a) 63 $\boxed{\text{sin}}$ $\longrightarrow$ 0.891   (b) 38.7 $\boxed{\text{cos}}$ $\longrightarrow$ 0.780   (c) 57.49 $\boxed{\text{tan}}$ $\longrightarrow$ 1.569

(d) 36.7 $\boxed{\text{sin}}$ $\longrightarrow$ 0.598

■ ■ ■

If you look at your calculator, you will not find keys to represent the secant, cosecant, or cotangent functions. To find the value of these trigonometric functions on the calculator, we must use the reciprocal relationships of the cosine, sine, and tangent, respectively. We use the $\boxed{1/x}$ key on the calculator to compute this reciprocal relationship rapidly.

▦ EXAMPLE 2   Find each of the following to three decimal places:

(a) sec 16.75°   (b) cot 43.3°   (c) csc 74.28°

Solution   Be sure your calculator is in the degree mode.

(a) Since the secant is the reciprocal of the cosine function, we can write

$$\sec 16.75° = \frac{1}{\cos 16.75°}$$

We calculate this expression as

16.75 $\boxed{\text{cos}}$ $\boxed{1/x}$ $\longrightarrow$ 1.044

(b) Since the cotangent is the reciprocal of the tangent function,

$$\cot 43.3° = \frac{1}{\tan 43.3°}$$

43.3 $\boxed{\text{tan}}$ $\boxed{1/x}$ $\longrightarrow$ 1.061

(c) Since the cosecant is the reciprocal of the sine function,

$$\csc 74.28° = \frac{1}{\sin 74.28°}$$

74.28 $\boxed{\text{sin}}$ $\boxed{1/x}$ $\longrightarrow$ 1.039

■ ■ ■

Determining the Angle from the Function Value

To find an angle given its trigonometric function value, we must reverse the process we have just used. For example, to find θ given that sin θ = 0.4384, we must find an angle θ whose sine value is 0.4384. In doing this we are actually using **inverse trigonometric ratios.** We will discuss these ratios in more detail in Chapter 13. For our discussion here, we need to recognize and use the notation. Inverse trigonometric functions are denoted as arcsin θ or $\sin^{-1} \theta$. On the graphing calculator, inverse trigonometric functions are denoted with the -1 exponent. We use the $\boxed{\text{shift}}$ key to access the inverse trig functions.

NOTE ✦ For the trig functions, the -1 exponent on the calculator is used to denote the inverse trigonometric functions, not a negative exponent.

▦ **EXAMPLE 3** Find θ to the nearest tenth of a degree for each of the following:

(a) $\sin \theta = 0.3684$ (b) $\cos \theta = 0.7983$ (c) $\tan \theta = 2.865$

Solution Be sure your calculator is in the degree mode.

(a) 0.3684 ⟨inv⟩ ⟨sin⟩ ⟶ 21.6 $21.6° \approx \theta$

(b) 0.7983 ⟨inv⟩ ⟨cos⟩ ⟶ 37.0 $37.0° \approx \theta$

(c) 2.865 ⟨inv⟩ ⟨tan⟩ ⟶ 70.8 $70.8° \approx \theta$ ■ ■ ■

▦ **EXAMPLE 4** Find θ to the nearest tenth of a degree for the following:

(a) $\cos \theta = 0.5436$ (b) $\tan \theta = 0.3685$

Solution

(a) 0.5436 ⟨inv⟩ ⟨cos⟩ ⟶ 57.071 $57.1° \approx \theta$

(b) 0.3685 ⟨inv⟩ ⟨tan⟩ ⟶ 20.23 $20.2° \approx \theta$ ■ ■ ■

▦ **EXAMPLE 5** Find θ to the nearest tenth of a degree for each of the following:

(a) $\sec \theta = 3.465$ (b) $\cot \theta = 0.1679$ (c) $\csc \theta = 2.693$

Solution

(a) $\sec \theta = \dfrac{1}{\cos \theta} = 3.465$

$\cos \theta = \dfrac{1}{3.465}$

3.465 ⟨1/x⟩ ⟨inv⟩ ⟨cos⟩ ⟶ 73.2 $73.2° \approx \theta$

(b) $\cot \theta = \dfrac{1}{\tan \theta} = 0.1679$

$\tan \theta = \dfrac{1}{0.1679}$

0.1679 ⟨1/x⟩ ⟨inv⟩ ⟨tan⟩ ⟶ 80.5 $80.5° \approx \theta$

(c) $\csc \theta = \dfrac{1}{\sin \theta} = 2.693$

$\sin \theta = \dfrac{1}{2.693}$

2.693 ⟨1/x⟩ ⟨inv⟩ ⟨sin⟩ ⟶ 21.8 $21.8° \approx \theta$ ■ ■ ■

CAUTION ✦ To avoid errors, it is best to write down the reciprocal trigonometric relationships and then perform the operations on your calculator.

▦ **EXAMPLE 6** The range R of a projectile fired at an angle of inclination θ with an initial speed V_o is given by

$$R = \frac{2\, V_o^2 \sin \theta \cos \theta}{g}$$

where g is the acceleration due to gravity (32 ft/s^2). Find the range if a tennis ball is fired at an angle of 48° with an initial speed of 110 ft/s.

Solution   From the information again, $\theta = 48°$ and $V_o = 110$ ft/s. Substituting these values into the formula gives

$$R = \frac{2(110 \text{ ft/s})^2 \sin 48° \cos 48°}{32 \text{ ft/s}^2}$$

$$R = 376 \text{ ft}$$

2 $\boxed{\times}$ 110 $\boxed{x^2}$ $\boxed{\times}$ 48 $\boxed{\sin}$ $\boxed{\times}$ 48 $\boxed{\cos}$ $\boxed{\div}$ 32 $\boxed{=}$ ⟶ 376.05   ∎∎∎

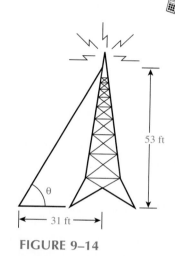

EXAMPLE 7   A guide wire is used to support a 53-ft radio antenna. The wire is anchored 31 ft from the base of the antenna, as shown in Figure 9–14. To find the angle the wire makes with the ground, we must solve the expression cot $\theta = 0.5849$. Find θ to the nearest tenth degree.

Solution   Since $\cot \theta = \dfrac{1}{\tan \theta} = 0.5849$,

$$\tan \theta = \frac{1}{0.5849}$$

$$\theta = 59.7°$$

0.5849 $\boxed{\text{2nd}}$ $\boxed{1/x}$ $\boxed{\text{inv}}$ $\boxed{\tan}$ ⟶ 59.6766   ∎∎∎

53 ft

31 ft

FIGURE 9–14

9–3 EXERCISES

Find the value of the given trigonometric function. Round to four decimal places.

1. csc 37.6°

2. tan 78.56°

3. cos 84°

4. sec 18.6°

5. sin 73.47°

6. sin 19.3°

7. tan 65.7°

8. csc 27.8°

9. sec 31.56°

10. cot 51.35°

11. cot 28.9°

12. cos 44°

13. tan 17.4°

14. sec 23°

15. csc 72°

16. cot 83.76°

17. sec 48.2°

18. cos 32.56°

19. cot 48.7°

20. sin 67.9°

21. cos 36.3°

22. sec 18.7°

23. sin 84.5°

24. sin 58.2°

25. csc 61.65°

26. cot 10.4°

27. sec 76.3°

28. csc 29.1°

Find θ to the nearest tenth of a degree.

29. csc $\theta = 1.259$

30. tan $\theta = 2.268$

31. cos $\theta = 0.378$

32. csc $\theta = 2.631$

33. cot $\theta = 3.76$

34. cos $\theta = 0.968$

35. tan $\theta = 2.526$

36. cot $\theta = 0.67$

37. sec $\theta = 2.87$

38. cot $\theta = 1.37$

39. sec $\theta = 1.169$

40. cos $\theta = 0.421$

41. tan $\theta = 3.617$

42. sin $\theta = 0.861$

43. csc $\theta = 4.815$

44. sec $\theta = 1.973$

45. tan $\theta = 6.739$

46. cos $\theta = 0.391$

47. cot $\theta = 3.785$

48. sec $\theta = 2.78$

49. sin $\theta = 0.765$

50. cot $\theta = 2.538$

51. tan $\theta = 1.328$

52. cot $\theta = 3.614$

53. csc $\theta = 1.38$

54. sec $\theta = 3.482$

55. cos $\theta = 0.540$

56. cos $\theta = 0.382$

M 57. The torque required to move the load up the thread of a screw is given by

$$T = \frac{FD}{2}\left(\frac{\cos A \tan B + f}{\cos A - f \tan B}\right)$$

Find T if $F = 500$ lb, $D = 1.2$ in., $f = 0.15$, $A = 12\,°$, and $B = 2.5°$. Round to tenths.

M 58. The axial force required for a cone brake is given by

$$F = \frac{T(\sin A + f \cos A)}{fR}$$

Find F if $T = 75$ ft-lb, $A = 12°$, $f = 0.25$, and $R = 0.4$ ft.

E 59. For a leading power factor, the voltage induced in an armature is given by

$$E = \sqrt{(V \cos \theta + IR)^2 + (V \sin \theta - IX)^2}$$

Find the induced voltage if $V = 500$ V, $I = 200$ A, $R = 0.1\ \Omega$, $X = 1.2\ \Omega$, and $\theta = 53°$. Round to tenths.

P 60. The index of refraction is defined as

$$n = \frac{\sin i}{\sin r}$$

where i = angle of incidence and r = angle of refraction. Find the index of refraction if $r = 40°$ and $i = 34.7°$.

C 61. Highway curves are banked at an angle θ. The banking angle for a car making a turn of radius r at velocity v is given by $\tan \theta = \dfrac{v^2}{gr}$ where $g = 32$ ft/s^2 or 9.8 m/s^2. Find the banking angle for a car moving 70 km/h to go around a curve 366 m in radius.

9–4

THE RIGHT TRIANGLE

We can model many applied problems by using a right triangle. Trigonometry is extremely useful in solving for unknown angles and unknown sides of a right triangle. Solving a right triangle means determining the measure of all sides and angles of the triangle. In this section we will discuss techniques for solving a right triangle.

Significant Digits

Significant digits and rounding are discussed in Appendix A–11. Two significant digits in the side measurement of a right triangle correspond to the nearest degree in the angle measurement; three significant digits correspond to the nearest tenth degree or multiple of 10′; and four significant digits correspond to the nearest hundredth degree or nearest minute. This information is summarized in the following box. You will need to remember these rules when computing your answers.

Significant Digits in Side Measurement	Accuracy of Degree Measurement
1 or 2	1°
3	0.1° or 10′
4	0.01° or 1′

▦ **EXAMPLE 1**   Find angle A shown in Figure 9–15.

Solution   Recall that each of the six trigonometric functions utilizes one of the acute angles and two sides of the right triangle. To find angle A, we set up a trigonometric ratio using the angle and two known sides of the triangle. As shown in Figure 9–15, we are given the side opposite the angle and the hypotenuse. Therefore, we use the sine function.

$$\sin A = \frac{\text{opposite}}{\text{hypotenuse}}$$

$$\sin A = \frac{17.2}{25.6}$$

$$\sin A \approx 0.672$$

$$A = 42.2°$$

$$17.2 \boxed{\div} 25.6 \boxed{=} \boxed{\text{inv}} \boxed{\text{sin}} \longrightarrow 42.21$$

∎∎∎

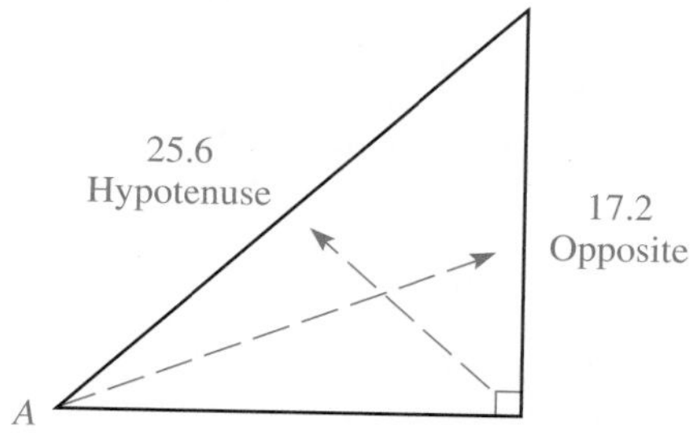

FIGURE 9–15

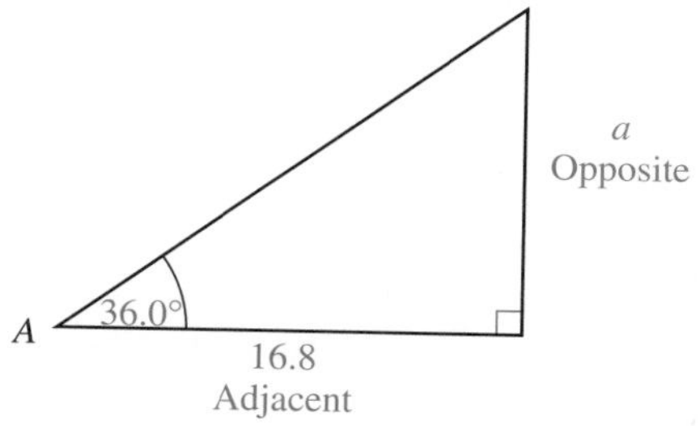

FIGURE 9–16

REMEMBER ✦ The opposite and adjacent sides are relative to the angle involved. The opposite side to one of the acute angles is the adjacent side relative to the other acute angle.

⊞ **EXAMPLE 2** Find side a in the triangle shown in Figure 9–16 if $A = 36.0°$ and $b = 16.8$. Round a to the correct number of significant digits.

Solution We set up a trigonometric ratio involving side a, a known angle, and a known side.

$$\tan A = \frac{\text{opposite}}{\text{adjacent}}$$

$$\tan 36.0° = \frac{a}{16.8}$$

$$a = 16.8(\tan 36.0°)$$

$$a = 12.2$$

$$16.8 \boxed{\times} 36 \boxed{\text{tan}} \boxed{=} \longrightarrow 12.205914$$

∎∎∎

NOTE ✦ There is usually more than one correct way to solve a right triangle. For example, the cotangent function could have been used in Example 2.

⊞ **EXAMPLE 3** Solve the triangle given in Figure 9–17.

Solution We are given ∠ A and the side adjacent to it. We can use the tangent function to find a, the opposite side.

$$\tan A = \frac{\text{opposite}}{\text{adjacent}}$$

$$\tan 37.4° = \frac{a}{21.3}$$

$$a \approx 16.3 \text{ ft}$$

$$21.3 \boxed{\times} 37.4 \boxed{\text{tan}} \boxed{=} \longrightarrow 16.2851$$

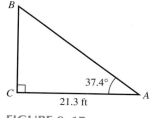

FIGURE 9–17

We can use the cosine function to find c, the hypotenuse.

$$\cos A = \frac{\text{adjacent}}{\text{hypotenuse}}$$

$$\cos 37.4° = \frac{21.3}{c}$$

$$c = 26.8 \text{ ft}$$

$$21.3 \boxed{\div} 37.4 \boxed{\cos} \boxed{=} \longrightarrow 26.8122$$

Since the sum of the angles of a triangle is 180°,

$$B = 180° - 90° - 37.4°$$

$$B = 52.6°$$

CAUTION ✦ When solving right triangles for missing sides or angles, try to use only information given in the problem rather than values that you have calculated. ■■■

Applications

We can use the trigonometric functions to find information in applied problems. The remainder of this section deals with methods of solving applied problems.

▦ **EXAMPLE 4** The angle of elevation from an observer on the ground to the top of a building is 37.4°. Find the height of the building if the observer is 240 ft from the base of the building.

Solution The **angle of elevation** is the *angle formed by the horizontal and the line of sight* of an observer. As shown in Figure 9–18, the angle of elevation is measured above the horizontal. Figure 9–19 gives a visual representation of the verbal statement.

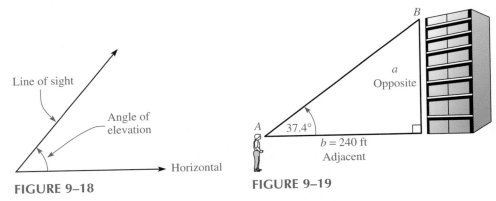

FIGURE 9–18

FIGURE 9–19

Next, we set up a trigonometric ratio and solve for a, the height.

$$\tan A = \frac{a}{b}$$

$$\tan 37.4° = \frac{a}{240}$$

$$a = 240(\tan 37.4°)$$

$$a \approx 183 \text{ ft}$$

$$240 \boxed{\times} 37.4 \boxed{\tan} \boxed{=} \longrightarrow 183.49389$$

■■■

Solving Applied Problems

1. Read the verbal statement.

2. Make a drawing to depict the verbal statement. Label all known quantities and check your figure against the verbal statement.

3. Set up a trigonometric ratio containing the unknown and two known quantities, and then solve for the unknown.

4. Check your solution to see if it is reasonable.

EXAMPLE 5 A person in a plane flying at an altitude of 850.0 ft sees a boat on the lake. If the angle of depression is 25.00°, find the straight-line distance from the plane to the boat.

Solution The **angle of depression** is measured from the horizontal down to the line of sight of an observer. As shown in Figure 9–20, the angle of depression is measured below the horizontal. A drawing of the verbal statement is shown in Figure 9–21.

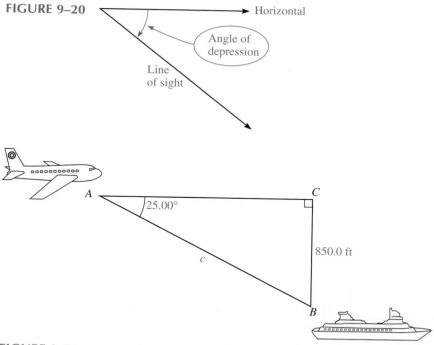

FIGURE 9–20

FIGURE 9–21

$$\sin 25.00° = \frac{850}{c}$$

$$c = \frac{850}{\sin 25.00°}$$

$$c = 2,011 \text{ ft}$$

850 $\boxed{\div}$ 25 $\boxed{\sin}$ $\boxed{=}$ ⟶ 2011.27135

■ ■ ■

EXAMPLE 6 From the top of a 45 ft lighthouse on the Florida coastline, a Coast Guard auxiliary watcher sights a boat in trouble. If the angle of depression of the boat is 8°, how far is the boat from the coastline? Refer to Figure 9–22.

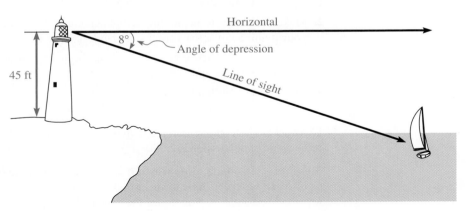

FIGURE 9–22

Solution From the information given, we can draw the right triangle shown in Figure 9–23. From the triangle we know an angle and the side opposite it, and we want to find the side adjacent to this angle.

$$\tan 8° = \frac{45}{x}$$

$$x = \frac{45}{\tan 8°}$$

$$x = 320 \text{ ft}$$

$$45 \boxed{\div} 8 \boxed{\tan} \boxed{=} \longrightarrow 320.1916$$

FIGURE 9–23

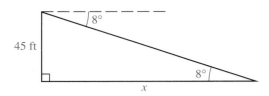

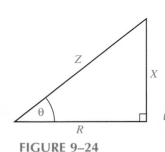

FIGURE 9–24

EXAMPLE 7 **Inductive reactance** X is a *measure of the retardation of current flow* by the capacitance and inductance. **Impedance** Z, measured in ohms Ω, is a *measure of the retardation of current flow* by all components of a circuit. Reactance, impedance, and total resistance R form a right triangle, as shown in Figure 9–24. The angle θ is called the **phase angle.** In an ac circuit with an inductive reactance of 50 Ω and a total resistance of 73 Ω, find the phase angle and impedance.

Solution Figure 9–25 gives a visual representation of the verbal statement. We can solve for impedance Z by using the Pythagorean theorem.

$$Z^2 = (50)^2 + (73)^2$$

$$Z^2 = 7{,}829$$

$$Z = 88 \ \Omega$$

FIGURE 9–25

We use the tangent function to solve for the phase angle θ.

$$\tan \theta = \frac{50}{73}$$

$$34° = \theta$$

50 $\boxed{\div}$ 73 $\boxed{=}$ $\boxed{\text{inv}}$ $\boxed{\text{tan}}$ $\longrightarrow$ 34

■ ■ ■

EXAMPLE 8   A manometer is used in fluid mechanics to measure the change in pressure, ΔP, relative to a change in elevation h. The relationship is given by the formula

$$\Delta P = \gamma h$$

where γ is the specific gravity of the liquid. The inclined well-type manometer shown in Figure 9–26 indicates the movement of gage fluid level L as pressure is applied to the well. The gage fluid has a specific gravity of 0.850 N/m³ and $L = 110$ mm. Calculate the change in pressure. Your answer should be in units of pascals.

FIGURE 9–26

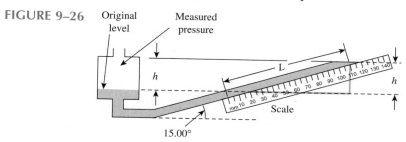

Original level   Measured pressure

h

L

h

mm 10 20 30 40 50 60 70 80 90 100 110 120 130 140

Scale

15.00°

Solution   To calculate the change in pressure, we must know h. From the right triangle in Figure 9–26, where $L = 110$ mm = 0.110 m

$$\sin \theta = \frac{h}{L}$$

$$L \sin \theta = h$$

$$0.110 \sin 15° = h$$

$$0.0285 \text{ m} = h$$

The change in pressure is

$$\Delta P = \gamma h$$

$$\Delta P = \left(0.850 \ \frac{N}{m^3}\right)(0.0285 \text{ m})$$

$$\Delta P = 0.0242 \ \frac{N}{m^2}$$

Pressure is normally measured in pascals (N/m²). The change in pressure is 0.0242 pascals.

■ ■ ■

EXAMPLE 9   One method used in surveying to determine the difference in elevation between two points is called trigonometric leveling. It is commonly used in topographic work and over very rugged terrain. From Figure 9–27 the difference in elevation between points A and B is determined by measuring the inclined distance between them using a tape and measuring the angle of elevation. Distance AB equals distance DC because $DABC$ is a parallelogram and CE represents the difference in elevation between points A and B. If the angle of elevation is 37.6° and $DC = 17.7$ ft, find the difference in elevation.

FIGURE 9–27

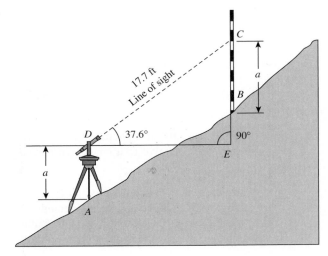

Solution   From the information given, we know the hypotenuse and an angle, and we want to find the opposite side. Therefore,

$$\sin 37.6° = \frac{EC}{17.7}$$

$$17.7 \sin 37.6° = EC$$

$$10.8 \text{ ft} = EC$$

■■■

9–4 EXERCISES

Solve the accompanying right triangle having the given parts. Express your answer to the same degree of accuracy as that given in the problem.

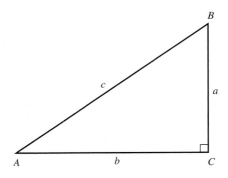

1. $a = 6.8$ m, $B = 43.8°$

2. $c = 16$ ft, $a = 9.8$ ft

3. $A = 36°43'$, $c = 39.5$ cm

4. $b = 123.6$ yd, $c = 196.83$ yd

5. $B = 14.5°$, $a = 16$ m

6. $A = 36°18'$, $c = 27.8$ ft

7. $A = 65°37'$, $b = 48.63$ cm

8. $B = 76.8°$, $a = 14.56$ yd

9. $c = 64$ ft, $a = 38$ ft

10. $A = 49.25°$, $c = 89.67$ m

11. $b = 43.3$ ft, $a = 29.8$ ft

12. $B = 57°14'$, $b = 78.8$ cm

13. From the roof of a building 80 ft high, the angle of depression to an object on the street is 58°. How far from the base of the building is the object?

E 14. The phase angle of an ac circuit is 28°, and the inductive reactance is 38 Ω. Find the impedance. (See Example 7.)

A 15. The average rise in elevation of a mountain road is 11.4°. What is the change in altitude (vertical) if the road length is 16.8 mi?

16. A telephone pole is to be supported by two guy wires. How much total wire is needed if the telephone pole is 40 ft high and the angle between the ground and the wire is 58°?

17. Find the length of a ramp if it makes an angle of 23° with the ground and the loading platform is 4.1 ft above the ground.

A 18. A house is 30 ft wide and the roof has an angle of inclination of 23°. Find the length *L* of the rafters. (See Figure 9–28.)

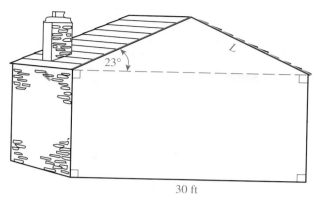

FIGURE 9–28

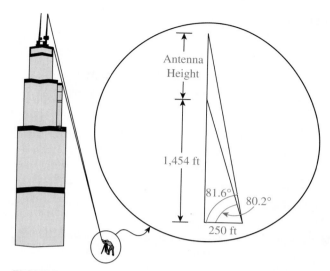

FIGURE 9–30

C 19. A surveyor wishes to determine the width of a river that is too wide to measure directly. She sights a point B on the opposite side of the river from where she stands at point C. Then she measures 100 yd from point C to point A, thus forming a right triangle. Sighting from point A to point B, she determines that angle A is 38°24′. Find the width of the river. See Figure 9–29.

M 22. A machine part is indicated in Figure 9–31. Find the distance between the center of the holes represented by lengths x and y.

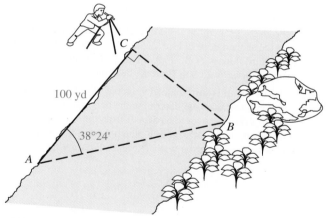

FIGURE 9–29

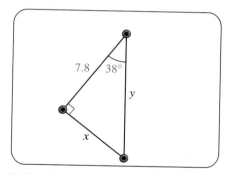

FIGURE 9–31

A 20. On a blueprint the walls of a rectangular room are 5.4 cm and 8.6 cm. Find the angle between the shorter wall and the diagonal across the room.

C 21. In 1973 the Sears Tower in Chicago was opened as the tallest building in the world. If the height to the top of the roof is 1,454 ft and the angle of elevation of an observer 250 ft from its base to the top of the roof is 80.2° and to the top of the antenna is 81.6°, how tall is the antenna? (Refer to Figure 9–30.) Ignore the height of the observer.

B 23. Radiation therapy is used to treat a tumor 7.62 cm below the surface of the skin. However, the radiologist must aim the radiation source at an angle to avoid the lungs, and he moves the radiation source 10.54 cm horizontally. What is the angle, relative to the patient's skin, at which the radiologist must aim the radiation source to hit the tumor? How far does the beam travel through the body before reaching the tumor?

M 24. For Figure 9–32, find the measure of angle B of the tip in the center of the drill bit.

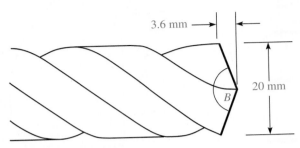

FIGURE 9–32

M 25. A grain conveyor belt used to store corn in a silo is 177 ft long. If the barn opening is 50 ft from the ground, at what angle relative to the ground will the conveyor belt sit?

C 26. The angle of elevation from point C through point B to point A is $12°18'$, and the distances are as given in Figure 9–33. Find the vertical and horizontal distance from A to C.

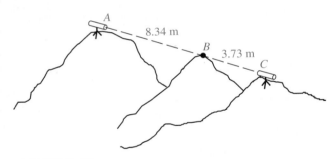

FIGURE 9–33

C 27. A plot plan requires the length of a property line to be 315.6 ft; however, a tree obstructs direct measurement. From the property line, point C, the surveyor turns the transit perpendicular and measures 54.70 ft to point A. The surveyor sets up the transit at point A and sights back to point B. Determine the measure of angle A and the distance c to locate B correctly. Refer to Figure 9–34.

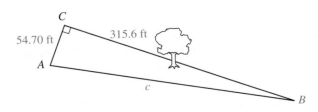

FIGURE 9–34

P 28. If light strikes a reflecting surface, the angle of incidence i is equal to the angle of reflection r. If a light beam has an angle of incidence equal to $42.3°$, find the distance y from the surface of the plane to a point on the beam if the horizontal distance x is 16.34 in. See Figure 9–35.

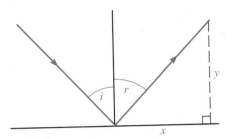

FIGURE 9–35

C 29. An observer in a balloon notes that the angle of depression to the top of a building is $35.8°$ and the angle of depression to the bottom of the building is $67.3°$. If the balloon is 59.4 ft horizontally from the base of the building, find the height of the balloon.

C 30. From the ground the angle of elevation from an observer to the second floor is $11.3°$, while the angle of elevation to the third floor is $21.8°$. Find the distance between floors if the observer is 25.0 m from the building.

A 31. A castle wall is surrounded by a moat of unknown width. An architect wants to determine the height of the wall. From a point known to be 50.0 ft from the wall, the angle of elevation is $28.4°$. Ignoring the height of the architect, how high is the castle wall? (Refer to Figure 9–36.)

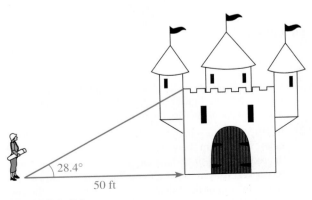

FIGURE 9–36

P 32. A radar station is tracking a missile as shown in Figure 9–37. The angle of elevation to the missile is 76.8° and the line of sight distance to the missile is 25.6 mi. Determine the height of the missile.

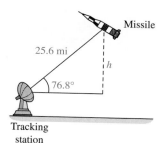

FIGURE 9–37

C 33. A civil engineering student is given the sketch of a survey shown in Figure 9–38. Find distance x, which represents the road frontage for a commercial lot.

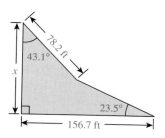

FIGURE 9–38

34. Barbara and Fred are planning to elope. To make sure that nothing goes wrong, Fred wants to calculate the length of the ladder so that it comes near the bottom of Barbara's bedroom window. If Barbara's window is 12 ft from the ground and if Fred places the ladder at a 60° angle with the ground, what length of ladder should Fred bring? Refer to Figure 9–39. Instead of measuring the 60° angle, how far from the base of the house should Fred place the ladder?

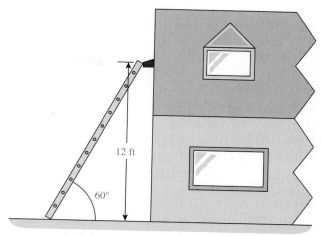

FIGURE 9–39

35. A flagpole was broken during a hurricane and needs to be replaced; however, no one knows its original length. If the broken portion makes an angle with the ground of 58° and the distance from the broken tip to the ground is 15 ft, how tall was the pole?

P 36. A rocket is supported by a cable attached with exploding bolts, as shown in Figure 9–40. If the cable is attached to the rocket at a point 60 m high and to the ground 35 m from the base of the rocket, what angle does the cable make with the ground and how long is the cable?

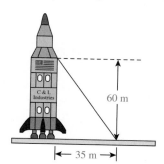

FIGURE 9–40

P 37. In bad weather airplanes use an automatic landing system that locks onto the plane and guides it to the runway. If lock-on occurs when the plane is 6.4 km (slant range) from the runway at an altitude of 1.2 km, what is the angle of descent?

C 38. To measure the width of a river, a surveyor chooses a point C on the riverbank directly across from a tree A. She then measures 75 ft from C to a point B on the same riverbank and uses a theodolite to measure angle CBA, which is 48°. How wide is the river? (Refer to Figure 9–41.)

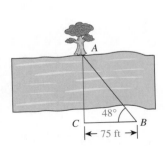

FIGURE 9–41

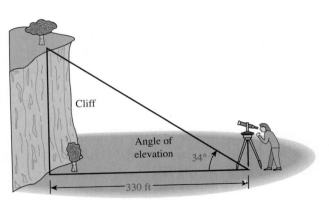

FIGURE 9–42

C 39. In the area of remote sensing, aerial photographs are used to measure inaccessible objects. If a satellite photograph showed the shadow of a crater on the moon to be 0.126 in. and the angle of the sun to the horizontal was 51.8° when the photograph was taken, how deep is the crater? (The scale of the map is 1 in. = 3,000 ft.)

C 40. To find the approximate height of an inaccessible cliff, Maria sets up her theodolite 330 ft from the base of the cliff and measures the angle of elevation to the top of the cliff to be 34°. Neglecting the height of the theodolite, how high is the cliff? (Refer to Figure 9–42.)

CHAPTER SUMMARY

Summary of Terms

adjacent side (p. 336)

angle (p. 332)

angle of depression (p. 351)

angle of elevation (p. 350)

cofunctions (p. 344)

complementary angles (p. 344)

coterminal (p. 333)

hypotenuse (p. 335)

impedance (p. 352)

inductive reactance (p. 352)

initial side (p. 332)

inverse trigonometric ratio (p. 345)

minutes (p. 332)

opposite side (p. 335)

phase angle (p. 352)

quadrantal angle (p. 334)

radius (p. 335)

seconds (p. 332)

standard position (p. 334)

terminal side (p. 332)

trigonometric functions (pp. 332, 336)

trigonometry (p. 332)

vertex (p. 332)

Summary of formulas

$$\sin \theta = \frac{\text{opposite}}{\text{hypotenuse}} = \frac{y}{r}$$

$$\cos \theta = \frac{\text{adjacent}}{\text{hypotenuse}} = \frac{x}{r}$$

$$\tan \theta = \frac{\text{opposite}}{\text{adjacent}} = \frac{y}{x}$$

$$\csc \theta = \frac{\text{hypotenuse}}{\text{opposite}} = \frac{r}{y}$$

$$\sec \theta = \frac{\text{hypotenuse}}{\text{adjacent}} = \frac{r}{x}$$

$$\cot \theta = \frac{\text{adjacent}}{\text{opposite}} = \frac{x}{y}$$

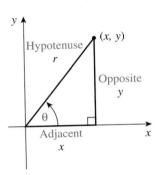

🖩 CHAPTER REVIEW

Section 9–1

Convert the following angles measured in degrees, minutes, and seconds to the nearest tenth of a degree.

1. 67°43′18″

2. 134°06′27″

3. 274°43′12″

4. 324°15′23″

5. 218°47′58″

6. 157°37′10″

Convert the following angles measured in degrees to angles measured to the nearest minute.

7. 16.78°

8. 131.6°

9. 318.15°

10. 249.6°

11. 108.43°

12. 68.27°

Draw the following angles in standard position, determine the quadrant in which the terminal side lies, and find two angles (one positive and one negative) coterminal with the given angle.

13. 147.8°

14. 67.6°

15. 290.3°

16. 205.6°

17. 314.63°

18. 147.5°

Section 9–2

Each of the following is a point on the terminal side of an angle in standard position. Determine the value to four decimal places of all six trigonometric functions of the angle.

19. (6, 12)

20. (12, 3)

21. (10, 15)

22. (4, 7)

23. ($\sqrt{3}$, 2)

24. (4, $\sqrt{5}$)

Use the given trigonometric function to find the value of the indicated trigonometric function. Give your answer to four decimal places. (Do not find θ.)

25. sin θ = 9/10; find tan θ and sec θ.

26. csc θ = 15/7; find sec θ and tan θ.

27. tan θ = 2.76; find sin θ and cos θ.

28. cos θ = 6/13; find sin θ and cot θ.

29. sec θ = 1.7864; find cos θ and tan θ.

30. cot θ = 4.86; find tan θ and sin θ.,

Section 9–3

Using a calculator, find the value of each trigonometric function rounded to three decimal places.

31. cos 16.84°

32. tan 21.8°

33. csc 67.8°

34. cos 47.6°

35. sec 72.77°

36. sin 26.9°

Find θ to the nearest tenth of a degree.

37. tan θ = 3.7864

38. csc θ = 2.7684

39. sec θ = 1.6327

40. cos θ = 0.3798

41. cot θ = 2.5378

42. sin θ = 0.9156

Section 9–4

Given right triangle *ABC* with the right angle at *C*, solve the triangle from the given information.

43. *b* = 7.3 yd, *A* = 56.7°

44. *B* = 49°37′, *c* = 67.8 ft

45. *c* = 71.8 m, *b* = 56.3 m

46. *a* = 123 cm, *c* = 256 cm

47. *A* = 53°48′, *a* = 98.3 km

48. The average rise in elevation of a mountain road is 14°. What is the change in altitude (vertical) if the road length is 4 km?

49. From the top of a 50-ft building, the angle of depression to a car on the ground is 37°. Find the horizontal distance from the base of the building to the car.

50. A cable is stretched from the top of a pole to a point 20 ft from the base of the pole. If the cable makes an angle of 60° with the ground, how tall is the pole?

🖩 CHAPTER TEST

The number in parentheses refers to the appropriate learning objective given at the beginning of the chapter.

1. Given csc θ = 10/7, use the calculator to find cot θ to four decimal places. (4)

2. Given right triangle *ABC* shown in Figure 9–43, find *a* if *b* = 56.8 and *A* = 42.5°. (7)

3. If (4, 9) is a point on the terminal side of angle *A* in standard position, then find cos *A* and cot *A* using a calculator. (3)

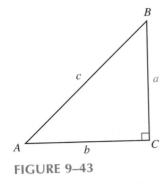

FIGURE 9–43

4. Convert 127°14′48″ into an angle measured to the nearest tenth of a degree. (1)

5. Using a calculator, find angle B to the nearest tenth of a degree if cot B = 2.2673. (6)

6. Using a calculator, evaluate csc 73.77°. (5)

7. Draw a 156° angle and determine the measure of two angles coterminal with it. (2)

8. A cathedral ceiling connects a wall 8 ft high with a wall 12 ft high. If the horizontal distance between the two walls is 20 ft, what is the angle of the rise of the ceiling relative to the horizontal? (8)

9. Given that (8, 5) is a point on the terminal side of angle A in standard position, find csc A using a calculator. (3)

10. Given that cot B = 0.6544, find sec B. (4)

11. Convert 124.83° to an angle measured in degrees and minutes. (1)

12. From a helicopter 600 ft high, the angle of depression to a house on the ground is 50°. Find the straight-line distance from the helicopter to the house. (8)

13. Find sec 24.3°. (5)

14. Given right triangle ABC shown in Figure 9–44, find c if A = 38.7° and a = 73.5. (7)

FIGURE 9–44

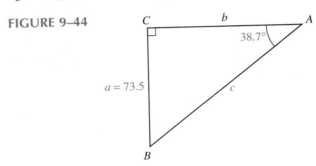

15. Calculate angle B to the nearest tenth of a degree if csc B = 2.0371. (6)

16. If sin A = 0.8437, find sec A. (4)

GROUP ACTIVITY

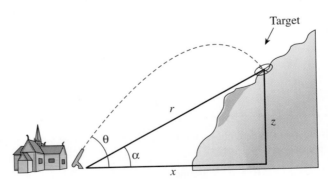

You are hired by a ski patrol unit in Banff, Canada, to help determine the correct angle θ for the avalanche cannon to be aimed to hit a point on the mountain. You know that the equation for the projectile path (assuming no air resistance) is given by:

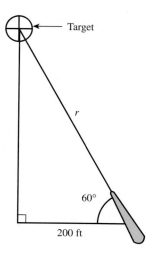

Target

r

60°

200 ft

$$z = \frac{V_{z0}}{V_{x0}}x - \frac{1}{2}\frac{g}{V_{x0}^2}x^2 \qquad (1)$$

where $V_{x0} = V_0 \cos\theta$ and $V_{z0} = V_0 \sin\theta$, θ is the angle that the cannon makes with the horizontal, g is the acceleration due to gravity (32 ft/sec²), V_0 is the muzzle velocity of the projectile, and z is the vertical height at a horizontal position x from the cannon. You also know that the range from the cannon to the target point can be found by using triangulation. This is accomplished by measuring the angle to the target from the cannon, then moving to one side and taking another angular measurement with respect to the line between the two points of measurement.

In later chapters you will see how to calculate the distance to a point when you cannot create a right triangle during the measurement process. Here we will assume that you can create a right triangle that is made up of the distance from the cannon to the target, the distance from the second measurement point to the target, and the distance between the two measurement points. The distance between the two points is 200 ft and the angle from the cannon to the target is 60°.

1. Find the distance to the target r.

2. The elevation (angle between the first measurement point and the horizontal α) was measured to be 30°. Find z. You are told that the muzzle velocity of the cannon is 300 ft/s.

3. Fill in equation 1 with the information that you now have. Keep the expressions in terms of variables. For extra credit see if you can show how to get to the following expression. (*Hint:* You will have to use three trigonometric identities: $\tan\theta = \sin\theta/\cos\theta$; $\sec\theta = 1/\cos\theta$; and $\sec^2\theta = \tan^2\theta + 1$.)

$$z = x\tan\theta - \frac{1}{2}\frac{g}{V_0^2}x^2(\tan^2\theta + 1) \qquad (2)$$

4. Rearrange equation 2 into the form for a quadratic equation in $\tan\theta$ ($\tan^2\theta - K\tan\theta + P = 0$). What are the values for K and P? Use the quadratic formula to solve for θ.

5. Discuss your results. Why are there two answers? If the velocity out of the cannon is equal to 100 ft/s, what happens to the solution? Discuss the significance of the solution for 100 ft/s. Can you find a minimum velocity that will still have a real solution?

SOLUTION TO CHAPTER INTRODUCTION

Figure 9–45 depicts the information given. From the two triangles, we find that we have two unknowns and must solve a system of equations. (Review Chapter 6.)

$$\tan 14.2° = \frac{x}{500 + y}$$

and

$$\tan 47.7° = \frac{x}{y}$$

FIGURE 9–45

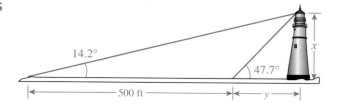

14.2°

47.7°

x

500 ft

y

Solving the second equation for y and substituting the result into the first equation gives

$$y = \frac{x}{\tan 47.7°}$$

$$\tan 14.2° = \frac{x}{500 + \dfrac{x}{\tan 47.7°}}$$

Solving for x, we have

$$\tan 14.2° \left(500 + \frac{x}{\tan 47.7°}\right) = x$$

$$500 \tan 14.2° + x\frac{\tan 14.2°}{\tan 47.7°} = x$$

$$126.5 + 0.2302x = x$$

$$126.5 = 0.7698x$$

$$164.3 = x$$

The lighthouse is approximately 164.3 ft high.

Y ou are a space scientist trying to find the linear velocity of the antenna on a geosynchronous satellite that is orbiting the earth with orbital radius of 26,259 miles. At this altitude, the satellite and the earth beneath it have the same angular velocity of $\pi/12$ radians/hour. (The solution to this problem is given at the end of this chapter.)

In the previous chapter we discussed the trigonometric functions of a first quadrant angle. In this chapter, we will discuss the trigonometric functions for obtuse angles and angles measured in radians.

Learning Objectives

After you complete this chapter, you should be able to

1. Determine the sign of a trigonometric function in any of the four quadrants, or if given sign(s) of trigonometric function(s), determine the quadrant(s) (Section 10–1).

2. Determine the six trigonometric functions from a point on the terminal side of an angle in any quadrant (Section 10–1).

3. Determine the reference angle for an angle whose terminal side lies in any quadrant (Section 10–2).

4. Find the value of the trigonometric functions for any given angle (Section 10–2).

5. Determine the measure of an angle in any quadrant given its trigonometric value (Section 10–2).

6. Convert angles measured in degrees to radians, and vice versa (Section 10–3).

7. Find the value of a trigonometric function given an angle measured in radians, and vice versa (Section 10–3).

8. Apply radian measure to finding the length of a circular arc, the area of a circular sector, and angular and linear velocity (Section 10–4).

Chapter 10

Trigonometric Functions of Any Angle

10–1

SIGNS OF THE TRIGONOMETRIC FUNCTIONS

Our discussion of trigonometry in Chapter 9 was limited to angles in the first quadrant. In this section we will discuss the sign of the trigonometric functions in the remaining three quadrants. Before we discuss the sign of the trigonometric functions, let us review several concepts discussed earlier.

First, we must be able to determine the quadrant in which the terminal side of an angle in standard position lies. For a given angle θ, the measurement boundaries for each quadrant are as summarized in Figure 10–1.

FIGURE 10–1

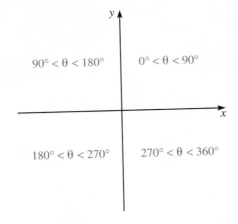

Second, we must know the definition of each trigonometric function to determine its sign. Therefore, the definitions of the trigonometric functions are repeated in the following box.

Definitions of the Trigonometric Functions

$$\sin \theta = \frac{y}{r} \qquad \csc \theta = \frac{r}{y}$$

$$\cos \theta = \frac{x}{r} \qquad \sec \theta = \frac{r}{x}$$

$$\tan \theta = \frac{y}{x} \qquad \cot \theta = \frac{x}{y}$$

Because r is always positive, the signs of the various trigonometric functions are determined by the sign of x and y. Recall from Chapter 9 that all the trigonometric functions are positive in quadrant I. Now, let us discuss the signs in the other quadrants.

REMEMBER ✦ A second, third, or fourth quadrant angle means that the terminal side of the angle lies in the second, third, or fourth quadrant respectively.

Quadrant II

Figure 10–2 shows a quadrant II angle and its resulting right triangle. From the figure we can see that y and r are both positive and x is negative. From this information, we can determine that only functions that involve x will be negative in quadrant II. This information is summarized as follows:

Positive	Negative
$\sin\theta\left(\dfrac{y}{r}=\dfrac{+}{+}=+\right)$	$\cos\theta\left(\dfrac{x}{r}=\dfrac{-}{+}=-\right)$
$\csc\theta\left(\dfrac{r}{y}=\dfrac{+}{+}=+\right)$	$\sec\theta\left(\dfrac{r}{x}=\dfrac{+}{-}=-\right)$
	$\tan\theta\left(\dfrac{y}{x}=\dfrac{+}{-}=-\right)$
	$\cot\theta\left(\dfrac{x}{y}=\dfrac{-}{+}=-\right)$

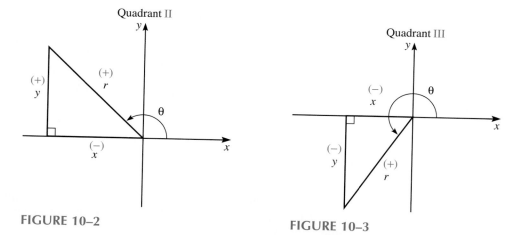

FIGURE 10–2   FIGURE 10–3

Quadrant III

Figure 10–3 shows a quadrant III angle and its resulting triangle. From the figure we can see that both x and y are negative and r is positive. The only positive functions in quadrant III involve both x and y. This information is summarized as follows:

Positive	Negative
$\tan\theta\left(\dfrac{y}{x}=\dfrac{-}{-}=+\right)$	$\sin\theta\left(\dfrac{y}{r}=\dfrac{-}{+}=-\right)$
$\cot\theta\left(\dfrac{x}{y}=\dfrac{-}{-}=+\right)$	$\cos\theta\left(\dfrac{x}{r}=\dfrac{-}{+}=-\right)$
	$\sec\theta\left(\dfrac{r}{x}=\dfrac{+}{-}=-\right)$
	$\csc\theta\left(\dfrac{r}{y}=\dfrac{+}{-}=-\right)$

Quadrant IV

Figure 10–4 shows a quadrant IV angle and its resulting triangle. From the figure we can see that x and r are both positive and y is negative. Therefore, the negative functions in quadrant IV involve y. This information is summarized as follows:

Positive	Negative
cos θ	sin θ
sec θ	csc θ
	tan θ
	cot θ

FIGURE 10–4

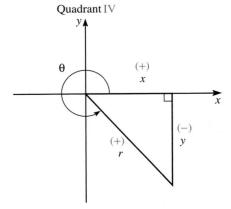

The sign of each trigonometric function in each quadrant is given in Table 10–1.

The graphic in Figure 10–5 summarizes the quadrants where each trigonometric function is positive.

TABLE 10–1
Signs of the Trigonometric Functions

Function	Quadrant			
	I	II	III	IV
Sine	+	+	−	−
Cosine	+	−	−	+
Tangent	+	−	+	−
Cosecant	+	+	−	−
Secant	+	−	−	+
Cotangent	+	−	+	−

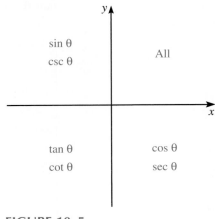

FIGURE 10–5

EXAMPLE 1 Determine the sign of each of the following trigonometric functions:

(a) tan 146° (b) cos 256° (c) sec 313°
(d) sin 218° (e) csc 94° (f) cot 190°
(g) sin(−230°) (h) tan 463° (i) sec(−108°)

Solution

(a) An angle of 146° has its terminal side in quadrant II because $90° \angle 146° \angle 180°$. Since both r and y are positive and x is negative, the tangent is negative.

$$\tan 146° = \frac{y}{x} = \frac{(+)}{(-)} = (-)$$

(b) An angle of 256° terminates in quadrant III, where both x and y are negative and r is positive. Therefore, the cosine is negative.

$$\cos 256° = \frac{x}{r} = \frac{(-)}{(+)} = (-)$$

(c) An angle of 313° is a quadrant IV angle, where both x and r are positive and y is negative. Thus,

$$\sec 313° = \frac{r}{x} = \frac{(+)}{(+)} = (+)$$

(d) $\sin 218° = \dfrac{y}{r} = \dfrac{(-)}{(+)} = (-)$

(e) $\csc 94° = \dfrac{r}{y} = \dfrac{(+)}{(+)} = (+)$

(f) $\cot 190° = \dfrac{x}{y} = \dfrac{(-)}{(-)} = (+)$

(g) Since an angle of −230° is rotated clockwise from the positive x axis, −230° has its terminal side in quadrant II. The angle −230° is coterminal with an angle of 130°. Therefore,

$$\sin(-230°) = \sin(130°) = \frac{y}{r} = \frac{(+)}{(+)} = (+)$$

(h) An angle of 463° involves revolving around the circle once and terminating in quadrant II at 103°. Thus,

$$\tan 463° = \tan 103° = \frac{y}{x} = \frac{(+)}{(-)} = (-)$$

(i) $\sec(-108°) = \sec(252°) = \dfrac{r}{x} = \dfrac{(+)}{(-)} = (-)$ ■ ■ ■

EXAMPLE 2 Determine the quadrant(s) in which the following conditions are true:

(a) tan θ is positive.
(b) sin θ and cos θ are both positive.
(c) csc θ is negative and cot θ is positive.
(d) sec θ and cos θ are both positive.

Solution
(a) I and III   (b) I   (c) III   (d) I and IV   ■■■

EXAMPLE 3   Determine the six trigonometric functions of an angle whose terminal side passes through the point $(-3, -8)$. Round to three decimal places.

Solution   Figure 10–6 represents the right triangle determined by this point. First, we use the Pythagorean theorem to determine the length of r.

$$r = \sqrt{x^2 + y^2} = \sqrt{(-3)^2 + (-8)^2} = \sqrt{73}$$
$$r \approx 8.544$$

FIGURE 10–6

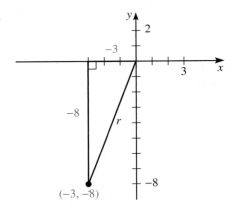

Substituting the values $x = -3$, $y = -8$, and $r = 8.544$ into the trigonometric functions gives the following values:

$$\sin \theta = \frac{y}{r} = \frac{-8}{8.544} \approx -0.936 \qquad \csc \theta = \frac{r}{y} = \frac{8.544}{-8} \approx -1.068$$

$$\cos \theta = \frac{x}{r} = \frac{-3}{8.544} \approx -0.351 \qquad \sec \theta = \frac{r}{x} = \frac{8.544}{-3} \approx -2.848$$

$$\tan \theta = \frac{y}{x} = \frac{-8}{-3} \approx 2.667 \qquad \cot \theta = \frac{x}{y} = \frac{-3}{-8} \approx 0.375 \qquad ■■■$$

10–1 EXERCISES

Determine the sign of the following trigonometric expressions without using a calculator.

1. $\sin 186°$

2. $\sin 256°$

3. $\cos 231°$

4. $\tan 318°$

5. $\cot 125°$

6. $\sin 278°$

7. $\csc 294°$

8. $\cot 97.3°$

9. $\sec 98°$

10. $\csc 108°$

11. $\tan 158.3°$

12. $\cos 263.5°$

13. $\sin 315°$

14. $\sec 184.3°$

15. $\cos 245°$

16. $4.6 \csc 358°$

17. $(\tan 138.6°)^3$

18. $\cos 294.8°$

19. $\cot 196.9°$

20. $\sin 177.9°$

21. $\sqrt{\csc 175.0°}$

22. $(\tan 145°)^2$

23. $3.6 \sin 190.6°$

24. $\cos 520°$

25. $\sec(-550°)$

26. $\dfrac{\csc 596°}{4.82}$

27. $\tan(-624°)$

28. $\cot^2(-195°)$

29. $\cos(-203°)$

30. $[\cot(-286.3°)]^{1/3}$

Find the quadrant(s) where the following conditions are true.

31. $\cos \theta$ is positive.

32. $\csc \theta$ is negative.

33. cot θ is negative.

34. sin θ is positive.

35. tan θ is negative.

36. sec θ is negative.

37. tan θ is negative, sin θ is negative.

38. csc θ is positive, cot θ is negative.

39. cos θ is negative, tan θ is positive.

40. sec θ is negative, csc θ is negative.

41. cot θ is negative, cos θ is positive.

42. cos θ is positive, sin θ is negative.

43. sin θ is positive, csc θ is positive.

44. sin θ is positive, tan θ is negative.

45. sec θ is negative, sin θ is negative.

46. tan θ is negative, cot θ is negative.

Using a calculator, find the six trigonometric functions for an angle θ whose terminal side passes through the point given. Round your answer to four decimal places.

47. (−4, 9)

48. (10, −3)

49. (6, −9)

50. (6, −10)

51. (−7, −2)

52. (−6, 4)

53. (−9, −4)

54. (−1, 6)

55. (−5, 11)

56. (−6, 9)

57. (7, −5)

58. (−3, −6)

59. (8, −2)

60. (−3, −4)

CS 61. A computer program determines that an angle θ has a sine value of 0.5 and a cosine value of 0.866 (rounded). What is the tangent value of this angle?

10–2

TRIGONOMETRIC FUNCTIONS OF ANY ANGLE

In the previous section we discussed the sign of the trigonometric functions in each quadrant. In this section we will discuss finding the value of a trigonometric function in each quadrant and finding an angle given its trigonometric value. We begin the discussion with finding the trigonometric value of obtuse angles (angles greater than 90°).

EXAMPLE 1   Using your calculator, find the value of the following trigonometric functions rounded to four decimal places:

(a) sin 218.6°   (b) tan 108.34°   (c) cos 313.3°

(d) sec 176.4°   (e) cot 346.73°

Solution   Using the same calculator keystrokes as discussed earlier gives the following values:

(a) 218.6 [sin] ⟶ −0.6239

(b) 108.34 [tan] ⟶ −3.0167

(c) 313.3 [cos] ⟶ 0.6858

(d) 176.4 [cos] [1/x] ⟶ −1.0020

(e) 346.73 [tan] [1/x] ⟶ −4.2402

■ ■ ■

From the results in Example 1, we can see that the calculator assigns the correct sign for the trigonometric functions in any quadrant. However, if we attempt to find an angle whose trigonometric value is given, the calculator may not give the desired angle. When we discuss the inverse trigonometric functions, you will understand why this is possible. This occurrence makes it necessary to discuss reference angles.

Reference Angle

In general, the numerical value of a trigonometric function in any quadrant is equivalent to the same function of a first quadrant angle, called the reference angle. The **reference angle** θ_R of any angle θ is the *positive angle formed by the terminal side of the angle and* *the nearest x axis.* The reference angle θ_R for a given angle θ in each of the quadrants is shown in Figure 10–7.

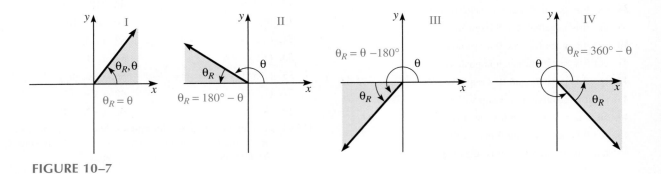

FIGURE 10–7

CAUTION ✦ The reference angle is always the *positive angle* measured between the terminal side of the angle and the nearest *x axis.*

A summary of how to calculate the reference angle θ_R from a given angle θ is given in the following box.

Reference Angle θ_R for a Given Angle θ

Quadrant I	$\theta_R = \theta$
Quadrant II	$\theta_R = 180° - \theta$
Quadrant III	$\theta_R = \theta - 180°$
Quadrant IV	$\theta_R = 360° - \theta$

EXAMPLE 2 Find the reference angle θ_R for each of the following given angles θ:

(a) 246°   (b) 136°   (c) 318.4°   (d) 168°37′

Solution

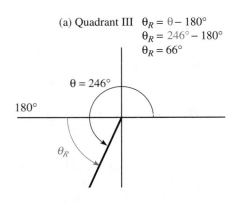

(a) Quadrant III $\theta_R = \theta - 180°$
$\theta_R = 246° - 180°$
$\theta_R = 66°$

(b) Quadrant II $\theta_R = 180° - \theta$
$\theta_R = 180° - 136°$
$\theta_R = 44°$

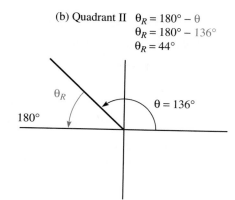

(c) Quadrant IV $\theta_R = 360° - \theta$
$\theta_R = 360° - 318.4°$
$\theta_R = 41.6°$

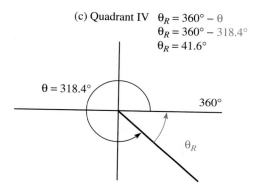

(d) Quadrant II $\theta_R = 180° - \theta$
$\theta_R = 180° - 168°37'$
$\theta_R = 179°60' - 168°37'$
$\theta_R = 11°23'$

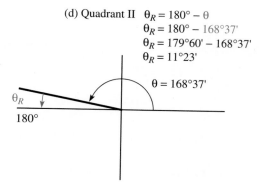

■ ■ ■

REMEMBER ✦ Reference angles are always determined using 180° or 360°.

Finding Angles from a Trigonometric Value

Recall from the previous chapter that we used the |inv| key to find a first quadrant angle given its trigonometric value. However, when we extend this process to the remaining quadrants, there should be two answers for each angle θ. For example, if cos θ = −0.6587, then θ is both a second quadrant and a third quadrant angle because the cosine function is negative in both quadrants. Since the calculator does not give both angles, we use the absolute value of the function to yield a reference angle and then use the sign of the function to determine the two values of θ. This process is illustrated in the next examples.

EXAMPLE 3 Find all angles θ, in degrees, such that $\tan \theta = -0.5051$.

Solution First, we find θ_R given that

$$\tan \theta_R = |\tan \theta|$$
$$\tan \theta_R = |-0.5051|$$
$$\tan \theta_R = 0.5051$$
$$\theta_R \approx 26.8°$$

Second, we place the reference angle θ_R in quadrants II and IV because the tangent is negative in both quadrants. Third, we calculate θ using the reference angle.

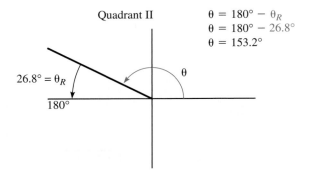

Quadrant II

$$\theta = 180° - \theta_R$$
$$\theta = 180° - 26.8°$$
$$\theta = 153.2°$$

$26.8° = \theta_R$

$180°$

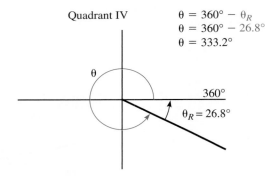

Quadrant IV

$$\theta = 360° - \theta_R$$
$$\theta = 360° - 26.8°$$
$$\theta = 333.2°$$

$360°$

$\theta_R = 26.8°$

Therefore, $\tan \theta = -0.5051$ results in $\theta = 153.2°$ and $333.2°$.

EXAMPLE 4 Find $\sec \theta$ when $\tan \theta = -1.6255$ and $\sin \theta < 0$.

Solution Since both $\tan \theta$ and $\sin \theta$ are negative, θ must be located in quadrant IV. Finding the reference angle gives

$$\tan \theta_R = |-1.6255|$$
$$\tan \theta_R = 1.6255$$
$$\theta_R \approx 58.4°$$

$$1.6255 \; \boxed{\text{inv}} \; \boxed{\text{tan}} \longrightarrow 58.4$$

Then we find θ in quadrant IV given the reference angle θ_R is 58.4°.

$$\theta = 360° - \theta_R$$
$$\theta = 360° - 58.4°$$
$$\theta = 301.6°$$

Last, we find the secant of 301.6°.

$$\sec 301.6° = \frac{1}{\cos 301.6°}$$

$$\sec \theta \approx 1.9084$$

301.6 $\boxed{\cos}$ $\boxed{1/x}$ $\longrightarrow$ 1.9084 ■■■

■ **EXAMPLE 5** Find θ if $(-8, 3)$ is a point on the terminal side of the angle. Round to tenths of a degree.

Solution First, we find the reference angle θ_R using the absolute value of the tangent function.

$$\tan \theta_R = \left| \frac{3}{-8} \right|$$

$$\theta_R \approx 20.6°$$

3 $\boxed{\div}$ 8 $\boxed{=}$ $\boxed{\text{inv}}$ $\boxed{\tan}$ $\longrightarrow$ 20.6

Second, we find θ in quadrant II (because $x < 0$, $y > 0$) using the reference angle θ_R.

$$\theta_R = 180° - \theta$$
$$\theta = 180° - \theta_R$$
$$\theta = 180° - 20.6°$$
$$\theta = 159.4°$$

■■■

Quadrantal Angles

In Section 9–3, we discussed the value of the trigonometric functions for the quadrantal angles 0° and 90°. Once again, if we let $r = 1$, the values of x, y, and r at 180° and 270° are as shown in Figure 10–8. The values of the trigonometric functions for the quadrantal angles are summarized in Table 10–2.

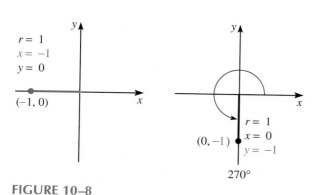

$r = 1$
$x = -1$
$y = 0$
$(-1, 0)$

$r = 1$
$x = 0$
$y = -1$
$(0, -1)$
270°

FIGURE 10–8

TABLE 10–2
Values of the Trigonometric Functions

Function	Quadrantal Angles			
	0°	90°	180°	270°
Sine	0	1	0	−1
Cosine	1	0	−1	0
Tangent	0	undef.	0	undef.
Cosecant	undef.	1	undef.	−1
Secant	1	undef.	−1	undef.
Cotangent	undef.	0	undef.	0

10–2 EXERCISES

Use a calculator to find the value of each of the following. Round your answer to thousandths.

1. cos 320°

2. tan 114.6°

3. sec 175°

4. $\sqrt{\sec 68.5°}$

5. sin 238°

6. csc 76.7°

7. cot 101.6°

8. cos 278.3°

9. tan 286°

10. sin 186.5°

11. cot 254.4°

12. 3(tan 124°) + 5

13. tan 173.65°

14. (cot 295.36°)²

15. sec 85.73°

16. (csc 160.9°)^{1/2}

17. sin 284.8°

18. $\sqrt{\sin 139.5°}$

19. cos 330.3°

20. cot 99.27°

21. cot 786°

22. 8 cos 194°18′

23. sin 418.65°

24. [sec(−286.56°)]^{1/3}

25. csc(−146°)

26. sec(−684°)

27. tan(−512°)

28. sin 156.4°

Determine the quadrant in which the terminal side of the given angle lies, and determine its reference angle.

29. 113.6°

30. 218.75°

31. 97.4°

32. 195.6°

33. 326.73°

34. 94.8°

35. 83.8°

36. 301.57°

37. 281.6°

38. 283.65°

39. 137.43°

40. 74.13°

41. 656°

42. 136.6°

43. 174.1°

44. 834°

45. 410.4°

46. 324.7°

47. 200.6°

48. 263.2°

Find all θ in degrees, 0° ≤ θ < 360°, such that the following statements are true. Round your answer to the nearest hundredth degree for Exercises 49–60 and to the nearest minute for Exercises 61–68.

49. sin θ = 0.8151

50. cos θ = −0.4099

51. tan θ = −0.9036

52. sec θ = −1.0255

53. tan θ = 2.2148

54. cot θ = −0.9163

55. cos θ = −0.9956

56. csc θ = −1.7305

57. sec θ = −1.3809

58. sin θ = 0.4020

59. csc θ = −1.0209

60. cot θ = 1.2174

61. sin θ = −0.9593

62. sec θ = −2.1855

63. cos θ = 0.6807

64. sin θ = 0.9990

65. cot θ = 0.6720

66. tan θ = −0.5914

67. sec θ = −1.1754

68. cos θ = −0.9752

Given the following conditions, determine the function value.

69. Find sin θ given sec θ = 3.6727 and tan θ < 0.

70. Find tan θ given sin θ = −0.3804 and cos θ < 0.

71. Find csc θ given tan θ = −0.0332 and sec θ > 0.

72. Find cos θ given tan θ = 0.8973 and cot θ > 0.

73. Find sec θ given cot θ = −0.3385 and sin θ < 0.

74. Find cot θ given cos θ = −0.5195 and csc θ > 0.

Find the angle, to the nearest tenth degree, whose terminal side passes through the point given.

75. (−6, −1)

76. (−5, −7)

77. (7, −3)

78. (−8, −12)

79. (−9, 12)

80. (−6, 1)

81. (4, −5)

82. (−1, −4)

P 83. A ray of light passing through an optical wedge is shifted by amount e, given by

$$e = \frac{t \sin(i - R)}{\cos R}$$

where i = angle of rotation, R = angle of reflection, and t = thickness of the wedge. Find the ray shift if t = 9 in., i = 56°, and R = 18°.

P 84. For a ray of light entering a prism, if the angle of incidence is in air, the following relationship is true:

$$n_R \sin R = \sin i$$

where i = angle of incidence, n_R = refractive index of the medium in which reflection takes place, and R = angle of refraction. Find the angle of refraction R if the angle of incidence is 66° and the refraction index is 1.375.

P 85. For objects projected with velocity v at an angle A above the horizontal, the time necessary to reach maximum height is given by

$$t = \frac{v \sin A}{g}$$

If t = 2 s, v = 30 m/s, and g = 9.8 m/s², find angle A.

P 86. In optics, the critical angle i_c is the incident angle that results in a 90° angle of refraction. This relationship is given by

$$\sin i_c = \frac{n_R}{n_i}$$

where n_R = refractive index of the refraction medium and n_i = refractive index of the incidence medium. Find the critical angle for a ray passing through a sheet of glass and refracted into air, given that n_R = 1 and n_i = 1.57.

E 87. The power P absorbed in an ac circuit is given by $P = IV \cos \theta$ where I = current, V = voltage, and θ = phase angle. Find the power P absorbed in a 2-A, 120-V circuit with a phase angle of 71°.

CS 88. To write a program that will determine points on a circle, you want point p with coordinates (x, y) to be put into a format in terms of $\cos \theta$ and $\sin \theta$ given a length z from the center of the circle. What equations give you this relationship for any point p (x, y)?

10–3

RADIAN MEASURE

Definitions

Until now, our discussion of angles has focused on the degree as the unit of angular measure. However, some technical problems require angles to be measured in radians. Given an angle whose vertex is at the center of the circle, a **radian** (rad) is defined to be the angle that intercepts an arc equal to the radius of the circle, as shown in Figure 10–9.

FIGURE 10–9

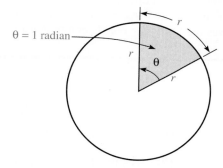
$\theta = 1$ radian

NOTE ✦ Radians are actually dimensionless units of angular measure. Therefore, an angle with the units omitted can be assumed to be measured in radians.

The quadrantal angles in radians are given in the following box.

Quadrantal Angles	
$90° = \dfrac{\pi}{2} \approx 1.57$ rad	$270° = \dfrac{3\pi}{2} \approx 4.71$ rad
$180° = \pi \approx 3.14$ rad	$360° = 2\pi \approx 6.28$ rad

EXAMPLE 1 Determine the quadrant in which the terminal side of the angle lies:

(a) $\dfrac{6\pi}{5}$ (b) 2.68 rad (c) $\dfrac{5\pi}{3}$

(d) 0.673 rad (e) $\dfrac{5\pi}{4}$ (f) -1.374 rad

Solution

(a) $\dfrac{6\pi}{5} = 1.2\pi$, which is greater than π but less than $\dfrac{3\pi}{2}$. Therefore, $\dfrac{6\pi}{5}$ is in quadrant III.

(b) 2.68 radians is greater than 1.57 but less than 3.14 rad. Therefore, 2.68 rad is in quadrant II.

(c) $\dfrac{5\pi}{3} = 1.67\pi$ radians, which is greater than $\dfrac{3\pi}{2} = 1.5\pi$ but less than 2π radians.

Therefore, $\dfrac{5\pi}{3}$ is in quadrant IV.

(d) I   (e) III   (f) IV   ∎∎∎

Radian and Degree Conversion

To convert an angle measured in degrees to radians, or vice versa, use the following proportion:

$$\frac{\text{radians}}{\text{degrees}} = \frac{\text{radians}}{\text{degrees}}$$

Remember proportions from our discussion in Chapter 1. Since a complete circle of 360° contains 2π radians, π radians equals 180°, substituting this value into the proportion gives the following conversion proportion.

Converting Degrees and Radians

$$\frac{\pi}{180°} = \frac{\text{radians}}{\text{degrees}}$$

Table 10–3 lists the degree and corresponding radian measures for some commonly used angles.

TABLE 10–3

Degrees	0°	30°	45°	60°	90°	120°	135°	150°	180°	210°	225°	240°	270°	300°	315°	330°	360°
Radians	0	$\dfrac{\pi}{6}$	$\dfrac{\pi}{4}$	$\dfrac{\pi}{3}$	$\dfrac{\pi}{2}$	$\dfrac{2\pi}{3}$	$\dfrac{3\pi}{4}$	$\dfrac{5\pi}{6}$	π	$\dfrac{7\pi}{6}$	$\dfrac{5\pi}{4}$	$\dfrac{4\pi}{3}$	$\dfrac{3\pi}{2}$	$\dfrac{5\pi}{3}$	$\dfrac{7\pi}{4}$	$\dfrac{11\pi}{6}$	2π

EXAMPLE 2

(a) Convert 240° to radians. Leave your answer as a rational multiple of π.

(b) Convert $\dfrac{7\pi}{6}$ rad to degrees.

Solution

(a) We take the basic proportion given above, substitute 240 in the degrees location and x in the radians location, and solve for x.

$$\frac{\pi}{180°} = \frac{x}{240°}$$

$$240°\pi = 180°x$$

$$\frac{240°\pi}{180°} = x$$

$$\frac{4\pi}{3} = x$$

(b) If we take the basic proportion and substitute $\dfrac{7\pi}{6}$ in the radians location and x in the degrees location, we have

$$\frac{\pi}{180°} = \frac{7\pi/6}{x}$$

$$\pi x = 180° \frac{7\pi}{6}$$

$$x = \frac{\overset{30}{\cancel{180}}(7\cancel{\pi})}{6\cancel{\pi}}$$

$$x = 210°$$

∎ ∎ ∎

Finding Function Values

We find the function value of an angle measured in radians just as we did in the previous sections when using angles measured in degrees. Make sure the calculator is in the radian mode, enter the radian measure of the angle, and press the desired trigonometric function key.

EXAMPLE 3 Find each of the following trigonometric values rounded to four decimal places:

(a) sin 3.68   (b) tan 1.25   (c) sec 2.15

(d) $\cos \dfrac{3\pi}{4}$ (e) $\csc \dfrac{8\pi}{5}$ (f) $\cot \dfrac{7\pi}{6}$

Solution **Make sure that your calculator is in the radian mode.** Then use the following keystrokes:

(a) 3.68 $\boxed{\sin}$ $\longrightarrow$ -0.5128

(b) 1.25 $\boxed{\tan}$ $\longrightarrow$ 3.0096

(c) 2.15 $\boxed{\cos}$ $\boxed{1/x}$ $\longrightarrow$ -1.8270

(d) 3 $\boxed{\times}$ $\boxed{\pi}$ $\boxed{\div}$ 4 $\boxed{=}$ $\boxed{\cos}$ $\longrightarrow$ -0.7071

(e) 8 $\boxed{\times}$ $\boxed{\pi}$ $\boxed{\div}$ 5 $\boxed{=}$ $\boxed{\sin}$ $\boxed{1/x}$ $\longrightarrow$ -1.0515

(f) 7 $\boxed{\times}$ $\boxed{\pi}$ $\boxed{\div}$ 6 $\boxed{=}$ $\boxed{\tan}$ $\boxed{1/x}$ $\longrightarrow$ 1.7321

∎ ∎ ∎

Finding an Angle from a Function Value

To find an angle given its trigonometric value, we use a reference angle just as we did in the previous section. Figure 10–10 illustrates the reference angle θ_R for an angle θ measured in radians.

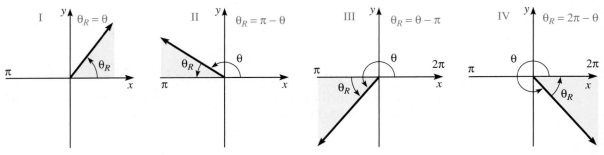

FIGURE 10–10

⊞ **EXAMPLE 4** Find all values of θ, $0 \leq \theta < 2\pi$, in radians to hundredths such that each of the following is true:

(a) $\sin \theta = -0.8634$ (b) $\csc \theta = 3.236$
(c) $\cot \theta = 0.3839$ (d) $\cos \theta = -0.8786$

Solution

(a) First, we determine the reference angle θ_R using the absolute value of the trigonometric function.

$$0.8634 \boxed{\text{inv}} \boxed{\text{sin}} \longrightarrow 1.04$$

Then we find θ in both quadrants III and IV, where the sine function is negative.

Quadrant III $\theta_R = \theta - \pi$ or $\theta = \theta_R + \pi$ $\theta = 4.18$
Quadrant IV $\theta_R = 2\pi - \theta$ or $\theta = 2\pi - \theta_R$ $\theta = 5.24$

Therefore, $\theta = 4.18$ and $\theta = 5.24$

(b) First, we find the reference angle θ_R by pressing the following keys:

$$3.236 \boxed{1/x} \boxed{\text{inv}} \boxed{\text{sin}} \longrightarrow 0.31$$

Remember that the cosecant is the reciprocal of the sine, thus the $1/x$ above. Then, using the reference angle $\theta_R = 0.31$ radian, we find θ in quadrants I and II where the sine function is positive.

Quadrant I $\theta_R = \theta$ so $\theta = 0.31$
Quadrant II $\theta_R = \pi - \theta$ so $\theta = \pi - \theta_R$ $\theta = 2.83$

Therefore, $\theta = 0.31$ and $\theta = 2.83$.

(c) To determine the reference angle θ_R, press the following keys:

$$0.3839 \boxed{1/x} \boxed{\text{inv}} \boxed{\text{tan}} \longrightarrow 1.20$$

Next, calculate θ in quadrants I and III.

Quadrant I $\theta_R = \theta$ so $\theta = 1.20$
Quadrant III $\theta_R = \theta - \pi$ so $\theta = \theta_R + \pi$ $\theta = 4.34$

Therefore, $\theta = 1.20$ and $\theta = 4.34$.

(d) To find the reference angle θ_R, press the following keys:

$$0.8786 \;\boxed{\text{inv}}\; \boxed{\text{cos}} \longrightarrow 0.50$$

Since the cosine function is negative in quadrants II and III, $\theta = 2.64$ and $\theta = 3.64$.

■ ■ ■

CAUTION ✦ Be careful in performing operations with the cotangent, secant, and cosecant functions. Write the reciprocal relationships and solve before using the calculator. Many errors result from pressing calculator keys before analyzing the situation.

EXAMPLE 5 The velocity v of an object of mass m attached to a spring that undergoes simple harmonic motion is given by

$$v = A\sqrt{\frac{k}{m}}\cos\sqrt{\frac{k}{m}}t$$

where k is the spring constant and A is the maximum distance the object moves in t seconds. Find the velocity of a 27.4-g object if it moves 8.00 cm after 0.150 s if $k = 375$ g/s^2.

Solution Substituting $A = 8.00$ cm, $k = 375$ g/s^2, $m = 27.4$ g, and $t = 0.150$ s, we have

$$v = 8.00 \text{ cm} \sqrt{\frac{375 \text{ g/s}^2}{27.4 \text{ g}}}\cos\sqrt{\frac{375 \text{ g/s}^2}{27.4 \text{ g}}}(0.150 \text{ s})$$

$$v = 8.00 \text{ cm} \left(\frac{3.70}{\text{s}}\right)\cos\left[\left(\frac{3.70}{\text{s}}\right)(0.150 \text{ s})\right]$$

$$v = 29.6 \text{ cm/s}\cos 0.555$$

$$v = 29.6 \text{ cm/s } (0.850)$$

$$v = 25.2 \text{ cm/s}$$

10–3 EXERCISES

Determine the quadrant in which the terminal side of each angle (measured in radians) lies.

1. 2.13

2. 5.34

3. $\dfrac{4\pi}{3}$

4. 0.86

5. −3.65

6. $\dfrac{\pi}{8}$

7. $-\dfrac{7\pi}{3}$

8. 3.29

Convert each angle measured in degrees to radian measure. Leave your answer as a rational multiple of π.

9. 68°

10. 256°

11. 318°

12. 146°

13. 330°

14. 187°

15. −280°

16. −159°

Convert each angle measured in degrees to decimal radians. Round your answer to hundredths.

17. 46.8°

18. 315.67°

19. 98.17°

20. 221.6°

21. 194.9°

22. −256°18′

23. −324.7°

24. 136.8°

Convert each angle measured in radians to degrees. Round your answer to hundredths.

25. 1.64

26. 3.57

27. 5.63

28. $\dfrac{3\pi}{4}$

29. $\dfrac{7\pi}{3}$

30. 1.15

31. $-\dfrac{11\pi}{6}$

32. -6.15

Find the value of the following functions. Round your answer to thousandths.

33. sin 2.86

34. sec 5.34

35. $\sqrt[3]{\tan 4.37}$

36. csc 3.29

37. $(\cot 1.17)^{1/2}$

38. $\cos^2 2.17$

39. sec 0.73

40. sin 1.69

41. csc 4.18

42. sin 3.59

43. $\cot \dfrac{5\pi}{4}$

44. $\cos \dfrac{5\pi}{6}$

45. $\cos \dfrac{3\pi}{8}$

46. $\tan \dfrac{5\pi}{3}$

47. $\csc \dfrac{7\pi}{6}$

48. $\tan \dfrac{3\pi}{4}$

Find all values of θ, $0 \le \theta < 2\pi$, to the nearest hundredth radian, that satisfy the following conditions.

49. $\cos \theta = 0.8345$

50. $\tan \theta = -2.673$

51. $\sin \theta = -0.3576$

52. $\cot \theta = -3.875$

53. $\csc \theta = 2.53$

54. $\cos \theta = 0.5641$

55. $\sec \theta = -1.635$

56. $\sin \theta = 0.335$

57. $\sec \theta = -2.973$

58. $\sin \theta = 0.8937$

59. $\cot \theta = -3.75$

60. $\tan \theta = 0.8634$

61. The area K of a regular polygon with n sides of length s is given by

$$K = \frac{1}{4} ns^2 \cot \frac{\pi}{n}$$

Find the area of an octagon ($n = 8$) whose sides are 3.4 m long.

62. The radius r of a circle inscribed in a regular polygon with n sides of length s is given by

$$r = \frac{s}{2} \cot \frac{\pi}{n}$$

Find the radius of a circle inscribed in a hexagon ($n = 6$) whose sides are 8 in. long.

P 63. A weight attached to a spring is pulled down 2 in. and released, thereby oscillating with simple harmonic motion and taking 0.12 s to return to its lowest position. This relationship is given by

$$y = 2 \cos \left(\frac{50\pi}{3} t \right)$$

Find the position y of the weight after 0.03 s.

E 64. The relationship between sinusoidal voltage V and time t is given by $V = V_0 \sin(2\pi ft)$, where V_0 is peak voltage and f is frequency. If the peak voltage is 163 V and the frequency is 60 Hz, find the voltage at time $t = 0.003$ s.

CS 65. Explain why a computer program designed to solve math problems shows an error message when you convert $\pi/2$ radians to θ degrees and attempt to find the tan value of this angle.

10–4

APPLICATIONS OF RADIAN MEASURE

In this section we will discuss three applications of radian measure: the length of a circular arc, the area of a circular sector, and angular velocity. Each of these applications requires that the angles be measured in radians.

Length of a Circular Arc

If we let s represent the length of an arc, then the length of an arc for a complete circle is given by the expression $2\pi r$. The expression 2π represents a central angle for the entire circle; however, we are attempting to find the arc length s for a given central angle θ that is less than 2π. The resulting expression for the length of a circular arc is as follows:

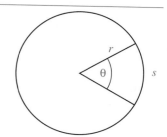

Length of a Circular Arc

$$s = r\theta$$

where

    s = length of the arc

    r = radius of the circle

    θ = measure of the central angle in radians

EXAMPLE 1   Find the length of the arc of a circle whose radius is 4.3 in. and whose central angle is $\dfrac{3\pi}{4}$.

Solution   Using the given values of $r = 4.3$ in. and $\theta = \dfrac{3\pi}{4}$, we solve for s by substituting into the formula as follows:

$$s = r\theta$$

$$s = (4.3 \text{ in.})\left(\frac{3\pi}{4}\right)$$

$$s \approx 10 \text{ in.}$$

■ ■ ■

EXAMPLE 2   A tree 250 ft from an observer on the ground intercepts an angle of 7.46° (see Figure 10–11). Find the approximate height of the tree.

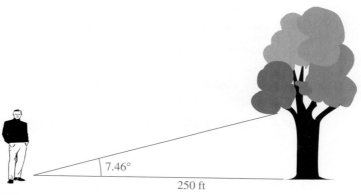

7.46°

250 ft

FIGURE 10–11

Solution   The height of the tree is approximated by the length of the intercepted arc. The length of the arc is given by $s = r\,\theta$ but θ must be in radians. Converting 7.46° to radians gives

$$\frac{\pi}{180°} = \frac{x}{7.46°}$$

$$x = \frac{7.46\pi}{180}$$

$$x = 0.130$$

The height of the tree is given by

$$s = (250 \text{ ft})\,(0.130)$$

$$s = 32.6 \text{ ft}$$

■ ■ ■

Area of a Sector

Radian measure is also used in finding the area of a sector of a circle. The area A of a sector is proportional to its central angle θ.

$$\frac{\text{area of sector } A}{\text{area of circle}} = \frac{\text{central angle of sector}}{\text{central angle of circle}}$$

$$\frac{A}{\pi r^2} = \frac{\theta}{2\pi}$$

Solving this equation for A gives the following formula for the area of a circular sector.

Area of a Sector

$$A = \frac{r^2\theta}{2}$$

where

> A = area of the sector
> r = radius of the circle
> θ = measure of the central angle in radians

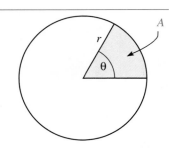

 EXAMPLE 3   Find the area of the sector of a circle whose central angle is 2.34 radians and whose radius is 3.15 ft.

Solution   Substituting $\theta = 2.34$ and $r = 3.15$ into the formula gives

$$A = \frac{(3.15 \text{ ft})^2(2.34)}{2}$$

$$A \approx 11.6 \text{ ft}^2$$

3.15 $\boxed{x^2}$ $\boxed{\times}$ 2.34 $\boxed{=}$ $\boxed{\div}$ 2 $\boxed{=}$ $\longrightarrow$ 11.61

■ ■ ■

EXAMPLE 4 Part of a baseball field and diamond is to be planted with grass seed. Determine the area of the field if the angle between the first and third baselines is 93° and the radius of the sector to be seeded is 205 ft from home plate.

Solution Figure 10–12 illustrates the verbal statement. We are asked to find the area of a sector of the circle. First, however, we must convert 93° to radians.

$$\frac{93°}{\theta} = \frac{180°}{\pi}$$

$$\theta \approx 1.62$$

FIGURE 10–12

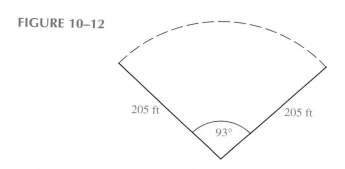

205 ft   205 ft

93°

Then we find the area of the sector.

$$A = \frac{(205)^2(1.62)}{2}$$

$$A \approx 34{,}000 \text{ ft}^2$$

93 $\times$ π $\div$ 180 $=$ $\times$ 205 x^2 $=$ $\div$ 2 $=$ ⟶ 34106.57   ■■■

Linear and Angular Velocity

Average velocity v is given by the expression

$$v = \frac{s}{t}$$

where s is distance and t is elapsed time. If an object is traveling in a circular path at a constant speed, the distance traveled equals the length of the arc. If we take the formula for arc length and divide both sides by t, we have

$$s = \theta r$$

$$\frac{s}{t} = \frac{\theta}{t} r$$

$$v = \omega r$$

where $v =$ **average linear velocity** and $\omega =$ **average angular velocity** ($\omega = \theta/t$).

Angular Motion

$$v = \omega r$$

where

ω = angular velocity measured in radians per unit of time

v = linear velocity

r = radius of the circle

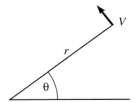

EXAMPLE 5   An object is moving along a circular path of radius 8.1 m with an angular velocity of 6.4 rad/s. Find the linear velocity of the object.

Solution   Substituting the quantities $r = 8.1$ m and $\omega = 6.4$ rad/s into the formula gives

$$v = \omega r$$
$$v = (6.4 \text{ rad/s})(8.1 \text{ m})$$
$$v \approx 52 \text{ m/s}$$

■■■

EXAMPLE 6   A truck with 30.0-in. diameter tires is traveling at 55.0 mi/h. What is the angular velocity of the tires in rad/s?

Solution   Since we are asked to find the angular velocity ω, we must rearrange the formula.

$$v = \omega r$$

$$\omega = \frac{v}{r}$$

Then we substitute $r = 15$ in. (diameter $= 30$ in.) and $v = 55$ mi/h into the formula.

$$\omega = \frac{55 \text{ mi/h}}{15 \text{ in.}}$$

However, the units of measure do not give rad/s. To obtain the required units of rad/s, we must convert 55 mi/h into in./s. This process is as follows:

$$\omega = \frac{\dfrac{55 \text{ mi}}{h} \cdot \dfrac{5,280 \text{ ft}}{1 \text{ mi}} \cdot \dfrac{12 \text{ in.}}{1 \text{ ft}} \cdot \dfrac{1 \text{ h}}{3,600 \text{ s}}}{15 \text{ in.}}$$

$$\omega \approx 64.5 \text{ rad/s}$$

■■■

▦ 10–4 EXERCISES

For a circle with radius r, diameter d, central angle θ, and arc length s, find the unknown quantity.

1. $r = 8.31$ ft, $\theta = \dfrac{\pi}{3}$, $s = ?$

2. $r = 18.4$ in., $\theta = \dfrac{2\pi}{7}$, $s = ?$

3. $s = 10.6$ cm, $d = 24$ cm, $\theta = ?$

4. $d = 38$ in., $\theta = 110°$, $s = ?$

5. $s = 34.8$ yd, $\theta = 240°$, $r = ?$

6. $r = 21$ m, $s = 35$ m, $\theta = ?$

For a sector with area A, central angle θ, and circle with radius r, find the unknown quantity.

7. $r = 27.2$ cm, $\theta = \dfrac{7\pi}{8}$, $A = ?$

8. $r = 6.83$ ft, $\theta = \dfrac{5\pi}{6}$, $A = ?$

9. $r = 16$ m, $\theta = 220°$, $A = ?$

10. $r = 6.7$ yd, $\theta = 330°$, $A = ?$

11. $A = 486$ ft^2, $r = 18$ ft, $\theta = ?$

12. $A = 56.7$ cm^2, $r = 6.5$ ft, $\theta = ?$

13. $A = 283.5$ in.^{2}, $\theta = 110°$, $r = ?$

14. $A = 687.4$ m^2, $\theta = 240°$, $r = ?$

P 15. A pendulum 37.4 in. long swings through an angle θ, which subtends an arc length of 23 in. Determine angle θ in degrees. (*Hint:* Find θ in radians, then convert.)

P 16. Determine the length of a pendulum if it swings through an angle of 15° and intercepts an arc of 6.8 in.

P 17. A flywheel rotates at an angular velocity of 4.83 rad/s. Determine the linear velocity, in feet per second, of a point on the rim if the radius of the flywheel is 1.66 ft.

18. Two concentric circles have radii of 18 ft and 23 ft. Using Figure 10–13, find the shaded area.

FIGURE 10–13

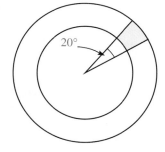

19. An isosceles triangle is set in a circle of radius 24 ft, as shown in Figure 10–14. Determine the measure of angle θ (in degrees) if the base intercepts an arc 28 ft long.

FIGURE 10–14

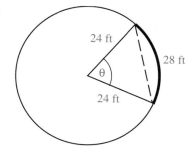

P 20. A car with 24-in.-diameter tires is traveling at 35 mi/h. What is the angular velocity of the tires in rad/s? (See Example 6.)

21. In a circle of radius 12 ft, the arc length of a sector is 18 ft. What is the area of the sector?

M 22. A circular sector whose central angle is 190° is cut from a circular piece of sheet metal of diameter 218 mm. A cone is then formed by joining the radii of the circular sector. Find the lateral surface area of the cone. (*Hint:* Remember its original shape.)

P 23. A satellite is in a circular orbit 210 mi above the equator of the earth. How many miles must the satellite move to change its longitude by 73°? (Assume that the radius of the earth at the equator is 3,960 mi.)

P 24. A windmill has a blade 20 ft in diameter that rotates at 18 r/min. What is the linear velocity in ft/s of a point on the end of the blade?

CS 25. A computer simulation for a car manufacturing company puts a car on a circular path at a constant speed of 70 mi/h. What is the angular velocity of this car in rad/s if the circular path had a diameter of 0.2 mi?

E 26. A microlens array is made up of microscopic lenslets tens to hundreds of micrometers in diameter. Since microlens arrays can be made extremely inexpensively they are a low-cost option used in line-scan CCD (charged-coupled device) cameras to collect light from a collimated (parallel) beam and focus it to an array of small, bright points at the focal plane. The numerical aperture (*NA*) for a lens in air is

$$NA = \sin \frac{a}{2f}$$

where a is the lens' diameter and f is its focal length. Find the numerical aperture of a microlens array lenslet 50 μm in diameter with a focal length of 0.02 mm.

C 27. Find the height of a microwave tower 914 ft away that intercepts an angle of 4.1°.

C 28. The GPS (global positioning system) consists of 24 satellites in a circular orbit of the earth at an altitude of 11,000 mi. Each satellite orbits the earth every 12 hours. If the radius of the earth is 3,960 mi, determine the linear speed of each satellite in miles/second.

M 29. An 8,000-rpm motor is directly connected to a 10-in.-diameter circular saw blade. What is the linear velocity of the saw's teeth in ft/min?

CS 30. The radius of the magnetic media in a 3.5-inch diskette is 1.7 inches. Find the linear velocity (inches/minute) of a point on the edge of the disk if it rotates at 360 revolutions per minute.

CHAPTER SUMMARY

Summary of Terms

average angular velocity (p. 385)

average linear velocity (p. 385)

average velocity (p. 385)

radian (p. 377)

reference angle (p. 371)

Summary of Formulas

$s = r\theta$ length of circular arc

$A = \dfrac{r^2\theta}{2}$ area of sector

$v = \omega r$ angular motion

CHAPTER REVIEW

Section 10–1

Without using a calculator, determine the signs of the following trigonometric functions.

1. sec 218°
2. cos 112°
3. tan(−334°)
4. −3 sin(−136°)
5. $\sqrt{\cot 68.4°}$
6. (csc 148.67°)³
7. sec 286.2°
8. csc 154.8°
9. cot 73.6°

Find the quadrant(s) where the following condition(s) are true.

10. tan θ is negative, cos θ is positive.
11. sec θ is positive, cot θ is positive.
12. sec θ is negative, cot θ is positive.
13. cot θ is negative, cos θ is negative.
14. csc θ is positive, tan θ is negative.
15. sin θ is negative, cos θ is positive.

Find the six trigonometric functions for an angle θ whose terminal side passes through the point given.

16. (−8, 3)
17. (7, −6)
18. (−10, −7)
19. (18, −11)
20. (−4, −7)
21. (−9, 5)

Section 10–2

Use a calculator to find the value of the following trigonometric functions. Round your answer to four decimal places.

22. csc 174.8°
23. $\sqrt{\sec 218.8°}$
24. sin 347.4°
25. −8 sec(−248°)
26. cos(−118°)
27. (cot 146.8°)³

Find the reference angle for the following angles.

28. 131.6°
29. 247.3°
30. 456°
31. 316.7°
32. 347.4°
33. 114.7°

Find θ, 0° ≤ θ < 360°, from the given information. Round your answer to the nearest tenth degree.

34. sin θ = −0.5519 and cos θ < 0
35. cot θ = 0.3739 and sin θ > 0
36. sec θ = 5.0159 and tan θ > 0
37. cos θ = −0.2351 and csc θ < 0
38. tan θ = 0.4411 and sec θ > 0
39. sin θ = −0.5008 and cot θ < 0

Section 10–3

Convert each angle measured in degrees to radians. Express your answer as a rational multiple of π.

40. 210°
41. 120°
42. 135°
43. 330°
44. 218°
45. 310°

Convert each angle measured in degrees to decimal radians. Round your answer to hundredths.

46. 247.9°
47. 127.56°
48. 158.73°
49. 318.45°
50. 334.3°
51. 214.5°

Convert each angle measured in radians to the nearest tenth degree.

52. $\dfrac{7\pi}{6}$
53. 2.756
54. 1.238
55. $\dfrac{7\pi}{4}$
56. −3.687
57. −5.147

Find the value of the following functions measured in radians.

58. $\cot \dfrac{7\pi}{4}$
59. csc 4.78
60. sec 1.73
61. $\tan \dfrac{3\pi}{4}$
62. cos 8.643
63. sin 9.576

Find all θ, 0 ≤ θ < 2π, that satisfy the following conditions.

64. sin θ = −0.7478

65. $\cot \theta = -1.7811$

66. $\csc \theta = 1.5442$

67. $\cos \theta = -0.0733$ and $\tan \theta < 0$

68. $\cot \theta = 1.0769$ and $\sin \theta > 0$

Section 10–4

For a circle with radius r, central angle θ, and arc length s, find the unknown quantity.

69. $s = 18.6$ m, $\theta = 126°$, $r = ?$

70. $\theta = 1.74$, $s = 2.86$ km, $r = ?$

71. $r = 14$ in., $s = 31$ in., $\theta = ?$

72. $r = 36.7$ ft, $\theta = 56°$, $s = ?$

For a sector with area A, central angle θ, and radius r, find the unknown quantity.

73. $A = 684$ in.^{2}, $\theta = 5.74$, $r = ?$

74. $A = 24.7$ m^2, $r = 11.8$ m, $\theta = ?$

75. $r = 23.5$ yd, $\theta = 256°$, $A = ?$

76. $r = 15.7$ ft, $\theta = 327°$, $A = ?$

77. Two concentric circles have radii of 15 m and 29 m. Using a central angle of 84°, find the area of the sector between the two circles.

78. A circular sector whose central angle is 215° is cut from a circular piece of sheet metal of diameter 18 in. If a cone is formed from the sector by bringing the radii together, find the surface area of the cone.

79. The propeller of a plane is 3.2 m in diameter and rotates at 2,500 r/min. What is the linear velocity in m/min of a point on the end of the blade?

M 80. The armature of a dynamo is 17.4 cm in radius and rotates at 1,100 r/min. What is the linear velocity of a point on the rim of the armature?

CHAPTER TEST

The number(s) in parentheses refers to the appropriate learning objective(s) at the beginning of the chapter.

1. Use a calculator to find csc 236.3°. (4)

2. The point $(-4, -7)$ is on the terminal side of angle θ. Determine $\sin \theta$ and $\cot \theta$. (2)

3. Find the length of the arc subtended by a circle of radius 12 ft if the central angle is 38°. (8)

4. Find $\csc (7\pi/6)$. (7)

5. Determine all θ, $0° \le \theta < 360°$, such that $\csc \theta = -1.1964$ and $\tan \theta < 0$. Give your answer to the nearest tenth degree. (3, 5)

6. For a circle with radius 15 in. and arc length 18 in., find the measure of the central angle to the nearest tenth degree. (1)

7. Determine the central angle of a pendulum 18 ft long if it swings through an arc 26.8 ft. (6)

8. Determine the quadrant(s) where $\cos \theta$ is negative and $\tan \theta$ is positive. (8)

9. Convert 3.87 radians to the nearest tenth degree. (8)

10. A car with 24-in.-diameter tires is traveling at 45 mi/h. What is the angular velocity, in rad/s, of a point on the rim of one of the tires? (3)

11. In a circle of radius 20 m, the arc length of a sector is 14 m. What is the area of this sector? (8)

12. Determine the reference angle of 218.6°. (8)

13. Find the central angle of a circle of radius 14 in. if the area of the sector is 36 in.^{2}. (3, 7)

14. Find all θ, $0 \le \theta < 2\pi$, such that $\sec \theta = -1.2776$ and $\tan \theta < 0$. (8)

15. Convert 140° to radians. Leave your answer as a rational multiple of π. (6)

16. Find all θ, $0 \le \theta < 2\pi$, such that $\csc \theta = -1.2877$ and $\tan \theta < 0$. (3, 7)

17. A conical tent is made from a circular piece of canvas 22 ft in diameter by removing a central angle of 218°. What is the surface area of the tent made from the larger piece? (8)

GROUP ACTIVITY

Thunder in the Valley Munitions Factory has hired you as a consultant to review test reports concerning projectile range calculations. From an old theoretical mechanics book you learn the following information. If air resistance is ignored, the projectile path is given by:

$$z = \frac{V_{z0}}{V_{x0}}x - \frac{1}{2}\frac{g}{V_{x0}^2}x^2 \tag{1}$$

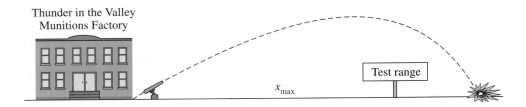

Thunder in the Valley
Munitions Factory

Test range

x_{max}

where V_{z0} and V_{x0} are the initial velocities of the projectile in the x and z directions. These are given by $V_{z0} = V\sin\alpha$, and $V_{x0} = V\cos\alpha$, where α is the elevation angle above the horizon. The variable g is the acceleration due to gravity (32 ft/s^2) and x is the horizontal distance from the cannon. The maximum range occurs when $z = 0$ and the maximum height occurs when the projectile is halfway to its maximum range. The resultant equations are given by:

$$x_{max} = 2\frac{V_{z0}V_{x0}}{g} \quad \text{and} \quad z_{max} = \frac{V_{z0}^2}{2g} \tag{2}$$

But you need to take into account the effect of air resistance. Some additional research reveals the following equation for the projectile's vertical velocity, as a function of time t, and height as a function of its horizontal range x.

$$V_z = \left(\frac{mg}{b} + V_{z0}\right)e^{-bt/m} - \frac{mg}{b} \tag{3}$$

$$z = \left(\frac{mg}{bV_{x0}} + \frac{V_{z0}}{V_{x0}}\right)x - \frac{m^2g}{b^2}\ln\left(\frac{mV_{x0}}{mV_{x0} - bx}\right) \tag{4}$$

The new variables are as follows: $b = k\rho A$, where $k = 2$ ft/s is a constant that is a function of the projectile's coefficient of drag, ρ is the density of the air in slugs/ft^3, and A is the cross-sectional area of the projectile in ft^2. The variable m is the projectile's mass in slugs.

1. As time increases, equation 3 will approach a constant value known as the terminal velocity. Discuss how you would find this terminal velocity. What is it?

2. Plot the curve given by equation 4. Discuss its shape. Does the maximum height of the projectile occur at the same place as it does for the no-air-resistance case (see equation 1)?

3. Solving the set of equations that lead to equations 3 and 4 for the maximum range is not as simple as it was in the first case. However, a solution can be reached by successive approximations. This gives the following range equation when the effect of air resistance is small:

$$x_{max} \cong \frac{2V_{x0}V_{z0}}{g} - \frac{8}{3}\frac{bV_{z0}^2 V_{x0}}{mg^2} \tag{5}$$

Assuming a balsa-wood ball with a cross-sectional area of $A = 1$ ft^2, and a mass given by $m = (0.21$ slug/ft$^3)(4/3 \, \pi \, (0.56$ ft$)^3) = 0.160$ slugs, with an initial velocity of 700 ft/s at an elevation of $\alpha = 30°$ and an air density at 20° C of $\rho = 2.33 \times 10^{-3}$ slug/ft^3, calculate x_{max}. How close is this estimate to your graph? Why are they different? Repeat the process using a steel ball (11.43 slugs). Is the estimate better? Try a velocity of 600 ft/s. Is the estimate better? Why or why not?

4. The report that you are reviewing said that the calculated estimated maximum range was 6,466 ft. You conclude that they took the data at a different temperature and did not record it. Given the following list of air densities at different temperatures, find the one that matches the recorded answer the closest: Discuss other factors that could lead to different answers. $\rho_{-50} = 3.07 \times 10^{-3}$, $\rho_{-30} = 2.82 \times 10^{-3}$, $\rho_{-10} = 2.60 \times 10^{-3}$, $\rho_{+30} = 2.26 \times 10^{-3}$ slugs/ft^3.

5. Program a calculator to find the maximum range given different elevation angles. Discuss how to find the angle that will give you the maximum range. Discuss what would happen if the projectile achieved an altitude where the air density was less than that assumed here. Would the elevation angle for the maximum range still be the same?

SOLUTION TO CHAPTER INTRODUCTION

Figure 10–15 depicts the problem outlined in the chapter introduction. We must calculate the linear velocity v given that the angular velocity is $\pi/12$ rad/h and the radius is 26,259 miles. Completing the calculation gives a linear velocity of

$$v = \omega r$$

$$v = (\pi/12)(26{,}259)$$

$$v = 6{,}875 \text{ mi/h}$$

FIGURE 10–15

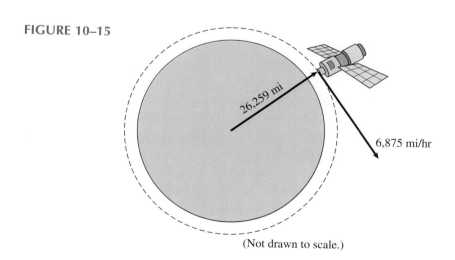

26,259 mi

6,875 mi/hr

(Not drawn to scale.)

Y
ou are a civil engineer responsible for surveying the land where a new municipal bridge will be built. You use indirect measurement and trigonometry to determine the distance from the bridge deck to the pilings at the river. (The solution to this problem is given at the end of the chapter.)

A problem of this type involves the vector quantities displacement and velocity. Also, if you make a drawing of this problem, as shown on the facing page, an oblique triangle results. Therefore, to solve a problem of this kind, you must know how to use vectors and how to solve oblique triangles; both topics are discussed in this chapter.

Learning Objectives

After you complete this chapter, you should be able to

1. Find the resultant of a given set of vectors (Section 11–1).

2. Use vectors to solve problems in science and technology (Section 11–2).

3. Apply the Law of Sines to solve oblique triangles (Section 11–3).

4. Apply the Law of Cosines to solve oblique triangles (Section 11–4).

5. Apply the Law of Sines and Law of Cosines to solve technical problems (Sections 11–3 and 11–4).

Chapter 11

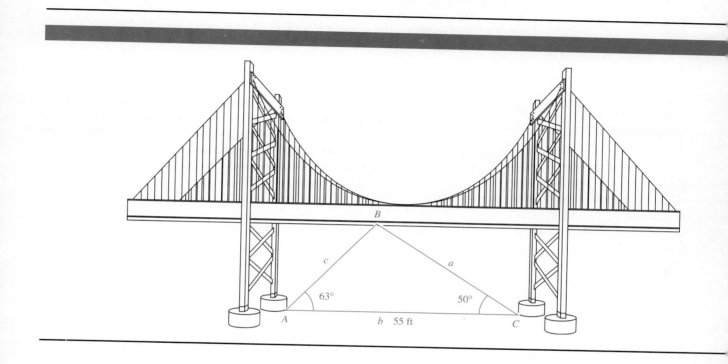

Vectors and Oblique Triangles

11–1

VECTORS

In technical areas some quantities called **scalars** can be specified using *only a number* called the **magnitude.** For example, the length of a table, the speed of a car, and the value of a trigonometric function all represent scalar quantities. However, there are other quantities, called **vectors,** that must be specified by both magnitude and direction. A car traveling 35 mi/h south is an example of the vector quantity called velocity. In the text a vector quantity will be represented in **boldface** type.

Geometric Representation

Vector quantities are represented visually in the rectangular coordinate plane by a directed line segment, as shown in Figure 11–1. The *beginning point* O, positioned at the origin, is called the **initial point** or **tail** of the vector, and the point *S* is called the **terminal point** or **head** of the vector. The direction of the arrow represents the direction of the vector, and the length of the line segment represents the magnitude of the vector.

FIGURE 11–1

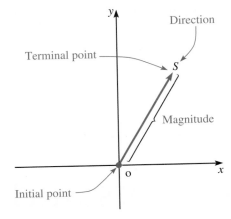

Addition of Vectors

The *sum of two or more vectors,* represented as **A + B,** is called their **resultant.** The resultant of several vectors is a single vector that has the same physical effect as the original vectors. Since the resultant is a vector, both its magnitude and its direction must be specified. However, before we discuss vector addition, we must discuss vector components.

Vector Components

If we draw a vector **A** as shown in Figure 11–2 and draw lines from the tail of the vector parallel to the *x* and *y* axes, the *two new vectors* are called the **components** of vector **A.** The original vector **A** is the sum of its *x* and *y* components, expressed by

$$\mathbf{A} = \mathbf{A}_x + \mathbf{A}_y$$

FIGURE 11–2

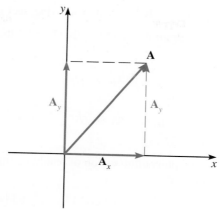

The *process of finding the components* of a vector is called **resolving the vector into its components.** If you are given the magnitude A and direction θ of vector **A,** the calculations for the components are as summarized in the following box.

Components of a Vector

Using the definition of the sine and cosine functions, we have

$$\cos \theta = \frac{A_x}{A} \qquad \text{or} \qquad A_x = A \cos \theta$$

$$\sin \theta = \frac{A_y}{A} \qquad \text{or} \qquad A_y = A \sin \theta$$

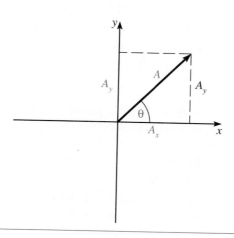

EXAMPLE 1 Resolve the following vectors into components:

(a) $s = 56$ m; $\theta = 67°$ (b) $F = 83$ lb; $\theta = 158°$

Solution

(a) $s_x = s \cos \theta = 56 \cos 67°$ 56 ⊠ 67 [cos] [≈] ⟶ 22

$s_x \approx 22$ m

$s_y = s \sin \theta = 56 \sin 67°$ 56 ⊠ 67 [sin] [≈] ⟶ 52

$s_y \approx 52$ m

A graphical illustration of resolving this vector into its components is given in Figure 11–3(a):

(b) $F_x = F \cos \theta = 83 \cos 158°$ 83 ⊠ 158 [cos] [≈] ⟶ −77

$F_x \approx -77$ lb

$F_y = F \sin \theta = 83 \sin 158°$ 83 ⊠ 158 [sin] [≈] ⟶ 31

$F_y \approx 31$ lb

A graphical illustration of resolving this vector into its components is given in Figure 11–3(b).

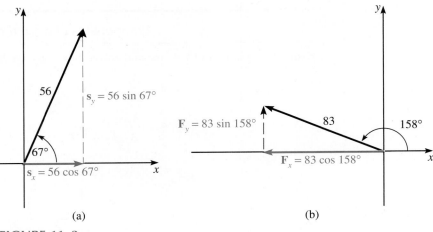

(a)   (b)

FIGURE 11–3   ▪▪▪

Vector Resultants Algebraically

Now we are in a position to add vectors algebraically. This process is illustrated in the next three examples.

EXAMPLE 2 Find the resultant of two forces, **A** and **B,** where **A** = 63 lb east and **B** = 47 lb south.

Solution Vectors **A** and **B** are shown in Figure 11–4. To find the magnitude of the resultant represented by R in the figure, we can use the Pythagorean theorem.

$$R^2 = A^2 + B^2$$
$$R = \sqrt{A^2 + B^2}$$
$$R = \sqrt{(63)^2 + (47)^2}$$
$$R \approx 79 \text{ lb}$$

FIGURE 11-4

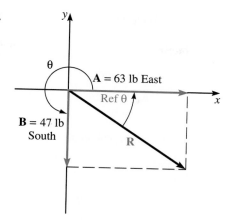

To find the direction of the resultant, we find the reference angle, labeled Ref θ in Figure 11-4, and then calculate θ,

$$\tan(\text{Ref } \theta) = \left| \frac{-47}{63} \right|$$

$$\text{Ref } \theta \approx 37°$$

Since θ is a fourth quadrant angle, we calculate

$$\theta = 360° - \text{Ref } \theta$$

$$\theta = 360° - 37°$$

$$\theta = 323°$$

The resultant of vectors **A** and **B** is 79 lb at 323° from standard position.　　　■■■

NOTE ◆ To make checking your answer easier, θ should always be an angle in standard position.

Vector Resultants by Components

Since the vectors in Example 2 were perpendicular, it was not necessary to resolve them into components. However, any vector that does not lie on the *x* or *y* axis must be resolved into components. The following box summarizes vector resultants, and the next example illustrates this procedure.

Resultant of Vectors

Magnitude:　　　$R = \sqrt{R_x^2 + R_y^2}$

Direction:　　$\tan(\text{Ref } \theta) = \left| \dfrac{R_y}{R_x} \right|$

where

R_x = sum of all *x* components

R_y = sum of all *y* components

▨ **EXAMPLE 3** Find the resultant of two displacements, **A** = 48 m at 63° and **B** = 73 m at 118°.

Solution Figure 11–5 shows the two vectors and their respective components. Using the formulas to find the x and y components of the vector gives

$$A_x = 48 \cos 63° \approx 22 \qquad\qquad A_y = 48 \sin 63° \approx 43$$
$$B_x = 73 \cos 118° \approx -34 \qquad\qquad B_y = 73 \sin 118° \approx 64$$

We use the formula to calculate the magnitude and direction of the resultant.

$$R = \sqrt{R_x^2 + R_y^2}$$
$$R \approx \sqrt{(22 - 34)^2 + (43 + 64)^2}$$
$$R \approx \sqrt{(-12)^2 + (107)^2}$$
$$R \approx 110 \text{ m}$$

$$\tan(\text{Ref } \theta) = \left| \frac{R_y}{R_x} \right|$$

$$\tan(\text{Ref } \theta) \approx \left| \frac{107}{-12} \right|$$

$$\text{Ref } \theta \approx 83.6°$$

Since R_y is positive and R_x is negative, θ is a second quadrant angle. Figure 11–6 shows R_x, R_y, R, and θ.

$$\theta = 180° - \text{Ref } \theta$$
$$\theta \approx 180° - 83.6°$$
$$\theta \approx 96°$$

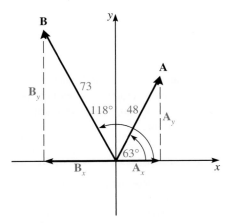

FIGURE 11–5

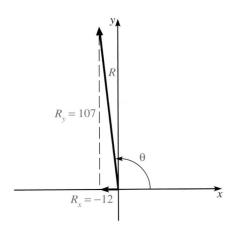

FIGURE 11–6

The resultant of vectors of **A** and **B** is 110 m at 96°. ■ ■ ■

A table proves useful in finding the resultant of more than two vectors. The table approach is shown in the next example.

EXAMPLE 4 Find the resultant of the following vectors:

$$\mathbf{A} = 38 \text{ N at } 218° \qquad \mathbf{B} = 56 \text{ N at } 48° \qquad \mathbf{C} = 23 \text{ N at } 108°$$

Solution We fill in a table showing the x and y components for each vector and the sum of these components.

Vector	x Component	y Component
A	$38 \cos 218° \approx -29.944$	$38 \sin 218° \approx -23.395$
B	$56 \cos 48° \approx 37.471$	$56 \sin 48° \approx 41.616$
C	$23 \cos 108° \approx -7.107$	$23 \sin 108° \approx 21.874$
Sum	$R_x \approx 0.420$	$R_y \approx 40.095$

The calculation for the magnitude of the resultant gives

$$R = \sqrt{R_x^2 + R_y^2}$$
$$R \approx \sqrt{(0.420)^2 + (40.095)^2}$$
$$R \approx \sqrt{1,607.785425}$$
$$R \approx 40$$

The direction of the resultant is given by

$$\tan (\text{Ref } \theta) = \left| \frac{R_y}{R_x} \right|$$

$$\tan (\text{Ref } \theta) \approx \left| \frac{40.095}{0.420} \right|$$

$$\text{Ref } \theta \approx 89°$$

Since R_x and R_y are both positive, θ is a quadrant I angle.

$$\theta \approx 89°$$

The resultant of **A, B,** and **C** is approximately 40 N at 89°. ▪▪▪

Resultants Using a Calculator

Most calculators convert between rectangular coordinates (x, y) and polar coordinates (R, θ). We discussed rectangular coordinates and graphing in Chapter 4, and we will discuss polar coordinates in more detail in Chapter 14. The following example illustrates how to use the calculator to fill in a table of components and calculate the resultant.

EXAMPLE 5 Find the resultant of the following vectors:

$$\mathbf{A} = 8.6 \text{ m/s at } 53° \qquad \mathbf{B} = 9.4 \text{ m/s at } 121° \qquad \mathbf{C} = 3.7 \text{ m/s at } 215°$$

Solution First we calculate the x and y components for each vector using the following keystrokes. Consult your calculator owner's manual for details. The results are shown in the table that follows:

Vector A: 8.6 $\boxed{x \leq y}$ 53 $\boxed{2\text{nd}}$ $\boxed{P \leftrightarrow R}$ ⟶ 6.8683 (y component)

$\boxed{x \geq y}$ ⟶ 5.1756 (x component)

Vector **B**: 9.4 $\boxed{x \gtrless y}$ 121 $\boxed{2^{nd}}$ $\boxed{P \leftrightarrow R}$ ────▶ 8.0574 $\boxed{x \gtrless y}$ ────▶ −4.8414

Vector **C**: 3.7 $\boxed{x \gtrless y}$ 215 $\boxed{2^{nd}}$ $\boxed{P \leftrightarrow R}$ ────▶ −2.1222 $\boxed{x \gtrless y}$ ────▶ 3.0309

Vector	x Component	y Component
A	5.1756	6.8683
B	−4.8414	8.0574
C	−3.0309	−2.1222
Sum	−2.7147	12.8035

The next step is to calculate R and θ given R_x and R_y. The calculator keystrokes follow.

2.7147 $\boxed{+/-}$ $\boxed{x \gtrless y}$ 12.8035 $\boxed{inv}$ $\boxed{2^{nd}}$ $\boxed{P \leftrightarrow R}$ ────▶ 101.97102 (θ)

$\boxed{x \gtrless y}$ ────▶ 13.08813 (R)

The resultant of the vectors is 13 m/s at 102°. ■■■

Application

EXAMPLE 6   A state trooper investigating an accident pushes a wheel tape to measure the skid marks as shown in Figure 11–7. If he applies a force of 8 lb at an angle of 52° with the ground, find the vertical and horizontal force.

FIGURE 11–7

Solution   Figure 11–8 shows the vector diagram.

The *x* component is given by

$$x = R \cos \theta$$
$$x = 8 \cos 308°$$
$$x = 5 \text{ lb}$$

FIGURE 11–8

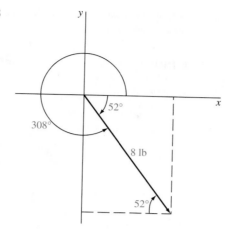

The *y* component is given by

$$y = R \sin \theta$$
$$y = 8 \sin 308°$$
$$y = -6 \text{ lb}$$

■■■

11–1 EXERCISES

Find the resultant of the following vectors. Give the direction as an angle in standard position.

1. 20 mi east; 17 mi east
2. 49 lb south; 36 lb north
3. 254 m south; 189 m north
4. 158 yd east; 176 yd west
5. 53.4 ft/s north; 39.5 ft/s west
6. 17.8 mi east; 28.3 mi north
7. 36 lb north; 24 lb west
8. 7.6 m/s east; 13.4 m/s south
9. 217 N north; 309 N west
10. 9.3 m/s² east; 14.8 m/s² north
11. 18.4 mi south; 30.5 mi east
12. 156.7 lb west; 204.5 lb south
13. 29 N east; 16 N north
14. 37.4 m east; 26.9 m south
15. 32.6 ft/s² south; 21.5 ft/s² west

Find the x and y components of the following vectors.

16. 148 N at 35.0°
17. 118 ft/s at 73.5°
18. 18.8 m/s² at 212.0°
19. 75.4 yd at 189.1°
20. 78.3 N at 142.6°
21. 124 yd at 285.3°
22. 38.9 m/s at 243.6°
23. 66.9 m at 195.8°
24. 315 lb at 312.6°
25. 418 mi at 156.7°

Find the resultant of the following vectors. Give the direction as an angle in standard position.

26. **A** = 16.7 lb at 126.4°; **B** = 31.8 lb at 78.0°
27. **A** = 14.8 ft/s at 243.5°; **B** = 29.5 ft/s at 118.7°
28. **A** = 119 m at 113.8°; **B** = 79.3 m at 212.6°
29. **A** = 48.6 yd at 82.0°; **B** = 63.5 yd at 108.4°
30. **A** = 27.8 N at 217.5°; **B** = 40.6 N at 126.8°
31. **A** = 63 N at 125°; **B** = 112 N at 281°
32. **A** = 212.4 lb at 246.80°; **B** = 186.5 lb at 301.60°
33. **A** = 82.1 m at 108°; **B** = 61.8 m east; **C** = 59.2 m at 51.2°
34. **A** = 24 ft/s at 62°; **B** = 35.6 ft/s at 284°; **C** = 41.9 ft/s at 125°
35. **A** = 46 N at 136°; **B** = 73 N at −73°; **C** = 52 N at 314°
36. **A** = 18 cm at −60°; **B** = 38 cm at 98°; **C** = 59 cm at 203°
37. **A** = 384 mi at 123.0°; **B** = 239 mi at 207.0°; **C** = 315 mi at 246.0°
38. **A** = 14.8 m/s at 78.3°; **B** = 37.4 m/s at 163.5°; **C** = 10.9 m/s at 309.7°
39. **A** = 486 N at 197.0°; **B** = 217 N at 318.0; **C** = 524 N at −153.0°
40. **A** = 316 ft at −122.0°; **B** = 214 ft at 326.0°; **C** = 190 ft at 118.0°

CS 41. If a 17-in. vector is drawn at a 40° angle from the lower left corner of a computer screen to the upper right corner, what is the area of the screen?

P 42. A cable supporting a transmission tower exerts a force of 516 N at an angle of 46.1° with the horizontal (see Figure 11–9). Find the horizontal and vertical components.

P 43. A ball is thrown with an initial velocity of 75 ft/s at an angle of 38° with the horizontal (see Figure 11–10). Find the vertical and horizontal components of the velocity.

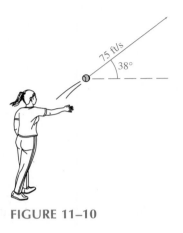

FIGURE 11–10

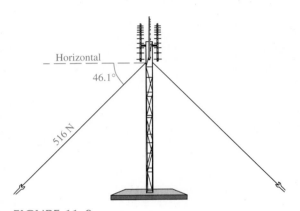

FIGURE 11–9

11–2

APPLICATIONS OF VECTORS

Vectors have numerous applications in science and technology. In this section we will discuss vectors as they apply to velocity, displacement, systems of forces, and inclined planes.

Velocity

Velocity is defined to be the rate at which displacement changes with respect to time. The next example shows an application of vectors to velocity.

EXAMPLE 1 As shown in Figure 11–11, a plane flies at a rate of 300 mi/h 60° north of east with respect to the air. If the velocity of the wind is 100 mi/h in a direction 23° south of east, what is the velocity of the plane with respect to the ground?

FIGURE 11–11

Solution As shown in Figure 11–12, we are finding the resultant of two vectors: the air velocity of the plane and the velocity of the wind. The calculations for the resultant are shown in the accompanying table.

Vector	x Component	y Component
Plane	300 cos 60° = 150	300 sin 60° = 259.81
Wind	100 cos(−23°) = 92.05	100 sin(−23°) = −39.07
Sum	R_x = 242.05	R_y = 220.74

FIGURE 11–12

The x component of the resultant is 242.05, and the y component is 220.74. The calculations for the magnitude and direction of the resultant are as follows:

$$R = \sqrt{R_x^2 + R_y^2}$$
$$R \approx \sqrt{(242.05)^2 + (220.74)^2}$$
$$R \approx 330$$

$$\tan(\text{Ref } \theta) = \left| \frac{R_y}{R_x} \right|$$

$$\tan(\text{Ref } \theta) \approx \left| \frac{220.74}{242.05} \right|$$

$$\text{Ref } \theta \approx 42°$$

Since R_x and R_y are both positive, $\theta = 42°$. The ground velocity of the plane is 330 mi/h at 42° north of east.

■ ■ ■

Displacement

Displacement is a vector quantity that describes an object's change in position. This application is illustrated in the next example.

EXAMPLE 2 A surveyor leaves her transit and walks 20 m 23° north of east, then 10 m west, and finally 6.0 m 56° south of west. How far is she from her starting position?

Solution We must find the resultant of the following three vectors, shown in Figure 11–13:

A = 20 m 23° north of east   **B** = 10 m west   **C** = 6.0 m 56° south of west

The accompanying table summarizes the x and y components of these vectors.

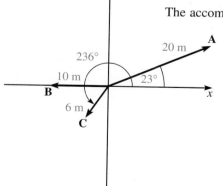

Vector	x Component	y Component
A	20 cos 23° ≈ 18.4	20 sin 23° ≈ 7.8
B	10 cos 180° = −10	10 sin 180° = 0
C	6.0 cos 236° ≈ −3.4	6.0 sin 236° ≈ −5.0
Sum	R_x ≈ 5	R_y ≈ 2.8

FIGURE 11–13

Next, we calculate the magnitude and direction for the resultant of the vectors.

$$R = \sqrt{R_x^2 + R_y^2}$$
$$R \approx \sqrt{(5)^2 + (2.8)^2}$$
$$R \approx 5.7$$

$$\tan (\text{Ref } \theta) = \left| \frac{R_y}{R_x} \right|$$

$$\tan (\text{Ref } \theta) \approx \frac{2.8}{5}$$

$$\text{Ref } \theta \approx 29°$$

Since R_y is positive and R_x is positive, θ is in quadrant I. Therefore, $\theta = 29°$. The surveyor is 5.7 m at 29° from her starting position.

Vector **A**: 20 $\boxed{x \gtrless y}$ 23 $\boxed{2^{nd}}$ $\boxed{P \leftrightarrow R}$ $\longrightarrow$ 7.8146

$\boxed{x \gtrless y}$ $\longrightarrow$ 18.4100

Vector **B**: 10 $\boxed{x \gtrless y}$ 180 $\boxed{2^{nd}}$ $\boxed{P \leftrightarrow R}$ $\longrightarrow$ 0

$\boxed{x \gtrless y}$ $\longrightarrow$ -10

Vector **C**: 6.0 $\boxed{x \gtrless y}$ 236 $\boxed{2^{nd}}$ $\boxed{P \leftrightarrow R}$ $\longrightarrow$ -4.9742

$\boxed{x \gtrless y}$ $\longrightarrow$ -3.3552

Resultant: 5 $\boxed{x \gtrless y}$ 2.8 $\boxed{inv}$ $\boxed{2^{nd}}$ $\boxed{P \leftrightarrow R}$ $\longrightarrow$ 29.2488

$\boxed{x \gtrless y}$ $\longrightarrow$ 5.7306   ■ ■ ■

EXAMPLE 3   Two tugboats are towing a loaded barge. Find the resultant force along the axis of the barge if the tension in the two ropes is 3,500 lb and the angle between the ropes is 48°. Refer to Figure 11–14.

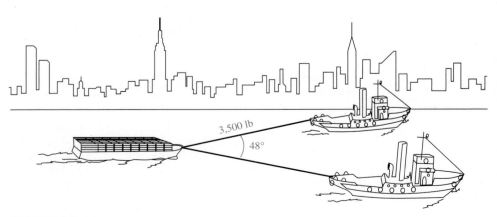

FIGURE 11–14

Solution   We are asked to find the resultant of two vectors of 3,500 lb at 24°, one above the *x*-axis and one below, as shown in Figure 11–15. The *x* and *y* components are shown in the following table.

$$R = \sqrt{(6{,}394.8)^2 + (0)^2}$$
$$R = 6{,}395$$

FIGURE 11–15

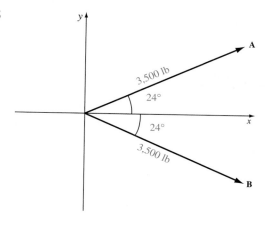

	x component	*y* component
A	3,197.4	1,423.6
B	3,197.4	−1,423.6
Total	6,394.8	0

The resultant force is 6,395 lb along the axis of the boat.   ∎∎∎

Force

Force is a vector quantity (push or pull) that tends to cause motion. According to the First Condition of Equilibrium, if an object is at rest or moving at a constant velocity, the resultant of all the forces acting on that body is zero. Therefore, the sum of all vectors along the *x* axis is zero, and the sum of all vectors along the *y* axis is zero. The next four examples illustrate the application of vectors to a system of forces.

EXAMPLE 4   A sign weighing 100 lb is hung from the ceiling by cables that make 43° and 62° angles with the ceiling, as shown in Figure 11–16. Determine the tension in each of the cables.

FIGURE 11–16

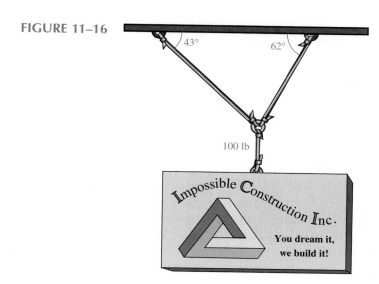

Solution First, we draw Figure 11–17, depicting the verbal statement. Second, we resolve the vectors into components, as shown in Figure 11–18. Third, we apply the First Condition of Equilibrium to yield the following equations:

$$\text{resultant of vectors along } x \text{ axis} = 0$$
$$A \cos 43° - B \cos 62° = 0$$
$$\text{resultant of vectors along } y \text{ axis} = 0$$
$$A \sin 43° + B \sin 62° + (-100) = 0$$

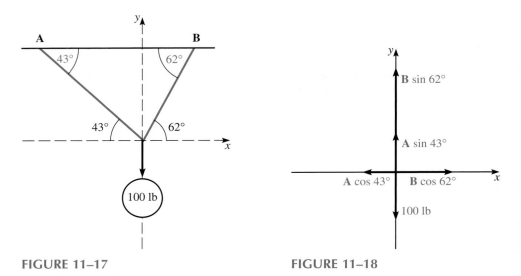

FIGURE 11–17   **FIGURE 11–18**

Using substitution to solve the system of equations for A and B gives

$$A = \frac{B \cos 62°}{\cos 43°}$$

$$A \approx 0.64B$$

$$0.64B \sin 43° + B \sin 62° - 100 = 0$$

$$0.44B + 0.88B = 100$$

$$1.32B = 100$$

$$B \approx 76 \text{ lb}$$

$$A = 0.64B = 0.64(76) \approx 49 \text{ lb}$$

The cable at the 43° angle has a tension of 49 lb, and the cable at the 62° angle has a tension of 76 lb. ■■■

Inclined Plane

A second application using a system of forces involves an object moving at a constant speed or at rest on an inclined plane. Figure 11–19 shows the forces acting in such a system. The normal force **N,** acting perpendicular to the plane, is the force exerted by the inclined plane on the object. The weight **W** of the object, acting vertically downward, is the force exerted by the object on the inclined plane. The frictional force **F** acts parallel to the inclined plane in the opposite direction of any possible movement. For ease in solving inclined plane problems, the x axis is drawn parallel to the inclined plane, and the y axis is drawn perpendicular to the inclined plane.

FIGURE 11–19

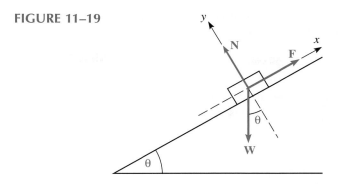

▣ **EXAMPLE 5** A 56-lb television is being loaded onto a truck using a ramp that makes an angle of 30° with the horizontal, as shown in Figure 11–20. Find the frictional and normal forces.

FIGURE 11–20

Solution Figure 11–21 shows the inclined plane with the pertinent information labeled. If we place the *x* and *y* axes in the manner described previously and rotate them, we obtain the system of forces shown in Figure 11–22. Using similar triangles, we find that the angle of the inclined plane equals the angle between the weight and the *y* axis. Resolving

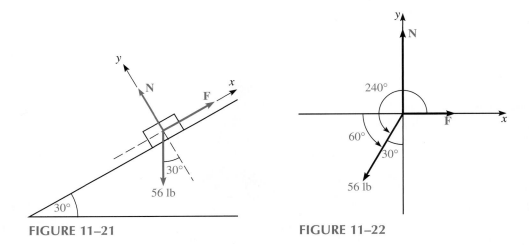

FIGURE 11–21   **FIGURE 11–22**

the vectors into components and applying the First Condition of Equilibrium gives the following equations:

$$56 \cos 60° = \mathbf{F}$$

$$56 \sin 60° = \mathbf{N}$$

Solving the above equations for **F** and **N** gives the frictional force as 28 lb and the normal force as 48 lb.

$$56 \boxed{\times} 60 \boxed{\cos} \boxed{=} \longrightarrow 28$$

$$56 \boxed{\times} 60 \boxed{\sin} \boxed{\approx} \longrightarrow 48 \qquad \blacksquare\,\blacksquare\,\blacksquare$$

EXAMPLE 6   A loading car like the one shown in Figure 11–23 is at rest on a track at a 15° angle. The weight of the car and its load is 3,500 lb and it is applied at a point 28 in. from the track, halfway between the two axles. The car is held by a cable attached 28 in. from the track. Find the tension on the cable and the force applied to each wheel.

Solution   A free-body diagram of the forces acting on the car is shown in Figure 11–24. The *x* axis was chosen parallel to the track and the *y* axis perpendicular to the track. The 3,500-lb weight is resolved into its components.

$$W_x = 3{,}500 \cos 15° \approx 3{,}380 \text{ lb}$$

$$W_y = 3{,}500 \sin 15° \approx 906 \text{ lb}$$

If we choose the pivot point at *A*, we obtain the following equation

$$906(30) \approx F_2(60)$$

$$453 \text{ lb} \approx F_2$$

Similarly, if we choose *B* as the pivot point, we obtain the following equation.

$$F_1(60) \approx 906(30)$$

$$F_1 \approx 453 \text{ lb}$$

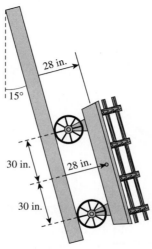

FIGURE 11–23

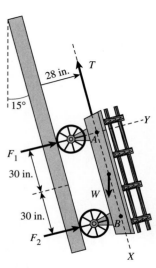

FIGURE 11–24

Applying the First Condition of Equilibrium gives

$$T = W_x = 3{,}380 \text{ lb}$$

■■■

EXAMPLE 7   A robotic arm weighing 140 lb pulls a 500-lb safe over a pulley using a rope as shown in Figure 11–25. Find the force and torque at the robot's elbow joint.

FIGURE 11–25

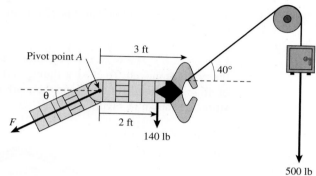

Solution   First, a free body diagram is made. (See Figure 11–26.) The First Condition of Equilibrium produces the following equations:

$$F \sin \theta + 140 = 500 \sin 40°$$
$$F \cos \theta = 500 \cos 40°$$

Using the Second Condition of Equilibrium, with the pivot point at *A*, produces:

$$\text{Torque} = (3)(500 \sin 40°) - (2)(140)$$

Using the first equation, adjusted as follows:

$$F \sin \theta = 500 \sin 40° - 140$$

and the second equation ($F \cos \theta = 500 \cos 40°$) to find the ratio of $\dfrac{F \sin \theta}{F \cos \theta}$ produces

$$\frac{F \sin \theta}{F \cos \theta} = \frac{500 \sin 40° - 140}{500 \cos 40°}$$

or

$$\tan \theta = \frac{500 \sin 40° - 140}{500 \cos 40°} \approx 0.4735$$

FIGURE 11–26

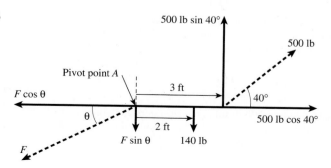

which yields $\theta \approx 25.34°$. Then using the second equation to solve for F yields:

$$F = \frac{500 \cos 40°}{\cos 25.34°} \approx 423.83 \text{ lb}$$

Use the third equation for torque:

$$\text{Torque} = (3)(500 \sin 40°) - (2)(140)$$
$$\approx 684.18 \text{ ft-lb}$$

The force on the robotic arm is 424 lb and the torque is 684 ft-lb counterclockwise. ■ ■ ■

EXAMPLE 8   The work done by a force is the product of the magnitude of the force and the distance moved. To close a sliding barn door, Mike pulls on a rope with constant force of 45 lb at an angle of 62°. Find the work done by moving the door 10 ft to its closed position, as shown in Figure 11–27.

FIGURE 11–27

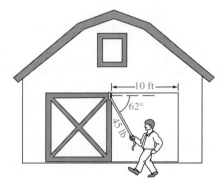

Solution   The magnitude of the force applied in the direction of the distance moved is the x component given by 45 lb cos 62°. The work done is given by

$$w = 45 \text{ lb} \cos 62°(10 \text{ ft})$$
$$w = 211 \text{ ft-lb}$$ ■ ■ ■

11–2 EXERCISES

1. Jane leaves the bakery, walks 8 blocks east, turns and walks 15 blocks at 45°, and then walks 10 blocks north. What is Jane's displacement from her initial position?

2. A boat travels horizontally across a river. If the boat travels 8 mi/h in still water and the river current travels 2 mi/h, find the velocity of the boat with respect to the water.

3. A plane is headed 37° north of east at a velocity of 457 mi/h with respect to the air. If the wind is from the northwest at 56 mi/h, what is the resultant velocity of the plane?

P 4. Movers are pushing a 200-lb sofa up a 20° ramp, applying a force of 150 lb at an angle of 35° with the horizontal. Determine the frictional force if the piano is moving at a constant velocity.

P 5. Two forces, one of 186 N and the other of 137 N, act on an object at right angles to each other. Find the magnitude of the resultant of these forces.

6. A jet flies 685 km south and then turns and flies 946 km at 58.4° west of south. What is the plane's displacement from its starting point?

P 7. A 20-lb ball is attached to a rope that hangs vertically from the ceiling. If the ball is pulled horizontally by a force **F** such that the rope makes a 35° angle with the vertical, find the force **F** and the tension in the rope.

M 8. A force of 15 lb is exerted on a bolt, as shown in Figure 11–28. Determine the horizontal and vertical components of the force.

FIGURE 11–28

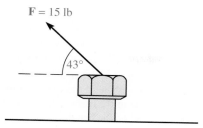

M 9. A collar that slides on a vertical rod is acted on by three forces, as shown in Figure 11–29. The direction of **F** may vary. Determine the direction of **F** so that the resultant of the forces is horizontal if the magnitude of **F** is 200 lb.

FIGURE 11–29

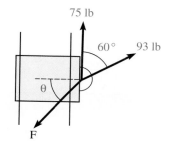

P 10. Two cables are tied together at point C and loaded as shown in Figure 11–30. Determine the tension in cables AC and BC.

FIGURE 11–30

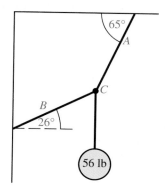

11. Frank wants to swim to a point directly across the river from his starting point. If he swims at a rate of 2.4 mi/h and the rate of the current is 1.5 mi/h, at what angle with respect to the shore should he swim?

P 12. A 4,800-N crate lying between two buildings is being lifted onto a truck. The crate is supported by a vertical cable joined by two ropes A and B that pass over pulleys attached to the buildings. From Figure 11–31, determine the tension in ropes A and B.

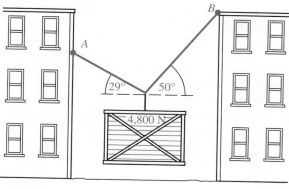

FIGURE 11–31

13. Four forces act on a screw, as shown in Figure 11–32. Determine the resultant force.

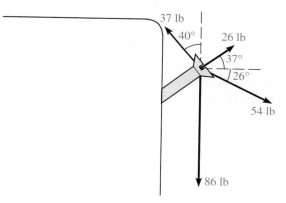

FIGURE 11–32

14. An airplane heads 37° north of west at a speed of 418 mi/h with respect to the air. If a wind blows at 38 mi/h from a direction 18° north of east, what is the velocity of the plane with respect to the ground?

P 15. Two cables are tied together at point C, as shown in Figure 11–33. Determine the tension in cables AC and BC. (You will have to use right triangle trigonometry to determine the necessary angles.)

FIGURE 11–33

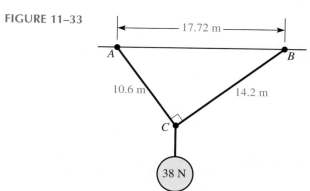

16. If an object weighing 75 N sits on a ramp inclined at 32° with the horizontal, what is the force of friction exerted on the object by the surface of the ramp?

17. A plane flies 27.8° south of east at a velocity of 380 mi/h with respect to the air. If the wind is blowing at 85.0 mi/h from a direction of 46.5° north of west, what is the velocity of the plane with respect to the ground?

18. Two ropes are tied together at point C, as shown in Figure 11–34. The maximum permissible tension in each rope is 678 lb. What is the maximum force **F** that can be applied without breaking the ropes if the direction of this force is 20°?

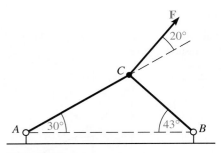

FIGURE 11–34

19. A sign weighing 250 lb is supported by two cables hung from the ceiling. One cable makes an angle of 48° with the ceiling, and the other cable makes an angle of 63° with the ceiling. Find the tension in each cable.

20. An object is dropped from a plane moving at 110 m/s at an angle of 15° below the horizontal. If the vertical velocity of the object as a function of time is given by $v_y = -110 \sin 15° - 9.8t$, what are the velocity and the direction of the object after 3 s?

21. Two cables with known tensions are attached to a telephone pole at point C. A third cable AC used as a guy wire is also attached to the pole at C. Determine the tension in AC so that the resultant of the three forces will be vertical. See Figure 11–35.

FIGURE 11–35

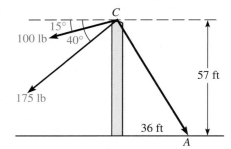

P 22. A 170-lb person sits in a hammock whose ropes each make different angles of 30° and 60° with the vertical. What is the tension in each rope?

P 23. A rock weighing 160 lb is suspended by a 15-ft rope fastened at the top to a beam. If the rock is held out 10 ft from the vertical by a horizontal force, find the magnitude of the horizontal force and the tension in the rope.

P 24. A roller is pulled up an inclined plane using a handle kept parallel to the plane. If the weight of the roller is 220 lb and the angle of the inclined plane is 35°, determine the force exerted on the handle.

A 25. A uniform plate girder weighs 8,000 lb. It is held by two crane cables, as shown in Figure 11–36. If the girder is held in a horizontal position, determine the tension in cables A and B and the angle α.

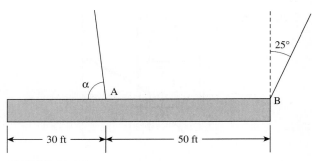

FIGURE 11–36

P 26. The collar slides freely on the horizontal smooth rod shown in Figure 11–37. Determine the magnitude of force F required to maintain equilibrium.

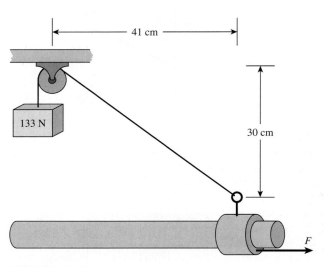

FIGURE 11–37

P 27. A 3,000-lb car is unloaded from a ship using a cable attached to a crane, as shown in Figure 11–38. A rope is attached to the car and pulled by a dock worker. Find the tension in the rope.

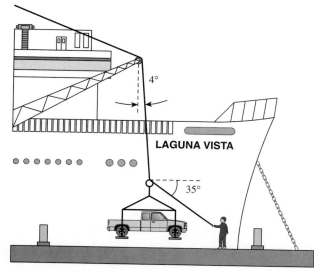

4°

LAGUNA VISTA

35°

FIGURE 11–38

M 28. Sharon is loading a delivery truck using a cart. The loading dock makes an angle of 17° with the horizontal. If the force Sharon applies to the cart must have a 250-N component acting up the dock, find the magnitude of the force. Refer to Figure 11–39.

FIGURE 11–39

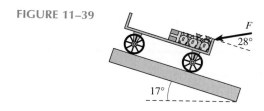

F

28°

17°

CS 29. Engineers use CAD to construct mathematical interpretations of an airplane in three-dimensional space. Once an airplane is made in this manner, it can be subjected to simulations. In one such simulation, a mathematical wind tunnel is used to apply force, via virtual wind, against the plane. If simulated wind blows at 100 mi/h 30° south of east and a plane travels 70 mi/h 70° north of east, what is the velocity of the plane with respect to the ground?

P 30. Carlos and Eduardo are moving a piano by pushing and pulling it along the floor, as shown in Figure 11–40. What is the resultant of these forces?

75 lb

15°

67 lb

10°

FIGURE 11–40

P 31. Jerry is pushing a cabinet, as shown in Figure 11–41. If Jerry is applying a force of 60 lb at an angle of 60° to move the cabinet 8 ft, how much work is done?

FIGURE 11–41

60°

60 lb

P 32. Mary Jane pulls a wagon, as shown in Figure 11–42. If she applies a force of 25 lb to move the wagon 36 ft, what work is done?

FIGURE 11–42

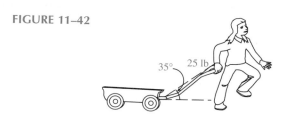

35° 25 lb

11–3

THE LAW OF SINES

In previous sections we have seen how to use trigonometry to solve right triangles. However, situations arise where it is necessary to solve an **oblique triangle**, a triangle without a right angle. We can use the Law of Sines and the Law of Cosines to solve oblique triangles, and the remainder of this chapter is devoted to these two topics.

Derivation of Law of Sines

We begin the discussion of solving oblique triangles by deriving the Law of Sines. Let ABC be an oblique triangle with sides a, b, and c opposite their respective angles, as shown in Figure 11–43. If an altitude h is drawn to the base, we can write the following relationships:

$$\sin A = \frac{h}{b} \qquad \sin B = \frac{h}{a}$$

$$h = b \sin A \qquad h = a \sin B$$

FIGURE 11–43

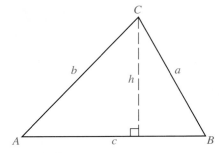

Equating the two expressions for h gives

$$b \sin A = a \sin B$$

Dividing both sides of the equation by $\sin A \sin B$ gives the following relationship:

$$\frac{b}{\sin B} = \frac{a}{\sin A}$$

Similarly, if we draw an altitude from angle A to side a, we can derive the following expression:

$$\frac{c}{\sin C} = \frac{b}{\sin B}$$

Combining these two results gives the Law of Sines, summarized as follows.

Law of Sines

In any triangle ABC,

$$\frac{a}{\sin A} = \frac{b}{\sin B} = \frac{c}{\sin C}$$

The length of a side is proportional to the sine of the angle opposite that side.

We can use the Law of Sines to solve the following two cases of triangles:

■ *Case 1*. Two angles and one side are given.
■ *Case 2*. Two sides and the angle opposite one of the sides are given.

A discussion of each of these cases follows.

Case 1

Example 1 illustrates the case where two angles and one side are given.

EXAMPLE 1   Given $A = 38°$, $B = 62°$, and $b = 16$, solve triangle ABC shown in Figure 11–44.

FIGURE 11–44

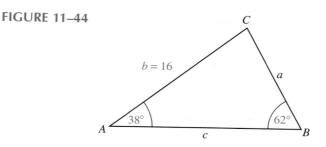

Solution   From the figure, we can see that we must solve for angle C, side a, and side c. Since we know that the sum of the angles of a triangle is 180°, we can solve for C using the following equation:

$$180° = A + B + C$$
$$C = 180° - 38° - 62°$$
$$C = 80°$$

To solve for side a, we can use the Law of Sines. The calculation is as follows:

This ratio contains the unknown. → $\dfrac{a}{\sin A} = \dfrac{b}{\sin B}$ ← Both terms of this ratio are known.

$$\frac{a}{\sin 38°} = \frac{16}{\sin 62°}$$

$$a = \frac{16 \sin 38°}{\sin 62°}$$

$$a \approx 11$$

$$16 \boxed{\times} 38 \boxed{\sin} \boxed{\div} 62 \boxed{\sin} \boxed{=} \longrightarrow 11.1565$$

To solve for side c, we can use the Law of Sines again.

$$\frac{c}{\sin C} = \frac{b}{\sin B}$$

$$\frac{c}{\sin 80°} = \frac{16}{\sin 62°}$$

$$c = \frac{16 \sin 80°}{\sin 62°}$$

$$c \approx 18$$

$$16 \boxed{\times} 80 \boxed{\sin} \boxed{\div} 62 \boxed{\sin} \boxed{=} \longrightarrow 17.8458$$

Sides a and c are 11 and 18, respectively, and C is 80°.

■ ■ ■

CAUTION ✦ A figure is helpful in solving oblique triangles; however, the triangles are not drawn to scale. Do not rely only on the relative size in the figure for information.

Case 2

When two sides and the angle opposite one of them is given, there may be no, one, or two solutions to the triangle. For this reason, Case 2 is called the **ambiguous case.** The following chart summarizes the possible cases.

Ambiguous Case

Given a, b, A of the oblique triangle ABC,

No solution

One solution

Two solutions

EXAMPLE 2   Given $c = 15$, $b = 23$, and $C = 42°$, solve triangle ABC shown in Figure 11–45.

FIGURE 11–45

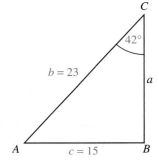

Solution   From the figure, we must first solve for angle B using the Law of Sines.

$$\frac{b}{\sin B} = \frac{c}{\sin C}$$

$$\frac{23}{\sin B} = \frac{15}{\sin 42°}$$

$$\sin B = \frac{23 \sin 42°}{15}$$

$$\sin B \approx 1.03$$

Since the value of the sine of an angle is never greater than 1, no triangle can be formed with the dimensions given here.   ■ ■ ■

REMEMBER ✦ Always set up your ratio using information given in the problem rather than quantities you found.

EXAMPLE 3 Given $a = 21$, $b = 18$, and $A = 38°$, solve triangle ABC shown in Figure 11–46.

FIGURE 11–46

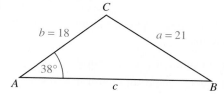

Solution From the figure, we can see that we should first solve for angle B using the Law of Sines.

$$\frac{a}{\sin A} = \frac{b}{\sin B}$$

$$\frac{21}{\sin 38°} = \frac{18}{\sin B}$$

$$\sin B = \frac{18 \sin 38°}{21}$$

$$\sin B \approx 0.5277$$

$$B \approx 32°$$

We have found a reference angle of $32°$ for angle B, but the sine function is positive in the first two quadrants. Therefore, there are two possibilities for the measure of angle B: $32°$ or $148°$. Since we cannot determine which is true, we must consider both possibilities.

If $B = 32°$,

$$C = 180° - 38° - 32°$$

$$C = 110°$$

$$\frac{a}{\sin A} = \frac{c}{\sin C}$$

$$\frac{21}{\sin 38°} = \frac{c}{\sin 110°}$$

$$c = \frac{21 \sin 110°}{\sin 38°}$$

$$c \approx 32$$

If $B = 148°$,

$$C = 180° - 38° - 148°$$

$$C = -6°$$

Since the angles of a triangle are not measured in negative numbers, the solution to the triangle is $B = 32°$, $C = 110°$, and $c = 32$. ∎∎∎

Examples 2 and 3 show that we can determine the number of situations to the ambiguous case by working through the problem. However, the number of solutions can also be determined from the given information. These results are summarized below.

Solutions to the Ambiguous Case

Given sides a and b and angle A of oblique triangle ABC, the following solutions are possible:

- no solution if $a < b \sin A$
- right triangle solution if $a = b \sin A$
- two solutions if $b \sin A < a < b$
- one solution if $a > b$ or $A > 90°$

Applications

EXAMPLE 4 Two forest ranger stations A and B are 48 miles apart. The bearing from A to B is 70° east of north. A ranger in each tower spots a fire. The fire's bearing from A and B is 33° east of north and 14° west of north, respectively. Using Figure 11–47, find the distance from the fire to each tower.

FIGURE 11–47

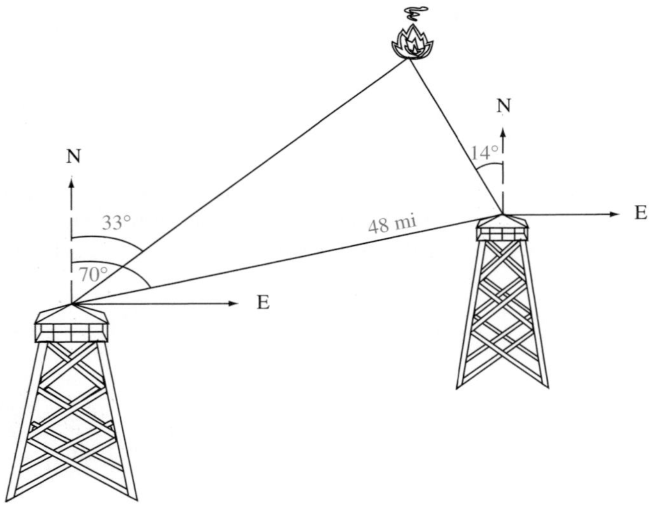

Solution Using the fact that the two dashed lines in Figure 11–47 are parallel, the angles within the triangle are $\angle A = 37°$, $\angle B = 96°$, and $\angle C = 47°$. Applying the Law of Sines to find $\overline{CB}$ and $\overline{CA}$ using x and y, respectively, gives

$$\frac{C}{\sin C} = \frac{x}{\sin A}$$

$$\frac{48}{\sin 47°} = \frac{x}{\sin 37°}$$

$$x \approx 39 \text{ miles}$$

and

$$\frac{C}{\sin C} = \frac{y}{\sin B}$$

$$\frac{48}{\sin 47°} = \frac{y}{\sin 96°}$$

$$y \approx 65 \text{ miles}$$

■ ■ ■

CAUTION ✦ When you are given two sides and the angle opposite one of them, remember to consider the possibility of two solutions. Neglecting to do so could be costly. For example, consider a navigator who considers only an angle of 47° and neglects the 133° angle. The plane may land at the wrong airport.

EXAMPLE 5 The Leaning Tower of Pisa was originally approximately 56 m high. If a surveyor trying to calculate the lean of the tower walks 72 m from the center base of the tower, the angle of elevation to the top is 40°. Find the lean of the Tower of Pisa.

Solution From Figure 11–48 we can use the Law of Sines to find angle C.

$$\frac{c}{\sin C} = \frac{b}{\sin B}$$

$$\frac{72}{\sin C} = \frac{56}{\sin 40°}$$

$$\sin C = 0.8264$$

$$C \approx 56°$$

Using the fact that the sum of the angles of a triangle is 180° gives

$$A + B + C = 180°$$

$$A = 180° - B - C$$

$$A \approx 180° - 40° - 56°$$

$$A \approx 84°$$

Since the angle of lean is the difference between vertical and A, the angle of lean is approximately 6°.

FIGURE 11–48

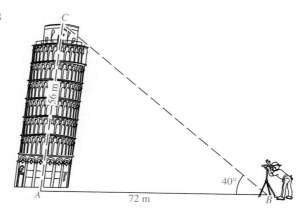

■ ■ ■

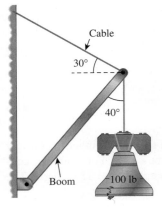

FIGURE 11–49

EXAMPLE 6 In the previous sections of this chapter, we discussed vectors using free-body diagrams to picture the physical situation. You could also use a force triangle. A 100-lb bell is supported by a tied boom, as shown in Figure 11–49. Use a force triangle to find the force F in the boom and the tension T in the cable.

Solution Since we are familiar with drawing a free-body diagram, let's draw one for this system of forces. The free-body diagram is shown in Figure 11–50.

To draw the force triangle, we begin with a force that is known in magnitude and direction. We then draw the lines of action for the two remaining forces, placing one through the tail and the other through the head of the known force. Be sure to maintain the angle between the forces. The force triangle is shown in Figure 11–51. We solve for F and T using the Law of Sines, as shown below.

$$\frac{100}{\sin 80°} = \frac{T}{\sin 40°} = \frac{F}{\sin 60°}$$

$$T = \frac{100 \sin 40°}{\sin 80°} \approx 65 \text{ lb}$$

$$F = \frac{100 \sin 60°}{\sin 80°} \approx 88 \text{ lb}$$

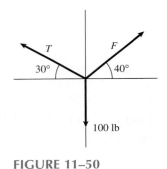

FIGURE 11–50

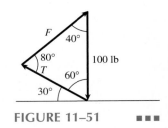

FIGURE 11–51 ■■■

11–3 EXERCISES

Solve triangle *ABC* given the information below.

1. $a = 15$; $C = 53°$; $B = 100°$

2. $a = 6.3$; $b = 4.8$; $A = 43°$

3. $a = 13$; $c = 10$; $A = 43°$

4. $b = 8.7$; $C = 53°$; $A = 39°$

5. $B = 67°$; $C = 70°$; $c = 23$

6. $b = 18$; $c = 25$; $C = 49°$

7. $a = 7.5$; $B = 58°$; $b = 9.0$

8. $A = 48°$; $C = 63°$; $a = 40$

9. $B = 42.8°$, $c = 36.7$; $C = 57.9°$

10. $C = 36.7°$; $a = 56.3$; $c = 39.8$

11. $A = 95°$; $C = 40°$; $b = 15$

12. $a = 39.1$; $C = 95.3°$; $c = 46.8$

13. $C = 82°$; $a = 38$; $B = 37°$

14. $A = 76.2°$; $b = 49.3$; $C = 37.8°$

15. $c = 14.8$; $b = 20.1$; $C = 30.2°$

16. $B = 63.4°$; $c = 14.8$; $A = 48.3°$

17. $C = 45.3°$; $b = 56.2$; $B = 72.5°$

18. $a = 31$; $c = 42$; $A = 23°$

19. $b = 25.3$; $C = 67.6°$; $c = 16.8$

20. $b = 18.6$; $B = 68.1°$; $A = 23.4°$

21. A pilot wants to fly to a city 46° east of north from her starting position. If the wind is blowing from the east at 25 mi/h and the plane flies at 400 mi/h with respect to the air, find the heading the pilot should take.

22. From a helicopter flying due west, the bearing of a stationary ship is 28° west of south at 3 P.M. and 15° west of south at 3:30 P.M. If the ground speed of the helicopter is 400 km/h, find the distance from the helicopter to the ship at 3 P.M.

C 23. A surveyor needs to determine the distance from *C* to *A* in Figure 11–52 to estimate the area of the pond. Using the information from the plat, find this length.

FIGURE 11–52

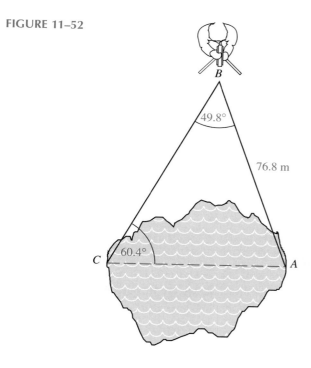

P 24. A sign is supported from the side of a building by two steel struts, as shown in Figure 11–53. Find the length of the struts from the information given in the figure.

FIGURE 11–53

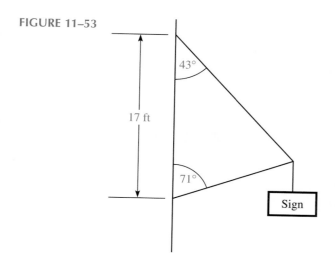

25. A ship leaves port and travels due south. At a certain point, the ship turns 24.6° north of east and travels on this course for 65.1 mi to a point of 83.4 mi from port. Determine the distance from the port to where the ship changed course.

C 26. An antenna 50 ft high is placed on the top of a building. The angle of elevation from a point on the ground to the top of the antenna is 63.7°, and the angle of elevation to the bottom of the antenna is 54.3°. Find the distance from the point on the ground to the top of the antenna.

27. An object is sighted from a helicopter at an angle of depression of 58°. After the helicopter has traveled 7.0 km, the angle of depression to the same side of the object is 64°. Determine the distance at that point from the helicopter to the object.

C 28. Town *B* is 28.3° south of west of town *A*. Determine the heading of a plane traveling from *A* to *B* if the wind is 38 mi/h from the east and the plane's speed is 400 mi/h with respect to air.

C 29. A telephone pole standing on level ground makes an angle of 80° with the horizontal. The pole is supported by a 30-ft prop whose base is 28 ft from the base of the pole. Find the angle made by the prop and the horizontal.

C 30. A 50-m-high antenna is mounted vertically on ground that slopes 13°. If two cables are attached to the top of the antenna at an angle of 53° with the horizontal, find the length of the cables.

CS 31. You are designing an algorithm that will be implemented in programming language C++ to give the vital statistics of a triangle when given certain information about the triangle. One menu selection is for the user to input "Angle, Angle, Side Opposite First Angle" information. If you choose this menu selection and input the following information "17, 71, 12," what other information about the triangle will the computer program produce?

P 32. A crate of 450 N is supported by two cables, as shown in Figure 11–54. Use a force triangle to find the tension in the cables.

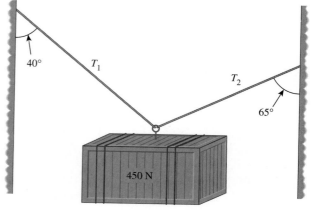

FIGURE 11–54

P 33. A 250-lb weight is supported by a cable that is attached to a wall and runs over a strut, as shown in Figure 11–55. Use a force triangle to find the force on the strut and the tension in the cable.

A 34. In the roof truss shown in Figure 11–56, find the length of support *BC*.

M 35. The toggle vise shown in Figure 11–57 is used to compress a block of wood. If *AB* is 62 cm, *BC* is 48 cm, and ∠*ACB* is 108°, find the distance between *A* and *C*.

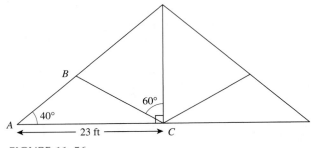

FIGURE 11–56

FIGURE 11–55

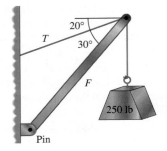

FIGURE 11–57

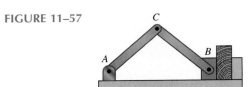

11–4

THE LAW OF COSINES

The second method of solving oblique triangles is the Law of Cosines. To derive the Law of Cosines, we are given triangle *ABC* with altitude *h*, shown in Figure 11–58. The altitude divides side *AB* into two parts: *x* and *c* − *x*. Using the Pythagorean theorem for each triangle gives

$$a^2 = h^2 + x^2 \qquad b^2 = (c - x)^2 + h^2$$

FIGURE 11–58

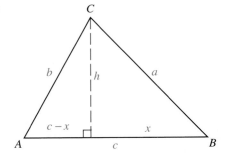

Solving each of these equations for h^2 gives

$$h^2 = a^2 - x^2 \qquad h^2 = b^2 - (c - x)^2$$

Equating the two expressions for h^2 gives

$$a^2 - x^2 = b^2 - (c - x)^2$$
$$a^2 - x^2 = b^2 - c^2 + 2cx - x^2$$

Solving the equation for b^2 gives

$$b^2 = a^2 + c^2 - 2cx$$

From Figure 11–58, we can see that the following expression is true:

$$\cos B = \frac{x}{a}$$

$$x = a \cos B$$

Substituting this expression for x gives one form of the Law of Cosines.

$$b^2 = a^2 + c^2 - 2ac \cos B$$

Using the same method and drawing altitudes to sides CB and AC gives similar results. The Law of Cosines is summarized as follows.

Law of Cosines

For oblique triangle ABC,

$$a^2 = b^2 + c^2 - 2bc \cos A$$
$$b^2 = a^2 + c^2 - 2ac \cos B$$
$$c^2 = a^2 + b^2 - 2ab \cos C$$

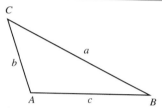

We can use the Law of Cosines to solve the following two additional cases of unknowns of an oblique triangle:

- *Case 3.* Two sides and the angle between those sides are given.
- *Case 4.* Three sides are given.

A discussion of each of these cases follows.

Case 3

The next example illustrates the case where we are given two sides and the included angle.

EXAMPLE 1 Solve triangle ABC if $a = 18.4$, $c = 26.3$, and $B = 47.9°$.

Solution First, we make a drawing of the oblique triangle as shown in Figure 11–59. We can use the Law of Cosines to solve for side b.

$$b^2 = a^2 + c^2 - 2ac \cos B$$
$$b^2 = (18.4)^2 + (26.3)^2 - 2(18.4)(26.3)\cos 47.9°$$
$$b^2 \approx 381.4$$
$$b \approx 19.5$$

FIGURE 11–59

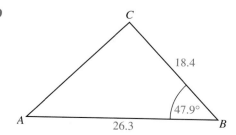

At this point, we can use either the Law of Cosines or the Law of Sines to solve for angles A and C. Since an angle is easier to find using the Law of Sines, we use it to solve for A.

$$\frac{a}{\sin A} = \frac{b}{\sin B}$$

$$\frac{18.4}{\sin A} = \frac{19.5}{\sin 47.9°}$$

$$\sin A = \frac{18.4 \sin 47.9°}{19.5}$$

$$\sin A \approx 0.7001$$

$$A \approx 44.4°$$

Then we can solve for angle C using the fact that the sum of the angles of a triangle is 180°. This calculation gives

$$C = 180° - A - B$$

$$C = 180° - 44.4° - 47.9°$$

$$C = 87.7° \qquad \blacksquare\blacksquare\blacksquare$$

Case 4

The next example illustrates solving an oblique triangle given three sides.

EXAMPLE 2 Solve triangle ABC given that $a = 17.8$, $b = 21.9$, and $c = 23.2$.

Solution First, we draw the oblique triangle ABC as shown in Figure 11–60. From the figure, we can see that we are given three sides of the triangle. We use the Law of Cosines to solve for angle A.

$$a^2 = b^2 + c^2 - 2bc \cos A$$

$$\cos A = \frac{a^2 - b^2 - c^2}{-2bc}$$

$$\cos A = \frac{(17.8)^2 - (21.9)^2 - (23.2)^2}{-2(21.9)(23.2)}$$

$$\cos A \approx 0.6899$$

$$A \approx 46.4°$$

FIGURE 11–60

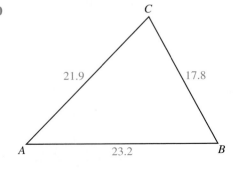

We can use the Law of Sines to solve for angle B.

$$\frac{b}{\sin B} = \frac{a}{\sin A}$$

$$\frac{21.9}{\sin B} = \frac{17.8}{\sin 46.4°}$$

$$\sin B \approx 0.8910$$

$$B \approx 63.0°$$

Then we solve for angle C using the fact that the sum of the angles of a triangle is 180°.

$$C = 180° - A - B$$

$$C = 180° - 46.4° - 63.0°$$

$$C = 70.6°$$

The following chart summarizes when to use the Law of Sines and the Law of Cosines.

Given	Use
Two angles and a side 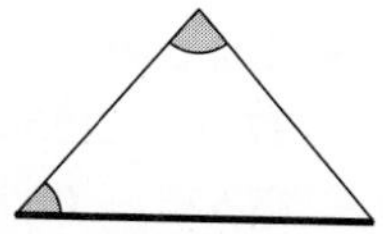	Law of Sines
Two sides and angle opposite one of them: (ambiguous case) 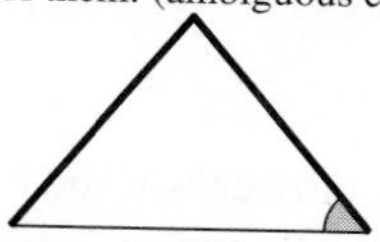	Law of Sines
Two sides and angle between them 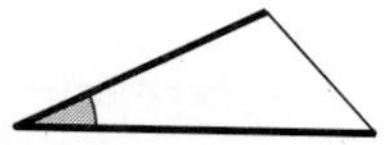	Law of Cosines
Three sides 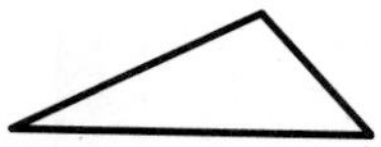	Law of Cosines

Applications

The Law of Cosines has many uses in solving oblique triangles resulting from applied problems. The next three examples illustrate this application.

EXAMPLE 3   An airplane flies from city A to city B, a distance of 275 miles, then turns through an angle of 43° and flies to city C, a distance of 250 miles. Using Figure 11–61, find the distance from city A to city C.

FIGURE 11–61

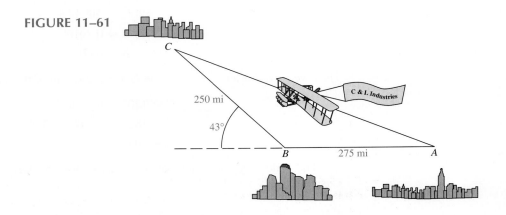

Solution   From the figure we can determine $\angle ABC$ inside the triangle to be 137°. Applying the Law of Cosines gives

$$b^2 = a^2 + c^2 - 2ac \cos B$$
$$b^2 = (250)^2 + (275)^2 - 2(250)275 \cos 137°$$
$$b^2 \approx 62{,}500 + 75{,}625 - (-100{,}561)$$
$$b^2 \approx 238{,}686$$
$$b \approx 489 \text{ miles}$$

■ ■ ■

EXAMPLE 4   A 75-ft vertical radio tower is to be erected on the side of a hill that makes an angle of 9° with the horizontal as shown in Figure 11–62. Find the length of two guy wires that will be anchored 80 ft uphill and downhill from the base of the tower.

FIGURE 11–62

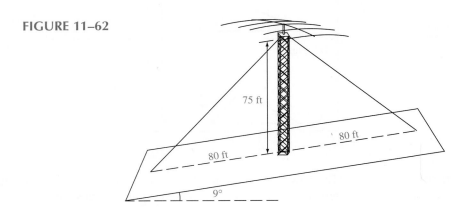

Solution From the information given in the problem statement and Figure 11–62, we can draw Figure 11–63 letting x and y represent the unknowns.

$$x^2 = (80)^2 + (75)^2 - 2(80)(75)\cos 99°$$
$$x^2 \approx 6{,}400 + 5{,}625 - (-1{,}877)$$
$$x^2 \approx 13{,}902$$
$$x \approx 118 \text{ ft}$$

Similarly,

$$y^2 = (80)^2 + (75)^2 - 2(80)(75)\cos 81°$$
$$y^2 \approx 6{,}400 + 5{,}625 - 1{,}877$$
$$y^2 \approx 10{,}148$$
$$y \approx 101 \text{ ft}$$

The two guy wires are 118 ft and 101 ft long.

FIGURE 11–63

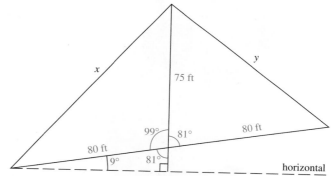

EXAMPLE 5 To program the movement of a robot, a position in Cartesian coordinates (x, y) must be translated into different coordinate systems depending on the type of robot. The jointed-arm robot, as shown in Figure 11–64, translates x, y coordinates into three angles of rotation: the base angle of rotation γ, the complete shoulder angle (preliminary shoulder angle of rotation α plus the elevation angle β), and an elbow angle ϕ. The robot's home reference point (origin) is taken as the center of its horizontal working plane. The angle γ would be equal to zero and the arm would be fully retracted. The coordinate system is shown in Figure 11–65.

Solution The following calculations find the four rotation angles, like α, β, ϕ, and γ. The following dimensions are given for the arm segments:

    Length of the upper arm $(A) = 3.0$ ft

    Length of the forearm $(B) = 4.0$ ft

    Pedestal height $= 13.0$ in.

It should be noted that the robot's maximum reach, R, is determined by the sum of the two arm segments $(A + B)$. The sum forms a sphere, with the center located at the lower pivot point. The robot can reach any point inside this sphere, assuming that it can rotate

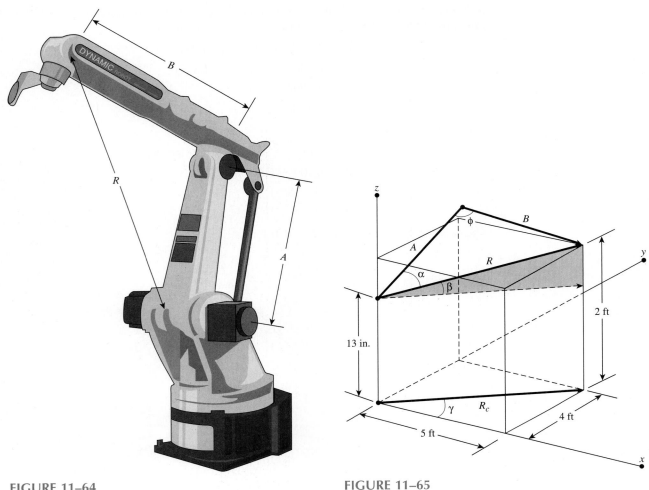

FIGURE 11–64

FIGURE 11–65

a full 360° about the z axis. The point that must be reached is located at $x = 5$ ft, $y = 4$ ft, $z = 2.0$ ft. The magnitude component of R in the x-y plane (R_c) is given by:

$$R_c = \sqrt{x^2 + y^2}$$
$$R_c = \sqrt{(5.0)^2 + (4.0)^2}$$
$$R_c \approx 6.4 \text{ ft}$$

The height of the robot's arm above the horizontal x-y plane is equal to the target height minus the pedestal height (24.0 in. − 13.0 in.), which allows the calculation of angle β, given by:

$$\sin \beta = \frac{24.0 \text{ in.} - 13.0 \text{ in.}}{(6.4 \text{ ft})(12.0 \text{ in./ft})} = 0.143$$

$$\beta = \sin^{-1}(0.143) \approx 8.2°$$

The angle γ is found through the ratio of the y coordinate to the x coordinate, given by:

$$\tan \gamma = \frac{y}{x} = \frac{4.0 \text{ ft}}{5.0 \text{ ft}} = 0.8$$

$$\gamma = \tan^{-1}(0.8) \approx 38.6°$$

If you believe that your calculation for R_c is correct, then you could have used the sine or cosine function to find γ.

Now that we know three sides of the triangle, we can use the Law of Cosines to find α.

$$\cos \alpha = \frac{A^2 + R^2 - B^2}{-2AR}$$

$$= \frac{(3)^2 + (6.4)^2 - (4)^2}{-2(3)(6.4)}$$

$$\alpha \approx 152.2°$$

We can use the Law of Cosines or Law of Sines to find ϕ. We use the Law of Sines because it is simpler.

$$\frac{R}{\sin \phi} = \frac{B}{\sin \alpha}$$

$$\frac{6.4}{\sin \phi} = \frac{4}{\sin (152.2)}$$

$$\phi \approx 48.3°$$

The final command sent to the robot would be: Move the complete shoulder angle $\alpha + \beta = 160.4°$, the base angle of rotation $\gamma = 38.6°$, and an elbow angle $\phi = 48.3°$. ■ ■ ■

11–4 EXERCISES

Solve triangle *ABC* given the information below.

1. $a = 17.6$; $b = 20.3$; $c = 27.2$

2. $b = 48$; $c = 56$; $A = 72°$

3. $a = 34$; $b = 26$; $C = 43°$

4. $a = 28.9$; $b = 47.3$; $c = 30.6$

5. $b = 16.8$; $c = 29.4$; $A = 73.8°$

6. $a = 78.6$; $c = 110.4$; $B = 56.7°$

7. $a = 63$; $b = 74$; $c = 58$

8. $a = 8.10$; $b = 17.8$; $c = 24.2$

9. $a = 23.7$; $b = 16.8$; $c = 11.7$

10. $b = 16$; $c = 25$; $A = 98°$

11. $a = 469$; $b = 342$; $c = 769$

12. $b = 238$; $a = 173$; $C = 83.0°$

13. $a = 68$; $c = 49$; $B = 102°$

14. $a = 14.8$; $b = 32.0$; $c = 27.5$

15. $C = 75.3°$; $a = 97.5$; $b = 109$

16. $B = 102.30°$; $c = 123.8$; $a = 175.8$

17. $a = 293$; $b = 436$; $c = 158$

18. $a = 36.8$; $b = 74.9$; $c = 44.6$

19. $A = 108°$; $b = 58$; $c = 73$

20. $a = 749$; $C = 86°$; $b = 486$

CS 21. A computer program uses the Law of Cosines to determine the angles of a triangle given the lengths of the three sides. A student inputs the lengths of the sides of an isosceles triangle as 6, 6, 13. Explain why the computer will give an "Error" message.

P 22. From a point on level ground directly between two telephone poles, cables are attached to the top of each pole. One cable is 74.8 ft long, and the other is 66.7 ft long. If the angle of intersection between the two cables is 103.6°, find the distance between the poles.

A 23. A tract of land in the shape of a parallelogram has sides 101 m and 123 m. If the shortest diagonal is 146 m, find the measure of the angles of the parallelogram.

P 24. A cruise ship heads north from port for 20 mi and then turns to a heading of 18° east of south and travels another 12 mi. At this point, how far is the ship from port?

A 25. A triangular piece of land is bounded by 186 ft of freeway fence, 135 ft of brick wall, and 215 ft of concrete median. Find the angle between the fence and the median.

P 26. Two airplanes are 805 km apart. The planes are flying toward the same airport, which is 322 km from one plane and 513 km from the other. Find the angle at which their paths intersect at the airport.

P 27. A boat moving at a speed of 8.3 knots with respect to the water heads directly southeast. If a current develops with a speed of 2.3 knots from 18° south of west, what will be the actual heading and speed of the boat?

P 28. A 50-ft antenna is placed on the top edge of a building. The distance from the top of the antenna to the base of a tree on the ground is 210 ft, and the distance from the bottom of the antenna to the same ground point is 185 ft. Find the angle of depression from the top of the antenna to the base of the tree.

P 29. A 150-ft ham radio antenna is installed in a vertical position on a hillside that is 18° to the horizontal. A 200-ft guy wire is attached to the top of the antenna and is anchored downhill. How far from the base of the antenna is the wire anchored if you measure along the ground lines.

M 30. A solar collector is placed on a roof that makes an angle of 21° to the horizontal. If the upper end of the collector is supported by a vertical brace, as shown in Figure 11–66, how long is the collector?

A 31. The two sides of a cathedral ceiling meet at an angle of 100°. If the distance to opposite walls is 12 ft and 15 ft, what length beam is needed to join the walls?

A 32. The truss consisting of two isosceles triangles shown in Figure 11–67 is used in constructing a steel bridge. Find the angles at which the sides meet.

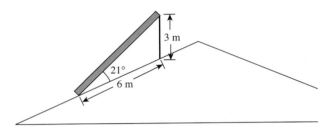

FIGURE 11–66

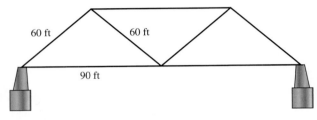

FIGURE 11–67

CHAPTER SUMMARY

Summary of Terms

ambiguous case (p. 416)

components (p. 394)

head (p. 394)

initial point (p. 394)

magnitude (p. 394)

oblique triangle (p. 413)

resolving into components (p. 395)

resultant (p. 394)

scalars (p. 394)

tail (p. 394)

terminal point (p. 394)

vectors (p. 394)

Summary of Formulas

$$\left.\begin{array}{l} A_x = A \cos \theta \\ A_y = A \sin \theta \end{array}\right\} \text{components of a vector}$$

$$\left.\begin{array}{ll} R = \sqrt{R_x^2 + R_y^2} & \text{magnitude} \\ \\ \tan(\text{Ref } \theta) = \left|\dfrac{R_y}{R_x}\right| & \text{direction} \end{array}\right\} \begin{array}{l}\text{resultant of} \\ \text{vectors}\end{array}$$

$$\frac{a}{\sin A} = \frac{b}{\sin B} = \frac{c}{\sin C} \quad \text{Law of Sines}$$

$$\left.\begin{array}{l} a^2 = b^2 + c^2 - 2bc \cos A \\ b^2 = a^2 + c^2 - 2ac \cos B \\ c^2 = a^2 + b^2 - 2ab \cos C \end{array}\right\} \text{Law of Cosines}$$

CHAPTER REVIEW

Section 11–1

Find the resultant of the following vectors. Give the angle in standard position.

1. 186 lb south; 265 lb west

2. 283 N east; 165 N south

3. 15 m/s at 118°; 29 m/s at 67°

4. 45 ft/s² at 193°; 72 ft/s² at 72°

5. 258 N at 72°; 197 N at 200°; 118 N at 156°

6. 96 yd at 283°; 150 yd at 147°; 100 yd at 61°

7. 18 lb at 136°; 38 lb at 310°; 10 lb at 66°

8. 86 m at 98°; 130 m at 172°; 59 m at 310°

Find the x and y components of the following vectors.

9. 291 lb at 221°

10. 17.84 ft at 198.3°

11. 486.7 km at 58.4°

12. 312.5 m/s at 127.6°

Section 11–2

13. Luke starts at the bicycle shop and walks 5 blocks north, then 8 blocks east, and finally 11 blocks southeast. What is Luke's final displacement from the bicycle shop?

14. A force of 218 lb is exerted on a screw, as shown in Figure 11–68. Determine the horizontal and vertical components of the force.

FIGURE 11–68

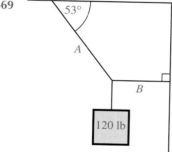

15. A 120-lb crate is supported, as shown in Figure 11–69. Find the tension in cables A and B.

FIGURE 11–69

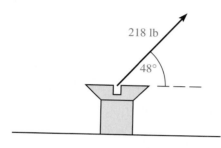

16. An object is dropped from a plane moving at 224 m/s at an angle of 23° above the horizontal. The vertical velocity of the object as a function of time is $v_y = 87.5 - 9.8t$. What are the velocity and the direction of the object after 12 s?

17. If an object weighing 253 lb rests on a ramp inclined at 18°, determine the friction force and the normal force.

18. Two ropes are tied together at point C, as shown in Figure 11–70. If the maximum tension in each rope without breaking the ropes is 756 lb, what is the maximum force **F** that can be applied? What should the direction of the force be?

19. An airplane heads 32° north of west at a velocity of 410 mi/h with respect to the air. If the wind is blowing at 63

FIGURE 11–70

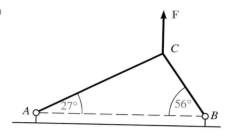

mi/h from 16° north of east, what is the velocity of the plane with respect to the ground?

20. An 89-N object attached to a cable hangs vertically from the ceiling. A horizontal force **F** is applied to the object such that the angle of the cable with the vertical is 22°. Find the force **F** and the tension **T** in the cable.

Section 11–3

Solve triangle ABC given the information below.

21. $a = 73.8$; $C = 16.2°$; $B = 93.1°$

22. $B = 72.4°$; $a = 124$; $b = 230$

23. $b = 14$; $C = 57°$; $c = 26$

24. $a = 18.7$; $b = 30.4$; $A = 39.7°$

25. $a = 56.8$; $A = 67.1°$; $b = 23.7$

26. $b = 39.6$; $c = 56.7$; $C = 102.1°$

27. A pilot wants to fly to a city that is 57° south of west from his current position. If the wind is blowing from the south at 56 mi/h and the plane flies at 350 mi/h with respect to the air, what should the heading of the plane be?

28. A cruise ship leaves port and travels due east. At a certain location, the ship turns to a course 47.5° north of east and travels on this course for 125.3 mi to a point 381.4 mi from port. Find the distance from port to where the ship changed course.

29. A telephone pole standing on level ground makes an angle of 85° with the horizontal. The pole is supported by a 56-ft prop placed 34 ft from the base of the pole. Find the angle made by the prop and the ground.

Section 11–4

Solve triangle ABC using the information given.

30. $a = 139$; $b = 110$; $c = 76.0$

31. $a = 257$; $B = 39.0°$; $c = 310$

32. $b = 10.8$; $c = 7.30$; $A = 63.1°$

33. $a = 18.4$; $b = 12.5$; $c = 21.8$

34. $a = 57.8$; $b = 63.4$; $c = 43.4$

35. $b = 186$; $C = 53.6°$; $a = 237$

36. Two boats are 176 mi apart. The boats are sailing toward the same port, which is 128 mi from one ship and 239 mi from the other ship. Find the angle at which the ships' paths intersect at the port.

37. Cables are attached to the top of two telephone poles and anchored at a point on level ground between the poles. One cable is 17 m long, and the other is 21 m long. If the angle of intersection between the two cables is 86°, find the distance between the poles.

38. A pilot flies north from an airport for 350 mi and then turns 36.1° west of north and travels another 275 mi. At this point, how far is the plane from the airport?

39. A small tract of land in the shape of a parallelogram has sides 25 yd and 27 yd long. If one diagonal is 34 yd, find the measure of the angle of the two sides of the parallelogram.

CHAPTER TEST

The number(s) in parentheses refers to the appropriate learning objective(s) given at the beginning of the chapter.

1. Find the resultant of each of the following vectors: (1)

       125 lb at 43.1° north of east   217 lb at 56.2° south of west

2. From a position 6 ft above the ground, an observer finds the angle of elevation to the top of a building to be 87°. From the same position the angle of elevation to a window 40 ft below the top is 63°. Find the distance from the observer to the top of the building. (5)

3. In triangle ABC, $A = 16°$, $c = 24.6$, and $B = 93°$. Find side a. (3)

4. In triangle ABC, $a = 173$, $b = 158$, and $c = 163$. Find angle C. (4)

5. A force of 56 N is applied to the top of a bolt at an angle of 48° with the horizontal. Find the vertical component of this force. (2)

6. Movers are pushing a 125-lb stereo cabinet up an 18° ramp, applying a force of 80 lb at an angle of 30° above the horizontal. If the cabinet is moving at a constant velocity, what is the frictional force of the ramp on the cabinet? (5)

7. Find the resultant of each of the following vectors: (1)

      16.8 m east   37.4 m at 30.0° south of west   21.3 m west

8. A vertical telephone pole is placed on a hill that makes an angle of 12° with the horizontal. Guy wires are attached to the pole 21 ft from the ground at points 12 ft from the base of the pole. Determine the length of each of the guy wires. (5)

9. In triangle ABC, $a = 16.8$, $b = 23.6$, and $B = 56.1°$. Solve for side c. (3)

10. Solve triangle ABC if $a = 86.2$, $b = 118$, and $c = 99.3$. (3, 4)

11. In triangle ABC, $a = 176$, $B = 97.0°$, $c = 198$. Find side b. (4)

12. An airplane pilot wants to fly from his starting point to his destination, which is 63° north of west from his current position. If the wind is blowing from the east at 43 mi/h and the plane flies at 475 mi/h with respect to the air, what heading should the pilot take? (5)

13. A boat travels across a river. If the boat travels 10 mi/h in still water and the current flows at 5 mi/h, what is the velocity of the boat relative to the shore? (2)

14. Find the resultant of each of the following vectors: (1)

      18 m/s at 63°   31 m/s at 174°   56 m/s at −38°

15. Solve triangle ABC if $A = 46°$, $B = 93°$, and $a = 29$. (3)

GROUP ACTIVITY

You have been hired by Linda's Derricks to analyze a load situation. The derrick is made up of a two-part boom. The lower part has a length of 40 ft, while the upper part is only 20 ft in length. The center of gravity for the lower boom is ¾ of the way up the boom (located 30 ft from the hinge), and the center of gravity for the upper boom is located at its center (10 ft from the top). The angle that the lower boom makes with the horizontal α is 60°. The angle that the upper boom makes with the horizontal β is 20°. The main cable is at an angle θ_1 of 30° with the horizontal. The angle that the top cable makes with the horizontal θ_2 is 20°.

1. Write the equations for the vertical force exerted by the wall V, the horizontal force exerted by the wall H, and the equation for T_1.

2. Solve your equations for T_1, V, and H.

3. Write the equation for T_2.

4. Solve for T_2 using your equation. What is T_3 equal to?

5. A 2,000-lb pterodactyl is flying in the area and spots the derrick. It wants to attempt a landing on the boom (not the wires). Where could it land (distance up the boom) and not break the cable if the cable tension limit is 322.1×10^3 pounds?

SOLUTION TO CHAPTER INTRODUCTION

We must find distances a and c given two angles and the side between them. Using the Law of Sines gives

$$\frac{a}{\sin A} = \frac{b}{\sin B}$$

$$\frac{a}{\sin 63°} = \frac{55}{\sin 67°}$$

$$a \approx 53 \text{ ft}$$

$$\frac{c}{\sin C} = \frac{b}{\sin B}$$

$$\frac{c}{\sin 50°} = \frac{55}{\sin 67°}$$

$$c \approx 46 \text{ ft}$$

The distances from the deck of the bridge to the two pilings are 53 ft and 46 ft.

T

he Coast Guard has used a buoy equipped with a downward looking acoustic sensor to plot the depth of the Canaveral Harbor entrance. They give you the following data and you are required to model the water depth using a cosine function. (The solution to this problem is given at the end of the chapter.)

Sound, water, and light waves are all examples of periodic waves; that is, the shape of the wave repeats at regular intervals. For example, a water particle moving down a stretched string will trace out a pattern duplicating the graph of a sine or cosine function. Another application of trigonometric graphs occurs in the screen display of an oscilloscope, which graphically displays an electrical signal. For example, you can graph a sinusoidal voltage and a cosine voltage on the x and y axes of an oscilloscope screen to display the resulting voltage. In this chapter we will discuss the basic shape of graphs of the trigonometric functions and alterations to the graph resulting from changes in amplitude, period, and phase angle.

Learning Objectives

After you complete this chapter, you should be able to

1. Identify the amplitude, period, and phase angle of a sine or cosine trigonometric function (Sections 12–1, 12–2, and 12–3).

2. Sketch the graph of a sine or cosine function using its amplitude, period, and phase angle (Sections 12–1, 12–2, and 12–3).

3. Apply the concept of amplitude, period, and phase shift to simple harmonic motion and ac circuits (Section 12–4).

4. Graph the secant, cosecant, cotangent, and tangent functions (Section 12–5).

5. Use addition of ordinates to graph composite functions (Section 12–6).

6. Graph parametric equations (Section 12–6).

time (t)	midnight	2	4	6	8	10	noon
depth (d in ft)	−0.85	0.85	1.7	0.85	−0.85	−1.7	−0.85

Chapter 12

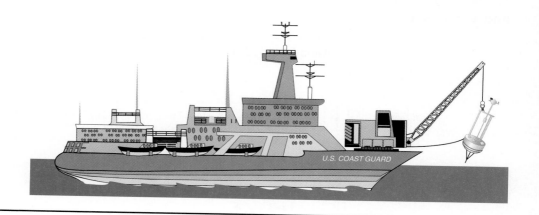

Graphs of the
Trigonometric Functions

12–1

GRAPHS OF SINE AND COSINE FUNCTIONS: AMPLITUDE

In this chapter we discuss the graphs of the six trigonometric functions. In graphing a trigonometric function such as $y = \sin x$, we use the same techniques that we used in graphing an algebraic function such as $2x - 5y = 7$ in Chapter 4. Namely, we fill in a table of values by choosing values for x, substituting each value into the function, and obtaining the corresponding value for y. However in graphing the trigonometric functions, the value chosen for x will be an angle, normally expressed in radians. By using radians, both the angle and the trigonometric function of the angle are expressed as real numbers.

Graph of $y = \sin x$ and $y = \cos x$

EXAMPLE 1 Graph $y = \sin x$.

Solution We begin by filling in the accompanying table of values. We can use special angles such as $\pi/4$, $\pi/3$, and $\pi/6$, or the quadrantal angles, or a calculator. The graph is shown in Figure 12–1.

x	0	$\pi/4$	$\pi/2$	$3\pi/4$	π	$5\pi/4$	$3\pi/2$	$7\pi/4$	2π	$5\pi/2$	3π
y	0	0.7	1	0.7	0	-0.7	-1	-0.7	0	1	0

FIGURE 12–1

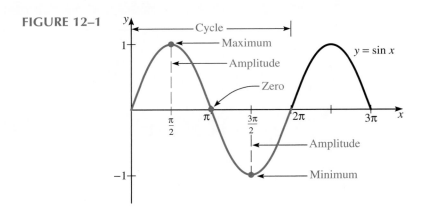

Two important aspects of the graph of $y = \sin x$ should be noted. First, the function is **periodic;** that is, the curve repeats itself at a regular interval. The **period** of the sine function is 2π, so the graph of the sine function looks exactly the same every 2π units, as shown by the color portion of the curve in Figure 12–1. The graph of the function through one period is called a **cycle.** Second, the amplitude of the sine function is 1. The **amplitude** represents the maximum variation of the curve from the x axis and is shown by the dashed line in Figure 12–1. Other properties of the sine function are summarized in the following box.

Properties of the Sine Function Graph

1. The graph crosses the x axis at the initial point, endpoint, and midpoint of a cycle. These points are called the **zeros of the function** (see Figure 12–1).

2. A maximum occurs midway between the first two zeros in a cycle, and a minimum occurs midway between the last two zeros. The maximum and minimum values are determined by the amplitude (see Figure 12–1).

EXAMPLE 2   Graph $y = \cos x$.

Solution   Using the same technique, we fill in the accompanying table of values for $y = \cos x$. The graph of $y = \cos x$ is given in Figure 12–2.

x	0	$\pi/4$	$\pi/2$	$3\pi/4$	π	$5\pi/4$	$3\pi/2$	$7\pi/4$	2π	$5\pi/2$	3π
y	1	0.7	0	-0.7	-1	-0.7	0	0.7	1	0	-1

FIGURE 12–2

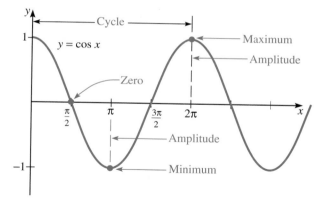

Notice that the graph of the cosine function in Figure 12–2 has the same basic shape as $y = \sin x$, called **sinusoidal.** It is also periodic, with period 2π, and it also has an amplitude of 1. Additional properties of the cosine function are summarized below.

Properties of the Cosine Function Graph

1. A maximum occurs at the initial point and the endpoint of a cycle. A minimum occurs midway in the cycle. The maximum and minimum values are determined by the amplitude (see Figure 12–2).

2. The zeros occur midway between a maximum and a minimum (see Figure 12–2).

LEARNING HINT ◆ In the remainder of this chapter, we will discuss how changes in amplitude, period, and phase angle affect the graph of the trigonometric functions. It is extremely important that you learn the properties of the sine and cosine functions outlined in this section.

Graph of $y = a \sin x$ and $y = a \cos x$

Next, let us examine the effect of a constant multiplier on the sine and cosine functions.

EXAMPLE 3   Graph $y = 3 \sin x$ through one period.

Solution   We will fill in a table of values just as we did previously, but a calculator will be helpful here. Figure 12–3 shows the graph of $y = 3 \sin x$. For comparison, the dotted line represents the graph of $y = \sin x$.

x	0	$\pi/4$	$\pi/2$	$3\pi/4$	π	$5\pi/4$	$3\pi/2$	$7\pi/4$	2π
y	0	2.1	3	2.1	0	-2.1	-3	-2.1	0

FIGURE 12–3

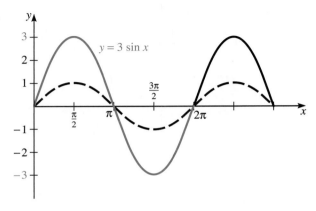

Notice that in the graph in Figure 12–3, the zeros of $y = 3 \sin x$ are the same as the zeros of $y = \sin x$. The difference in the two graphs occurs in their amplitude. Therefore, *multiplying the sine or cosine function by a constant changes only the amplitude* or causes a vertical stretch or compression of the basic graph. The graphs of $y = a \sin x$ or $y = a \cos x$ will have a maximum of $|a|$ and a minimum of $- |a|$. With this knowledge we can sketch the graph of the sine and cosine functions, and we do not have to fill in a table of values. This process is illustrated in the next example.

EXAMPLE 4   Sketch the graph of $y = 4 \cos x$ through one period.

Solution   We can sketch the graph of this function without a table of values because we know the basic properties of the cosine graph, and we know that the constant multiplier of 4 results in an amplitude change. To summarize, we know the following:

1. A maximum of 4 occurs at 0 and 2π, which are endpoints of the period.

2. A minimum of -4 occurs midway between the endpoints, which is π.

3. Zeros occur midway between each maximum and minimum at $\pi/2$ and $3\pi/2$.

The graph of $y = 4 \cos x$ is given in Figure 12–4.

FIGURE 12–4

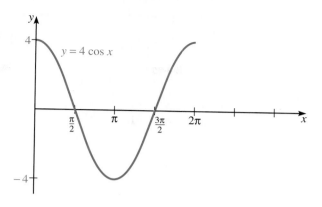

$y = 4 \cos x$

Graph of
$y = -a \sin x$ and
$y = -a \cos x$

In this section we discuss how a negative sign placed before the multiplier affects the graph.

EXAMPLE 5   Graph $y = -2 \sin x$ through one period.

Solution   We know from our previous discussion that the 2 in front of the sine function results in an amplitude change that affects the maximum and minimum values. Since we do not know the effect of the negative sign on the graph, we fill in a table of values. The graph is shown in Figure 12–5. For comparison, the dotted line represents the graph of $y = 2 \sin x$.

x	0	$\pi/4$	$\pi/2$	$3\pi/4$	π	$5\pi/4$	$3\pi/2$	$7\pi/4$	2π
y	0	-1.4	-2	-1.4	0	1.4	2	1.4	0

FIGURE 12–5

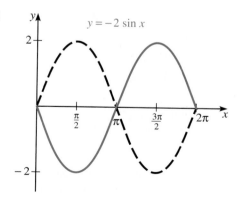

$y = -2 \sin x$

Notice that in Figure 12–5, the *negative sign in front of the amplitude has the effect of inverting or flipping the graph above the x axis.* The locations along the *x* axis where the maximum and minimum values occur are interchanged, but the zeros remain in the same location.

Summary

$$y = a \sin x$$

or

$$y = a \cos x$$

$|a|$ = amplitude.

A negative sign before a inverts the graph.

EXAMPLE 6 Sketch the graph of $y = -3 \cos x$ through one period.

Solution As in Example 4, we can sketch the graph of this function without a table of values, because we know the following:

1. The maximum of 3 and minimum of -3 occur at the midpoint and endpoint of the period, respectively.
2. The locations along the x axis where the maximum and minimum occur have been interchanged due to the negative sign, the midpoint and endpoints, respectively.
3. Zeros occur midway between each maximum and minimum at $\pi/2$ and $3\pi/2$.

The graph of $y = -3 \cos x$ is shown in Figure 12–6.

FIGURE 12–6

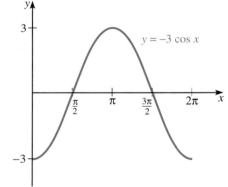

■■■

LEARNING HINT ✦ In sketching the graph of the sine and cosine functions, list the known information as in Example 6. Then use the calculator to check several points on the graph.

Graphing Calculator

The sine and cosine functions can be easily graphed using a graphing calculator. We can use our knowledge of the characteristics of the graph to set the viewing rectangle.

EXAMPLE 7 Using a graphing calculator, graph $y = -0.5 \sin x$.

Solution From our knowledge of how this graph should appear, we would set the y maximum to 0.5 and the y minimum to -0.5. For a graph of one period, we would set the

x minimum to 0 and the x maximum to 2π (6.28). We would also expect to see the graph inverted due to the negative sign. The viewing rectangle and the resulting graph are shown in Figure 12–7.

FIGURE 12–7

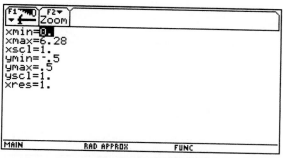

Viewing rectangle

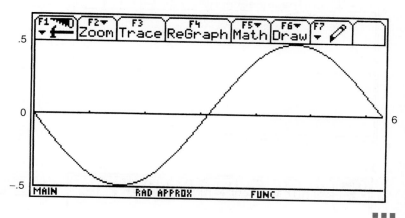

12–1 EXERCISES

Sketch the graph of each function through one period. You may want to use a graphing calculator to check your graphs.

1. $y = -2 \cos x$

2. $y = 2 \sin x$

11. $y = -8 \cos x$

12. $y = 6 \sin x$

3. $y = -4 \sin x$

4. $y = -\cos x$

13. $y = -5 \sin x$

14. $y = 3.9 \cos x$

5. $y = 6 \cos x$

6. $y = 2.5 \sin x$

15. $y = 0.5 \cos x$

16. $y = -0.8 \cos x$

7. $y = -3 \sin x$

8. $y = -4.8 \cos x$

17. $y = -1.3 \sin x$

18. $y = 1.5 \cos x$

9. $y = 1.8 \sin x$

10. $y = 1.5 \sin x$

19. $y = 4 \sin x$

Determine the equation for the following graphs.

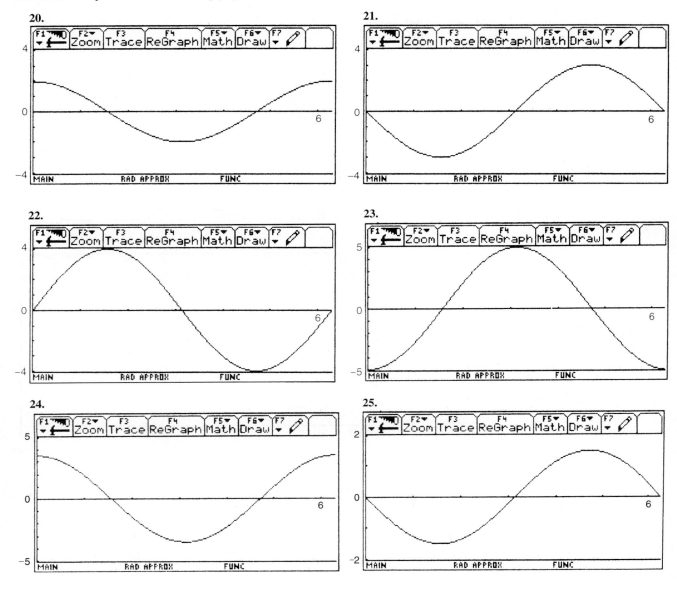

12-2

GRAPHS OF SINE AND COSINE FUNCTIONS: PERIOD

Graph of $y = \sin bx$ or $y = \cos bx$

In the previous section, we found that the graph of $y = a \sin x$ or $y = a \cos x$ each have a period of 2π. However, such is not the case for all sine and cosine functions, as you will see in this section.

EXAMPLE 1   Graph $y = \cos 2x$.

Solution   Since we are uncertain about this graph, we fill in a table just as in the previous section. The graph of $y = \cos 2x$ is shown in Figure 12–8. The dotted line represents the graph of $y = \cos x$.

x	0	π/4	π/2	3π/4	π	5π/4	3π/2	7π/4	2π
y	1	0	−1	0	1	0	−1	0	1

FIGURE 12–8

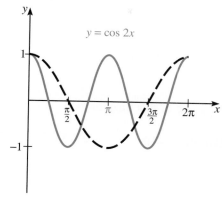

$y = \cos 2x$

Comparing the two graphs in Figure 12–8 shows that the period of $y = \cos 2x$ is π instead of 2π. For graphs of the form $y = a \sin bx$ or $y = a \cos bx$, the period is given by $2\pi/b$. The zeros, minimum, and maximum occur at the same location relative to the period. For example, the maximum and minimum still occur at the endpoints and midpoint, respectively, of the period, but these locations are 0, $\pi/2$, and π for the graph of $y = \cos 2x$.

EXAMPLE 2   Sketch the graph of $y = \sin \dfrac{x}{2}$ through one period.

Solution   This function undergoes a period change given by

$$\frac{2\pi}{1/2} = 4\pi$$

In graphing this function, we know the following:

1. The period is 4π.

2. The amplitude is 1.

3. The zeros occur at the endpoints and midpoint of the period, which are 0, 2π, and 4π.

4. The maximum and minimum occur midway between the zeros, at π and 3π.

The graph is shown in Figure 12–9.

FIGURE 12–9

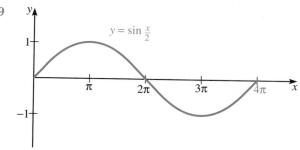

$y = \sin \dfrac{x}{2}$

EXAMPLE 3 Using the graphing calculator, graph $y = \cos \pi x$.

Solution From our knowledge of how this graph should appear, we set the viewing rectangle with the following:

$$x\text{-min} = 0$$

$$x\text{-max} = \text{one period} = \frac{2\pi}{\pi} = 2$$

$$y\text{-min} = -1$$

$$y\text{-max} = 1$$

The graph of this function is shown in Figure 12–10.

FIGURE 12–10

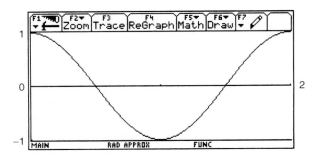

LEARNING HINT ✦ Sketching the graph for the sine and cosine functions will be easier if you list the known information, such as period and amplitude, before sketching the graph.

Graph of
$y = a \sin bx$ **or**
$y = a \cos bx$

EXAMPLE 4 Graph $y = 2 \cos 2\pi x$ through one period.

Solution From previous discussions, we know the following:

1. The amplitude is 2.
2. The period is $2\pi/2\pi = 1$.
3. The maximum occurs at the endpoints of the period, which are 0 and 1.
4. The minimum occurs at the midpoint of the period, 0.5.
5. The zeros occur midway between each maximum and minimum, 0.25 and 0.75.

The graph is shown in Figure 12–11. Since the period is 1 radian, the x axis is marked in numerical values of radians, rather than in units of π.

FIGURE 12–11

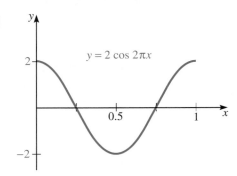

Summary

$$y = a \sin bx$$

or

$$y = a \cos bx$$

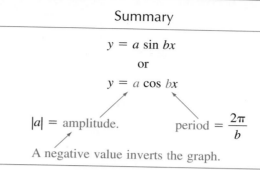

$|a|$ = amplitude.　　period $= \dfrac{2\pi}{b}$

A negative value inverts the graph.

EXAMPLE 5　Sketch the graph of $y = -3 \sin \dfrac{2}{3}x$.

Solution　For this graph, we know the following:

1. The amplitude is 3.

2. Due to the negative sign in front of the amplitude, the graph is inverted; the locations of the maximum and minimum are interchanged.

3. The period is $\dfrac{2\pi}{2/3} = 3\pi$.

4. The zeros occur at the endpoints and midpoint of the period, which are 0, 1.5π, and 3π.

5. The maximum and minimum values occur midway between the zeros at 2.25π and 0.75π, respectively.

The graph of $y = -3 \sin \dfrac{2}{3}x$ is shown in Figure 12–12.

FIGURE 12–12

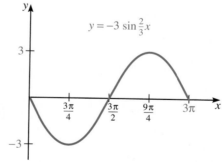

$$y = -3 \sin \tfrac{2}{3}x$$

EXAMPLE 6　Determine the equation for the graph shown in Figure 12–13.

Solution　From the graph we first identify it as a sine function. We can also determine that the amplitude is 3. Note that the graph is not inverted, so a is positive. From the graph, we can also determine that the period is $\pi/2$. To calculate the value of b, we solve

$$\frac{2\pi}{b} = \text{new period} = \frac{\pi}{2}$$

$$\frac{2\pi}{b} = \frac{\pi}{2}$$

FIGURE 12–13

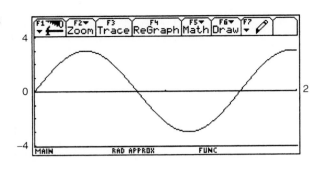

$$\pi b = 4\pi$$
$$b = 4$$

Substituting the values of a and b, the equation for the graph is $y = 3 \sin 4x$. ■■■

LEARNING HINT ✦ To aid in graphing, write the properties for the sine and cosine functions, along with the summary on amplitude and period changes, on an index card. Then use this information while you do the exercises.

12–2 EXERCISES

State the amplitude and period for each function, and sketch its graph through one period.

1. $y = \sin 2x$

2. $y = \cos 3x$

3. $y = \sin 3\pi x$

4. $y = -\cos \dfrac{x}{4}$

5. $y = \cos \dfrac{x}{4}$

6. $y = \sin 2x$

7. $y = 2 \sin 3\pi x$

8. $y = 1.5 \cos 2\pi x$

9. $y = 3 \cos \dfrac{x}{4}$

10. $y = 3.7 \sin \dfrac{x}{4}$

11. $y = -2 \sin 3\pi x$

12. $y = -5 \sin 6\pi x$

13. $y = -3 \cos \dfrac{x}{4}$

14. $y = 1.5 \cos \dfrac{3x}{4}$

15. $y = -\dfrac{1}{2} \sin \dfrac{x}{2}$

16. $y = \dfrac{1}{2} \sin 6x$

17. $y = 4 \cos 5\pi x$

18. $y = 3 \cos \dfrac{4x}{3}$

19. $y = -3 \sin \dfrac{3x}{2}$

20. $y = -2 \cos 6\pi x$

Determine the equation for the following graphs.

21.

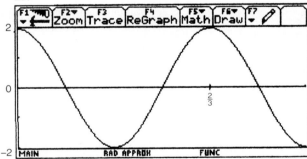

22.

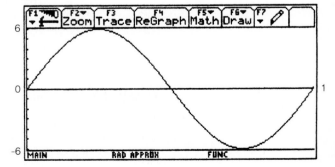

23.

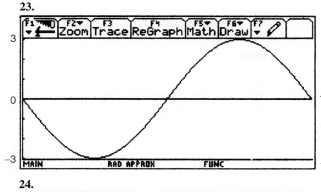

24.

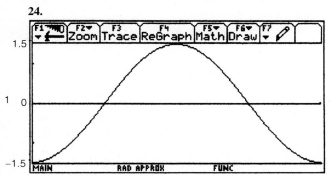

M 25. Linear displacement x of an oscillating object as a function of time t can be written as

$$x = 3 \sin 2\pi t$$

Graph this function through two cycles.

E 26. The current I in an ac circuit can be given by $I = 6 \cos 120\pi t$. Sketch the graph of I as a function of t for $0 \leq t \leq 0.1$ s.

M 27. The displacement of an object oscillating at the end of a spring is given by $y = 8 \cos 6\pi t$. Sketch the graph of y as a function of t for $0 \leq t \leq 1$.

E 28. Under certain conditions the voltage in an ac circuit is given by $V = 100 \sin \pi t$. Graph V as a function of t for $0 \leq t \leq 0.5$ s.

12–3

GRAPHS OF SINE AND COSINE FUNCTIONS: PHASE SHIFT

If you study Figures 12–1 and 12–2, you can see that the graphs would be identical if the graph of the sine function were shifted $\pi/2$ units to the left. For this reason, the cosine function is said to *lead* the sine by $\pi/2$. This horizontal movement along the x axis is called a **phase shift.**

Graph of
$y = a \sin(bx + c)$ **and**
$y = a \cos(bx + c)$

EXAMPLE 1 Graph $y = \cos(x + \pi)$ through one period.

Solution In this particular example a and b equal one. We fill in the accompanying table of values. To calculate y when $x = \pi/2$, press the following calculator keys:

$$\boxed{\pi} \; \boxed{\div} \; \boxed{2} \; \boxed{+} \; \boxed{\pi} \; \boxed{=} \; \boxed{\cos} \longrightarrow 0$$

We know that the period of this function is 2π and the amplitude is 1.

x	0	$\pi/2$	π	$3\pi/2$	2π
y	-1	0	1	0	-1

Figure 12–14 shows the graph of $y = \cos(x + \pi)$. Notice that the *graph of $y = \cos x$ has just been moved along the x axis π units to the left.* For comparison, the dotted line represents the graph of $y = \cos x$.

FIGURE 12–14

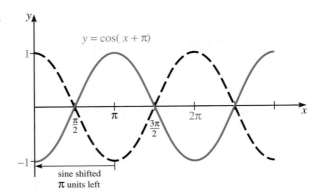

$y = \cos(x + \pi)$

sine shifted
π units left

■ ■ ■

For an equation of the form $y = a \sin(bx + c)$ or $y = a \cos(bx + c)$, the **phase shift** is given by c/b, and the *graph is shifted along the x axis in the opposite direction of the sign in front of c*. The angle c is called the **phase angle.** Notice in Example 1 that the phase shift and the phase angle are the same because $b = 1$.

LEARNING HINT ✦ Students frequently have difficulty in determining the direction of the phase shift. The x axis is positive to the right of zero and negative to the left. The graph shifts along the x axis in the opposite direction of the sign in front of the phase angle. For example, $y = \sin(x - \pi)$ shifts the graph π units to the right.

EXAMPLE 2   Sketch the graph of $y = 3 \cos(x - \pi)$.

Solution   From our previous discussion, we know the following:

1. The amplitude is 3.
2. The period is $2\pi/b = 2\pi/1 = 2\pi$.
3. The phase shift is $c/b = \pi/1 = \pi$ units along the x axis and to the right.
4. The graph has a maximum at the endpoints and a minimum at the midpoint of its cycle.
5. The graph has a zero between each maximum and minimum.

To sketch the graph, we *begin the curve at π* and draw a cosine function with an amplitude of 3 and a period of 2π. Then we extend the graph back to the y axis. Figure 12–15 gives the graph of $y = 3 \cos(x - \pi)$.

FIGURE 12–15

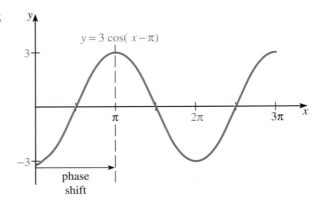

$y = 3 \cos(x - \pi)$

phase
shift

■ ■ ■

EXAMPLE 3   Sketch $y = -2 \sin\left(2x + \dfrac{\pi}{2}\right)$ through one cycle.

Solution   We know the following about the graph of this function:

1. The curve is inverted; the locations of the maximum and minimum along the x axis are reversed.

2. The amplitude is $|-2| = 2$.

3. The period is $\dfrac{2\pi}{b} = 2\pi/2 = \pi$.

4. The phase shift is $\dfrac{c}{b} = \dfrac{\pi/2}{2} = \dfrac{\pi}{4}$ units to the left because c is positive.

5. The zeros occur at the endpoints, $-\pi/4$, and $-\pi/4 + \pi = 3\pi/4$, and at the midpoint, $(-\pi/4 + 3\pi/4)/2 = \pi/4$, of the period.

6. Since the graph is inverted, the minimum is midway between the first two zeros, $-\pi/4$ and $\pi/4$ at 0, and the maximum is midway between the last two zeros, $\pi/4$ and $3\pi/4$ at $\pi/2$.

To sketch the graph, *begin at* $-\pi/4$ and extend the graph to the right so that one cycle appears to the right of the y axis. Figure 12–16 shows the graph of this function.

FIGURE 12–16

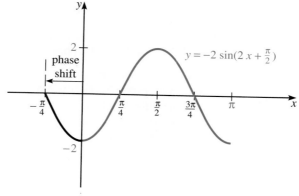

■ ■ ■

Summary

$$y = a \sin(bx + c)$$

or

$$y = a \cos(bx + c)$$

amplitude $= |a|$ period $= \dfrac{2\pi}{b}$ phase shift $= \dfrac{c}{b}$

A negative sign inverts the graph.

EXAMPLE 4   In an ac circuit with only a constant inductance, the voltage is given by

$$V = V_m \sin\left(2\pi ft + \dfrac{\pi}{2}\right)$$

where f = the frequency of the current, V_m = maximum voltage (amplitude), and t = time. In a 60-Hz system with a maximum voltage of 12 V, the equation becomes

$$V = 12 \sin(120\pi t + 0.5\pi)$$

Graph the voltage as a function of time.

Solution   From the equation, we know that the amplitude is 12; the period, which is the reciprocal of the frequency, is 1/60 s; and the phase shift is 1/240 s. The graph is shown in Figure 12–17.

FIGURE 12–17

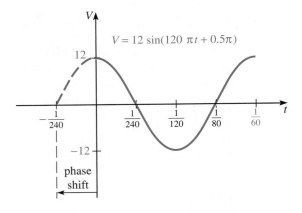

Equations of Sine and Cosine Functions

EXAMPLE 5   Determine the equation of an inverted cosine function with an amplitude of 3, a period of 3π, and a phase shift of $3\pi/2$ units to the left along the x axis.

Solution   To write the equation of this function, we must substitute for a, b, and c in the equation $y = a \cos(bx + c)$ where $|a|$ represents amplitude, b represents period change, and c represents the phase angle. From the given information, $|a| = 3$. To find the value of b, we know that $2\pi/b$ gives the period for the cosine function. Since the period of this function is 3π, we have

$$\frac{2\pi}{b} = 3\pi$$

$$b = \frac{2}{3}$$

To find the value of c, we know that c/b represents the phase shift and $b = 2/3$. Since the phase shift is $3\pi/2$, we have

$$\frac{c}{b} = \frac{3\pi}{2}$$

$$\frac{c}{2/3} = \frac{3\pi}{2}$$

$$c = \pi$$

Since the phase shift is left, the sign in front of c is positive. Also, since the cosine function is inverted, there is a negative sign in front of the amplitude. The equation of the function is

$$y = -3 \cos\left(\frac{2}{3}x + \pi\right)$$

You can check your answer by comparing the amplitude, period, and phase shift of this function with the original information.

■■■

Applications

EXAMPLE 6 Under certain conditions, the angular displacement s of a point at the end of an oscillating pendulum is given by

$$s = 3 \cos(\pi t - \pi)$$

Determine the amplitude, period, and phase shift of this function, and sketch its graph for $0 \le t \le 3$.

Solution From the equation we can determine that $a = 3$, $b = \pi$, and $c = -\pi$. Using this information, we obtain

$$\text{amplitude} = 3$$

$$\text{period} = \frac{2\pi}{\pi} = 2$$

$$\text{phase shift} = \frac{c}{b} = \frac{-\pi}{\pi} = -1 \text{ unit right}$$

The graph is shown in Figure 12–18.

FIGURE 12–18

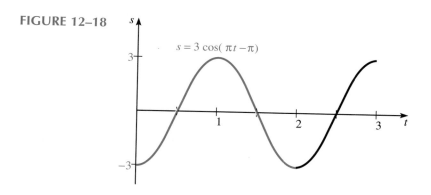

$$s = 3 \cos(\pi t - \pi)$$

■■■

When a conductor is rotated through a magnetic flux, a voltage is created. The number of revolutions per second that the conductor is rotated is called the frequency of the voltage. The unit for frequency is the hertz (Hz). The voltage has a sinusoidal wave form. The time it takes for one complete cycle of the conductor through the magnetic flux is one period of the resulting sine wave. Therefore, the period is the reciprocal of the frequency. Each time the conductor rotates through one cycle, it completes one revolution, 2π radians. The angular velocity is the product of 2π and the frequency ($\omega t = 2\pi f t$).

In a purely resistive ac circuit, the voltage and current equal zero at the same time. They also reach their peak values at the same time. When this occurs, the voltage and current are said to be in phase. Whenever an ac circuit contains an inductance or capacitance, the current and voltage are out of phase. When the phase angle is known, the instantaneous current and voltage can be calculated. In a parallel circuit, the voltage is the reference, and the current either leads or lags the voltage. In a series circuit, the current is the reference, and the voltage either leads or lags the current.

EXAMPLE 7 Compute the voltage and current in a parallel circuit 15 μs after t_0 if the peak current is 25 mA, the peak voltage is 25 V, the frequency is 9 kHz, and the voltage leads the current by π/4.

Solution The equation for voltage, which is the reference, is

$$V = V_{peak} \sin 2\pi ft$$

where $V_{peak} = 25$ V, $f = 9$ kHz, and $t = 15$ μs. Substituting these values gives

$$V = 25 \text{ V} \sin \left[2\pi \left(9 \times 10^3 \text{ Hz} \right) \left(15 \times 10^{-6} \text{ s} \right) \right]$$

$$V = 25 \text{ V} \sin 0.8482$$

$$V \approx 19 \text{ V}$$

The current lags the voltage by 45° (π/4). The equation for the current is

$$I = I_{peak} \sin \left(2\pi ft - \frac{\pi}{4} \right)$$

where $I_{peak} = 25$ mA, $f = 9$ kHz, and $t = 15$ μs. The current is

$$I = 25 \text{ mA} \sin \left[2\pi \left(9 \times 10^3 \text{ Hz} \right) \left(15 \times 10^{-6} \text{ s} \right) - \frac{\pi}{4} \right]$$

$$I \approx 25 \text{ mA} \sin 0.0628$$

$$I \approx 1.6 \text{ mA}$$

■ ■ ■

12–3 EXERCISES

State the amplitude, period, and phase shift of the following functions. Sketch the graph through one cycle.

1. $y = \cos\left(x + \dfrac{\pi}{4}\right)$

2. $y = \sin\left(x - \dfrac{3\pi}{2}\right)$

3. $y = -\sin\left(x - \dfrac{\pi}{3}\right)$

4. $y = \cos\left(x + \dfrac{3\pi}{4}\right)$

5. $y = 2 \cos(x + 1)$

6. $y = -\sin(x - 2)$

7. $y = -\cos(2x - \pi)$

8. $y = \sin(3\pi x - \pi)$

9. $y = \sin(3x + 2\pi)$

10. $y = -\cos\left(2x + \dfrac{\pi}{3}\right)$

11. $y = 2 \cos\left(x + \dfrac{\pi}{4}\right)$

12. $y = 3 \sin(x - 2\pi)$

13. $y = -3 \sin(x + 3)$

14. $y = -\cos\left(x + \dfrac{3\pi}{4}\right)$

15. $y = 2 \sin(2\pi x + 1)$

16. $y = 6 \sin(3x - 1)$

17. $y = -\cos(3\pi x - 2)$

18. $y = 3.1 \sin(2x - \pi)$

19. $y = 4 \sin\left(6\pi x + \dfrac{1}{2}\right)$

20. $y = -3 \cos(3\pi x + 1)$

21. $y = -2.5 \cos(3x - 2\pi)$

22. $y = 1.4 \cos(4x - 3\pi)$

23. $y = \sin(2\pi x - 5\pi)$

24. $y = -5 \sin(3x - 6\pi)$

25. Write the equation of a cosine function with an amplitude of 3 and period of 4.

26. Write the equation of a sine function with an amplitude of 2, period of 2π, and a phase shift of π/3 units to the right.

27. Write the equation of an inverted sine function with amplitude π, frequency 1/6 (frequency is the reciprocal of period), and phase shift π/3 units left.

28. Write the equation of a cosine function with amplitude 1/2, period 1/3, and phase shift 2/3 unit right.

29. Write the equation of a cosine function with amplitude 4, period π, and phase shift π/3 units right.

30. Write the equation of an inverted sine function with amplitude 6, frequency 1/2 (frequency is the reciprocal of period), and phase shift 3π/4 units left.

E 31. In a certain ac circuit, the current I is given by

$$I = 5.3 \cos\left(120\pi t + \frac{\pi}{2} \right)$$

Graph this function through one cycle.

M 32. The angular displacement of a point at the end of a pendulum is given by

$$s = A \cos(\omega t + \theta)$$

Sketch the graph of s versus t when $A = 3.4$, $\omega = 5.6$ rad/s, and $\theta = \pi/6$ for $0 \le t \le 2$.

P 33. The equation of a certain sinusoidal wave on a string is given by

$$y = 4 \sin\left(2\pi t - \frac{\pi}{2}\right)$$

Graph this function through one cycle.

M 34. The displacement y of a point on the rim of a rotating gear is given by

$$y = 8 \cos\left(t + \frac{\pi}{2}\right)$$

Graph this function through one cycle.

E 35. Using Example 7 as a guide, find the current and voltage in a series circuit (use current as a reference) if $I_{peak} = 30$ mA, $V_{peak} = 20$ V, $f = 100$ Hz, $t = 2.5$ ms, and the current leads the voltage by $\pi/6$. The equations for a series circuit are

$$I = I_{peak} \sin 2\pi ft$$

$$V = V_{peak} \sin\left(2\pi ft + \theta\right)$$

E 36. Find the current and voltage in a series circuit given $I_{peak} = 1.8$ mA, $V_{peak} = 12$ V, $f = 16$ kHz, $t = 40$ μs, and current leads voltage by $\pi/3$.

E 37. Find the current and voltage in a parallel circuit given $I_{peak} = 4$ mA, $V_{peak} = 10$ V, $f = 20$ kHz, $t = 15$ μs, and voltage leads current by $\pi/4$.

12–4

APPLICATIONS

Trigonometric functions are very useful in describing periodic phenomena, such as wave movement, oscillation of mechanical systems, and current flow in a circuit. In this section we will discuss two such applications: simple harmonic motion and alternating current.

Simple Harmonic Motion

When a spring with an object attached to the end is stretched and released, the object does not simply return to its equilibrium or starting position. The object actually oscillates above and below the equilibrium point. The oscillation of this system is called *simple harmonic motion*. The displacement y of an object in simple harmonic motion is given by the following equation.

Simple Harmonic Motion

$$y = a \sin(2\pi ft + \theta)$$

where

$a = $ amplitude of motion

$f = $ frequency $= \dfrac{1}{\text{period}}$

$= $ number of oscillations per unit of time

$t = $ time

$\theta = $ phase angle

EXAMPLE 1 A system vibrating in simple harmonic motion is given by the equation

$$y = 2 \sin\left(6\pi t + \frac{\pi}{2}\right)$$

where $y = $ displacement in inches and $t = $ time in seconds.

(a) Find the amplitude, frequency, period, and phase shift for this function.

(b) Determine the displacement of the object 3 s after release.

(c) Determine when the object will first reach its maximum negative displacement.

Solution

(a) Determine the amplitude, frequency, period, and phase shift just as in the previous sections.

$$y = 2 \sin\left(6\pi t + \frac{\pi}{2}\right)$$

$$\text{amplitude} = 2$$

$$\text{frequency} = \frac{1}{\text{period}} = \frac{1}{\dfrac{2\pi}{6\pi}} = 6\pi/2\pi$$

$$f = 3$$

$$\text{period} = \frac{1}{\text{frequency}} = \frac{2\pi}{6\pi} = \frac{1}{3}$$

$$\text{phase shift} = \frac{\pi/2}{6\pi} = \frac{1}{12}$$

(b) To determine the displacement y at 3 s, substitute $t = 3$ into the equation and solve for y.

$$y = 2 \sin\left(6\pi(3) + \frac{\pi}{2}\right)$$

$$y = 2 \sin\left(18\pi + \frac{\pi}{2}\right)$$

$$y = 2 \sin \frac{37\pi}{2}$$

$$y \approx 2 \text{ in., its maximum height}$$

(c) The first maximum negative displacement for the regular sine function $y = a \sin(x)$ occurs at $x = 3\pi/2$. Therefore, the first time the object reaches its maximum negative displacement occurs when the angle equals $3\pi/2$. Thus,

$$6\pi t + \frac{\pi}{2} = \frac{3\pi}{2}$$

Solving for t gives

$$6\pi t = \pi$$
$$t \approx 0.17 \text{ s}$$

■ ■ ■

EXAMPLE 2 A wooden block attached to the end of a spring is pulled down 10 cm and released, oscillating with simple harmonic motion and taking 3 s to return to its point of release. The phase angle is zero.

(a) Determine the period, frequency, and amplitude of the motion.

(b) Write an equation to describe the block's displacement.

(c) Determine the position of the block 1.25 s after release.

Solution

(a) The maximum displacement or amplitude is 10 cm. Since the block returns to its starting point in 3 s, its period is 3 s. Frequency, which is 1/period, equals 1/3.

(b) The general form of the equation is

$$y = a \sin(2\pi f t + \theta)$$

We know that $\theta = 0$, and we just found that

$$a = 10$$

$$f = \frac{1}{3}$$

Therefore, the equation is

$$y = 10 \sin \frac{2}{3}\pi t$$

(c) We can find the position of the block 1.25 s after release by substituting $t = 1.25$ into the above equation.

$$y = 10 \sin\left(\frac{2}{3}\pi(1.25)\right)$$

$$y \approx 10 \sin(2.618)$$

$$y \approx 5 \text{ cm}$$

■ ■ ■

EXAMPLE 3   The water level at Trident Pier, Florida is measured by a buoy using a downward looking acoustic sensor. The rise and fall of the water level due to the tide can be modeled using a sinusoidal curve. (The model of tidal flow actually is a composite of several sinusoidal curves, but has been simplified for this problem.) The following data was accumulated.

t (time)	midnight	2	4	6	8	10	noon
y (depth in ft)	1	2	1	−1	−2	−1	1

Use a cosine function to model this data.

Solution   The function would be modeled by finding a, b, and c for the equation

$$y = a \cos(bt - c)$$

The amplitude is given by

$$a = \frac{1}{2}(\text{high} - \text{low}) = \frac{1}{2}(2 - (-2)) = 2$$

The period is

$$P = 2(\text{low time} - \text{high time}) = 2(8 - 2) = 12$$

which means

$$b = \frac{2\pi}{P} = \frac{2\pi}{12} \approx 0.524$$

Because high tide occurs 2 hours after midnight, $c/b = 2$.

$$\frac{c}{b} = 2$$

$$\frac{c}{0.524} = 2$$

$$c \approx 1.05$$

The data can be modeled by

$$y = 2 \cos(0.524t - 1.05)$$ ■ ■ ■

Alternating Current

Alternating current (ac) is generated when a coil of wire rotates in a magnetic field. In an ac circuit, the current I as a function of time t is given by the following equation.

Sinusoidal Current

$$I = I_m \sin(2\pi ft + \theta)$$

where

$$I_m = \text{peak current}$$
$$f = \text{frequency}$$
$$\theta = \text{phase angle}$$

Similarly, voltage V in an ac circuit is given by the following equation.

Sinusoidal Voltage

$$V = V_0 \sin(2\pi ft + \theta)$$

where

$$V_0 = \text{peak voltage}$$
$$f = \text{frequency}$$
$$\theta = \text{phase angle}$$

NOTE ✦ The applications in this section use the sine function; however, the cosine function could just as easily be used because $\sin \theta = \cos(90° - \theta)$.

EXAMPLE 4 An alternating current has a peak current of 3 A and a frequency of 50 Hz. If the phase angle is zero, write the equation of current as a function of time, find the period, and find the current at $t = 0.1$ s.

Solution To write the equation, we need values of I_m, f, and θ. From the verbal statement, we know that $I_m = 3$, $f = 50$, and $\theta = 0$. Therefore, the equation of current as a function of time is

$$I = I_m \sin(2\pi ft + \theta)$$
$$I = 3 \sin 100\pi t$$

The period is the reciprocal of the frequency; therefore,

$$\text{period} = \frac{1}{50} \text{ s}$$

We find the current at time 0.1 s by substituting $t = 0.1$ into the equation.

$$I = 3 \sin(100\pi(0.1))$$
$$I = 3 \sin 10\pi = 0$$

■■■

EXAMPLE 5 A certain alternating voltage has a peak voltage of 220 V and a frequency of 60 Hz. Write an equation for the voltage V if the phase angle is $\pi/2$. Also, determine the period and the voltage when $t = 0.03$ s. Graph the function through one cycle.

Solution The general equation is given by

$$V = V_0\sin(2\pi ft + \theta)$$

Since $V_0 = 220, f = 60$, and $\theta = \pi/2$, the equation becomes

$$V = 220 \sin\left(120\pi t + \frac{\pi}{2}\right)$$

The period is the reciprocal of the frequency; therefore,

$$\text{period} = \frac{1}{60} \text{ s}$$

The voltage when $t = 0.03$ s is given by

$$V = 220 \sin\left(120\pi(0.03) + \frac{\pi}{2}\right)$$
$$V \approx 68.0 \text{ V}$$

The graph through one cycle is given in Figure 12–19.

■■■

Fourier Series

In calculus the sine and cosine functions are used to approximate a Fourier series. Historically, the values of $\sin \theta$ or $\log x$ were calculated using an infinite series as an approximation. Although computers and calculators make this unnecessary today, the

FIGURE 12–19

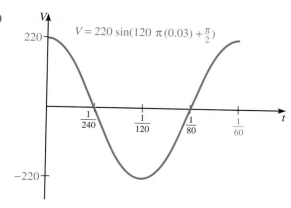
$$V = 220 \sin(120\pi(0.03) + \frac{\pi}{2})$$

Fourier series is used extensively in electrical engineering, as you will see when you study calculus.

🖩 **EXAMPLE 6**   A half-wave rectifier circuit converts ac input to pulsating dc on every positive half-cycle. Figure 12–20 shows the initial sinusoidal voltage and the resulting output voltage. The Fourier series for the half-wave rectifier is

$$y = \frac{1}{\pi} - \frac{3}{2\pi} \cos 2x - \frac{2}{15\pi} \cos 4x + \cdots + \frac{1}{2} \sin x$$

Using a graphing calculator, graph $y = \sin x$ for $0 \le x \le 3.14$ and the resulting Fourier series, adding a term with each successive graph.

FIGURE 12–20

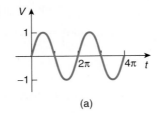
(a)

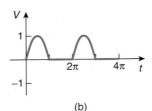
(b)

Solution   The resulting graphs are shown in Figure 12–21a, b, and c. Note that the Fourier series is only an approximation to the curve, but it approximates the curve more closely as we add terms. The half-wave rectifier circuit may actually be seen on an oscilloscope screen, thus allowing a technician to troubleshoot and analyze the electrical current.

12–4 EXERCISES

For the equations of systems vibrating in simple harmonic motion in Exercises 1–6, do the following:

(a) Find the amplitude, period, and phase angle.

(b) Determine the displacement y of the object after 1.5 s.

(c) Determine when the object will reach its maximum positive displacement.

(d) Sketch the graph through one cycle.

1. $y = 1.4 \sin\left(6\pi t - \dfrac{\pi}{2}\right)$

2. $y = 3 \sin\left(2\pi t + \dfrac{\pi}{4}\right)$

3. $y = 2 \sin\left(3\pi t + \dfrac{\pi}{2}\right)$

4. $y = 2.8 \sin\left(4\pi t - \dfrac{\pi}{3}\right)$

5. $y = 3.1 \sin(\pi t - \pi)$

6. $y = 2.5 \sin\left(\dfrac{\pi t}{2} + \dfrac{\pi}{4}\right)$

Write an equation describing the simple harmonic motion of each system.

7. period = 2 s; amplitude = 4 in.; phase angle = $\pi/3$

8. period = 0.5 s; amplitude = 1.2 m; phase angle = $\pi/2$

9. period = 1.3 s; amplitude = 6 cm; phase angle = $-\pi/4$

10. period = 0.75 s; amplitude = 2.1 ft; phase angle = $-5\pi/6$

11. period = 4 s; amplitude = 2 yd; phase angle = π

12. period = 1.25 s; amplitude = 0.75 m; phase angle = $-\pi$

Sketch two cycles of $I = I_m \sin(2\pi ft + \theta)$ for the following.

13. $I_m = 4.6$ A; $f = 60$ Hz; $\theta = -\pi/2$

14. $I_m = 7.4$ A; $f = 50$ Hz; $\theta = \pi/6$

Sketch two cycles of $V = V_0 \sin(2\pi ft + \theta)$ for the following.

15. $V_0 = 65$ V; $f = 450$ Hz; $\theta = \pi/4$

16. $V_0 = 100$ V; $f = 60$ Hz; $\theta = -\pi$

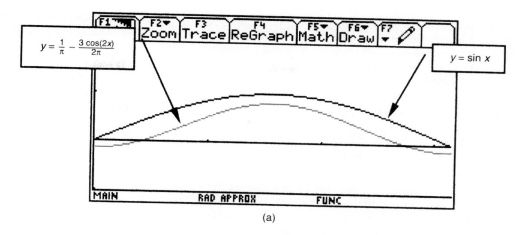

(a)

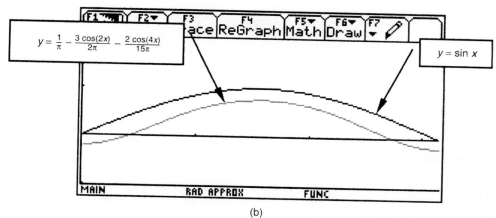

(b)

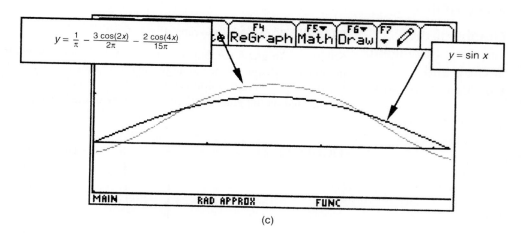

(c)

FIGURE 12–21

M 17. A weight at the end of a spring vibrates in simple harmonic motion. The weight moves from its starting position to a maximum height of 18 cm and completes a cycle in 1.5 s. Assuming the phase angle is zero, write an equation to describe the position of the weight as a function of time, and find its position at 0.65 s.

E 18. The voltage in an ac circuit is given by the equation

$$V = 240 \sin(60\pi t - \pi)$$

Find the amplitude, period, and phase shift of this function, and sketch its graph through two cycles.

E 19. A certain alternating voltage has a peak voltage of 135 V and a frequency of 50 Hz. Assuming the phase angle is zero, write an equation for the voltage, find the amplitude and period, and graph the function through two cycles.

E 20. In an ac circuit the current I is given by

$$I = 4.8 \sin\left(60\pi t - \frac{\pi}{2}\right)$$

Find the amplitude, period, and phase shift of this function, and graph the function through two cycles.

M 21. A point on the end of a tuning fork moves in simple harmonic motion given by $y = a \sin 2\pi f t$. Find the frequency for middle C if the period is 0.0037879.

M 22. A harbor buoy oscillates in simple harmonic motion. Assuming the phase angle is zero, write the equation that describes the motion of the buoy if it moves 5 ft from its low point to its high point, and it returns to its high point every 15 seconds. (At $t = 0$, the buoy is at its high point.)

E 23. An electrical circuit known as a full-wave rectifier has a Fourier series given by

$$y = \frac{2}{\pi} + \frac{4}{3\pi} \cos 2x - \frac{4}{15\pi} \cos 4x + \frac{4}{35\pi} \cos 6x + \cdots$$

Graph $y = |\cos x|$ for $0 \le x \le 3.14$ and the Fourier series on the same graph.

12–5

GRAPHS OF THE OTHER TRIGONOMETRIC FUNCTIONS

Thus far in this chapter, we have discussed changes in amplitude, period, and phase angle for the sine and cosine functions only. In this section we will briefly consider the graphs of the remaining trigonometric functions. After developing the basic shape of each function, we will sketch the graph of these functions with changes in amplitude, period, and phase angle.

Graphs of the Tangent, Cotangent, Secant, and Cosecant Functions

EXAMPLE 1 Sketch the graph of $y = \tan x$.

Solution To graph the tangent function, we fill in the accompanying table of values using our knowledge of special angles and using the calculator. Note from the table of values that the tangent function is undefined at $\pi/2$ and $3\pi/2$ due to division by zero. This is represented on the graph of Figure 12–22 by dashed vertical lines at $\pi/2$ and $3\pi/2$. These

FIGURE 12–22

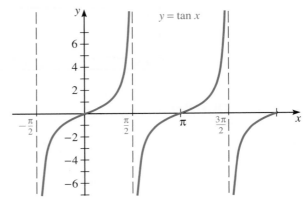

dashed lines are called **asymptotes,** which are straight lines that approach but never touch a curve. Figure 12–22 gives the graph of $y = \tan x$.

x	0	$\pi/4$	$\pi/2$	$3\pi/4$	π	$5\pi/4$	$3\pi/2$	$7\pi/4$	2π
y	0	1	undef.	−1	0	1	undef.	−1	0

To graph the remaining trigonometric functions, we can use the reciprocal relationships.

$$\cot x = \frac{1}{\tan x} \qquad \csc x = \frac{1}{\sin x} \qquad \sec x = \frac{1}{\cos x}$$

We can fill in a table of values using values from the corresponding reciprocal function. Figure 12–23 shows the graph of $y = \cot x$. Figures 12–24 and 12–25 show the graphs of $y = \sec x$ and $y = \csc x$, respectively.

FIGURE 12–23

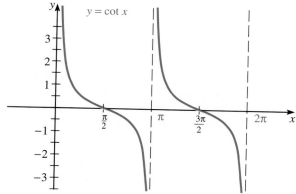

FIGURE 12–24

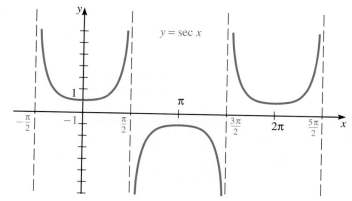

FIGURE 12–25

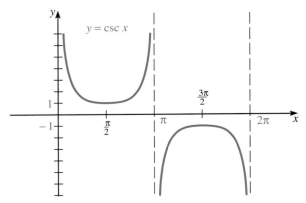

■ ■ ■

CAUTION ✦ If you use a calculator to find some of the values, you can get an error message when no error exists. For example, if you calculate cot $\pi/2$ using a calculator, you strike the following keys:

$$\boxed{\pi} \ \boxed{\div} \ 2 \ \boxed{=} \ \boxed{\tan} \ \boxed{1/x}$$

However, after you strike the $\boxed{\tan}$ key, you will get an error message because the tangent of $\pi/2$ is undefined. However, cot $\pi/2 = 0$. For the quadrantal angles, it is probably best not to use a calculator.

The period of both the tangent and the cotangent functions is π, while the period of both the secant and the cosecant functions is 2π. None of these functions has an amplitude because there is no maximum distance from the x axis to the graph. Sketching the graph of the secant, cosecant, tangent, or cotangent using only amplitude, period, and phase angle is more difficult than graphing the sine and cosine. For this reason, it is best to fill in a table of values when graphing these functions. The period and amplitude are summarized in the following table.

	period	amplitude
tangent	π	undefined
cotangent	π	undefined
secant	2π	undefined
cosecant	2π	undefined

LEARNING HINT ✦ Be sure to learn the basic shape and location of asymptotes for these functions. It will make graphing much easier.

Changes in Amplitude, Period, and Phase Angle

EXAMPLE 2   Sketch the graph of $y = 2 \tan x$ through one cycle.

Solution   Since the period of this tangent function is π, the table of values need not extend beyond π. The graph of $y = 2 \tan x$ is shown in Figure 12–26. The dotted line represents the graph of $y = \tan x$. Notice that the 2 multiplier alters the slope of the graph.

x	0	$\pi/4$	$\pi/2$	$3\pi/4$	π
y	0	2	undef.	-2	0

FIGURE 12–26

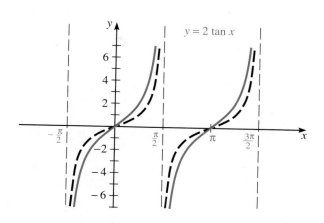

EXAMPLE 3   Graph $y = \sec(2x - \pi)$ through one cycle.

Solution   Examining the equation of this function reveals a period of π and a phase shift of $\pi/2$ units to the right. We fill in the accompanying table of values using a calculator. The graph of $y = \sec(2x - \pi)$ is given in Figure 12–27.

x	0	$\pi/8$	$\pi/4$	$3\pi/8$	$\pi/2$	$5\pi/8$	$3\pi/4$	$7\pi/8$	π
y	-1	-1.41	undef.	$+1.41$	1	$+1.41$	undef.	-1.41	-1

FIGURE 12–27

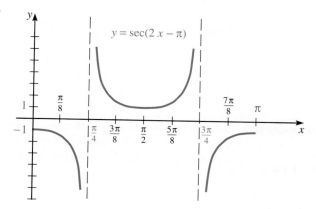

EXAMPLE 4   Graph $y = -3 \cot(x + \dfrac{\pi}{2})$ through one cycle.

Solution The graph of this function is inverted; the slope of the graph is altered; the period is π; and the graph is shifted $\pi/2$ units to the left. The table of values is given below, and the graph is shown in Figure 12–28.

x	0	$\pi/8$	$\pi/4$	$3\pi/8$	$\pi/2$	$5\pi/8$	$3\pi/4$	$7\pi/8$	π
y	0	1.2	3	7.2	undef.	-7.2	-3	-1.2	0

FIGURE 12–28

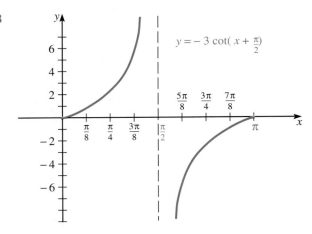

$$y = -3\cot\left(x + \tfrac{\pi}{2}\right)$$

12–5 EXERCISES

Sketch the graph of each function through one cycle.

1. $y = 3\tan x$

2. $y = -\csc x$

3. $y = -2\sec x$

4. $y = 4\cot x$

5. $y = \tan\left(x - \dfrac{\pi}{2}\right)$

6. $y = \sec(x + \pi)$

7. $y = \csc(x + \pi)$

8. $y = \tan(3x - \pi)$

9. $y = 4\tan\left(x - \dfrac{\pi}{2}\right)$

10. $y = \csc\left(2x - \dfrac{\pi}{4}\right)$

11. $y = -\sec(2\pi x - \pi)$

12. $y = -\cot\dfrac{x}{3}$

13. $y = -\cot\left(x + \dfrac{\pi}{4}\right)$

14. $y = 3\sec 2x$

15. $y = 2\tan\left(\dfrac{x}{2} - \dfrac{\pi}{2}\right)$

16. $y = -2\tan(2\pi x - 3\pi)$

17. $y = -\sec\left(2x - \dfrac{\pi}{2}\right)$

18. $y = \csc(4x - \pi)$

19. $y = \csc\left(\dfrac{x}{2} + \pi\right)$

20. $y = -4\cot\left(\pi x + \dfrac{\pi}{2}\right)$

12–6

COMPOSITE TRIGONOMETRIC CURVES

In the previous sections of this chapter, we dealt with graphs of one function at any given time. However, many applications deal with a combination of several functions. In this section we will discuss graphing composite functions and parametric equations.

Addition of Ordinates

A **composite function** is the sum or difference of several functions. One method of graphing composite functions is by **addition of ordinates,** whereby we graph each of the functions separately and then graph the composite function by adding the y values (ordinates) for selected points. This procedure is demonstrated in the next two examples.

EXAMPLE 1   Sketch the graph of $y = 2 + \sin x$.

Solution   This function is the sum of the simpler functions $y = 2$ and $y = \sin x$. We can sketch the graph by adding 2 to the important values of $y = \sin x$. The following table of values shows this result, and Figure 12–29 shows the three curves on the same graph.

x	0	$\dfrac{\pi}{2}$	π	$\dfrac{3\pi}{2}$	2π
$\sin x$	0	1	0	-1	0
$2 + \sin x$	2	3	2	1	2

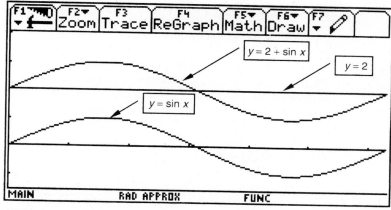

FIGURE 12–29

EXAMPLE 2   Graph $y = x + \sin x$.

Solution   First, graph the functions $y = x$ and $y = \sin x$ on the same set of coordinate axes, as shown in Figure 12–30. Then locate selected points along the x axis. Any x coordinate can be used; however, let us choose points on the x axis where either function has a maximum, minimum, or zero. These locations are denoted with dotted vertical lines in Figure 12–30. Next, add the y coordinate of the two graphs at each of these points, and place a point at the new y coordinate. Finally, draw a curve through these points. The graph of $y = x + \sin x$ is shown as the color curve in Figure 12–30.

FIGURE 12–30

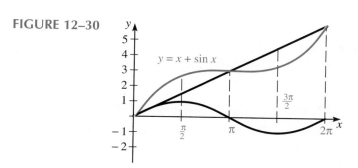

Graphing by Addition of Ordinates

To graph composite functions using addition of ordinates, follow these steps:

1. Graph each function on the same set of axes.
2. Locate relevant points on the x axis, such as the maximum, minimum, or zero of either function, and draw dotted vertical lines through these points.
3. Add the y coordinate of each function at the points chosen in Step 2, and place a point at the location of the new y coordinate.
4. Complete the graph by drawing a smooth curve through the points from Step 3.

EXAMPLE 3   Graph $y = \cos 2x - 3 \sin x$.

Solution   First, graph the functions $y = \cos 2x$ and $y = -3 \sin x$ on the same set of axes as shown in Figure 12–31. Second, locate points on the x axis where a maximum, minimum, or zero occurs, and draw a vertical line through these points: 0, $\pi/2$, π, $3\pi/2$, and 2π (Figure 12–31). Next, place *dots* at points that represent the sum of the y coordinates at these points. Last, draw a smooth curve through these points. The graph of $y = \cos 2x - 3 \sin x$ is shown as the color curve in Figure 12–31.

FIGURE 12–31

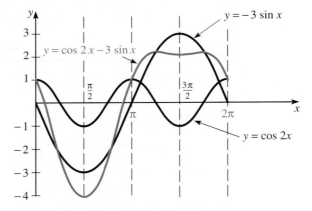

NOTE ✦ If each function is periodic, the period of the composite function is the least common multiple of the period of the individual functions. For example, $y = \cos 2x - 3 \sin x$ has a period of 2π because the period of $\cos 2x$ is π and the period of $-3 \sin x$ is 2π.

CAUTION ✦ When adding ordinates, be careful in adding negative numbers. The y coordinate resulting from adding 3 and -8 is -5.

EXAMPLE 4   Use the graphing calculator to graph $y = \sin 3x + \cos \pi x$.

Solution   The graph of this function will be a combination of $y = \sin 3x$ and $y = \cos \pi x$. The viewing rectangle and the resulting graph are shown in Figure 12–32.

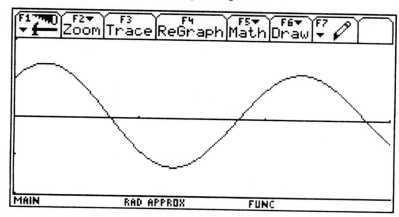

FIGURE 12–32

■■■

Parametric Equations

The equations $y = 3m$ and $x = 2m + 11$ are called **parametric equations** *because the variables x and y are defined in terms of the same third variable m,* called a **parameter.** *When parametric equations describe simple harmonic motion,* the resulting graph is called a **Lissajous figure.**

One application of parametric equations is the oscilloscope. An oscilloscope converts an electrical signal into a graphic display. If we have two voltage signals represented by equations of the form

$$V = V_0\sin(2\pi ft + \theta)$$

we can compare them by associating one signal with the x axis and the other signal with the y axis. For given values of time, we can plot values for x and y. If the two signals are equal in frequency and are in-phase, the resulting figure is a straight line. If the signals are equal in frequency but are out-of-phase, the figure will be a circle or an ellipse. A difference in frequency between the two signals results in a more complex figure.

EXAMPLE 5 Graph $x = 3 \sin t$ and $y = \cos(t - \pi)$.

Solution Since these functions are equal in frequency but are out-of-phase, we can expect the graph to be a circle or an ellipse. We fill in the accompanying table of values by choosing values for t and substituting them into each equation to find the resulting value for x and y. Then we plot the x and y coordinates and join the points with a smooth curve. The graph is shown in Figure 12–33.

t	0	$\pi/4$	$\pi/2$	$3\pi/4$	π	$5\pi/4$	$3\pi/2$	$7\pi/4$	2π
x	0	2.1	3	2.1	0	-2.1	-3	-2.1	0
y	-1	-0.7	0	0.7	1	0.7	0	-0.7	-1

FIGURE 12–33

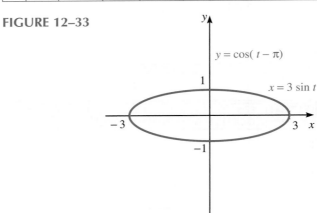

LEARNING HINT ✦ If you arrange the table of values in order such that t, x, and y represent the rows, you will make fewer graphing errors because the (x, y) ordered pairs are in the correct order.

EXAMPLE 6   Graph $x = 2\sin(t - \pi/2)$ and $y = \cos(2t + \pi)$.

Solution   Since these functions are out-of-phase and have a different frequency, we can expect their graph to be somewhat more complex than in the previous example. However, the procedure used to graph these equations is the same as in Example 5. The table of values is given below, and the graph is shown in Figure 12–34. This graph is a portion of a parabola. Notice that the points repeat in reverse order after $t = \pi$.

t	0	$\pi/4$	$\pi/2$	$3\pi/4$	π	$5\pi/4$	$3\pi/2$	$7\pi/4$	2π
x	-2	-1.4	0	1.4	2	1.4	0	-1.4	-2
y	-1	0	1	0	-1	0	1	0	-1

FIGURE 12–34

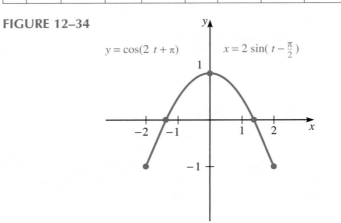

EXAMPLE 7 Use a graphing calculator to graph the Lissajous figure given by $x = 2 \sin 2\pi t$ and $y = \cos\left(\pi t + \dfrac{3\pi}{2}\right)$.

Solution To graph a parametric equation on most graphing calculators, you must choose "parametric" for the graph under MODE. This allows you to enter an x and y equation, both in terms of t. The viewing rectangle and the resulting graph are shown in Figure 12–35.

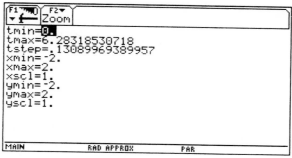

Viewing rectangle

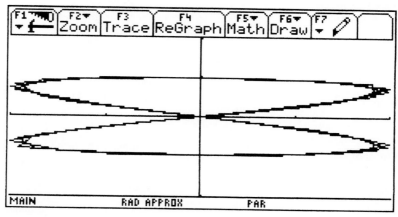

FIGURE 12–35

■ ■ ■

12–6 EXERCISES

Graph the following composite functions by using addition of ordinates. The graphing calculator may be used to check your work.

1. $y = 2x + \sin x$

2. $y = \dfrac{x}{2} + \cos x$

3. $y = x - \cos x$

4. $y = -x + 2 \sin x$

5. $y = (3x - 1) + \sin 2x$

6. $y = x - 3 \cos 2x$

7. $y = \dfrac{x}{2} - 3 \sin x$

8. $y = \dfrac{x}{4} + 2 \sin x$

9. $y = -4x + \cos\left(x - \dfrac{\pi}{2}\right)$

10. $y = 3x + \sin\left(x + \dfrac{\pi}{2}\right)$

11. $y = x^2 - \sin(2x - \pi)$

12. $y = x^2 + 3 \cos\left(x - \dfrac{\pi}{4}\right)$

13. $y = 3 \cos 2x + \sin(x + \pi)$

14. $y = \cos x - 4 \sin\dfrac{x}{2}$

15. $y = \cos\left(x + \dfrac{\pi}{2}\right) + 3 \cos x$

16. $y = -2 \sin x + \sin\dfrac{x}{3}$

17. $y = \sin x - 2 \cos x$

18. $y = \cos x + \sin\left(2x - \dfrac{\pi}{4}\right)$

19. $y = 2 \sin x + 3 \cos\dfrac{x}{2}$

20. $y = \sin\dfrac{x}{2} - 2 \cos\left(x - \dfrac{\pi}{2}\right)$

Graph the following parametric equations. The graphing calculator may be used to check your work.

21. $x = \sin t;\ y = 2 \cos t$

22. $x = 2 \sin t;\ y = -3 \sin t$

23. $x = \sin\left(t + \dfrac{\pi}{4}\right);\ y = \cos\left(t - \dfrac{\pi}{2}\right)$

24. $x = 2 \cos\dfrac{t}{3};\ y = -3 \sin t$

25. $x = 1.5 \cos 4t;\ y = \sin\left(\dfrac{t}{2} + \dfrac{\pi}{4}\right)$

26. $x = 2 \sin(t + \pi);\ y = 3 \sin\left(2t - \dfrac{\pi}{2}\right)$

27. $x = \sin\left(\dfrac{t}{2} - \pi\right);\ y = -\cos 3t$

28. $x = \cos 3t;\ y = 3 \sin t$

E 29. The current in a certain ac circuit is given by the expression

$$i = 1.5 \sin(\pi t) + 3 \cos\left(t - \dfrac{\pi}{2}\right)$$

Graph the current as a function of time.

E 30. The voltage in a certain ac circuit is given by

$$V = 30 \cos \pi t + 60 \cos 40\pi t$$

Graph this relationship.

P 31. The x and y components of a certain object shot 43° above the horizontal are given by $x = 20t \cos 43°$ and $y = 20t \sin 43° - 16t^2$. Graph these parametric equations.

E 32. The first two terms in a certain Fourier series are given by

$$y = \cos \pi x + \dfrac{1}{9} \cos 3\pi x$$

Sketch the graph.

P 33. Under certain conditions, the path of a pendulum is given by $x = 4 \sin \pi t$ and $y = \cos 2\pi t$. Plot the path of the pendulum.

CHAPTER SUMMARY

Summary of Terms

addition of ordinates (p. 466)

amplitude (p. 438)

asymptotes (p. 463)

composite function (p. 466)

cycle (p. 438)

Lissajous figure (p. 469)

parameter (p. 469)

parametric equation (p. 469)

period (p. 438)

periodic (p. 438)

phase angle (p. 450)

phase shift (p. 449)

sinusoidal (p. 439)

zeros of a function (p. 439)

Summary of Formulas

$y = a \sin(bx + c)$ or $y = a \cos(bx + c)$

where

$$\text{amplitude} = |a|$$

$$\text{period} = \dfrac{2\pi}{b}$$

$$\text{phase shift} = \dfrac{c}{b}$$

CHAPTER REVIEW

Section 12–1

Sketch the graph of each function through one period.

1. $y = -\sin x$

2. $y = 2 \cos x$

3. $y = 1.8 \cos x$

4. $y = -3 \cos x$

5. $y = -5 \sin x$

6. $y = 2.4 \sin x$

Section 12–2

State the amplitude and period for each function, and sketch the graph through one period.

7. $y = \sin 2x$

8. $y = \cos\dfrac{x}{2}$

9. $y = 3 \cos\dfrac{x}{4}$

10. $y = -2 \sin 2\pi x$

11. $y = 6 \sin 4x$

12. $y = 3 \cos \dfrac{x}{3}$

13. $y = 1.6 \cos 3\pi x$

14. $y = -4 \cos 6\pi x$

15. The linear displacement x of an object oscillating at the end of a spring is given by $x = 4 \cos \pi t$. Graph this function through two cycles.

16. The voltage in a certain ac circuit is given by $v = 210 \sin 40\pi t$. Graph v as a function of t for $0 \le t \le 0.1$ s.

Section 12–3

Give the amplitude, period, and phase shift for the following functions, and sketch the graph through one cycle.

17. $y = 2 \sin\left(x + \dfrac{\pi}{4}\right)$

18. $y = \cos\left(3x - \dfrac{\pi}{2}\right)$

19. $y = \cos\left(2x - \dfrac{\pi}{2}\right)$

20. $y = 1.4 \sin\left(x + \dfrac{\pi}{4}\right)$

21. $y = -3 \sin(2\pi x - 3)$

22. $y = -4 \cos\left(\dfrac{x}{2} + \pi\right)$

23. $y = 4 \cos\left(\dfrac{x}{3} + \dfrac{\pi}{2}\right)$

24. $y = 6 \sin(\pi x - 3\pi)$

25. Find the equation of a cosine function with an amplitude of 2, a frequency of π, and a phase shift of $\pi/2$ units right.

26. Find the equation of a sine function with an amplitude of 1.4, a frequency of 4, and a phase shift of $\pi/3$ units left.

Section 12–4

The equations in Problems 27–30 represent a system vibrating in simple harmonic motion.

(a) Find the amplitude, period, and phase shift.

(b) Determine the displacement y after 2.4 s.

(c) Determine when the object will reach maximum positive displacement.

(d) Sketch the graph through one cycle.

27. $y = 8 \sin\left(t - \dfrac{\pi}{4}\right)$

28. $y = 6 \sin(3\pi t + 2\pi)$

29. $y = 3.1 \sin(2t + \pi)$

30. $y = 2 \sin\left(\dfrac{t}{3} - \dfrac{\pi}{2}\right)$

E 31. The voltage in an ac circuit is given by $V = 60 \sin (30t - \pi)$. Find the amplitude, period, and phase shift of this function, and sketch the graph through two cycles.

Section 12–5

Sketch the graph of the following functions through one cycle.

32. $y = -\csc 2x$

33. $y = 4 \tan x$

34. $y = -3 \cot x$

35. $y = -4 \sec x$

36. $y = \tan\left(\dfrac{x}{2} - \pi\right)$

37. $y = \cot(\pi x - 3\pi)$

38. $y = 2 \cot\left(x + \dfrac{\pi}{2}\right)$

39. $y = -2 \csc(x + \pi)$

40. $y = -\sec(2\pi x - \pi)$

41. $y = 3 \sec\left(\dfrac{x}{3} - \dfrac{\pi}{3}\right)$

Section 12–6

Use addition of ordinates to graph the following functions.

42. $y = 2x + 3 \cos x$

43. $y = x - 4 \sin 2\pi x$

44. $y = x^2 - 4 \sin x$

45. $y = \cos\left(\pi x + \dfrac{\pi}{2}\right) - x^2$

46. $y = \sin 2x + 3 \cos x$

47. $y = 3 \cos x - \sin \dfrac{x}{2}$

48. $y = 2 \cos\left(x - \dfrac{\pi}{2}\right) + 3 \sin(x + \pi)$

49. $y = \sin\left(\dfrac{x}{2} + \pi\right) + 4 \cos\left(x - \dfrac{\pi}{4}\right)$

Graph the following parametric equations.

50. $x = 3 \sin t; \ y = \cos\left(t + \dfrac{\pi}{2}\right)$

51. $x = -\sin 2t; \ y = 2 \sin 2t$

52. $x = \cos\left(\dfrac{t}{2} - 2\pi\right); \ y = 4 \sin 3t$

53. $x = \sin\left(t - \dfrac{\pi}{3}\right); \ y = \cos\left(t + \dfrac{\pi}{2}\right)$

54. $x = \cos 4t; \ y = -4 \sin(t + 1)$

CHAPTER TEST

The number(s) in parentheses refers to the appropriate learning objective(s) given at the beginning of the chapter.

1. Sketch the graph of $y = -2 \cos\left(x + \dfrac{\pi}{2}\right)$. (2)

2. Graph $y = 2x + 2 \sin x$ by addition of ordinates. (5)

3. Write an equation describing the simple harmonic motion of an object if the period is 0.8 s, the amplitude is 2 in., and the phase angle is $\pi/3$ radians. (3)

4. For the following trigonometric function, identify the amplitude, period, and phase shift:

$$y = 1.7 \sin\left(3\pi x - \frac{\pi}{3}\right)$$

(1)

5. Sketch the graph of $y = 3 \sin 4x$ through one cycle.

(2)

6. Sketch the graph of $y = -2 \tan x$ through one cycle.

(4)

7. Sketch the graph of $y = -\sin(x - \pi)$ through one cycle.

(2)

8. An alternating current has an amplitude of 2.4 A and a frequency of 60 Hz. If the phase angle is zero, write the equation of current as a function of time, and find the current at $t = 1.8$ s.

(3)

9. Sketch the graph of $y = \cot\left(x + \frac{\pi}{2}\right)$ through one cycle.

(4)

10. Graph $x = \sin 2t$ and $y = \cos\left(t - \frac{\pi}{2}\right)$.

(6)

11. For $y = -3 \cos(\pi x - \pi)$, identify the amplitude, period, and phase shift.

(1)

12. The following two voltages are applied to the horizontal and vertical planes of an oscilloscope:

$$v_1 = 60 \sin\left(100\pi t - \frac{\pi}{2}\right) \qquad \text{and} \qquad v_2 = 100 \cos 120\pi t$$

(6)

Sketch the figure that would appear on the screen.

13. Write an equation for sinusoidal voltage if the peak voltage is 115 V, the frequency is 50 Hz, and the phase angle is $-\pi/6$. Graph this function through one cycle.

(2, 3)

14. Sketch the graph of $y = -2 \csc(x + \pi)$ through one cycle.

(4)

GROUP ACTIVITY

Mountain Circuits Corp.

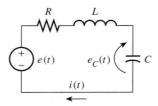

You have been hired by Mountain Circuits Corp. to determine an equation for the current $i(t)$. After a review of some electrical engineering books, you learn that the solution for the circuit involves finding the complementary function of a linear differential equation with constant coefficients. The complete solution is made up of only the transient response because the steady state solution is zero (a dc current cannot flow through a capacitor). A method of finding the complementary function is to let $i(t) = e^{pt}$ where p must satisfy the equation:

$$Lp^2 + Rp + \frac{1}{C} = 0$$

(1)

1. Using the quadratic formula, find the two roots for p. What conditions will produce a complex root?

2. If the roots may be written as $p_1, p_2 = -\alpha \pm j\beta$, where j is equal to the square root of -1, what are the values of α and β in terms of L, R, and C?

3. Assuming:

$$\frac{1}{LC} > \frac{R^2}{4L^2}$$

The complete solution becomes:

$$i(t) = K_1 e^{(-\alpha + j\beta)t} + K_2 e^{(-\alpha - j\beta)t} \tag{2}$$

The constants K_1 and K_2 are found from the initial conditions of the circuit. If $i(t) = 0$ at $t = 0$, find the relationship between K_1 and K_2.

4. The change in current with respect to time is written as:

$$\frac{di}{dt} = K_1 e^{(-\alpha + j\beta)t}(-\alpha + j\beta) + K_2 e^{(-\alpha - j\beta)t}(-\alpha - j\beta) \tag{3}$$

If the circuit is initially at rest (no charge on the capacitor), then the change in current with respect to time (at $t = 0$) is $1/L$. Using this information and the relationship between K_1 and K_2 found earlier, find a value for K_1 using Equation 3.

5. Using the boundary value information (solutions for K_1 and K_2), show that Equation 3 may be written as:

$$i(t) = \frac{1}{\beta L} e^{-\alpha t} \sin \beta t \tag{4}$$

You will need the identity:

$$\sin \beta t = \frac{e^{j\beta t} - e^{-j\beta t}}{2j}$$

6. What determines the period of the oscillations? Write an expression for the period in terms of L, C, and R.

7. Plot $i(t)$ (Equation 4) when $L = 5$ H, $C = 0.00625$ F, and $R = 10\ \Omega$. Does your calculation for the period agree with your plot? *Note*: The term βt is in radians. Therefore, you may need to convert to degrees before taking the sine of βt.

8. The solution that you have examined is a condition known as underdamped oscillation. The other two conditions are critically damped and overdamped. What do you think are the prerequisites for these conditions? You do not need to solve the equations.

SOLUTION TO CHAPTER INTRODUCTION

To model the data provided, we must determine a, b, and c in the equation

$$d = a \cos(bt - c)$$

The amplitude is given by

$$a = \frac{1}{2} (\text{high} - \text{low}) = \frac{1}{2} (1.7 - (-1.7)) = 1.7$$

The period is given by

$$P = 2(\text{low time} - \text{high time}) = 2(10 - 4) = 12$$

which means

$$b = \frac{2\pi}{P} = \frac{2\pi}{12} \approx 0.524$$

Since high tide occurs 4 hours after midnight, $c/b = 4$ and $c = 2.096$. The data can be modeled by

$$d = 1.7 \cos(0.524t - 2.096)$$

Y

ou have contracted your services to a photographer taking a picture of a 4-foot watercolor painting setting on an easel. The camera lens is 1 ft below the lower edge of the painting. The angle y subtended by the lens x feet from the painting is given by

$$y = \arctan \frac{4x}{x^2 + 5} \quad x > 0$$

The photographer wants you to determine the distance she can set up the camera from the painting so that angle y is a maximum. (The solution to this problem is given at the end of the chapter.)

Electronic signals can be monitored on an oscilloscope and described mathematically by equations using trigonometric functions. Also, sound waves from a stereo can be represented by sinusoidal functions. In this chapter we will discuss trigonometric identities, that is, formulas that can be substituted into a trigonometric expression to transform and simplify it. The trigonometric identities are useful in solving a trigonometric equation, another topic of this chapter.

Learning Objectives

After you complete this chapter, you should be able to

1. Prove that a given trigonometric expression is an identity (Section 13–1).

2. Apply the formula for the sum or difference of two angles to simplify a trigonometric expression and to verify an identity (Section 13–2).

3. Apply the double- and half-angle formulas to simplify a trigonometric expression and to verify an identity (Section 13–3).

4. Write a trigonometric expression as a single term by using the appropriate sum, difference, half- or double-angle formula (Sections 13–2 and 13–3).

5. Solve a trigonometric equation (Section 13–4).

6. Use the inverse trigonometric functions to evaluate an expression (Section 13–5).

Chapter 13

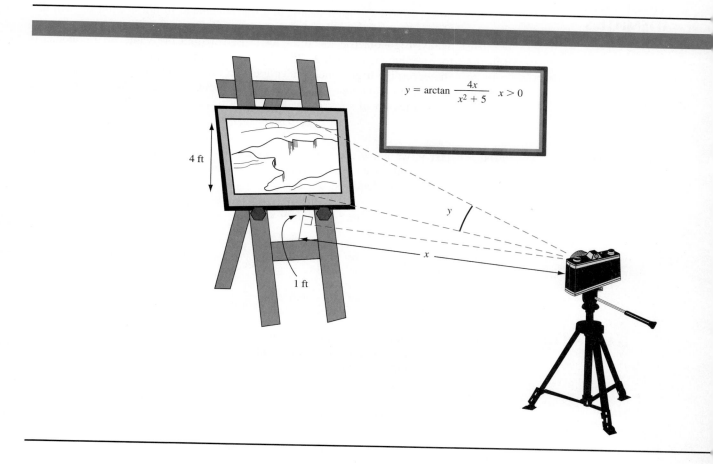

$$y = \arctan \frac{4x}{x^2 + 5} \quad x > 0$$

4 ft

1 ft

y

x

Trigonometric Equations and Identities

13–1

BASIC TRIGONOMETRIC IDENTITIES

An **identity** is a relationship that is *true for all permissible values* of the variable(s). For example, the expression $3(x + 1) + 8 = x + 11 + 2x$ is an algebraic identity because any real number substituted for x results in a true statement. In this section we will discuss identities as they relate to trigonometry.

Historically, trigonometric identities were used to find the value of trigonometric functions for specific angles such as $75°$. With the use of a calculator, this is no longer necessary; however, the trigonometric identities are useful in other applications, such as solving trigonometric equations and Laplace transforms in calculus. We will begin the discussion with a review of some of the basic identities discussed earlier.

Basic Identities

From Chapter 9, we know that certain trigonometric functions are reciprocals: sine and cosecant, cosine and secant, and tangent and cotangent. Rearranging these relationships leads to the following identities:

$$\csc\theta = \frac{1}{\sin\theta} \qquad \text{or} \qquad \sin\theta = \frac{1}{\csc\theta} \qquad \text{or} \qquad \sin\theta\csc\theta = 1$$

$$\sec\theta = \frac{1}{\cos\theta} \qquad \text{or} \qquad \cos\theta = \frac{1}{\sec\theta} \qquad \text{or} \qquad \cos\theta\sec\theta = 1$$

$$\cot\theta = \frac{1}{\tan\theta} \qquad \text{or} \qquad \tan\theta = \frac{1}{\cot\theta} \qquad \text{or} \qquad \tan\theta\cot\theta = 1$$

LEARNING HINT ✦ Notice the different forms that these reciprocal identities can take. It is extremely important that you be able to identify the different forms of any given identity. This requires great familiarity with each one.

In Chapter 9, we also defined the trigonometric functions in terms of x, y, and r. These relationships were given as follows:

$$\sin\theta = \frac{y}{r} \qquad \csc\theta = \frac{r}{y}$$

$$\cos\theta = \frac{x}{r} \qquad \sec\theta = \frac{r}{x}$$

$$\tan\theta = \frac{y}{x} \qquad \cot\theta = \frac{x}{y}$$

If we take the definition for the tangent function and divide the numerator and denominator by r, we obtain the following additional identity:

$$\tan\theta = \frac{y}{x} = \frac{y/r}{x/r} = \frac{\sin\theta}{\cos\theta}$$

Since the cotangent function is the reciprocal of the tangent, the following relationship is also true:

$$\cot\theta = \frac{\cos\theta}{\sin\theta}$$

Several additional identities result from applying the Pythagorean theorem.

$$x^2 + y^2 = r^2 \qquad \text{Pythagorean theorem}$$

$$\frac{x^2}{r^2} + \frac{y^2}{r^2} = \frac{r^2}{r^2} \qquad \text{divide by } r^2$$

$$\left(\frac{x}{r}\right)^2 + \left(\frac{y}{r}\right)^2 = 1 \qquad \text{simplify}$$

$$(\cos \theta)^2 + (\sin \theta)^2 = 1 \qquad \text{substitute}$$

$$\cos^2 \theta + \sin^2 \theta = 1 \qquad \text{simplify}$$

NOTE ✦ The expressions $\cos^2 \theta$ and $(\cos \theta)^2$ are identical; $\cos^2 \theta$ is just easier to write.

Similarly, by applying the Pythagorean theorem, we can derive the following identities:

$$x^2 + y^2 = r^2 \qquad \text{Pythagorean theorem}$$

$$\frac{x^2}{y^2} + \frac{y^2}{y^2} = \frac{r^2}{y^2} \qquad \text{divide by } y^2$$

$$\left(\frac{x}{y}\right)^2 + 1 = \left(\frac{r}{y}\right)^2 \qquad \text{simplify}$$

$$\cot^2 \theta + 1 = \csc^2 \theta \qquad \text{substitute}$$

and

$$x^2 + y^2 = r^2 \qquad \text{Pythagorean theorem}$$

$$\frac{x^2}{x^2} + \frac{y^2}{x^2} = \frac{r^2}{x^2} \qquad \text{divide by } x^2$$

$$1 + \left(\frac{y}{x}\right)^2 = \left(\frac{r}{x}\right)^2 \qquad \text{simplify}$$

$$1 + \tan^2 \theta = \sec^2 \theta \qquad \text{substitute}$$

The eight basic identities derived so far are summarized below.

Basic Trigonometric Identities

$$\sec \theta = \frac{1}{\cos \theta} \qquad\qquad \tan \theta = \frac{\sin \theta}{\cos \theta}$$

$$\csc \theta = \frac{1}{\sin \theta} \qquad\qquad \cot \theta = \frac{\cos \theta}{\sin \theta}$$

$$\cot \theta = \frac{1}{\tan \theta} \qquad \sin^2 \theta + \cos^2 \theta = 1$$

$$1 + \tan^2 \theta = \sec^2 \theta \qquad \cot^2 \theta + 1 = \csc^2 \theta$$

EXAMPLE 1 Using $\theta = 45°$, verify the following identity:

$$\tan \theta = \frac{\sin \theta \cos \theta}{1 - \sin^2 \theta}$$

Solution To verify the identity for a specific value, substitute the value 45° for θ and prove the equality as follows:

$$\tan 45° = \frac{\sin 45° \cos 45°}{1 - \sin^2 45°}$$

$$1 = \frac{\left(\frac{\sqrt{2}}{2}\right)\left(\frac{\sqrt{2}}{2}\right)}{1 - \left(\frac{\sqrt{2}}{2}\right)^2}$$

$$1 = \frac{1/2}{1/2}$$

$$1 = 1 \qquad \blacksquare\blacksquare\blacksquare$$

Proving an Identity

To prove that an expression is an identity, we must prove the equality of the equation for all permissible values of the variable. It would be impossible to substitute every permissible value to prove an identity. Therefore, we prove an identity by transforming one member of the equation into the other.

EXAMPLE 2 Prove the following identity:

$$\cot x = \frac{\sec x \cos x}{\tan x}$$

Solution To prove this identity, we must show that the two sides of the equation are identical for all permissible values of x. To do so, we choose the more complicated side of the equation, substitute trigonometric identities, and perform algebraic operations to simplify the expression until the two sides of the equation are identical. To transform the right member into cot x, we can substitute 1 for sec x cos x.

$$\cot x = \frac{\sec x \cos x}{\tan x}$$

$$\cot x = \frac{1}{\tan x} \qquad \text{substitute}$$

$$\cot x = \cot x \qquad \text{substitute} \qquad \blacksquare\blacksquare\blacksquare$$

LEARNING HINT ✦ There is no set procedure for proving an identity; however, some general guidelines are given in the following box. There may be several correct ways to prove an identity, and it may take several false starts before you find a correct way. However, the more you practice, the easier the proofs become.

Guidelines for Proving Identities

1. Substitute trigonometric identities into the more complicated side of the identity until it is identical to the less complicated side (see Example 3).

2. When working with the more complicated side of an identity, always be aware of the expression you are working toward. If the less complicated side involves only one

trigonometric function, then substitute in terms of that function on the other side of the identity. Another useful transformation is to multiply or divide the numerator and denominator of a fraction by the same quantity (see Example 5).

3. Perform any algebraic operations to simplify an expression. This step may involve finding a common denominator to add fractions, factoring a polynomial, or multiplying polynomials. If one side of the identity has a single term in the denominator and several terms in the numerator, write each term in the numerator over the denominator and simplify (see Examples 4 and 6).

4. When in doubt, transform the more complicated member to sine and cosine functions. This technique may not be the easiest one, but it will usually give acceptable results (See Example 3).

EXAMPLE 3   Prove the following identity:

$$\frac{\sin x \sec x}{\sin^2 x - \tan^2 x} = -\frac{\cos x}{\sin^3 x}$$

Solution   The left side of the equation appears more complicated; therefore, we will simplify it. Since the right side involves only the sine and cosine functions, let us substitute the sine and cosine functions on the left side of the equation.

$$\frac{\sin x \sec x}{\sin^2 x - \tan^2 x} = -\frac{\cos x}{\sin^3 x}$$

$$\frac{\sin x \, \dfrac{1}{\cos x}}{\sin^2 x - \dfrac{\sin^2 x}{\cos^2 x}} = -\frac{\cos x}{\sin^3 x}$$

After substituting, we perform normal algebraic simplification by multiplying the fractions in the numerator and adding the fractions in the denominator.

$$\frac{\dfrac{\sin x}{\cos x}}{\dfrac{\sin^2 x \cos^2 x - \sin^2 x}{\cos^2 x}} = -\frac{\cos x}{\sin^3 x}$$

$$\frac{\sin x}{\cos x} \cdot \frac{\cos^2 x}{\sin^2 x(\cos^2 x - 1)} = -\frac{\cos x}{\sin^3 x} \qquad \text{divide and factor}$$

$$\frac{\cos x}{\sin x(\cos^2 x - 1)} = -\frac{\cos x}{\sin^3 x} \qquad \text{simplify}$$

$$\frac{\cos x}{\sin x(-\sin^2 x)} = -\frac{\cos x}{\sin^3 x} \qquad \text{substitute}$$

$$-\frac{\cos x}{\sin^3 x} = -\frac{\cos x}{\sin^3 x} \qquad \text{simplify}$$

■■■

CAUTION ✦ When proving that an expression is an identity, be very careful *not* to handle the identity as if it were an equation. This means you *cannot* perform operations such as adding the same expression to both sides or dividing both sides by the same expression. These operations apply only to an equation where the statement is known to be true; an identity must be proven to be true.

EXAMPLE 4 Prove the following identity:

$$\frac{\sin x}{1 + \cos x} + \frac{1 + \cos x}{\sin x} = 2 \csc x$$

Solution Since the left side of the equation appears to be more complicated, we simplify by adding the fractions.

$$\frac{\sin x}{1 + \cos x} + \frac{1 + \cos x}{\sin x} = 2 \csc x$$

$$\frac{\sin^2 x + (1 + \cos x)^2}{\sin x(1 + \cos x)} = 2 \csc x \qquad \text{add}$$

$$\frac{\sin^2 x + 1 + 2 \cos x + \cos^2 x}{\sin x(1 + \cos x)} = 2 \csc x \qquad \text{square the binomial}$$

$$\frac{(1 - \cos^2 x) + 1 + 2 \cos x + \cos^2 x}{\sin x(1 + \cos x)} = 2 \csc x \qquad \text{substitute for } \sin^2 x$$

$$\frac{2 + 2 \cos x}{\sin x(1 + \cos x)} = 2 \csc x \qquad \text{simplify}$$

$$\frac{2(1 + \cos x)}{\sin x(1 + \cos x)} = 2 \csc x \qquad \text{factor}$$

$$\frac{2}{\sin x} = 2 \csc x \qquad \text{simplify (reduce)}$$

$$2\left(\frac{1}{\sin x}\right) = 2 \csc x \qquad \text{simplify}$$

$$2 \csc x = 2 \csc x \qquad \text{substitute} \qquad \blacksquare\blacksquare\blacksquare$$

Graphing Calculator

A graphing calculator can be used to verify that an expression is an identity. Using Example 4, you would input each side of the equation as a graph. You input

$$y = \frac{\sin x}{1 + \cos x} + \frac{1 + \cos x}{\sin x}$$

as one equation to graph and

$$y = 2 \csc x$$

as the other. However, a calculator recognizes only the sine, cosine, and tangent functions, so the second graph must be entered using $\csc x = 1/\sin x$. The second equation input to the graphing calculator is

$$y = \frac{2}{\sin x}$$

Then you graph the two equations on the same axis. If a single curve results, the original expression is an identity. Figure 13–1 shows the graph resulting from this process for Example 4.

FIGURE 13–1

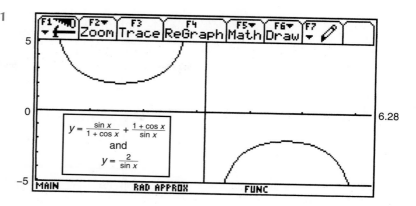

EXAMPLE 5   Verify the following identity:

$$\tan x = \tan x \csc^2 x - \cot x$$

Solution   Since the right hand of the equation appears to be more complicated, we transform it to look like the left side. Notice that the left side of the equation contains only the tangent function; therefore, the substitutions that we make on the right side should all lead toward the tangent.

$$\tan x = \tan x \csc^2 x - \cot x$$
$$\tan x = \tan x (1 + \cot^2 x) - \cot x \qquad \text{substitute}$$
$$\tan x = \tan x + \tan x \cot^2 x - \cot x \qquad \text{multiply}$$
$$\tan x = \tan x + \tan x \left(\frac{1}{\tan^2 x} \right) - \frac{1}{\tan x} \qquad \text{substitute}$$
$$\tan x = \tan x + \frac{1}{\tan x} - \frac{1}{\tan x} \qquad \text{simplify}$$
$$\tan x = \tan x$$

■ ■ ■

EXAMPLE 6   Verify the following identity:

$$\sec^2 x + \tan^2 x = \sec^4 x - \tan^4 x$$

Solution   Figure 13–2 shows the graph resulting from entering

$$y = \frac{1}{(\cos x)^2} + (\tan x)^2$$

and

$$y = \frac{1}{(\cos x)^4} - (\tan x)^4$$

and pressing graph. The result shows that the expression is an identity.

FIGURE 13–2

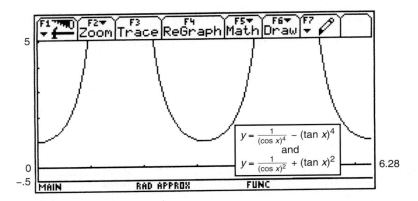

Since the right side of the equation appears to be more complicated, we begin by factoring it.

$$\sec^2 x + \tan^2 x = \sec^4 x - \tan^4 x$$
$$\sec^2 x + \tan^2 x = (\sec^2 x - \tan^2 x)(\sec^2 x + \tan^2 x) \qquad \text{factor}$$
$$\sec^2 x + \tan^2 x = 1(\sec^2 x + \tan^2 x) \qquad \text{substitute}$$
$$\sec^2 x + \tan^2 x = \sec^2 x + \tan^2 x \qquad \text{simplify} \qquad ■■■$$

13–1 EXERCISES

Prove the following identities. Use the graphing calculator to verify that the expression is an identity.

1. $\cos x \csc x = \cot x$

2. $\dfrac{\tan x}{\sin x} = \sec x$

3. $\sin x \cos x \csc x \sec x = 1$

4. $\dfrac{\cot x}{\cos x} = \csc x$

5. $\dfrac{\sin^2 A}{1 - \cos A} = 1 + \cos A$

6. $\cos x(1 + \tan^2 x) = \sec x$

7. $\dfrac{\tan x - 1}{1 - \cot x} = \tan x$

8. $\dfrac{\cos A}{1 - \sin A} = \dfrac{1 + \sin A}{\cos A}$

9. $\sin x + \cot x \cos x = \csc x$

10. $(\cot x + \csc x)(\cot x - \csc x) = -1$

11. $\cot A(\tan A + \cot A) = \csc^2 A$

12. $(\sec x - \csc x)(\sin x + \cos x) = \tan x - \cot x$

13. $\dfrac{\cos x}{\sin^2 x - 1} = -\sec x$

14. $\dfrac{\tan^4 x - 1}{\sec^2 x} = \tan^2 x - 1$

15. $\cos x(\tan^2 x + 1) = \sec x$

16. $\csc x \cot x \cos x = \cot^2 x$

17. $\tan x = \sec x \csc x - \cot x$

18. $\sin A \tan A + \cos A = \sec A$

19. $\dfrac{2 - \sin^2 A}{\cos A} = \sec A + \cos A$

20. $\cos x(\sec x - \cos x) = \sin^2 x$

21. $\cos^2 x \csc x - \csc x = -\sin x$

22. $\dfrac{1}{\csc A - \cot A} = \csc A + \cot A$

23. $\csc A + \tan A + \cot A = \dfrac{\cos A + 1}{\sin A \cos A}$

24. $\dfrac{4 \csc^4 x - 4 \cot^4 x}{\csc^2 x + \cot^2 x} = 4$

25. $\dfrac{1 + \cot x}{1 - \cot x} = \dfrac{1 + \tan x}{\tan x - 1}$

26. $\dfrac{\cot x}{\csc x + 1} = \dfrac{\csc x - 1}{\cot x}$

27. $(\sin A - \cos A)(\sin A + \cos A) = \cos^2 A(\tan^2 A - 1)$

28. $\dfrac{\sin x - \tan x}{\sin x \tan x} = \dfrac{\cos x - 1}{\sin x}$

29. $\dfrac{\cot x + \cos x}{1 + \sin x} = \cot x$

30. $\cos^2 x - \sin^2 x = 1 - 2 \sin^2 x$

31. $\dfrac{1 + \sin y}{\cos y} + \dfrac{\cos y}{1 + \sin y} = 2 \sec y$

32. $\csc x \cot x = \dfrac{1 + \cot^2 x}{\sec x}$

33. $\dfrac{\sin x}{1 + \cos x} = \csc x - \cot x$

34. $\dfrac{\cot x - \cos x}{\cot x} = \dfrac{\cos^2 x}{1 + \sin x}$

13–2

THE SUM OR DIFFERENCE OF TWO ANGLES

The trigonometric identities in the previous section dealt with functions of one angle. In this section we will consider trigonometric functions that are the sum or difference of two angles. Let us begin by deriving an expression for sin $(A + B)$.

Deriving the Sum and Difference Formulas

In Figure 13–3, angle A is in standard position, while angle B has its initial side as the terminal side of angle A. The angle $A + B$ is an angle in standard position. From a point P on the terminal side of angle $A + B$, perpendicular lines are drawn to the x axis at point T and to the terminal side of angle A at point R. Then perpendicular lines are drawn from R to the x axis at point U and from R to line PT at point Q. Angle TPR is equal to angle A (similar triangles). **Similar triangles** have equal corresponding angles and proportional corresponding sides.

FIGURE 13–3

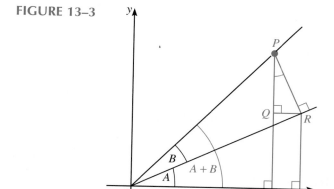

From Figure 13–3,

$$\sin (A + B) = \frac{PT}{OP} = \frac{TQ + QP}{OP} = \frac{RU}{OP} + \frac{QP}{OP}$$

From triangle ORU,

$$RU = OR \sin A$$

and from triangle PQR,

$$QP = PR \cos A$$

Substituting gives

$$\sin (A + B) = \frac{OR \sin A}{OP} + \frac{PR \cos A}{OP}$$

$$\sin (A + B) = \sin A \frac{OR}{OP} + \cos A \frac{PR}{OP}$$

But, in triangle OPR,

$$\frac{OR}{OP} = \cos B \qquad \text{and} \qquad \frac{PR}{OP} = \sin B$$

Substituting gives the final result.

$$\sin (A + B) = \sin A \cos B + \cos A \sin B$$

Similarly,

$$\cos (A + B) = \cos A \cos B - \sin A \sin B$$

To derive the formula for the difference of two angles, we substitute $-B$ for angle B in the sum formula as follows:

$$\sin (A + (-B)) = \sin A \cos (-B) + \cos A \sin (-B)$$

Since $\sin (-B) = -\sin B$ and $\cos (-B) = \cos B$, we have

$$\sin (A - B) = \sin A \cos B - \cos A \sin B$$

Similarly,

$$\cos (A - B) = \cos A \cos B + \sin A \sin B$$

To derive the formula for the tangent of the sum or difference of two angles, we use the fact that $\tan C = \sin C/\cos C$. Thus,

$$\tan (A \pm B) = \frac{\sin (A \pm B)}{\cos (A \pm B)} = \frac{\sin A \cos B \pm \cos A \sin B}{\cos A \cos B \mp \sin A \sin B}$$

Dividing the numerator and denominator by $\cos A \cos B$ gives

$$\tan (A \pm B) = \frac{\tan A \pm \tan B}{1 \mp \tan A \tan B}$$

The formulas for the sum and difference of two angles are summarized below.

Sum or Difference of Two Angles

$$\sin (A \pm B) = \sin A \cos B \pm \cos A \sin B$$
$$\cos (A \pm B) = \cos A \cos B \mp \sin A \sin B$$

$$\tan (A \pm B) = \frac{\tan A \pm \tan B}{1 \mp \tan A \tan B}$$

NOTE ✦ The two signs given in the formula above denote the signs for two formulas, one for sum and one for difference. In each formula, pair the upper signs together and pair the lower signs together.

Applying the Sum and Difference Formulas

EXAMPLE 1 Without using a calculator, find exact values for sin 15° and cos 75°.

Solution To find exact values for sin 15° and cos 75°, we can express 15° and 75° in terms of the special angles 0°, 30°, 45°, 60°, or 90°, because we know exact values for these angles. We can write sin 15° as sin (45° − 30°) and cos 75° as cos (30° + 45°). Applying the appropriate formula and simplifying the result gives

$$\sin 15° = \sin(45° - 30°) = \sin 45° \cos 30° - \cos 45° \sin 30°$$

$$\sin 15° = \frac{\sqrt{2}}{2}\left(\frac{\sqrt{3}}{2}\right) - \frac{\sqrt{2}}{2}\left(\frac{1}{2}\right)$$

$$\sin 15° = \frac{\sqrt{6}}{4} - \frac{\sqrt{2}}{4}$$

$$\sin 15° = \frac{\sqrt{6} - \sqrt{2}}{4}$$

$$\cos 75° = \cos(30° + 45°) = \cos 30° \cos 45° - \sin 30° \sin 45°$$

$$\cos 75° = \frac{\sqrt{3}}{2}\left(\frac{\sqrt{2}}{2}\right) - \frac{1}{2}\left(\frac{\sqrt{2}}{2}\right)$$

$$\cos 75° = \frac{\sqrt{6} - \sqrt{2}}{4}$$

■ ■ ■

CAUTION ✦ Remember that sin $(A + B) \neq$ sin $A +$ sin B. This holds true for similar sum and difference formulas.

EXAMPLE 2 We are given that sin $A = 12/13$, A is in quadrant I, cos $B = -4/5$, and B is in quadrant II.

(a) Find sin $(A + B)$.
(b) Find cos $(A + B)$.
(c) Determine the quadrant in which the terminal side of $(A + B)$ lies.

Solution To find sin $(A + B)$ and cos $(A + B)$, we need to have values for sin A, cos A, sin B, and cos B. Angles A and B are shown in Figure 13–4. From the information given,

FIGURE 13–4

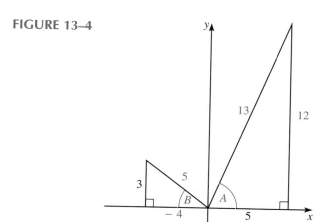

we can apply the Pythagorean theorem to find the remaining side of each triangle and the required values for the sine and cosine functions. These values are

$$\sin A = \frac{12}{13} \qquad \sin B = \frac{3}{5}$$

$$\cos A = \frac{5}{13} \qquad \cos B = -\frac{4}{5}$$

We substitute into the formula to obtain the following:

(a) $\sin (A + B) = \sin A \cos B + \cos A \sin B$

$$\sin (A + B) = \frac{12}{13}\left(-\frac{4}{5}\right) + \frac{5}{13}\left(\frac{3}{5}\right)$$

$$\sin (A + B) = \frac{-48 + 15}{65}$$

$$\sin (A + B) = -\frac{33}{65}$$

(b) $\cos (A + B) = \cos A \cos B - \sin A \sin B$

$$\cos (A + B) = \frac{5}{13}\left(-\frac{4}{5}\right) - \frac{12}{13}\left(\frac{3}{5}\right)$$

$$\cos (A + B) = \frac{-20 - 36}{65}$$

$$\cos (A + B) = -\frac{56}{65}$$

(c) Finally, since $\sin (A + B)$ and $\cos (A + B)$ are both negative, the terminal side of angle $A + B$ lies in quadrant III, where the sine and cosine functions are both negative.

■ ■ ■

EXAMPLE 3　Reduce the following expressions to a single term:

(a) $\cos 3A \cos 2A - \sin 3A \sin 2A$

(b) $\dfrac{\tan 5A - \tan A}{1 + \tan 5A \tan A}$

Solution

(a) This expression fits the expanded form of $\cos (A + B)$. Therefore, we can write

$$\cos 3A \cos 2A - \sin 3A \sin 2A = \cos (3A + 2A) = \cos 5A$$

(b) Since this expression fits the expanded form of $\tan (A - B)$,

$$\frac{\tan 5A - \tan A}{1 + \tan 5A \tan A} = \tan (5A - A) = \tan 4A$$

■ ■ ■

Proving an Identity　　We use the same techniques developed in the previous section to prove an identity. We have just added the formulas for the sum or difference of two angles to the list of possible trigonometric substitutions that we can make.

EXAMPLE 4 Verify the identity $\sin (A + B) \sin (A - B) = \sin^2 A - \sin^2 B$.

Solution Since the left side of the equation appears to be more complicated, simplify it by expanding $\sin (A + B)$ and $\sin (A - B)$.

$$\sin (A + B) \qquad\qquad \sin (A - B) \qquad\qquad = \sin^2 A - \sin^2 B$$

$$(\sin A \cos B + \cos A \sin B) (\sin A \cos B - \cos A \sin B) = \sin^2 A - \sin^2 B$$

Because the left side is the product of the sum and difference of a binomial, multiplying gives

$$\sin^2 A \cos^2 B - \cos^2 A \sin^2 B = \sin^2 A - \sin^2 B$$

Since the right side of the equation contains only the sine function, let us substitute for $\cos^2 B$ and $\cos^2 A$.

$$\sin^2 A(1 - \sin^2 B) - (1 - \sin^2 A)\sin^2 B = \sin^2 A - \sin^2 B$$

Multiplying and simplifying the result gives

$$\sin^2 A - \sin^2 A \sin^2 B - \sin^2 B + \sin^2 A \sin^2 B = \sin^2 A - \sin^2 B$$

$$\sin^2 A - \sin^2 B = \sin^2 A - \sin^2 B \quad \blacksquare\blacksquare\blacksquare$$

EXAMPLE 5 Prove the following identities:

(a) $\tan \left(\dfrac{\pi}{4} - x \right) = \dfrac{1 - \tan x}{1 + \tan x}$

(b) $\dfrac{\sin (x + y)}{\sin x \cos y} = 1 + \cot x \tan y$

Solution

(a) To prove this identity, expand the left side of the equation using the formula for $\tan (A - B)$.

$$\tan \left(\frac{\pi}{4} - x \right) = \frac{1 - \tan x}{1 + \tan x}$$

$$\frac{\tan \pi/4 - \tan x}{1 + \tan \pi/4 \tan x} = \frac{1 - \tan x}{1 + \tan x}$$

Substituting $\tan \pi/4 = 1$ gives the final result.

$$\frac{1 - \tan x}{1 + \tan x} = \frac{1 - \tan x}{1 + \tan x}$$

(b) To prove this identity, let us expand the left side of the equation using the formula for $\sin (A + B)$ and simplify the result.

$$\frac{\sin (x + y)}{\sin x \cos y} = 1 + \cot x \tan y$$

$$\frac{\sin x \cos y + \cos x \sin y}{\sin x \cos y} = 1 + \cot x \tan y$$

$$\frac{\sin x \cos y}{\sin x \cos y} + \frac{\cos x \sin y}{\sin x \cos y} = 1 + \cot x \tan y$$

$$1 + \cot x \tan y = 1 + \cot x \tan y \qquad \blacksquare\blacksquare\blacksquare$$

Application

EXAMPLE 6   A type of interference in connection with musical instruments is interference between two waves traveling in opposite directions. A standing wave occurs when two waves of the same amplitude and frequency travel in opposite direcctions. A wave $y_1 = A \sin(kx - \omega t)$ traveling to the right and a wave of the same frequency and amplitude $y_2 = A \sin(kx + \omega t)$ traveling to the left are superimposed to give a resultant $y = y_1 + y_2$. Find an expression for the resultant y.

Solution   To find y we add the two waves.

$$y = y_1 + y_2$$

$$y = A \sin(kx - \omega t) + A \sin(kx + \omega t)$$

Then, we use the formula for the sum and difference of angles.

$$y = A[\sin kx \cos \omega t - \cos kx \sin \omega t] + A[\sin kx \cos \omega t + \cos kx \sin \omega t]$$

$$y = 2A \sin kx \cos \omega t$$

■ ■ ■

13–2 EXERCISES

Use the sum or difference formula to find the exact value of the given function.

1. $\sin 105°$

2. $\tan 15°$

3. $\cos 15°$

4. $\sin 75°$

5. $\cos 195°$

6. $\tan 75°$

7. $\cos 165°$

8. $\tan 285°$

9. Given that $\cos A = 12/13$ and A is in quadrant I and that $\sin B = 15/17$ and B is in quadrant II, find the following:

(a) $\cos(A + B)$ (b) $\tan(A + B)$

(c) quadrant of $A + B$

10. Given that $\tan A = -3/4$ and A is in quadrant II and that $\cos B = 5/13$ and B is in quadrant I, find the following:

(a) $\sin(A - B)$ (b) $\cos(A - B)$

(c) quadrant of $A - B$

11. Given that $\sin A = 15/17$ and A is in quadrant II and that $\tan B = 12/5$ and B is in quadrant III, find the following:

(a) $\sin(A - B)$ (b) $\cos(A - B)$

(c) quadrant of $A - B$

12. Given that $\cos A = 7/25$ and A is in quadrant I and that $\sin B = 12/13$ and B is in quadrant II, find the following:

(a) $\tan(A + B)$ (b) $\sin(A + B)$

(c) quadrant of $A + B$

Reduce each expression to a single term.

13. $\sin 75° \cos 40° - \cos 75° \sin 40°$

14. $\cos 48° \cos 36° + \sin 48° \sin 36°$

15. $\cos A \cos 5A - \sin A \sin 5A$

16. $\sin 3x \cos y + \cos 3x \sin y$

17. $\sin A \sin 3A - \cos A \cos 3A$

18. $\cos(x - y) \cos x - \sin(x - y) \sin x$

19. $\dfrac{\tan(A + B) - \tan A}{1 + \tan(A + B) \tan A}$

20. $\sin(C - D) \cos D + \cos(C - D) \sin D$

Verify the following identities.

21. $\cos\left(\dfrac{\pi}{2} + x\right) = -\sin x$

22. $\sin\left(\dfrac{\pi}{3} - x\right) = \dfrac{\sqrt{3} \cos x - \sin x}{2}$

23. $\tan(\pi + x) = \tan x$

24. $\sin(30° + x) = \dfrac{\cos x + \sqrt{3} \sin x}{2}$

25. $\dfrac{\sin 5A}{\csc 2A} + \dfrac{\cos 5A}{\sec 2A} = \cos 3A$

26. $\cos(A + B) - \cos(A - B) = -2 \sin A \sin B$

27. $\cos(A + B) \cos(A - B) = \cos^2 A - \sin^2 B$

28. $\sin 2A = 2 \sin A \cos A$ (*Hint:* $2A = A + A$.)

29. $\sin(A + B) - \sin(A - B) = 2 \cos A \sin B$

30. $\csc(A + B) = \dfrac{\csc A \csc B}{\cot B + \cot A}$

31. $\cot(A + B) = \dfrac{\cot A \cot B - 1}{\cot A + \cot B}$

32. $\sin (A + B + C) = \sin A \cos B \cos C$
$$+ \cos A \sin B \cos C$$
$$+ \cos A \cos B \sin C$$
$$- \sin A \sin B \sin C$$

P 33. The expression for the displacement y of an object in simple harmonic motion can be written

$$y = a \sin 2\pi ft \cos B + a \cos 2\pi ft \sin B$$

Write the expression on the right side of the equation as a single term.

P 34. A spring vibrating in simple harmonic motion described by

$$y_1 = A_1 \cos \left(wt + \frac{\pi}{2} \right)$$

is subjected to another motion described by

$$y_2 = A_2 \cos (wt - \pi)$$

Show that the resultant motion is

$$y_1 + y_2 = - A_2 \cos wt - A_1 \sin wt$$

P 35. The displacement y of a wave traveling through some liquid can be given by

$$y = 12 \sin \left(\frac{\pi t}{3} - \frac{3\pi}{2} \right)$$

Expand and simplify this expression.

P 36. The displacement of light passing through glass can be written as

$$d = \frac{t \sin (A - B)}{\cos B}$$

Simplify this expression.

P 37. Using Example 6, find the resultant of two waves $y_1 = B \cos (kx - \omega t)$ and $y_2 = B \cos (kx + \omega t)$.

13–3

DOUBLE-ANGLE AND HALF-ANGLE IDENTITIES

Derivation of the Double-Angle Identity

Using the formulas for the sum of two angles, we can derive the formulas for a double angle. The derivations for $\sin 2A$, $\cos 2A$, and $\tan 2A$ follow.

$$\sin 2A = \sin (A + A)$$
$$\sin 2A = \sin A \cos A + \sin A \cos A$$
$$\sin 2A = 2 \sin A \cos A$$

$$\cos 2A = \cos (A + A)$$
$$\cos 2A = \cos A \cos A - \sin A \sin A$$
$$\cos 2A = \cos^2 A - \sin^2 A$$
$$\cos 2A = (1 - \sin^2 A) - \sin^2 A$$
$$\cos 2A = 1 - 2 \sin^2 A$$
$$\cos 2A = \cos^2 A - (1 - \cos^2 A)$$
$$\cos 2A = 2 \cos^2 A - 1$$

$$\tan 2A = \tan (A + A)$$
$$\tan 2A = \frac{\tan A + \tan A}{1 - \tan A \tan A}$$
$$\tan 2A = \frac{2 \tan A}{1 - \tan^2 A}$$

The double-angle formulas are summarized in the following box.

Double-Angle Formulas

$$\sin 2A = 2 \sin A \cos A$$
$$\cos 2A = \cos^2 A - \sin^2 A$$
$$\cos 2A = 1 - 2 \sin^2 A$$
$$\cos 2A = 2 \cos^2 A - 1$$
$$\tan 2A = \frac{2 \tan A}{1 - \tan^2 A}$$

EXAMPLE 1   Use a double-angle formula to find the exact value of $\cos 2A$, $\sin 2A$, and $\tan 2A$ if $\sin A = 8/17$ and the terminal side of angle A lies in quadrant I.

Solution   To fill in the required double-angle formulas, we need to know the value of $\sin A$, $\cos A$, and $\tan A$. We can determine these values using the Pythagorean theorem, as shown in Figure 13–5. You may also recall from Chapter 9 that 8, 15, 17 are dimensions for a right triangle.

$$\sin A = \frac{8}{17} \qquad \cos A = \frac{15}{17} \qquad \tan A = \frac{8}{15}$$

Calculating $\sin 2A$, $\cos 2A$, and $\tan 2A$ gives the following:

$$\sin 2A = 2 \sin A \cos A$$
$$\sin 2A = 2\left(\frac{8}{17}\right)\left(\frac{15}{17}\right)$$
$$\sin 2A = \frac{240}{289}$$
$$\cos 2A = \cos^2 A - \sin^2 A$$
$$\cos 2A = \left(\frac{15}{17}\right)^2 - \left(\frac{8}{17}\right)^2$$

FIGURE 13–5

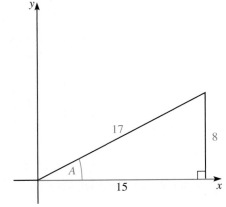

$$\cos 2A = \frac{225 - 64}{289}$$

$$\cos 2A = \frac{161}{289}$$

$$\tan 2A = \frac{2 \tan A}{1 - \tan^2 A}$$

$$\tan 2A = \frac{2(8/15)}{1 - (8/15)^2}$$

$$\tan 2A = \frac{16}{15} \cdot \frac{225}{161}$$

$$\tan 2A = \frac{240}{161}$$

Since sin $2A$, cos $2A$, and tan $2A$ are all positive, the terminal side of angle $2A$ lies in quadrant I. We check by calculating angle A using sin $A = 8/17$ and multiplying the resulting angle by 2.

$$\sin A = \frac{8}{17} \qquad A \approx 28° \qquad 2A \approx 56°$$

■ ■ ■

EXAMPLE 2 Write $4 \sin^2 2x - 2$ as a single expression.

Solution This expression appears to be one of the identities for cos $2A$: $1 - 2 \sin^2 A$. However, we need to arrange it in the same order as the identity.

$$-2 + 4 \sin^2 2x$$

Next, factor out -2.

$$-2(1 - 2 \sin^2 2x)$$

The expression in parentheses is equivalent to cos $2A$ where $A = 2x$. The expression is simplified to

$$-2 \cos 4x$$

■ ■ ■

EXAMPLE 3 Prove the identity $\cos x(1 - \cos 2x) = \sin x \sin 2x$.

Solution Since the left side of the equation appears to be more complicated, simplify it. Also, because the right side of the equation contains only the sine function, all substitutions on the left side will work toward that end.

$$\cos x(1 - \cos 2x) = \sin x \sin 2x$$
$$\cos x - \cos x \cos 2x = \sin x \sin 2x \qquad \text{multiply}$$
$$\cos x - \cos x(1 - 2\sin^2 x) = \sin x \sin 2x \qquad \text{substitute}$$
$$\cos x - \cos x + 2\sin^2 x \cos x = \sin x \sin 2x \qquad \text{multiply}$$
$$2\sin^2 x \cos x = \sin x \sin 2x \qquad \text{simplify}$$
$$\sin x(2\sin x \cos x) = \sin x \sin 2x \qquad \text{simplify}$$
$$\sin x \sin 2x = \sin x \sin 2x \qquad \text{substitute}$$

■ ■ ■

Half-Angle Formulas   Using the double-angle formulas, we can derive expressions for sin $A/2$, cos $A/2$, and tan $A/2$. To derive cos $A/2$, we solve for cos B in the identity cos $2B = 2\cos^2 B - 1$.

$$\cos 2B = 2\cos^2 B - 1$$

$$\frac{\cos 2B + 1}{2} = \cos^2 B$$

$$\pm\sqrt{\frac{\cos 2B + 1}{2}} = \cos B$$

This formula is true for all angles; therefore, substituting $A/2$ for B gives the identity

$$\cos\frac{A}{2} = \pm\sqrt{\frac{1 + \cos A}{2}}$$

The sign in front of the radical is determined by the quadrant in which the terminal side of angle $A/2$ lies. In a similar manner, use the identity cos $2B = 1 - \sin^2 B$, solve for sin B, and substitute $B = A/2$ to obtain the half-angle formula for the sine function, as follows:

$$\cos 2B = 1 - 2\sin^2 B$$

$$\frac{1 - \cos 2B}{2} = \sin^2 B$$

$$\pm\sqrt{\frac{1 - \cos 2B}{2}} = \sin B$$

Substituting $\dfrac{A}{2} = B$ gives

$$\sin\frac{A}{2} = \pm\sqrt{\frac{1 - \cos A}{2}}$$

We can find tan $A/2$ using the identity

$$\tan\frac{A}{2} = \frac{\sin A/2}{\cos A/2}$$

$$\tan\frac{A}{2} = \frac{\pm\sqrt{\dfrac{1 - \cos A}{2}}}{\pm\sqrt{\dfrac{1 + \cos A}{2}}}$$

$$\tan\frac{A}{2} = \pm\sqrt{\frac{1 - \cos A}{1 + \cos A}}$$

If we multiply the numerator and denominator by $\sqrt{1 - \cos A}$ and simplify, the result is

$$\tan\frac{A}{2} = \frac{1 - \cos A}{\sin A}$$

If we multiply the numerator and denominator by $\sqrt{1 + \cos A}$, the result is

$$\tan\frac{A}{2} = \frac{\sin A}{1 + \cos A}$$

The half-angle formulas are summarized below.

Half-Angle Formulas

$$\sin \frac{A}{2} = \pm \sqrt{\frac{1 - \cos A}{2}}$$

$$\cos \frac{A}{2} = \pm \sqrt{\frac{1 + \cos A}{2}}$$

$$\tan \frac{A}{2} = \frac{1 - \cos A}{\sin A} = \frac{\sin A}{1 + \cos A}$$

EXAMPLE 4 Use the half-angle formula to find an exact value of $\cos \pi/8$.

Solution Since $\pi/8$ is half of $\pi/4$, we can use the following half-angle formula where $A = \pi/4$:

$$\cos \frac{A}{2} = \pm \sqrt{\frac{1 + \cos A}{2}}$$

$$\cos \frac{\pi}{8} = \pm \sqrt{\frac{1 + \cos \frac{\pi}{4}}{2}}$$

$$\cos \frac{\pi}{8} = + \sqrt{\frac{1 + \frac{\sqrt{2}}{2}}{2}}$$

$$\cos \frac{\pi}{8} = \frac{\sqrt{2 + \sqrt{2}}}{2}$$

Since $\pi/8$ is in quadrant I, we use the plus sign before the radical. ■■■

EXAMPLE 5 Given that $\sin A = -12/13$ and A is in quadrant III, find the following:

(a) $\cos \dfrac{A}{2}$ (b) $\tan \dfrac{A}{2}$

Solution To find $\cos A/2$ and $\tan A/2$, we need to know the value of $\cos A$, which we obtain using the Pythagorean theorem.

$$\cos A = -\frac{5}{13}$$

(a) $\cos \dfrac{A}{2} = \pm \sqrt{\dfrac{1 + \cos A}{2}}$

$$\cos \frac{A}{2} = \pm \sqrt{\frac{1 + \frac{-5}{13}}{2}}$$

$$\cos \frac{A}{2} = \pm \sqrt{\frac{4}{13}}$$

$$\cos \frac{A}{2} = -\frac{2}{13}\sqrt{13}$$

Since angle A is in quadrant III, $180° < A < 270°$, $A/2$ is in quadrant II, and $90° < A/2 < 135°$. Because the cosine is negative in quadrant II, we place a negative sign in front of the answer.

(b) $\tan \dfrac{A}{2} = \dfrac{1 - \cos A}{\sin A}$

$\tan \dfrac{A}{2} = \dfrac{1 + \dfrac{5}{13}}{\dfrac{-12}{13}}$

$\tan \dfrac{A}{2} = -\dfrac{3}{2}$

Notice that in Example 5, the sign of $\tan A/2$ is determined by the formula (answer). Only the sine and cosine functions require determining the correct sign.

EXAMPLE 6   Verify the following identity:

$$\cos^2 \frac{B}{2} = \frac{\sec B + 1}{2 \sec B}$$

Solution   To verify this identity, let us start by substituting for $\cos^2 B/2$.

$$\cos^2 \frac{B}{2} = \frac{\sec B + 1}{2 \sec B}$$

$$\left(\pm \sqrt{\frac{1 + \cos B}{2}}\right)^2 = \frac{\sec B + 1}{2 \sec B} \qquad \text{substitute}$$

$$\frac{1 + \cos B}{2} = \frac{\sec B + 1}{2 \sec B} \qquad \text{simplify}$$

If we compare the first term of the numerator on each side of the equation, we know that we must transform 1 into $\sec B$. However, $\sec B$ equals $1/\cos B$. Thus, we should divide the numerator and denominator of the left side of the equation by $\cos B$ as follows:

$$\frac{\dfrac{1}{\cos B} + \dfrac{\cos B}{\cos B}}{2\left(\dfrac{1}{\cos B}\right)} = \frac{\sec B + 1}{2 \sec B}$$

$$\frac{\sec B + 1}{2 \sec B} = \frac{\sec B + 1}{2 \sec B}$$

CAUTION ✦ Be very careful in using the double- and half-angle formulas. Determine the necessary value of A, $2A$, and $A/2$ before substituting into the formula.

Application

EXAMPLE 7   The index of refraction n of a prism whose angle is A and whose angle of minimum deviation is D is given by

$$n = \frac{\sin\left(\dfrac{A + D}{2}\right)}{\sin \dfrac{A}{2}} \qquad n > 0$$

Simplify this expression.

Solution   We begin by substituting for $\sin\left(\dfrac{A + D}{2}\right)$ and $\sin\left(\dfrac{A}{2}\right)$.

$$n = \frac{\sin\left(\dfrac{A + D}{2}\right)}{\sin\dfrac{A}{2}}$$

$$n = \frac{\sqrt{\dfrac{1 - \cos(A + D)}{2}}}{\sqrt{\dfrac{1 - \cos A}{2}}}$$

$$n = \sqrt{\frac{1 - \cos(A + D)}{1 - \cos A}}$$

$$n = \sqrt{\frac{1 - \cos A \cos D + \sin A \sin D}{1 - \cos A}}$$

■ ■ ■

13–3 EXERCISES

Use double- or half-angle formulas to find exact values for the following functions.

1. $\sin 15°$

2. $\tan 67.5°$

3. $\cos 105°$

4. $\cos 165°$

5. $\tan 120°$

6. $\sin 112.5°$

7. If $\sin A = 8/17$ and A is in quadrant I, find $\cos 2A$ and $\tan A/2$.

8. If $\tan A = -40/9$ and A is in quadrant II, find $\sin 2A$ and $\cos A/2$.

9. If $\cos A = -24/25$ and $180° < A < 270°$, find $\tan A/2$ and $\sin A/2$.

10. If $\sin A = 24/25$ and $90° < A < 180°$, find $\sin A/2$ and $\tan 2A$.

Write each expression as a single term.

11. $4 \sin 2A \cos 2A$

12. $\dfrac{2 \tan 3B}{1 - \tan^2 3B}$

13. $4 \sin^2 x \cos^2 x$

14. $16 \sin^2 A - 8$

15. $-\sqrt{\dfrac{1 + \cos 210°}{2}}$

16. $\dfrac{1 - \cos 70°}{\sin 70°}$

Prove the following identities.

17. $\cos 3x = \cos x(1 - 4 \sin^2 x)$

18. $\cos 2x = \dfrac{1 - \tan^2 x}{1 + \tan^2 x}$

19. $\tan B \sin 2B + \cos 2B = 1$

20. $\cot 2A = \dfrac{\cot^2 A - 1}{2 \cot A}$

21. $\tan \dfrac{A}{2} = \csc A - \cot A$

22. $\dfrac{\tan 2x}{\sin 2x} = \sec 2x$

23. $\sec 2C = \dfrac{\tan C + \cot C}{\cot C - \tan C}$

24. $(\sin x + \cos x)^2 = 1 + \sin 2x$

25. $\sin 2A = \dfrac{2 \tan A}{1 + \tan^2 A}$

26. $\tan \dfrac{A}{2}(\cos A + 1) = \sin A$

27. $\cos^2 A - \sin^2 A = \cos 2A$

28. $\sec \dfrac{x}{2} = \pm\sqrt{\dfrac{2}{1 + \cos x}}$

29. $\left(\cos \dfrac{x}{2} - \sin \dfrac{x}{2}\right)^2 = 1 - \sin x$

30. $\dfrac{\cos 2x - \cos x}{\cos x - 1} = 2 \cos x + 1$

31. $1 = 2 \cos^2 \dfrac{B}{2} - \cos B$

32. $\cot \dfrac{B}{2} = \csc B + \cot B$

P 33. The index of refraction n is given by

$$n = \frac{\sin \dfrac{(A + D)}{2}}{\sin \dfrac{A}{2}}$$

where A = angle of prism and D = angle of minimum deviation. Simplify this expression using cot, sin, and cos.

P 34. The horizontal range y of an object projected with velocity v at an angle A above the horizontal is given by $y = vt \cos A$, but

$$t = \frac{2v \sin A}{g}$$

Substitute for t and simplify the result.

E 35. In an ac circuit containing reactance, the instantaneous power is given by

$$P = V_{pk} I_{pk} \cos \omega t \sin \omega t$$

Show that P can also be written as

$$P = \frac{V_{pk} I_{pk}}{2} \sin 2\omega t$$

13–4

TRIGONOMETRIC EQUATIONS

So far in this chapter, we have verified identities, which are equations that are true for all permissible values of the variable. In this section we will solve **conditional equations,** that is, equations that are true only for specific values of the variable. Solving a trigonometric equation means the same thing as solving an algebraic equation, namely, finding values of the variable that satisfy the equation. To solve trigonometric equations, we use the same techniques used in solving algebraic equations such as isolating the variable, collecting like terms, and factoring. In addition to these techniques, we will also simplify the equation by substituting trigonometric identities.

EXAMPLE 1 Solve the equation $2 \cos A + 1 = 0$ for $0 \leq A < 2\pi$.

Solution Since this is a linear trigonometric equation, we use the techniques of solving an algebraic linear equation as follows:

$$2 \cos A + 1 = 0$$
$$2 \cos A = -1 \qquad \text{transpose}$$
$$\cos A = -\frac{1}{2} \qquad \text{divide}$$

To find angle A, we must determine all angles $0 \leq A < 2\pi$ whose cosine value equals 1/2. Recall from our previous discussion of the 30°–60°–90° triangle that the reference angle is $\pi/3$. Since the cosine is negative in quadrants II and III,

$$A = \frac{2\pi}{3} \qquad \text{and} \qquad A = \frac{4\pi}{3}$$

After checking, the solutions are $2\pi/3$ and $4\pi/3$. ∎ ∎ ∎

CAUTION ✦ Be sure to find *all* angles that satisfy the equation. You may have a tendency to stop with one solution.

EXAMPLE 2 Solve the equation $\sin x - 2 \sin^2 x = 0$ for $0 \leq x < 2\pi$.

Solution Since this is a quadratic trigonometric equation, we can solve it by factoring.

$$\sin x - 2 \sin^2 x = 0$$
$$\sin x(1 - 2 \sin x) = 0$$

Next, we set each factor equal to zero and solve for x.

$$\sin x = 0 \qquad\qquad 1 - 2 \sin x = 0$$

$$x = 0 \quad \text{and} \quad x = \pi \qquad\qquad -2 \sin x = -1$$

$$\sin x = \frac{1}{2}$$

$$x = \frac{\pi}{6} \quad \text{and} \quad x = \frac{5\pi}{6}$$

After checking, the solutions are $x = 0$, $\pi/6$, $5\pi/6$, and π. ∎∎∎

Guidelines for Solving Trigonometric Equations

1. If the equation contains one function of a single angle, use algebraic techniques to solve for the angle (see Example 1).

2. Solve a quadratic equation containing a single function of the same angle by factoring, if possible. Otherwise, use the quadratic formula (see Example 2).

3. If the equation contains several functions of the same angle, substitute trigonometric identities to obtain a single function (see Example 3).

4. If the equation contains several angles, substitute trigonometric identities to obtain a function of a single angle (see Example 4).

EXAMPLE 3 Solve $2 \sin^2 A - \cos^2 A = 0$ for $0 \le A < 2\pi$.

Solution Since this equation involves two functions of the same angle, substitute trigonometric identities to obtain one trigonometric function.

$$2 \sin^2 A - \cos^2 A = 0$$

$$2 \sin^2 A - (1 - \sin^2 A) = 0 \qquad \text{substitute}$$

$$2 \sin^2 A - 1 + \sin^2 A = 0 \qquad \text{simplify}$$

$$3 \sin^2 A - 1 = 0 \qquad \text{simplify}$$

$$3 \sin^2 A = 1 \qquad \text{transpose}$$

$$\sin^2 A = \frac{1}{3} \qquad \text{divide}$$

$$\sin A = \pm\frac{\sqrt{3}}{3} \qquad \text{square root}$$

Since A is not one of the special angles, we use the calculator to find the reference angle, $A = 0.6155$ radian. Since $\sin A$ is both positive and negative, we place this angle in all four quadrants: $A = 0.6155$, 2.5261, 3.7571, and 5.6677. After these values have been checked, the solutions are 0.6155, 2.5261, 3.7571, and 5.6677. ∎∎∎

The graphing calculator can also be used to solve trigonometric equations. As in the case of algebraic equations, the solution to a trigonometric equation is the x coordinate of the point where the graph crosses the x axis. To solve the equation in Example 3 using the graphing calculator, enter the equation $y = 2 \sin^2 x - \cos x$. Since the graphing calculator expects variables of x and y for the graph, x was substituted for A. The viewing rectangle and the resulting graph are shown in Figure 13–6. In the viewing rectangle, notice that xmax = 6.28 or 2π, so all solutions in the desired range are shown.

FIGURE 13–6

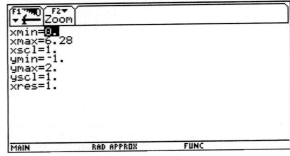

Viewing rectangle

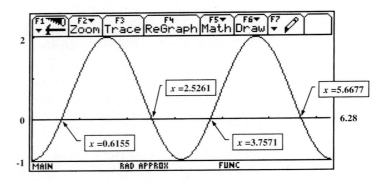

■ ■ ■

EXAMPLE 4   Solve the equation cos $2t$ + 3 sin t = 2 for $0 \le t < 2\pi$.

Solution   Since this equation involves more than one function and will not factor, we should make a trigonometric substitution. We want an expression in terms of a single angle, $2t$ or t, and a single trigonometric function, cosine or sine. Substituting $1 - 2 \sin^2 t$ for cos $2t$ gives

$$\cos 2t + 3 \sin t = 2$$
$$(1 - 2 \sin^2 t) + 3 \sin t = 2 \qquad \text{substitute}$$
$$-2 \sin^2 t + 3 \sin t - 1 = 0 \qquad \text{transpose}$$
$$2 \sin^2 t - 3 \sin t + 1 = 0 \qquad \text{multiply by } -1$$
$$(2 \sin t - 1)(\sin t - 1) = 0 \qquad \text{factor}$$

$$2 \sin t - 1 = 0 \qquad\qquad \sin t - 1 = 0 \qquad \text{set factors equal to zero}$$
$$2 \sin t = 1 \qquad\qquad\qquad \sin t = 1$$

$$\sin t = \frac{1}{2} \qquad\qquad\qquad t = \frac{\pi}{2}$$

$$t = \frac{\pi}{6} \qquad \text{and} \qquad t = \frac{5\pi}{6}$$

After checking, the solutions are $t = \pi/6$, $\pi/2$, and $5\pi/6$.   ■ ■ ■

FIGURE 13–7

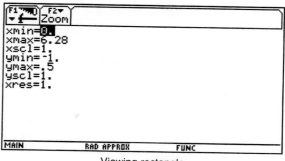

Viewing rectangle

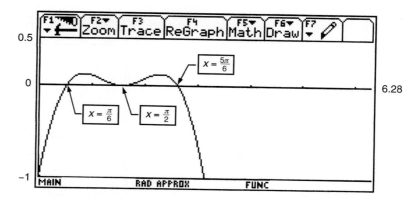

To solve the equation in Example 4 using the graphing calculator, enter the equation $y = \cos 2t + 3 \sin t - 2$. Notice that in the graphing calculator solution, unlike the algebraic solution, you do not need to make any trigonometric substitutions. The viewing rectangle and graph are shown in Figure 13–7.

EXAMPLE 5   Solve the equation $\cos 5x \cos 3x + \sin 5x \sin 3x = -1$ for $0 \le x < 2\pi$.

Solution   Since the left side of this equation reduces to $\cos (5x - 3x) = \cos 2x$, this equation becomes

$$\cos 2x = -1$$

Substituting for $\cos 2x$ and solving gives

$$2 \cos^2 x - 1 = -1 \qquad \text{substitute}$$
$$2 \cos^2 x = 0 \qquad \text{transpose and add}$$
$$\cos^2 x = 0 \qquad \text{divide}$$
$$\cos x = 0 \qquad \text{take square root of both sides}$$
$$x = \frac{\pi}{2} \qquad \text{and} \qquad x = \frac{3\pi}{2}$$

The graphing calculator viewing rectangle and graph are shown in Figure 13–8.

FIGURE 13–8

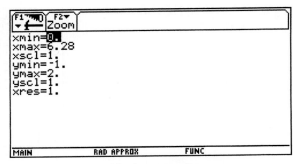

Viewing rectangle

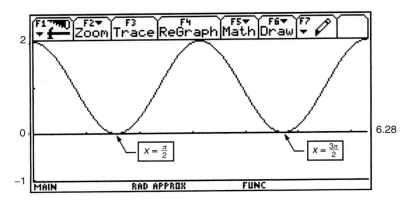

EXAMPLE 6   Solve the equation $\sin A/2 = 1 - \cos A$ for $0 \leq A < 2\pi$.

Solution   First, we substitute for $\sin A/2$.

$$\pm\sqrt{\frac{1 - \cos A}{2}} = 1 - \cos A$$

Next, we remove the radical by squaring both sides of the equation.

$$\frac{1 - \cos A}{2} = (1 - \cos A)^2$$

$$\frac{1 - \cos A}{2} = 1 - 2\cos A + \cos^2 A \qquad \text{remove parentheses}$$

$$1 - \cos A = 2 - 4\cos A + 2\cos^2 A \qquad \text{remove fractions}$$

$$0 = 1 - 3\cos A + 2\cos^2 A \qquad \text{simplify}$$

$$0 = (1 - \cos A)(1 - 2\cos A) \qquad \text{factor}$$

$$1 - \cos A = 0 \qquad\qquad 1 - 2\cos A = 0 \qquad \text{solve}$$

$$\cos A = 1 \qquad\qquad\qquad \cos A = \frac{1}{2}$$

$$A = 0 \qquad A = \frac{\pi}{3} \qquad \text{and} \qquad A = \frac{5\pi}{3}$$

FIGURE 13–9

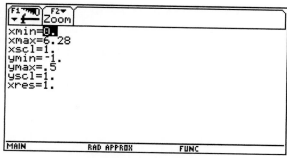

Viewing rectangle

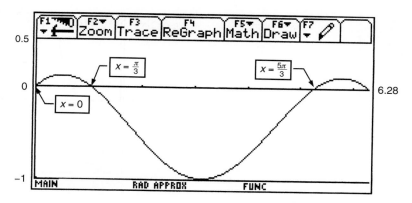

Since we squared both sides of the equation, we must check each value by substituting into the original equation. The solutions are $A = 0$, $\pi/3$, and $5\pi/3$. The graphing calculator solution is shown in Figure 13–9.

∎ ∎ ∎

EXAMPLE 7 Solve the equation $-\cot x + 3 \tan x = 2$ for $0 \leq x < 2\pi$.

Solution We begin by substituting $\cot x = 1/\tan x$.

$$-\cot x + 3 \tan x = 2$$

$$-\frac{1}{\tan x} + 3 \tan x = 2 \qquad \text{substitute}$$

$$-1 + 3 \tan^2 x = 2 \tan x \qquad \text{remove fraction}$$

$$3 \tan^2 x - 2 \tan x - 1 = 0 \qquad \text{transpose}$$

$$(3 \tan x + 1)(\tan x - 1) = 0 \qquad \text{factor}$$

$$3 \tan x + 1 = 0 \qquad\qquad \tan x - 1 = 0 \qquad \text{solve}$$

$$3 \tan x = -1 \qquad\qquad\quad \tan x = 1$$

$$\tan x = \frac{-1}{3}$$

$$x = 2.8 \qquad \text{and} \qquad x = 6.0 \qquad x = \frac{\pi}{4} \qquad \text{and} \qquad x = \frac{5\pi}{4}$$

After the above values have been checked for x, the solutions are $x = \pi/4$, 2.8, $5\pi/4$, and 6.0.

FIGURE 13–10

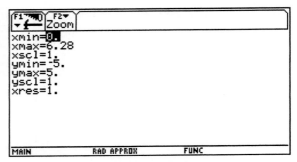

Viewing rectangle

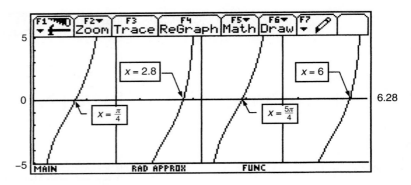

To determine the graphing calculator solution, we must substitute $1/\tan x$ for $\cot x$. The equation entered is $y = -\dfrac{1}{\tan x} + 3 \tan x - 2$. The viewing rectangle and the graph are shown in Figure 13–10. ∎ ∎ ∎

Application

EXAMPLE 8 The range R of a projectile propelled at an angle θ with the horizontal at velocity $\angle C$ is given by

$$R = \frac{2v^2 \cos \theta \sin \theta}{g}$$

Bobo the Clown is to be shot from a cannon at a velocity of 100 ft/s with the intention of falling into a net 150 ft away. (See Figure 13–11.) Determine the angle of the cannon needed so that Bobo makes the net.

150 ft

FIGURE 13–11

Solution   Since we are using the English system, $g = 32$ ft/s². The equation we must solve is

$$150 \text{ ft} = \frac{2(100 \text{ ft/s})^2 \cos \theta \sin \theta}{32 \text{ ft/s}^2}$$

$$150 \text{ ft} = \frac{(100 \text{ ft/s})^2 \cdot \sin 2\theta}{32 \text{ ft/s}^2}$$

$$\frac{150(32)}{(100)^2} = \sin 2\theta$$

$$0.4800 = \sin 2\theta$$

$$14° \approx \theta \qquad \blacksquare\blacksquare\blacksquare$$

13–4 EXERCISES

Solve the following equations for $0 \leq x < 2\pi$. Use a calculator only when x is not one of the special angles. You may use a graphing calculator to check your solution.

1. $(\sqrt{3} \sin x - 1)(2 \cos x + 1) = 0$

2. $(\sqrt{3} \tan x + 1)(\tan x + 1) = 0$

3. $\csc x + 1 = 0$

4. $\cot x - 1 = 0$

5. $4 \sin^2 x = 1$

6. $\cot^2 x - 1 = 0$

7. $\tan^2 x + 6 \tan x + 7 = 0$

8. $\sin x \csc x - \csc x = 0$

9. $\cot^2 x = \cot x$

10. $\csc^2 x - 5 \cot x + 3 = 0$

11. $\sin^2 x = 1 - \cos x$

12. $\sin x = \sin \dfrac{x}{2}$

13. $\cot^2 x - 1 = \csc x$

14. $\cos 2x - \cos x = 0$

15. $3 \csc x \sec x = \sec x$

16. $2 \sin^2 x = \cos 2x - 2$

17. $\cos 2x + \sin x = 1$

18. $\dfrac{\tan 3x + \tan x}{1 - \tan 3x \tan x} = -1$

19. $5 \cos^2 x = 4 \cos x + 1$

20. $\sin x - 2 \sin^3 x = 0$

21. $\sin 4x \cos x + \cos 4x \sin x = -1$

22. $\cos 2x - 2 \cos^2 x = 0$

23. $\tan^2 x = 1 + \sec x$

24. $\dfrac{1 - \cos x}{\sin x} = \sqrt{3}$

25. $2 \cos 3x = 1$

26. $2 \sin 5x = \sqrt{3}$

27. $\sin 2x = \cos x$

28. $\cos x \cos 3x - \sin x \sin 3x = -1$

29. $\cot^2 x - \cot x = 0$

30. $2 \cos^2 3x = 1$

P 31. The angle of a projectile shot at 86 m/s with a range of 500 m is given by $7{,}396 \tan A = 2{,}450 \sec^2 A$. Find angle A.

P 32. The vertical displacement y of an object at the end of a spring is given by $y = 3 \cos t + \sin 2t$ where t represents time. Determine the least amount of time required for the displacement to be zero.

P 33. The height from the water level of a point on the rim of a waterwheel is given by

$$y = 3 + 4 \cos \left(t - \frac{\pi}{2} \right)$$

At what times, t, is the point at water level $(y = 0)$?

P 34. The displacement x of a piston is given by $x = 3 \sin A + \sin 2A$. Determine the angle A where the displacement is zero.

B 35. Jim has paid a consultant to determine that the monthly sales S of his pocket PC is given by

$$S = 68.7 + 40.31 \sin \frac{\pi t}{8}$$

where t is the time in months since January $(t = 1)$. Determine the first month when sales will be 100.

P 36. A baseball leaves a bat at an angle θ with the horizontal and initial velocity $v = 125$ ft/s. The ball is caught by an outfielder 275 ft away. Find θ if the range r of the baseball is given by $r = \dfrac{1}{32} v \sin^2 2\theta$.

P 37. Snell's Law of Refraction states that as a light ray passes from one medium (air, water) to another, the ratio of the sine of the angle of incidence θ_i to the sine of the angle of refraction θ_r is a constant n, called the index of refraction. The formula is

$$\frac{\sin \theta_i}{\sin \theta_r} = n$$

Find the angle of refraction if the angle of incidence is 50° and the index of refraction of light passing through water is 1.33.

13–5

THE INVERSE TRIGONOMETRIC FUNCTIONS

In Chapter 7, we discussed exponential and logarithmic functions. They are inverse functions because $y = b^x$ is equivalent to $x = \log_b y$. In this section we will discuss inverse functions of another type, inverse trigonometric functions. Although we did not refer to them as inverse trigonometric functions, we discussed them briefly in Chapter 9 when we solved for an angle given two sides of a right triangle. For example, to find angle A given that $\sin A = 16/19$, we pressed the $\boxed{\text{inv}}$ and then the $\boxed{\text{sin}}$ key on the calculator. The condition $\sin A = 16/19$ can also be written as

$$\sin^{-1}\left(\frac{16}{19}\right) = A \qquad \text{or} \qquad \arcsin \frac{16}{19} = A$$

where the -1 exponent or the "arc" prefix is used to denote an inverse trigonometric function. We will begin the discussion of the inverse trigonometric functions with their graphs.

Graphs of the Inverse Trigonometric Functions

The strict definition of the inverse of a function requires that each value chosen for x result in a single value for y. A function must pass the vertical line list. This means that a vertical line drawn through the graph must touch the graph exactly once. In Figure 13–12(b) you can see that a vertical line through $x = 0$ (the y axis) would touch the graph at least three times. To define the trigonometric functions uniquely, we restrict the domain. This is shown on the respective graphs in second color.

In general, to find the inverse of any function, we interchange x and y in the equation and solve for y. We implement this procedure in graphing by interchanging the x and y coordinates of each ordered pair, as shown in the next example.

EXAMPLE 1 Graph $y = \sin x$ and $y = \arcsin x$.

Solution We know how to graph $y = \sin x$ from the previous chapter. The table of values for $y = \sin x$ is as follows:

x	$-\pi/2$	0	$\pi/2$	π	$3\pi/2$
y	-1	0	1	0	-1

To form the table of values of $y = \arcsin x$, we interchange the x and y coordinates in the table of values given above for $y = \sin x$. Thus, we obtain the following table:

x	-1	0	1	0	-1
y	$-\pi/2$	0	$\pi/2$	π	$3\pi/2$

The graphs of $y = \sin x$ and $y = \arcsin x$ are shown in Figure 13–12(a) and (b), respectively.

The graphs of $y = \cos x$ and $y = \arccos x$ are shown in Figure 13–13(a) and (b), respectively, and the graphs of $y = \tan x$ and $y = \arctan x$ are shown in Figure 13–14(a)

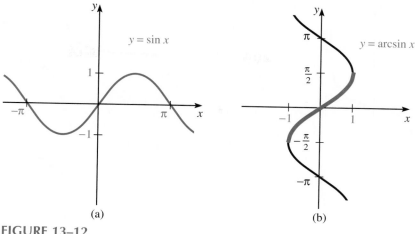

(a)   (b)

FIGURE 13–12

and (b), respectively. Notice that in Figures 13–12, 13–13, and 13–14, the graphs for arc-sin, arccos, and arctan do not represent a function because a given x can be paired with more than one y coordinate. To define the inverse trigonometric functions, we must restrict the range, or the values, of y. These restrictions are denoted by the colored line on the appropriate graph in Figures 13–12, 13–13, and 13–14. The domain and range for each of the inverse trigonometric functions are summarized in Table 13–1.

NOTE ✦ To denote the function form of the trigonometric inverses, we use an initial capital letter, such as Arcsin or Sin^{-1}.

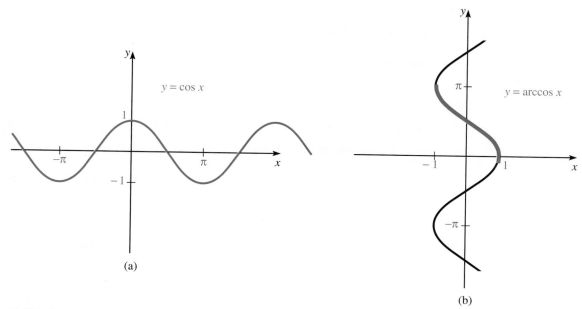

FIGURE 13–13

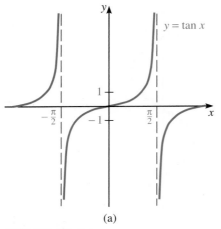

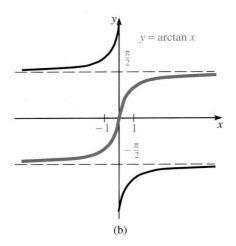

(a)   (b)

FIGURE 13–14

TABLE 13–1

Function	Domain	Range	Quadrants
$y = \text{Arcsin } x$	$-1 \leq x \leq 1$	$-\pi/2 \leq y \leq \pi/2$	I, IV
$y = \text{Arccos } x$	$-1 \leq x \leq 1$	$0 \leq y \leq \pi$	I, II
$y = \text{Arctan } x$	All reals	$-\pi/2 < y < \pi/2$	I, IV
$y = \text{Arcsec } x$	$x \leq -1, x \geq 1$	$0 \leq y \leq \pi, y \neq \dfrac{\pi}{2}$	I, II
$y = \text{Arccsc } x$	$x \leq -1, x \geq 1$	$-\pi/2 \leq y \leq \pi/2, y \neq 0$	I, IV
$y = \text{Arccot } x$	All reals	$0 < y < \pi$	I, II

EXAMPLE 2   Find the following:

(a) $y = \text{Arcsin } \dfrac{\sqrt{3}}{2}$ 　　　　　(b) $y = \text{Arctan } -\dfrac{\sqrt{3}}{3}$ 　　　　　(c) $y = \text{Arccos } (0.348)$

Solution

(a) The result of an inverse trigonometric function is an angle; therefore, the expression $y = \text{Arcsin } \dfrac{\sqrt{3}}{2}$ requires finding an angle y whose sine function equals $\dfrac{\sqrt{3}}{2}$ over the interval $\dfrac{-\pi}{2}$ to $\dfrac{\pi}{2}$. We know from previous chapters that the reference angle for y is $\dfrac{\pi}{3}$. Since the Arcsin function is positive in quadrant I, $y = \dfrac{\pi}{3}$.

(b) The expression $y = \text{Arctan } \dfrac{-\sqrt{3}}{3}$ requires finding an angle y whose tangent function equals $-\dfrac{\sqrt{3}}{3}$. The reference angle for y is $\dfrac{\pi}{6}$. Since the Arctangent function is negative in quadrant IV, angle y is $\dfrac{-\pi}{6}$.

(c) The expression $y = \text{Arccos } (0.348)$ requires that we find the angle whose cosine function equals 0.348. We will need to use the calculator because y is not one of the special angles. The calculator keystrokes are

$$0.348 \boxed{\text{inv}} \boxed{\text{cos}} \rightarrow 1.215$$
$$y = 1.215$$

When performing this operation, be sure your calculator is in the radian mode. ■■■

EXAMPLE 3 Evaluate the following expressions:

(a) $\sin (\text{Arctan } \sqrt{3})$ (b) $\cos (\text{Arcsin } -1/2)$ (c) $\tan (\text{Arcsin } 0.876)$

Solution

(a) If we analyze the expression $\sin (\text{Arctan } \sqrt{3})$, the expression in parentheses (Arctan $\sqrt{3}$) results in an angle for which we calculate the sine value.

$$\sin (\text{Arctan } \sqrt{3}) = \sin \frac{\pi}{3}$$

$$\sin (\text{Arctan } \sqrt{3}) = \frac{\sqrt{3}}{2}$$

(b) To evaluate $\cos (\text{Arcsin } -1/2)$, we determine an angle in quadrant I or IV whose sine is $-1/2$. Then we calculate the cosine value of that angle.

$$\cos \left(\text{Arcsin } \frac{-1}{2} \right) = \cos \left(-\frac{\pi}{6} \right)$$

$$\cos \left(\text{Arcsin } \frac{-1}{2} \right) = \frac{\sqrt{3}}{2}$$

(c) Begin by determining an angle whose sine value is 0.876. You will need to use the calculator because this is not one of the special angles. Using this angle, calculate the value of the tangent.

$$\tan (\text{Arcsin } 0.876) = \tan (1.0675)$$
$$\tan (\text{Arcsin } 0.876) = 1.8163$$
$$0.876 \boxed{\text{inv}} \boxed{\text{sin}} \boxed{\text{tan}} \rightarrow 1.8163$$

■■■

EXAMPLE 4 Without using a calculator, evaluate the following expressions:

(a) $\tan \left(\text{Arcsin } \frac{3}{5} \right)$ (b) $\cos \left(\text{Arctan } \frac{-8}{15} \right)$

Solution

FIGURE 13–15

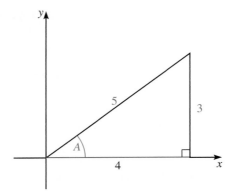

(a) We must find an angle A whose sine value is 3/5, and then we must calculate its tangent value. It is not necessary to know the actual measure of angle A, Figure 13–15 shows an angle A whose sine value is 3/5. Applying the Pythagorean theorem to find the length of the remaining side gives

$$5^2 = x^2 + 3^2$$
$$25 - 9 = x^2$$
$$16 = x^2$$
$$\pm 4 = x$$

After finding the remaining side, we find that $\tan A = 3/4$. Therefore,

$$\tan \left(\text{Arcsin } \frac{3}{5} \right) = \frac{3}{4}$$

(b) Similarly, we can find an angle B whose tangent value is $-8/15$. Since the Arctangent function is negative in quadrant IV, angle B is in quadrant IV, as shown in Figure 13–16. Using the Pythagorean theorem, we find the hypotenuse of the triangle to be 17. Then we find the cosine value for angle B to be 15/17.

$$\cos \left(\text{Arctan } \frac{-8}{15} \right) = \frac{15}{17}$$

■ ■ ■

NOTE ✦ You may also recall from Chapter 9 that 3, 4, 5 and 8, 15, 17 are dimensions for a right triangle.

FIGURE 13–16

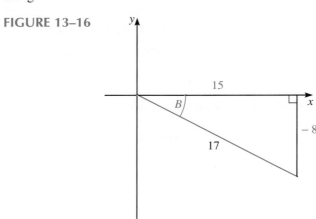

CHAPTER REVIEW

Section 13–1

Prove the following identities.

1. $\dfrac{\sin A}{1 - \cos A} = \dfrac{1 + \cos A}{\sin A}$

2. $\dfrac{\sec x - 1}{1 - \cos x} = \sec x$

3. $(\sin A + \cos A)(\sin A - \cos A) = \sin^2 A - \cos^2 A$

4. $\csc B + \sin B = \dfrac{2 - \cos^2 B}{\sin B}$

5. $\dfrac{1 - \cos^2 B}{1 + \cot^2 B} = \sin^4 B$

6. $\dfrac{\sin x - \tan x}{\sin x \tan x} = \dfrac{\cos x - 1}{\sin x}$

7. $\dfrac{\sin x}{1 - \cos x} - \dfrac{1}{\sin x} = \cot x$

8. $\dfrac{\sec x + \csc x}{1 + \cot x} = \sec x$

9. $\dfrac{\cot A \cos A}{\cot A - \cos A} = \dfrac{\cos A}{1 - \sin A}$

10. $\sec A \csc A - \tan A = \cot A$

11. $\dfrac{\tan x}{\sec x + 1} = \dfrac{\sec x - 1}{\tan x}$

12. $1 - \cot B \sin B \cos B = \sin^2 B$

Section 13–2

13. Given that $\tan A = -12/5$ and A is in quadrant II and that $\cos B = 9/41$ and B is in quadrant I, find the following:
 (a) $\cos (A - B)$ **(b)** $\tan (A + B)$ **(c)** $\sin (A - B)$
 (d) quadrant of $(A - B)$

14. Given that $\sin A = 8/17$, $\cos B = 12/13$, and both A and B are in quadrant I, find the following:
 (a) $\sin (A + B)$ **(b)** $\cos (A - B)$ **(c)** $\tan (A + B)$
 (d) quadrant of $(A + B)$

Reduce each expression to a single term.

15. $\sin (C - D) \cos B + \cos (C - D) \sin B$

16. $\cos 3A \cos A - \sin 3A \sin A$

17. $\dfrac{\tan 3x + \tan 2x}{1 - \tan 3x \tan 2x}$

18. $\sin 56° \cos 10° - \cos 56° \sin 10°$

Verify the following identities.

19. $\sin \left(\dfrac{\pi}{2} - x \right) = \cos x$

20. $\tan \left(x + \dfrac{\pi}{4} \right) = \dfrac{\tan x + 1}{1 - \tan x}$

21. $\sin \left(x + \dfrac{\pi}{2} \right) - \cos \left(x + \dfrac{\pi}{2} \right) = \cos x + \sin x$

22. $\dfrac{\sin (x - y)}{\sin x \cos y} = 1 - \cot x \tan y$

23. $\dfrac{\sin A}{\sin 3A} + \dfrac{\cos A}{\cos 3A} = \dfrac{\sin 4A}{\sin 3A \cos 3A}$

24. $\dfrac{\sin (A + B)}{\cos (A - B)} = \dfrac{\tan A + \tan B}{1 + \tan A \tan B}$

25. $\sin (A - \pi) + \cos (A + \pi) = -\sin A - \cos A$

Section 13–3

Write each expression as a single term.

26. $\sqrt{\dfrac{1 - \cos 9A}{2}}$

27. $8 - 16 \sin^2 B$

28. $10 \sin 6A \cos 6A$

29. $\dfrac{1 - \cos 5B}{\sin 5B}$

30. If $\sin A = 40/41$ and A is in quadrant II, find $\tan \dfrac{A}{2}$ and $\cos 2A$.

31. If $\tan B = -8/15$ and $90° < B < 180°$, find $\sin \dfrac{B}{2}$ and $\cos 2B$.

Verify the following identities.

32. $\sin \dfrac{A}{2} \cos \dfrac{A}{2} = \dfrac{\sin A}{2}$

33. $\sin 4x = 4 \sin x \cos x \cos 2x$

34. $\dfrac{\csc^2 x}{\csc 2x} = 2 \cot x$

35. $\sin x \sin 2x = \cos x (1 - \cos 2x)$

36. $1 - 8 \sin^2 x \cos^2 x = \cos 4x$

37. $\sin 2x = \tan x(1 + \cos 2x)$

38. $\tan \left(x + \dfrac{x}{2} \right) = \dfrac{\sin 2x + \sin x}{\cos 2x + \cos x}$

Section 13–4

Solve the following equations for $0 \le x < 2\pi$. Use a calculator only when x is not one of the special angles.

39. $\sqrt{3} \sin x = 1$ **40.** $\tan x = -1$

41. $6 \cos^2 x + \cos x = 2$

42. $\cos \dfrac{x}{2} = \cos x + 1$

43. $\tan^2 x - \tan x = 0$

44. $4 \sin^2 x = 3 - 8 \cos x$

45. $\sin^2 x + 5 \sin x = 6$

46. $\sin 3x \cos x - \cos 3x \sin x = 0$

47. $5 \cos^2 x - \sin^2 x = 0$

48. $3 \cos 2x + 2 = 0$

Section 13–5

Evaluate the following without using a calculator.

49. $\sin \left(\text{Arccos} -\dfrac{15}{17} \right)$

50. $\cos \left(\text{Arctan} \dfrac{4}{5} \right)$

51. $\tan \left(\text{Arcsin} -\dfrac{40}{41} \right)$

52. $\sec \left(\text{Arcsin} -\dfrac{12}{13} \right)$

Give an algebraic expression for each of the following.

53. $\sin (\text{Arctan } x)$

54. $\csc (\text{Arcsin } -2x)$

55. $\sin (\pi - \text{Arccos } x)$

56. $\cos \left(\text{Arcsec} \dfrac{2}{x} \right)$

57. $\cos \left(\dfrac{1}{2} \text{Arcsin } x \right)$

58. $\sin (2 \text{ Arctan } x)$

CHAPTER TEST

The number(s) in parentheses refers to the appropriate learning objective(s) given at the beginning of the chapter.

1. If $\cos A = -9/41$ and $90° < A < 180°$, find $\tan A/2$ and $\sin 2A$. (3)

2. Solve the following equation for $0 \leq x < 2\pi$: (4, 5)

$$5 \cos 2x + 3 = 0$$

3. Prove the identity (1)

$$\frac{1 + \cos^2 A}{\sin^2 A} + 1 = 2 \csc^2 A$$

4. Find an algebraic expression for the following: (6)

$$\cos \left(\frac{1}{2} \text{Arcsin } 3x \right)$$

5. Reduce the following to an expression containing a single term: (2)

$$\cos (A + B) \cos (C - D) - \sin (A + B) \sin (C - D)$$

6. Verify the following identity: (2)

$$\cos (x - y) \cos (x + y) = \cos^2 x - \sin^2 y$$

7. Evaluate the following without using a calculator. Express the answer in radians. (6)

$$\sec \left(\text{Arcsin} -\frac{12}{13} \right)$$

8. Prove the following identity: (1)

$$\frac{1 + \cot^2 x}{\cot^2 x} = \sec^2 x$$

9. Solve the following equation for $0 \leq y < 2\pi$: (5)

$$6 \sin^2 A - \sin A = 1$$

10. Verify the identity (3)

$$\csc^2 \frac{A}{2} = \frac{2 \sec A}{\sec A - 1}$$

P 11. The angle A of a projectile shot at 25 m/s with a range of 625 m is given by 625 $\tan A = 306.25 \sec^2 A$. Solve for A. (5)

12. Reduce the following to a single term: (4)

$$\sin \frac{A}{4} \cos \frac{A}{4}$$

13. Verify the identity (1)

$$\frac{\cot x \cos x}{\cot x - \cos x} = \frac{\cos x}{1 - \sin x}$$

14. Verify the following identity: (3)

$$\frac{\cot^2 x - 1}{2 \cot x} = \cot 2x$$

P 15. A spring vibrating in harmonic motion is subjected to two forces. The motion of each is described as (2)

$$y_1 = A_1 \sin \left(x + \frac{\pi}{4}\right) \qquad \text{and} \qquad y_2 = A_2 \cos \left(x + \frac{\pi}{4}\right)$$

Find a simplified expression for the resultant, $y_1 + y_2$.

GROUP ACTIVITY

Harriet's **E**ver-ready **R**easonable **O**rganization performs several different tasks for different customers. One customer required the accurate calculation of the area of stainless-steel triangles. The extremely accurate measurement of angles was found to be too difficult for Harriet's organization. You have learned of her difficulties and offer your services to develop a formula for the area of a triangle that uses only the length of its three sides, which can be measured with a high degree of accuracy.

1. If (x, y) is any point on the unit circle, then by using the Pythagorean theorem (see Chapter 5) show that $x^2 + y^2 = \mathbf{1}$.

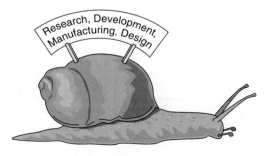

Or be a HERO and use
Harriet's Ever-ready Reasonable Organization!

2. Using the definitions for sin θ and cos θ in terms of x and y, show that $\sin^2 \theta = 1 - \cos^2 \theta$.

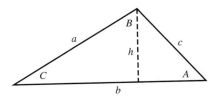

3. Discuss how to demonstrate that the area of any triangle with sides a, b, c and angles A, B, C is given by:

$$\text{area} = \tfrac{1}{2} ab \sin C$$

4. Using the equation for the area of a triangle, show that the area may also be written in terms of only the triangle sides a, b, and c. In other words, show:

$$\text{area} = \sqrt{s(s - a)(s - b)(s - c)}$$

where $s = \tfrac{1}{2}(a + b + c)$. (*Hint:* You will also need to use the law of cosines.) The area of a triangle in terms of its three sides first appeared around 200 A.D. in the writings of Hero of Alexandra.

SOLUTION TO THE CHAPTER INTRODUCTION

The easiest way to determine the distance from the painting that gives the maximum angle subtended by the lens is to use the graphing calculator to graph the function and TRACE to find the maximum. The graph is shown in Figure 13–21. Note that the angle mode is set to degrees. The maximum angle is approximately 42° and the distance is approximately 2.2 ft.

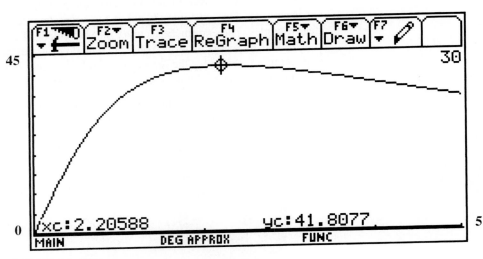

FIGURE 13–21

Y ou are an electrical engineer responsible for designing a tuning circuit for a radio station. The circuit is to contain a variable capacitor, a 0.0020-H coil, and a 20-Ω resistor. You must calculate the capacitance necessary to tune the station to a broadcast frequency of 101 MHz. (The solution to this problem is given at the end of the chapter.)

Complex numbers, discussed in this chapter, apply to electronics. For example, one of these applications is resonant frequency as it relates to electrical systems. However, the concept of resonance also relates to mechanical systems. If mechanical impulses are applied to a system at the proper frequency, the system will resonate or vibrate. A historical note of such an occurrence is the collapse of the Tacoma Narrows Bridge in 1940 due to oscillations caused by a 42-mi/h wind. The height of the oscillations or vibrations increased to the point that the main span of the bridge broke apart and fell into Puget Sound.

Learning Objectives

After you complete this chapter, you should be able to:

1. Add, subtract, multiply, and divide imaginary numbers (Section 14–1).

2. Simplify powers of the imaginary number j (Section 14–1).

3. Add, subtract, multiply, and divide complex numbers in rectangular form (Section 14–2).

4. Convert complex numbers among rectangular, polar, and exponential forms (Section 14–3).

5. Use polar and exponential forms to find the product, quotient, power, and roots of complex numbers (Section 14–4).

6. Apply the concept of complex numbers to ac circuits (Section 14–5).

Chapter 14

Complex Numbers

14–1

INTRODUCTION TO COMPLEX NUMBERS

We briefly discussed complex numbers in Chapter 3 when we took the square root of a negative number. We also used complex numbers in solving quadratic equations in Chapter 5. In this chapter we will discuss operations with complex numbers in rectangular, polar, and exponential forms and applications of complex numbers to electronics.

A **complex number** is *any number that can be written as a + bj where a and b represent real numbers.* In mathematics, complex numbers are represented as $a + bi$. Since i is used to represent electrical current $a + bj$ is used in engineering applications. Because we will discuss applications of complex numbers to engineering, we will use the $a + bj$ representation.

The **real part** of the complex number is *represented by the letter a*, and the **imaginary part** is *represented by bj*. A number *consisting only of the imaginary part* is called a **pure imaginary number**. A complex number *written as a + bj* is said to be in **rectangular form**. Complex numbers can also be expressed in polar and exponential forms, but our discussion in the next two sections will use the rectangular form.

Rectangular Form of a Complex Number

$a + bj$ where a, b are real numbers

Operations with Imaginary Numbers

We perform algebraic operations with imaginary numbers in much the same way as in previous chapters with algebraic expressions. We only need to convert to j notation before performing the algebraic operation.

EXAMPLE 1 Perform the following indicated operation:

$$(\sqrt{-6})^2$$

Solution Before squaring this number, we must write the expression in terms of j.

$$\sqrt{-6} = \sqrt{-1 \cdot 6} = \sqrt{-1} \cdot \sqrt{6} = j\sqrt{6}$$

Therefore, this expression can be written as follows:

$$(j\sqrt{6})^2$$

NOTE ✦ The j is written in front of the radical to avoid confusion about whether the j is under the radical or not.

Once we have the expression in terms of j, we square it just as we have done previously.

$$(j\sqrt{6})^2 = (j^2)(\sqrt{6})^2 = j^2(6)$$

However, $j^2 = -1$. Therefore,

$$(\sqrt{-6})^2 = -6 \qquad\blacksquare\blacksquare\blacksquare$$

CAUTION ✦ If a number is expressed as the square root of a negative, always write the number in terms of j before performing any algebraic operation.

EXAMPLE 2 Multiply the following complex numbers:

$$\sqrt{-2} \cdot \sqrt{-32}$$

Solution Before multiplying the two radicals, we must express each negative number under the radical in terms of j.

$$\sqrt{-2} \cdot \sqrt{-32}$$
$$j\sqrt{2} \cdot j\sqrt{32}$$

Then we multiply and simplify the result.

$$j^2\sqrt{64}$$
$$(-1)(8)$$
$$-8$$

■■■

EXAMPLE 3   Write $\sqrt{-27}$ in simplified form.

Solution   First, let us write the radical in terms of j as follows:

$$j\sqrt{27}$$

Second, we simplify the radical by removing $\sqrt{9} = 3$. Thus, the simplified form is

$$3j\sqrt{3}$$

■■■

Powers of j

Many problems involve powers of j greater than two, making simplification of higher powers of j necessary. Powers of j are cyclic, meaning they repeat in a fixed pattern. Look at the following powers of j and note the pattern of the answers:

$$j^1 = j \qquad\qquad j^5 = j^4 \cdot j = j$$
$$j^2 = -1 \qquad\qquad j^6 = j^4 \cdot j^2 = -1$$
$$j^3 = j^2 \cdot j = -j \qquad\qquad j^7 = j^6 \cdot j = -j$$
$$j^4 = j^2 \cdot j^2 = 1 \qquad\qquad j^8 = j^4 \cdot j^4 = 1$$

Notice that the powers of j repeat in the sequence j, -1, $-j$, and 1.

Simplifying Powers of j

To simplify a power of j, perform the following steps:

1. Because powers of j have a cycle of 4, divide the exponent by 4.
2. Pair the remainder from Step 1 with the corresponding exponent below, and find the resulting value.

$$j^0 = 1$$
$$j^1 = j$$
$$j^2 = -1$$
$$j^3 = -j$$

CAUTION ✦ In simplifying any expression involving j, never leave a power of j greater than one.

EXAMPLE 4 Simplify j^{43}.

Solution We *divide 43 by 4* and obtain a remainder of 3. The preceding box tells us that $j^3 = -j$. Therefore,

$$j^{43} = -j$$

■ ■ ■

EXAMPLE 5 Simplify $6j^{24} - 5j^{18}$.

Solution First, we simplify the power of j, and then we subtract, if possible.

$$6j^{24} - 5j^{18}$$
$$6j^0 - 5j^2$$
$$6(1) - 5(-1)$$
$$6 + 5$$
$$11$$

■ ■ ■

Equality of Complex Numbers

Two complex numbers are equal if their real parts are equal and their imaginary parts are equal. For example, $3 + 8j$ is equal to $x + yj$ only if $x = 3$ and $y = 8$. We can use this principle of equality to solve equations with complex numbers by equating real parts and by equating imaginary parts. The next two examples illustrate this process.

EXAMPLE 6 Determine x and y such that $2x + 5yj = 10 + 15j$.

Solution According to the definition of equality, these two complex numbers are equal if

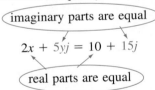

Therefore,

$$2x = 10 \quad \text{and} \quad 5yj = 15j$$

Thus,

$$x = 5 \quad \text{and} \quad 5y = 15$$
$$y = 3$$

■ ■ ■

EXAMPLE 7 Solve the following for x and y:

$$(3 + x) - 5j = 2x + (4y + 3)j$$

Solution To determine the value of x and y, we must be able to equate the real parts and the imaginary parts of these complex numbers. Therefore, we must be able to separate the two complex numbers into their real and imaginary parts.

$$(3 + x) - 5j = 2x + (4y + 3)j$$

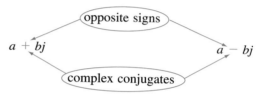

Equating real parts and imaginary parts and solving for x and y gives

$$3 + x = 2x \qquad \text{and} \qquad -5j = (4y + 3)j$$
$$3 = x \qquad\qquad\qquad -5 = 4y + 3$$
$$-5 - 3 = 4y$$
$$-2 = y$$

∎∎∎

Complex Conjugates

The **conjugate** of the complex number $a + bj$ is the complex number $a - bj$. To determine the complex conjugate of a number, *reverse the sign of the imaginary part.*

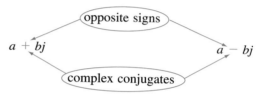

EXAMPLE 8   Determine the complex conjugate of the following complex numbers:

(a) $3 - 7j$ (b) $-5 + 11j$ (c) $8j$ (d) -5

Solution

(a) We obtain the complex conjugate of $3 - 7j$ by reversing the sign of the imaginary part: $3 + 7j$.
(b) To determine the complex conjugate of $-5 + 11j$, we reverse the sign of the imaginary part to obtain $-5 - 11j$.
(c) To find the complex conjugate of $8j$, first, we write the number in $a + bj$ form $(0 + 8j)$ and then reverse the sign of the imaginary part: $0 - 8j$. Thus, the complex conjugate of $8j$ is $-8j$.
(d) Similarly, if we write -5 in $a + bj$ form, we have $-5 + 0j$. Since the imaginary part is zero, reversing its sign does not affect the number. Therefore, the complex conjugate of -5 is -5.

∎∎∎

CAUTION ✦ Be sure to reverse only the sign of the imaginary part to obtain a complex conjugate.

14–1 EXERCISES

Write each expression in terms of j.

1. $\sqrt{-64}$

2. $\sqrt{-36}$

3. $\sqrt{-8}$

4. $\sqrt{-24}$

5. $\sqrt{-72}$

6. $-\sqrt{-0.16}$

7. $-\sqrt{-40}$

8. $\sqrt{-16/25}$

9. $\sqrt{-8/49}$

Perform the indicated operation and simplify.

10. $\sqrt{-8} \cdot \sqrt{-4}$

11. $\sqrt{-3} \cdot \sqrt{-12}$

12. $\sqrt{-5} \cdot \sqrt{-8}$

13. $\sqrt{-9} \cdot \sqrt{6}$

14. $\sqrt{-20} \cdot \sqrt{5}$

15. $(3\sqrt{-6})^2$

16. $(7\sqrt{-3})^2$

17. $-2\sqrt{-16}$

Simplify the following powers of j.

18. j^{15}

19. j^{36}

20. j^{49}

21. j^{127}

22. j^{86}

23. j^{74}

24. j^{264}

25. j^{246}

26. $j^{14} + 3j^{39}$

27. $7j^{56} - j^{34}$

28. $6j^{60} - 7j^{82}$

29. $8j^{59} + 5j^{97}$

Solve for x and y.

30. $2x + 3yj = 10 + 18j$

31. $7 + 6yj = 14x - 42j$

32. $4j - 10 = 2x + yj - 8$

33. $10 + 3yj + 8j = 2(x + 11j)$

34. $3(x + 5j) = 7yj - 6 + 8yj$

35. $19 - 8yj + 2x = 15 - 2yj + 18j$

36. $x + 7yj - 4 = 5 + 21j + 2x$

37. $2x + 10j - 4yj = 8j - 16 + 2yj$

38. $4(x - yj) = 8 + 2yj + 10j$

39. $7x - 5yj + 18j = 16 - 3yj - 4j + 9x$

Determine the conjugate of the following complex numbers.

40. $2 + 7j$

41. $8 - 5j$

42. $-5 - 11j$

43. $-7 + 9j$

44. $-3j$

45. $-2j$

46. 8

47. -10

48. What type of number results from (a) the sum and (b) the product of a complex number and its conjugate?

E 49. The impedance Z in a resistance-inductance-capacitance *(RLC)* series ac circuit is given by

$$Z = R + jX_L - jX_C$$

Determine an expression for the impedance if $R = 44\ \Omega$, $X_L = 36\ \Omega$, and $X_C = 50\ \Omega$.

CS 50. Computer scientists often use the modulo operator to solve various problems in mathematics. Modular arithmetic is denoted by x mod y, which is equal to the remainder when x is divided by y. Knowing this, use the modulo operator to restate the algorithm for simplifying a power of j.

14–2

OPERATIONS IN RECTANGULAR FORM

In this section we will discuss addition, subtraction, multiplication, and division of complex numbers in a rectangular form, $a + bj$. Remember that you must write the imaginary part of a complex number in terms of j before performing any operation. Also, remember that the answer should not contain any power of j greater than 1.

Addition and Subtraction

To add or subtract complex numbers, combine the real parts and combine the imaginary parts.

EXAMPLE 1 Perform the following indicated operations:

(a) $(-3 + 5j) + (6 - 7j)$

(b) $(-10 + \sqrt{-64}) - (-3 - \sqrt{-49})$

Solution

(a) To add these complex numbers, we combine the real parts, -3 and 6, and we combine the imaginary parts, $5j$ and $-7j$.

real parts / imaginary parts

$$(-3 + 5j) + (6 - 7j) = \overline{(-3 + 6)} + \overline{(5j - 7j)}$$
$$= 3 - 2j$$

(b) First, we must change the complex numbers to rectangular $a + bj$ form:

$$(-10 + \sqrt{-64}) - (-3 - \sqrt{-49}) = (-10 + j\sqrt{64}) - (-3 - j\sqrt{49})$$
$$= (-10 + 8j) - (-3 - 7j)$$

Subtracting the two complex numbers gives

$$(-10 + 8j) - (-3 - 7j) = -10 + 8j + 3 + 7j$$
$$= (-10 + 3) + (8j + 7j)$$
$$= -7 + 15j$$

■ ■ ■

Graphical Solution

Addition and subtraction of complex numbers in rectangular form can also be done graphically. Complex numbers are graphed in the **complex plane** where the horizontal axis represents the real part of the complex number and the vertical axis represents the imaginary part. The complex number $8 - 6j$ is represented by the ordered pair $(8, -6)$. The graph of $8 - 6j$ is shown in Figure 14–1.

FIGURE 14–1

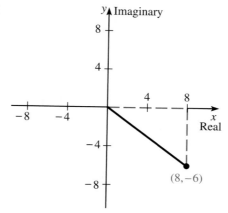

Graphical addition and subtraction of complex numbers involve the same technique as graphically finding the resultant of vectors (shown in Chapter 11). The next two examples illustrate this process.

EXAMPLE 2   Find $(-2 + 3j) + (5 + 4j)$ graphically.

Solution   First, we represent graphically the complex numbers $-2 + 3j$ and $5 + 4j$ as the ordered pairs $(-2, 3)$ and $(5, 4)$, respectively. Then we complete a parallelogram and read the ordered pair for the resultant as shown in Figure 14–2. The ordered pair for the resultant is $(3, 7)$, which is equivalent to $3 + 7j$. You can check this result by adding the two complex numbers algebraically.

$$(-2 + 3j) + (5 + 4j) = (-2 + 5) + (3 + 4)j = 3 + 7j$$

FIGURE 14–2

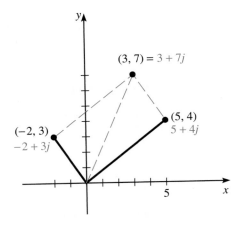

EXAMPLE 3   Subtract $6 - 5j$ from $4 + 9j$ graphically.

Solution   Subtracting $6 - 5j$ is equivalent to adding $-6 + 5j$. Following the same procedure as in Example 2, we add $4 + 9j$ and $-6 + 5j$. The ordered pair for the difference is $(-2, 14)$, which is equivalent to the complex number $-2 + 14j$, as shown in Figure 14–3.

FIGURE 14–3

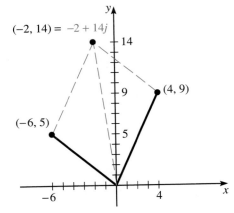

Multiplication

After converting to $a + bj$ form, we multiply complex numbers using the same techniques used to multiply polynomials, such as the distributive property or FOIL. Then we simplify powers of j and combine like terms.

EXAMPLE 4   Multiply the following:

(a) $(3 + 5j)(8 - 7j)$ (b) $(2 + 3\sqrt{-3})(5 - \sqrt{-3})$

Solution

(a) Since the numbers are written in $a + bj$ form, we *multiply the binomials* using FOIL.

$$(3 + 5j)(8 - 7j) = 24 - 21j + 40j - 35j^2$$

Making the substitution $j^2 = -1$ and combining like terms gives

$$(3 + 5j)(8 - 7j) = 24 - 21j + 40j - 35(-1)$$
$$(3 + 5j)(8 - 7j) = 24 - 21j + 40j + 35$$
$$(3 + 5j)(8 - 7j) = 59 + 19j$$

(b) First, we must convert the complex numbers to $a + bj$ form.

$$(2 + 3\sqrt{-3})(5 - \sqrt{-3}) = (2 + 3j\sqrt{3})(5 - j\sqrt{3})$$

Next, we multiply the binomials using FOIL.

$$(2 + 3j\sqrt{3})(5 - j\sqrt{3}) = 10 - 2j\sqrt{3} + 15j\sqrt{3} - 3j^2\sqrt{9}$$

Last, we substitute $j^2 = -1$, simplify radicals, and combine like terms.

$$(2 + 3j\sqrt{3})(5 - j\sqrt{3}) = 10 - 2j\sqrt{3} + 15j\sqrt{3} - 3(-1)(3)$$
$$(2 + 3j\sqrt{3})(5 - j\sqrt{3}) = 10 - 2j\sqrt{3} + 15j\sqrt{3} + 9$$
$$(2 + 3j\sqrt{3})(5 - j\sqrt{3}) = (10 + 9) + (-2j + 15j)\sqrt{3}$$
$$(2 + 3j\sqrt{3})(5 - j\sqrt{3}) = 19 + 13j\sqrt{3}$$

■ ■ ■

Division

Dividing complex numbers is similar to dividing radical expressions (Chapter 3) because j is a radical. Divide complex numbers by rationalizing the denominator. If you remember that $j = \sqrt{-1}$, you can understand why rationalizing the denominator is necessary.

EXAMPLE 5   Divide the following complex numbers:

(a) $\dfrac{4 + \sqrt{-6}}{\sqrt{-1}}$ (b) $\dfrac{3 - 2j}{5 + 7j}$

Solution

(a) First, we *convert the complex number to $a + bj$ form.*

$$\frac{4 + \sqrt{-6}}{\sqrt{-1}} = \frac{4 + j\sqrt{6}}{j}$$

Then we *rationalize the denominator* by multiplying the numerator and denominator by the conjugate of j, which is $-j$.

$$\frac{4 + \sqrt{-6}}{\sqrt{-1}} = \frac{4 + j\sqrt{6}}{j} \cdot \frac{-j}{-j}$$

$$\frac{4 + \sqrt{-6}}{\sqrt{-1}} = \frac{-4j - j^2\sqrt{6}}{-j^2}$$

Finally, we substitute $j^2 = -1$ and *simplify* the result.

$$\frac{4 + \sqrt{-6}}{\sqrt{-1}} = \frac{-4j - (-1)\sqrt{6}}{-(-1)}$$

$$\frac{4 + \sqrt{-6}}{\sqrt{-1}} = \frac{\sqrt{6} - 4j}{1}$$

$$\frac{4 + \sqrt{-6}}{\sqrt{-1}} = \sqrt{6} - 4j$$

(b) Since this complex number is given in $a + bj$ form, we multiply the numerator and denominator by the conjugate of $5 + 7j$, which is $5 - 7j$.

$$\frac{3 - 2j}{5 + 7j} = \frac{(3 - 2j)}{(5 + 7j)} \cdot \frac{(5 - 7j)}{(5 - 7j)}$$

Multiplying the numerators and denominators using FOIL gives

$$\frac{3 - 2j}{5 + 7j} = \frac{15 - 21j - 10j + 14j^2}{25 - 35j + 35j - 49j^2}$$

Substituting $j^2 = -1$ and combining like terms gives

$$\frac{3 - 2j}{5 + 7j} = \frac{15 - 21j - 10j + 14(-1)}{25 - 49(-1)}$$

$$\frac{3 - 2j}{5 + 7j} = \frac{15 - 21j - 10j - 14}{25 + 49}$$

$$\frac{3 - 2j}{5 + 7j} = \frac{1 - 31j}{74}$$

■■■

Graphing Calculator

The graphing calculator can be used to perform computations with complex numbers. On the TI-92 press [MODE], go down to Complex Format, choose Rectangular, and press [ENTER]. This allows you to enter complex numbers in rectangular form, $a + bj$. Note that the calculator will use $a + bi$ as the form of the complex number. The calculator keystrokes for addition, multiplication, and division of complex numbers are shown in Example 6 using the information given in Examples 3, 4, and 5.

EXAMPLE 6 Use the graphing calculator to perform the following operations.

(a) $(4 + 9j) - (6 - 5j)$ (b) $(3 + 5j)(8 - 7j)$ (c) $\dfrac{3 - 2j}{5 + 7j}$

Solution

(a) Press $(4 + 9i) - (6 - 5i)$ and [ENTER].
    Note that i is used on the calculator rather than j.

FIGURE 14–4

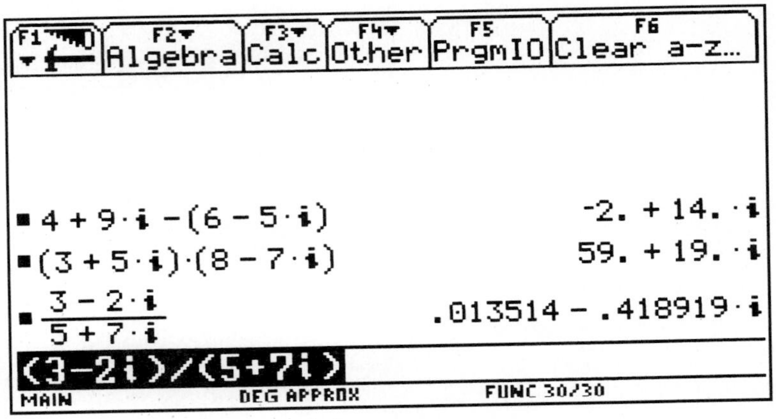

(b) Press $(3 + 5i)(8 - 7i)$ and ENTER.

(c) Press $(3 - 2i) \div (5 + 7i)$ and ENTER.

Figure 14–4 shows a screen print of the results.   ■■■

Application

When alternating current, ac, passes through a series *RLC* (resistance, inductance, capacitance) circuit, the magnitude of the current is determined by the total opposition to its flow represented by the total resistance, reactance of the inductor, and reactance of the capacitor. This opposition is called impedance. Impedance, represented by *Z*, is a vector quantity that can be written as a complex number. Ohm's law may be stated as $Z = V/I$ ohms.

EXAMPLE 7 Determine the impedance in an ac series circuit with a current of $5 + 0j$ A and source voltage of $V = 11.2 - 4.1j$ V.

Solution Since $Z = V/I$, we can calculate the impedance as follows.

$$Z = \frac{V}{I}$$

$$Z = \frac{11.2 - 4.1j}{5 + 0j}$$

$$Z = 2.2 - 0.8j \ \Omega$$

■■■

14–2 EXERCISES

Perform the indicated operations. Leave your answer in $a + bj$ form. You may use a graphing calculator to check your answers.

1. $(7 - 5j) + (-4 - 2j)$
2. $(8 - 16j) + (7 + 3j)$
3. $(-8 + 11j) - (6 - 5j)$
4. $(14 + 8j) + (17 - 9j)$
5. $(7 - 5j) - (10 - 13j)$
6. $(-12 + j) - (19j + 6)$
7. $(3 + 6j) - (8j - 5) + (-2 + 3j)$
8. $(2 + j) + (7j - 8) - (16 + 5j)$
9. $(1 - 4j) + (-9 + 4j) - (15 - 9j)$
10. $(j - 2) + (-5 + 7j) + (8j - 13)$
11. $(4 - 5j)(3 + 4j)$
12. $(\sqrt{-4} + 8) - (16 + \sqrt{-25})$
13. $(-5 + \sqrt{-36}) - (3 + \sqrt{-9})$
14. $(-9 + \sqrt{-49}) + (16 - \sqrt{-100})$
15. $(9 + 7j)(3 - 2j)$
16. $(18 - \sqrt{-36}) - (\sqrt{-64} + 10)$
17. $\sqrt{-9}(\sqrt{-4} + \sqrt{-81})$
18. $\sqrt{-64}(\sqrt{25} - \sqrt{-81})$
19. $5j(3 - 6j)$
20. $-3j(7 + 10j)$

21. $(8 + 7j)(3 + 2j)$
22. $(4 + \sqrt{-1})(\sqrt{-9} + 6)$
23. $(\sqrt{-25} + 11)(3 + \sqrt{-64})$
24. $(\sqrt{-16} + 5)(9 - \sqrt{-25})$
25. $(8 - 3j)(7 + 4j)$
26. $(8 - 7j)(4 - 3j)$
27. $(2 + 5j)^2$
28. $(7 - 3j)^2$
29. $(7 - 2j)(7 + 2j)$
30. $(8 - 5j)(8 + 5j)$
31. $(-6 - 3j)^2$
32. $(-7 + 9j)^2$
33. $(2 - 3j)^3$
34. $(1 - 2j)^3$

35. $\dfrac{6 + 7j}{2j}$
36. $\dfrac{5 - 6j}{4j}$

37. $\dfrac{3 - \sqrt{-4}}{\sqrt{-9}}$
38. $\dfrac{\sqrt{-9} + 6}{\sqrt{-16}}$

39. $\dfrac{2 + \sqrt{-36}}{\sqrt{-25}}$
40. $\dfrac{9 + \sqrt{-25}}{\sqrt{-4}}$

41. $\dfrac{8 - 5j}{j}$
42. $\dfrac{9 + 11j}{2j}$

43. $\dfrac{3 - 2j}{1 + 3j}$
44. $\dfrac{6 - 10j}{j + 5}$

45. $\dfrac{-3 + 9j}{2 - 5j}$

46. $\dfrac{-8 + 6j}{3j - 10}$

47. $\dfrac{5 - 4j}{7 - 2j}$

48. $\dfrac{9j - 5}{4j - 1}$

49. $\dfrac{-8 + 5j}{6 - j}$

50. $\dfrac{1 + j}{1 - j}$

Perform the indicated operations graphically.

51. $(-3 + 2j) + (-5j - 8)$

52. $(3 + 5j) + (-6 + 4j)$

53. $(4j - 3) - (7 + 2j)$

54. $(7 - 8j) - (-4 - 3j)$

55. $(5 + 6j) + (5 - 6j)$

56. $(-2 + 5j) + (-2 - 5j)$

E 57. Determine the impedance in an ac series circuit with a current $2 + 0j$ A and source voltage of $87 + 50j$ V.

E 58. Determine the impedance in an ac series circuit with current of $60 + 0j$ mA and an effective voltage of $4.5 - 7.8j$ V.

E 59. Determine the impedance in an ac series circuit with a current of $7.8 + 0j$ mA and an effective voltage of $0 - 110j$ V.

14–3

POLAR FORM AND EXPONENTIAL FORM

In the previous two sections, we discussed complex numbers written in rectangular form: $a + bj$. However, in this section we will discuss two additional forms of complex numbers: polar and exponential. Certain operations, such as finding roots, are performed more easily when the complex number is written in exponential or polar form.

Polar Form

In the previous section we briefly discussed graphing complex numbers. If we take the complex number $x + yj$, plot it on the complex plane, and drop a perpendicular line to the real axis, we obtain a right triangle as shown in Figure 14–5. Recall that this procedure is the same one used in Chapter 11 for vectors. By forming the right triangle, we have the hypotenuse of the triangle, labeled r, and an angle θ in standard position. From our previous knowledge of trigonometry, we know that

$$x = r \cos \theta \qquad \tan \theta = \frac{y}{x}$$

$$y = r \sin \theta \qquad r = \sqrt{x^2 + y^2}$$

FIGURE 14–5

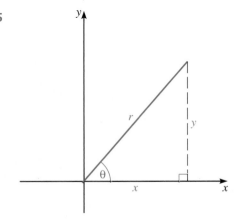

If we take the rectangular form of the complex number $x + yj$, substitute for x and y, and simplify, we obtain the polar form of the complex number.

$$x + yj = (r \cos \theta) + (r \sin \theta)j \qquad \text{substitute}$$
$$= r(\cos \theta + j \sin \theta) \qquad \text{factor}$$

NOTE ✦ The j is placed in front of $\sin \theta$ so as not to confuse what is the angular measure. It is j times $\sin \theta$, not sine of the product of θ and j.

Polar Form of a Complex Number

The polar form of the complex number $x + yj$ is

$$r(\cos \theta + j \sin \theta)$$

where

$$r = \sqrt{x^2 + y^2} \qquad\qquad x = r \cos \theta$$

$$\tan \theta = \frac{y}{x} \qquad\qquad y = r \sin \theta$$

Another notation for polar form is $r\angle\theta$. The length r is called the **magnitude** of the complex number, and the angle θ is called the **argument.**

⊞ **EXAMPLE 1**   Represent $6 + 7j$ graphically and determine its polar form.

Solution   We plot the ordered pair $(6, 7)$ on the complex plane and draw a vector from the origin to the point as shown in Figure 14–6. To convert $6 + 7j$ to polar form, we must determine r and θ.

$$r = \sqrt{x^2 + y^2} \qquad\qquad \tan \theta = \frac{y}{x}$$

$$r = \sqrt{36 + 49} \qquad\qquad \tan \theta = \frac{7}{6}$$

$$r \approx 9.22 \qquad\qquad \theta \approx 49.4°$$

FIGURE 14–6

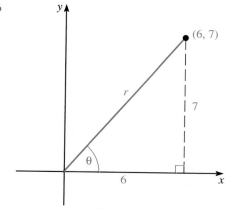

The polar form of $6 + 7j$ is $9.22(\cos 49.4° + j \sin 49.4°)$, or $9.22\angle 49.4°$.

$$6\;\boxed{x^2}\;\boxed{+}\;7\;\boxed{x^2}\;\boxed{=}\;\boxed{\sqrt{\;\;}} \longrightarrow 9.22$$
$$7\;\boxed{\div}\;6\;\boxed{=}\;\boxed{\text{inv}}\;\boxed{\text{tan}} \longrightarrow 49.4$$

■ ■ ■

NOTE ✦ Be sure that your calculator is in the degree mode.

⊞ **EXAMPLE 2** Represent $-5 - 8j$ graphically and find its polar form.

Solution To represent $-5 - 8j$ graphically, plot the ordered pair $(-5, -8)$ and draw a vector from the origin to the point as shown in Figure 14–7. To find its polar form, we must determine r and θ. We can find r just as we did in Example 1; however, we should find a reference angle θ_R to determine θ because it is not a first quadrant angle.

$$r = \sqrt{x^2 + y^2}$$
$$r = \sqrt{(-5)^2 + (-8)^2} \qquad\qquad \tan \theta_R = \frac{|-8|}{|-5|}$$
$$r = \sqrt{89} \qquad\qquad\qquad \theta_R \approx 57.99°$$
$$r \approx 9.43 \qquad\qquad \theta = 180° + \theta_R \approx 237.99°$$

The polar form of $-5 - 8j$ is $9.43(\cos 237.99° + j \sin 237.99°)$, or $9.43\angle 237.99°$.

$$5\;\boxed{+/-}\;\boxed{x^2}\;\boxed{+}\;8\;\boxed{+/-}\;\boxed{x^2}\;\boxed{=}\;\boxed{\sqrt{\;\;}} \longrightarrow 9.43$$
$$8\;\boxed{\div}\;5\;\boxed{=}\;\boxed{\text{inv}}\;\boxed{\text{tan}}\;57.99\;\boxed{+}\;180\;\boxed{=} \longrightarrow 237.99$$

FIGURE 14–7

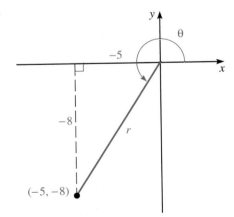

■ ■ ■

CAUTION ✦ Do not rely on your calculator to determine the value of θ. Calculate the reference angle and then determine θ according to the quadrant in which the complex number lies.

⊞ **EXAMPLE 3** Represent $7.8(\cos 136.8° + j \sin 136.8°)$ graphically and express the complex number in rectangular form.

Solution   The graph of 7.8(cos 136.8° + j sin 136.8°) is given in Figure 14–8, where 7.8 is the length of the vector from the origin and 136.8° is the angle this vector makes with the positive x axis. To convert this complex number from polar to rectangular form, we must determine x and y. We know that $r = 7.8$ and $\theta = 136.8°$. Thus,

$$x = r \cos \theta \qquad \text{and} \qquad y = r \sin \theta$$
$$x = 7.8 \cos 136.8° \qquad\qquad\qquad y = 7.8 \sin 136.8°$$
$$x \approx -5.7 \qquad\qquad\qquad\qquad y \approx 5.3$$

7.8 $\boxed{\times}$ 136.8 $\boxed{\cos}$ $\boxed{=}$ $\longrightarrow$ -5.69

7.8 $\boxed{\times}$ 136.8 $\boxed{\sin}$ $\boxed{=}$ $\longrightarrow$ 5.34

FIGURE 14–8

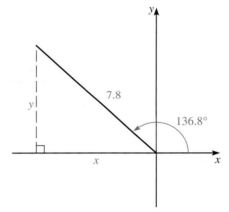

The rectangular form of 7.8 (cos 136.8° + j sin 136.8°) is $-5.7 + 5.3j$. From Figure 14–8 we can verify that this answer is reasonable because the point $(-5.7, 5.3)$ would be placed in quadrant II. ∎∎∎

Application

In a series ac circuit consisting of a resistor, an inductor, and a capacitor, the impedance Z is given by

$$Z = R + jX_L - jX_C$$

where X_L is the inductive reactance and X_C is the capacitive reactance. In constructing an impedance diagram, resistance is plotted along the positive real axis, inductive reactance is plotted along the positive imaginary axis, and capacitive reactance is plotted along the negative imaginary axis.

EXAMPLE 4   For a resistance of 15 Ω and a capacitive reactance of 10 Ω, express the impedance in rectangular form, polar form, and as an impedance diagram.

Solution $R = 15$ Ω and $X_C = 10$ Ω. The impedance in rectangular form is

$$Z = 15 - 10j \ \Omega$$

The polar form is given by

$$r = \sqrt{(15)^2 + (-10)^2}$$
$$r = \sqrt{325}$$
$$r \approx 18.0$$
$$\theta = \tan^{-1}\left(\frac{-10}{15}\right)$$
$$\theta = -33.7°$$

The polar form is $18\angle -34°$. The impedance diagram is shown in Figure 14–9.　■■■

FIGURE 14–9

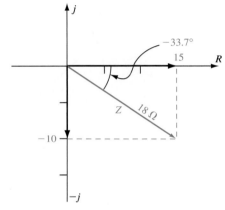

Exponential Form

The exponential form of a complex number $x + yj$ is given below.

Exponential Form of a Complex Number

The exponential form of a complex number $x + yj$ is

$$re^{j\theta}$$

where θ is in radians and

$$r = \sqrt{x^2 + y^2} \qquad \tan \theta = \frac{y}{x}$$

Recall that e is the base for natural logarithms, discussed in Chapter 8. To convert a complex number in polar form to exponential form, convert θ from degrees to radians.

EXAMPLE 5 Convert $3 - 2j$ to polar and exponential forms.

Solution To convert $3 - 2j$ into polar form, we must determine r and θ. Once again, we should find a reference angle θ_R and then calculate θ.

$$r = \sqrt{(3)^2 + (-2)^2} \qquad \tan \theta_R = \frac{2}{3}$$

$$r \approx 3.61 \qquad\qquad \theta_R \approx 33.69°$$

$$\theta \approx 326.31°$$

$$3 \boxed{x^2} \boxed{+} 2 \boxed{+/-} \boxed{x^2} \boxed{=} \boxed{\sqrt{}} \longrightarrow 3.61$$

$$2 \boxed{\div} 3 \boxed{=} \boxed{\text{inv}} \boxed{\tan} \; 33.69$$

$$\boxed{\text{STO}} \; 360 \boxed{-} \boxed{\text{RCL}} \boxed{=} \longrightarrow 326.31$$

The polar form of $3 - 2j$ is

$$3.61(\cos 326.31° + j \sin 326.31°)$$

To find the exponential form, we convert θ from degrees to radians ($326.31° \approx 5.70$ radians). The exponential form of $3 - 2j$ is

$$3.61e^{5.70j} \qquad\qquad\qquad ■■■$$

CAUTION ✦ Be sure that your angle is in radians for exponential form.

EXAMPLE 6 Convert $5.83e^{4.76j}$ to polar and rectangular forms.

Solution From $5.83e^{4.76j}$, we know that $r = 5.83$ and $\theta = 4.76$ radians. To convert from exponential form to polar form, we must determine $r(\cos \theta + j \sin \theta)$, where θ is in degrees. Since $r = 5.83$, we must convert 4.76 radians to 272.73°. Some calculators have a button to convert between radians and degrees. The polar form of $5.83e^{4.76j}$ is

$$5.83(\cos 272.73° + j \sin 272.73°)$$

To convert to rectangular form, we calculate $x = r \cos \theta$ and $y = r \sin \theta$. The rectangular form of $5.83e^{4.76j}$ is

$$x = 5.83 \cos 272.73° \qquad \text{and} \qquad y = 5.83 \sin 272.73°$$

$$0.277 - 5.823j$$

$$5.83 \boxed{\times} 272.73 \boxed{\cos} \boxed{=} \longrightarrow 0.2777$$

$$5.83 \boxed{\times} 272.73 \boxed{\sin} \boxed{=} \longrightarrow -5.823 \qquad ■■■$$

EXAMPLE 7 In an ac circuit, the current is represented by the complex number $2.5 - 4.8j$ amperes. Write this number in polar and exponential forms, and determine the magnitude of the current.

Solution First, convert from rectangular to polar form. For polar form, we must determine r and θ in degrees.

$$r = \sqrt{x^2 + y^2} \qquad\qquad \tan\theta_R = \left|\frac{y}{x}\right| = \frac{4.8}{2.5}$$

$$r = \sqrt{(2.5)^2 + (-4.8)^2} \qquad\qquad \theta_R \approx 62.5°$$
$$r \approx 5.4 \qquad\qquad\qquad \theta \approx 297.5° \quad \text{(quadrant IV)}$$

2.5 $\boxed{x^2}$ $\boxed{+}$ 4.8 $\boxed{+/-}$ $\boxed{x^2}$ $\boxed{=}$ $\boxed{\sqrt{\ \ }}$ $\longrightarrow$ 5.4

4.8 $\boxed{\div}$ 2.5 $\boxed{=}$ $\boxed{\text{inv}}$ $\boxed{\text{tan}}$ 62.5

$\boxed{\text{STO}}$ 360 $\boxed{-}$ $\boxed{\text{RCL}}$ $\boxed{=}$ $\longrightarrow$ 297.5

The polar form of $2.5 - 4.8j$ is

$$5.4(\cos 297.5° + j \sin 297.5°)$$

To obtain the exponential form of this complex number, we convert the angle in polar
form from degrees to radians. The exponential form of $2.5 - 4.8j$ is

$$5.4e^{5.2j}$$

The magnitude of the current is determined by r, which is 5.4 A. ∎∎∎

The different forms of a complex number are summarized below.

Forms of a Complex Number

Rectangular:

$$x + yj$$

Polar:

$$r(\cos\theta + j\sin\theta)$$

where θ is usually expressed in degrees.

Exponential:

$$re^{j\theta}$$

where θ is in radians and

$$x = r\cos\theta \qquad\qquad y = r\sin\theta$$

$$r = \sqrt{x^2 + y^2} \qquad \tan\theta = \frac{y}{x}$$

14–3 EXERCISES

Represent each complex number graphically and express each number in polar and exponential forms.

1. $-2 + 3j$	**2.** $5 - 6j$	**7.** $6 - 13j$	**8.** $-8 + 7j$
3. $-7 - 10j$	**4.** $5 - 11j$	**9.** $-7.5 + 10.1j$	**10.** $-6.9 - 15j$
5. $6 + 9j$	**6.** $3 + 2j$	**11.** $-5.7 - 9.3j$	**12.** $-1.5 + 3.8j$

13. $8j$

14. $-15j$

15. 15

16. -10

Express each complex number in rectangular form.

17. $6(\cos 300° + j \sin 300°)$

18. $10(\cos 120° + j \sin 120°)$

19. $258\angle 100°$

20. $4.7\angle 228°$

21. $7.2\angle -56.1°$

22. $75\angle -38°$

23. $137(\cos 180° + j \sin 180°)$

24. $9.3(\cos 201° + j \sin 201°)$

25. $7e^{1.8j}$

26. $4.8e^{4.1j}$

27. $163e^{2.6j}$

28. $0.328e^{5.4j}$

29. $28.6e^{1.31j}$

30. $274e^{2.88j}$

31. $0.739e^{1.1j}$

32. $56e^{4.9j}$

Express each complex number in polar and rectangular forms.

33. $6.75e^{3.9j}$

34. $11.6e^{4.1j}$

35. $175e^{1.8j}$

36. $238e^{5.2j}$

37. $26.9e^{2.5j}$

38. $67.4e^{0.75j}$

E 39. The current in an ac circuit is given by the complex number $3.6 - 5.3j$. Write this number in polar and exponential forms.

E 40. Determine the magnitude of a voltage in an ac circuit represented by $V = 108.6 + 58.7j$ volts.

CS 41. Computer screens and graphing calculators allow us to draw equations in action, which in turn can spark further explorations and applications. Draw a geometric interpretation of a complex number and its conjugate. What is their relationship?

E 42. In an *RL* series circuit with a resistance of 10 Ω and an inductive reactance of 8 Ω, express the impedance in rectangular form, polar form, exponential form, and as an impedance diagram.

E 43. In an *RC* series circuit with a resistance of 20 Ω and a capacitive reactance of 18 Ω, express the impedance in rectangular form, polar form, exponential form, and as an impedance diagram.

14-4

PRODUCTS, QUOTIENTS, POWERS, AND ROOTS OF COMPLEX NUMBERS

In Section 14-2, we discussed the product and quotient of complex numbers in rectangular form. However, these operations and others are done more easily when the complex numbers are written in alternate forms. In this section we will discuss the product, quotient, power, and roots of complex numbers written using polar and exponential forms.

Multiplication

The derivation of the rules for multiplying complex numbers in exponential or polar form follows. When multiplying complex numbers in exponential form, we apply the Laws of Exponents to obtain

$$r_1 e^{j\theta_1} \cdot r_2 e^{j\theta_2} = r_1 r_2 e^{j(\theta_1 + \theta_2)}$$

Using this information to express the product of complex numbers in polar form, we have

$$r_1 e^{j\theta_1} \cdot r_2 e^{j\theta_2} = r_1(\cos \theta_1 + j \sin \theta_1) r_2(\cos \theta_2 + j \sin \theta_2)$$

If we expand the expression for the product of the exponential forms above, we have

$$r_1 r_2 e^{j(\theta_1 + \theta_2)} = r_1 r_2 [\cos(\theta_1 + \theta_2) + j \sin(\theta_1 + \theta_2)]$$

Since the two exponential expressions are equal, the two polar expressions are also equal. Thus,

$$r_1(\cos \theta_1 + j \sin \theta_1) \cdot r_2(\cos \theta_2 + j \sin \theta_2) = r_1 r_2 [\cos (\theta_1 + \theta_2) + j \sin (\theta_1 + \theta_2)]$$

Multiplication of Complex Numbers

Exponential form:

$$r_1 e^{j\theta_1} \cdot r_2 e^{j\theta_2} = r_1 r_2 e^{j(\theta_1 + \theta_2)}$$

Polar form:

$$r_1(\cos \theta_1 + j \sin \theta_1) \cdot r_2(\cos \theta_2 + j \sin \theta_2) = r_1 r_2[\cos (\theta_1 + \theta_2) + j \sin (\theta_1 + \theta_2)]$$

To multiply complex numbers in exponential or polar form, multiply magnitudes and add angles.

EXAMPLE 1   Using polar form, multiply $2 - 7j$ and $-5 + 9j$.

Solution   First, we convert each number to polar form.
    For $2 - 7j$:

$$r_1 = \sqrt{2^2 + (-7)^2} \qquad \tan \theta_R = \frac{7}{2}$$

$$r_1 = \sqrt{53} \qquad\qquad \theta_R \approx 74.1°$$
$$r_1 \approx 7.28 \qquad\qquad \theta_1 \approx 285.9°$$

$$7.28(\cos 285.9° + j \sin 285.9°)$$

$$2 \boxed{x^2} \boxed{+} 7 \boxed{+/-} \boxed{x^2} \boxed{\sqrt{}} \longrightarrow 7.28$$

$$7 \boxed{\div} 2 \boxed{=} \boxed{\text{inv}} \boxed{\text{tan}} 74.1 \boxed{\textbf{STO}} 360 \boxed{-} \boxed{\textbf{RCL}} \boxed{=} \longrightarrow 285.9$$

For $-5 + 9j$:

$$r_2 = \sqrt{(-5)^2 + 9^2} \qquad \tan \theta_R = \frac{9}{5}$$

$$r_2 = \sqrt{106} \qquad\qquad \theta_R \approx 60.9°$$
$$r_2 \approx 10.3 \qquad\qquad \theta_2 \approx 119.1°$$

$$10.3(\cos 119.1° + j \sin 119.1°)$$

$$5 \boxed{+/-} \boxed{x^2} \boxed{+} 9 \boxed{x^2} \boxed{=} \boxed{\sqrt{}} \longrightarrow 10.3$$

$$9 \boxed{\div} 5 \boxed{=} \boxed{\text{inv}} \boxed{\text{tan}} 60.9 \boxed{\textbf{STO}} 180 \boxed{-} \boxed{\textbf{RCL}} \boxed{=} \longrightarrow 119.1$$

Multiplying the two complex numbers in polar form gives

$$7.28(\cos 285.9° + j \sin 285.9°) \cdot 10.3(\cos 119.1° + j \sin 119.1°)$$

add angles

multiply magnitudes

$$= (7.28)(10.3)[\cos (285.9° + 119.1°) + j \sin (285.9° + 119.1°)]$$
$$= 75.0(\cos 405.0° + j \sin 405.0°)$$

We do not leave an angular measure greater than 360° or 2π radians; therefore, the product is

$$75.0(\cos 45.0° + j \sin 45.0°)$$

■ ■ ■

EXAMPLE 2 Using exponential form, multiply $3 + 8j$ and $-6 - 5j$.

Solution First, convert each number to exponential form. Be sure your calculator is in radian mode.

For $3 + 8j$:

$$r_1 = \sqrt{9 + 64} \qquad \tan \theta_R = \frac{8}{3}$$

$$r_1 \approx 8.5 \qquad \theta_R \approx 1.2$$

$$\theta_1 \approx 1.2$$

$$8.5e^{1.2j}$$

For $-6 - 5j$:

$$r_2 = \sqrt{36 + 25} \qquad \tan \theta_R = \frac{5}{6}$$

$$r_2 \approx 7.8 \qquad \theta_R \approx 0.69$$

$$\theta_2 \approx 3.8 \text{ (Q II angle)}$$

$$7.8e^{3.8j}$$

Multiplying the complex numbers in exponential form gives

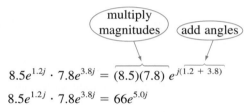

$$8.5e^{1.2j} \cdot 7.8e^{3.8j} = \overbrace{(8.5)(7.8)}\ e^{j(1.2\ +\ 3.8)}$$

$$8.5e^{1.2j} \cdot 7.8e^{3.8j} = 66e^{5.0j}$$

You can check your answer by multiplying the rectangular forms and converting the result to exponential form. ∎ ∎ ∎

Division

Similarly, we can derive the following formulas for division of complex numbers. The rules are given below.

Division of Complex Numbers

Exponential form:

$$r_1 e^{j\theta_1} \div r_2 e^{j\theta_2} = \frac{r_1}{r_2}e^{j(\theta_1 - \theta_2)}$$

Polar form:

$$\frac{r_1(\cos \theta_1 + j \sin \theta_1)}{r_2(\cos \theta_2 + j \sin \theta_2)} = \frac{r_1}{r_2}[\cos (\theta_1 - \theta_2) + j \sin (\theta_1 - \theta_2)]$$

To divide complex numbers in exponential or polar form, divide magnitudes and subtract angles.

EXAMPLE 3   Using polar form, divide the first complex number in Example 1 by the second.

Solution   From Example 1, the polar forms of these two numbers are 7.28∠285.9° and 10.3∠119.1°, respectively. Dividing these numbers gives

$$\frac{7.28(\cos 285.9° + j \sin 285.9°)}{10.3(\cos 119.1° + j \sin 119.1°)}$$

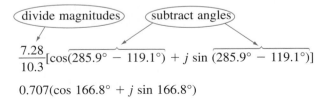

$$\frac{7.28}{10.3}[\cos(285.9° - 119.1°) + j \sin (285.9° - 119.1°)]$$

$$0.707(\cos 166.8° + j \sin 166.8°)$$ ∎∎∎

EXAMPLE 4   Using exponential form, divide the first number in Example 2 by the second number.

Solution   The exponential forms from Example 2 are $8.5e^{1.2j}$ and $7.8e^{3.8j}$. Dividing gives

$$\frac{8.5e^{1.2j}}{7.8e^{3.8j}} = \frac{8.5}{7.8}e^{j(1.2 - 3.8)}$$

$$\frac{8.5e^{1.2j}}{7.8e^{3.8j}} \approx 1.1e^{-2.6j}$$

Since we want a nonnegative angle less than 2π, add -2.6 to 2π. The quotient is

$$1.1e^{3.7j}$$ ∎∎∎

Powers of Complex Numbers

Exponential and polar forms are helpful in raising a complex number to a power. The rule for the power of a complex number in exponential form is based on the Law of Exponents. *The formula for the nth power of a complex number in polar form* is called **DeMoivre's theorem.** The formula for the power of a complex number written in polar and exponential forms is summarized in the following box.

Powers of Complex Numbers

Exponential form:

$$(re^{j\theta})^n = r^n e^{jn\theta}$$

Polar form:

$$[r(\cos \theta + j \sin \theta)]^n = r^n(\cos n\theta + j \sin n\theta)$$

EXAMPLE 5   Using polar and exponential forms, find $(4 - 6j)^3$.

Solution   First *converting* $4 - 6j$ *to polar form* gives

$$r = \sqrt{(4)^2 + (-6)^2} \qquad \tan \theta_R = \frac{6}{4}$$

$$r = \sqrt{52} \qquad\qquad \theta_R = 56.3°$$
$$r \approx 7.21 \qquad\qquad \theta \approx 303.7°$$

$$[7.21(\cos 303.7° + j \sin 303.7°)]^3$$

Applying DeMoivre's theorem gives

$$(7.21)^3[\cos 3(303.7°) + j \sin 3(303.7°)]$$
$$375(\cos 191.1° + j \sin 191.1°)$$

$$191.1° = 911.1° - 720°$$

To find $(4 - 6j)^3$ using exponential form, we must *convert* $4 - 6j$ *to exponential form*. We must convert $303.7°$ to radians.

$$7.21e^{5.30j}$$

Applying the formula gives

$$(7.21e^{5.30j})^3 = (7.21)^3 e^{3(5.30j)} = 375e^{15.9j}$$

Converting the angle to less than 2π gives

$$375e^{3.33j}$$

■ ■ ■

Roots of Complex Numbers

DeMoivre's theorem was originally intended for positive integral powers of n; however, you can also use it to find rational roots of a complex number. Every complex number has n distinct nth roots.

EXAMPLE 6   Find all possible roots for $\sqrt{-4 + 7j}$.

Solution   First, we *convert* $-4 + 7j$ *to polar form*.

$$-4 + 7j = 8.1\angle 119.7°$$
$$[8.1(\cos 119.7° + j \sin 119.7°)]^{1/2}$$

Next, we *apply DeMoivre's theorem*.

$$\sqrt{8.1}\left(\cos \frac{119.7°}{2} + j \sin \frac{119.7°}{2}\right)$$

$$2.8(\cos 59.9° + j \sin 59.9°)$$

We should have two square roots. Since two complex numbers are equal if their magnitudes are equal and the angles are coterminal, we can find the second square root by adding $360°$ to the original angle.

$$\sqrt{8.1}\left(\cos \frac{119.7° + 360°}{2} + j \sin \frac{119.7° + 360°}{2}\right)$$

$$2.8(\cos 239.9° + j \sin 239.9°)$$

The two square roots of $-4 + 7j$ are

$$2.8(\cos 59.9° + j \sin 59.9°)$$
$$2.8(\cos 239.9° + j \sin 239.9°)$$

■ ■ ■

Roots of Complex Numbers

$$\sqrt[n]{x + yj} = \sqrt[n]{r}\left(\cos \frac{\theta + 360k°}{n} + j \sin \frac{\theta + 360k°}{n}\right)$$

where

$$k = 0, 1, 2, \ldots, (n - 1)$$

EXAMPLE 7 Find all values for the fourth root of -1.

Solution Converting -1 to polar form gives

$$1(\cos 180° + j \sin 180°)$$

Evaluating $\sqrt[4]{-1}$ gives

$$\sqrt[4]{1}\left(\cos \frac{180° + 360k°}{4} + j \sin \frac{180° + 360k°}{4}\right)$$

where $k = 0, 1, 2,$ and 3; there are four roots.

$k = 0$	$1(\cos 45° + j \sin 45°) = 0.71 + 0.71j$
$k = 1$	$1(\cos 135° + j \sin 135°) = -0.71 + 0.71j$
$k = 2$	$1(\cos 225° + j \sin 225°) = -0.71 - 0.71j$
$k = 3$	$1(\cos 315° + j \sin 315°) = 0.71 - 0.71j$

■ ■ ■

CAUTION ✦ When you are finding roots of a complex number, the number of answers should equal the index of the root. For example, the cube root of a complex number has three answers. Also, the complex number should be in polar form before you find powers and roots.

14–4 EXERCISES

Perform the indicated operations. Leave your answer in exponential form.

1. $4e^{2j} \cdot 7e^{3j}$

2. $2.8e^{1.3j} \cdot 3.4e^{4.6j}$

3. $8.2e^{3.2j} \cdot 6.1e^{2.9j}$

4. $7e^{3j} \cdot 6e^{j}$

5. $(5 - j)(2 + 3j)$

6. $(2 - 7j)(3 + j)$

7. $\dfrac{8e^{6j}}{2e^{j}}$

8. $\dfrac{16e^{4j}}{10e^{3j}}$

9. $\dfrac{10.4e^{5.3j}}{2.5e^{6.4j}}$

10. $\dfrac{34.8e^{2.7j}}{14.7e^{5.9j}}$

11. $\dfrac{9 + 6j}{3 - 4j}$

12. $\dfrac{11 - 8j}{7 + 4j}$

13. $(2 - 5j)^3$

14. $(1 + j)^4$

Perform the indicated operations. Leave your answer in polar form.

15. $(14\angle 146.8°)(8\angle 287.4°)$

16. $(7.8\angle 318.1°)(11.3\angle 78.4°)$

17. $16e^{4.8j} \cdot 12e^{6.1j}$

18. $5.8e^{4.5j} \cdot 2.3e^{3.2j}$

19. $\dfrac{6(\cos 120° + j \sin 120°)}{20(\cos 75° + j \sin 75°)}$

20. $\dfrac{15(\cos 216° + j \sin 216°)}{7(\cos 108° + j \sin 108°)}$

21. $\dfrac{7.8e^{1.1j}}{3.2e^{4.6j}}$

22. $\dfrac{2.5e^{6.2j}}{1.4e^{3.5j}}$

23. $(6\angle 147°)^3$

24. $(2\angle 214°)^3$

25. $(4 + 7j)^4$

26. $(8 - 5j)^2$

Find all possible roots.

27. $\sqrt[3]{1 - j}$

28. $\sqrt[4]{2 - 5j}$

29. $\sqrt{6(\cos 312° - j \sin 312°)}$

30. $\sqrt[3]{2(\cos 120° + j \sin 120°)}$

31. $\sqrt[4]{j}$

32. $\sqrt[3]{-j}$

33. $\sqrt[6]{1}$

34. $\sqrt[4]{1}$

35. $\sqrt{2 + 4j}$

36. $\sqrt{3 - 5j}$

E 37. In an ac circuit the voltage V is the product of the current I and the impedance Z. If $I = 6\angle 40°$ A and $Z = 4\angle 75°$ Ω, find the voltage.

E 38. In an ac circuit the impedance Z is the quotient of the voltage V and the current I. If the voltage is $12 \angle 16.2°$ V and the current is $4.2 \angle 14.1°$ A, find the impedance.

E 39. In an ac circuit impedances $Z_1 = 3 + 2j$ and $Z_2 = 1 + 6j$ are connected in parallel. Find the equivalent impedance Z given by

$$Z = \frac{Z_1 Z_2}{Z_1 + Z_2}$$

E 40. Find the equivalent impedance in Exercise 39 if $Z_1 = 8 + 7j$ and $Z_2 = 10 - 2j$.

CS 41. You are using a math program to compute all the possible answers to a problem assigned in class. Two of the answers have the letter j in them, indicating a complex solution. The correlation between these answers can be studied using the following rules for complex numbers:

(1) $(w + xj) + (y + zj) = (w + y) + (x + z)j$

(2) $(w + xj)(y + zj) = (wy - xz) + (wz + xy)j$

When do these rules for complex numbers reduce to ordinary addition and multiplication of real numbers?

14–5

APPLICATIONS

In this section we will discuss the application of complex numbers to voltage and current in an alternating current (ac) circuit. We will begin with a discussion of the basic elements of a circuit.

Resistance is the opposition to the flow of current. It is denoted by R, measured in ohms Ω, and represented by ⋀⋀ in a circuit diagram. A **capacitor** consists of two closely spaced conductors that store an electrical charge. **Capacitance** is denoted by C, measured in farads F, and represented by ⊣⊢ in a circuit diagram. An **inductor** is a coil of wire that induces current. **Inductance** is denoted by L, measured in henries H, and represented by ⌒⌒⌒ in a circuit diagram. **Reactance,** denoted by X, is the effective resistance of any component of a circuit. Since both capacitors and inductors resist the flow of current, we can define capacitive reactance and inductive reactance by the following equations.

Capacitive Reactance

$$X_C = \frac{1}{2\pi f C}$$

Inductive Reactance

$$X_L = 2\pi f L$$

where

$$f = \text{frequency of current}$$
$$C = \text{capacitance}$$
$$L = \text{inductance}$$

In a parallel ac circuit with multiple frequencies, an inductor and a resistor together act as a "high-pass filter" to allow passage of only high-frequency currents. Low-frequency currents are diverted through the inductor. Similarly, a "low-pass filter" allows only low-frequency currents to pass.

The voltage V across any component is given by $V = IX$, where I represents current in amperes and X represents reactance in ohms. From this information, the following equations give the voltage across a resistor, a capacitor, and an inductor.

Voltage Drop Across a Circuit Component

Resistor:

$$V_R = IR$$

Capacitor:

$$V_C = IX_C$$

Inductor:

$$V_L = IX_L$$

EXAMPLE 1 The ac circuit shown in Figure 14–10 contains a 2.6-Ω resistor, a 0.03-H inductor, and a 0.00002-F (20-μF) capacitor. If a 9-V, 1,000-Hz power supply is connected across the circuit, determine the capacitive and inductive reactances.

FIGURE 14–10

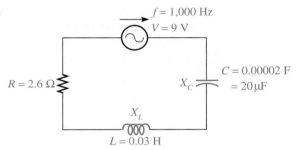

Solution From the statement of the problem, we know that $R = 2.6\ \Omega$, $L = 0.03$ H, $C = 0.00002$ F, $V = 9$ V, and $f = 1,000$ Hz.

Calculating the capacitive reactance X_C gives

$$X_C = \frac{1}{2\pi fC}$$

$$X_C = \frac{1}{2\pi(1,000)(0.00002)}$$

$$X_C \approx 8\ \Omega$$

$2\ \boxed{\times}\ \boxed{\pi}\ \boxed{\times}\ 1000\ \boxed{\times}\ 0.00002\ \boxed{=}\ \boxed{1/x}\ \longrightarrow\ 7.96$

Calculating X_L gives

$$X_L = 2\pi fL$$
$$X_L = 2\pi(1,000)(0.03)$$
$$X_L \approx 190\ \Omega$$

$2\ \boxed{\times}\ \boxed{\pi}\ \boxed{\times}\ 1000\ \boxed{\times}\ 0.03\ \boxed{=}\ \longrightarrow\ 188.5$

■ ■ ■

To determine the voltage across a combination of resistors, inductors, and capacitors, we must account for the reactance and the phase of the voltage across individual components. A polar diagram shows the effective voltage across a resistor, capacitor, and inductor as vectors, with the phase differences appearing as angles. Since the voltage across a resistor is in-phase with the current, a real number R can be used to represent resistance. Graphically, this relationship is illustrated in Figure 14–11(a). Since the voltage across an inductor leads the current by $90°$, the inductive reactance is represented by the positive, pure imaginary number $X_L j$ as shown in Figure 14–11(b). Because the voltage across a capacitor lags $90°$ behind the current, it can be represented by the negative, pure imaginary number $-X_C j$ as shown in Figure 14–11(c).

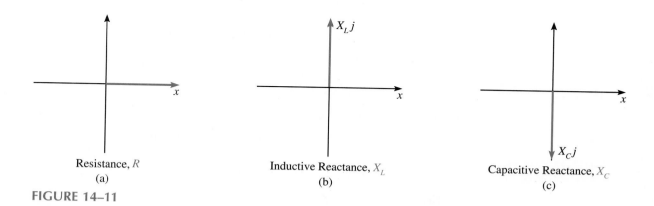

Resistance, R
(a)

Inductive Reactance, X_L
(b)

Capacitive Reactance, X_C
(c)

FIGURE 14–11

The total voltage V across a circuit containing a resistor, capacitor, and inductor is the sum of the individual voltages, represented by the following equation:

$$V = V_R + V_L + V_C$$
$$V = IR + IX_L j - IX_C j$$
$$V = I[R + j(X_L - X_C)]$$

Impedance Z is the effective resistance of the entire circuit and is given by $R + j(X_L - X_C)$. Making the substitution gives the following equation.

Impedance

$$V = IZ$$

where

$$Z = R + j(X_L - X_C)$$

The magnitude of Z, denoted $|Z|$, is

$$|Z| = \sqrt{R^2 + (X_L - X_C)^2}$$

and the phase angle θ between the current and voltage is given by

$$\tan \theta = \frac{X_L - X_C}{R}$$

▦ **EXAMPLE 2** For the circuit shown in Figure 14–12, assume that $X_L = 95.0\ \Omega$, $X_C = 103\ \Omega$, $R = 2.70\ \Omega$, and $V = 12.0$ V. Calculate the following:

(a) phase angle

(b) magnitude of the impedance

(c) current in the circuit

(d) voltage across each circuit component

FIGURE 14–12

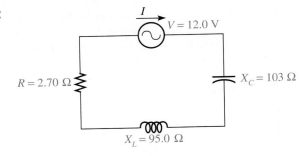

Solution

(a) The phase angle is given by

$$\tan \theta = \frac{X_L - X_C}{R} = \frac{95.0 - 103}{2.70}$$

$$\theta \approx -71.4°$$

The negative sign on the phase angle means that the voltage lags the current by 71.4°.

(b) The magnitude of the impedance is given by

$$|Z| = \sqrt{R^2 + (X_L - X_C)^2}$$

$$|Z| \approx 8.44\ \Omega$$

2.7 $\boxed{x^2}$ $\boxed{+}$ $\boxed{(}$ 95 $\boxed{-}$ 103 $\boxed{)}$ $\boxed{x^2}$ $\boxed{=}$ $\boxed{\sqrt{}}$ $\boxed{\text{STO}}$ ⟶ 8.44

(c) The current is given by

$$I = \frac{V}{|Z|} = \frac{12.0}{8.44}$$

$$I \approx 1.42\ \text{A}$$

12 $\boxed{\div}$ $\boxed{\text{RCL}}$ $\boxed{=}$ $\boxed{\text{STO}}$ ⟶ 1.42

(d)

$$V_R = IR = (1.42)(2.70)$$ $\boxed{\text{RCL}}$ $\boxed{\times}$ 2.7 $\boxed{=}$ ⟶ 3.83 V

$$V_L = IX_L = (1.42)(95.0)$$ $\boxed{\text{RCL}}$ $\boxed{\times}$ 95 $\boxed{=}$ ⟶ 135 V

$$V_C = IX_C = (1.42)(103)$$ $\boxed{\text{RCL}}$ $\boxed{\times}$ 103 $\boxed{=}$ ⟶ 146 V ■■■

▦ **EXAMPLE 3** An electric motor with an inductance of 0.020 H and a resistance of 18 Ω is connected to a 110-V, 60-Hz power source.

(a) Calculate the magnitude of the impedance of the motor.

(b) Calculate the current through the motor.

(c) If a 0.00042-F (420-µF) capacitor is added to the circuit, how much current will the circuit draw?

(d) Find the phase angle in part (c).

Solution

(a) To calculate the magnitude of the impedance, we must know R, X_L, and X_C. From the information given, $R = 18 \ \Omega$, $X_C = 0$ since there is no capacitor in the circuit, and $X_L = 2\pi fL = 2\pi(60)(0.02) \approx 7.5 \ \Omega$. Therefore, the magnitude of the impedance is

$$|Z| = \sqrt{R^2 + (X_L - X_C)^2} = \sqrt{(18)^2 + (7.5)^2}$$
$$|Z| \approx 20 \ \Omega$$

18 $\boxed{x^2}$ $\boxed{+}$ 7.5 $\boxed{x^2}$ $\boxed{=}$ $\boxed{\sqrt{}}$ $\longrightarrow$ 19.5

(b) To calculate the current through the motor, we solve the equation $V = IZ$ for I and substitute.

$$I = \frac{V}{|Z|} = \frac{110}{20}$$
$$I = 5.5 \text{ A}$$

(c) If a capacitor is added to the circuit, its capacitive reactance is

$$X_C = \frac{1}{2\pi fC} = \frac{1}{2\pi(60)(0.00042)}$$
$$X_C \approx 6.3 \ \Omega$$

2 $\boxed{\times}$ $\boxed{\pi}$ $\boxed{\times}$ 60 $\boxed{\times}$ 0.00042 $\boxed{=}$ $\boxed{1/x}$ $\longrightarrow$ 6.3

Calculating the magnitude of the impedance gives

$$|Z| = \sqrt{R^2 + (X_L - X_C)^2}$$
$$|Z| \approx 18 \ \Omega$$

18 $\boxed{x^2}$ $\boxed{+}$ $\boxed{(}$ 7.5 $\boxed{-}$ 6.3 $\boxed{)}$ $\boxed{x^2}$ $\boxed{=}$ $\boxed{\sqrt{}}$ $\longrightarrow$ 18

The resulting current is

$$I = \frac{V}{|Z|} = \frac{110}{18}$$
$$I \approx 6.1 \text{ A}$$

(d) The phase angle between the voltage and current is given by

$$\tan \theta = \frac{X_L - X_C}{R} = \frac{7.5 - 6.3}{18}$$
$$\theta \approx 3.8°$$

The voltage leads the current by 3.8°. ∎ ∎ ∎

Complex Current

In the discussion so far, the phase angle of the current has been taken as zero. However, if an arbitrary angle is chosen, current, voltage, and impedance must be treated as complex numbers.

EXAMPLE 4 In a particular circuit the current is given as $4 - j$ amperes, and the impedance is given as $5 + 3j$ ohms. Determine the magnitude of the voltage across the circuit.

Solution We use the formula $V = IZ$ to calculate the voltage. Multiplying gives

$$V = (4 - j)(5 + 3j) = 20 + 12j - 5j - 3j^2$$
$$V = 23 + 7j \text{ V}$$

The magnitude of the voltage is

$$|V| = \sqrt{(23)^2 + (7)^2}$$
$$|V| \approx 24 \text{ V} \qquad \blacksquare\blacksquare\blacksquare$$

Resonance

Since both inductive and capacitive reactance depend on the frequency of the voltage, impedance also depends on frequency. When inductive reactance equals capacitive reactance, the equation for impedance becomes $|Z| = R$. When this is true, the impedance is a minimum, and the current is a maximum. The frequency f_0 at which inductive reactance and capacitive reactance are equal is called the **resonant frequency.** The formula for f_0 given below is the result of setting $X_L = X_C$ and solving for f.

Resonant Frequency

$$f_0 = \frac{1}{2\pi\sqrt{LC}} \text{ Hz}$$

EXAMPLE 5 Find the resonant frequency of a circuit containing an inductor of 0.0300 H, a resistor of 7.00 Ω, a capacitor of 0.0000190 F (19 μF), and a voltage source of 9.00 V.

Solution The resonant frequency is given by

$$f_0 = \frac{1}{2\pi\sqrt{LC}} = \frac{1}{2\pi\sqrt{(0.03)(0.000019)}}$$
$$f_0 \approx 211 \text{ Hz}$$

$$2 \boxed{\times} \boxed{\pi} \boxed{\times} \boxed{(} 0.03 \boxed{\times} 0.000019 \boxed{)} \boxed{\sqrt{}} \boxed{=} \boxed{1/x} \longrightarrow 211 \qquad \blacksquare\blacksquare\blacksquare$$

The resonant circuit described in Example 5 is fundamental to the operation of different electrical systems. The frequency response of this circuit is shown in Figure 14–13.

FIGURE 14–13

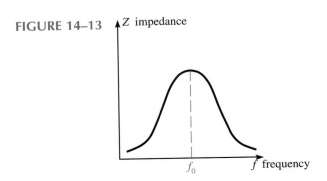

Radio and television receivers use this type of circuit. When the receiver is tuned to a particular station, it is set at or near the resonant frequency f_0. As a result, stations transmitting at frequencies to the left or right of f_0 do not interfere with the station's broadcast.

EXAMPLE 6   The tuning circuit of a television contains a variable capacitor, a 0.0030-H coil, and a 15-Ω resistor. Calculate the capacitance necessary to tune the circuit to a television station broadcasting at 60×10^6 Hz.

Solution   To solve for the capacitance, use the formula for resonant frequency.

$$f_0 = \frac{1}{2\pi\sqrt{LC}}$$

Next, solve for C.

$$f_0^2 = \frac{1}{4\pi^2 LC}$$

$$4f_0^2\pi^2 LC = 1$$

$$C = \frac{1}{4f_0^2\pi^2 L}$$

Then substitute values for L and f_0.

$$C = \frac{1}{4(60 \times 10^6)^2\pi^2(0.0030)}$$

$$C = 2.3 \times 10^{-15} \text{ F}$$

$4\ \boxed{\times}\ 60000000\ \boxed{x^2}\ \boxed{\times}\ \boxed{\pi}\ \boxed{x^2}\ \boxed{\times}\ 0.003\ \boxed{=}\ \boxed{1/x} \longrightarrow 2.3\ {-15}$ ■ ■ ■

14–5 EXERCISES

E 1. An ac circuit contains a 0.036-H inductor, a 0.0000080-F capacitor, and a 7.5-Ω resistor. If a 9.0-V, 60-Hz power source is connected across the circuit, find the following:

  (a) capacitive reactance   **(b)** inductive reactance

  (c) magnitude of the impedance   **(d)** phase angle between the current and voltage

  (e) current   **(f)** voltage across the resistor

  (g) voltage across the capacitor   **(h)** voltage across the inductor

E 2. Answer parts (a)–(h) from Exercise 1 if the circuit contains a 0.0540-H inductor, a 0.00000600-F capacitor, a 12.6-Ω resistor, and a 110-V, 60.0-Hz power source.

E 3. A current of 0.15 A flows through a 0.31-H inductor that is connected to a 115-V ac source. What is the frequency of the source?

E 4. A 0.00000007-F capacitor is connected to a 9-V, 60-Hz power source. Find the following:

  (a) capacitive reactance   **(b)** current flow

E 5. In a circuit, $R = 7.0\ \Omega$, $X_C = 13\ \Omega$, and $X_L = 8.0\ \Omega$. Find the magnitude of the impedance and the phase angle between the current and voltage.

E 6. Calculate the magnitude of the voltage in a circuit if the complex current is $3.00 - 4.00j$ and the complex impedance is $5.00 + 7.00j$.

E 7. If the complex current in a particular circuit is given by $2.00 + 5.00j$ and the complex impedance is given by $7.00 - 6.00j$, find the magnitude of the voltage.

E 8. The voltage across a circuit is given by $8.00 - 3.00j$, and the current is $9.00 + 5.00j$. Find the magnitude of the impedance.

E 9. The voltage across a particular circuit is given by $1.00 + 8.00j$, and the impedance is given by $7.00 - 3.00j$. Find the magnitude of the current.

E 10. A 0.0000060-F (6-μF) capacitor, a 0.060-H inductor, and an 8.0-Ω resistor are connected with a 28-V, 400-Hz power source. Find the following:

  (a) impedance of the circuit   **(b)** current that flows through it

E 11. Find the resonant frequency of the circuit in Exercise 10.

E 12. A 0.0000090-F (9-μF) capacitor, a 0.090-H inductor, and a 45-Ω resistor are connected across a 115-V, 60-Hz power line. Find the following:
(a) current in the circuit
(b) voltage drop across each element of the circuit

E 13. The tuning circuit of a radio contains an 8.5-Ω resistor, a 0.018-H coil, and a variable capacitor. Calculate the capacitance required to tune the radio to a station broadcasting at 7.3×10^5 Hz (730 kHz).

E 14. The antenna circuit of a radio contains a 0.20-H coil, a variable capacitor, and a 48-Ω resistor. Find the capacitance required for resonance if the station transmits at 6.4×10^5 Hz (640 kHz).

E 15. The power P dissipated in an ac circuit is given by $P = IV \cos \theta$ where V = voltage, I = current, and θ = phase angle between the current and voltage. Find the power dissipated in the circuit in Exercise 10.

E 16. Find the power dissipated by the circuit described in Exercise 12.

CS 17. Digital circuits are the foundation of our current computer's logic system. Imaginary numbers can be used to represent values that correspond to voltage relationships across circuits. Why can these values be used?

E 18. In a parallel ac circuit, admittance Y (in siemens) is the reciprocal of impedance Z. Find the admittance for an impedance of $9.0 - 3.0j$ Ω.

E 19. Using the information from Exercise 18, find the admittance for an impedance of $12 - 6j$ Ω.

E 20. The susceptance B (in siemens) in an ac circuit is the reciprocal of the inductive or capacitive reactance. If the inductive reactance is $2.7 \angle 90°$, find the susceptance.

CHAPTER SUMMARY

Summary of Terms

argument (p. 533)

capacitance (p. 545)

capacitor (p. 545)

complex number (p. 522)

complex plane (p. 527)

conjugate (p. 525)

DeMoivre's theorem (p. 542)

imaginary part (p. 522)

impedance (p. 547)

inductance (p. 545)

inductor (p. 545)

magnitude (p. 533)

pure imaginary number (p. 522)

reactance (p. 545)

real part (p. 522)

rectangular form (p. 522)

resistance (p. 545)

resonant frequency (p. 550)

Summary of Formulas

$re^{j\theta}$ where θ is in radians exponential form of complex numbers

$r_1 e^{j\theta_1} \cdot r_2 e^{j\theta_2} = r_1 r_2 e^{j(\theta_1 + \theta_2)}$ exponential form

$r_1(\cos \theta_1 + j \sin \theta_1) \cdot r_2(\cos \theta_2 + j \sin \theta_2) =$

$r_1 r_2[(\cos(\theta_1 + \theta_2) + j \sin(\theta_1 + \theta_2)]$ polar form
$\left.\right\}$ multiplication of complex numbers

$r_1 e^{j\theta_1} \div r_2 e^{j\theta_2} = \dfrac{r_1}{r_2} e^{j(\theta_1 - \theta_2)}$ exponential form

$\dfrac{r_1(\cos \theta_1 + j \sin \theta_1)}{r_2(\cos \theta_2 + j \sin \theta_2)} = \dfrac{r_1}{r_2}[\cos(\theta_1 - \theta_2) + j \sin(\theta_1 - \theta_2)]$ polar form $\left.\right\}$ division of complex numbers

$(re^{j\theta})^n = r^n e^{jn\theta}$ exponential form

$[r(\cos \theta + j \sin \theta)]^n = r^n(\cos n\theta + j \sin n\theta)$ polar form $\left.\right\}$ powers of complex numbers

$\sqrt[n]{x + yj} = \sqrt[n]{r}\left(\cos \dfrac{\theta + 360k°}{n} + j \sin \dfrac{\theta + 360k°}{n}\right)$ roots of complex numbers

$r = \sqrt{x^2 + y^2}$ $\tan \theta = \dfrac{y}{x}$ $x = r \cos \theta$ $y = r \sin \theta$

CHAPTER REVIEW

 Section 14–1

Perform the indicated operation and simplify.

1. $\sqrt{-8} \cdot \sqrt{3}$
2. $\sqrt{-54} \cdot \sqrt{-9}$
3. $(6\sqrt{-5})^2$
4. $\sqrt{-16} + \sqrt{-9}$
5. $\sqrt{-4/25} + (3\sqrt{-7})^2$
6. $j^{14} + \sqrt{-49}$
7. $\sqrt{-4/9} + j^{25}$

Simplify the following powers of j.

8. j^{786}
9. j^{127}
10. j^{39}
11. j^{472}
12. $j^{34} + j^{66}$
13. $3j^{102} - j^{40}$
14. $10j^{109} + 7j^{213}$
15. $5j^{415} + j^{33}$

 Section 14–2

Perform the indicated operations. Leave your answer in $a + bj$ form.

16. $(-2 + 3j) + (5j - 6)$
17. $(16 - 7j) + (-2 - 15j)$
18. $(10 - \sqrt{-25}) - (\sqrt{-16} + 7)$
19. $(\sqrt{-114} + 9) - (4 + \sqrt{-49})$
20. $(4 - 9j)(4 + 9j)$
21. $(2 - 5j)(2 + 5j)$
22. $(\sqrt{-9} + 6)(7 - \sqrt{-81})$
23. $(4 + \sqrt{-25})(11 - \sqrt{-64})$
24. $(8 + 7j)(5 - 4j)$
25. $(3 - 5j)^3$
26. $(2 + 9j)^3$
27. $\dfrac{10 - \sqrt{-49}}{-j}$
28. $\dfrac{5 - \sqrt{-18}}{j}$
29. $\dfrac{2j - 1}{9 - j}$
30. $\dfrac{8 + j}{3 - 5j}$

 Section 14–3

Express each complex number in polar and exponential forms.

31. $-7 + 8j$
32. $13 - 10j$
33. $-14 - 8j$
34. $6.8 - 7.9j$
35. $8j$
36. 12
37. 14
38. $-5j$

Express each complex number in rectangular form.

39. $14\angle 276°$
40. $25\angle -62°$
41. $1.7(\cos 214° + j \sin 214°)$
42. $14.2(\cos 170° + j \sin 170°)$
43. $23e^{5.7j}$
44. $1.7e^{3.4j}$
45. $17.8e^{2j}$
46. $47.3e^{1.1j}$

 Section 14–4

Perform the indicated operations. Leave your answer in exponential form.

47. $(1 - j)^2$
48. $(2 + 3j)^3$
49. $3e^{7j} \cdot 5e^{1.7j}$
50. $6.7e^{1.2j} \cdot 3e^{4.9j}$
51. $\dfrac{6.9e^{2.7j}}{2.1e^{3.4j}}$
52. $\dfrac{11.8e^{5.7j}}{6.6e^{3.9j}}$

Perform the indicated operation. Leave your answer in polar form ($0 \le \theta < 360.0°$).

53. $(6.5\angle 112°)(9.3\angle 318°)$
54. $(7.9e^{5.8j})(1.6e^{3.7j})$
55. $\dfrac{8.7e^{1.8j}}{16.4e^{2.9j}}$
56. $\dfrac{8(\cos 78.4° + j \sin 78.4°)}{10(\cos 218.1° + j \sin 218.1°)}$
57. $(5 - 6j)^2$
58. $(7 + 9j)^2$

Find all possible roots.

59. $\sqrt[3]{2 - 5j}$
60. $\sqrt{3 + 7j}$
61. $\sqrt[4]{1 + 2j}$
62. $\sqrt{16(\cos 240° + j \sin 240°)}$
63. $\sqrt[6]{j}$
64. $\sqrt[3]{1.8(\cos 78° + j \sin 78°)}$
65. $\sqrt[5]{-1}$
66. $\sqrt[4]{-1}$

Section 14–5

E For Problems 67–74, an ac circuit contains a 0.043-H inductor, a 0.0000060-F (6-μF) capacitor, and a 9.3-Ω resistor. A 12-V, 50-Hz power source is connected across the circuit.

67. Find the capacitive reactance.
68. Find the inductive reactance.
69. Find the magnitude of the impedance.
70. Find the phase angle between the current and voltage.
71. Find the current.
72. Find the voltage across the resistor.
73. Find the voltage across the capacitor.
74. Find the voltage across the inductor.
75. Calculate the magnitude of the voltage in a circuit if the complex current is $6.00 + 11.00j$ and the complex impedance is $7.00 - 7.33j$.
76. The voltage across a circuit is given by $10.00 - 2.00j$, and the current is given by $1.00 + 7.00j$. Find the magnitude of the impedance.

77. A television tuner circuit contains a 0.15-μH (1.5 $\times$ 10^{-7}H) coil, a variable capacitor, and a 52-Ω resistor. Find the capacitance required for resonance if the station transmits at 6.4 $\times$ 10^7 Hz.

78. A 0.038-H inductor is connected to a 110-V, 60-Hz power source. Find the following:
    (a) inductive reactance
    (b) current flow

CHAPTER TEST

The number(s) in parentheses refers to the appropriate learning objective(s) given at the beginning of the chapter.

1. Add $\sqrt{-16} + \sqrt{-9} + \sqrt{-81}$.   (1)

2. Convert $8.6\angle218.4°$ to exponential form.   (4)

3. An ac circuit contains a 0.015-H inductor, a 4.2-Ω resistor, and a 0.0000040-F capacitor. A 115-V, 60-Hz power source is connected across the circuit. Find the inductive and capacitive reactance.   (6)

4. Simplify the power of j in j^{94}.   (2)

5. Find all possible roots for $\sqrt[3]{2 + 5j}$.   (5)

6. Multiply $(3 - 7j)(6 + 9j)$.   (3)

7. Using exponential form, divide the following and leave your answer in rectangular form: (4, 5)

$$\frac{8e^{2.7j}}{10e^{1.4j}}$$

8. Use DeMoivre's theorem to evaluate $(6 - 8j)^4$.   (5)

9. Multiply and simplify $\sqrt{-18} \cdot \sqrt{-40}$.   (1)

10. In a given circuit, $R = 6.3$ Ω, $X_C = 11.0$ Ω, and $X_L = 8.6$ Ω. Find the magnitude of the impedance and the phase angle between the current and voltage.   (6)

11. Simplify the power of j in j^{986}.   (2)

12. Express $7 - 10j$ in polar and exponential forms.   (4)

13. A complex current in a particular circuit is $6.00 - 11.00j$ and the complex impedance is $9.00 + 7.00j$. What is the magnitude of the voltage?   (6)

14. Add $\sqrt{-9/16} + \sqrt{-49}$.   (1)

15. Using polar form multiply the following and express your answer in exponential form: (4, 5)

$$[1.9(\cos 218° + j \sin 218°)][3.4(\cos 314° + j \sin 314°)]$$

16. Express $7.4e^{3.6j}$ in polar and rectangular forms.   (4)

17. An electric motor has an inductance of 0.040 H and a resistance of 15.4 Ω. If the motor is connected to a 115-V, 60-Hz power source, find the current through the motor.   (6)

GROUP ACTIVITY

Puffin Circuit Designs

You have been hired by Puffin Circuit Designs to determine the steady-state current (final current attained) in the middle branch of the circuit shown in Figure 14–14 when $e_1(t) =$ 10 cos $(2t - 30°)$ and $e_2(t) = 5$ cos $(2t + 45°)$. After a visit to the library, you discover that if only the steady-state parts are of interest, then the circuit may be solved in terms of the complex amplitudes. Figure 14–16 is an equivalent circuit where the complex amplitudes shown in Figure 14–15 were used to define the elements. When an excitation (voltage or current) is applied that has the form Ce^{st}, the proportionality constants between the complex amplitudes of the voltage and the current (sL and $1/sC$) in an inductor or capacitor are called the component's impedance. It plays a similar role, as does the resistance (R) in a resistor.

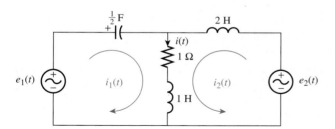

FIGURE 14–14   Original circuit.

FIGURE 14–15   Relationship between the complex amplitudes of the voltage and current in linear elements.

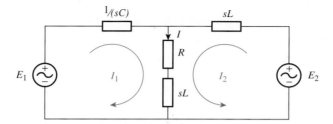

FIGURE 14–16   Circuit using definitions of Figure 14–15.

There are at least three classes of circuit problems with excitations that can be written in the form Ce^{st}: (1) $Ae^{\kappa t}$, where both A and κ are real; (2) A cos $(\omega t + \phi)$, where A, ω, and ϕ are all real; and (3) $Ae^{\kappa t}$ cos $(\omega t + \phi)$, where A, ω, ϕ, and κ are all real. The steady-state solution may be found by using the same process for all three classes of problems.

The problem here is a class 2 problem. To solve this problem, you need to rewrite the inputs in the form $e(t) = Ce^{st}$, define values for C and s, write the circuit equations using Figure 14–16, solve the set of equations for I_1 and I_2, and solve for $I = I_1 + I_2$.

1. Discuss how to show $A \cos (\omega t + \phi) = Re[Ae^{j\phi}e^{j\omega t}]$.

2. Write this equation in the form Ce^{st} by defining what C and s are equal to in terms of A, ϕ, and ω.

3. Write $e_1(t)$ and $e_2(t)$ in the form Ce^{st} by defining what C and s are equal to in terms of the constant terms (10 and 5), t, and the phase angles ($-30°$ and $45°$).

4. Using the definitions for s, from the previous step, replace each element in the circuit of Figure 14–14 with its impedance.

5. Using Kirchhoff's voltage law (the algebraic sum of all voltage rises must be equal to the sum of all voltage drops along a closed path), show that the resulting set of equations for the circuit in Figure 14–16 is given by

$$E_1 - (-j + 1 + j2)I_1 + (1 + j2)I_2 = 0$$
$$E_2 - (1 + j2)I_1 + (j4 + 1 + j2)I_2 = 0$$

6. Solve the two sets of equations for I_1 and I_2.

7. Show that I is given by

$$I = I_1 + I_2 = 3.55 - j10.22 = 10.82e^{-j70.8°}$$

Therefore, $i(t) = 10.82e^{-j70.8°}e^{j2t}$. For the original excitations, $i(t) = 10.82 \cos (2t - 70.8°)$.

8. What do you think the result would be if the excitations were $e_1(t) = 10 \sin (2t - 30°)$ and $e_2(t) = 5 \sin (2t + 45°)$? *Hint:* Convert the sine functions to cosine functions ($\cos \theta = \sin (\theta + 90°)$) and proceed as before.

9. Develop an algorithm to convert between the various forms of complex notations (for example, between exponential, polar, and rectangular). Decide what process you will use. Develop a good human–machine interface (HMI) for the user to interface with your algorithm. Try to convert your algorithm into a working program on a PC or programmable calculator.

SOLUTION TO CHAPTER INTRODUCTION

To solve for the capacitance, use the following formula, solve for C, and substitute the values given:

$$f_0 = \frac{1}{2\pi \sqrt{LC}}$$

$$f_0^2 = \frac{1}{4\pi^2 LC}$$

$$4f_0^2\pi^2 LC = 1$$

$$C = \frac{1}{4f_0^2\pi^2 L}$$

$$C = \frac{1}{4(101 \times 10^6)^2\pi^2(0.0020)}$$

$$C = 1.2 \times 10^{-15} \text{ F}$$

Y

ou are responsible for installing a satellite dish and receiver. A satellite dish is a paraboloid of revolution. This figure is formed by rotating a parabola around its axis. Rays entering the dish along lines parallel to the axis are reflected through the focus. The receiver should be placed at the parabola's focus. If the satellite dish is 8 ft across and 1.3 ft deep, where should the receiver be placed in order to pick up incoming signals? (The solution to this problem is given at the end of the chapter.)

A problem like this requires a knowledge of the equation of an ellipse, one of the conic sections to be discussed in this chapter. In addition to the ellipse, we will discuss the straight line, circle, parabola, and hyperbola.

Learning Objectives

After you complete this chapter, you should be able to

1. Determine the equation of a line from given information (Section 15–1).

2. Determine the slope, x intercept, and y intercept of an equation, and use this information to graph the line (Section 15–1).

3. Find the center and radius of a circle given the equation in standard or general form (Section 15–2).

4. Find the equation of a circle given pertinent information (Section 15–2).

5. Find the focus, directrix, and vertex of a parabola from its equation, and sketch the graph of the parabola (Section 15–3).

6. Find the equation of a parabola from given information (Section 15–3).

7. Find the center, vertices, foci, and endpoints of the minor axis of an ellipse, and sketch the graph of the ellipse (Section 15–4).

8. Find the equation of an ellipse from given information (Section 15–4).

9. Find the center, vertices, foci, and endpoints of the conjugate axis, and the slope of the asymptotes of a hyperbola, and sketch the graph of the hyperbola (Section 15–5).

10. Find the equation of a hyperbola from given information (Section 15–5).

11. Given a general second-degree equation, determine the type of conic section, determine pertinent information about the conic section, and sketch its graph (Section 15–6).

12. Apply the concepts of analytic geometry to technical problems (Section 15–6).

Chapter 15

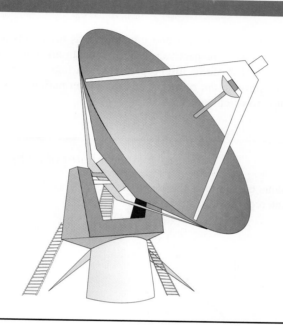

Analytic Geometry

15–1

THE STRAIGHT LINE

In this section we will discuss how to find the equation for a line from given information, which may include such quantities as slope, x intercept, y intercept, or points through which the line passes. We can use two forms for the equation of a line: the slope–intercept form or the point–slope form. We will start the discussion with the slope–intercept form.

Slope–Intercept Form

In Chapter 4, we defined the x and y intercepts as the points at which the graph crosses the x axis and y axis, respectively. The intercepts are represented as the ordered pairs $(x, 0)$ and $(0, y)$. The **slope–intercept form** of a line uses *the slope and the y intercept to determine uniquely the equation of a line*. You may want to review the discussion of slope and how to calculate it in Section 4–4.

Slope–Intercept Form of a Line

If m represents the slope of the line and b represents the y coordinate of the y intercept, the slope–intercept form is given by

$$y = mx + b$$

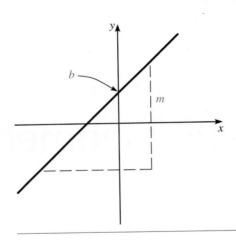

The **general form of a line** is given by

$$Ax + By + C = 0$$

where A, B, and C represent integers, and A and B are not both 0.

NOTE ✦ In this section you should always place the equation of the straight line in general form, which may require rearranging the equation. Remember that by algebraic convention, we usually do not leave fractional coefficients or leave the first term negative.

EXAMPLE 1 Determine the equation of the straight line with slope -3 and y intercept 5. Place your answer in general form.

Solution To use the slope–intercept form of a line, we must know the slope of the line and the y intercept. Here we are given both of these quantities. Therefore, we substitute $m = -3$ and $b = 5$ into the slope–intercept formula.

$$y = mx + b$$
$$y = -3x + 5$$

Then we must arrange the equation in general form, $Ax + By + C = 0$.

$$3x + y - 5 = 0 \qquad \blacksquare\blacksquare\blacksquare$$

EXAMPLE 2 Determine the equation of the line having slope 2/3 and passing through the point $(0, -4)$.

Solution We are given the slope of the line and the y intercept, written in point form. Therefore, $m = 2/3$ and $b = -4$. Substituting into the slope–intercept form and rearranging the equation to the general form gives

$$y = mx + b$$
$$y = \frac{2}{3}x - 4$$
$$3y = 2x - 12$$
$$-2x + 3y + 12 = 0$$
$$2x - 3y - 12 = 0 \qquad \blacksquare\blacksquare\blacksquare$$

Graphs

The slope–intercept form of a line is also very useful in graphing the line, as illustrated in the next example.

EXAMPLE 3 Find the slope, y intercept, and x intercept for the equation $2x - 5y = 8$, and use this information to graph the line.

Solution To determine the slope and the y intercept from the equation, we rearrange the equation in the slope–intercept form by solving the equation for y.

$$2x - 5y = 8$$
$$-5y = -2x + 8$$
$$y = \frac{-2x + 8}{-5}$$
$$y = \underbrace{\frac{2}{5}}_{\text{slope}}x - \underbrace{\frac{8}{5}}_{y \text{ intercept}}$$

To find the x intercept, we substitute $y = 0$ into the equation and solve for x.

$$2x - 5y = 8$$
$$2x - 5(0) = 8$$
$$2x = 8$$
$$x = 4$$

To graph this equation using the slope and the y intercept, start by placing a point on the y axis at the intercept. Then, from that point, plot the slope to determine a second point.

$$m = \frac{\text{rise}}{\text{run}} = \frac{\text{up or down depending on sign (in } y \text{ direction)}}{\text{left or right depending on sign (in } x \text{ direction)}}$$

In this case, since the slope is 2/5, we go up 2 units and right 5 units. Place the point and draw a line. Figure 15–1 shows the graph of $2x - 5y = 8$. We can verify that this line is correct by verifying that the x intercept is 4. ∎∎∎

FIGURE 15–1

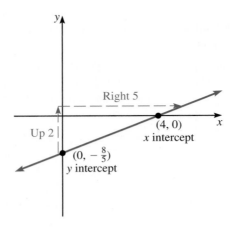

Point–Slope Form

The second form of a linear equation that we discuss in this section is called **point–slope,** defined as follows.

Point–Slope Form of a Line

If m is the slope of the line and (x_1, y_1) represents a point on the line, the equation of the line can be written as

$$y - y_1 = m(x - x_1)$$

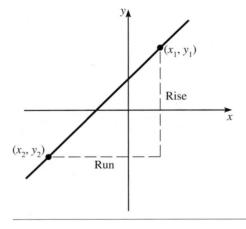

EXAMPLE 4   Determine the equation of a line having slope -3 and crossing the x axis at 4.

Solution   We are given the slope of the line and the x intercept. Since we are given the x intercept, it is easier to use the point–slope form, which requires that we write the x intercept as a point: (4, 0). Substituting $m = -3$ and $(x_1, y_1) = (4, 0)$ gives

$$y - y_1 = m(x - x_1)$$
$$y - 0 = -3(x - 4)$$
$$y = -3x + 12$$

Arranging the equation in general form, $Ax + By + C = 0$, gives

$$3x + y - 12 = 0$$

■■■

EXAMPLE 5   Find the equation of the line passing through $(-6, 3)$ and $(5, 4)$.

Solution   First, we must calculate the slope of the line. Let $(x_1, y_1) = (-6, 3)$ and $(x_2, y_2) = (5, 4)$. Then

$$m = \frac{y_2 - y_1}{x_2 - x_1}$$

$$m = \frac{4 - 3}{5 - (-6)} = \frac{1}{11}$$

Then we substitute $m = 1/11$ and (5, 4) as (x_1, y_1) into the point–slope formula.

NOTE ✦ Using either point will result in the same equation.
Substituting into the point–slope equation and rearranging to general form gives

$$y - y_1 = m(x - x_1)$$

$$y - 4 = \frac{1}{11}(x - 5) \qquad \text{substitute}$$

$$11y - 44 = x - 5 \qquad \text{eliminate fractions}$$
$$-x + 11y - 39 = 0 \qquad \text{transpose}$$
$$x - 11y + 39 = 0 \qquad \text{multiply by } -1$$

■■■

Horizontal, Vertical, and Perpendicular Lines

It is easier to write the equation for a horizontal or vertical line by inspection rather than by using the point–slope formula. The following box provides a summary to aid in this process.

Horizontal Line

Equation:	$y = b$
Slope:	0
Recognition:	Ordered pairs have the same y coordinate.

$$(x_1, b) \qquad (x_2, b)$$

Vertical Line

Equation: $x = a$

Slope: undefined

Recognition: Ordered pairs have the same x coordinate.

$$(a, y_1) \qquad (a, y_2)$$

EXAMPLE 6 Find the equation of each line satisfying the following conditions:

(a) a line having 0 slope and passing through $(-2, 6)$
(b) a line passing through $(-7, 4)$ and $(-7, 2)$

Solution

(a) Since the *slope is 0,* this is a horizontal line with equation $y = b$. From the ordered pair we can see that the y coordinate is 6. Therefore, the equation is

$$y = 6$$

(b) Since the ordered pairs have the *same x coordinate,* this is a vertical line with equation $x = a$, where a is the x coordinate of the ordered pairs. The equation is

$$x = -7 \qquad \blacksquare\blacksquare\blacksquare$$

Recall from Chapter 4 that perpendicular lines, except vertical and horizontal lines, have negative reciprocal slopes. The next example illustrates finding the equation of a line perpendicular to a given line.

EXAMPLE 7 Find the equation of the line perpendicular to $3x + y = 6$ and passing through $(2, -3)$.

Solution A line perpendicular to $3x + y = 6$ will have a slope that is the negative reciprocal of the slope of this line. The slope of this line is (arrange the equation in slope–intercept form):

$$3x + y = 6$$
$$y = -3x + 6$$

$$\overset{\nearrow}{\boxed{\text{slope}}}$$

The slope of the perpendicular line is 1/3, and the line passes through the point $(2, -3)$. Substituting this information into the point–slope form and arranging the result in the general form gives

$$y - y_1 = m(x - x_1)$$

$$y - (-3) = \frac{1}{3}(x - 2)$$

$$3y + 9 = x - 2$$
$$-x + 3y + 11 = 0$$
$$x - 3y - 11 = 0 \qquad \blacksquare\blacksquare\blacksquare$$

Applications

EXAMPLE 8   The resistance of a certain electrical circuit component increases by 0.015 Ω for every increase of 1°C. The resistance at 0°C is found to be 1.5 Ω. Find the equation for resistance R in terms of temperature T, and use the equation to find the resistance when the temperature is 30°C.

Solution   Since we are asked to find resistance in terms of temperature, R is the dependent variable and the equation takes the following form:

$$R = mT + b$$

This form is known and used because the change in resistance is constant for a given change in temperature. To determine the slope, recall that slope is defined to be

$$m = \frac{\text{change in } R}{\text{change in } T}$$

However, we were told that resistance changes 0.015 Ω for each 1°C change in temperature. Thus,

$$m = \frac{0.015 \ \Omega}{1°C} = 0.015 \ \Omega/°C$$

Then we determine the R intercept (comparable to a y intercept). We were told that the resistance is 1.5 Ω at 0°C. Therefore, $b = 1.5 \ \Omega$. The resulting equation is given by

$$R = 0.015T + 1.5$$

To find the resistance at 30°C, we substitute $T = 30°$ into the equation.

$$R = 0.015T + 1.5$$
$$R = 0.015(30) + 1.5$$
$$R = 1.95$$

At 30°C, the resistance is 1.95 Ω. ■■■

EXAMPLE 9   The length L of a spring varies linearly with the force F applied. If a force of 2.0 lb produces a length of 18 in., and a force of 5.0 lb produces a length of 24 in., find the equation relating length and force. Leave your answer in slope–intercept form.

Solution   Since force is the independent variable, we are asked to find the equation relating the ordered pairs (2, 18) and (5, 24). First, we must find the slope of the line passing through these two points.

$$m = \frac{y_2 - y_1}{x_2 - x_1} = \frac{24 - 18}{5 - 2} = 2.0 \text{ in-lb}$$

Then we use the point–slope form of a line with $m = 2$ and $(F_1, L_1) = (2, 18)$. The equation is

$$L - L_1 = m(F - F_1)$$
$$L - 18 = 2(F - 2)$$
$$L = 2.0F + 14$$

■■■

EXAMPLE 10 A computer systems consultant charges a flat fee of $250 and $75/hr. Find an equation for the cost C (in dollars) and the time t (in hours).

Solution The cost is the sum of the fixed cost (flat fee) and the variable cost (cost per hour). Therefore,

$$C = 250 + 75t$$

■ ■ ■

15–1 EXERCISES

From the given information determine the equation of the line. Leave your answer in general form, $Ax + By + C = 0$, with A, B, and C integers, and $A \geq 0$.

1. Slope $= -2$; y intercept $= -1$.

2. Slope $= -3/4$; y intercept $= -1$.

3. Slope $= 4$; x intercept $= 6$.

4. Slope $= -1$; passing through $(4, 6)$.

5. Passing through $(-2, 4)$ and $(3, 1)$.

6. 0 slope; passing through $(-8, -3)$.

7. Slope $= -2$; passing through $(3, -1)$.

8. Passing through $(5, -7)$ and $(-3, -4)$.

9. Undefined slope; passing through $(4, -7)$.

10. Parallel to line $y = -4$ and passing through $(-7, 0)$.

11. Passing through $(-1, 5)$ and $(6, -3)$.

12. Slope $= -4$, x intercept $= -5$.

13. Horizontal line; passing through $(7, -3)$.

14. Passing through $(6, 9)$ and $(-4, 8)$.

15. Parallel to line $x = -2$ and passing through $(3, 1)$. (*Hint:* Parallel lines have equal slopes.)

16. Perpendicular to line $y = 6$ and passing through $(-8, 4)$.

17. Parallel to the line $y = 3x - 6$ and passing through $(-1, 6)$.

18. Parallel to the line $3x + 2y = 10$ and passing through $(8, -4)$.

19. Perpendicular to the line $2x - y = 7$ and passing through $(-1, -4)$.

20. Perpendicular to the line $y = 4x - 7$ and passing through $(4, 3)$.

Find the slope, y intercept, and x intercept for the following equations. Use the slope and y intercept to graph the line.

21. $3x - 4y = 12$

22. $5x + 8y = 16$

23. $x = -8$

24. $x + 6y = 10$

25. $4x - 6y + 15 = 0$

26. $y = 5$

By arranging each equation in slope–intercept form, determine whether the following lines are parallel, perpendicular, or neither.

27. $2x + y = 9$ and $2y - x = 14$

28. $4x + 3y = 6$ and $10 + 6x - 8y = 0$

29. $7x - 5y + 8 = 0$ and $y = 7x + 12$

30. $3x + 2y = 5$ and $6y = 3x + 19$

31. $3y - 2x + 8 = 0$ and $4x - 9 = 6y$

32. $x + 5y - 7 = 0$ and $5x + y + 10 = 0$

B 33. Fixed cost is the cost of operation no matter how many items are manufactured, while variable cost is the cost of manufacturing each item. Total cost is the sum of fixed and variable costs. Let x represent the number of units manufactured per day and C represent total cost. If fixed cost is $800 per day and variable cost is $3 per unit, write an equation relating total cost to units manufactured.

P 34. The length of a heated object varies with temperature. A metal rod was found to be 12.4 m at 0°C and 12.5 m at 20.0°C. Write an equation for length as a function of temperature.

P 35. The pressure P on an object under water varies linearly with depth D. Find the equation relating pressure and depth if the pressure at 55 ft is 38 lb/in^2 and the pressure at 28 ft is 26 lb/in^2.

E 36. The electric resistance of a resistor increases 0.03 Ω for each 1°C increase in temperature. If its resistance at 0°C is 2 Ω, find the equation relating resistance to temperature.

P 37. The average velocity v of an object is the ratio of change in displacement s and change in time t. Find the equation relating displacement and time if the average velocity is 20 ft/s and $s = 15$ ft when $t = 0$ s.

P 38. The length of a spring varies linearly with the force applied. If a spring that is initially 12 in. long measures 16 in. when a force of 4.0 lb is applied, find the equation relating length and force.

E 39. The resistance of a circuit component increases 0.075 Ω for each 1.0°C increase in temperature. If the resistance is 3.5 Ω at 0°C, write the equation relating resistance to temperature.

P 40. The acceleration a of an object is the change in velocity v divided by the change in time t. Find the equation relating velocity and time if the acceleration is 24 m/s^2 and $v = 14$ m/s when $t = 0$ s.

B 41. A computer manufacturer incurs a fixed cost of $125 for a board plus an additional cost of $35 for installation. Write an equation relating the total cost C for installing n boards.

B 42. The profit per share for Rick's Rafts was $1.57 in 1995 and $2.35 in 1999. Write an equation for the profit per share in terms of the year and use it to predict the profit per share in 2003.

15–2

THE CIRCLE

The **circle** is defined as the set of points in a plane that are equidistant from a fixed point. *The fixed point* is called the **center** of the circle, and the distance from *the center to a point on the circle* is called the **radius**. If (h, k) represents the center of the circle and r represents the radius, the standard equation of a circle is as follows.

Standard Form of the Circle

If r represents the radius of a circle and (h, k) represents the center, then

$$(x - h)^2 + (y - k)^2 = r^2$$

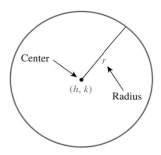

EXAMPLE 1 Determine the center, radius, and graph of the circle represented by the following equation:

$$(x - 4)^2 + (y + 3)^2 = 5$$

Solution   Since the equation is in standard form, the center is $(4, -3)$ and the radius is $\sqrt{5}$. To graph the circle, place a point at $(4, -3)$ to represent the center. Then draw a circle $\sqrt{5} \approx 2.2$ units from the center. The graph is shown in Figure 15–2.

FIGURE 15–2

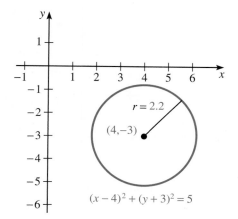

■■■

General Form

The **general form of the equation of a circle** results from simplifying its standard form.

General Form of the Equation of a Circle

$$x^2 + y^2 + Ax + By + C = 0$$

where A, B, and C represent constants.

The next two examples illustrate the procedure for finding the center and the radius of a circle from its equation in general form.

EXAMPLE 2   Determine the center, radius, and graph for the following circle:

$$x^2 + y^2 - 2x - 6y + 6 = 0$$

Solution   To determine the center and radius, we must arrange this equation in standard form by completing the square on the x and y terms. First, we group the x and y terms and transpose the constant.

$$x^2 - 2x + y^2 - 6y = -6$$

Then we complete the square by adding the square of half the linear term to both sides of the equation.

$$\left(\frac{2}{2} = 1^2\right) \qquad \left(\frac{6}{2} = 3^2\right)$$

$$(x^2 - 2x + 1) + (y^2 - 6y + 9) = -6 + 1 + 9$$

Next, we simplify the result:

$$(x - 1)^2 + (y - 3)^2 = 4$$

The center is (1, 3) and the radius is 2. Figure 15–3 shows the graph of the circle.

FIGURE 15–3

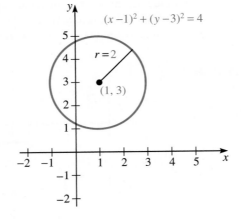

EXAMPLE 3   Find the center, radius, and graph of the following circle:

$$3x^2 + 3y^2 + 6x - 12y = 8$$

Solution Unlike the previous equation, the quadratic terms have a coefficient of 3. The standard form of the equation for a circle requires that the coefficient of the squared terms be 1. Therefore, before completing the square, we must divide the equation by 3.

$$3x^2 + 3y^2 + 6x - 12y = 8$$

$$x^2 + y^2 + 2x - 4y = \frac{8}{3}$$

Then we complete the square as in the previous example.

$$(x^2 + 2x + 1) + (y^2 - 4y + 4) = \frac{8}{3} + 1 + 4$$

$$(x + 1)^2 + (y - 2)^2 = \frac{23}{3}$$

The center is $(-1, 2)$ and the radius is $\sqrt{23/3} \approx 2.8$. The graph is shown in Figure 15–4.

FIGURE 15–4

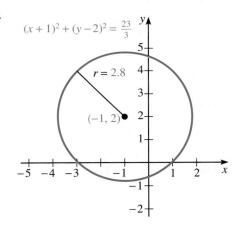

$(x + 1)^2 + (y - 2)^2 = \frac{23}{3}$

$r = 2.8$

$(-1, 2)$

■ ■ ■

CAUTION ✦ When determining the center of a circle, be careful of the signs. The standard form is $(x - h)^2 + (y - k)^2 = r^2$. Thus, the coordinates for the center of the circle have signs opposite those given in the standard form.

Finding the Equation of a Circle

Until now we have found the radius and center of a circle from its equation. However, we must be able to determine the equation of a circle from its geometric information. The next two examples illustrate this process.

EXAMPLE 4 Determine the equation of a circle whose center is at $(-2, 3)$ and whose radius is 5. Leave your answer in standard form.

Solution To find the equation of this circle, we substitute the coordinates for the center and the value of the radius into the standard form for the equation of a circle.

$$(x - h)^2 + (y - k)^2 = r^2$$

$$h = -2 \qquad k = 3 \qquad r = 5$$

The equation is

$$(x + 2)^2 + (y - 3)^2 = 25$$

∎∎∎

EXAMPLE 5 Determine the equation of a circle having its center at $(3, -1)$ and passing through $(-1, 2)$. Leave your answer in standard form.

Solution We are given the center but must determine the radius of the circle. The radius is defined as the distance from the center to any point on the circle, represented by the distance from $(3, -1)$ to $(-1, 2)$. Thus, using the distance formula gives

$$r = \sqrt{(x_2 - x_1)^2 + (y_2 - y_1)^2}$$
$$r = \sqrt{(3 + 1)^2 + (-1 - 2)^2}$$
$$r = \sqrt{25}$$
$$r = 5$$

We substitute into the standard form, $(x - h)^2 + (y - k)^2 = r^2$, the values $h = 3$, $k = -1$, and $r = 5$. The equation is

$$(x - 3)^2 + (y + 1)^2 = 25$$

∎∎∎

Application

EXAMPLE 6 An archway is composed of a semicircle atop a rectangle. If the rectangular portion is drawn to scale on the rectangular coordinate system, the four vertices of the rectangle are $(3, 2)$, $(9, 2)$, $(3, 12)$, and $(9, 12)$. Find the equation of the circle containing the semicircle top. Use the standard form for your answer.

FIGURE 15–5

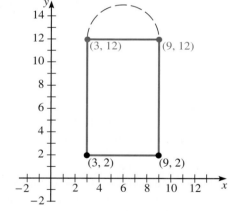

Solution Figure 15–5 shows the rectangle. The center of the circle is represented by $(6, 12)$, which is the midpoint of the line representing the top of the rectangle.

$$x_{mid} = \frac{3 + 9}{2} = \frac{12}{2} = 6$$

$$y_{mid} = \frac{12 + 12}{2} = \frac{24}{2} = 12$$

$$center = (6, 12)$$

The radius is half the diameter represented by the distance between the vertices.

$$r = \frac{9 - 3}{2} = \frac{6}{2} = 3$$

The equation of the circle is given by

$$(x - 6)^2 + (y - 12)^2 = 9$$

∎

15–2 EXERCISES

Find the radius and center of the circle and sketch its graph.

1. $x^2 + y^2 = 9$ **2.** $x^2 + y^2 = 16$

3. $x^2 + y^2 = 49$

4. $(x + 4)^2 + (y - 3)^2 = 4$

5. $(x - 1)^2 + y^2 = \sqrt{3}$ **6.** $x^2 + (y + 5)^2 = \sqrt{2}$

7. $(x - 3)^2 + (y + 2)^2 = 8$

8. $x^2 + y^2 - 16y - 2x - 8 = 0$

9. $x^2 - 12x + y^2 + 2y + 28 = 0$

10. $x^2 + y^2 - 5x + 12y = 4$

11. $2x^2 + 2y^2 + 10x - 4y - 6 = 0$

12. $3x^2 + 3y^2 + 15y - 9x = 0$

13. $3x^2 + 3y^2 + 6x - 8y + 5 = 0$

14. $4x^2 + 4y^2 - 8y + 12x + 7 = 0$

Determine the standard form of the equation for the circle from the given information.

15. Center at $(0, 0)$; radius of 9.

16. Center at the origin; radius of 7.

17. Center at $(-3, -1)$; radius of 3.

18. Center at $(6, 10)$; radius of 4.

19. Center at $(0, -4)$; radius of $\sqrt{6}$.

20. Center at $(6, 1)$; radius of $\sqrt{9}$.

21. Center at $(-8, 5)$; passing through $(2, 3)$.

22. Center at $(3, 6)$; passing through $(1, 4)$.

23. Center at $(6, -4)$; passing through $(3, 2)$.

24. Center at $(-7, -6)$; passing through $(-4, 0)$.

25. Center at $(3, 4)$; tangent to x axis.

26. Center at $(-5, 4)$; tangent to y axis.

27. A diameter is the segment from $(-6, -4)$ to $(3, 5)$.

28. A diameter is the segment from $(7, -3)$ to $(-4, 5)$.

P 29. Centripetal force F is given by

$$F = \frac{mv^2}{R}$$

where m = mass of an object moving with speed v in a circular path of radius R. A particle moves in the xy plane, and its center is $(-4, 6)$. Find the equation of the circular path traveled by the particle, which has mass 0.0004 kg and moves at 3.1 m/s under a centripetal force of 1,200 N.

M 30. A draftsperson is drawing a friction drive as shown in Figure 15–6. Determine the equation for each circle if the origin is at the center of the larger circle and the radii of the circles are 9 cm and 4 cm.

FIGURE 15–6

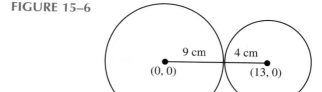

E 31. The impedance Z in an ac circuit is given by

$$Z^2 = R^2 + (X_L - X_C)^2$$

where R = resistance, X_L = inductive reactance, and X_C = capacitive reactance. Graph the relationship between resistance and inductive reactance if $Z = 30\ \Omega$ and $X_C = 50\ \Omega$.

15–3

▬▬▬▬▬▬

THE PARABOLA

Definition

The **parabola** is defined as the set of points in a plane equidistant from a given point and a given line. *The given point is called the **focus**, and the given line is called the **directrix**. The line through the focus and perpendicular to the directrix is called the **axis** of the* parabola.

The point of intersection of the axis and the parabola is called the **vertex.** Figure 15–7 illustrates this terminology.

FIGURE 15–7

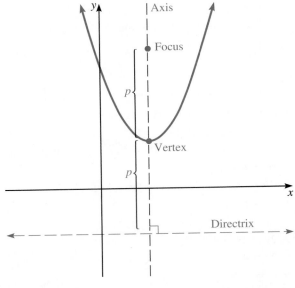

Geometry

The parabola can be placed so that its axis is a vertical line or a horizontal line. A parabola with a vertical axis is given by the following equation:

$$(x - h)^2 = 4p(y - k)$$

where (h, k) represents the coordinate of the vertex, and p represents the directed distance along the parabola's axis from the vertex to the focus. If $4p$ is a positive number, the parabola opens upward; if it is a negative value, the parabola opens downward. Figure 15–8(a) and (b) illustrates this terminology.

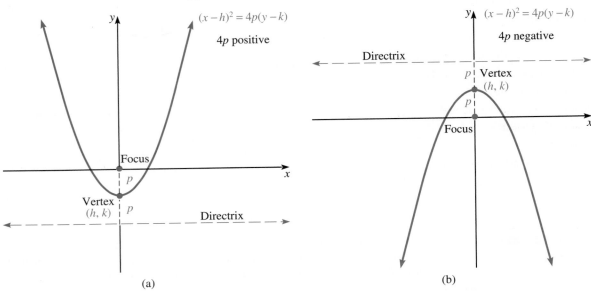

(a)

(b)

FIGURE 15–8

NOTE ✦ The focus is *always* located inside the parabola, and the directrix is *always* a line outside the parabola.

LEARNING HINT ✦ Rather than trying to memorize formulas, we will find p in each case and count from the vertex to determine the focus and directrix.

Similarly, a parabola whose axis is a horizontal line is represented by the following equation:

$$(y - k)^2 = 4p(x - h)$$

A negative value for $4p$ means that the parabola opens left, and a positive value means that the parabola opens right. Figure 15–9(a) and (b) illustrates this type of parabola.

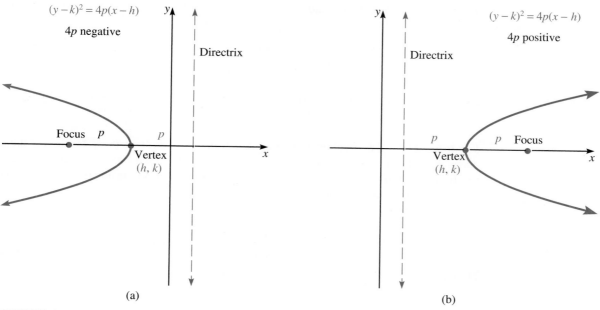

(a)

(b)

FIGURE 15–9

Finding the Focus, Directrix, and Vertex

EXAMPLE 1 Determine the vertex, directrix, focus, and graph of the parabola given by

$$x^2 = 16y$$

Solution The standard equation for a parabola of this form is

$$(x - h)^2 = 4p(y - k)$$

Writing the given equation in this form yields

$$(x - 0)^2 = 16(y - 0)$$

From this form of the equation, we can determine that the vertex is $(0, 0)$ and $4p = 16$ or $p = 4$. Let us determine the remaining information from the graph shown in Figure 15–10. Since the value of $4p$ is positive, the graph opens upward. The focus is 4 units

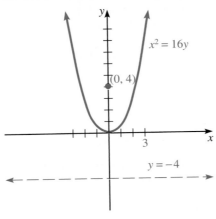

FIGURE 15–10

from the vertex inside the parabola, and the directrix is 4 units from the vertex outside the parabola. The vertex, focus, and directrix are summarized as follows:

$$\text{vertex:} \quad (0, 0)$$
$$\text{focus:} \quad (0, 4)$$
$$\text{directrix:} \quad y = -4$$

■ ■ ■

EXAMPLE 2 Determine the vertex, directrix, focus, and graph of the parabola given by

$$(x + 3)^2 = -4(y - 1)$$

Solution Since the equation is in standard form, the vertex is $(-3, 1)$; the parabola opens down because $4p$ is negative; and $p = -1$. We can determine the focus and directrix from the graph as shown in Figure 15–11. The focus is 1 unit inside the parabola, and the directrix is 1 unit outside the parabola.

$$\text{vertex:} \quad (-3, 1)$$
$$\text{focus:} \quad (-3, 0)$$
$$\text{directrix:} \quad y = 2$$

■ ■ ■

FIGURE 15–11

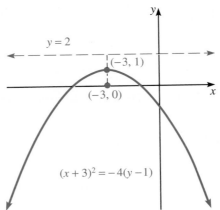

EXAMPLE 3 Determine the vertex, directrix, focus, and graph of the parabola given by

$$y^2 + 8y = 12x - 40$$

Solution Since this equation is not in standard form, we must complete the square.

$$(y^2 + 8y + 16) = 12x - 40 + 16 \quad \text{complete the square}$$
$$(y + 4)^2 = 12x - 24 \quad \text{simplify}$$
$$(y + 4)^2 = 12(x - 2) \quad \text{factor}$$

The vertex is $(2, -4)$, the parabola opens to the right, and $4p = 12$ or $p = 3$. The focus, vertex, and directrix are as follows:

$$\text{vertex:} \quad (2, -4)$$
$$\text{focus:} \quad (5, -4)$$
$$\text{directrix:} \quad x = -1$$

The graph is shown in Figure 15–12. ■■■

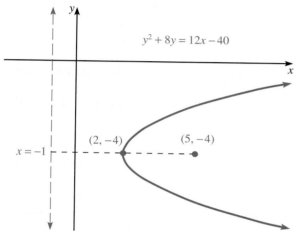

FIGURE 15–12

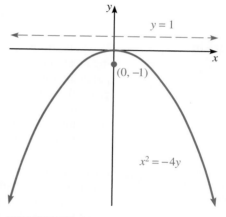

FIGURE 15–13

Finding the Equation of a Parabola

EXAMPLE 4 Find the equation of a parabola whose directrix is the line $y = 1$ and whose vertex is the point $(0, 0)$.

Solution To determine the equation of a parabola, we must fill in the value of p and (h, k). From the information given, we can sketch the graph shown in Figure 15–13. Because the directrix is a horizontal line and the vertex is $(0, 0)$, we know the equation of the parabola is of the form $x^2 = 4py$. We also know that p is the directed distance along the parabola's axis from the vertex to the focus; therefore, $p = -1$. The equation of the parabola is

$$x^2 = -4y$$ ■■■

EXAMPLE 5 Find the equation of the parabola whose directrix is $x = 5$ and whose focus is $(1, -2)$.

Solution Figure 15–14 shows a graph of the parabola. We know that the focus is located on the axis of the parabola, and the vertex is located on the axis of the parabola

FIGURE 15–14

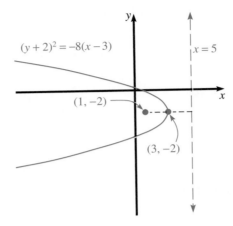

halfway between the focus and the directrix. Therefore, the coordinates of the vertex are $(3, -2)$, and the value of p is -2. The equation of the parabola is given by

$$(y + 2)^2 = 4(-2)(x - 3) \qquad (y + 2)^2 = -8(x - 3) \qquad \blacksquare\blacksquare\blacksquare$$

Application

Parabolas occur in several different applications. A parabolic reflector is formed by rotating a parabola about its axis of symmetry. Any light rays emanating from the focus of the parabolic reflector will reflect off the mirror in lines parallel to the axis of symmetry. This principle is used in flashlights, searchlights, and some automobile headlights.

Conversely, light rays emanating from a distant source so they are parallel to the axis of symmetry of the parabolic reflector will be reflected to a single point at the focus. This principle is used in certain solar energy devices, satellite dishes, and reflecting telescopes.

EXAMPLE 6 A satellite dish is shaped like a parabolic reflector. The signals from the satellite strike the surface of the dish and are reflected to a single point, where the receiver is located. If the dish is 18 in. across at its opening and is 12 in. deep at its center, where should the receiver be placed?

Solution The form of the equation of the parabola is $x^2 = 4py$. Since $(9, 12)$ is a point on the parabola,

$$x^2 = 4py$$
$$(9)^2 = 4p(12)$$
$$81 = 48p$$
$$1.7 \approx p \qquad \blacksquare\blacksquare\blacksquare$$

The receiver should be located 1.7 in. from the base of the dish along the axis of symmetry.

EXAMPLE 7 A cable used to support a swinging bridge approximates the shape of a parabola. Determine the equation of the parabola if the length of the bridge is 100 ft and the vertical distance from where the cable is attached to the bridge to the lowest point of the cable is 20 ft. Assume that the origin is at the lowest point of the cable.

Solution Figure 15–15 gives a sketch of the parabola. Since we can assume that the origin is at the lowest point, the vertex is $(0, 0)$, and the general form of the equation is $x^2 = 4py$. Next, we must determine the value of p. From the information given, the points $(50, 20)$ and $(-50, 20)$ are on the curve. Substituting the point $(50, 20)$ to determine p gives

$$x^2 = 4py$$
$$(50)^2 = 4p(20)$$
$$2{,}500 = 80p$$
$$31.25 = p$$

The equation for the parabola is

$$x^2 = 4py$$
$$x^2 = 125y$$

FIGURE 15–15

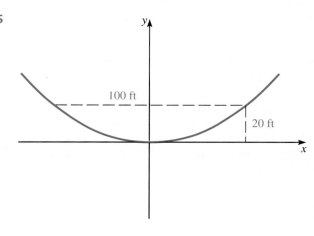

100 ft

20 ft

15–3 EXERCISES

Find the coordinates of the vertex and focus, determine the equation of the directrix, and sketch the graph of the parabola given by each of the following equations.

1. $y^2 = 8x$

2. $y^2 = 12x$

3. $x^2 = 12y$

4. $x^2 = 2y$

5. $y^2 = -16x$

6. $y^2 = -20x$

7. $x^2 = -10y$

8. $x^2 = -y$

9. $(y - 3)^2 = -8x$

10. $(y + 5)^2 = 6x$

11. $(x + 2)^2 = 12(y - 3)$

12. $(x - 3)^2 = -8(y - 2)$

13. $(y - 6)^2 = -2(x + 4)$

14. $(y + 2)^2 = 3(x - 5)$

15. $(x - 7)^2 = -16(y + 3)$

16. $(x + 8)^2 = 4(y + 1)$

17. $x^2 + 2x = -y + 3$

18. $x^2 - 4y + 8x = 0$

19. $y^2 + 10y = 2x - 15$

20. $y^2 - 12y + 8x - 20 = 0$

21. $3x^2 + 6x - 4y + 15 = 0$

22. $2x^2 + 8x + 4y - 20 = 0$

Find the equation of the parabola from the given geometric properties.

23. Vertex (0, 0); directrix $y = 4$.

24. Focus (1, 8); directrix $y = -2$.

25. Vertex (3, 2); focus (3, 5).

26. Directrix $x = 4$; vertex (6, 3).

27. Directrix $x = -3$; focus (1, 4).

28. Vertex (−4, 6); focus (−4, 5).

29. Vertex (0, −3), passes through (4, 7), and directrix is parallel to the x axis.

30. Directrix $x = -3$; focus (−1, −2).

31. Vertex (3, −6); focus (3, −2).

32. Vertex (3, −2), passes through (1, 6); axis parallel to y axis.

P 33. The height s of a projectile is given by

$$s = v(\sin \theta)t + \frac{at^2}{2}$$

If $v = 50$ m/s, $\theta = 30°$, and $a = -9.8$ m/s^2, write the equation in standard form.

P 34. A cable hangs in a parabolic curve between two poles 70 ft apart. At a distance 20 ft from each pole, the cable is 4 ft above its lowest point. Find the height of the cable 45 ft from each pole.

P 35. The stopping distance y of a car traveling at v mi/h is given by

$$y = v + \frac{v^2}{20}$$

Arrange this equation in standard form and plot its graph.

P 36. If a light beam passes through the focus of a parabolic reflector, the light rays will be reflected on a line parallel to the axis of the parabola. A light beam passes through the focus of a parabolic reflector with its cross-section represented by $y^2 = 18x$ and strikes the reflector at (0.5, 3). Find the equation of the reflected line.

E 37. Heat H in a resistor R develops at a rate given by $H = Ri^2$ where i = current. Graph H versus i if $R = 10\ \Omega$.

E 38. A shipping lane is determined by points equidistant from a lighthouse and the opposite shoreline. Determine the equation for the shipping lane if the lighthouse is 3 mi from the shoreline.

E 39. A satellite dish is shaped like a parabolic reflector. The satellite signals hit the surface of the dish and reflect to a single point, where the receiver is located. If the dish is 9 ft across and is 4 ft deep at its center, where should the receiver be located?

E 40. A cable television company owns a satellite dish in the shape of a parabolic reflector. Find the location of the receiver if the dish is 10 ft across and 4 ft deep at its center.

P 41. A reflecting telescope contains a mirror shaped like a parabolic reflector. If the mirror is 6 ft across at its opening and is 2 ft deep, where should the light be focused?

M 42. A simply supported beam is 12 ft long and has a load placed at its center. The deflection of the beam at its center is 2 inches. Given that the shape of the deflected beam is a parabola, find its equation.

P 43. An astronomer constructs a reflecting telescope with a parabolic mirror at one end of a hollow tube forming the image at the other end of the tube. If the mirror has a radius of 8 inches and a depth of 0.0013 inches, find the position of the image.

E 44. The parabolic headlight on a car is 8 inches in diameter and 5 inches deep. Where should the lightbulb be placed so the reflected rays are parallel?

C 45. The cables of a suspension bridge are in the shape of a parabola. The towers supporting the cable are 800 ft apart and 75 ft high. If the cables are at a height of 20 ft midway between the towers, what is the height of the cable at a point 40 ft from a tower?

15–4

THE ELLIPSE

Definition

The **ellipse** is defined as *the set of points in the plane the sum of whose distance from two fixed points is constant. The two fixed points* are called the **foci** (plural of **focus**) of the ellipse. The **center** of the ellipse is *the midpoint of the line segment joining the foci. The line segment whose endpoints are on the ellipse and passing through the foci* is called the **major axis.** The major axis is also the longer axis. *The endpoints of the major axis* are called the **vertices** of the ellipse. *The line segment whose endpoints are on the ellipse and passing through the center perpendicular to the major axis* is called the **minor axis.** *The endpoints of the minor axis* are simply called the **endpoints of the minor axis.** This terminology is illustrated in Figure 15–16.

Geometry

An ellipse may have its major axis horizontal or vertical. The equation of *an ellipse elongated horizontally* is given by

$$\frac{(x - h)^2}{a^2} + \frac{(y - k)^2}{b^2} = 1 \qquad a^2 > b^2$$

where (h, k) represent the coordinates of the center, a is the distance from the center to the vertices, and b is the distance from the center to the endpoints of the minor axis. The distance from the center to either focus is represented by c where

$$c = \sqrt{a^2 - b^2}$$

FIGURE 15–16

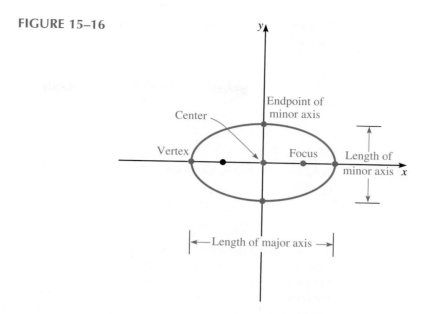

Notice that the foci are always located on the major axis. Figure 15–17 illustrates the terminology for this ellipse.

Similarly, *an ellipse elongated vertically* is represented by the equation

$$\frac{(x-h)^2}{b^2} + \frac{(y-k)^2}{a^2} = 1$$

where h, k, a, and b represent the same quantities as above. The major axis, vertices, and foci are located on a vertical line. Figure 15–18 illustrates the terminology for this ellipse.

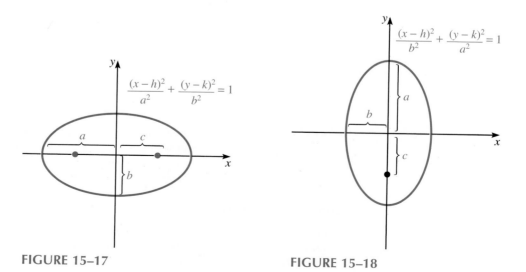

FIGURE 15–17   **FIGURE 15–18**

LEARNING HINT ✦ For an equation of an ellipse in standard form, a^2 is the larger number in the denominator. Furthermore, the term under which a^2 appears determines the direction of elongation.

EXAMPLE 1 Find the coordinates for the center, vertices, endpoints of the minor axis, and foci of the ellipse represented by

$$16x^2 + 25y^2 = 400$$

Also sketch the graph.

Solution To begin, we place this equation in standard form by dividing the equation by 400.

$$\frac{x^2}{25} + \frac{y^2}{16} = 1 \qquad \text{or} \qquad \frac{(x-0)^2}{25} + \frac{(y-0)^2}{16} = 1$$

The center is at the origin. Since the larger denominator is under x^2, the major axis is horizontal. To find the vertices, move $\sqrt{25} = 5$ units horizontally from the center. To find the endpoints of the minor axis, move $\sqrt{16} = 4$ units vertically from the center. To find the coordinates of the foci, move $\sqrt{25 - 16} = 3$ units horizontally from the center. Figure 15–19 gives the graph. The results are summarized as follows:

$$\begin{aligned}
\text{center:} &\quad (0, 0) \\
\text{vertices:} &\quad (-5, 0) \text{ and } (5, 0) \\
\text{endpoints of minor axis:} &\quad (0, 4) \text{ and } (0, -4) \\
\text{foci:} &\quad (3, 0) \text{ and } (-3, 0)
\end{aligned}$$

FIGURE 15–19

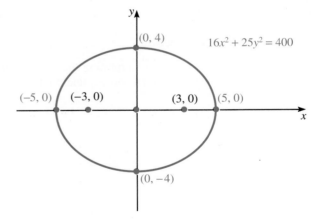

EXAMPLE 2 Find the coordinates for the center, vertices, endpoints of the minor axis, and foci of the ellipse represented by

$$\frac{(x-3)^2}{4} + \frac{(y+1)^2}{49} = 1$$

Also sketch the graph.

Solution Since the equation is in standard form, we can determine that the center is the point $(3, -1)$. Also, because the larger denominator is under the y term, the major axis is vertical. To determine the vertices, move $\sqrt{49} = 7$ units vertically from the center. To de-

FIGURE 15–20

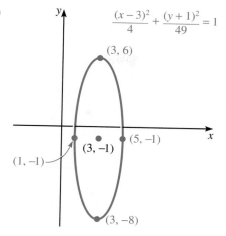

$$\frac{(x-3)^2}{4} + \frac{(y+1)^2}{49} = 1$$

termine the endpoints of the minor axis, move $\sqrt{4} = 2$ units horizontally from the center. To determine the coordinates of the foci, move $\sqrt{49 - 4} = \sqrt{45} \approx 6.7$ units vertically from the center. The graph is shown in Figure 15–20, and the results are summarized as follows:

center: $(3, -1)$

vertices: $(3, 6)$ and $(3, -8)$

endpoints of minor axis: $(1, -1)$ and $(5, -1)$

foci: $(3, 5.7)$ and $(3, -7.7)$

EXAMPLE 3   Find the coordinates for the center, vertices, endpoints of the minor axis, and foci of the ellipse represented by the equation

$$2x^2 + 3y^2 + 8x - 18y = 1$$

Also sketch the graph.

Solution   First, we begin by arranging the equation in standard form and completing the square.

$$(2x^2 + 8x) + (3y^2 - 18y) = 1 \qquad \text{group terms}$$
$$2(x^2 + 4x) + 3(y^2 - 6y) = 1 \qquad \text{factor}$$
$$2(x^2 + 4x + 4) + 3(y^2 - 6y + 9) = 1 + 8 + 27 \qquad \text{complete the square}$$
$$2(x + 2)^2 + 3(y - 3)^2 = 36 \qquad \text{factor}$$
$$\frac{(x + 2)^2}{18} + \frac{(y - 3)^2}{12} = 1 \qquad \text{divide}$$

The graph is shown in Figure 15–21. Summarizing the information gives

$$a = \sqrt{18} \approx 4.2 \qquad b = \sqrt{12} \approx 3.5 \qquad c = \sqrt{18 - 12} \approx 2.4$$

center: $(-2, 3)$

vertices: $(2.2, 3)$ and $(-6.2, 3)$

endpoints of minor axis: $(-2, 6.5)$ and $(-2, -0.5)$

foci: $(0.4, 3)$ and $(-4.4, 3)$

FIGURE 15–21

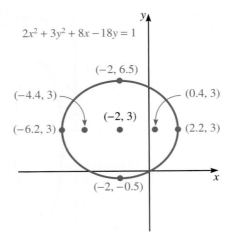

$$2x^2 + 3y^2 + 8x - 18y = 1$$

Finding the Equation of an Ellipse

Since we know the geometric properties of the ellipse, we can use them to find its equation. The next two examples illustrate this process.

EXAMPLE 4 Determine the equation of an ellipse centered at the origin if one focus is at (0, 3) and the endpoints of the minor axis are at (2, 0) and (−2, 0). Figure 15–22 shows this information.

FIGURE 15–22

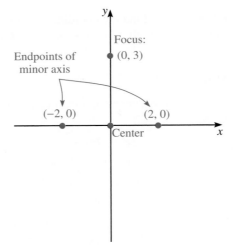

Solution To determine the equation, we must know the values of h, k, a^2, and b^2. Because the endpoints of the minor axis are (2, 0) and (−2, 0), the center is midway, at (0, 0), and $h = 0$, $k = 0$. We also know that $b = 2$ and $b^2 = 4$. One focus at (0, 3) tells us that $c = 3$, or $c^2 = 9$. Then $c^2 = a^2 - b^2$ and $c^2 + b^2 = a^2$. Thus

$$a^2 = 9 + 4$$
$$a^2 = 13$$

The major axis is in the y direction; therefore, the equation is

$$\frac{(x - 0)^2}{4} + \frac{(y - 0)^2}{13} = 1$$

where h points to 0 in $(x - 0)^2$, k points to 0 in $(y - 0)^2$, b^2 points to 4, and a^2 points to 13.

■ ■ ■

EXAMPLE 5 Determine the equation of an ellipse if the foci are at $(-6, 3)$ and $(4, 3)$ and the length of the major axis is 14. Figure 15–23 shows the given information.

FIGURE 15–23

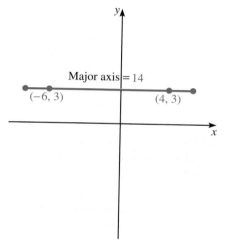

Major axis $= 14$
$(-6, 3)$ $(4, 3)$

Solution Once again we must determine h, k, a^2, and b^2. The center of the ellipse is located midway between the foci. The center is

$$\left(\frac{-6 + 4}{2}, \frac{3 + 3}{2}\right) = (-1, 3)$$

Therefore, $h = -1$ and $k = 3$. Since c represents the distance from the center $(-1, 3)$ to each focus $(-6, 3)$, $c = 5$ and $c^2 = 25$. Because the length of the major axis is 14, $2a = 14$, $a = 7$, and $a^2 = 49$. From this information we can determine b^2.

$$c^2 = a^2 - b^2$$
$$b^2 = a^2 - c^2$$
$$b^2 = 49 - 25$$
$$b^2 = 24$$

From the location of the foci, we can determine that the major axis is horizontal; therefore, the equation is

$$\frac{(x - (-1))^2}{49} + \frac{(y - 3)^2}{24} = 1$$

$$\frac{(x + 1)^2}{49} + \frac{(y - 3)^2}{24} = 1$$

■ ■ ■

Applications

Applications of ellipses are found in many areas of science and engineering. The orbit of the planets around the sun are elliptical, with the sun's position as one focus. Bridges are often shaped as semielliptical arches. Ellipses are also used in the construction of the ceiling of a whispering gallery. In a whispering gallery, a person standing at one focus can whisper and be heard by another person at the other focus.

EXAMPLE 6 The ceiling of a whispering gallery relative to the position of two people is 60 ft long and 25 ft high at its highest point. Where should the two people stand to hear a whisper?

Solution We set up a rectangular coordinate system so the center of the ellipse is at the origin and the major axis is along the *x* axis. The equation is

$$\frac{x^2}{a^2} + \frac{y^2}{b^2} = 1$$

where $a = 60$ and $b = 25$. Since

$$c^2 = a^2 - b^2 = 3,600 - 625 = 2,975$$

$$c \approx 54.5$$

The people should be located approximately 54.5 ft from the center.

EXAMPLE 7 A semielliptical arch is used in the construction of a shopping center. If the span of the archway is 150 ft and the maximum height is 20 ft, determine the equation of the archway.

Solution Figure 15–24 shows the archway. For convenience let us place the origin at the center of the semiellipse. From the drawing we can determine that $a = 75$ ft and $b = 20$ ft. Therefore, the equation of the ellipse is

$$\frac{x^2}{5,625} + \frac{y^2}{400} = 1$$

FIGURE 15–24

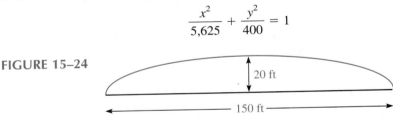

20 ft

150 ft

15–4 EXERCISES

Determine the coordinates for the center, vertices, endpoints of the minor axis, and foci of the ellipse. Round to tenths, if necessary. Sketch the graph.

1. $\dfrac{x^2}{25} + \dfrac{y^2}{9} = 1$

2. $\dfrac{x^2}{49} + \dfrac{y^2}{25} = 1$

8. $\dfrac{(x-3)^2}{64} + \dfrac{(y-2)^2}{25} = 1$

3. $\dfrac{x^2}{16} + \dfrac{y^2}{36} = 1$

4. $\dfrac{x^2}{16} + \dfrac{y^2}{81} = 1$

9. $\dfrac{(x+1)^2}{36} + \dfrac{y^2}{81} = 1$

5. $\dfrac{x^2}{10} + \dfrac{y^2}{6} = 1$

6. $\dfrac{x^2}{4} + \dfrac{y^2}{36} = 1$

10. $\dfrac{(y+6)^2}{8} + \dfrac{(x-4)^2}{16} = 1$

7. $\dfrac{(x+2)^2}{4} + \dfrac{(y-6)^2}{25} = 1$

11. $\dfrac{(y-1)^2}{4} + \dfrac{(x-3)^2}{12} = 1$

12. $\dfrac{(x - 2)^2}{10} + \dfrac{(y + 7)^2}{36} = 1$

13. $\dfrac{(x + 8)^2}{30} + \dfrac{(y + 4)^2}{12} = 1$

14. $3x^2 + 7y^2 = 21$

15. $16x^2 + 36y^2 - 144 = 0$

16. $9x^2 + 25y^2 = 225$

17. $x^2 + 8x + 3y^2 - 12y - 1 = 0$

18. $y^2 + 5x^2 - 20x = 20$

19. $4x^2 + 7y^2 + 14y + 24x = 75$

20. $8x^2 - 32x + 2y^2 - 12y - 23 = 0$

21. $6x^2 + y^2 + 24x = 12$

22. $x^2 + 6y^2 - 24y - 40 = 0$

Find the equation of the ellipse from the given geometric properties.

23. Center at origin; endpoint of major axis (0, 7); and focus at (0, 3).

24. Center at origin; endpoint of minor axis (5, 0); and focus at $(0, \sqrt{12})$.

25. Center at (4, −3); focus at (3, −3); and length of minor axis is 4.

26. Center at (1, 5); length of minor axis is 6; length of major axis is 10; and major axis is horizontal.

27. Foci at (−7, −3) and (1, −3); length of minor axis is 6.

28. Vertices at (3, −1) and (3, 7); focus at (3, 1).

29. Vertices at (−3, −8) and (−3, 4); length of minor axis is 10.

30. Endpoints of the minor axis at (4, 10) and (4, 7); length of major axis is 16.

P 31. A satellite orbits the earth in an elliptical path. The maximum altitude of the satellite is 325 mi, the minimum altitude is 130 mi, and the earth's center is at one focus. Determine the equation of the satellite's path. (The radius of the earth is 3,960 mi.)

C 32. A bridge in the shape of a semiellipse spans 80 ft and is 20 ft high at its maximum. What is the maximum height of a truck that can cross the bridge at a point 30 ft from the center?

M 33. A semielliptical arch for a railroad tunnel has a major axis of 150 ft and a height at the center of 50 ft. Find the equation for the elliptical tunnel.

P 34. Determine the positions where two people should stand if a whispering gallery is 50 ft long and 20 ft high at its highest point.

15–5

THE HYPERBOLA

Definition

The **hyperbola** is defined as *the set of points in a plane the difference of whose distance from two fixed points is a constant. The two fixed points are called the* **foci.** *The midpoint of the line segment joining the foci is called the* **center** *of the hyperbola. The line segment whose endpoints are on the hyperbola and if extended would join the foci is called the* **transverse axis.** *The points where the hyperbola intersects the transverse axis are called the* **vertices** *of the hyperbola. The line segment passing through the center and perpendicular to the transverse axis is called the* **conjugate axis.** These definitions are illustrated in Figure 15–25.

Geometry

A hyperbola can have a horizontal or a vertical transverse axis. The equation of a *hyperbola with a horizontal transverse axis is given by*

$$\dfrac{(x - h)^2}{a^2} - \dfrac{(y - k)^2}{b^2} = 1$$

where the point (h, k) is the center, a represents the distance from the center to a vertex, and b represents the distance from the center to an endpoint of the conjugate axis. The slopes of the asymptotes are given by

$$m = \pm \dfrac{b}{a}$$

FIGURE 15–25

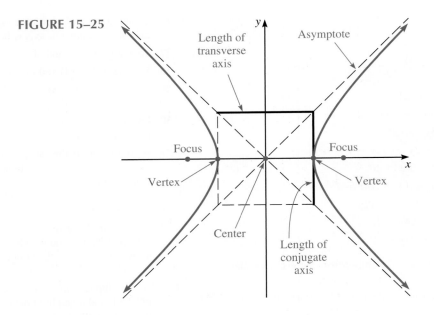

The foci are a distance c from the center where $c = \sqrt{a^2 + b^2}$ along the transverse axis. Figure 15–26 illustrates this terminology.

The equation of a *hyperbola whose transverse axis is vertical* is given by

$$\frac{(y - k)^2}{a^2} - \frac{(x - h)^2}{b^2} = 1$$

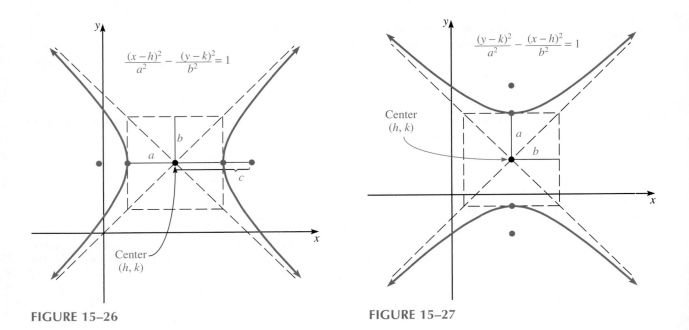

FIGURE 15–26

FIGURE 15–27

where h, k, a, and b represent the same quantities as above. The slopes of the asymptotes are given by

$$m = \pm \frac{a}{b}$$

Figure 15–27 illustrates this hyperbola.

LEARNING HINT ✦ If the equation is in standard form, the direction of opening of the hyperbola is determined by which term is positive. If the y term is positive, the hyperbola opens upward and downward. If the x term is positive, the hyperbola opens left and right. The letter a^2 always appears in the denominator of the positive term (and a^2 need *not* be larger than b^2).

To graph a hyperbola, locate the center. Use the distance a to locate the vertices, and use the distance b to locate the endpoints of the conjugate axis. Draw a rectangle through the vertices and endpoints. The asymptotes pass through the center and opposite corners of this rectangle. Finally, draw the hyperbola starting at the vertices and extending toward the asymptotes.

EXAMPLE 1   Determine the foci, center, vertices, endpoints of the conjugate axis, and slopes of the asymptotes of the following hyperbola:

$$\frac{x^2}{16} - \frac{y^2}{25} = 1$$

Also sketch the graph.

Solution   Since the equation is in standard form, the center is at $(0, 0)$, $a = \sqrt{16} = 4$, $b = \sqrt{25} = 5$, and the transverse axis is horizontal. Figure 15–28 shows the graph of this hyperbola, and the results are summarized as follows:

FIGURE 15–28

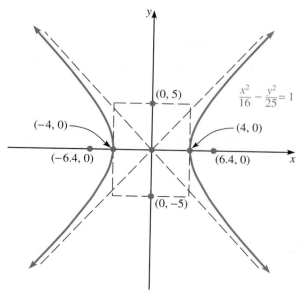

$$\begin{aligned}
\text{center:} \quad & (0, 0) \\
\text{vertices:} \quad & (4, 0) \text{ and } (-4, 0) \\
\text{endpoints of conjugate axis:} \quad & (0, 5) \text{ and } (0, -5) \\
\text{foci:} \quad & (6.4, 0) \text{ and } (-6.4, 0) \\
& (c = \sqrt{a^2 + b^2} = \sqrt{16 + 25} \approx 6.4)
\end{aligned}$$

$$\text{slope of asymptotes:} \quad m = \pm\frac{5}{4} \qquad \blacksquare\blacksquare\blacksquare$$

EXAMPLE 2 Determine the foci, center, vertices, endpoints of the conjugate axis, and slopes of the asymptotes of the following hyperbola:

$$\frac{(y + 1)^2}{36} - \frac{(x - 3)^2}{81} = 1$$

Also sketch the graph.

Solution Since this equation is in standard form, we know that the center is $(3, -1)$, $a = 6$, $b = 9$, and the transverse axis is vertical. The distance from the center to the foci is given by $c = \sqrt{36 + 81} = \sqrt{117} \approx 10.8$. Figure 15–29 shows the graph, and the results are summarized as follows:

$$\begin{aligned}
\text{center:} \quad & (3, -1) \\
\text{vertices:} \quad & (3, 5) \text{ and } (3, -7) \\
\text{endpoints of conjugate axis:} \quad & (12, -1) \text{ and } (-6, -1) \\
\text{foci:} \quad & (3, 9.8) \text{ and } (3, -11.8)
\end{aligned}$$

$$\text{slope of asymptotes:} \quad m = \pm\frac{2}{3} \qquad \blacksquare\blacksquare\blacksquare$$

FIGURE 15–29

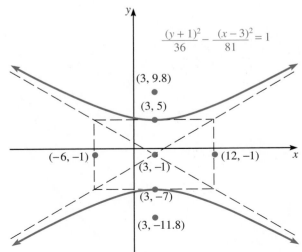

EXAMPLE 3 Determine the foci, center, vertices, endpoints of the conjugate axis, and equations of the asymptotes of the hyperbola given by

$$4y^2 - 24y - x^2 + 6x = 37$$

Also sketch the graph.

Solution First, we must arrange the equation in standard form by completing the square.

$$4y^2 - 24y - x^2 + 6x = 37$$
$$4(y^2 - 6y) - (x^2 - 6x) = 37$$
$$4(y^2 - 6y + 9) - (x^2 - 6x + 9) = 37 + 36 - 9$$
$$4(y - 3)^2 - (x - 3)^2 = 64$$
$$\frac{(y - 3)^2}{16} - \frac{(x - 3)^2}{64} = 1$$

From the equation in standard form, we know that the center is $(3, 3)$, $a = 4$, $b = 8$, and the transverse axis is vertical. Figure 15–30 shows the graph, and the results are summarized as follows:

$$\text{center:} \quad (3, 3)$$
$$\text{vertices:} \quad (3, 7) \text{ and } (3, -1)$$
$$\text{endpoints of conjugate axis:} \quad (11, 3) \text{ and } (-5, 3)$$
$$\text{foci:} \quad (3, 11.9) \text{ and } (3, -5.9)$$
$$c = \sqrt{a^2 + b^2} \approx 8.9$$
$$\text{slopes of asymptotes:} \quad m = \pm\frac{1}{2}$$

■■■

Finding the Equation of a Hyperbola

EXAMPLE 4 Find the equation of the hyperbola with its center at the origin, one focus at $(5, 0)$, and one vertex at $(-3, 0)$. This information is shown in Figure 15–31.

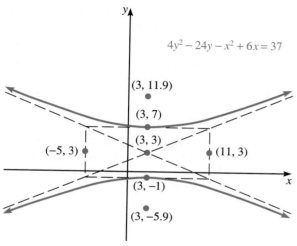

$4y^2 - 24y - x^2 + 6x = 37$

FIGURE 15–30

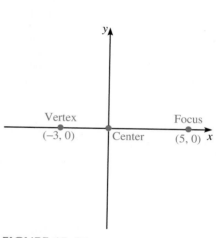

FIGURE 15–31

Solution To find the equation of the hyperbola, we must fill in values for h, k, a^2, and b^2. From the information given, we know that $h = 0$ and $k = 0$. We also know that $a = 3$ and $c = 5$. From this information we can determine b^2.

$$c^2 = a^2 + b^2$$
$$b^2 = c^2 - a^2$$
$$b^2 = 25 - 9$$
$$b^2 = 16$$

From the location of the vertex and focus, we can also determine that the transverse axis is horizontal (the x term is positive). The equation is

$$\frac{x^2}{9} - \frac{y^2}{16} = 1$$

■■■

EXAMPLE 5 Determine the equation of the hyperbola if its center is at $(-4, 2)$, its vertex is at $(-4, 7)$, and the slope of an asymptote is $5/2$.

Solution To determine the equation, we must find h, k, a^2, and b^2. Since the center is $(-4, 2)$, $h = -4$ and $k = 2$. Since the vertex is located at $(-4, 7)$, the transverse axis is a vertical line (the y term in the equation is positive). From the equation of the asymptote, we know that $b = 2$ and $a = 5$. The equation is

$$\frac{(y - 2)^2}{25} + \frac{(x + 4)^2}{4} = 1$$

■■■

Applications

The LOng RAnge Navigation (LORAN) system uses a master radio sending station and a secondary sending station widely separated from the master station to send signals to aircraft and ships. The difference in the arrival terms of the signals from the two sending stations to the aircraft or ship is a constant on a hyperbola, with the sending stations at the foci.

EXAMPLE 6 Two LORAN stations are 300 mi apart. If a ship records a time difference of 0.0005 s between the signals, determine where the ship would reach shore if it continues to follow the hyperbola corresponding to this time difference.

Solution We set up a rectangular coordinate system with the two stations on the x axis and the origin midway between them. The ship lies on a hyperbola whose foci are the locations of the two stations. Since the time difference is 0.0005 s and the speed of sound is 186,000 mi/s, the distance from the ship to each station is

$$d = rt = (186{,}000 \text{ mi/s}) (0.0005 \text{ s}) = 93 \text{ mi}$$

The difference of the distances from the ship to each station, 93 mi, equals $2a$. Therefore, $a = 46.5$ and the vertex of the hyperbola is $(46.5, 0)$. Since the focus is at $(150, 0)$, the ship would reach shore $150 - 46.5 = 103.5$ mi from the master station. ■■■

EXAMPLE 7 The impedance of an ac circuit is given by

$$Z^2 = R^2 + (X_L - X_C)^2$$

where R = resistance, X_L = inductive reactance, and X_C = capacitive reactance. Graph the relationship between Z and R if $(X_L - X_C) = 20\ \Omega$. Let Z be the independent variable.

Solution   Since $(X_L - X_C) = 20$, the equation becomes

$$Z^2 = R^2 + 400$$
$$Z^2 - R^2 = 400$$
$$\frac{Z^2}{400} - \frac{R^2}{400} = 1$$

Since $a^2 = 400$, $a = 20$, $b^2 = 400$, $b = 20$, $c = \sqrt{a^2 + b^2}$, $c \approx 28.3$, and because Z is the independent variable, the transverse axis is horizontal. The graph is shown in Figure 15–32. ∎ ∎ ∎

FIGURE 15–32

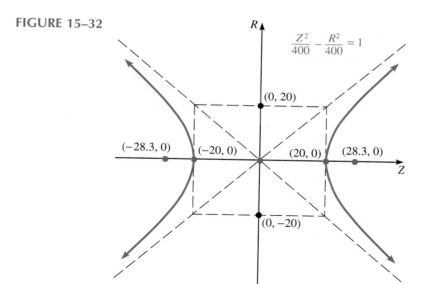

Hyperbolas of the Form $xy = k$

The hyperbolas discussed thus far are quadratic in form. However, there is another type of equation that produces a hyperbola. An equation of the form $xy = k$, where k is some nonzero constant, produces a hyperbola. The x and y axes serve as the asymptotes. Also, if k is positive, the hyperbola lies in quadrants I and III, and if k is negative, the hyperbola lies in quadrants II and IV. The vertices are given by $(\sqrt{|k|}, \sqrt{|k|})$ placed in the appropriate quadrant with appropriate signs. The foci are given by $(\sqrt{|2k|}, \sqrt{|2k|})$ placed in the appropriate quadrant with appropriate signs. The transverse axis is $y = x$ if $k > 0$, and $y = -x$ if $k < 0$.

EXAMPLE 8   Graph $xy = -4$.

Solution   Since $k < 0$, the hyperbola is located in quadrants II and IV. The transverse axis is the line $y = -x$. The vertices are located at $(-2, 2)$ and $(2, -2)$, and the foci are located at $(-2.8, 2.8)$ and $(2.8, -2.8)$. The graph is shown in Figure 15–33.

FIGURE 15–33

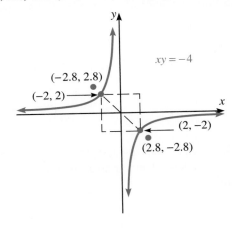

15–5 EXERCISES

For the following hyperbolas, determine the coordinates of the center, vertices, foci, and endpoints of the conjugate axis. Find the slopes of the asymptotes, and sketch the graph. Round to tenths if necessary.

1. $\dfrac{x^2}{24} - \dfrac{y^2}{4} = 1$

2. $\dfrac{x^2}{36} - \dfrac{y^2}{9} = 1$

3. $\dfrac{y^2}{49} - \dfrac{x^2}{16} = 1$

4. $\dfrac{y^2}{81} - \dfrac{x^2}{64} = 1$

5. $\dfrac{x^2}{64} - \dfrac{y^2}{25} = 1$

6. $\dfrac{y^2}{9} - \dfrac{x^2}{49} = 1$

7. $\dfrac{(y + 1)^2}{16} - \dfrac{(x - 2)^2}{81} = 1$

8. $\dfrac{(x + 4)^2}{49} - \dfrac{(y + 3)^2}{36} = 1$

9. $\dfrac{(x + 2)^2}{100} - \dfrac{(y - 3)^2}{49} = 1$

10. $\dfrac{(y - 6)^2}{36} - \dfrac{(x - 4)^2}{81} = 1$

11. $\dfrac{(y - 7)^2}{20} - \dfrac{(x - 3)^2}{16} = 1$

12. $\dfrac{(x + 9)^2}{36} - \dfrac{(y - 5)^2}{81} = 1$

13. $3x^2 - 12x - y^2 + 10y = 22$

14. $x^2 + 12x - 4y^2 - 8y = 8$

15. $2y^2 - x^2 + 12y - 14x = 25$

16. $y^2 - 5x^2 + 10x + 18y = -12$

17. $y^2 - 4x^2 + 16y - 24x = 8$

18. $3x^2 - 18x - 2y^2 + 16y = 1$

19. $5x^2 - 20x - 3y^2 + 12y = 7$

20. $2x^2 - 7y^2 + 16x - 14y = 67$

From the given information, find the equation for the hyperbola in standard form.

21. Center $(0, 0)$; focus $(-5, 0)$; and slope of asymptote 3/4.

22. Center $(0, 0)$; vertex $(-8, 0)$; and focus $(-11, 0)$.

23. Center $(5, 3)$; slope of asymptote 7/3; and focus $(10, 3)$.

24. Center $(-2, 6)$; vertex $(-2, 3)$; and endpoint of conjugate axis $(3, 6)$.

25. Vertices $(0, 4)$ and $(6, 4)$; slope of asymptote 8/3.

26. Foci $(-6, -4)$ and $(2, -4)$; endpoint of conjugate axis $(-2, -1)$.

27. Center $(-1, -5)$; vertex $(-1, -2)$; and length of conjugate axis is 10.

28. Center $(5, 2)$; focus $(0, 2)$; and length of conjugate axis is 6.

E 29. Sketch the relationship between impedance and resistance if $X_L - X_C = 14\ \Omega$ (see Example 7).

P 30. Boyle's law states that for a constant temperature, the product of the pressure P and volume V of an ideal gas is a constant k ($PV = k$). Graph the relationship between P and V if $k = 1{,}800$. (V is the independent variable.)

E 31. The area of a triangle is $A = bh/2$. Sketch the relationship between the base and height of a triangle whose area is 250 ft². (Let b be the independent variable.)

E 32. The volume V of a conduit 2.5 m long is given by $V = 2.5\pi(R^2 - r^2)$ where R = radius of the outside of the conduit, and r = inside radius. Sketch a graph of the relationship between the radii when the volume is fixed at 60π m^3. (Let R be the independent variable.)

C 33. Two LORAN stations are situated 200 mi apart. A ship records a time difference of 0.0004 s. Where would the ship reach shore if it follows the hyperbola corresponding to this time difference?

C 34. Two LORAN stations are 100 mi apart. A ship records a time difference of 0.0001 s. Where would the ship reach shore if it follows the hyperbola corresponding to this time difference?

15–6

THE SECOND-DEGREE EQUATION

Thus far in this chapter we have discussed the equations for the circle, parabola, ellipse, and hyperbola. The equation for these conic sections are special cases of the more general second-degree equation. The general form of a second-degree equation is given by

$$Ax^2 + Bxy + Cy^2 + Dx + Ey + F = 0$$

The type of conic section is determined by coefficients A, B, and C of this equation. These results are summarized below.

The Second-Degree Equation

$$Ax^2 + Bxy + Cy^2 + Dx + Ey + F = 0$$

1. Parabola:

$B = 0$; $A = 0$ or $C = 0$ (quadratic in only x or y)

2. Circle:

$B = 0$, $A = C$ (quadratic in both x and y, and the coefficients of the quadratic terms are the same)

3. Ellipse:

$B = 0$, $A \neq C$ but they have the same sign (quadratic in both x and y; the coefficients of the quadratic terms are different, but they have the same sign)

4. Hyperbola:

$B = 0$, A and C have opposite signs (quadratic in both x and y, but they have different signs)

5. Hyperbola:

$A = C = D = E = 0$ but $B \neq 0$ (not a quadratic equation but has the xy term)

The flowchart in Figure 15–34 will help you determine the type of conic section that is described by a quadratic function.

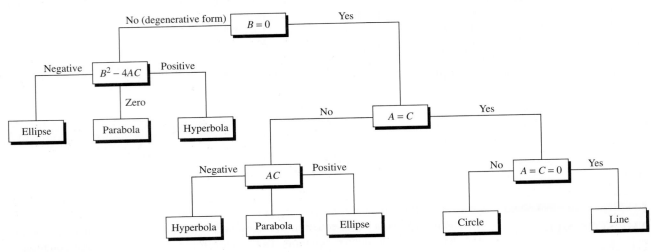

FIGURE 15–34

EXAMPLE 1   Identify the curve represented by the following equation:

$$3x^2 - 6x - y^2 + 8y = 1$$

Determine the appropriate information for the figure and sketch its graph.

Solution   To determine the type of figure, let us arrange the equation in the form of the general second-degree equation.

$$3x^2 - y^2 - 6x + 8y - 1 = 0$$

$$A \qquad C$$

Because $A = 3$, $B = 0$, and $C = -1$, this equation represents a hyperbola. To determine the pertinent information and sketch the graph, arrange the equation in standard form.

$$3x^2 - 6x - y^2 + 8y = 1$$
$$3(x^2 - 2x) - (y^2 - 8y) = 1$$
$$3(x^2 - 2x + 1) - (y^2 - 8y + 16) = 1 + 3 - 16$$
$$3(x - 1)^2 - (y - 4)^2 = -12$$
$$-\frac{(x - 1)^2}{4} + \frac{(y - 4)^2}{12} = 1$$
$$\frac{(y - 4)^2}{12} - \frac{(x - 1)^2}{4} = 1$$
$$a = \sqrt{12} \approx 3.5$$
$$c^2 = a^2 + b^2$$
$$c^2 = 12 + 4$$
$$c = 4$$

center: (1, 4)

vertices: (1, 7.5) and (1, 0.5)

endpoints of conjugate axis: (−1, 4) and (3, 4)

foci: (1, 8) and (1, 0)

slope of asymptotes: $m = \pm\dfrac{a}{b} = \pm\dfrac{\sqrt{12}}{2} = \pm\sqrt{3} \approx \pm 1.7$

The graph is shown in Figure 15–35.

FIGURE 15–35

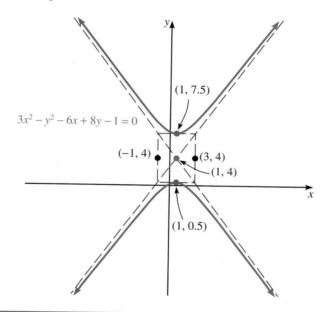

$3x^2 - y^2 - 6x + 8y - 1 = 0$

Standard Equations of the Conics

$(y - k)^2 = 4p(x - h)$

$(x - h)^2 = 4p(y - k)$

$\left.\begin{array}{c}\\ \\\end{array}\right\}$ parabola, vertex at (h, k)

$(x - h)^2 + (y - k)^2 = r^2$ circle, center (h, k)

$\dfrac{(x - h)^2}{a^2} + \dfrac{(y - k)^2}{b^2} = 1$ ellipse, center (h, k)

$\dfrac{(x - h)^2}{a^2} - \dfrac{(y - k)^2}{b^2} = 1$ hyperbola, center (h, k)

EXAMPLE 2 Identify the curve represented by the equation

$$x^2 - 3y - 8x = 2$$

Determine the appropriate information for the figure and sketch its graph.

Solution First, we arrange the equation in general second-degree form.

$$x^2 - 8x - 3y - 2 = 0$$

FIGURE 15–36

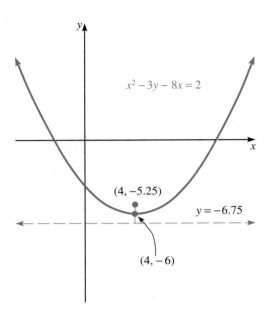

Since $B = 0$ and $B^2 - 4AC = 0$, this equation represents a parabola. Second, we arrange the equation in standard form.

$$(x^2 - 8x) = 3y + 2$$
$$(x^2 - 8x + 16) = 3y + 2 + 16$$
$$(x - 4)^2 = 3(y + 6)$$
$$\text{vertex:} \quad (4, -6)$$
$$\text{focus:} \quad (4, -5.25)$$
$$\text{directrix:} \quad y = -6.75$$

The graph is given in Figure 15–36. ■ ■ ■

EXAMPLE 3 Identify the curve represented by the following equation:

$$4x^2 + 8x - 24y + 4y^2 = 5$$

Determine the appropriate information for the figure and sketch its graph.

Solution Arranging the equation in general second-degree form gives

$$4x^2 + 4y^2 + 8x - 24y - 5 = 0$$
$$\underset{A}{\uparrow} \quad \underset{C}{\uparrow}$$

Since $B = 0$ and $A = C = 0$, the equation represents a circle. To determine the pertinent information, we arrange the equation in standard form.

$$4x^2 + 4y^2 + 8x - 24y - 5 = 0$$
$$4x^2 + 8x + 4y^2 - 24y = 5$$
$$4(x^2 + 2x) + 4(y^2 - 6y) = 5$$
$$4(x^2 + 2x + 1) + 4(y^2 - 6y + 9) = 5 + 36 + 4$$
$$4(x + 1)^2 + 4(y - 3)^2 = 45$$
$$(x + 1)^2 + (y - 3)^2 = 11.25$$

The center is $(-1, 3)$ and the radius is 3.4. Figure 15–37 shows the graph of this circle.

FIGURE 15–37

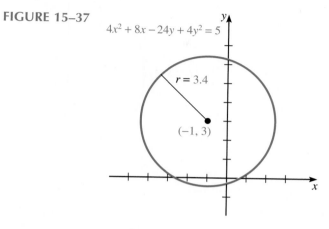

$4x^2 + 8x - 24y + 4y^2 = 5$

$r = 3.4$

$(-1, 3)$

EXAMPLE 4 Identify the curve represented by the following equation:

$$9x^2 + 36x + 4y^2 - 16y = 20$$

Determine the appropriate information for the figure and sketch its graph.

Solution Arranging the equation in general second-degree form gives

$$9x^2 + 4y^2 + 36x - 16y - 20 = 0$$

Since $B = 0$, $A \neq C$, and AC is positive, this equation represents an ellipse. Next, we arrange the equation in standard form for an ellipse.

$$9x^2 + 36x + 4y^2 - 16y = 20$$
$$9(x^2 + 4x) + 4(y^2 - 4y) = 20$$
$$9(x^2 + 4x + 4) + 4(y^2 - 4y + 4) = 20 + 36 + 16$$
$$9(x + 2)^2 + 4(y - 2)^2 = 72$$
$$\frac{(x + 2)^2}{8} + \frac{(y - 2)^2}{18} = 1$$

$$a = \sqrt{18} \approx 4.2$$
$$b = \sqrt{8} \approx 2.8$$
$$c = \sqrt{a^2 - b^2} = \sqrt{10} \approx 3.2$$

center: $(-2, 2)$

vertices: $(-2, 6.2)$ and $(-2, -2.2)$

endpoints of minor axis: $(-4.8, 2)$ and $(0.8, 2)$

foci: $(-2, 5.2)$ and $(-2, -1.2)$

Figure 15–38 shows the graph.

FIGURE 15–38

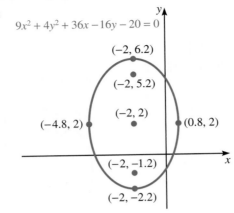

$9x^2 + 4y^2 + 36x - 16y - 20 = 0$

(−2, 6.2)

(−2, 5.2)

(−2, 2)

(−4.8, 2)   (0.8, 2)

(−2, −1.2)

(−2, −2.2)

15–6 EXERCISES

Identify the following equations as representing a circle, parabola, ellipse, or hyperbola.

1. $3x^2 - 4y + 6x = 12$

2. $8x + 3y^2 - 6y + 4x^2 = 20$

3. $4x^2 + 10x - y^2 + 6y - 13 = 0$

4. $x^2 - y^2 + 3x - 12y + 10 = 0$

5. $6x + 2y^2 - 4y + 2x^2 = 6$

6. $3x(x - 1) = 4y + 3x - 7$

7. $4x + 14 = 3y^2 + 15y$

8. $6y^2 + 8x - 18y + 6x^2 = 0$

9. $5y^2 + 8 - 10x + 6y = 2 - 4x^2$

10. $4x + 18 - y^2 + 6y = -1$

11. $3 - 7x^2 + 21y + y^2 - 14x = 0$

12. $3x - x^2 + 7y - 4y^2 + 18 = 0$

13. $4x - 3y + x^2 - 10 + y^2 = 7$

14. $5x^2 + 5y^2 = 1$

15. $9x^2 + 4y^2 - 36x + 12y = -3$

16. $xy + 8 = 0$

Identify the curve represented by each of the given equations. Identify the information pertinent to the curve and sketch its graph.

17. $x^2 + 2y - 6x = -17$

18. $x^2 + 6x + y^2 - 10y = 43$

19. $4x^2 + 6y^2 - 8x + 36y = -26$

20. $y^2 - 6x + 8y + 58 = 0$

21. $8x + y^2 + 2y + x^2 + 13 = 0$

22. $49y^2 - 16x^2 + 196y - 160x = 988$

23. $3 - xy = 0$

24. $4x^2 + 25y^2 - 24x + 150y + 161 = 0$

25. $25x^2 - 54y - 9y^2 = 306$

26. $2x^2 + 2y^2 - 2x + 20y + 18 = 0$

27. The length of a room is 9 ft more than the width. Express the volume of the room as a function of the length if the room is 8 ft high. Identify the curve represented by this equation.

28. The height of a triangle is three more than five times the base. Express the area of the triangle as a function of the base, and identify the resulting curve.

29. The two legs of a right triangle are represented by x and y, and the hypotenuse is $y + 1$. Identify the curve represented by the equation relating x and y.

CHAPTER SUMMARY

Summary of Terms

axis (p. 571)

center (pp. 567, 578, 585)

circle (p. 567)

conjugate axis (p. 585)

directrix (p. 571)

ellipse (p. 578)

endpoints of minor axis (p. 578)

focus/foci (pp. 571, 578, 585)

general form of a circle (p. 568)

general form of a line (p. 560)

hyperbola (p. 585)

major axis (p. 578)

minor axis (p. 578)

parabola (p. 571)

point–slope form (p. 562)

radius (p. 567)

slope–intercept form (p. 560)

transverse axis (p. 585)

vertex/vertices (pp. 571, 578, 585)

$$\left.\begin{array}{l} (x - h)^2 = 4p(y - k) \\ (y - k)^2 = 4p(x - h) \end{array}\right\} \text{ parabola}$$

$$\left.\begin{array}{l} \dfrac{(x - h)^2}{a^2} + \dfrac{(y - k)^3}{b^2} = 1 \\[2mm] \dfrac{(x - h)^2}{b^2} + \dfrac{(y - k)^2}{a^2} = 1 \end{array}\right\} \text{ ellipse}$$

$$\left.\begin{array}{l} \dfrac{(x - h)^2}{a^2} - \dfrac{(y - k)^2}{b^2} = 1 \\[2mm] \dfrac{(y - k)^2}{a^2} - \dfrac{(x - h)^2}{b^2} = 1 \end{array}\right\} \text{ hyperbola}$$

Summary of Formulas

$y = mx + b$ slope–intercept form

$y - y_1 = m(x - x_1)$ point–slope form

$(x - h)^2 + (y - k)^2 = r^2$ circle

$Ax^2 + Bxy + Cy^2 + Dx + Ey + F = 0$
second-degree equation

CHAPTER REVIEW

Section 15–1

Using the information given, find the equation of the line. Place your answer in general form, $Ax + By + C = 0$.

1. Slope $= -3/4$; passes through $(-1/2, 6)$.
2. Slope $= -4$; x intercept $= 5$.
3. Passing through $(9, -5)$ and $(-4, -3)$.
4. Vertical line; passes through $(-8, -7)$.
5. Undefined slope; passes through $(-3, 4)$.
6. Passing through $(6, -2)$ and $(1, 5)$.
7. Horizontal line through $(6, -8)$.
8. Perpendicular to $y = 3x + 8$; passing through $(1, 4)$.
9. Parallel to $3x - 2y = 6$; passing through $(-3, -5)$.
10. Slope $= 0$; y intercept $= -2$.
11. Perpendicular to $x - 5y = 10$; passing through $(7, -4)$.
12. Parallel to $4x - y = 5$; passing through $(6, -1)$.

Find the slope, x intercept, and y intercept of the following lines. Use this information to graph the line.

13. $7x - 4y = 10$
14. $2x + y = 10$
15. $2x + 3y + 6 = 0$
16. $3y - 4x = 6$
17. $x - 3y = 9$
18. $5x - 2y + 15 = 0$

E 19. The resistance of a resistor is found to increase 0.02 Ω for each 1°C increase in temperature. Write an equation to relate resistance and temperature if the resistance at 0°C is 2.3 Ω.

P 20. The pressure in a lake is measured to be 28.5 lb/in^2 at a depth of 32 ft and 39 lb/in^2 at 56 ft. Find the linear equation for pressure as a function of depth. (Let depth be the independent variable.)

P 21. The acceleration of an object is the change in velocity divided by the change in time. Find the linear equation relating velocity and time if the acceleration is 10 m/s^2 and the velocity is 3.5 m/s where $t = 0$ s.

Section 15–2

Find the radius and center of the circle and sketch its graph.

22. $(x - 3)^2 + y^2 = 16$
23. $(y + 3)^2 + (x + 4)^2 = 4$
24. $(x - 1)^2 + (y + 2)^2 = \sqrt{7}$
25. $(x + 1)^2 + (y - 6)^2 = \sqrt{6}$
26. $2x^2 + 2y^2 - 4y - 6 = 0$
27. $3x^2 + 3y^2 - 9x + 15y = 18$

Using the given information, find the equation, in standard form, of the circle.

28. Center at $(4, -2)$; radius of 2.
29. Center at $(-2, -4)$; radius of $\sqrt{3}$.
30. Center at $(4, 0)$; passing through $(6, 5)$.
31. Center at $(6, -5)$; tangent to the x axis.
32. Diameter is a segment from $(-4, 6)$ to $(2, -4)$.
33. Diameter is a segment from $(-7, 4)$ to $(3, -6)$.
34. The legs of a right triangle are represented by x and y, and the hypotenuse is 9 ft. Write an equation relating x and y, and sketch the graph of the equation.

E 35. The impedance Z in an ac circuit is given by

$$Z^2 = R^2 + (X_L - X_C)^2$$

where R = resistance, X_L = inductive reactance, and X_C = capacitive reactance. Graph the relationship between resistance and inductive reactance when $Z = 25\ \Omega$ and $X_C = 40\ \Omega$.

Section 15–3

For the given equations of parabolas, find the coordinates of the vertex and focus, determine the equation of the directrix, and sketch the graph.

36. $3(y + 1) = x^2$

37. $y^2 = -2(x - 3)$

38. $y^2 + x + 4y = -12$

39. $x^2 + 3y - 4x + 13 = 0$

40. $x^2 - x - 8y = 3$

41. $x^2 + 8x + 2y + 10 = 0$

Find the equation of the parabola from its given geometric properties.

42. Vertex at $(-1, 4)$; focus $(-1, 7)$.

43. Vertex at $(-2, 0)$; focus $(1, 0)$.

44. Directrix $x = 2$; focus $(4, -6)$.

45. Directrix $y = -3$; focus $(4, 0)$.

46. Vertex at $(1, 3)$; passes through $(4, 6)$; and axis parallel to the y axis.

C 47. A cable suspended between two bridge supports that are 200 ft apart approximates the shape of a parabola. The vertical distance from where the cable is attached to its lowest point is 35 ft. Assuming that the origin is at the lowest point, find the equation representing the shape of the cable.

E 48. Heat H in a resistor of resistance R and current i develops at a rate given by $H = Ri^2$. Graph H as a function of i if $R = 25\ \Omega$.

Section 15–4

Determine the coordinates for the center, vertices, endpoints of the minor axis, and foci of the given ellipse. Sketch the graph.

49. $\dfrac{(x + 1)^2}{36} + \dfrac{(y - 3)^2}{49} = 1$

50. $\dfrac{(x - 2)^2}{16} + \dfrac{(y - 5)^2}{9} = 1$

51. $81x^2 + 25y^2 - 300y = 1{,}800$

52. $20x^2 + 14y^2 - 40x - 112y = 46$

53. $9y^2 + 24x + 18y + 4x^2 + 44 = 0$

54. $16y^2 + 9x^2 - 18x = 135$

Determine the equation of the ellipse from the given geometric properties.

55. Center at $(5, 1)$; focus at $(5, -4)$; and endpoint of minor axis at $(8, 1)$.

56. Endpoints of minor axis at $(3, -4)$ and $(3, 2)$; focus at $(7, -1)$.

57. Vertices at $(-4, 2)$ and $(4, 2)$; length of minor axis is 6.

58. Center at $(5, 3)$; foci at $(2, 3)$ and $(8, 3)$; and length of minor axis is 4.

P 59. A satellite orbits the earth in an elliptical orbit. The maximum height is 400 mi, the minimum height is 175 mi, and the earth's center is at one focus. Determine the equation of the satellite's path. (The radius of the earth is approximately 3,955 mi.)

C 60. A covered bridge in the shape of a semiellipse spans 150 ft and is 50 ft high at its maximum. What is the height of the bridge 60 ft from the center?

Section 15–5

Determine the coordinates of the center, vertices, endpoints of the conjugate axis, and foci for each hyperbola. Sketch the graph.

61. $\dfrac{(x + 3)^2}{16} - \dfrac{(y - 1)^2}{4} = 1$

62. $\dfrac{(x - 2)^2}{49} - \dfrac{(y - 3)^2}{9} = 1$

63. $\dfrac{(y + 4)^2}{25} - \dfrac{(x + 3)^2}{36} = 1$

64. $\dfrac{(y - 8)^2}{81} - \dfrac{(x + 5)^2}{36} = 1$

65. $5y^2 + 40 = 4x^2 + 40y$

66. $25x^2 + 325 = 36y^2 + 350x$

Determine the equation of the hyperbola from its geometric properties.

67. Center $(0, 0)$; vertex $(0, 3)$; and focus $(0, -5)$.

68. Center $(3, -4)$; focus $(3, 2)$, and length of conjugate axis $= 8$.

69. Vertices $(5, 7)$ and $(-1, 7)$; length of conjugate axis $= 10$.

70. Foci $(-1, -3)$ and $(-7, -3)$; length of transverse axis $= 4$.

71. Boyle's law states that for a constant temperature, the product of pressure P and volume V of an ideal gas is a constant k $(PV = k)$. Graph the relationship between P and V if $k = 400$.

72. The area of a rectangle is $A = LW$. Sketch the relationship between length L and width W if the area is fixed at $200\ \text{m}^2$.

Section 15–6

Identify the curve represented by the following equations. Identify the information pertinent to the curve and sketch its graph.

73. $2x^2 - 4x + 2y^2 + 24y = -69$

74. $4y^2 + 9x^2 + 144x - 36y + 576 = 0$

75. $x^2 = -4y + 8x - 44$

76. $\dfrac{(y+7)^2}{81} - \dfrac{(x-1)^2}{64} = 1$

77. $3(x-3)^2 + (y+1)^2 = 18$

78. $3y^2 - 30y + 3x^2 = -67$

79. $\dfrac{(x-3)^2}{8} - \dfrac{(y+1)^2}{25} = 1$

80. $3y^2 - 6x + 6y + 3 = 0$

81. $16y^2 + 32y + 25x^2 + 150x = 159$

82. $xy = -7$

CHAPTER TEST

The number in parentheses refers to the appropriate learning objective given at the beginning of the chapter.

1. Determine the slope, y intercept, and x intercept of $2x - 10y = 9$. Sketch the graph. **(2)**

2. Find the equation of a circle with center at $(2, -5)$ and passing through $(3, -1)$. **(4)**

3. Determine the center, foci, vertices, endpoints of conjugate axis, and slopes of the asymptotes of the following hyperbola **(9)**

$$\frac{(x-1)^2}{16} - \frac{(y+3)^2}{4} = 1$$

Also sketch the graph.

E 4. The impedance Z in an ac circuit is given by **(12)**

$$Z^2 = R^2 + (X_L - X_C)^2$$

where R = resistance, X_L = inductive reactance, and X_C = capacitive reactance. Graph the relationship between resistance and inductive reactance when $Z = 36\ \Omega$ and $X_C = 20\ \Omega$.

5. Find the center and radius of the circle given by **(3)**

$$(x-4)^2 + (y+8)^2 = 25$$

6. Determine the vertex, focus, and directrix of the following parabola: **(5)**

$$(x+3)^2 = -8(y+6)$$

Also sketch its graph.

P 7. A satellite orbits the earth in an elliptical orbit. The maximum height is 200 mi, the minimum height is 110 mi, and the earth's center is at one focus. Find the equation of the satellite's path. (Assume that the radius of the earth = 3,960 mi.) **(12)**

8. Find the equation of the ellipse with its center at $(-7, 4)$, its vertex at $(-7, 1)$, and its minor axis of length 4. **(8)**

9. Determine the equation of a parabola whose focus is $(4, -5)$ and whose directrix is $x = -2$. **(6)**

P 10. The pressure in a lake is measured to be 22.5 lb/in² at a depth of 18 ft and 36.33 lb/in² at 50 ft. If the relationship between pressure and depth is linear, find the equation relating the variables. **(12)**

11. Determine the equation of the straight line passing through $(-7, 6)$ with a slope of -3. **(1)**

12. Identify the conic section represented by the following equation: **(11)**

$$4x^2 + 9y^2 - 32x + 144y + 631 = 0$$

13. Determine the coordinates for the center, vertices, foci, and endpoints of the minor
axis of the ellipse represented by **(7)**

$$16y^2 + 9x^2 - 64y + 54x + 1 = 0$$

Also sketch the graph.

14. Find the equation of a hyperbola with its center at $(6, -3)$, its vertex at $(2, -3)$, and
the length of its conjugate axis 12. **(10)**

GROUP ACTIVITY

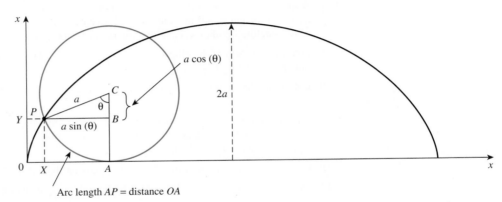

You have been hired by Quickest Descent Emergency Exit Inc. to determine the parametric equation for a cycloid and to determine what the equation $7.3x^2 + 4xy + 5y^2 = 9$
represents. After a trip to the library you find that a point on a wheel rolling along the x
axis without slipping can generate a cycloid. The point, P, on the wheel traces out a line
as indicated in Figure 15–39. The curve is called a cycloid. The segment of the circle

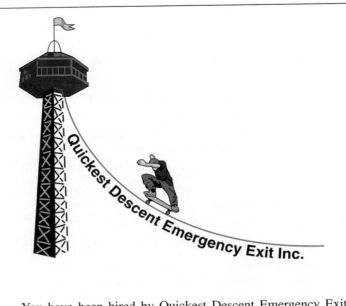

Arc length AP = distance OA

FIGURE 15–39 Cycloid of one cycle $(0 \rightarrow \pi)$

(*AP*) is given by $a\theta$, where θ is in radians. This arc is also equal to the distance along the *x* axis *OA*.

1. Show that an expression for *x* is given by $x = a(\theta - \sin\theta)$.
2. Discuss how to write a similar expression for *y*. What is that expression?

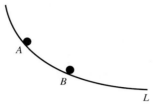

FIGURE 15–40 Two objects located on a cycloid curve. Both will reach *L* at the same time.

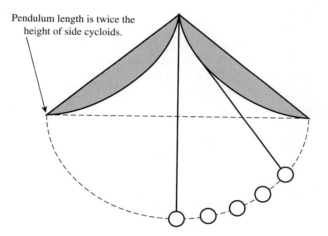

Pendulum length is twice the height of side cycloids.

FIGURE 15–41 Construction of Huygens pendulum clock

The period of a pendulum is not strictly independent of the amplitude of its swing. The equation that is normally derived assumes small displacements from its equilibrium position. Christian Huygens (1629–1695), a Dutch mathematician, physicist, and astronomer, was interested in finding the curve that a pendulum bob could traverse such that its period was independent of its displacement. The cycloid curve produces a pendulum whose period is exactly the same whether the bob traversed a long or a short arc. A flexible pendulum constrained by a cycloid on either side generates the cycloid path for the pendulum. He also proved that no matter what point along the cycloid curve that a mass starts from (e.g., *A*, *B*) it would reach the lowest point *L* in the same time (given by $\pi\sqrt{a/g}$ s, where *g* is the acceleration due to gravity in ft/s^2). This problem is known as the tautochrone problem (*tautochrone* means "equal time" in Greek). Another interesting problem is to find the path of quickest descent if a particle is constrained to follow some nonvertical path acted on only by gravity. The solution to this problem (known as the brachistochrone problem) is again an inverted cycloid. This discovery is attributed to many famous mathematicians (e.g., Johann Bernoulli and Blaise Pascal) and was a significant step in creating the branch of mathematics known as the calculus of variations.

Rotation of Axis

After another trip to the library you learn that you can determine the properties of the graph $7.3x^2 + 4xy + 5y^2 = 9$ through a clever process of rotating the axis. When rotating the axis, the origin remains fixed while the x axis and y axis are rotated through an angle θ to a new position (denoted by x' and y'). You want to find a way to express X and Y in terms of X', Y', and θ.

1. From Figure 15–42, $X' = r \cos \alpha$ and $Y' = r \sin \alpha$. Write similar terms for X and Y in terms of α and θ.

2. Discuss how to derive the expressions for X and Y in terms of X' and Y'. What are these expressions?

3. Rotate the graph of $9x^2 + 25y^2 = 400$ through an angle of 45 degrees. What is the resulting equation? Sketch the graph.

4. Earlier you learned how the translation of the axis affected the equation for the conic sections. Discuss how you would derive the translation of axis equations (X and Y in terms of X' and Y' under translation only). What are these equations? Discuss how the rotation of the axis results can be combined with the translation of axis results. What are these equations?

5. Using the second-degree equation $Ax^2 + Bxy + Cy^2 + Dx + Ey + F = 0$ representing the graph of a circle, parabola, ellipse, and hyperbola, discuss a method for eliminating the xy term (B must be equal to zero) by rotating the axis through some angle θ. Verify that this equation is given by $\tan 2\theta = B/(A - C)$ where $A \neq C$.

6. Determine how to transform $7.3x^2 + 4xy + 5y^2 = 9$ into an equation that does not have an xy term. Through what angle do you need to translate the axis? What is the final equation? Graph the equation and determine its properties (circle, parabola, ellipse, or hyperbola).

7. Graph several conic sections on a piece of graph paper or via the computer and exchange them with other members of your team. Each team member is to identify the graph (circle, parabola, ellipse, or hyperbola) and determine its properties. Write an equation for the graph.

FIGURE 15–42

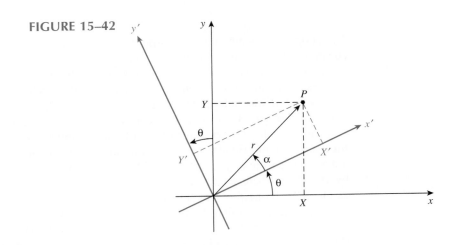

SOLUTION TO CHAPTER INTRODUCTION

If we make a drawing of the satellite dish, placing the vertex at the origin, we would have Figure 15–43. The equation for this parabola is of the form

$$y^2 = 4px$$

From the information given, the point (1.3, 4) is on the parabola. Substituting into the equation, we can solve for p.

$$(4)^2 = 4p(1.3)$$
$$16 = 5.2p$$
$$3.1 = p$$

The focus is 3.1 ft from the vertex, which is where the receiver should be placed.

FIGURE 15–43

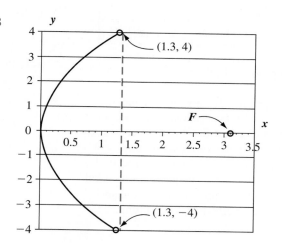

A

s a biologist, you measure the number n of bacteria in a petri dish as a function of time t. You know that the function is defined by an equation of the form $n = me^{0.2t} + b$. Using the following data, you are to predict the number of bacteria at the end of 18 hours:

t (h)	0	1	2	3	4	5
n (thousands)	19.2	21	23	25	28	32

(The solution to this problem is given at the end of the chapter.)

Problems of this type involve a statistical method called *least squares* whereby experimental data are used to generate an equation that best fits the curve representing the relationship between the variables. In turn, you can use the equation to approximate other values. In this chapter we will discuss methods of displaying empirical data, measures of central tendency, measures of dispersion, and the least squares method of curve fitting.

Learning Objectives

After you complete this chapter, you should be able to

1. Organize data into a frequency distribution and give a graphical representation using a histogram or a frequency polygon (Section 16–1).

2. Calculate the arithmetic mean, median, and mode from empirical data (Section 16–2).

3. Calculate the range and standard deviation from empirical data (Section 16–3).

4. Use the mean and standard deviation to calculate the variation of data values that should fall within one or two standard deviations of the mean (Section 16–3).

5. Use the least squares method to fit empirical data to an equation (Section 16–4).

Chapter 16

Introduction to Statistics and Empirical Curve Fitting

16–1

FREQUENCY DISTRIBUTIONS

In this section we will discuss methods of organizing and displaying statistical data. First, we will discuss the frequency distribution as a method of organizing the data. Then we will discuss the histogram and the frequency polygon as means to represent the data graphically.

Frequency Distribution

When a statistical experiment is conducted, repeated readings of measurements, or **empirical data,** are taken. Since the number of measurements can be quite large, we need some method of organizing the data. One such organizational technique, called a **frequency distribution,** involves a *table used to tabulate the number of occurrences* of a particular measurement. The next example illustrates how to create a frequency distribution.

EXAMPLE 1 The following data concern the number of cars passing through a chosen intersection in 1-minute intervals for a 15-minute period:

6	8	15	12	9
15	14	10	6	6
8	13	7	10	9

Construct a frequency distribution for these data.

Solution To construct a frequency distribution, we arrange the data values in ascending order and tabulate the number of times each value occurs. The frequency distribution for these data is given in the following table:

Number of Cars	6	7	8	9	10	11	12	13	14	15
Frequency	3	1	2	2	2	0	1	1	1	2

■ ■ ■

When data contain a large number of observations with a wide range of values, it is often helpful to organize the data into equal intervals or groups often called **classes.** There are several rules that you must follow when grouping data:

■ All intervals must be of equal width and cannot overlap.
■ A data value can belong to only one interval.

If we chose different interval widths, the frequency distribution would look different. There is no single correct rule for choosing interval widths for grouped data. However, if the interval width is too small or too large, it may be difficult to determine trends or patterns in the data. As a general rule, construct between 5 and 12 intervals. To determine an approximate width of the intervals, subtract the smallest data value from the largest, and divide the result by the number of intervals. This process is illustrated in the next example.

EXAMPLE 2 A quality control engineer conducted a survey of faulty integrated circuits. She randomly selected groups of 100 integrated circuits and recorded the numbers that were faulty in each group as follows:

9	21	16	12	31	27	46	18	7
36	17	29	41	6	26	35	39	11
23	38	45	10	30	32	8		

Construct a frequency distribution using grouped data.

Solution   First, we divide the range of values into nine equal intervals. The approximate width of each interval is $(46 - 6)/9 \approx 4$. Then we determine the frequency distribution of each interval. This information is summarized in the following table:

Interval	4–8	9–13	14–18	19–23	24–28	29–33	34–38	39–43	44–48
Frequency	3	4	3	2	2	4	3	2	2

■■■

Making a Frequency Distribution

1. Find the largest and smallest number in the data to find the range of values.

2. Divide the range into class intervals, usually between 5 and 12. Determine the lower and upper class limits for each interval.

3. Tally the number of observations for each class interval.

Graphic Representation

The graph of a frequency distribution makes recognition of a pattern or trend in the data easier to detect. One such graphical representation is the **histogram**, *a bar graph in which the data values are represented on the horizontal axis and the frequency is represented on the vertical axis.*

EXAMPLE 3   The following grades were recorded on a mathematics test:

84	67	73	81	94	57	78	71
81	53	62	91	77	64	89	43
98	76	78	69	93	80	51	70

Draw a histogram to depict the distribution of the grades.

Solution   First, we will divide the grades into equal intervals. The approximate interval width for six intervals is $(98 - 43)/6 \approx 9$. The frequency distribution is given in the following table:

Interval	40–49	50–59	60–69	70–79	80–89	90–99
Frequency	1	3	4	7	5	4

Next, we construct a histogram using rectangles whose width represents the length of the data interval and whose height represents the frequency of that interval. In the case of grouped data, generally the horizontal axis is labeled with the midpoint of the interval. Recall from a previous chapter that the midpoint of the interval is half the difference in interval endpoints. The histogram for this data set is shown in Figure 16–1.

FIGURE 16–1

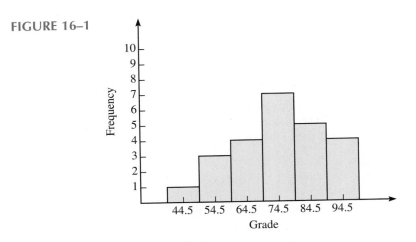

■ ■ ■

EXAMPLE 4   The graphing calculator can be used to graph data into a histogram. Bob did a survey of college students on the number of hours per week they spend using a computer. The data is summarized below. Use a graphing calculator to graph a histogram.

> 2, 10, 8, 21, 15, 40, 6, 8, 15, 25, 16, 48, 18, 15, 8, 31, 14, 22, 18, 3, 14, 16, 18, 27, 32, 35, 7, 10, 18, 23

Solution   On the TI-92, press ⬚APPS⬚, ⬚6⬚, and ⬚ENTER⬚. This puts you into the Data/Matrix editor and allows you to input the data. After the data has been entered, press ⬚F2⬚, Plot Setup. Press ⬚F4⬚ to check the plot you are entering. Then press ⬚F1⬚ to define the plot. Select Histogram, the correct cell column, and the histogram bucket width. In this case, we chose $(48 - 2)/9 \approx 5$. Then press ⬚ENTER⬚ and ⬚◆⬚, ⬚Graph⬚. The histogram will be displayed. You may need to press ⬚F2⬚, Zoom, to get a view. You can press ⬚F3⬚, Trace, and use the arrow keys to move along the graph. The information at the bottom of the screen will display the upper and lower limits on the class and the number of entries. From the histogram shown in Figure 16–2, you can see that most students, 9, spent 15 to 20 hours per week on the computer.

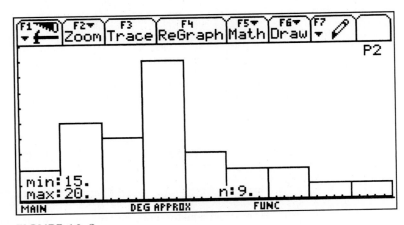

FIGURE 16–2

■ ■ ■

A second method of graphically representing a frequency distribution is called a **frequency polygon,** *a broken-line graph with the data value on the horizontal axis and the* *frequency of its occurrence on the vertical axis.* To construct a frequency polygon, place a point above the data group at the appropriate frequency. Then connect successive points with straight line segments. Since a polygon is a closed figure, always add an additional data group with a frequency of zero on each end of the horizontal axis.

EXAMPLE 5   Draw a frequency polygon for the data given in Example 3.

Solution   You can construct a frequency polygon from a histogram by joining the midpoints of the tops of the rectangles with line segments. The frequency polygon for this histogram is shown in Figure 16–3.

FIGURE 16–3

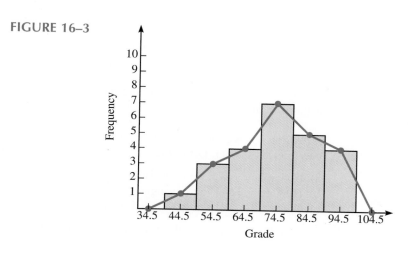

NOTE ✦ As shown in the next example, you do not have to draw a histogram before drawing a frequency polygon.

EXAMPLE 6   In an effort to monitor fuel consumption in compact cars, an auto manufacturer asked 150 car owners to report their mpg results. Using the following grouped data, construct a frequency polygon.

Miles per Gallon	25.0–25.4	25.5–25.9	26.0–26.4	26.5–26.9	27.0–27.4	27.5–27.9
Frequency	6	9	10	16	15	17
Miles per Gallon	28.0–28.4	28.5–28.9	29.0–29.4	29.5–29.9	30.0–30.4	
Frequency	26	20	18	8	5	

Solution   To construct the frequency polygon, add the two additional data groups of 24.5–24.9 and 30.5–30.9, each with a frequency of zero. Then label the midpoint of each data group on the horizontal axis, and place a dot at the corresponding frequency.

Connect successive points with a line segment as shown in Figure 16–4.

FIGURE 16–4

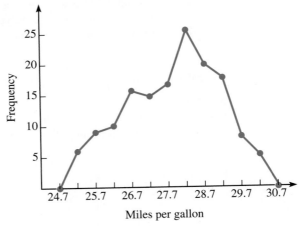

Often, it is necessary to interpret data organized in a histogram. This process is illustrated in the next example.

EXAMPLE 7 A group of steel shafts were weighed (in ounces). Figure 16–5 is a histogram of these data. Using the histogram, answer the following:

(a) How many shafts were measured?
(b) How many steel shafts weighed 1.85 oz?
(c) How many steel shafts weighed more than 1.95 oz?

(Exact values are used on the horizontal axis rather than the midpoint of an interval.)

FIGURE 16–5

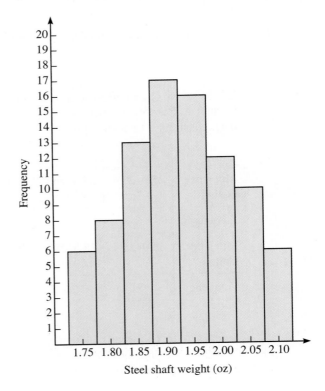

Solution

(a) From the histogram, we can determine the number of steel shafts measured by total-ing the frequency.

$$\text{number measured} = 6 + 8 + 13 + 17 + 16 + 12 + 10 + 6$$
$$= 88$$

Eighty-eight steel shafts were measured.

(b) To determine the number of shafts that weighed 1.85 oz, we find 1.85 on the horizontal axis; then we move vertically to the top of the rectangle and horizontally to read the frequency. There are 13 steel shafts that weigh 1.85 oz.

(c) The number of shafts that weighed more than 1.95 oz is the total of the frequencies in groups 2.00, 2.05, and 2.10 oz: $12 + 10 + 6$. Twenty-eight of the shafts weigh more than 1.95 oz.

∎∎∎

Another method of analyzing data is to use the **cumulative frequency.** This represents the sum of frequencies in a "less than" or "more than" class. In Example 7, the number of steel shafts that weighed more than 1.95 oz is an example of cumulative frequency. If we restrict the discussion to all values less than a given class limit, the graph of the cumulative frequency is called an ogive.

EXAMPLE 8 Construct the cumulative frequency for the test scores given in Example 3.

Test Score	Cumulative Frequency
less than 49	1
less than 59	4
less than 69	8
less than 79	15
less than 89	20
less than 99	24

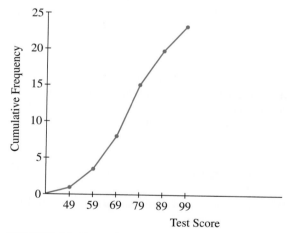

FIGURE 16–6

Solution

The accompanying ogive is shown in Figure 16–6. From the graph, it is easy to see that 8 students made less than 69 on the test.

∎∎∎

16–1 EXERCISES

Use data sets *A–F* for Problems 1–18.

A Grades on a math test: 86, 65, 74, 93, 88, 72, 86, 81, 74, 88, 72, 65, 69, 74, 97

B Number of defective diodes in groups of 100: 3, 10, 8, 6, 14, 10, 13, 8, 5, 18, 8, 7, 12, 3, 14, 7, 12, 6, 8, 15

C Number of hours worked in a week by technicians at C Corporation: 43, 35, 41, 56, 50, 48, 42, 40, 38, 46, 40, 35, 50, 46, 40, 48, 43

D Age of employees at M Corporation: 30, 45, 56, 47, 25, 47, 30, 32, 27, 33, 63, 42, 35, 39, 25, 35, 56, 42, 33, 42, 45, 39

E The IQ of math students at Little Rock School: 86, 110, 95, 134, 100, 89, 97, 110, 98, 80, 100, 83, 147, 110, 84, 98, 100, 120, 134, 80

F Starting salary for electrical engineers (in thousands): 18, 23, 15, 17, 20, 18, 21, 17, 22, 18, 23, 19, 21, 18, 20, 23, 19, 17

Construct a frequency distribution for the indicated data set.

1. Set A   2. Set B
3. Set C   4. Set D
5. Set E   6. Set F

Construct a histogram for the indicated data set.

7. Set A   8. Set B
9. Set C   10. Set D
11. Set E   12. Set F

Construct a frequency polygon for the indicated data set.

13. Set A   14. Set B
15. Set C   16. Set D
17. Set E   18. Set F

19. Construct a frequency distribution from the histogram shown in Figure 16–7. (Exact values are used on the horizontal axis rather than the midpoint of the interval because the data are not grouped.)

20. From the histogram in Figure 16–7, how many factories were surveyed?

21. From the histogram in Figure 16–7, how many employees were counted?

22. From the histogram in Figure 16–7, how many factories employed 800 people?

23. Construct a frequency distribution from the histogram in Figure 16–8.

24. From the histogram in Figure 16–8, how many students took the test?

25. From the histogram in Figure 16–8, how many students scored in the interval for 80 or above on the test?

26. Using the information in Figure 16–7, construct a cumulative frequency and ogive.

27. Using the information in Figure 16–8, construct a cumulative frequency and ogive.

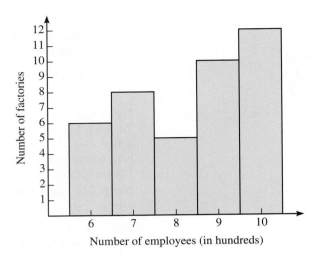

FIGURE 16–7

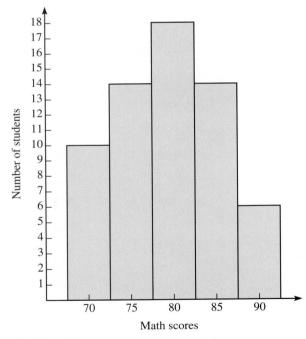

FIGURE 16–8

16–2

MEASURES OF CENTRAL TENDENCY

Although frequency distributions and graphs provide useful information about data, it is often helpful to obtain numbers that are representative of the distribution of the data. These numbers are called **measures of central tendency.** In this section we will discuss three measures of central tendency: mean, median, and mode.

Mean

The most frequently used measure of central tendency is the **arithmetic mean,** often called the **average.** To calculate the arithmetic mean $\bar{x}$ of a group data, *add the data values and divide this total by the number of values.* (Some textbooks use the Greek letter mu, μ, to denote the arithmetic mean.) The following formula summarizes this calculation.

Arithmetic Mean

Given data values $x_1, x_2, x_3, \cdots, x_n$, the arithmetic mean $\bar{x}$ is given by

$$\bar{x} = \frac{x_1 + x_2 + x_3 + \cdots + x_n}{n} = \frac{\sum_{i=1}^{n} x_i}{n}$$

where the Greek letter Σ (upper case sigma) denotes a sum. The notation $\sum_{i=1}^{n}$ means to add $x_1, x_2, \cdots$ through x_n. The variable n equals the number of values in the data set.

EXAMPLE 1   The monthly electrical energy consumption (in kilowatt-hours) of 8 families with the same size house was found to be

618   597   715   674   703   637   657   586

Find the average monthly energy consumption of these families.

Solution   The average or arithmetic mean is given by

$$\bar{x} = \frac{\sum_{i=1}^{n} x_i}{n} = \frac{\sum_{i=1}^{8} x_i}{8} = \frac{618 + 597 + 715 + 674 + 703 + 637 + 657 + 586}{8}$$

$$\bar{x} = \frac{5{,}187}{8}$$

$$\bar{x} \approx 648$$

■ ■ ■

We can also calculate the arithmetic mean given a frequency distribution. The formula is given below, and its application is illustrated in the next example.

Arithmetic Mean from a Frequency Distribution

$$\bar{x} = \frac{\displaystyle\sum_{i=1}^{n} f_i x_i}{\displaystyle\sum_{i=1}^{n} f_i}$$

where x_i = a data value and f_i = frequency of that value.

EXAMPLE 2   A quality control engineer counted the number of defective light bulbs in lots of 100. From the following frequency distribution, determine the average number of defective bulbs per 100:

Number of Defective Bulbs	3	5	6	8	4
Frequency	6	2	1	2	4

Solution   Substituting into the formula for the arithmetic mean from a frequency distribution gives

$$\bar{x} = \frac{\displaystyle\sum_{i=1}^{n} f_i x_i}{\displaystyle\sum_{i=1}^{n} f_i}$$

$$\bar{x} = \frac{3(6) + 5(2) + 6(1) + 8(2) + 4(4)}{6 + 2 + 1 + 2 + 4}$$

$$\bar{x} = 4.4$$

On average, there are approximately 4 defective bulbs per hundred.   ■ ■ ■

If the frequency distribution is given in terms of intervals rather than exact values, substitute the midpoint of each interval for x_i into the mean formula.

EXAMPLE 3   Find the mean for the following grouped data:

x	10–14	15–19	20–24	25–29	30–34
Frequency	6	8	12	7	4

Solution   To find the mean, we use a frequency distribution to substitute into the mean formula. Since the data are grouped, use the midpoint of each interval.

$$\bar{x} = \frac{\sum\limits_{i=1}^{n} f_i x_i}{\sum\limits_{i=1}^{n} f_i}$$

midpoint
10–14　　midpoint　midpoint
　　　　　20–24　25–29
　midpoint
　15–19　　　　　　　　midpoint
　　　　　　　　　　　30–34

$$\bar{x} = \frac{12(6) + 17(8) + 22(12) + 27(7) + 32(4)}{6 + 8 + 12 + 7 + 4}$$

$$\bar{x} \approx 21$$

∎∎∎

EXAMPLE 4   Using the information in Example 6 of Section 16–1, find the average fuel consumption of the cars.

Solution   Since the data are grouped, we use the midpoint of each interval as the data value and substitute into the formula for the arithmetic mean from a frequency distribution.

$$\bar{x} = [(25.2)(6) + (25.7)(9) + (26.2)(10) + (26.7)(16) + (27.2)(15)$$
$$+ (27.7)(17) + (28.2)(26) + (28.7)(20) + (29.2)(18) + (29.7)(8)$$
$$+ (30.2)(5)] \div [6 + 9 + 10 + 16 + 15 + 17 + 26 + 20$$
$$+ 18 + 8 + 5]$$

$$\bar{x} \approx 27.8$$

The average fuel consumption of these compact cars is 27.8 mpg.

∎∎∎

Median

A second measure of central tendency is the **median,** which is *the middle number when the data are arranged in numerical order in an array.* If the number N of data items is odd, count $N/2$ (rounded up) units from either end of the array to reach the median.

EXAMPLE 5   Find the median of the following beginning salaries of 9 computer technicians:

$19,160 $20,283 $20,476 $20,537 $21,571

$21,963 $22,419 $22,849 $23,473

Solution   To find the median salary, count $9/2 \approx 5$ from either end of the array. The median salary is $21,571 since there are four salaries higher and four salaries lower than it.

four data values below median　　　median

$19,160 $20,283 $20,476 $20,537 $21,571

four data values above median

$21,963 $22,419 $22,849 $23,473

∎∎∎

The median of an even number of data values is the mean of the middle two values. Count $N/2$ units from each end of the array of numbers, and find the mean of the two values. The next example illustrates how to calculate the median in this case.

EXAMPLE 6   The following numbers represent the age of Mr. Carter's employees:

37   21   41   32   18   40   28   47

Find the median of these data values.

Solution   First, arrange the data values in ascending order. To find the median value, count $8/2 = 4$ values from each end and take the mean of the two values. The median is the mean of the fourth and fifth values.

$$\text{median} = \text{mean of these values}$$

$$
\underbrace{18 \quad 21 \quad 28 \quad 32}_{4} \quad \overset{\nearrow \nwarrow}{} \quad \underbrace{37 \quad 40 \quad 41 \quad 47}_{4}
$$

$$\text{median} = \frac{32 + 37}{2} = \frac{69}{2} = 34.5$$

The median age of these employees is 34.5.   ■ ■ ■

The median divides the data into two equal parts. In the same way, the **quartiles, Q_1, Q_2, and Q_3,** divide the data into four equal parts. Quartiles provide us an easy way to graphically see how the data is distributed. You will often see standardized test scores, such as GRE, reporting your score and quartile. If you are in the 90th quartile, 90% of the people taking the test scored less than you.

Finding Quartiles

---

1. Arrange the data in ascending order.

2. Q_2 is the median. It divides the data into two equal parts.

3. Q_1 is the median of the lower half of the data.

4. Q_3 is the median of the upper half of the data.

EXAMPLE 7   Determine the quartiles for the following data: 16, 8, 12, 21, 7, 18, 8, 36, 10, 27, 17, 20, 16.

Solution   We begin by arranging the data in ascending order.

$$7, 8, 8, 10, 12, 16, 16, 17, 18, 20, 21, 27, 36$$

Next, we find the median, Q_2, as 13 numbers/2 $\approx$ 7. The median is 16. The lower half of the data is

$$7, 8, 8, 10, 12, 16$$

The median Q_1 of the lower half is $\dfrac{8 + 10}{2} = 9$.

The upper half of the data is

$$17, 18, 20, 21, 27, 36$$

The median Q_3 of the upper half is $\dfrac{20 + 21}{2} = 20.5$. ▪▪▪

Once you have determined the quartiles, you can construct a **box plot.** This can also be done on a graphing calculator, as shown in the next example.

EXAMPLE 8 Using the data from Example 7, construct a box plot.

Solution Press $\boxed{\text{APPS}}$, $\boxed{6}$, and $\boxed{\text{ENTER}}$. Input the data in one of the columns. When that is complete press $\boxed{\text{F2}}$. Check the plot number you want. Then, press $\boxed{\text{F1}}$ to define the plot. Choose Box Plot, the correct column number, and press $\boxed{\text{ENTER}}$. Then, press $\boxed{\blacklozenge}$ and $\boxed{\text{Graph}}$. The box plot is graphed. If you press $\boxed{\text{F3}}$, Trace, you can see the values for Q_1, Q_2, and Q_3. The lines on each side show the minimum and maximum data values. The box plot is shown in Figure 16–9.

FIGURE 16–9

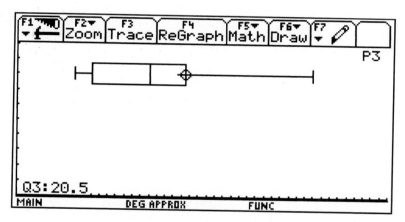

Mode

A third measure of central tendency, rarely used in technical applications, is the **mode,** *the data value that occurs most frequently.* There may be more than one mode for a given data set if several data values have the same maximum frequency. The mode is used primarily in quality control as an inspection method for determining central tendency. It gives a quick and approximate measure of the central tendency and is used to describe the most typical value of a distribution.

EXAMPLE 9 The following data result from a motion-time study on the time duration of a task:

6 8 10 7 6 11 7 6 12
17 9 8 6 14 11 2 8 10

Find the mode for this data set.

Solution To determine the mode, we construct the following frequency distribution:

Data	2	6	7	8	9	10	11	12	14	17
Frequency	1	4	2	3	1	2	2	1	1	1

From the frequency distribution, the mode is 6 because it occurs most frequently, four times.   ■ ■ ■

16–2 EXERCISES

For the data sets in Problems 1–14, find the mean, median, and mode.

1. 16, 21, 18, 15, 26, 40, 13, 29, 16, 14, 20, 19, 27

2. 3, 14, 8, 7, 2, 5, 6, 8, 3, 4, 9

3. 73, 86, 93, 60, 73, 84, 82, 97

4. 205, 310, 846, 350

5. 3, 3, 3, 6, 8, 8, 10, 10, 10, 12, 15

6. 750, 310, 946, 1,150, 1,365, 310

7. 75, 95, 80, 60, 65, 75, 90, 75, 85

8. 4, 4, 4, 9, 9, 10, 10, 10, 13, 13, 18, 20

9. 1, 360, 2,108, 50, 1,840, 780

10. 1, 16, 8, 12, 10, 9, 24, 2, 6, 10, 8

11. Number of defective integrated circuits per lot of 100: 6, 10, 14, 5, 6, 8, 10, 6

12. IQ of math students: 120, 86, 95, 100, 89, 115, 108, 90, 80, 100, 110, 130, 85, 99, 106, 95, 100, 103, 108, 96

13. Test scores for a math class: 73, 60, 85, 78, 85, 94, 73, 85, 77, 87, 82, 72, 85, 70, 82

14. Salary (in thousands) of civil engineers: 18, 21, 19, 24, 16, 28, 21

15. A math student needs an average of 75 on nine math exams to get a grade of C in the course. In his first 8 tests, Fred made 86, 64, 70, 77, 75, 69, 76, and 83. What must he make on the last test to make a grade of C in the course?

Ch 16. During the last 5 months a photo lab used the following amounts of a certain chemical: 56 lb, 68 lb, 32 lb, 47 lb, and 60 lb. Based on the average monthly consumption, how much chemical is needed for the next 12 months?

B 17. A technician must estimate the labor cost of repairing a computer. She knows that five similar jobs required 6.1 h, 3.3 h, 4.2 h, 5.1 h, and 7.2 h. Using the arithmetic mean of the previous jobs and an hourly rate of $35 per hour, determine how much the technician should estimate for labor cost.

B 18. A company has 46 employees. Fifteen earn $5.50 per hour, 12 earn $10.00 per hour, and 19 earn $14.50 per hour. Determine the mean earnings per hour.

B 19. The monthly utility bills for XYZ Corporation are as follows:

$96.25	$108.10	$74.12	$60.58	$87.34
$74.12	$136.45	$103.15	$94.83	$81.50
$96.25	$72.51			

Construct a frequency distribution and histogram for the data, and find the mean, median, and mode of this data set.

20. Using the data in Problem 19, determine the quartiles.

21. Using the information from Problem 20, construct a box plot.

16–3

MEASURES OF DISPERSION

In the previous section we discussed measures of central tendency, which are used to determine data values representative of a collection of data. However, it is also important to have some indication of the spread of the data values, known as **measures of dispersion** or **variation.** In this section we will discuss two measures of dispersion: range and standard deviation.

Range

Range is the measure of dispersion that is least used but easiest to calculate. Because it is not sensitive to variation within a data group or to variation from the mean, it is used infrequently. The **range** is defined as *the difference between the two extreme values in a data set.*

EXAMPLE 1   A machine is tooling a part that should be 26 cm long. A quality control engineer randomly selects 8 parts and finds their lengths (in centimeters) to be as follows:

24.7   26.8   23.2   27.3   25.8   24.9   26.8   26.2

Determine the range of these values.

Solution From the list of lengths, we determine that the largest value is 27.3 cm and the smallest value is 23.2 cm. Therefore,

$$\text{range} = \text{highest value} - \text{lowest value}$$
$$\text{range} = 27.3 - 23.2$$
$$\text{range} = 4.1$$

The range of this data set is 4.1 cm.

■■■

Standard Deviation

The most widely used measure of dispersion is the **standard deviation,** which is a measure of how much the data relate to the arithmetic mean. The standard deviation is given by the following formula.

Standard Deviation

$$s = \sqrt{\frac{\sum\limits_{i=1}^{n} (x_i - \bar{x})^2}{n}}$$

where $\bar{x}$ = mean of the set of n values $x_1, x_2, \ldots, x_n$. (Some textbooks use the Greek letter σ (lowercase sigma) to denote standard deviation.)

Using the formula above, we can calculate standard deviation by computing the following values successively:

■ Calculate the arithmetic mean $\bar{x}$ of the data values.
■ Subtract the mean from each data value.
■ Square each of these differences.
■ Calculate the arithmetic mean of the squared differences.
■ Calculate the square root of this arithmetic mean.

It is usually best to organize these calculations into a table, as illustrated in the next example.

EXAMPLE 2 The following values represent the number of defective diodes in lots of 100:

1   6   5   2   9   6   4   8   7   5

Find the standard deviation of these data values.

Solution First, we use the information in Table 16–1 to calculate the mean of the data values.

$$\bar{x} = \frac{\sum x}{n} = \frac{53}{10} = 5.3$$

TABLE 16–1

x	$x - \bar{x}$	$(x - \bar{x})^2$
1	−4.3	18.49
6	0.7	0.49
5	−0.3	0.09
2	−3.3	10.89
9	3.7	13.69
6	0.7	0.49
4	−1.3	1.69
8	2.7	7.29
7	1.7	2.89
5	−0.3	0.09
Σ 53		Σ 56.1

Second, we calculate the difference between the mean and the data values as shown in the second column of Table 16–1. Third, we square each entry in the second column and place the result in the third column of Table 16–1. Fourth, we find the arithmetic mean of these squared differences.

$$\frac{\Sigma (x - \bar{x})^2}{n} = \frac{56.1}{10} = 5.61$$

Last, the standard deviation is the square root of this mean.

$$s = \sqrt{5.61} \approx 2.4$$

The average number of defective diodes per 100 is 5 with a standard deviation of 2.

The graphing calculator will perform these calculations. For Example 2, perform the following steps using the TI-92. Press APPS, 6, and ENTER. This puts you into the Data/Matrix Editor so that you can enter the data. After the data values have been input, press F5. Define the calculation type as one variable, input the correct column for the data, and press ENTER twice. The following information will be displayed; mean of x values, sum

FIGURE 16–10

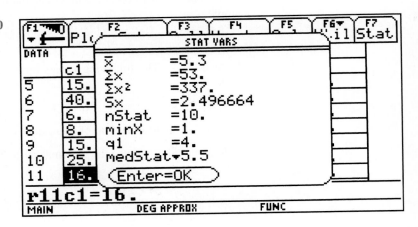

of x values, sum of x^2 values, sample standard deviation, number of data points, minimum of x values, Q_1, median, Q_3, and maximum of x values. The screen for this calculation is shown in Figure 16–10.

Standard Deviation from Grouped Data

To calculate the standard deviation from grouped data, use the following formula.

Standard Deviation from a Frequency Distribution

$$s = \sqrt{\frac{\sum\limits_{i=1}^{n} f_i(x_i - \overline{x})^2}{\sum\limits_{i=1}^{n} f_i}}$$

where $\overline{x}$ = mean of the data set $x_1, x_2, \ldots, x_n$ and f_i = frequency of data value x_i.

EXAMPLE 3 The first two columns in Table 16–2 show the test scores for a mathematics class. Using the frequency distribution, find the standard deviation.

Solution First, fill in the $f \cdot x$ column in Table 16–2 by multiplying the columns labeled x and f. Then calculate the mean as follows:

$$\overline{x} = \frac{\sum f \cdot x}{f} = \frac{1{,}968}{25} = 78.72$$

TABLE 16–2

x	f	$f \cdot x$	$x - \overline{x}$	$(x - \overline{x})^2$	$f(x - \overline{x})^2$
70	4	280	−8.72	76.04	304.16
73	6	438	−5.72	32.72	196.32
78	3	234	−0.72	0.52	1.56
80	5	400	1.28	1.64	8.20
88	7	616	9.28	86.12	602.84
	Σ 25	Σ 1,968			Σ 1,113.08

Next, fill in the remainder of the table. Last, use the equation and the information in Table 16–2 to calculate the standard deviation.

$$s = \sqrt{\frac{1{,}113.08}{25}} = 6.67$$

The average grade on the test is 79 with a standard deviation of 7 points. ∎

Alternate Formula for Standard Deviation

If you use a calculator to determine standard deviation, the following formula is easier.

Alternate Formula for Standard Deviation

$$s = \sqrt{\frac{\sum_{i=1}^{n} (x_i)^2}{n} - \left(\frac{\sum_{i=1}^{n} x_i}{n}\right)^2}$$

mean of the squares   square of the mean

▦ **EXAMPLE 4** The following measurements (in millimeters) represent the variation of a metal part from its specified length:

1.83   4.31   3.42   4.08   2.67   3.95

Find the standard deviation of these measurements.

Solution First, calculate the square of the mean, $\left(\dfrac{\sum x}{n}\right)^2$.

1.83 $\boxed{+}$ 4.31 $\boxed{+}$ 3.42 $\boxed{+}$ 4.08 $\boxed{+}$ 2.67 $\boxed{+}$

3.95 $\boxed{=}$ $\boxed{\div}$ 6 $\boxed{=}$ $\boxed{x^2}$ $\boxed{\text{STO}}$ $\longrightarrow$ 11.40187778

Second, calculate the mean of the squares, $\dfrac{\sum x^2}{n}$.

1.83 $\boxed{x^2}$ $\boxed{+}$ 4.31 $\boxed{x^2}$ $\boxed{+}$ 3.42 $\boxed{x^2}$ $\boxed{+}$ 4.08

$\boxed{x^2}$ $\boxed{+}$ 2.67 $\boxed{x^2}$ $\boxed{+}$ 3.95 $\boxed{x^2}$ $\boxed{=}$ $\boxed{\div}$ 6 $\boxed{=}$ $\longrightarrow$ 12.16653333

Third, subtract these values.

$\boxed{-}$ $\boxed{\text{RCL}}$ $\boxed{=}$ $\longrightarrow$ 0.764655550

Last, take the square root of the resulting value.

$\boxed{\sqrt{}}$ $\longrightarrow$ 0.874445853

The standard deviation is approximately 0.87 mm.   ■■■

NOTE ✦ Some calculators have special keys to aid in statistical calculations, such as $\boxed{\Sigma x}$, $\boxed{\Sigma x^2}$, and $\boxed{s}$ (standard deviation). Consult your owner's manual or instructions.

Significance of Standard Deviation

Thus far in this section, we have seen that the standard deviation is a measure of the dispersion or variation of a data set from the mean, and we have discussed how to calculate it. Now, we will discuss the significance of the standard deviation.

If we take enough experimental measurements, the curve of the data set usually approaches a normal, bell-shaped distribution as shown in Figure 16–11. Notice that for the normal distribution, approximately 68% of the measurements fall within one standard deviation of the mean, or within $\bar{x} - s$ and $\bar{x} + s$. Also, roughly 95% of the measurements fall within the two standard deviations of the mean, or within $\bar{x} - 2s$ and $\bar{x} + 2s$.

FIGURE 16–11

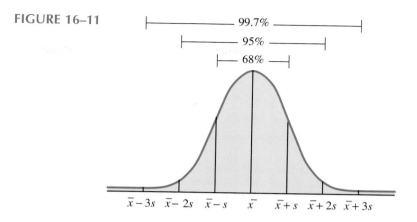

$$\bar{x} - 3s \quad \bar{x} - 2s \quad \bar{x} - s \quad \bar{x} \quad \bar{x} + s \quad \bar{x} + 2s \quad \bar{x} + 3s$$

EXAMPLE 5 For the data set in Example 3, calculate the interval of values within one standard deviation of the mean. Compare this result with the actual percentage of the data set that falls within this range.

Solution To find the interval of values within one standard deviation of the mean, we calculate $\bar{x} - s$ and $\bar{x} + s$. From the calculations in Example 3, $\bar{x} = 78.72$ and $s = 6.67$. Thus,

$$\bar{x} - s = 78.72 - 6.67 = 72.05$$
$$\bar{x} + s = 78.72 + 6.67 = 85.39$$

The interval of values within one standard deviation of the mean is 72.05 to 85.39. From the data set itself, we find that 14 of the 25 values, or 56%, fall within this interval.

■ ■ ■

16–3 EXERCISES

Use data sets *A–E* for Problems 1–10.

A Average daily temperature (°F) for February of major cities: 36°, 72°, 65°, 19°, 18°, 25°, 60°, 58°, 30°, 15°, 48°

B Age of employees at ABC Stores: 18, 26, 43, 32, 19, 54, 39, 29, 22, 47, 39, 21, 60, 34, 23, 25, 58, 48, 18, 53

C Test scores: 63, 87, 73, 98, 60, 78, 85, 81, 70, 98, 93, 90, 72, 93, 83, 68, 88, 81, 72, 79, 61, 65, 75, 63

D Yearly income of technicians (in thousands): 18, 28, 21, 29, 31, 23, 25, 19, 20, 28, 18, 30, 20, 26, 18, 27, 23

E IQ of math students: 86, 140, 100, 96, 103, 110, 115, 97, 90, 108, 89, 97, 100, 105, 120, 118, 136, 102, 109

Find the range and standard deviation of the indicated data set. You may want to use the graphing calculator to check your answers.

1. Set *A*

2. Set *B*

3. Set *C*

4. Set *D*

5. Set *E*

TABLE 16–3

Grade	Frequency
60	2
65	1
70	3
80	4
85	6
90	3
95	2
100	1

Using the indicated data set, determine the interval of values one standard deviation from the mean. Compare this value to the actual percentage of data values in this range.

6. Set *A*

7. Set *B*

8. Set *C*

9. Set *D*

10. Set *E*

11. Using the frequency distribution in Table 16–3, construct a histogram and find the mean, median, mode, and standard deviation.

12. Using the frequency distribution in Table 16–4, construct a frequency polygon and calculate the mean, median, mode, and standard deviation.

13. Calculate the standard deviation for Problem 4 of Exercises 16–1.

14. Calculate the standard deviation for Problem 12 of Exercises 16–2.

15. Calculate the standard deviation for Problem 13 of Exercises 16–2.

TABLE 16–4

Employee's Age	Frequency
18	2
25	5
28	2
36	6
39	4
42	3
48	3
56	1

16. Calculate the standard deviation for Problem 14 of Exercises 16–2.

16–4

EMPIRICAL CURVE FITTING

In this chapter we have discussed statistical methods as they relate to a single variable. We have discussed methods of tabulating and displaying data and of calculating measures of central tendency and dispersion for one variable. However, we know from our study of functions in Chapter 4 that a relationship may exist between two variables. In this section we will discuss a method of determining the relationship between two variables using empirical data, or measurements.

Many of the relationships between two variables can be derived mathematically and stated in a formula, such as $A = \pi r^2$ for the area of a circle. However, in technology we are often given the relationship between two variables in the form of measurements. This relationship is activated by systematically varying one variable while measuring the second variable. Due to measurement errors, these relationships do not give exact formulas. The method of finding the curve that best fits the data is called **empirical curve fitting.**

Before attempting to fit a curve, we must first determine the type of curve that best fits the data. We can determine the form of the curve by using prior knowledge of the relationship between the variables or by graphing the data set as ordered pairs on the Cartesian coordinate plane.

EXAMPLE 1 Jack repeatedly clocks the velocity of a moving car as a function of time as follows:

t	1	2	3	4	5	6
v	28.5	35	46	50	60	66

Plot the data points on the Cartesian plane and determine the apparent relationship between the variables.

Solution   The graph of the data points is given in Figure 16–12. From the graph, the relationship between the variables appears to be linear. From physics, we know that the

FIGURE 16–12

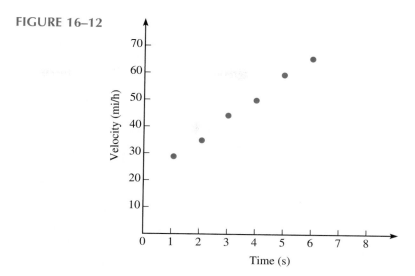

velocity v of an object as a function of time is given by $v = v_0 + at$, where v_0 = initial velocity, a = acceleration, and t = time. Both from the graph of the data and from prior knowledge, we should fit the data to a linear equation.

■ ■ ■

Least Squares Method

The data points from Example 1 and the equation representing that particular relationship, $v = 20 + 8t$ (we will discuss how to determine this equation later), are plotted on the same Cartesian plane in Figure 16–13. Note that the data points do not all fall on the graph. The difference between the y value of the data point and the y value of the curve is called the **deviation,** as illustrated in Figure 16–13.

The most commonly used technique of curve fitting is called the **least squares method.** The method is used to find the equation of the curve that minimizes the deviation.

FIGURE 16–13

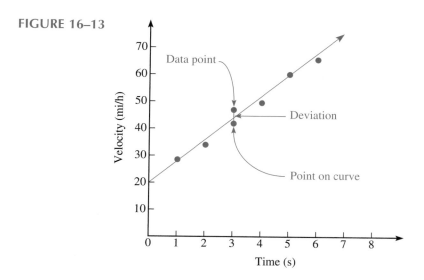

Linear Curve Fitting

We will begin our discussion of curve fitting with linear equations because they are easier. Also, fitting a nonlinear curve is based on the methods used in linear curve fitting.

The equation of a straight line can be expressed in slope–intercept form as $y = mx + b$ where m represents the slope of the line and b represents the y coordinate of the y intercept. Using advanced mathematics, we can derive the following formula to calculate the value of m and b.

Curve Fitting a Straight Line

$$y = mx + b$$

where

$$m = \frac{n \sum xy - \sum x \sum y}{n \sum x^2 - (\sum x)^2}$$

$$b = \frac{\sum x^2 \sum y - \sum x \sum xy}{n \sum x^2 - (\sum x)^2}$$

(x, y) represents the data points, and n is the number of data points.

 EXAMPLE 2   A teacher wants to be able to predict a student's grade on a test based on the number of hours of study. Assuming that the relationship between the variables is linear, find the least squares equation of the line that fits the data, and use it to predict the test score of a student who studies 4 hours.

x (hours)	8	6	5	7	8.3	7.5	6.3	8.7
y (grade)	91	72	62	81	94	86	76	98

Solution   Normally, we would plot the data to determine the nature of the relationship between the variables, but it is given to be linear. To find the equation of the line, we must calculate m and b using the formulas above. Table 16–5 provides the necessary information.

TABLE 16–5

x (Hours)	y (Grade)	xy	x^2
8	91	728	64
6	72	432	36
5	62	310	25
7	81	567	49
8.3	94	780.2	68.89
7.5	86	645	56.25
6.3	76	478.8	39.69
8.7	98	852.6	75.69
$\Sigma 56.8$	$\Sigma 660$	$\Sigma 4{,}793.6$	$\Sigma 414.52$

Using the formulas to calculate m and b gives

$$m = \frac{n\,\Sigma\,xy - \Sigma\,x\,\Sigma y}{n\,\Sigma\,x^2 - (\Sigma\,x)^2}$$

$$m = \frac{8(4{,}793.6) - 56.8(660)}{8(414.52) - (56.8)^2} \approx 9.57$$

$([8 \times 4793.6 - 56.8 \times 660)] \div$

$([8 \times 414.52 - 56.8 \, x^2)] = \longrightarrow 9.57$

$$b = \frac{\Sigma\,x^2\,\Sigma\,y - \Sigma\,x\,\Sigma\,xy}{n\,\Sigma\,x^2 - (\Sigma\,x)^2}$$

$$b = \frac{414.52(660) - 56.8(4{,}793.6)}{8(414.52) - (56.8)^2} \approx 14.5$$

$([414.52 \times 660 - 4793.6 \times 56.8)] \div$

$([8 \times 414.52 - 56.8 \, x^2)] = \longrightarrow 14.5$

The least squares equation of the straight line that best fits the data is

$$y = 9.57x + 14.5$$

Using this equation to predict the grade of a student who studies 4 hours gives

$$y = 9.57(4) + 14.5$$
$$y \approx 53$$

The graph in Figure 16–14 shows the least squares line and the data points.

FIGURE 16–14

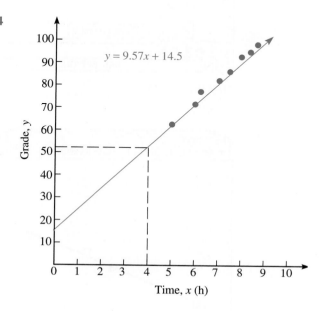

CAUTION ✦ You should always graph the data points and the least squares curve on the same axes to point out calculation errors.

Graphing Calculator

Graphing calculators allow you to enter data, compute statistics for the data, and graph the data and the least squares equation. The following example illustrates the procedure.

EXAMPLE 3 Using the data in Example 2, use the graphing calculator to calculate the least squares line and graph it with the data points.

Solution Remember to refer to the owner's manual for your graphing calculator for specific keystrokes. First, select FUNCTION from the MODE dialog box. Second, display the Data/Matrix Editor and enter the values for hours studied and the corresponding grade. Third, from the Calculate dialog box, choose Lin Reg (linear regression), assign x and y to the variables set up in the Data/Matrix Editor, and store the equation to $y|(x)$. After pressing Enter, the linear regression information is displayed. This result is shown in Figure 16–15. From this information, the equation for the linear regression line is $y = 9.572954x + 14.532028$. Fourth, define the plot using a scatter, box, and variable definition. Fifth, display the $y =$ Editor. You should see that Plot 1 and $y1$ are checked. Set the display style to Dot. Last, graph the data points and the linear regression line. This result is shown in Figure 16–16.

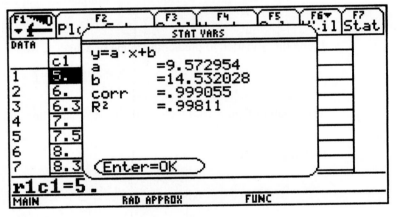

FIGURE 16–15

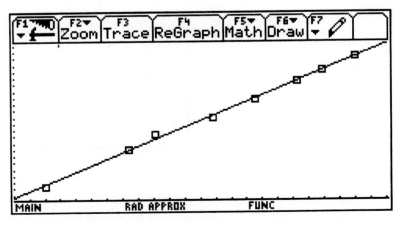

FIGURE 16–16

EXAMPLE 4   The following data resulted from attaching a weight to a spring and measuring the length of the spring:

(0, 4)   (2, 7.5)   (4, 11)   (1.2, 6)   (3, 9)   (5, 12)

Determine a linear least squares line that fits the data, and use it to determine the length of the spring when 8 oz is attached to the spring.

Solution   To determine the linear equation that best fits the data, we fill in Table 16–6.

TABLE 16–6

x (oz)	y (in.)	xy	x^2
0	4	0	0
1.2	6	7.2	1.44
2	7.5	15	4
3	9	27	9
4	11	44	16
5	12	60	25
$\Sigma 15.2$	$\Sigma 49.5$	$\Sigma 153.2$	$\Sigma 55.44$

Next, we substitute the required values from the table into the formulas for m and b.

$$m = \frac{n \Sigma xy - \Sigma x \Sigma y}{n \Sigma x^2 - (\Sigma x)^2}$$

$$m = \frac{6(153.2) - 15.2(49.5)}{6(55.44) - (15.2)^2}$$

$$m \approx 1.64$$

$$b = \frac{\Sigma x^2 \Sigma y - \Sigma x \Sigma xy}{n \Sigma x^2 - (\Sigma x)^2}$$

$$b = \frac{55.44(49.5) - 15.2(153.2)}{6(55.44) - (15.2)^2}$$

$$b \approx 4.09$$

The linear equation that best fits the data is

$$y = 1.64x + 4.09$$

To determine the length of the spring when an 8-oz object is attached, we substitute $x = 8$ into the equation.

$$y = 1.64(8) + 4.09$$
$$y = 17.21 \text{ in.}$$

Figure 16–17 shows the least squares equation and the data points.

FIGURE 16–17

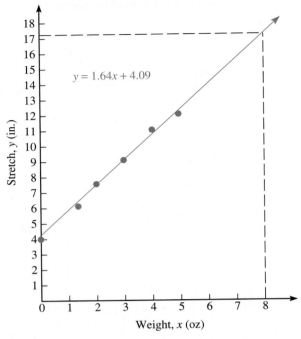

So far in this section, we have discussed only a linear relationship between two variables. However, from previous chapters we know that variables can also be related by nonlinear functions. If the graph of the data points indicates a nonlinear relationship $f(x)$, then we can express the least squares equation by rewriting the slope–intercept form in the following manner. Since nonlinear curves are harder to determine, you will be given the form of the equation for the least squares curve.

Nonlinear Curve Fitting

Nonlinear Least Squares Curve

$$y = m[f(x)] + b$$

where

$$m = \frac{n \, \Sigma \, ([f(x)]y) - \Sigma \, f(x) \, \Sigma \, y}{n \, \Sigma \, [f(x)]^2 - [\Sigma \, f(x)]^2}$$

$$b = \frac{\Sigma \, [f(x)]^2 \, \Sigma \, y - \Sigma \, f(x) \, \Sigma \, ([f(x)]y)}{n \, \Sigma \, [f(x)]^2 - [\Sigma \, f(x)]^2}$$

$f(x)$ is a function, m is the slope of the line, and b is the y coordinate of the y intercept.

EXAMPLE 5   Find the least square equation of the form $y = mx^2 + b$ for the following data:

x (s)	0	1	2	3	4	5
y (ft)	8.7	9.8	13.7	19.6	28	38.6

Then use the equation to find y when $x = 8$ s.

Solution   Using $f(x) = x^2$, we fill in Table 16–7.

TABLE 16–7

x	y	$f(x) = x^2$	$f(x)y$	$[f(x)]^2$
0	8.7	0	0	0
1	9.8	1	9.8	1
2	13.7	4	54.8	16
3	19.6	9	176.4	81
4	28	16	448	256
5	38.6	25	965	625
	$\Sigma 118.4$	$\Sigma 55$	$\Sigma 1,654.0$	$\Sigma 979$

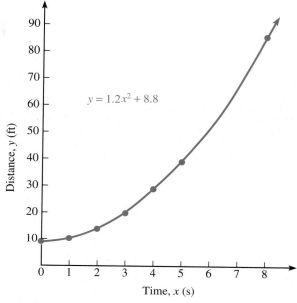

FIGURE 16–18

Next, we substitute into the formulas to calculate m and b.

$$m = \frac{n \, \Sigma \, ([f(x)]y) - \Sigma \, f(x) \, \Sigma \, y}{n \, \Sigma \, [f(x)]^2 - [\Sigma \, f(x)]^2}$$

$$m = \frac{6(1,654.0) - 55(118.4)}{6(979) - (55)^2}$$

$$m \approx 1.2$$

$$b = \frac{\Sigma \, [f(x)]^2 \, \Sigma \, y - \Sigma \, f(x) \, \Sigma \, ([f(x)]y)}{n \, \Sigma \, [f(x)]^2 - [\Sigma \, f(x)]^2}$$

$$b = \frac{979(118.4) - 55(1,654.0)}{6(979) - (55)^2}$$

$$b \approx 8.8$$

The least squares equation that best fits the data is given by

$$y = 1.2x^2 + 8.8$$

When $x = 8$ s,

$$y = 1.2(8^2) + 8.8$$
$$y \approx 85.6 \text{ ft}$$

Figure 16–18 shows the graph of the curve and the data points.

■ ■ ■

EXAMPLE 6 The temperature of a metal pipe was measured as it was heated, and the following data were obtained:

x (min)	1	2	3	4	5
y (°C)	77	90	107	131	165

Fit a curve to the data if the equation is of the form $y = me^{0.3x} + b$. Find y when the time is 8 min.

Solution Using $f(x) = e^{0.3x}$, we set up Table 16–8.

TABLE 16–8

x	y	$f(x) = e^{0.3x}$	$f(x)y$	$[f(x)]^2$
1	77	1.35	103.94	1.82
2	90	1.82	163.99	3.32
3	107	2.46	263.18	6.05
4	131	3.32	434.94	11.02
5	165	4.48	739.48	20.09
	$\Sigma 570$	$\Sigma 13.43$	$\Sigma 1,705.53$	$\Sigma 42.30$

From Table 16–8, we substitute the necessary information into the formulas for m and b.

$$m = \frac{n \, \Sigma \, ([f(x)]y) - \Sigma \, f(x) \, \Sigma \, y}{n \, \Sigma \, [f(x)]^2 - [\Sigma \, f(x)]^2}$$

$$m = \frac{5(1,705.53) - 13.43(570)}{5(42.30) - (13.43)^2}$$

$$m \approx 28.0$$

$$b = \frac{\Sigma \, [f(x)]^2 \, \Sigma \, y - \Sigma \, f(x) \, \Sigma \, ([f(x)]y)}{n \, \Sigma \, [f(x)]^2 - [\Sigma \, f(x)]^2}$$

$$b = \frac{42.30(570) - 13.43(1,705.53)}{5(42.30) - (13.43)^2}$$

$$b \approx 38.7$$

The least squares equation that best fits the data is

$$y = 28.0e^{0.3x} + 38.7$$

When $x = 8$ min, the temperature is

$$y = 28.0e^{0.3(8)} + 38.7 = 347.3°$$

Figure 16–19 shows a plot of the least squares curve and the data points.

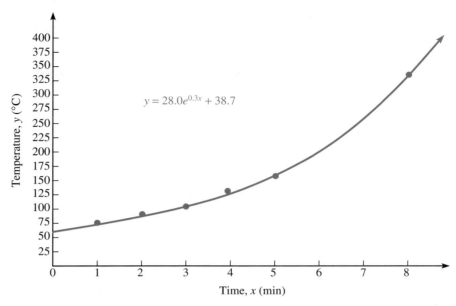

$$y = 28.0e^{0.3x} + 38.7$$

FIGURE 16–19

Graphing Calculator

The graphing calculator can also be used to analyze nonlinear curves, specifically cubic, exponential, natural logarithmic, power, quadratic, and quartic. The following example illustrates this process.

EXAMPLE 7 The unemployment rate of Florida (given in thousands) versus the United States for 1980–1999 is given on page 636.

Use the graphing calculator to model the unemployment as a quadratic and a regression curve, and determine which is the best fit for the data.

Solution We start by entering the data using the Data/Matrix Editor. Then, from the Calculate dialog box, choose QuadReg for a quadratic regression equation. The quadratic equation fitting this data is $y = 1.347471x^2 - 5.602328x + 106.504386$. Figure 16–20 displays the plot of the data points and the quadratic regression equation. We

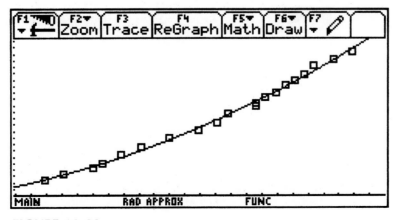

FIGURE 16–20

repeat this process by choosing LnReg as the regression line. The natural logarithm equation that best fits the data is $y = 18.120921 + 58.794528 \ln x$. Figure 16–21 displays the plot of the data points and the natural log regression equation. By inspection of the two graphs, you can determine the curve that best fits the data points: the quadratic curve. ■ ■ ■

Year	Florida	United States
1980	4.3	107.0
1981	4.5	108.7
1982	4.8	110.2
1983	4.9	111.5
1984	5.1	113.5
1985	5.3	115.5
1986	5.6	117.8
1987	5.9	119.9
1988	6.1	121.7
1989	6.2	123.9
1990	6.5	125.9
1991	6.5	126.4
1992	6.6	128.1
1993	6.7	129.1
1994	6.8	131.0
1995	6.9	132.3
1996	7.0	133.9
1997	7.1	136.2
1998	7.3	137.6
1999	7.5	139.3

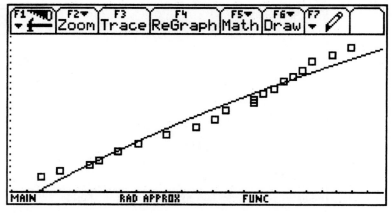

FIGURE 16–21

16–4 EXERCISES

Using the least squares method, find the equation of the straight line that best fits each data set. Plot the data points and the least squares equation on the same Cartesian plane.

1.

x	1	2	3	4	5	6
y	8	11	14	18	20	24

2.

x	1	2	4	6	9	11
y	4	10	22	35	53	66

3.

x	2	7	9	12	16	20
y	10	24	29	37	48	59

4.

x	1	3	5	6	9	11
y	9	12	16	18	24	27

Using the least squares method, find the best-fitting curve of the given form for each data set. Plot the data points and the curve on the same graph.

5.

x	1	2	3	4	5
y	2	13	32	57	91

$y = mx^2 + b$

6.

x	4	8	12	20	30
y	20	24	29	36	42

$y = m\sqrt{x} + b$

7.

x	0.1	0.2	0.3	0.4	0.5
y	10	11	12	13	15

$y = m(10^x) + b$

8.

x	1	3	5	8
y	16	85	599	11,930

$y = m(e^x) + b$

9. The displacement s of a pendulum bob from its equilibrium position as a function of time t was measured as follows:

t (s)	0	5.4	6.8	7.9	9.3
s (cm)	14.3	15.2	15.5	15.7	15.8

Find the least squares equation of the form $s = mt + b$, and determine the displacement when $t = 12$ s.

10. The growth n of bacteria in a petri dish as a function of time t is given in the following table:

t (h)	0.75	1.0	1.5	1.75	2.0	2.25
n (thousands)	18	21	27	33	38	46

Using the least squares method, fit the data to a curve of the form $n = m(e^t) + b$. How many bacteria will be in the dish at $t = 3.5$ h?

11. The displacement s of an object moving at a constant velocity as a function of time t was measured as follows:

t (s)	3	4	7	9	10	15	20
s (ft)	131	165	280	352	390	573	760

Find the least squares equation of the form $s = mt + b$, and find the displacement when $t = 12$ s.

12. The pressure P and volume V of a gas at a constant temperature were measured. Using the following data, find the least squares equation of the form

$$P = m\frac{1}{V} + b$$

V (cm^3)	20	30	35	40	50
P (kPa)	70	45	40	35	27

13. The resistance R of a wire at various temperatures T was as follows:

T (°F)	32	46	60	72	85	93
R (Ω)	620	867	1,120	1,333	1,570	1,710

Use the least squares method to find the line of the form $R = mT + b$. Use the equation to evaluate the resistance at 80°F.

14. The distance d an object rolls down an inclined plane as a function of time t was measured and is summarized in the table at the right:

Find the least squares equation of the form $d = mt^2 + b$, and find the distance when $t = 9$ s.

t (s)	1	2	3.4	5.6	7.2
d (cm)	13.9	23	48	110	170

CHAPTER SUMMARY

Summary of Terms

arithmetic mean (p. 615)

average (p. 615)

deviation (p. 627)

empirical curve fitting (p. 626)

empirical data (p. 608)

frequency distribution (p. 608)

frequency polygon (p. 610)

histogram (p. 609)

least squares method (p. 627)

measures of central tendency (p. 615)

measures of dispersion (p. 620)

median (p. 617)

mode (p. 619)

range (p. 620)

standard deviation (p. 621)

variation (p. 620)

Summary of Formulas

$$\bar{x} = \frac{x_1 + x_2 + x_3 + \cdots + x_n}{n} = \frac{\sum\limits_{i=1}^{n} x_i}{n} \qquad \text{arithmetic mean}$$

$$\bar{x} = \frac{\sum\limits_{i=1}^{n} f_i x_i}{\sum\limits_{i=1}^{n} f_i} \qquad \text{arithmetic mean from a frequency distribution}$$

$$s = \sqrt{\frac{\sum\limits_{i=1}^{n} (x_i - \bar{x})^2}{n}} \qquad \text{standard deviation}$$

$$s = \sqrt{\frac{\sum\limits_{i=1}^{n} f_i (x_i - \bar{x})^2}{\sum\limits_{i=1}^{n} f_i}} \qquad \text{standard deviation from a frequency distribution}$$

$$s = \sqrt{\left(\frac{\sum\limits_{i=1}^{n} (x_i)^2}{n}\right) - \left(\frac{\sum\limits_{i=1}^{n} x_i}{n}\right)^2} \qquad \text{standard deviation (alternate formula)}$$

$$y = mx + b \qquad \text{linear least squares}$$

where

$$m = \frac{n \sum xy - \sum x \sum y}{n \sum x^2 - (\sum x)^2}$$

$$b = \frac{\Sigma\, x^2 \,\Sigma\, y - \Sigma\, x \,\Sigma\, xy}{n\,\Sigma\, x^2 - (\Sigma\, x)^2}$$

$$y = m\,[f(x)] + b \qquad \text{nonlinear least squares}$$

where

$$m = \frac{n\,\Sigma\,([f(x)]y) - \Sigma\, f(x)\,\Sigma\, y}{n\,\Sigma\,[f(x)]^2 - [\Sigma\, f(x)]^2}$$

$$b = \frac{\Sigma\,[f(x)]^2\,\Sigma\, y - \Sigma\, f(x)\,\Sigma\,([f(x)]y)}{n\,\Sigma\,[f(x)]^2 - [\Sigma\, f(x)]^2}$$

CHAPTER REVIEW

Use data sets *A–D* as required.

A Age of employees at ABC Corporation: 27, 35, 18, 45, 27, 36, 18, 27, 57, 27, 18, 35, 63, 27

B Grades on algebra test: 68, 94, 83, 75, 71, 88, 62, 91, 83, 77, 64, 75, 94, 83, 81, 79, 86, 96

C Salary of civil engineers (in thousands): 19, 24, 30, 21, 24, 18, 30, 19, 27, 23, 21, 23, 24, 18, 32

D Number of defective diodes per package of 200: 6, 20, 14, 18, 7, 10, 11, 18, 6, 12, 24, 14, 18, 10, 13, 14, 18

Section 16–1

Construct a frequency distribution, histogram, and frequency polygon for the indicated data set.

1. Set *A*
2. Set *B*
3. Set *C*
4. Set *D*

Section 16–2

Determine the mean, median, and mode for the indicated data set.

5. Set *A*
6. Set *B*
7. Set *C*
8. Set *D*

9. Fred must estimate the cost of repairing a television. He knows that the parts will cost $36.50, but he must calculate the cost of labor. He knows that four similar jobs required 3.1 h, 4.7 h, 2.9 h, and 4.3 h. Using the arithmetic mean of the previous jobs and an hourly rate of $26 per hour, determine how much Fred should estimate for the cost of repairing the television.

10. Carlos scored 86, 94, 90, 81, and 85 on his 5 algebra tests. What must he score on his last test (the sixth) to earn an average of 90?

11. The monthly electric bills during the last year for the Smith family are as follows:

$50.27	$42.78	$67.34	$86.73	$94.86
$150.78	$126.81	$105.73	$93.70	$73.85
$62.38	$56.71			

If Mrs. Smith is attempting to make a budget for the coming year, based on the monthly average, what should she budget for the coming year?

12. A company has 35 employees: 5 are 26 years old, 3 are 30 years old, 7 are 33 years old, 7 are 46 years old, 5 are 50 years old, and 8 are 56 years old. Determine the mean age of the employees.

Section 16–3

Determine the range and standard deviation for the indicated data set.

13. Set *A*
14. Set *B*
15. Set *C*
16. Set *D*

Using the indicated data set, determine the range of values one standard deviation from the mean. Compare this value to the actual percentage of the data values within this range.

17. Set *A*
18. Set *B*
19. Set *C*
20. Set *D*

21. Using the frequency distribution in Table 16–9, calculate the standard deviation.

TABLE 16–9

Employee's Salary (thousands)	Frequency
18	2
23	8
27	10
29	7
30	4
32	1

22. Using the frequency distribution in Table 16–10, calculate the standard deviation.

TABLE 16–10

Grade	Frequency
61	1
64	1
68	2
74	3
77	10
81	6
83	8
88	5
90	3
94	2
96	2
98	1

Section 16–4

Using the least squares method, find the equation of the straight line that best fits each data set. Plot the data points and the least squares equation on the same Cartesian plane.

23.

x	2	4	6	9
y	30	45	62	85

24.

x	1	6	7	13
y	18	36	38	60

25.

x	2	5	9	12	15
y	22	36	50	60	74

26.

x	1	3	6	8
y	22	30	50	55

Using the least squares method, find the best-fitting curve of the given form for each data set. Plot the data points and the curve on the same graph.

27.

x	1	4	6	9
y	11	45	85	179

$y = mx^2 + b$

28.

x	1	2	3	4	5	6
y	18	40	77	130	190	274

$y = mx^2 + b$

29.

x	0.1	0.4	0.8	1.0	1.3	1.8	2.0
y	14	18	29	43	70	205	320

$y = m10^x + b$

30.

x	0.3	0.7	0.9	1.6	2.1
y	15	18	20	30	45

$y = me^x + b$

CHAPTER TEST

The number(s) in parentheses refers to the appropriate learning objective(s) given at the beginning of the chapter.

1. Calculate the mean, median, and mode for the following data set: (2)

18　23　16　27　18　36　21　23　18　21　25　16　27

2. Draw a frequency polygon for the following data set: (1)

250　220　286　200　250　197　200
190　250　200　186　186　250　197

3. Calculate the standard deviation for the following data set: (3)

25　16　29　16　34　10　15　19　24　21　28　32　12　16　24

4. Using the least squares method, determine the straight line that best fits the following data set: (5)

x	1	2	3	4	5
y	10	16	23	30	36

   Using the equation, calculate y when $x = 8$.

5. Organize the following data set into a frequency distribution, and draw a histogram. (1)

   2   6   3   8   11   3   14   7   8   2   3   14   8   2   9   18   7

6. Using the mean and standard deviation for the following data set, determine the range of values within one standard deviation of the mean: (4)

   16   28   40   10   18   25   36   14   28   41   26   27   19   25   38

7. KLZ Corporation employs 47 people. Ten earn $24,000 annually, 7 earn $28,500 annually, 9 earn $32,000 annually, 11 earn $36,000 annually, 6 earn $42,000 annually, and 4 earn $50,000 annually. Calculate the mean salary and the standard deviation. (2, 3)

8. Using the least squares method, determine the equation of the curve of the form $y = mx^2 + b$ that best fits the following data set: (5)

x	1	3	4	6	9
y	15	35	55	100	215

GROUP ACTIVITY

Dolphin Dan has hired you to perform an unusual task. It seems that Dan has a treasure map that was drawn by a rather mathematical pirate, Dusty Pete (name received from a freak flour-bin accident). It seems that the pirate did not want any ordinary person to find his gold. He did this by making reference to a particular geometric curve in his writings. You have to be able to use this curve to locate his gold successfully. However, the map was nearly destroyed during a freak galley accident involving some bootlegged tuna. In addition, the map has been passed around among many less intelligent bandits of the sea. Dan happened to find the map at a garage sale held by one of Dusty's great, great grandchildren. Originally the map had several numbers written along the edge. Most of the numbers were lost during the tuna incident. However, one of the pirates had memorized them and wrote what he remembered on the back of the map. Other clues were to be found on the front of the map. However, most of the data is unintelligible.

Dan tells you that Dusty was in love with mathematics. He was known to have affection for four curves in particular. The first is the "witch of Agnesi," which is named after Maria Gaetana Agnesi (1718–1799), a mathematician known for her work in differential calculus. She was a prodigy, having mastered several languages at an early age. At age 9 she published a Latin discourse in defense of higher education for women. The curve is called the witch of Agnesi not because she was thought to be a witch, but because the shape of the curve was called *aversiera*, which in Italian means "to turn." This word is also a slang short form for *avversiere,* which means "wife of the devil." A series of mistranslations over time finally set the name of the curve to the "witch of Agnesi." The second curve, the cissoid of Diocles (meaning "ivy-shaped"), was invented by Diocles in about 180 B.C. in connection with his attempt to duplicate the cube by geometric methods. The name first appears in the work of Geminus about 100 years later. Diocles was a contemporary of Nicomedes. He also wrote about burning mirrors, which proves the focal property of a parabolic mirror for the first time.

\multicolumn{4}{c}{Map Data}			
x	y	x	y
0.0	5.958	5.5	3.298
0.5	5.971	6.0	3.062
1.0	5.797	6.5	2.823
1.5	5.646	7.0	2.552
2.0	5.401	7.5	2.289
2.5	5.068	8.0	2.219
3.0	4.740	8.5	1.963
3.5	4.556	9.0	1.884
4.0	4.201	9.5	1.649
4.5	3.789	10.0	1.581
5.0	3.498		

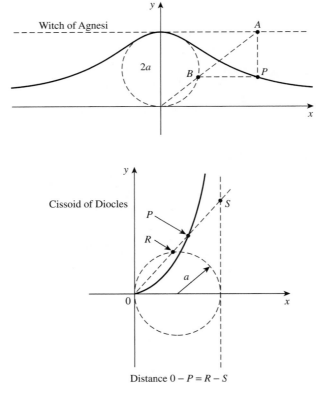

Distance $0 - P = R - S$

Largely ignored by later Greeks, the work had a large influence on the Arab mathematicians, in particular al'Haitam. The properties of parabolic mirrors were brought to the West around 1200. The equations and data, as found on the map, are as follows:

$$\textbf{Witch of Agnesi: } y = \frac{8a^3}{x^2 + 4a^2} \qquad a = 3$$

$$\textbf{Cissoid of Diocles: } y^2 = \frac{1}{16}\left(\frac{x^3}{2a - x}\right) \qquad a = 6$$

$$\textbf{Exponentials: } y = 6e^{-x/8} \qquad \text{and} \qquad y = e^{x/7} - 1$$

1. Plot the map data by any means that you want to use.

2. Plot each of the pirate's equations and determine which one is the closest fit. Three of the curves will come close to meeting at one point. What is the value of x and y at that point? What are the errors (deviation) between the three curves?

3. In previous sections, the curve-fitting process minimized the deviation between the raw data and the computed curve. This was called the least squares method. What other error measurements can you find? Discuss each in terms of the sensitivity to errors and the ease of implementation. What are the resulting error measurements between the map data and the curves?

4. What methods does your programmable calculator use for curve fitting and error measurements?

5. Dan wants to use your results to find the pirate's gold. Dusty's instructions read: "Move due east from the brimstone a distance, in meters, as determined by the curve of ancestral marvel where the vertical value is equal to a lonely number." You rationalize that the lonely number is equal to 1 and the vertical value is equal to the y axis of the graph. Find the x value of your chosen graph when $y = 1$. Since Dan knows where the brimstone is, your answer should lead him to the treasure. (Therefore, you should receive a very good tip.)

SOLUTION TO CHAPTER INTRODUCTION

To determine the equation for the curve, we must calculate m and b. First, we fill in Table 16–11.

TABLE 16–11

t	n	$f(t) = e^{0.2t}$	$[f(t)n]$	$[f(t)]^2$
0	19.2	1.00	19.20	1.00
1	21	1.22	25.65	1.49
2	23	1.49	34.31	2.23
3	25	1.82	45.55	3.32
4	28	2.23	62.32	4.95
5	32	2.72	86.99	7.39
	$\Sigma\ 148.2$	$\Sigma\ 10.48$	$\Sigma\ 274.02$	$\Sigma\ 20.38$

Next, we substitute into the formulas to calculate m and b.

$$m = \frac{N\,\Sigma\,([f(t)]n) - \Sigma\,f(t)\,\Sigma\,n}{N\,\Sigma\,[f(t)]^2 - [\Sigma\,f(t)]^2}$$

$$m = \frac{6(274.02) - 10.48(148.2)}{6(20.38) - (10.48)^2}$$

$$m \approx 7.31$$

$$b = \frac{\Sigma\,[f(t)]^2\,\Sigma\,n - \Sigma\,f(t)\,\Sigma\,([f(t)]n)}{N\,\Sigma\,[f(t)]^2 - [\Sigma\,f(t)]^2}$$

$$b = \frac{20.38(148.2) - 10.48(274.02)}{6(20.38) - (10.48)^2}$$

$$b \approx 11.9$$

The least squares equation for the growth of bacteria is

$$n = 7.31e^{0.2t} + 11.9$$

To predict the number of bacteria at 18 hours, we substitute 18 into the equation.

$$n = 7.31e^{0.2(18)} + 11.9$$

$$n = 279$$

At the end of 18 hours, there are 279,000 bacteria in the petri dish.

Y ou are offered a job as a technician at XYZ Corporation for a yearly salary of $23,480 with a promise of yearly raises of 8% for eight years. You are also offered a job with ABC Company for $26,700 with a promise of yearly raises of $1,500 for 8 years. Which job should you accept if your decision is based solely on your salary at the end of 8 years? (The answer to this problem is given at the end of the chapter.)

To determine your salaries at the end of 8 years, you must calculate the nth term of a geometric progression and of an arithmetic progression, both topics discussed in this chapter. We will also discuss finding the sum of an infinite geometric series and the binomial formula.

Learning Objectives

After you complete this chapter, you should be able to

1. Find the nth term of an arithmetic progression from given information (Section 17–1).

2. Determine the sum of the first n terms of an arithmetic progression (Section 17–1).

3. Find the nth term of a geometric progression from given information (Section 17–2).

4. Determine the sum of the first n terms of a geometric progression (Section 17–2).

5. Find the sum of an infinite geometric progression (Section 17–3).

6. Represent a repeating decimal in fractional form (Section 17–3).

7. Apply the binomial formula to raise a binomial to a given power (Section 17–4).

8. Apply Pascal's triangle to expand a binomial to a given power (Section 17–4).

9. Find a specified term in a binomial expansion (Section 17–4).

Chapter 17

Sequences, Series, and the Binomial Theorem

17–1

ARITHMETIC PROGRESSIONS

Definitions

A **sequence** is a collection of numbers in a certain order. For example, both of the following represent a sequence:

$$7, 8, 9 \quad \text{and} \quad 1, 7, 13, 19, \ldots$$

Each member of the sequence is called a **term.** *If the sequence has a last term,* it is called a **finite sequence,** as exemplified by the first sequence above. *Any sequence that is not finite* is called an **infinite sequence,** as exemplified by the second sequence above. The notation a_n denotes the nth term of the sequence; thus, a_3 in the infinite sequence above is 13, or $a_3 = 13$.

Arithmetic Progression

An **arithmetic progression** is a sequence of numbers in which the difference between successive terms is a constant, called the **common difference.** For example, the sequence of numbers 4, 11, 18, 25, . . . represents an arithmetic progression with a common difference of 7.

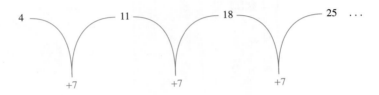

In general, if a_1 is the first term and d is the common difference of the sequence, you can generate an arithmetic progression as follows:

$$a_1 \quad a_1 + d \quad a_1 + 2d \quad a_1 + 3d \quad a_1 + 4d$$
$$a_1 \quad\quad a_2 \quad\quad a_3 \quad\quad a_4 \quad\quad a_5$$

Using this information, we can generate the following formula to find the nth term of an arithmetic progression.

---

nth Term of an Arithmetic Progression

The nth term of an arithmetic progression is given by

$$a_n = a_1 + (n - 1)d$$

where a_1 = the first term of the progression and d = the common difference.

EXAMPLE 1 Find the eighth term of the following arithmetic progression:

$$6, 10, 14, \ldots$$

Solution From the list of terms in the progression, we know that the first term $a_1 = 6$. We can determine the common difference d by subtracting any two consecutive terms. The common difference d is

$$d = 10 - 6 = 4$$

Furthermore, since we are asked to find the eighth term of the progression, $n = 8$. Substituting $a_1 = 6$, $d = 4$, and $n = 8$ into the formula gives

$$a_n = a_1 + (n - 1)d$$
$$a_8 = 6 + (8 - 1)4$$
$$a_8 = 6 + 7(4)$$
$$a_8 = 34$$

The eighth term of this arithmetic progression is 34.

∎∎∎

EXAMPLE 2 Determine how many numbers between 7 and 800 are divisible by 3.

Solution First, we must find the smallest number, a_1, and the largest number, a_n, that are in the range from 7 to 800 and that are divisible by 3. A number is divisible by 3 if the sum of its digits is divisible by 3. Therefore, we start with 7 and count upward, and then we start with 800 and count downward to find the smallest and largest numbers in this range divisible by 3. The smallest a_1 is 9, and the largest a_n is 798. Next, we substitute $a_n = 798$, $a_1 = 9$, and $d = 3$ into the formula and solve for n.

$$a_n = a_1 + (n - 1)d$$
$$798 = 9 + (n - 1)3$$
$$798 = 9 + 3n - 3$$
$$792 = 3n$$
$$264 = n$$

There are 264 numbers between 7 and 800 that are divisible by 3.

∎∎∎

EXAMPLE 3 An object moving in a line is given an initial velocity of 4.3 m/s and a constant acceleration of 1.2 m/s^2. How long will it take the object to reach a velocity of 14.8 m/s?

Solution We are told that $a_1 = 4.3$, $a_n = 14.8$, and $d = 1.2$, and we are asked to determine n. Substituting these values into the formula gives

$$a_n = a_1 + (n - 1)d$$
$$14.8 = 4.3 + (n - 1)1.2$$
$$14.8 = 4.3 - 1.2 + 1.2n$$
$$n \approx 9.8$$

After 9.8 s, the object is traveling 14.8 m/s.

14.8 $\boxed{-}$ 4.3 $\boxed{+}$ 1.2 $\boxed{=}$ $\boxed{\div}$ 1.2 $\boxed{=}$ ⟶ 9.75

∎∎∎

Sum of an Arithmetic Progression

You can calculate the sum S_n of the first n terms of an arithmetic progression manually by adding the terms. However, to develop a formula, write an expression for the sum of the first n terms forward and backward, and add the resulting two equations.

$$S_n = a_1 + (a_1 + d) + \cdots + (a_n - d) + a_n$$
$$\underline{S_n = a_n + (a_n - d) + \cdots + (a_1 + d) + a_1}$$
$$2S_n = (a_1 + a_n) + (a_1 + a_n) + \cdots + (a_1 + a_n) + (a_1 + a_n)$$

Since there are n terms of $(a_1 + a_n)$,

$$2S_n = n(a_1 + a_n)$$

Solving the equation for S_n gives the following formula.

Sum of First n Terms of an Arithmetic Progression

The sum of the first n terms of an arithmetic progression is given by

$$S_n = \frac{n(a_1 + a_n)}{2}$$

EXAMPLE 4 Find the sum of the positive even integers up to and including 250.

Solution From the information given, we know that $a_1 = 2$, $d = 2$, and $a_n = 250$. First, we must use the previous formula to determine the number n of even integers in this range.

$$a_n = a_1 + (n - 1)d$$
$$250 = 2 + (n - 1)2$$
$$250 = 2 + 2n - 2$$
$$n = 125$$

Substituting $a_1 = 2$, $a_n = 250$, and $n = 125$ into the sum formula gives

$$S_n = \frac{n(a_1 + a_n)}{2}$$

$$S_{125} = \frac{125(2 + 250)}{2}$$

$$S_{125} = 15{,}750$$

The sum of the positive, even integers through 250 is 15,750. ■■■

Application

EXAMPLE 5 In an integrated circuit with an initial current of 960 mA, the temperature in the components decreased from 25% to 22% to 19%. Assuming that each temperature decrease is caused by a decrease in the initial current, what is the value of the current at the fifth measurement?

Solution We know that $n = 5$, $a_1 = 25\%$, and $d = -3\%$. First, we must calculate a_5 before we can determine S_5.

$$a_n = a_1 + (n - 1)d$$
$$a_5 = 25\% + 4(-3\%)$$
$$a_5 = 13\%$$

Then we can calculate S_5.

$$S_n = \frac{n(a_1 + a_n)}{2}$$

$$S_5 = \frac{5(25 + 13)}{2}$$

$$S_5 = 95\%$$

Last, we calculate the value of the current at this point. Since 95% of the current is lost in heat, 5% of the 960-mA current remains.

$$\text{current} = (5\%)(960 \text{ mA}) = 48 \text{ mA} \qquad ■■■$$

Under certain circumstances we may need to solve for two unknowns in an arithmetic progression. Such a problem requires a system of two equations: one resulting from the a_n formula and the second from the S_n formula. The next example illustrates this technique.

EXAMPLE 6 A technician accepts a job to pay off his $12,000 college loan. If he pays $457.50 toward the loan the first month and increases his payment by $15 each month, how long will it take him to pay off the loan?

Solution We are told that $a_1 = \$457.5$, $S_n = \$12,000$, and $d = \$15$, and we must calculate n, which represents months. Therefore,

$$a_n = a_1 + (n - 1)d$$
$$a_n = 457.5 + (n - 1)15$$
$$a_n = 442.5 + 15n$$

Also,

$$S_n = \frac{n(a_1 + a_n)}{2}$$

$$12,000 = \frac{n(457.5 + a_n)}{2}$$

Then we solve the system of two equations by substitution.

$$12,000 = \frac{n[457.5 + (442.5 + 15n)]}{2}$$

$$24,000 = n(900 + 15n)$$
$$24,000 = 15n^2 + 900n$$
$$15n^2 + 900n - 24,000 = 0$$
$$n^2 + 60n - 1,600 = 0$$
$$(n + 80)(n - 20) = 0$$
$$n = -80 \qquad \text{and} \qquad n = 20$$

Since n must be positive, the technician will repay his loan in 20 months. ■■■

NOTE ✦ If you had difficulty with Example 6, review Chapter 5 on solving quadratic equations and Chapter 6 on solving systems of equations.

EXAMPLE 7   An auditorium has 45 rows of seats. If the first row contains 31 seats and each successive row contains two additional seats, what is the seating capacity of the auditorium?

Solution   The number of seats in the rows of the auditorium forms an arithmetic progression where the common difference, d, is 2. To find the total number of seats, we use the formula

$$S_n = \frac{n(a_1 + a_n)}{2}$$

where $n = 45$. Since the first row has 31 seats, $a_1 = 31$. To determine a_n, we solve

$$a_n = a_1 + (n - 1)d$$
$$a_{45} = 31 + (45 - 1)2$$
$$a_{45} = 119$$

Substituting into the sum formula above, we have

$$S_{45} = \frac{45(31 + 119)}{2}$$

$$S_{45} = 3{,}375$$

The auditorium contains 3,375 seats.   ■ ■ ■

17–1 EXERCISES

Find the indicated term of the following arithmetic progressions.

1. The fifth term of 3, 9, 15, . . .

2. The eighth term of $-5, -2, 1, . . .$

3. The tenth term of $8, \dfrac{15}{2}, 7, . . .$

4. The thirteenth term of 13, 7, 1, . . .

5. The twenty-fifth term of $-3, 4, 11, . . .$

6. The fifteenth term of $-2, -4, -6, . . .$

7. The ninth term where $a_1 = 7$ and $d = -3$

8. The eighteenth term where $a_1 = -4$ and $d = 6$

9. The twenty-third term where $a_1 = -15$ and $d = 4$

10. The thirtieth term where $a_1 = 10$ and $d = 0.5$

Find the sum of the first n terms of each arithmetic progression.

11. $n = 18$, $a_1 = -3$, and $a_{18} = 31$

12. $n = 9$, $a_1 = 6$, and $a_9 = -26$

13. $n = 20$, $a_1 = 4$, and $a_{20} = 14$

14. $n = 33$, $a_1 = -18$, and $a_{33} = -10$

15. $n = 15$, $a_1 = -6$, and $d = 3$

16. $n = 27$, $a_1 = 38$, and $d = -2$

Use the information given in the following problems to find the requested quantities.

17. If $a_1 = 3$, $d = -2$, and $a_n = -11$, find n and S_n.

18. If $a_1 = -18$, $d = 1/2$, and $a_n = -13$, find n and S_n.

19. If $a_1 = 5$ and $a_{18} = 56$, find d and S_{18}.

20. If $a_1 = -0.5$ and $a_{12} = -33.5$, find d and S_{12}.

21. If $a_1 = -5$, $n = 10$, and $d = 4$, find a_{10} and S_{10}.

22. If $a_1 = 16$, $n = 30$, and $d = -2$, find a_{30} and S_{30}.

23. If $a_1 = 10$, $a_n = 31$, and $S_n = 164$, find d and n.

24. If $a_1 = -8$, $a_n = 1$, and $S_n = -45.5$, find d and n.

Solve the following problems.

B 25. A technician accepts a job at a starting salary of $23,186 and receives a $1,500 increase each year for six years. Find the technician's salary during the sixth year and the total earnings for the 6-year period.

B 26. A computer depreciates $350 per year. If the computer is currently worth $7,860, what will its value be after 7 years?

B 27. A taxi driver charges $1.50 for the first mile and $0.60 for each additional mile. How much does the driver charge for a 15-mile trip?

P 28. An object falls 16 ft during the first second, 48 ft during the second second, and 80 ft during the third second. Find the distance the object falls in the sixth second and the total distance fallen in the first six seconds.

C 29. The population of Glenwood increases arithmetically from 75,230 to 125,280 in 8 years. Assuming that the growth rate is constant, find the yearly increase in population.

CS 30. A computer manufacturer reported sales of 1,860 units at the end of January. If sales are to increase 46 units per month, in what month will sales reach 2,320 units?

B 31. Fred's Sneaks sells $45,000 worth of sneakers in its first year of operation. If the goal is to increase sales by $12,000 each year for 13 years, what would be the total sales during the first 14 years of operation?

B 32. A computer costing $10,000 depreciates 10% the first year, 9.5% the second year, and 9.0% the third year. Determine the value of the computer after 8 years, assuming the percentages apply to the original cost.

B 33. Linda's Custom Frame Shop has an annual income of $12,000 during its first year. If the shop increases income by $8,000 each year for 5 years, what is the total income during the first 5 years?

A 34. A ceramic tile medallion is designed in the shape of a trapezoid 15 ft wide at the bottom and 8 ft wide at the top. Refer to Figure 17–1. If the tiles are 12 in. by 12 in. and each successive row will have one less tile than the previous row, how many tiles should the architect order?

B 35. Robert agrees to repay a loan of $15,000 at the rate of $1,000 per year plus 8% of the unpaid balance. Find the total interest paid.

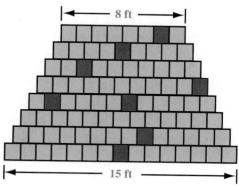

FIGURE 17–1

17–2

GEOMETRIC PROGRESSIONS

A **geometric progression** is a sequence in which each successive term is a *constant multiple*, called the **common ratio**, r, of the previous term. For example, the sequence 1, 3, 9, 27, . . . is a geometric progression with a common ratio of 3.

Definitions

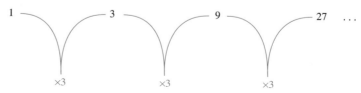

In general, if a_1 is the first term of a geometric progression with a common ratio r, the terms are given by

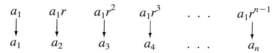

The nth term of a geometric progression is given by the following formula.

---

nth Term of a Geometric Progression

The nth term of a geometric progression is given by

$$a_n = a_1 r^{n-1}$$

where a_1 = the first term of the progression and r = the common ratio.

EXAMPLE 1   Find the sixth term of the following geometric progression:

$$4, 20, 100, \ldots$$

Solution   To find the sixth term a_6, we determine a_1, r, and n.

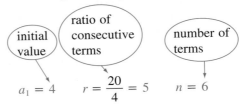

Substituting $a_1 = 4$, $r = 5$, and $n = 6$ into the formula gives

$$a_n = a_1 r^{n-1}$$
$$a_6 = 4(5)^{6-1}$$
$$a_6 = 12{,}500$$

$$4 \;\boxed{\times}\; 5 \;\boxed{y^x}\; \boxed{(}\; 6 \;\boxed{-}\; 1 \;\boxed{)}\; \boxed{=} \longrightarrow 12500 \qquad \blacksquare\blacksquare\blacksquare$$

Applications

EXAMPLE 2   Each layer of 3-in. insulation reduces fuel consumption (in kilowatts) by 8%. How many layers would reduce consumption from 948 kW to 894 kW?

Solution   From the information given, we know that

$$a_1 = 948 \qquad a_n = 894 \qquad r = 92\%$$

and we must solve for n. The common ratio between successive values is 92% since each layer reduces consumption by 8%. Substituting these values into the formula gives

$$a_n = a_1 r^{n-1}$$
$$894 = 948(0.92)^{n-1}$$

To solve for the exponent n, we use logarithms.

$$\log 894 = \log [948(0.92)^{n-1}]$$
$$\log 894 = \log 948 + (n - 1)\log 0.92$$
$$\log 894 = \log 948 + n \log 0.92 - \log 0.92$$

$$\frac{\log 894 - \log 948 + \log 0.92}{\log 0.92} = n$$

$$n \approx 1.7$$

Approximately two layers, 6 in. (1.7×3 in.), of insulation are needed.

$$894 \;\boxed{\log}\; \boxed{-}\; 948 \;\boxed{\log}\; \boxed{+}\; 0.92 \;\boxed{\log}\; \boxed{=}$$
$$\boxed{\div}\; 0.92 \;\boxed{\log}\; \boxed{=} \longrightarrow 1.7034 \qquad \blacksquare\blacksquare\blacksquare$$

NOTE ✦ If you have difficulty solving the logarithmic equation in Example 2, review Chapter 8 on exponential and logarithmic functions.

EXAMPLE 3   A radioactive product has a half-life of 5 years. If the radioactivity level is 56.0 microcuries after 23 years, determine the original level of radioactivity.

Solution We are told that $a_n = 56$ and $r = 0.5$. To determine a_1, we must calculate n, the number of half-lives elapsed.

$$n = \frac{23 \text{ years}}{5 \text{ years}} = 4.6$$

Therefore, including a_1, the number of terms n is 5.6. Substituting into the formula gives

$$a_n = a_1 r^{n-1}$$
$$56.0 = a_1(0.5)^{5.6-1}$$
$$a_1 \approx 1,360$$

$$56 \boxed{\div} \boxed{(} \boxed{0.5} \boxed{y^x} \boxed{4.6} \boxed{)} \boxed{=} \longrightarrow 1358.1 \qquad \blacksquare\blacksquare\blacksquare$$

Sum of a Geometric Progression

To derive the formula for the sum of the first n terms of a geometric progression, multiply the equation representing the sum by r, subtract this result from the original sum, and solve for S_n.

$$S_n = a_1 + a_1 r + a_1 r^2 + \cdots + a_1 r^{n-2} + a_1 r^{n-1} \qquad \text{sum of } n \text{ terms}$$
$$-rS_n = -(a_1 r + a_1 r^2 + \cdots + a_1 r^{n-2} + a_1 r^{n-1} + a_1 r^n) \quad \text{multiply by } r$$
$$S_n - rS_n = a_1 + 0 + 0 + 0 + \cdots + 0 + 0 - a_1 r^n \qquad \text{and subtract}$$
$$S_n(1 - r) = a_1 - a_1 r^n$$
$$S_n = \frac{a_1(1 - r^n)}{1 - r} \qquad r \neq 1$$

Sum of the First n Terms of a Geometric Progression

The sum of the first n terms of a geometric progression is given by

$$S_n = \frac{a_1(1 - r^n)}{1 - r} \qquad r \neq 1$$

where a_1 = the first term, r = the common ratio, and n = the number of terms in the progression.

 EXAMPLE 4 Find the sum of the first seven terms of the geometric progression

$$9, 6, 4, \ldots$$

Solution From the given information, we know that $a_1 = 9$, $r = \dfrac{6}{9} = \dfrac{2}{3}$, and $n = 7$. Substituting these values into the formula gives

$$S_n = \frac{a_1(1 - r^n)}{1 - r}$$

$$S_7 = \frac{9\left(1 - \left(\dfrac{2}{3}\right)^7\right)}{1 - \dfrac{2}{3}}$$

$$S_7 \approx 25.4$$

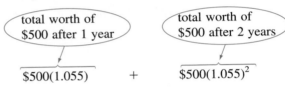

CAUTION ✦ In calculating an expression such as $1 - 0.5^6$, raise 0.5 to the sixth power before subtracting the result from 1.

Applications

EXAMPLE 5 Determine the total worth of a yearly $500 investment after 15 years if the interest rate is 5.5% compounded annually.

Solution At the end of the first year, the total worth is

principal   interest

$500   + $500(0.055)

500(1 + 0.055)

500(1.055)

At the end of the second year, the total worth is

total worth of   total worth of
$500 after 1 year $500 after 2 years

$500(1.055)   + $500(1.055)^2$

Continuing this pattern gives the following geometric progression:

$$500(1.055) + 500(1.055)^2 + 500(1.055)^3 + \cdots + 500(1.055)^{15}$$

or

$$500[(1.055) + (1.055)^2 + (1.055)^3 + \cdots + (1.055)^{15}]$$

Since the expression in brackets represents the sum of a geometric progression, we will substitute the following values into the formula:

$$a_1 = 1.055 \qquad r = \frac{(1.055)^2}{1.055} = 1.055 \qquad n = 15$$

$$S_n = \frac{a_1(1 - r^n)}{1 - r}$$

Substituting gives

$$S_{15} = \frac{1.055\,[1 - (1.055)^{15}]}{1 - 1.055}$$

$$S_{15} \approx 23.64$$

The total value of the investment is $\approx$$11,821 or

$$\$500(23.64) = \$11,820.57$$

EXAMPLE 6 At the beginning of each year, Rosa puts $2,000 into an IRA paying 9% interest compounded annually. How much money will be in the account in 12 years?

Solution   At the end of the first year, Rosa will have the interest added to her initial investments. Therefore, for the last (twelfth) $2,000 invested, the value will be

$$\$2,000(1 + 0.09) = \$2,000(1.09)$$

The next to the last $2,000 will have interest added twice or $2,000(1.09)(1.09) = $2,000(1.09)^2$. Similarly, the value of the first $2,000 investment will be $2,000(1.09)^{12}$ because the interest has been added twelve times. The sum of the sequence becomes

$$\$2,000(1.09) + \$2,000(1.09)^2 + \$2,000(1.09)^3 + \cdots \$2,000(1.09)^{12}$$

or

$$\$2,000[1.09 + (1.09)^2 + (1.09)^3 + \cdots (1.09)^{12}]$$

For the sequence in brackets $a_1 = 1.09$, $r = 1.09$, and $n = 12$.
The sum of the geometric series becomes

$$S_{12} = \frac{1.09[1 - (1.09)^{12}]}{1 - 1.09}$$

$$S_{12} = 21.95$$

The total value of the investment is $2,000(21.95) \approx \$43,910.00$.   ■ ■ ■

🖩 17–2 EXERCISES

Find the indicated term of each geometric progression.

1. The tenth term of 3, 6, 12, . . .

2. The fifth term of 24, 6, 1.5, 0.375, . . .

3. The ninth term of 8, −4, 2, . . .

4. The tenth term of −9, 7.5, −6.25, . . .

5. The eighth term of −2, −6, −18, . . .

6. The fourth term where $a_1 = 8$ and $r = \dfrac{1}{2}$

7. The sixth term where $a_1 = -10$ and $r = -4.1$

8. The eighth term where $a_1 = 72$ and $r = -3$

9. The eleventh term where $a_1 = 86$ and $r = 2.5$

10. The seventh term where $a_1 = 108$ and $r = -\dfrac{1}{3}$

Find the sum of the first n terms of each geometric progression.

11. −6, −3, −1.5, . . .; $n = 8$

12. 4, 12, 36, . . .; $n = 7$

13. 6, 9, 13.5, . . .; $n = 5$

14. $n = 5$, $a_1 = 7$, and $r = 3$

15. $n = 8$, $a_1 = -4$, and $r = -0.5$

16. $n = 6$, $a_1 = 60$, and $a_6 = 14,580$

17. $n = 7$, $a_1 = -0.5$, and $a_7 = -2,048$

Use the information given in the following problems to find the requested quantities for each geometric progression.

18. If $a_1 = 6$, $r = 3$, and $a_n = 486$, find n and S_n.

19. If $a_1 = \dfrac{1}{8}$ and $r = \dfrac{1}{2}$, find a_{10} and S_{10}.

20. If $r = 2$ and $S_5 = 558$, find a_1 and a_5.

21. If $a_1 = 9$, $r = \dfrac{1}{3}$, and $S_n = 13\dfrac{1}{3}$, find n and a_n.

22. If the seventh term is 8,748 and the common ratio is 3, find the first term and the sum of the first seven terms.

23. If the first term is 8, the common ratio is 3, and the nth term is 5,832, find n and the sum of the first n terms.

24. If the sum of the first five terms is −93 and the common ratio is 2, find the first term and the third term.

25. If the second term is $\dfrac{1}{2}$ and the seventh term is $\dfrac{1}{64}$, find the tenth term.

26. If the fourth term is 1,296 and the seventh term is 279,936, find the second term.

Solve the following problems.

B 27. An engineer accepts a job with a starting salary of $20,130 with an 8% raise each year for seven years. What is the technician's salary during the seventh year?

Bi 28. The number of bacteria in a culture increases by 30% each hour. If the culture contains 25,000 bacteria at 1:00 P.M., how many bacteria are present at 7:00 P.M.?

B 29. Joe invests $1,500 per year in a savings account for 9 years. Find the total value of the account if the interest rate of 6% is compounded annually.

B 30. A technician receives a job offer for $23,500 per year. If he receives a 5% raise each year, how many years will pass before his annual salary is $34,720.20?

B 31. Suppose you deposited $150 into an account at the beginning of every month for 20 years. If the account pays 9% interest compounded monthly what is your balance at the end of 20 years?

B 32. The annual sales for the Easy Chair Company from 1990 to 1998 can be modeled by $a_n = 145e^{0.143n}$, where $n = 0$, $1, 2 . . . 8$; n represents the year counted from 1990 (in 1990, $n = 0$); and a_n = annual sales (in millions of dollars). What are the total sales in this 8-year period?

B 33. A new piece of office equipment was purchased for $20,000. If the company depreciates the equipment at 15% per year, how much is the equipment worth after 5 years?

B 34. A deposit of $2,000 is made for an account that earns 7% interest compounded monthly. The balance in the account is given by

$$A_n = 2,000\left(1 + \frac{0.07}{12}\right)^n, \qquad n = 1, 2, 3 . . .$$

Find the balance in the account after 5 years by calculating the sixtieth term of the sequence.

M 35. A tracer dye is placed in a system to detect leaks. After an hour 7/8 of the dye is left. At the end of the second hour 7/8 of the remaining dye is left. If 200 ml of dye is injected, how much is left after 6 hours?

B 36. The enrollment at YSU doubled every 5 years from 1980 to 2000. If the enrollment was 6,000 in 1980, find the enrollment in 2000.

C 37. Historically the consumption of electricity in Greenwood has increased 4% per year. If Greenwood uses 1.5 billion kilowatt-hours now, what will Greenwood use 8 years from now?

P 38. A tennis ball is dropped from a height of 25 ft. If the ball rebounds 3/5 of its original height, how high will the ball rebound after the fourth bounce?

17–3

■■■■■■■■■■

INFINITE GEOMETRIC SERIES

In the previous section we found the sum of a finite number of terms in a geometric progression. In this section we will develop a formula for the sum of all the terms of an infinite geometric series.

Consider the following geometric progressions:

$$3, 6, 12, 24, . . . \qquad \text{where } r = 2$$

and

$$36, 18, 9, 4.5, . . . \qquad \text{where } r = \frac{1}{2}$$

The sum of the terms in the first geometric progression increases without bound because each successive term of the progression is larger. However, the sum of the terms of the second geometric progression approaches some limiting value because successive terms are smaller. From Table 17–1, you can see that as n, the number of terms, increases, the sum S_n for the second progression approaches 72.

TABLE 17–1

n	Series	S_n
3	$36 + 18 + 9$	63
4	$36 + 18 + 9 + 4.5$	67.5
5	$36 + 18 + 9 + 4.5 + 2.25$	69.75
6	$36 + 18 + 9 + 4.5 + 2.25 + 1.125$	70.875
11	$36 + 18 + 9 + 4.5 + \cdots + 0.0352$	71.965
15	$36 + 18 + 9 + \cdots + 0.0022$	71.9978
20	$36 + 18 + 9 + \cdots + 6.8665 \times 10^{-5}$	71.9999
30	$36 + 18 + 9 + \cdots + 6.7055 \times 10^{-8}$	71.9999999

If we compare the two geometric progressions given earlier, when $r > 1$, the sum of the terms increases without bound; but when $r < 1$, the sum of the terms approaches some limiting value. If we consider the formula for the sum of the first n terms of a geometric progression,

$$S_n = \frac{a_1(1 - r^n)}{1 - r}$$

we see that for $|r| < 1$, the expression r^n becomes smaller as the value of n becomes larger. In calculus, we say that

$$\lim_{n \to \infty} r^n = 0 \qquad \text{for } |r| < 1$$

This notation is read, "the limit of r^n as n increases without bound is zero." If we apply this knowledge about r^n for $|r| < 1$ to the sum formula, we obtain the following equation.

Sum of an Infinite Geometric Progression

The sum of an infinite geometric progression is given by

$$S = \frac{a_1}{1 - r} \qquad \text{for } |r| < 1$$

where $a_1 =$ the first term of the sequence and $r =$ the common ratio.

EXAMPLE 1   Find the sum of the following infinite geometric progression:

$$18, 6, 2, \ldots$$

Solution   From the progression, we know that $a_1 = 18$ and $r = \dfrac{6}{18} = \dfrac{1}{3}$.

The sum of the geometric progression is

$$S = \frac{a_1}{1 - r}$$

$$S = \frac{18}{1 - 1/3}$$

$$S = 27 \qquad\qquad ■ ■ ■$$

EXAMPLE 2   Find the sum of the following infinite geometric progression:

$$-8, 3, -1.125, \ldots$$

Solution   We know that $a_1 = -8$ and $r = \dfrac{3}{-8} = -\dfrac{3}{8}$. The sum is

$$S = \frac{-8}{1 - (-3/8)}$$

$$S = -\frac{64}{11} \approx -5.82 \qquad\qquad ■ ■ ■$$

Repeating Decimals   You can use a geometric progression to find the fractional equivalent of a repeating decimal. The next two examples illustrate the procedure.

EXAMPLE 3 Find the rational equivalent to the repeating decimal

$$0.\overline{39} = 0.393939 \ldots$$

Solution The repeating decimal $0.\overline{39}$ can be written as a series in the following manner:

$$0.\overline{39} = 0.39 + 0.0039 + 0.000039 + \cdots$$

The terms in this series form a geometric progression with

$$a_1 = 0.39 \quad \text{and} \quad r = \frac{0.0039}{0.39} = 0.01$$

Using the formula for the sum of an infinite geometric progression gives

$$S = \frac{0.39}{1 - 0.01} = \frac{0.39}{0.99} = \frac{39}{99} = \frac{13}{33}$$

The fractional equivalent of $0.\overline{39}$ is $\dfrac{13}{33}$. ■■■

EXAMPLE 4 Find the fractional equivalent of $0.6\overline{813}$.

Solution First, write the repeating decimal as a series in the following manner:

$$0.6\overline{813} = 0.6 + 0.0813 + 0.0000813 + \cdots$$

Second, determine a_1 and r for the terms that comprise the geometric progression.

$$a_1 = 0.0813 \quad \text{and} \quad r = 0.001$$

Third, find the sum of the infinite geometric progression.

$$S = \frac{0.0813}{1 - 0.001} = \frac{0.0813}{0.999} = \frac{813}{9,990}$$

The fractional form of the repeating decimal is

$$0.6 + S = \frac{3}{5} + \frac{813}{9,990} = \frac{6,807}{9,990} = \frac{2,269}{3,330}$$ ■■■

Application

⊞ **EXAMPLE 5** A ball attached to the end of an elastic band oscillates up and down. The total distance traveled during its initial oscillation is 7.8 cm, and the distance traveled during each successive oscillation is 68% of the previous distance. Find the total vertical distance the ball traveled before coming to rest.

Solution We know that $a_1 = 7.8$ and $r = 0.68$. The total vertical distance is given by

$$S = \frac{a_1}{1 - r}$$

$$S = \frac{7.8}{1 - 0.68}$$

$$S \approx 24 \text{ cm}$$

$$7.8 \boxed{÷} \boxed{(} \boxed{1} \boxed{-} \boxed{0.68} \boxed{)} \boxed{=} \longrightarrow 24.375$$ ■■■

EXAMPLE 6 Initially a clock pendulum swings through an arc of 12 inches (refer to Figure 17–2). On each subsequent swing, the length of the arc is 95% of the previous length. What is the length of the arc after 15 swings?

FIGURE 17–2

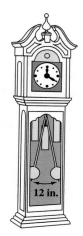

12 in.

Solution   The length of the first arc is 12 inches. The length of the second arc is

$$0.95(12 \text{ inches})$$

The length of the third swing is

$$0.95(\text{length of second swing})$$
$$0.95(0.95)(12)$$

In general, the length of the arc on swing *n* is given by

$$(0.95)^{n-1}(12)$$

To find the length of the arc after 15 swings, we evaluate

$$(0.95)^{15-1}(2)$$
$$5.9 \text{ inches}$$

■ ■ ■

17–3 EXERCISES

Find the sum of each infinite geometric progression.

1. $-8, 4, -2, \ldots$

2. $15, -10, \dfrac{20}{3}, \ldots$

3. $12, 4, \dfrac{4}{3}, \ldots$

4. $4.32, 3.6, 3.0, \ldots$

5. $42, 36, \dfrac{216}{7}, \ldots$

6. $54, 18, 6, \ldots$

7. $69.12, 28.8, 12, \ldots$

8. $18, 12, 8, \ldots$

9. $36, 24, 16, \ldots$

10. $63, 21, 7, \ldots$

11. $64, 16, 4, \ldots$

12. $25a^3, 5a, \dfrac{1}{a}, \ldots$

13. $14, -7\sqrt{2}, 7, \ldots$

14. $15, 5\sqrt{3}, 5, \ldots$

15. $6x^2, 3x, 1.5, \ldots$

Find the fractional equivalent for each repeating decimal.

16. $0.\overline{6}$

17. $0.\overline{9}$

18. $0.\overline{17}$

19. $0.8\overline{3}$

20. $0.13\overline{5}$

21. $0.42\overline{3}$

22. $0.3\overline{15}$

23. $3.\overline{738}$

24. $8.\overline{217}$

P 25. A pendulum swings through an arc of 36 in. On each successive swing, the pendulum covers an arc equal to 90% of the previous swing. Find the length of the arc on the sixth swing and the total distance the pendulum travels before coming to rest.

P 26. An object suspended from a spring oscillates up and down. The first oscillation is 25 cm, and each successive oscillation is 80% of the preceding one. Find the total distance the object travels before coming to rest.

P 27. A rubber ball rebounds 3/5 of its height. If it is initially 30 ft high, what total vertical distance does it travel before coming to rest?

P 28. An object decelerates such that it travels 45 m during the first second, 15 m during the second second, and 5 m during the third second. Determine the total distance the object travels before coming to rest.

17–4

THE BINOMIAL THEOREM

In previous algebraic operations, we have raised a binomial (an expression with two terms) to the second or third power, but rarely to any higher power. However, in more advanced mathematics such multiplication will be necessary. In this section we will develop a formula for raising a binomial to any power.

Expanding $(x + y)^n$

Let us expand the expression $(x + y)^n$ where $n = 0, 1, 2, 3, 4,$ and 5.

$$
\begin{aligned}
(x + y)^0 &= 1 \\
(x + y)^1 &= x + y \\
(x + y)^2 &= x^2 + 2xy + y^2 \\
(x + y)^3 &= x^3 + 3x^2y + 3xy^2 + y^3 \\
(x + y)^4 &= x^4 + 4x^3y + 6x^2y^2 + 4xy^3 + y^4 \\
(x + y)^5 &= x^5 + 5x^4y + 10x^3y^2 + 10x^2y^3 + 5xy^4 + y^5
\end{aligned}
$$

The process of multiplying $(x + y)$ times itself five times is quite lengthy. However, the multiplications above illustrate certain patterns that will prove helpful in defining a formula.

■ Each expansion of $(x + y)^n$ contains $n + 1$ terms. For example $(x + y)^4$ contains $n + 1 = 5$ terms.

■ The first term is always x^n, and the last term is always y^n. For example, when $n = 3$,

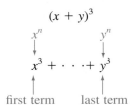

$$(x + y)^3$$

$$x^3 + \cdots + y^3$$

first term   last term

■ Proceeding from left to right, the powers of x decrease by 1 in each term, and the powers of y increase by 1. Moreover, the sum of the exponents in each term equals n. For example,

powers of x	4	3	2	1	0
$(x + y)^4 =$	x^4	$+ 4x^3y$	$+ 6x^2y^2$	$+ 4xy^3$	$+ y^4$
power of y	0	1	2	3	4
sum of exponents	4	4	4	4	4

■ The coefficients of the terms form a symmetrical, triangular pattern, called **Pascal's triangle.**

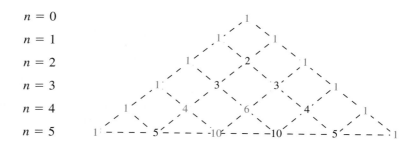

$n = 0$
$n = 1$
$n = 2$
$n = 3$
$n = 4$
$n = 5$

Notice that each row begins and ends with 1, and each coefficient is the sum of the two numbers above it on each side. For example,

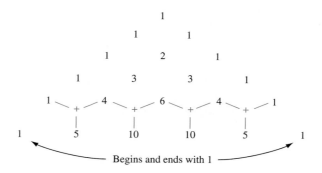

Begins and ends with 1

EXAMPLE 1   Using Pascal's triangle, find $(a + b)^6$.

Solution   First, we use Pascal's triangle to determine the coefficients. From our previous expansion, we know the coefficient for $n = 5$.

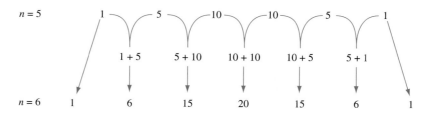

$n = 5$

1   5   10   10   5   1

1 + 5   5 + 10   10 + 10   10 + 5   5 + 1

$n = 6$ 1   6   15   20   15   6   1

Second, we determine the powers of a and b for each term.

descending powers of a

$$6 \quad 5 \quad 4 \quad 3 \quad 2 \quad 1 \quad 0$$
$$1a^6 + 6a^5b + 15a^4b^2 + 20a^3b^3 + 15^2b^4 + 6ab^5 + 1b^6$$
$$0 \quad 1 \quad 2 \quad 3 \quad 4 \quad 5 \quad 6$$

ascending powers of b

The result is

$$(a + b)^6 = a^6 + 6a^5b + 15a^4b^2 + 20a^3b^3 + 15a^2b^4 + 6ab^5 + b^6$$ ■■■

EXAMPLE 2   Expand $(4a - 3b)^4$.

Solution   From Pascal's triangle, the coefficients for $n = 4$ are

$$1 \qquad 4 \qquad 6 \qquad 4 \qquad 1$$

However, the terms that are substituted in descending and ascending order for x and y also contain a coefficient. To determine the expanded form, we *substitute 4a for x in descending powers and $-3b$ for y in ascending powers in each term.* The expanded form is

$$(4a - 3b)^4 = 1(4a)^4(-3b)^0 + 4(4a)^3(-3b)^1 + 6(4a)^2(-3b)^2$$
$$+ 4(4a)^1(-3b)^3 + 1(4a)^0(-3b)^4$$
$$(4a - 3b)^4 = 256a^4 - 768a^3b + 864a^2b^2 - 432ab^3 + 81b^4$$ ■■■

CAUTION ✦ When the binomial you are expanding contains coefficients, you must be very careful in applying Pascal's triangle.

The Factorial Function

Pascal's triangle is helpful in expanding smaller powers of a binomial, but it can be difficult for large powers. A formula has been developed to do this, but it uses the **factorial function,** defined in the following box. *The number n! is the product of the first n positive integers.*

The Factorial Function

The factorial function is given by

$$n! = n(n - 1)(n - 2)(n - 3) \ldots 3 \cdot 2 \cdot 1$$

where $n =$ any positive integer. By definition,

$$0! = 1$$
$$1! = 1$$

EXAMPLE 3   Calculate the following:

(a) 6!   (b) 15!

Solution

(a) $6! = 6 \cdot 5 \cdot 4 \cdot 3 \cdot 2 \cdot 1 = 720$
(b) $15! = 1.31 \div 10^{12}$

   15 $\boxed{x!}$ ⟶   1.307674368   12 ■■■

Another symbol useful in expanding a binomial is $\left(\dfrac{n}{j}\right)$, read "$n$ items taken j at a time." The symbol is defined as follows:

$$\left(\frac{n}{j}\right) = \frac{n!}{j! \, (n - j)!}$$

EXAMPLE 4   Evaluate the following:

(a) $\begin{pmatrix} 6 \\ 2 \end{pmatrix}$ (b) $\begin{pmatrix} 16 \\ 7 \end{pmatrix}$

Solution   (a) $\begin{pmatrix} 6 \\ 2 \end{pmatrix} = \dfrac{6!}{2!(6-2)!} = \dfrac{6!}{2!\,4!} = \dfrac{6 \cdot 5 \cdot 4!}{2 \cdot 1 \cdot 4!} = \dfrac{6 \cdot 5}{2 \cdot 1} = 15$

(b) $\begin{pmatrix} 16 \\ 7 \end{pmatrix} = \dfrac{16!}{7!\,(16-7)!} = \dfrac{16!}{7!\,9!} = \dfrac{16 \cdot 15 \cdot 14 \cdot 13 \cdot 12 \cdot 11 \cdot 10 \cdot 9!}{7!\,9!}$

$\begin{pmatrix} 16 \\ 7 \end{pmatrix} = \dfrac{16 \cdot 15 \cdot 14 \cdot 13 \cdot 12 \cdot 11 \cdot 10}{7!} = 11{,}440$ ■ ■ ■

NOTE ✦ By definition, $\begin{pmatrix} n \\ 0 \end{pmatrix} = 1$ and $\begin{pmatrix} n \\ n \end{pmatrix} = 1$.

The Binomial Theorem

Pascal's triangle is difficult to use in determining the coefficients for expanding a binomial such as $(x + 5)^{12}$ because, in this case, you need to produce twelve rows of Pascal's triangle. The Binomial Theorem can be applied to expanding a binomial.

Binomial Theorem

$$(x + a)^n = \begin{pmatrix} n \\ 0 \end{pmatrix} x^n + \begin{pmatrix} n \\ 1 \end{pmatrix} a x^{n-1} + \begin{pmatrix} n \\ 2 \end{pmatrix} a^2 x^{n-2} + \ldots$$

$$\begin{pmatrix} n \\ n \end{pmatrix} a^n x^0$$

where $n = $ any positive integer.

EXAMPLE 5   Use the Binomial Theorem to expand $(x + 3)^5$.

Solution   From the information given, $a = 3$ and $n = 5$.

$$(x + 3)^5 = \begin{pmatrix} 5 \\ 0 \end{pmatrix} x^5 + \begin{pmatrix} 5 \\ 1 \end{pmatrix} (3)x^4 + \begin{pmatrix} 5 \\ 2 \end{pmatrix} (3)^2 x^3 + \begin{pmatrix} 5 \\ 3 \end{pmatrix} (3)^3 x^2 + \begin{pmatrix} 5 \\ 4 \end{pmatrix} (3)^4 x + \begin{pmatrix} 5 \\ 5 \end{pmatrix} 3^5 x^0$$

$$(x + 3)^5 = x^5 + 5(3)x^4 + 10(9)x^3 + 10(27)x^2 + 5(81)x^4 + 243$$
$$(x + 3)^5 = x^5 + 15x^4 + 90x^3 + 27x^2 + 405x^4 + 243$$ ■ ■ ■

EXAMPLE 6   Use the Binomial Theorem to expand $(x + y)^8$.

Solution   We substitute $a = y$ and $n = 8$.

$$(x + y)^8 = \begin{pmatrix} 8 \\ 0 \end{pmatrix} x^8 + \begin{pmatrix} 8 \\ 1 \end{pmatrix} yx^7 + \begin{pmatrix} 8 \\ 2 \end{pmatrix} y^2 x^6 + \begin{pmatrix} 8 \\ 3 \end{pmatrix} y^3 x^5 + \begin{pmatrix} 8 \\ 4 \end{pmatrix} y^4 x^4 + \begin{pmatrix} 8 \\ 5 \end{pmatrix} y^5 x^3 + \begin{pmatrix} 8 \\ 6 \end{pmatrix} y^6 x^2$$

$$+ \begin{pmatrix} 8 \\ 7 \end{pmatrix} y^7 x + \begin{pmatrix} 8 \\ 8 \end{pmatrix} y^8 x^0$$

$$(x + y)^8 = x^8 + 8x^7 y + 28x^6 y^2 + 56x^5 y^3 + 70x^4 y^4 + 56x^3 y^5 + 28x^2 y^6 + 8xy^7 + y^8$$

■ ■ ■

EXAMPLE 7 Use the Binomial Theorem to expand $(5x - 2y)^4$.

Solution For the Binomial Theorem $x = 5x$, $a = (-2y)$, and $n = 4$.

$$(5x - 2y)^4 = \binom{4}{0}(5x)^4 + \binom{4}{1}(-2y)(5x)^3 + \binom{4}{2}(-2y)^2(5x)^2 + \binom{4}{3}(-2y)^3(5x) + \binom{4}{4}(-2y)^4$$

$$(5x - 2y)^4 = (5x)^4 + 4(-2y)(5x)^3 + 6(-2y)^2(5x)^2 + 4(-2y)^3(5x) + (-2y)^4$$

$$(5x - 2y)^4 = 625x^4 - 1{,}000x^3y + 600x^2y^2 - 160xy^3 + 16y^4$$ ■■■

Finding a Given Term

We can use the binomial formula to find a specific term of a binomial expansion without finding all the terms. The $(k + 1)$ term of the expansion $(x + y)^n$ is given by

$$\binom{n}{n - k} x^{(n - k)} y^k$$

EXAMPLE 8 Find the ninth term of the expansion of $(x + y)^{12}$.

Solution To find the ninth term of the expansion, we compute $k + 1 = 9$ or $k = 8$ and $n = 12$. The ninth term is

$$\binom{12}{12 - 8} x^{12 - 8} y^8$$

$$\binom{12}{4} x^4 y^8$$

$$\frac{12!}{4! \, 8!} x^4 y^8$$

$$495x^4 y^8$$ ■■■

17–4 EXERCISES

Write the expansion of each binomial.

1. $(a + 5)^4$

2. $(x - 7)^5$

3. $(m - 1)^7$

4. $(y - 3)^6$

5. $(2x - 5y)^3$

6. $(2b - 3)^4$

7. $(3y + x)^6$

8. $(4m + n)^5$

9. $(m + 3)^5$

10. $(d - 7)^4$

11. $(b - 6)^7$

12. $(y + 8)^5$

13. $(3x - 1)^4$

14. $(2m + 5)^5$

15. $(5m - 4)^6$

16. $(8x - 3y)^4$

17. $(x - 4y)^7$

18. $(7x + 2y)^3$

19. $(2a^2 - 3b)^4$

20. $(4x + 5y^2)^3$

Write the first four terms in the binomial expansion of each expression.

21. $(1 - x)^8$

22. $(1 + x)^7$

23. $(1 + x)^3$

24. $(1 - x)^2$

25. Find the seventh term of the expansion of $(y - 8)^{10}$.

26. Find the fourth term of the expansion of $(x + 7)^6$.

27. Find the x^3 term of the expansion of $(x - 2)^8$.

28. Find the term containing y^5 of the expansion of $(m - y)^7$.

CHAPTER SUMMARY

Summary of Terms

arithmetic progression (p. 648)

common difference (p. 648)

common ratio (p. 653)

factorial function (p. 664)

finite sequence (p. 648)

geometric progression (p. 653)

infinite sequence (p. 648)

Pascal's triangle (p. 663)

sequence (p. 648)

term (p. 648)

Summary of Formulas

$a_n = a_1 + (n - 1)d$ nth term of an arithmetic progression

$S_n = \dfrac{n(a_1 + a_n)}{2}$ sum of the first n terms of an arithmetic progression

$a_n = a_1 r^{n-1}$ nth term of a geometric progression

$S_n = \dfrac{a_1(1 - r^n)}{1 - r}$ sum of the first n terms of a geometric progression $r \neq 1$

$S = \dfrac{a_1}{1 - r}$ sum of an infinite geometric progression for $|r| < 1$

$n! = n(n - 1)(n - 2)(n - 3) \ldots 3 \cdot 2 \cdot 1$ factorial function

$(x + a)^n = \dbinom{n}{0}x^n + \dbinom{n}{1}ax^{n-1} + \dbinom{n}{2}a^2x^{n-2} + \cdots \dbinom{n}{n}a^n$ Binomial Theorem

$\dfrac{n(n - 1)(n - 2) \cdots (n - k + 1)}{k!}x^{(n-k)}y^k$; $(k + 1)$ term of $(x + y)^n$

CHAPTER REVIEW

Section 17–1

Find the indicated term of the arithmetic progression.

1. The eighth term of $-4, 2, 8, \ldots$

2. The fifteenth term of $-12, -9, -6, \ldots$

3. The twenty-third term of $48, 42, 36, \ldots$

4. The eighteenth term of $-1, -5, -9, \ldots$

Find the sum of the first n terms of each arithmetic progression.

5. $6, 9, 12, \ldots; n = 9$

6. $-10, -2, 6, 14, \ldots; n = 7$

7. $n = 10, a_1 = 3$, and $d = -2$

8. $n = 15, a_1 = -5$, and $d = 6$

9. The population of Orange Grove increased arithmetically from 25,000 to 40,000 in 12 years. What was each year's population growth?

10. Pablo accepts a job earning \$28,250 per year. If he receives a \$945 raise each year, what will his salary be at the end of 6 years?

Section 17–2

Find the indicated term of the geometric progression.

11. The fifth term of $-3, 9, -27, \ldots$

12. The sixth term of $64, -16, 4, \ldots$

13. The ninth term of $625, 125, 25, \ldots$

14. The thirteenth term of $6, 12, 24, \ldots$

Find the sum of the first n terms of the geometric progression.

15. $2, 6, 18, \ldots; n = 8$

16. $-3, 12, -48, \ldots; n = 5$

17. $n = 10, a_1 = -8$, and $r = -2$

18. $n = 12, a_1 = 4$, and $r = 0.5$

B 19. Rick accepts a job earning \$18,500 per year. Each year for 3 years, he receives an 8% pay raise. How much does he earn during the third year?

B 20. Manufacturing equipment currently worth \$60,000 depreciates 15% each year. Determine the value of the equipment after 8 years.

Section 17–3

Find the sum of each infinite geometric progression.

21. $\dfrac{1}{3}, \dfrac{1}{12}, \dfrac{1}{48}, \ldots$

22. 24, 6, 1.5, . . .

23. 18, 6, 2, . . .

24. $48x^3, 12x^2, 3x, \ldots$

Find the fractional equivalent for each repeating decimal.

25. $0.\overline{34}$

26. $0.\overline{67}$

27. $3.\overline{124}$

28. $7.3\overline{18}$

P 29. A rubber ball rebounds 45% of its height on successive bounces. If it is initially 50 ft high, what total distance does it travel before coming to rest?

P 30. An object suspended from an elastic band oscillates up and down. The first oscillation is 26 in., and each successive oscillation is 65% of the preceding one. Find the total distance the object travels.

Section 17–4

Write the expansion of each binomial.

31. $(x - z)^3$

32. $(m - n)^4$

33. $(2x - 3y)^4$

34. $(5a + b)^3$

35. $(a + 2b)^4$

36. $(3x - y)^5$

37. $(y^2 - 5z)^3$

38. $(m + 4n)^4$

39. Find the fifth term of the expansion of $(x - 6)^8$.

40. Find the fourth term of the expansion of $(y + 5)^9$.

41. Find the x^4 term of the expansion of $(x + 3)^7$

42. Find the y^5 term of the expansion of $(y - 3)^9$.

CHAPTER TEST

The number in parentheses refers to the appropriate learning objective given at the beginning of the chapter.

1. Find the sum of the following infinite geometric progression: (5)

$$80, 40, 20, \ldots$$

2. Find the thirty-fourth term of the following arithmetic progression: (1)

$$7, 13, 19, \ldots$$

B **3.** Paul accepts a job earning $20,000 a year. If he receives a 7% raise each year for 6 years, what will his salary be during the sixth year? (3)

4. Find the fractional form of the repeating decimal $0.28\overline{1}$. (6)

5. Apply the binomial formula to expand $(2x - y)^4$. (7)

6. Determine the sum of the first 12 terms of the following geometric progression: (4)

$$3, 9, 27, \ldots$$

7. Expand $(m - n)^4$. (8)

8. Find the tenth term of the following geometric progression: (3)

$$81, 27, 9, \ldots$$

9. Determine the sum of the first 26 terms of the following arithmetic progression: (2)

$$9, 16, 23, \ldots$$

10. Find the fractional equivalent of $3.\overline{27}$. (6)

11. Determine the sum of the first 19 terms of the following arithmetic progression: (2)

$$2, 11, 20, \ldots$$

12. Find the sum of the following infinite geometric progression: (5)

$$128, 16, 2, \ldots$$

13. Find the x^4 term in the expansion of $(x + 3)^8$. (9)

C 14. The population of Palm Coast increases 4% each year. If the current population is 46,280, determine what the population will be in 20 years. (3)

15. Find the sum of the first 11 terms of the following geometric progression: (4)

$$-1, 3, -9, 27, \ldots$$

16. Apply the binomial formula to expand $(2x - y)^5$. (7)

17. Find the fifth term of the expansion of $(3y + 8)^8$. (9)

18. Find the ninth term of the following geometric progression: (3)

$$32, 8, 2, \ldots$$

GROUP ACTIVITY

The Dream Job

Raccoon Solutions seems to have an outstanding employee relationship. Their medical plan is one of the best in the nation. They are a company that comes up with unique solutions to other corporations' problems. Ninety-five percent of the employees have advanced degrees in engineering and science. The opportunity to work with outstanding people and to work on different kinds of problems makes this the perfect place to work.

You were called in for an interview and offered a position with the other Raccoon employees. They promised you $120,000 a year for 6 years, plus a choice of one of the following options:

1. A bonus of $10,000 per year.

2. An annual increase of 5% per year (starts after the first year).

3. An annual increase of 6.5% per year (starts after the second year).

4. An annual increase of $9,000 per year (starts after the first year).

You have to tell Raccoon Solutions what your decision is soon.

1. Calculate what your payment would be for every year for the 6 years for each option.
2. Calculate the total that Raccoon Solutions will pay you over the 6 years for each option. What formulas can you use?
3. Identify which option will pay the most overall.
4. What circumstances might make other options a better choice, even though they may pay less money? For example, suppose you worked for Raccoon Solutions for 12 years.
5. What are the advantages of being paid more at the beginning of your career at Raccoon Solutions?
6. Are any of these options examples of arithmetic or geometric sequences?

SOLUTION TO CHAPTER INTRODUCTION

Your salary at XYZ Corporation represents a geometric progression. To determine your salary at the end of the eighth year, substitute $a_1 = 23{,}480$, $r = 1.08$, and $n = 8$ into the formula.

$$a_n = a_1 r^{n-1}$$
$$a_8 = (23{,}480)(1.08)^7$$
$$a_8 \approx \$40{,}240.59$$

To determine your salary at ABC Company, substitute $a_1 = 26{,}700$, $d = 1{,}500$, and $n = 8$ into the formula for an arithmetic progression.

$$a_n = a_1 + (n - 1)d$$
$$a_8 = 26{,}700 + 7(1{,}500)$$
$$a_8 = \$37{,}200$$

From the above calculation, you should accept the job at XYZ Corporation.

Appendix A

A Review of Basic Algebraic Concepts

A basic understanding of algebra is important to every student entering a technical field. This appendix reviews the algebraic concepts found in most elementary algebra courses. Topics include the following:

- Real numbers
- Laws and operations of real numbers
- Zero and the order of operations
- Exponents
- Scientific notation
- Roots and radicals
- Addition and subtraction of algebraic expressions
- Multiplying polynomials
- Polynomial division
- Percentages
- Significant digits
- Units and unit conversion
- Calculator
- Computer number systems

We will begin the review with a discussion of real numbers.

A–1

REAL NUMBERS

Imagine all the computations a technician must make daily to solve problems or finalize reports. Such calculations require a thorough understanding and working knowledge of the operations, rules, and symbols of real numbers.

Integers

Historically, the different types of numbers evolved from the need for a number to express a real-life situation. For example, counting the number of desks in a classroom or expressing how many items are contained in a shipment requires the use of numbers called **integers.** The counting numbers such as 1, 2, 3 are called **positive integers.** The

negative integers, such as -1, -2, -3, are often used to show a decrease in a quantity. **Zero** is also an integer, but it is neither positive nor negative. The group of numbers designated as integers includes the positive and negative integers and zero.

To represent part of a group or a part of a whole requires the use of a rational number. A **rational number** is any number in the form a/b (ratio), where a is an integer and b is nonzero integer. The numbers 3/5, $-1/2$, and $-15/4$ are examples of rational numbers. Since any integer can also be expressed in the form a/b where b equals 1, all integers are also rational numbers.

Rational Numbers

A rational number may be expressed as a decimal by dividing the denominator into the numerator. The result is either a terminating or repeating decimal.

EXAMPLE 1 Express the following rational numbers in decimal form:

(a) $\dfrac{4}{5}$ (b) $\dfrac{1}{3}$

Solution

(a) Dividing 5 into 4 gives 0.8, a terminating decimal.

(b) $\dfrac{1}{3} = 0.333$ . . . is a repeating decimal because the digit 3 repeats.   ■ ■ ■

When the length of the diagonal of a unit square (see Figure A–1) is computed, the result is the number $\sqrt{2}$, read "square root of two." Such numbers are called irrational numbers.

FIGURE A–1

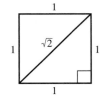

Irrational Numbers

Irrational numbers are numbers that cannot be expressed in the form of a rational number (a/b where a is an integer and b is a nonzero integer). In decimal form, irrational numbers neither terminate nor repeat. For instance, the $\sqrt{2}$ expressed in decimal form equals 1.4142 . . . . The decimal part of this number will never form a repetitive pattern nor will it terminate. For this reason, when decimal forms of irrational numbers are used in calculations, it is necessary to round to some approximate value.

 EXAMPLE 2   Using a calculator, express the following irrational numbers in decimal form:

(a) $\sqrt{8}$ (b) $\sqrt[3]{12}$

Solution

(a) $\sqrt{8} = 2.8284271$ . . . is a nonrepeating decimal.
(b) $\sqrt[3]{12} = 2.2894285$ . . . is a nonrepeating decimal.   ■ ■ ■

Irrational numbers are often encountered when finding the square root, or higher root, of certain numbers. The square root of any number that is not a perfect squared number is an irrational number. Two quantities often found in technical formulas, π and e, are both irrational numbers because they cannot be written as either terminating or repeating decimals.

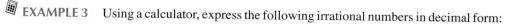 **EXAMPLE 3**   Using a calculator, express the following irrational numbers in decimal form:

(a) π (b) e

Solution

(a) Pressing the π key on your calculator gives 3.1415926 . . . , which is a nonrepeating decimal.
(b) Pressing 1 and e^x on your calculator gives 2.7182818 . . . , which is a nonrepeating decimal. ■■■

The majority of technical applications involve only rational and irrational numbers. The rational and irrational numbers are combined to make the real numbers. In this appendix when we speak of numbers, we mean real numbers unless stated otherwise.

EXAMPLE 4   Classify each of the following numbers as integer, rational, or irrational. A number may be in more than one category.

(a) $\dfrac{2}{3}$ (b) -8 (c) $-\sqrt{4}$ (d) 0.5

(e) $\sqrt{3}$ (f) $-\dfrac{\sqrt{5}}{8}$ (g) $\dfrac{\pi}{4}$ (h) $\dfrac{e}{2}$

Solution

(a) $\dfrac{2}{3}$ is a rational number because it is represented as a ratio.

(b) -8 is both an integer and a rational number.
(c) $-\sqrt{4} = -2$ is both an integer and a rational number.
(d) 0.5 is a rational number because it is expressed as a terminating decimal.
(e) $\sqrt{3}$ is an irrational number because $\sqrt{3} = 1.7320508$ . . . , a nonrepeating, nonterminating decimal.

(f) $-\dfrac{\sqrt{5}}{8}$ is an irrational number because it is a nonrepeating, nonterminating decimal.

(g) $\dfrac{\pi}{4}$ is an irrational number.

(h) $\dfrac{e}{2}$ is an irrational number. ■■■

Complex Numbers

Some applications in physics and electronics involve complex numbers. **Complex numbers** are numbers in the form $a + bj$ where a and b are real numbers and $j = \sqrt{-1}$. When working with quadratic equations, you will encounter the square root of a negative number, such as $\sqrt{-36}$. Such a number is called an **imaginary number** and is equivalent to $6j$ or the imaginary part of a complex number. Complex numbers are explained in greater detail in Chapters 3 and 5.

Real Number Line

The **real number line** is a pictorial representation of the real number system. Each real number can be associated with one and only one point on the line; conversely, each point on the line can be assigned only one real number.

The point associated with the real number zero is called the **origin.** It is customary to locate positive values to the right of zero and negative values to the left, as shown in Figure A–2.

FIGURE A–2

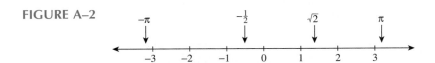

The relative position of two real numbers on the number line can be clarified using the equality symbol, $=$ (equal to), and the two inequality symbols, $<$ (less than) and $>$ (greater than).

EXAMPLE 5   For each of the following, place the appropriate symbol ($=$, $<$, or $>$) between the two numbers to make the statement true.

(a) $\dfrac{10}{2} \,\square\, 5$ (b) $-1.5 \,\square\, 0$ (c) $3 \,\square\, -2$

Solution

(a) $\dfrac{10}{2} = 5$ (Read 10 divided by 2 is equal to 5.) Since these values are equal, they would occupy the same position on the number line.

(b) $-1.5 < 0$ (Read -1.5 is less than 0.) Figure A–3 shows the position of these numbers on the number line.

(c) $3 > -2$ (Read 3 is greater than -2.) Figure A–4 shows the relative position of -2 and 3 on the number line.

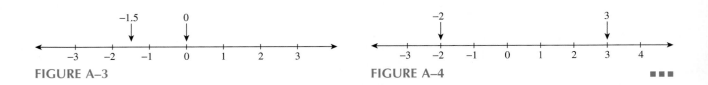

FIGURE A–3   FIGURE A–4   ■ ■ ■

The pictorial or graphical representation of the numbers in parts (b) and (c) in Example 5 demonstrates the following fact: Any number to the left of another number on the number line is less than that number; any number to the right of another number on the number line is greater than that number.

EXAMPLE 6   Decide if the following statements are true or false.

(a) $-5 < -2$ (b) $\sqrt{3} > 2$ (c) $\dfrac{1}{3} > -\dfrac{3}{5}$ (d) $-\dfrac{\pi}{2} < -0.5$

Solution

(a) True, because -5 is to the left of -2 (Figure A–5).
(b) False, because $\sqrt{3} \approx 1.7$ is not to the right of 2 (Figure A–6).
(c) True (see Figure A–7).
(d) True (see Figure A–8).

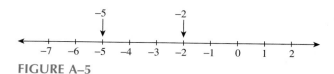

FIGURE A–5

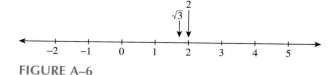

FIGURE A–6

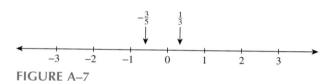

FIGURE A–7

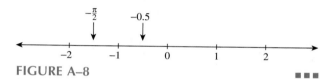

FIGURE A–8   ■ ■ ■

Absolute Value

When applying the rules for adding or subtracting real numbers, you need to know the meaning of absolute value. The **absolute value** of a number is the distance from the origin to the number. Since distance is always positive or zero, the absolute value of a number is always positive or zero. Two vertical bars | | placed around a number is the mathematical symbol used to indicate the absolute value of that number.

EXAMPLE 7   Find the absolute value of the following numbers:

(a) $|-3|$ (b) $\left|\dfrac{1}{2}\right|$ (c) $|-\sqrt{30}|$

Solution

(a) As shown in Figure A–9, the distance from the origin to -3 equals 3. Therefore, $|-3| = 3$.

(b) $\left|\dfrac{1}{2}\right| = \dfrac{1}{2}$

(c) $|-\sqrt{30}| = \sqrt{30}$ ■ ■ ■

FIGURE A–9

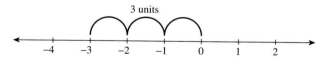

A–1 EXERCISES

Classify each of the following numbers as integer, rational number, or irrational number. A given number may be in more than one category.

1. $\dfrac{1}{3}$

2. $-\sqrt{4}$

3. 0.575

4. $\sqrt{12}$

5. -5

6. $\dfrac{\pi}{2}$

7. $-\sqrt{144}$

8. 0

9. $15\dfrac{1}{4}$

10. $-\dfrac{1}{8}$

Decide if the following are true or false.

11. $-\dfrac{1}{2} < -1$

12. $\pi > 3$

13. $-5 < -3$

14. $\dfrac{1}{8} = 0.128$

15. $\dfrac{5}{13} > \dfrac{4}{15}$

16. $0 > |-0.5|$

17. $13\dfrac{2}{3} > \sqrt{271}$

18. $-3.2 < -3.5$

19. $-|5| < 2$

20. $|-8| < |-4|$

21. Locate the following numbers on the same number line:

$$-\dfrac{1}{2}, \quad 1.3, \quad \sqrt{5}, \quad -3, \quad 4.5$$

22. Locate the following numbers on the same number line:

$$\sqrt{20}, \quad -2\dfrac{2}{3}, \quad \left|\dfrac{\pi}{3}\right|, \quad -1.45, \, 0.33$$

23. Locate the following numbers on the number line.

(a) The temperature when it drops to 10 degrees below zero.
(b) The length of a structural beam found to be $\sqrt{40}$ m.
(c) The circumference of a circle found to be π cm.
(d) A current source of 5 negative units.

A–2

LAWS AND OPERATIONS OF REAL NUMBERS

One main characteristic of algebra that distinguishes it from arithmetic is its use of literal numbers. Literal numbers are used extensively in algebra, and the laws pertaining to literal numbers must be defined explicitly. A **literal number** is a letter or other symbol used to represent a real number. Using literal numbers allows you to write a single expression to represent several different situations. For example, the perimeter of any square is the sum of all its sides. The formula $P = 4s$ describes the perimeter of any square no matter what the length of the sides.

When adding or multiplying literal numbers, understanding the laws of operations is essential.

Laws of Operations

Commutative Law
    Addition: $a + b = b + a$
    Multiplication: $a \cdot b = b \cdot a$

Associative Law
    Addition: $(a + b) + c = a + (b + c)$
    Multiplication: $(a \cdot b) \cdot c = a \cdot (b \cdot c)$

Distributive Law
                      $a(b + c) = a \cdot b + a \cdot c$
                      $(b + c)a = b \cdot a + c \cdot a$

Commutative Law

The commutative law states that the order in which you add or multiply two real numbers does not affect the sum or product. For example, the commutative law means that $5 + 3 = 3 + 5$ and $2 \cdot 8 = 8 \cdot 2$. However, the commutative law does not apply to the operations of subtraction or division.

Associative Law

Another important law is the associative law. The associative law states that the grouping of real numbers in addition or multiplication does not alter the results. For example, $(5 + 3) + 2 = 5 + (3 + 2)$ and $(2 \cdot 6) \cdot 8 = 2 \cdot (6 \cdot 8)$. Again, the associative law does not apply to the operations of subtraction or division.

Distributive Law

Many times in formulas and applications you need to multiply a number by the sum of two other numbers. The distributive law allows the expression of this product. Applying the distributive law gives $5(3 + 4) = 5 \cdot 3 + 5 \cdot 4$ and $(8 + 2)6 = 8 \cdot 6 + 2 \cdot 6$. The distributive law is used extensively in evaluating and solving equations and formulas.

EXAMPLE 1 For each of the following equations, indicate whether the commutative, associative, or distributive law is involved.

(a) $3.5 + 2.4 = 2.4 + 3.5$
(b) $1.002 \cdot 2 = 2 \cdot 1.002$
(c) $(6 + 24) + 8 = 6 + (24 + 8)$
(d) $(3 \cdot 5) \cdot 10 = 3 \cdot (5 \cdot 10)$
(e) $9(2 + 4) = 9 \cdot 2 + 9 \cdot 4$

Solution

(a) commutative because the order of the numbers is changed
(b) commutative because the order of the numbers is changed
(c) associative because the grouping of the numbers is changed
(d) associative because the grouping is changed
(e) distributive because multiplication is distributed over addition　■■■

Addition and Subtraction

The laws stated thus far are true for all real numbers and help to simplify operations with algebraic expressions. To clarify the four fundamental operations of addition, subtraction, multiplication, and division with signed numbers, more rules are needed. The following two rules are given for the operation of addition.

Adding Real Numbers

Rule 1 To add two real numbers with like signs, find the sum of their absolute values and attach their common sign to the sum.

Rule 2 To add two real numbers with unlike signs, find the difference of their absolute values and attach the sign of the number with the larger absolute value.

EXAMPLE 2 Perform the operation of addition.

(a) $3 + 8$　　(b) $(-4) + (-5)$　　(c) $(-10) + 3$
(d) $12 + (-8)$　　(e) $(-2.5) + (5.4)$

Solution

(a) $3 + 8 = 11$; since the numbers have the same sign, we add their absolute values and use their common sign, $+$.

(b) $(-4) + (-5) = -9$; both numbers have the same sign. Therefore, we add the numbers and use their common sign.

(c) $(-10) + (3) = -7$; the numbers have unlike signs. Therefore, we subtract their absolute values and use the sign of the larger.

(d) $12 + (-8) = 4$; the sum is positive because 12 is larger in absolute value.

(e) $(-2.5) + (5.4) = 2.9$; the sum is positive because 5.4 is larger in absolute value. ■■■

The operation of subtraction is not difficult if the following rule is applied carefully.

Subtracting Real Numbers

Rule 3   To subtract two real numbers, change the subtraction sign to an addition sign and change the sign of the subtrahend (the number to be subtracted). Then apply the rules of addition.

EXAMPLE 3   Perform the operation of subtraction.

(a) $18 - 20$ (b) $(-7) - 2$ (c) $21 - (-3)$
(d) $(-3.8) - (-12.9)$ (e) $(-25) - 10.5$

Solution

(a) $18 - 20 = 18 + (-20) = -2$; change to addition, change sign of subtrahend, apply rules of addition.

(b) $(-7) - 2 = -7 + (-2) = -9$

(c) $21 - (-3) = 21 + (+3) = 24$

(d) $(-3.8) - (-12.9) = (-3.8) + (+12.9) = 9.1$

(e) $(-25) - (10.5) = (-25) + (-10.5) = -35.5$ ■■■

Multiplication and Division

The next rule applies for multiplication or division of signed numbers.

Multiplication or Division of Real Numbers

Rule 4   The product or quotient of two real numbers with like signs is positive. The product or quotient of two real numbers with unlike signs is negative.

In other words, multiplying or dividing an even number of negative signs results in a positive answer; multiplying or dividing an odd number of negative signs results in a negative answer.

EXAMPLE 4 Perform the indicated operations.

(a) $5 + (-4)$ (b) $(-3)(-4)$ (c) $45 \div (-5)$
(d) $12 - (-4)$ (e) $(-9) + (-4)$ (f) $10 - 15$
(g) $(-2.5)(-1.8)(-3.6)$ (h) $-8.2 \div (-2)$

Solution

(a) 1   (b) 12   (c) -9 (d) 16
(e) -13 (f) -5 (g) -16.2 (h) 4.1   ■■■

A–2 EXERCISES

Perform the indicated operations.

1. $(-12) + 6$

2. $7 + 2$

3. $(-3) + (-14)$

4. $8 + (-12)$

5. $13 - (-5)$

6. $24 - 45$

7. $(-6) - 8$

8. $3(-4)$

9. $(-9)(6)$

10. $15 \div (-5)$

11. $(-28) \div (-7)$

12. $(-2.5) + 1$

13. $(-4.5)(3.1)$

14. $-100 \div 10$

15. $-10.5 - 2.5$

16. $(-9)(3)(-4)$

17. $10 - (-3) + 6$

18. $(-134.6) + 12.5 - 28.7$

19. $-5.3 + (-7.8) - (-1.3)$

20. $6 + (-8) - (-4) - 10$

For each of the following equations, identify the operational law being illustrated.

21. $5(x + y) = 5x + 5y$

22. $45(3 \cdot 8) = (45 \cdot 3)8$

23. $4 + 5 = 5 + 4$

24. $(9 + b)a = 9a + ba$

Solve the following problems.

25. In Denver, Colorado, the temperature at 3 P.M. was 12°C. By 10 P.M. it had dropped to -5°C. What is the degree difference between the two temperatures?

26. Temperature on the Celsius (C) scale is related to the Fahrenheit (F) scale by the formula $C = 5/9(F - 32)$.

Convert each of the following Fahrenheit temperatures to Celsius temperatures.

a. $F = -22$, find C.   b. $F = 77$, find C.
c. $F = -12.2$, find C.   d. $F = 48.9$, find C.

27. When a body is moved by a force in the line of action of the force, work is done. The formula for work (w) is $w =$ force (f) $\times$ distance (d).

a. $f = 100$ lb and $d = 3$ ft, find w.
b. $f = -60$ lb and $d = -6$ ft, find w.
c. $f = -22$ lb and $d = -2$ ft, find w.
d. $f = 34$ lb and $d = 12.5$ ft, find w.

28. Using the formula $F = 9C/5 + 32$, convert the following Celsius (C) temperatures to Fahrenheit (F) temperatures.

a. $C = 57$, find F.   b. $C = -2$, find F.
c. $C = -263$, find F.   d. $C = 145$, find F.

29. In many physics applications the sum of component vectors must be found. Find the sum of each of the following:

a. $\mathbf{A}_x = -24.0$ b. $\mathbf{B}_x = -283$
   $\mathbf{A}_y = 15.4$ $\mathbf{B}_y = -246$
c. $\mathbf{A}_x = 35.6$ d. $\mathbf{B}_x = -102.7$
   $\mathbf{A}_y = -10.8$ $\mathbf{B}_y = -20.3$

30. The algebraic sum of all currents at a point in a circuit must be zero. Currents toward a point are positive; currents away from it are negative. Find the sum of the currents for the circuit diagrams in Figures A–10 and A–11.

FIGURE A–10

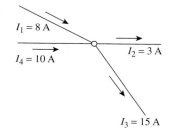

$I_1 = 8$ A
$I_4 = 10$ A
$I_2 = 3$ A
$I_3 = 15$ A

FIGURE A–11

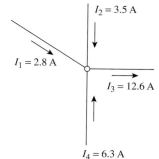

$I_2 = 3.5$ A
$I_1 = 2.8$ A
$I_3 = 12.6$ A
$I_4 = 6.3$ A

A–3

ZERO AND THE ORDER OF OPERATIONS

Zero is an important real number that is encountered quite often in algebraic and arithmetic operations. Also, the order of performing operations is an important topic in algebra and arithmetic. This section will present the rules and guidelines for both topics.

Operations with Zero

The four basic operations of addition, subtraction, multiplication, and division should be performed with zero as summarized below.

Operations with Zero

For any real number, the following statements are true:

Addition	$a + 0 = a$	and	$0 + a = a$
Subtraction	$a - 0 = a$	and	$0 - a = -a$
Multiplication	$a \cdot 0 = 0$	and	$0 \cdot a = 0$
Division	$0/a = 0$	and	$a/0 =$ undefined

As shown in the box, division by zero is undefined. The reason for this fact is given as follows: $6 \div 3 = 2$ because $6 = 3 \cdot 2$. Similarly, $0 \div 8 = 0$ because $0 = 8 \cdot 0$. However, applying this same reasoning to division by zero gives $3 \div 0 = x$, but we can find no real number x such that $0 \cdot x = 3$. Therefore, division by zero is undefined.

EXAMPLE 1 Evaluate each of the following:

(a) $0 - 5$ (b) $(-8) + 0$ (c) $4(0)$ (d) $\dfrac{0}{7}$ (e) $\dfrac{3}{0}$

Solution

(a) -5 by the subtraction rule
(b) -8 by the addition rule
(c) 0 by the multiplication rule
(d) 0 by the division rule
(e) undefined by the division rule

■ ■ ■

Order of Operations

Often in algebraic problems an expression or formula is encountered that involves a variety of operations and/or grouping symbols. The rules for order of operations are as follows.

Rules for Order of Operations

1. Simplify within grouping symbols.

2. Perform all multiplication and division, proceeding from left to right.

3. Perform all addition and subtraction, proceeding from left to right.

The first rule refers to grouping symbols. The most common symbols of grouping are parentheses ( ), brackets [ ], braces { }, and the fraction bar /. Use the rules to calculate the following example.

EXAMPLE 2   Calculate $10 - 3(5 + 1) \div 2 + 5$.

Solution   First, simplify within the parentheses.

$$10 - 3(5 + 1) \div 2 + 5 = 10 - 3(6) \div 2 + 5$$

Second, multiply and divide from left to right.

$$= 10 - (18) \div 2 + 5 \qquad \text{multiply}$$
$$= 10 - (9) + 5 \qquad \text{divide}$$

Third, add and subtract from left to right.

$$= (1) + 5 \qquad \text{subtract}$$
$$= 6 \qquad \text{add} \qquad \blacksquare\blacksquare\blacksquare$$

Within a given step from the previous box, remember to simplify from left to right. For example, in the expression $8 \div 2 \cdot 3 \div 6$, multiplication and division are done at the same time, moving from left to right as shown.

$$8 \div 2 \cdot 3 \div 6 = 4 \cdot 3 \div 6$$
$$= 12 \div 6$$
$$= 2$$

EXAMPLE 3   Calculate $12 \div 4 + 6(10 - 3) - 2$.

Solution   First, simplify within the parentheses.

$$12 \div 4 + 6(10 - 3) - 2 = 12 \div 4 + 6(7) - 2$$

Second, multiply and divide from left to right.

$$= (3) + 6(7) - 2 \qquad \text{divide}$$
$$= 3 + (42) - 2 \qquad \text{multiply}$$

Third, add and subtract from left to right.

$$= (45) - 2 \qquad \text{add}$$
$$= 43 \qquad \text{subtract} \qquad \blacksquare\blacksquare\blacksquare$$

Although the fraction bar is a grouping symbol, it is used to separate a numerator from a denominator denoting division. The next example illustrates the use of the fraction bar as well as the previous rules for order of operations.

EXAMPLE 4   Calculate $\dfrac{5 \cdot 2 - 3 \cdot 8}{10 - 3(4 - 2)}$.

Solution Simplify the numerator and then the denominator according to the rules for the order of operations.

$$\frac{5 \cdot 2 - 3 \cdot 8}{10 - 3(4 - 2)} = \frac{10 - 24}{10 - 3(4 - 2)} \qquad \text{multiply}$$

$$= \frac{-14}{10 - 3(4 - 2)} \qquad \text{subtract}$$

$$= \frac{-14}{10 - 3(2)} \qquad \text{parentheses}$$

$$= \frac{-14}{10 - 6} \qquad \text{multiply}$$

$$= \frac{-14}{4} \qquad \text{subtract}$$

$$= -3.5 \qquad \text{divide} \qquad \blacksquare\blacksquare\blacksquare$$

Try the problem in Example 4 with your calculator. Follow these steps.

$$\boxed{(}\ 5\ \boxed{\times}\ 2\ \boxed{-}\ 3\ \boxed{\times}\ 8\ \boxed{)}\ \boxed{\div}\ \boxed{(}\ 10\ \boxed{-}\ 3\ \boxed{\times}\ \boxed{(}\ 4\ \boxed{-}\ 2\ \boxed{)}\ \boxed{)}\ \boxed{=}$$

CAUTION ✦ When performing these operations on your calculator, remember to enter the parentheses.

EXAMPLE 5   Perform the following operations:

(a) $12 + 5(6 - 3 \cdot 4) - 7$ (b) $\dfrac{56 - 4(5.2 - 3.4)}{12.4 + 3(-1.2)}$

Solution

(a)
$$
\begin{aligned}
12 + 5(6 - 3 \cdot 4) - 7 &= 12 + 5(6 - 12) - 7 &&\text{multiply within parentheses} \\
&= 12 + 5(-6) - 7 &&\text{subtract within parentheses} \\
&= 12 - 30 - 7 &&\text{multiply} \\
&= -25 &&\text{subtract}
\end{aligned}
$$

(b)
$$\frac{56 - 4(5.2 - 3.4)}{12.4 + 3(-1.2)} = \frac{56 - 4(1.8)}{12.4 + 3(-1.2)} \qquad \text{parentheses}$$

$$= \frac{56 - 7.2}{12.4 - 3.6} \qquad \text{multiply}$$

$$= \frac{48.8}{8.8} \qquad \text{subtract}$$

$$= 5.5454 \ldots \qquad \text{divide} \qquad \blacksquare\blacksquare\blacksquare$$

A–3 EXERCISES

Perform the indicated operations.

1. $23 + 0$

2. $3(0)$

3. $0 - 12$

4. $6 - 0$

5. $0 - (-8)$

6. $\dfrac{0}{10}$

7. $\dfrac{-16}{0}$

8. $23 - 0$

9. $8 - 3(4)$

10. $3(0) + 2(-4)$

11. $12 \div 2 + 8$

12. $\dfrac{3(-6)}{2} - 3(-4)$

13. $\dfrac{-44}{11} + \dfrac{10 - 2}{0 - 4}$

14. $10 - (36 \div 6) + 3(-5)$

15. $\dfrac{(-12)(-2) - 7(3)}{2(5) + 9}$

16. $1.3 - 2(4.5) + (6.8)$

17. $\dfrac{0(-7) + 45 \div 9}{(-3.5)2 - 3(1.5)}$

18. $\dfrac{6.7 - 3.4(2.2) + 5.4}{(-45.1)(-3) - 23.4(-2)}$

19. $\dfrac{34.5}{(-.5)} + 12.4$

20. $34.5(-100) - 67(0) + 96 - (-4)$

Solve the following problems.

21. The perimeter of a rectangle with dimensions L and W can be found using the formula $P = 2L + 2W$.
  a. Find P, when $L = 40$ ft and $W = 8$ ft.
  b. Find P, when $L = 100$ m and $W = 25$ m.

22. The formula $v_2 = v_1 + a(t_2 - t_1)$ is often used in physics. Find v_2 when $a = 60$ ft/s^2, $t_1 = 2$ s, $t_2 = 4$ s, and $v_1 = 30$ ft/s.

23. Using the formula given in Exercise 22, find v_2 for $a = 12$ ft/s^2, $t_1 = 9$ s, $t_2 = 12$ s, and $v_1 = 25$ ft/s.

24. The area of a geometric figure called a trapezoid is found using the formula, $A = .5h(B + b)$. Find A if $h = 4.5$ in., $B = 5$ in., and $b = 3$ in.

25. Using the formula in Exercise 24, find A if $h = 12.6$ ft, $B = 10.5$ ft, and $b = 8.9$ ft.

A–4

EXPONENTS

One important process in algebra involves a way of expressing repeated multiplication by using exponents. This section will present the definitions and rules for this essential process.

Laws of Exponents

When two numbers are multiplied, each number is called a **factor.** For example, in the expression $2 \cdot 5 = 10$, 2 and 5 are both factors of the product 10. However, if one factor is multiplied repeatedly, such as $2 \cdot 2 \cdot 2 \cdot 2 = 16$, we need a shortcut notation for this. Exponential notation a^n represents a as the factor being repeated and n as the number of times it is multiplied. For example, $3 \cdot 3 \cdot 3 \cdot 3 \cdot 3 = 3^5$. In the expression a^n, the number a is called the **base,** the number n is called the **exponent,** and the expression is read as the "nth power of a."

EXAMPLE 1 Write the following multiplications in exponential form:

(a) $4 \cdot 4 \cdot 4$ (b) $(-1)(-1)(-1)(-1)$ (c) $x \cdot x$

Solution

(a) $4 \cdot 4 \cdot 4 = 4^3$ read "the third power of 4" or "four cubed"
(b) $(-1)(-1)(-1)(-1) = (-1)^4$ read "the fourth power of -1"
(c) $x \cdot x = x^2$ read "the second power of x" or "x squared" ■■■

NOTE ✦ A base written without an exponent is understood to have an exponent of 1. Thus, $x = x^1$.

EXAMPLE 2 Complete each of the following:

(a) Name the base and the exponent: $(-3)^4$ and x^5

(b) Write the following using exponents: $\dfrac{1}{2} \cdot \dfrac{1}{2} \cdot \dfrac{1}{2}$ and $a \cdot a$.

(c) Calculate: $\left(\dfrac{1}{4}\right)^2$, 5^3, $(-3)^3$, $(-3)^4$

(d) Write each value as a power of 10: 100,000 and 100

Solution

(a) -3 is the base and 4 is the exponent; x is the base and 5 is the exponent

(b) $\left(\dfrac{1}{2}\right)^3$; a^2

(c) $\left(\dfrac{1}{4}\right)\left(\dfrac{1}{4}\right) = \dfrac{1}{16}$

$$5^3 = 5 \cdot 5 \cdot 5 = 125$$
$$(-3)^3 = (-3)(-3)(-3) = -27$$
$$(-3)^4 = (-3)(-3)(-3)(-3) = 81$$

(d) 10^5; 10^2 ■ ■ ■

Rather than writing repeated factors to perform algebraic operations, the Laws of Exponents make computations involving exponents more efficient.

Laws of Exponents

For any real numbers a and b and any positive integers m and n, the following rules apply:

$$a^m \cdot a^n = a^{m+n} \qquad\qquad \text{[Rule 1–1]}$$

If the bases are the same, add exponents to multiply.

$$(ab)^m = a^m \, b^m \qquad\qquad \text{[Rule 1–2]}$$

The exponent applies to each factor.

$$(a^n)^m = a^{nm} \qquad\qquad \text{[Rule 1–3]}$$

Multiply exponents to raise to a power.

$$\dfrac{a^m}{a^n} = a^{m-n} \qquad \text{where } a \neq 0 \qquad \text{[Rule 1–4]}$$

If the bases are the same, subtract exponents to divide.

$$\left(\dfrac{a}{b}\right)^m = \dfrac{a^m}{b^m} \qquad \text{where } b \neq 0 \qquad \text{[Rule 1–5]}$$

The exponent applies to the numerator and denominator.

CAUTION ✦ Be careful in applying the Laws of Exponents. Rules 1–1 and 1–4 can be used only when the bases are the same.

EXAMPLE 3   Perform the indicated operations.

(a) $3^5 \cdot 3^2$ (b) $x^2 \cdot x^3 \cdot x^2$ (c) $\dfrac{2^8}{2^3}$ (d) $\dfrac{a^4}{a}$

Solution

(a) $3^5 \cdot 3^2 = 3^{5+2} = 3^7$ [Rule 1–1]

(b) $x^2 \cdot x^3 \cdot x^2 = x^{2+3+2} = x^7$ [Rule 1–1]

(c) $\dfrac{2^8}{2^3} = 2^{8-3} = 2^5$ [Rule 1–4]

(d) $\dfrac{a^4}{a} = a^{4-1} = a^3$ [Rule 1–4]   ■■■

Rules 1–2, 1–3, and 1–5 are the power laws. The use of parentheses in expressions where these rules apply is important to the simplification process.

EXAMPLE 4   Simplify the following:

(a) $(xy)^3$ (b) $(2^2)^3$ (c) $(x^3)^4$ (d) $\left(\dfrac{2}{5}\right)^3$ (e) $\left(\dfrac{x}{y}\right)^5$

Solution

(a) $(xy)^3 = x^3 y^3$ [Rule 1–2]

(b) $(2^2)^3 = 2^{2 \cdot 3} = 2^6$ [Rule 1–3]

(c) $(x^3)^4 = x^{3 \cdot 4} = x^{12}$ [Rule 1–3]

(d) $\left(\dfrac{2}{5}\right)^3 = \dfrac{2^3}{5^3}$ [Rule 1–5]

(e) $\left(\dfrac{x}{y}\right)^5 = \dfrac{x^5}{y^5}$ [Rule 1–5]   ■■■

Negative Exponents

Two special laws of exponents involve numbers with negative and zero exponents. The first is the law for negative exponents. Consider applying the division law [Rule 1–4] to the following example:

$$\frac{5^3}{5^5} = \frac{5 \cdot 5 \cdot 5}{5 \cdot 5 \cdot 5 \cdot 5 \cdot 5} = \frac{1}{5^2}$$

Law of Negative Exponents

For any real number a and integer m,

$$a^{-m} = \frac{1}{a^m} \qquad \text{(where } a \neq 0\text{)} \qquad \text{[Rule 1–6]}$$

EXAMPLE 5   Simplify and express each quotient with positive exponents:

(a) $\dfrac{5}{5^4}$ (b) $\dfrac{x^3}{x^7}$ (c) $\dfrac{t^5}{t^{12}}$

Solution

(a) $5^{1-4} = 5^{-3} = \dfrac{1}{5^3}$

(b) $x^{3-7} = x^{-4} = \dfrac{1}{x^4}$

(c) $t^{5-12} = t^{-7} = \dfrac{1}{t^7}$

■ ■ ■

All the laws of exponents given earlier apply to negative exponents also. You must pay particular attention to the rules for signed numbers to obtain the correct answer.

EXAMPLE 6   Perform the indicated operation.

(a) $6^{-2} \cdot 6^{-3}$ (b) $\dfrac{x^{-1}}{x^{-2}}$ (c) $\dfrac{(ab)^{-5}}{(a^2b^3)^{-1}}$

Solution

(a) $6^{-2} \cdot 6^{-3} = 6^{-2 + (-3)} = 6^{-5} = \dfrac{1}{6^5}$ [Rules 1–1, 1–6]

(b) $\dfrac{x^{-1}}{x^{-2}} = x^{-1 - (-2)} = x^{-1+2} = x$ [Rule 1–4]

(c) $\dfrac{(ab)^{-5}}{(a^2b^3)^{-1}} = \dfrac{a^{-5} \cdot b^{-5}}{a^{-2} \cdot b^{-3}} = a^{-5-(-2)}b^{-5-(-3)}$ [Rules 1–2, 1–4]

$= a^{-3}b^{-2} = \dfrac{1}{a^3b^2}$ [Rule 1–6]

■ ■ ■

Zero Exponent

The Law of Zero Exponent also involves the division law [Rule 1–4]. Consider the following example:

$$\frac{8^2}{8^2} = 8^{2-2} = 8^0$$

Since we know that any number divided by itself, except zero, equals 1, we can conclude that any nonzero number raised to the zero power is equal to 1.

Law of Zero Exponent

For any nonzero number a,

$$a^0 = 1 \qquad \text{[Rule 1–7]}$$

EXAMPLE 7   Evaluate the following:

(a) 10^0 (b) x^0 (c) $(-3)^0$

Solution

(a) $10^0 = 1$ [Rule 1–7]
(b) $x^0 = 1$ [Rule 1–7]
(c) $(-3)^0 = 1$ [Rule 1–7]   ∎∎∎

In the order of operation, exponential factors should be evaluated after simplifying within grouping symbols and before multiplying and dividing.

EXAMPLE 8   Calculate $5(2 + 4) - 4^2$.

Solution

$$5(2 + 4) - 4^2 = 5(2 + 4) - 16 \qquad \text{exponent}$$
$$= 5(6) - 16 \qquad \text{parentheses}$$
$$= 30 - 16 \qquad \text{multiply}$$
$$= 14 \qquad \text{subtract}$$ ∎∎∎

Before doing the exercises for this section, complete the following example.

EXAMPLE 9   Identify the base in each of the following exponential expressions:

(a) $4x^2$ (b) $-a^3$ (c) $(-3)^2$ (d) -3^2

Solution

(a) $4x^2$ means $4 \cdot x^2$; x is the base.
(b) $-a^3$ means $-1 \cdot a^3$; a is the base.
(c) $(-3)^2 = -3 \cdot -3 = 9$; -3 is the base.
(d) $-3^2 = -1 \cdot 3 \cdot 3 = -9$; 3 is the base.   ∎∎∎

A–4 EXERCISES

Simplify using the Laws of Exponents. Express results with positive exponents only.

1. $(-2)^5$

2. $3^0 \cdot 2^3$

3. $x \cdot x^3$

4. $(x^2)^3$

5. $\dfrac{10^5}{10^4}$

6. $\dfrac{8^2 \cdot 3^4}{8 \cdot 3^2}$

7. $\left(\dfrac{2}{5}\right)^2$

8. $\dfrac{x^5}{x^3}$

9. $\dfrac{2x^2}{x^5}$

10. $(2x^3)^3$

11. $\left(\dfrac{10}{y^2}\right)^3$

12. $(-2x)^5$

13. $\dfrac{(a^0 b^2)}{ab^6}$

14. $-8x^0$

15. $\dfrac{12s^0}{-6s^3}$

16. $(20x^2y)^{-1}$

17. $(-2)^4 \cdot x^{-1} \cdot x^3$

18. $\dfrac{x^{-3}}{x^4}$

19. $\dfrac{10^{-3} \cdot 10}{10^6 \cdot 10^{-4}}$

20. $\left(\dfrac{3a^2 b}{2ab^3}\right)^{-2}$

Evaluate the given expressions.

21. $3(-8)^2 - (-6)$

22. $8 \div (-2) - (-1)^3$

23. $(-5)^2 + (-3)(20) \div 10$

24. $10^3 - (-2)10^4$

Solve the given problems. Round to three decimal places when necessary.

25. The formula for the total surface area of a cylinder is $2\pi r^2 + 2\pi rh$. Find the total surface area of a cylinder with a diameter of 4 m and a height of 12 m.

26. If a body in space moves from location $d_0 = 500$ ft to a location $d = 2,800$ ft, with a velocity $v_0 = 245$ ft/s in time $t = 5$ s, what local gravity is operating? Use the formula
$$g = \frac{d - d_0 - v_0 t}{.5t^2}.$$

27. The volume of a sphere is $V = (4\pi r^3)/3$, where r is the radius. What is the volume of a sphere with a radius of 1.5 m?

28. Using the formula given in Exercise 27, find the volume of a sphere with a radius of 12 ft.

29. Prove that $(a + b)^2 \neq a^2 + b^2$ using $a = 3$ and $b = 2$.

30. The volume of a cube is $V = e^3$, where e is the length of a side. What is the volume of a cube whose sides are 7.2 cm long?

A–5

SCIENTIFIC NOTATION

Numbers such as 16,300,000,000 or 0.000000015 are often encountered in technical work. However, actually working with these numbers in this form is cumbersome or even impossible with some calculators. Scientific notation is a technique for writing such very large and very small numbers in a much more concise form.

A number is said to be in **scientific notation** when it is written as the product of a number between 1 and 10 and a power of 10. Stated formally, a number is expressed in scientific notation when it is in the form $N \times 10^k$ where $1 \leq N < 10$ and k is an integer.

To write a number in scientific notation, we first determine the factor N between 1 and 10. To do this we write the original number with a decimal point after the leftmost nonzero digit. Second, we must find the value of k. The value of k, the exponent of 10, is equal to the number of places the decimal point was moved from its original position. If the decimal point was moved to the left, k is positive. If the decimal point was moved to the right, k is negative.

EXAMPLE 1 Write the following numbers in scientific notation:

(a) 16,300,000,000 (b) 0.000000015

Solution

(a) $16,300,000,000 = 1.63 \times 10^{10}$ ($N = 1.63$ and $k = 10$)

(b) $0.000000015 = 1.5 \times 10^{-8}$ ($N = 1.5$ and $k = -8$) ∎

To change a number from scientific notation back to standard decimal notation, reverse the process given above.

EXAMPLE 2 Rewrite each of the following in standard form:

(a) 3.58×10^8 (b) 2.3×10^{-6}

Solution

(a) $3.58 \times 10^8 = 3.58000000$

$= 358,000,000$

(b) $2.3 \times 10^{-6} = 0000002.3$

$= 0.0000023$ ∎

Multiplication and Division

Multiplying and dividing very large or very small numbers is much easier when the numbers are expressed in scientific notation because the Laws of Exponents can be used.

EXAMPLE 3   Calculate $(123{,}000{,}000)(0.0000083)$.

Solution

Step 1:   Rewrite each number in scientific notation.
$$(1.23 \times 10^8)(8.3 \times 10^{-6})$$

Step 2:   Multiply the numbers 1.23×8.3.
$$1.23 \times 8.3 = 10.209$$

Step 3:   Use the Laws of Exponents on the powers of 10.
$$10^8 \times 10^{-6} = 10^{8-6} = 10^2$$

Step 4:   Combine the results of Steps 2 and 3.
$$10.209 \times 10^2 = 1{,}020.9$$ ■■■

EXAMPLE 4   Calculate $\dfrac{(0.000043)(9{,}000{,}000)}{(0.0004)}$.

Solution

Step 1: $\dfrac{(4.3 \times 10^{-5})(9.0 \times 10^6)}{(4.0 \times 10^{-4})}$ convert to scientific notation

Step 2: $\dfrac{4.3 \times 9.0}{4.0} = 9.675$ perform operation on numbers

Step 3: $\dfrac{10^{-5} \times 10^6}{10^{-4}} = 10^5$ perform operation on powers of 10

Step 4: $9.675 \times 10^5 = 967{,}500$ combine results of Steps 2 and 3   ■■■

A–5 EXERCISES

Rewrite each number in scientific notation.

1. 8,956,000

2. 0.000345

3. −450,000

4. 0.000000236

5. 6,900,000,000

6. 12

7. The normal red blood cell count is 5,000,000 per mm^3.

8. The thermal conductivity of wood is 0.00024 cal/cm·s.

Rewrite each number in standard decimal notation.

9. 1.23×10^7

10. -8.9×10^{-5}

11. 5.4×10^9

12. 6.7×10^{-8}

13. 2.0×10^{-3}

14. 4.5×10^{10}

15. The measure of the temperature at which hydrogen would become an ionized plasma is given by 1.05×10^5 K.

16. The radius of a typical atom is 10^{-8} cm.

Perform the following calculations by rewriting each number in scientific notation and using the Laws of Exponents to calculate the answers. Write each answer in scientific notation.

17. $\dfrac{(0.00000098)(42.4)}{2{,}300}$

18. $(89{,}000{,}000)(1{,}200{,}000)$

19. $(200)^4$

20. $(0.50200)(0.00023)(8{,}000{,}000)$

21. $\dfrac{6{,}300{,}000{,}000}{8{,}230{,}000}$

22. $\dfrac{(450,000)(0.004500)}{(3,400)(0.000002)}$

23. The sun exerts a force of about

$$40,000,000,000,000,000,000,000,000,000 \text{ N}$$

on the earth, which travels 958,000,000,000 m in its annual orbit of the sun. To compute how much work is done by the sun on the earth during the year the following

computation would be necessary: $(4.0 \times 10^{28})(9.58 \times 10^{11})$. Perform this computation and express your answer in scientific notation. (Final units are joules.)

24. An electron in a beam of a television tube has a mass of 9.1×10^{-31} kg. Its velocity is 30,000,000 m/s. Find its kinetic energy, *KE* by performing this computation: $0.5(9.1 \times 10^{-31})(3.0 \times 10^7)(3.0 \times 10^7)$. (Final units are joules.)

A–6

ROOTS AND RADICALS

To solve many equations and to evaluate various formulas, we sometimes find it necessary in algebra to take the "square root" of a value or of an expression. This section will explain square roots and other roots that are often found in technical applications.

Terminology

When explaining exponents in Section A–4, we discussed repeated multiplication of a factor *n* times. For example, $5 \cdot 5 = 5^2 = 25$ or $2 \cdot 2 \cdot 2 = 2^3 = 8$. Now let's consider this process starting with the product and working backward. Given the product of 25, what number multiplied twice gives 25? In other words, what is the second root or square root of 25? For the second example, we would ask what is the third root or cube root of 8? This process can be generalized by the following definition: The **nth root** of the number *a*, written $\sqrt[n]{a}$, is the number which, when used as a factor *n* times, gives *a* as the product. The symbol $\sqrt{}$ is called the **radical sign,** *a* is the **radicand,** and *n* is the **index.** When the index is not written, it is assumed to be 2, indicating square root.

EXAMPLE 1 Simplify the following:

(a) $\sqrt{36}$ (b) $\sqrt[3]{27}$ (c) $\sqrt[4]{16}$

Solution

(a) $\sqrt{36} = 6$ because $6 \cdot 6 = 6^2 = 36$
(b) $\sqrt[3]{27} = 3$ because $3 \cdot 3 \cdot 3 = 3^3 = 27$
(c) $\sqrt[4]{16} = 2$ because $2 \cdot 2 \cdot 2 \cdot 2 = 2^4 = 16$ ∎∎∎

Technically, $\sqrt{25}$ equals both $+5$ and -5 because $+5 \cdot +5 = 25$ and $-5 \cdot -5 = 25$. To avoid confusion with which root to use, we define $\sqrt[n]{a}$ as the **principal nth root** of *a*, which is $+a$. In this section we will always take the principal root.

EXAMPLE 2 Simplify the following:

(a) $\sqrt[4]{81}$ (b) $\sqrt[3]{-27}$ (c) $\sqrt{-16}$

Solution

(a) $\sqrt[4]{81} = 3$ because $(3)^4 = 81$
(b) $\sqrt[3]{-27} = -3$ because $(-3)^3 = -27$
(c) $\sqrt{-16} =$ no real solution since no real number squared gives -16 ∎∎∎

NOTE ✦ On a calculator, the $\boxed{\sqrt{x}}$ key is used for square roots and the $\boxed{\sqrt[y]{x}}$ or $\boxed{x^y}$ are usually used to find other roots. A calculator will always display the principal root, and most calculators will not accept a negative radicand, regardless of the index.

Simplifying Radicals

To simplify radicals, you need an important property of roots; that is,

$$\sqrt{ab} = \sqrt{a} \cdot \sqrt{b} \text{ where } a, b \geq 0$$

This property is usually necessary when the radicand does not have an integral root. For example, when you take the $\sqrt{8}$, there is no integral value that when squared will result in 8. The radical may be rewritten using the above property.

$$\sqrt{8} = \sqrt{(4)(2)} = \sqrt{4} \cdot \sqrt{2} = 2\sqrt{2}$$

The procedure to follow when simplifying a radical is to use the above rule to remove or "take out" the square root of any perfect square factor in the radicand. Try this as you simplify the following examples.

EXAMPLE 3 Simplify the following:

(a) $\sqrt{12}$ (b) $\sqrt{125}$ (c) $\sqrt{32}$

Solution

(a) $\sqrt{12} = \sqrt{(4)(3)} = \sqrt{4} \cdot \sqrt{3} = 2\sqrt{3}$
(b) $\sqrt{125} = \sqrt{(25)(5)} = \sqrt{25} \cdot \sqrt{5} = 5\sqrt{5}$
(c) $\sqrt{32} = \sqrt{(16)(2)} = \sqrt{16} \cdot \sqrt{2} = 4\sqrt{2}$ ■ ■ ■

Some radicals cannot be simplified, such as $\sqrt{2}$ and $\sqrt{13}$. These radicals have radicand values that cannot be factored so that one of the factors is a perfect square.

Simplification of radicals allows us to state an exact value of the radical. Of course, the calculator can always be used to find decimal approximations of radicals. Try taking the $\sqrt{5}$ on a calculator. As you will see, $\sqrt{5} = 2.236067977 \ldots$.

NOTE ✦ Recall that such numbers are classified as irrational numbers and an exact decimal value does not exist for these numbers.

Order of Operations

There are two important facts that should be stressed concerning radicals and the order of operations. The first has to do with the radical symbol; it is a grouping symbol and all operations under the radical should be performed before the root is extracted or simplified. The second important fact pertains to the order in which radicals should be evaluated when combined with other operations. Radicals are simplified in the same step as exponents and should be done before multiplication, division, addition, or subtraction.

EXAMPLE 4 Simplify the following:

(a) $\sqrt{7 + 3^2}$ (b) $\sqrt{25} - 3(2)^2 + 8$ (c) $-\sqrt{40}$ (d) $\sqrt[3]{-8} + 3^2$

Solution

(a) $\sqrt{7+9} = \sqrt{16} = 4$
(b) $5 - 3(4) + 8 = 5 - 12 + 8 = -7 + 8 = 1$
(c) $-\sqrt{(4)(10)} = -\sqrt{4} \cdot \sqrt{10} = -2\sqrt{10}$
(d) $-2 + 9 = 7$

■ ■ ■

A–6 EXERCISES

Simplify each expression.

1. $\sqrt{81}$ **2.** $\sqrt{144}$

3. $\sqrt[5]{32}$ **4.** $\sqrt[4]{16}$

5. $\sqrt[3]{8}$ **6.** $\sqrt{\dfrac{4}{25}}$

7. $\sqrt{\dfrac{16}{25}}$ **8.** $\sqrt[3]{\dfrac{-1}{8}}$

9. $\sqrt{\dfrac{1}{4}}$ **10.** $\sqrt[5]{-243}$

11. $\sqrt{32}$ **12.** $\sqrt{45}$
13. $\sqrt{2,500}$ **14.** $\sqrt[5]{-32}$

15. $\sqrt{24}$ **16.** $\sqrt{\dfrac{9}{49}}$

Evaluate the following using the order of operations rules.

17. $\sqrt{4 + 3(7)}$ **18.** $\sqrt{16} - 3(2)^2$

19. $\dfrac{\sqrt{100} - 8}{10}$ **20.** $\dfrac{\sqrt[3]{8}}{2}$

21. $\dfrac{\sqrt[3]{-1,000} - 3(-4)^3}{(-1)^3 + \sqrt{4}}$

Use your calculator to find the following roots to the nearest hundredth.

22. $\sqrt{45}$ **23.** $\sqrt{\dfrac{1}{2}}$

24. $\sqrt{15}$ **25.** $\sqrt{4.8}$
26. $\sqrt{13.001}$ **27.** $\sqrt[3]{12}$

28. $\sqrt[5]{95}$ **29.** $\sqrt[3]{2,400}$
30. $\sqrt{0.002}$

Solve each of the following. (Round all answers to hundredths.)

31. Find the length of the diagonal of the rectangle in Figure A–12 using the Pythagorean Theorem.

$$c = \sqrt{a^2 + b^2}$$

FIGURE A–12

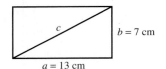

32. Find the length of the diagonal of a 4-meter square.

33. Find the length of a side (e) of a cube with a volume of 64 in³ ($V = e^3$).

34. This calculation is necessary for finding the length of an unknown side of an oblique triangle:

$$c = \sqrt{20.4^2 + 12.3^2 - 2(20.4)(12.3)(0.5)}$$

Find the unknown side, c.

35. This calculation is necessary for finding the magnitude of a vector when its two component vectors are known:

$$\mathbf{v} = \sqrt{2.2^2 + 4.3^2}$$

Find the magnitude of the vector, $\mathbf{v}$.

A–7

ADDITION AND SUBTRACTION OF ALGEBRAIC EXPRESSIONS

Thus far we have worked mostly with the algebraic operations of addition, subtraction, multiplication, division, and taking roots. In this section we will extend our rules for working with literal numbers to include algebraic expressions. The rules discussed previously also apply to operations with algebraic expressions.

Terminology

Algebraic expressions are composed of **terms** that are literal expressions separated by plus and minus signs. Since a term may contain several factors, be careful not to confuse terms with factors because the rules are not the same.

EXAMPLE 1 Identify the terms in the following expressions:

(a) $2x + 6xy - 4y$ (b) $a - \dfrac{2}{b}$

Solution

(a) $2x + 6xy - 4y$ is an algebraic expression with three terms: $2x$, $6xy$, and $-4y$.

(b) $a - \dfrac{2}{b}$ is an algebraic expression with two terms: a and $-\dfrac{2}{b}$ ■■■

Algebraic expressions are named according to the number of terms that they contain. An expression with only one term is called a **monomial** expression, a two-term expression is called a **binomial,** and a three-term expression is called a **trinomial.** Many expressions contain more than three terms, and these expressions are called **multinomial** expressions. In fact, any expression containing two or more terms may be called a multinomial.

EXAMPLE 2 Identify the following expressions by the number of terms.

(a) $2x$ (b) $8x - y$ (c) $2xy + 3x - 5$ (d) $-3a + b - 5c + 16$

Solution

(a) $2x$ is a monomial expression with one term $2x$.
(b) $8x - y$ is a binomial expression with terms $8x$ and $-y$.
(c) $2xy + 3x - 5$ is a trinomial expression.
(d) $-3a + b - 5c + 16$ is a multinomial expression containing 4 terms. ■■■

As seen in Example 2, a term can be composed of both literal and numerical factors. Any set of factors in a term is the **coefficient** of the remaining factors. For example, the expression $3x^2y$ contains three factors. The coefficient of y is $3x^2$; while the coefficient of 3 is x^2y. A numerical factor is called the **numerical coefficient** of the term. In our previous example, 3 is the numerical coefficient. When a numerical coefficient is not written, it is understood to be 1 or -1. Terms are called **like** if they differ only in their numerical coefficients. Like terms have exactly the same variables and corresponding variables have the same exponents. For example, in the expression $3x + 5a^2 - 2xy + 4x + a^2 + 4y$, the like terms are $3x$ and $4x$, $5a^2$ and a^2, but $-2xy$ and $4y$ are not like terms.

To simplify an algebraic expression, all like terms should be combined. This is accomplished by adding or subtracting the numerical coefficients and leaving the variable unchanged.

EXAMPLE 3 Simplify the following:

(a) $2x + 5x - x$ (b) $-5a^2 + a^2$

Solution

(a) $2x + 5x - x = (2 + 5 - 1)x = 6x$ (All three terms are like, so we combine the numerical coefficients and leave the variable x unchanged.)
(b) $-5a^2 + a^2 = (-5 + 1)a^2 = -4a^2$ ■■■

CAUTION ✦ Be careful when performing subtraction. Group the subtrahend in parentheses to perform the subtraction correctly.

EXAMPLE 4   Perform the indicated operation.

(a) $(2x + 5y) + (3x - 4y)$ (b) $(5a - b) - (4a + 6b - 8c)$

Solution

(a) $(2x + 5y) + (3x - 4y) = 2x + 5y + 3x - 4y$ remove parentheses
$$= (2 + 3)x + (5 - 4)y$$ combine like terms
$$= 5x + y$$

(b) $(5a - b) - (4a + 6b - 8c)$
$$= 5a - b - 4a - 6b + 8c$$ remove parentheses
$$= (5 - 4)a + (-1 - 6)b + 8c$$ combine like terms
$$= a - 7b + 8c$$ ■■■

When expressions are added, the parentheses may just be dropped and like terms combined. However, when subtracting one expression from another, additional steps must be taken. When a grouping symbol is preceded by a subtraction sign, you must change the sign of each term inside the symbol to remove the grouping symbol. Review the following example for clarification of this principle.

EXAMPLE 5   Subtract $(4x - 5y + 10) - (3x + 2y - 8)$.

Solution

Step 1:   Remove parentheses and change signs.
$$4x - 5y + 10 - 3x - 2y + 8$$

Step 2:   Group like terms together.
$$4x - 3x - 5y - 2y + 10 + 8$$

Step 3:   Combine like terms.
$$(4 - 3)x + (-5 - 2)y + (10 + 8)$$

Step 4:   Simplify the answer.
$$x - 7y + 18$$ ■■■

When you combine algebraic expressions, remember always to combine only like terms. Try the following example before starting the exercises.

EXAMPLE 6   Simplify the following:

(a) $10x - 5x + x$ (b) $x^2 + x$
(c) $(2a + 3b) + (6a + b)$ (d) $(2s - 4t) - (s + 7t)$
(e) $(5 - 3x + 8y) - (-3 + x)$

Solution

(a) $(10 - 5 + 1)x = 6x$

(b) $x^2 + x$; the two terms are not like and cannot be combined

(c) $2a + 6a + 3b + b = (2 + 6)a + (3 + 1)b$
$$= 8a + 4b$$

(d) $2s - 4t - (s + 7t) = 2s - s - 4t - 7t$
$$= (2 - 1)s + (-4 - 7)t$$
$$= s - 11t$$

(e) $5 - 3x + 8y + 3 - x = -3x - x + 8y + 5 + 3$
$$= (-3 - 1)x + 8y + 8$$
$$= -4x + 8y + 8$$

■ ■ ■

EXERCISE A–7

Simplify the following algebraic expressions.

1. $2x + 4x - x$

2. $3y - 5y + 6y$

3. $a + 3a - 2a$

4. $-2s + 3s - 5s$

5. $12p - 3p - 9p$

6. $5x^2 - 8x^2$

7. $9xz + 4xz - 2xz$

8. $2y - 5y + 4y + 2y + 4y$

9. $2a - 3b + a - 4b + 5$

10. $3t^2 - 2t + 4t^2$

Perform the indicated operations.

11. $(5x - 3y) + (2x + 8y)$

12. $(x^2 + 2y^2) + (3x^2 + y^2)$

13. $(a + 2b) - (2a - 3b)$

14. $(3t - 5s) - (-2t + 2s)$

15. $(x - 2y + 6) + (2x - y + 6)$

16. $(2x + y - 10) + (5x + 4y + 2)$

17. $(x^2 + y^2 + x - y) - (y + x - 2y^2)$

18. $(3 - 2xy + 2x^2 + y) - (xy + 5x^2 - 3y + 8)$

19. $3(2m + 5n) - 4(6m - 2n)$

20. $-2(x - y + 5) + 5(3x + 2y - 6)$

Solve the following problems.

21. In computing the slope of a roadway, a surveyor obtains the following expression:

$$(m - 2) + m - (2m + 5) - (1 - 4m)$$

Simplify the expression.

22. The expression $2(2w - 2) + 2w$ describes the perimeter of a certain rectangle. Simplify the expression.

23. The expression $0.144x + 0.1034(2,000 - x)$ represents the total interest earned from two investments. Simplify the expression.

24. Many times subscripted variables are used in expressions, such as $I_1 + 2I_2 - (3I_2 + 3)$. Simplify the expression.

25. In finding the current through an electric current circuit, an electrical engineering student is confronted with the expression:

$$(2I_1 + I_2) - (3I_1 - I_2) + (I_2 - I_1)$$

Simplify the expression.

A–8

MULTIPLYING POLYNOMIALS

The process of multiplying polynomials can be broken into three categories—multiplying by a monomial, multiplying binomials, and multiplying polynomials. Each of these categories is discussed in this section.

Multiplying by a Monomial

First, we will discuss multiplying monomials. Multiply the coefficients and then multiply variables by applying the Laws of Exponents. The next two examples illustrate this procedure.

EXAMPLE 1 Multiply $(2x^2y^3)(-6xy^4z^2)$.

Solution Multiplying the numerical coefficients gives $(2)(-6) = -12$. Next, we multiply the variables by adding exponents on like bases. This gives $x^2 \cdot x = x^3$; $y^3 \cdot y^4 = y^7$; and z^2 remains unchanged. The product is $-12x^3y^7z^2$. ■■■

EXAMPLE 2 Multiply $(-8m^2)(2m^3n^4)(-m^4n)$.

Solution First, we multiply the numerical coefficients.

$$(-8)(2)(-1) = 16$$

Next, we multiply the variables

$$(-8\underbrace{m^2)(2m^3n^4)(-m^4}n) = m^2(m^3)(m^4)$$
$$= m^{2+3+4}$$
$$= m^9$$
$$(-8m^2)(2m^3\underbrace{n^4)(-m^4}n) = n^4(n)$$
$$= n^{4+1}$$
$$= n^5$$

The product is $16m^9n^5$. ■■■

Next, we will multiply a polynomial by a monomial. We apply the Distributive Property to multiply the monomial and each term in the polynomial. The next two examples illustrate this procedure.

EXAMPLE 3 Multiply $2xy^2(6xy - 7x^2y^3 + 6)$.

Solution First, apply the Distributive Property by multiplying $2xy^2$ by each term of the polynomial inside the parentheses. Writing out this product gives

$$2xy^2(6xy - 7x^2y^3 + 6) \qquad \text{or} \qquad 2xy^2(6xy) + 2xy^2(-7x^2y^3) + 2xy^2(6)$$

Second, multiply monomial factors using the techniques given previously for monomial multiplication.

$$12x^2y^3 - 14x^3y^5 + 12xy^2$$

Since none of the terms are alike, the product is

$$12x^2y^3 - 14x^3y^5 + 12xy^2$$ ■■■

EXAMPLE 4 Multiply $-5a^2b^3c(4ac^3 + 9a^2bc^2)$.

Solution Multiply $-5a^2b^3c$ by each term of the polynomial. Writing out this product gives

$$(-5a^2b^3c)(4ac^3) + (-5a^2b^3c)(9a^2bc^2)$$

Multiplying monomial factors gives the product of

$$-20a^3b^3c^4 - 45a^4b^4c^3$$ ■■■

Multiplying Binomials

There are several techniques for multiplying binomials; however, the most frequently used method is called FOIL. FOIL is based on applying the Distributive Property to binomials, and the letters in FOIL represent the order in which you multiply terms (see Figure A–13). The letters in FOIL represent

- F: multiply the *first* terms in each set of parentheses
- O: multiply the *outside* terms in each set of parentheses
- I: multiply the *inside* terms in each set of parentheses
- L: multiply the *last* terms in each set of parentheses

Multiplying binomials using FOIL is also based on multiplying monomials. When you multiply the first elements in each set of parentheses, you are multiplying two monomial factors.

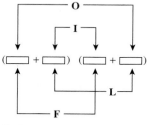

FIGURE A–13

EXAMPLE 5 Multiply $(7x - 8y)(4x - 5y)$.

Solution Figure A–14 shows the terms to multiply:

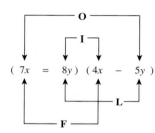

F gives $(7x)(4x) = 28x^2$
O gives $(7x)(-5y) = -35xy$
I gives $(-8y)(4x) = -32xy$
L gives $(-8y)(-5y) = 40y^2$

FIGURE A–14

FOIL gives $28x^2 - 35xy - 32xy + 40y^2$. After combining like terms, the product is $28x^2 - 67xy + 40y^2$. ■■■

Multiplying Polynomials

Since FOIL applies only to multiplying binomials, we must use another technique when any factor is not a binomial. We will use the Distributive Property to distribute each term of one polynomial over the second polynomial. The next example illustrates this process.

EXAMPLE 6 Multiply $(2a^3 - a^2 + 5a - 1)(a + 6)$.

Solution To multiply these polynomials, we multiply each term in the first polynomial by each term in the second polynomial. Since multiplying by the binomial is easier, we will distribute $a + 6$ over the polynomial $(2a^3 - a^2 + 5a - 1)$.

$$a(2a^3 - a^2 + 5a - 1) + 6(2a^3 - a^2 + 5a - 1)$$

Multiplying gives the following result:

$$a(2a^3 - a^2 + 5a - 1) + 6(2a^3 - a^2 + 5a - 1)$$
$$2a^4 - a^3 + 5a^2 - a + 12a^3 - 6a^2 + 30a - 6$$

Last, we combine like terms to yield the following product:

$$2a^4 + 11a^3 - a^2 + 29a - 6$$

■■■

A–8 EXERCISES

Find each of the following products.

1. $4a(3a^2b)$

2. $(-4xy)(-x^3y^2)$

3. $(-12s^5)(2s)$

4. $(-3x)(5x^8)(x^3)$

5. $2x^2(x^2 - 3y)$

6. $-4a^3(a^2 + a - 12)$

7. $-mn(mn - m + n)$

8. $5s^4t^2(-4s^3 + 6st)$

9. $x^4(x^5 + x^3 - x)$

10. $-23xy(2xy + 4)$

11. $(x - 3)(x + 4)$

12. $(y + 5)(y + 4)$

13. $(2a - 4b)(a - b)$

14. $(3x + 6)(2x - 4)$

15. $(3s^2 + t)(5s^2 - t)$

16. $(w - 3)(w + 3)$

17. $(2xy)^2(3x)^3$

18. $(-6a^3)^2(-3)$

19. $(2x + 3y)^2$

20. $(12a - 2b)^2$

21. $(x - y)(x^2 + xy - y^2)$

22. $(2s + t)(s^3 - 2s + 4)$

23. $(x - 3)(x^2 - 2x - 2)$

24. $(2x + y)(3x - y)(4x - 1)$

25. $(6y + 2)(2y - 4)(y + 1)$

26. $(x - 2)^3$

Solve the following problems.

27. In finding the maximum power in an electric circuit, the expression $(R + 2r)^2 - r(2R + r)$ might arise. Multiply and simplify this expression.

28. In determining the area of a special rectangular plot of land, the expression $(200x + 5)(100 + 3x)$ is found. Perform the indicated multiplication.

A–9

POLYNOMIAL DIVISION

Before discussing polynomial division, we need to review the Laws of Exponents that relate to division. These laws are summarized below.

$$\frac{a^m}{a^n} = a^{m-n} \quad \text{if } m > n$$

$$\frac{a^m}{a^n} = \frac{1}{a^{n-m}} \quad \text{if } m < n$$

Since we do not want negative exponents, always subtract the smaller exponent from the larger and write the result in the part of the fraction where the larger exponent appeared. Division of polynomials can be placed into two categories—division by a monomial and division by a polynomial.

Division by a Monomial

Division by a Monomial

To divide a polynomial by a monomial,

1. Divide each term of the polynomial by the monomial.

2. Divide or reduce the numerical coefficients.

3. Divide the variables using the Laws of Exponents.

EXAMPLE 1 Divide $\dfrac{24x^2y^3 - 9x^3y^2 + 27xy}{-3xy}$.

Solution   To divide a polynomial by a monomial, divide each term of the polynomial by the monomial. Writing each term of the polynomial over the monomial gives

$$\frac{24x^2y^3}{-3xy} - \frac{9x^3y^2}{-3xy} + \frac{27xy}{-3xy}$$

Dividing each term using monomial division gives the quotient as

$$-8xy^2 + 3x^2y - 9$$

∎

EXAMPLE 2   Divide $\dfrac{54x^2y^4 + 6x^5y^2 - 36xy^5}{9x^3y^2}$.

Solution   Divide $9x^3y^2$ into each term of the polynomial.

$$\frac{54x^2y^4}{9x^3y^2} + \frac{6x^5y^2}{9x^3y^2} - \frac{36xy^5}{9x^3y^2}$$

The quotient is

$$\frac{6y^2}{x} + \frac{2x^2}{3} - \frac{4y^3}{x^2}$$

∎

Division by a Polynomial

Division by a polynomial requires a different process than the one used for division by a monomial. Division by a polynomial is very similar to long division in arithmetic. As a matter of fact, you use the same procedure in dividing by a polynomial as you use for long division in arithmetic. Compare the examples from arithmetic and algebra that follow.

$$\begin{array}{r} 2 \\ 36{\overline{\smash{\big)}\,785}} \end{array}$$

divide 36 into 78; as a trial divisor divide 3 into 7

$$\begin{array}{r} x \phantom{{}-3x-10} \\ x+2{\overline{\smash{\big)}\,x^2-3x-10}} \end{array}$$

divide $x + 2$ into $x^2 - 3x$; as a trial divisor divide x into x^2

$$\begin{array}{r} 2 \\ 36{\overline{\smash{\big)}\,785}} \\ 72 \end{array}$$

multiply 36 by 2

$$\begin{array}{r} x \phantom{{}-3x-10} \\ x+2{\overline{\smash{\big)}\,x^2-3x-10}} \\ x^2 + 2x \end{array}$$

multiply $x + 2$ by x

$$\begin{array}{r} 2 \\ 36{\overline{\smash{\big)}\,785}} \\ -72 \\ \hline 65 \end{array}$$

subtract and bring down the next number

$$\begin{array}{r} x \phantom{{}-3x-10} \\ x+2{\overline{\smash{\big)}\,x^2 \ominus 3x-10}} \\ \ominus\, x^2 + 2x \\ \hline -5x - 10 \end{array}$$

subtract by changing signs on the bottom polynomial and add; bring down the next term

$$\begin{array}{r} 21 \\ 36{\overline{\smash{\big)}\,785}} \\ -72 \\ \hline 65 \\ -36 \\ \hline 29 \end{array} + \frac{29}{36}$$

repeat Steps 1, 2, and 3 until the last number has been used

$$\begin{array}{r} x - 5 \\ x+2{\overline{\smash{\big)}\,x^2 \ominus 3x-10}} \\ \ominus\, x^2 + 2x \\ \hline -5x \ominus 10 \\ \oplus\,{-5x} \ominus 10 \\ \hline 0 \end{array}$$

repeat Steps 1, 2, and 3 until the last term has been used

In algebraic long division as in arithmetic, write the remainder as a fraction, writing the remainder as the numerator of the fraction and the divisor as the denominator of the fraction. Always precede the remainder with an addition sign.

Polynomial Division

In dividing by a polynomial, you must perform the following steps before dividing:

1. Always arrange both polynomials in descending powers. This means place the variable with the largest exponent first, then the term with the second largest exponent next, and so on.

2. Place a zero numerical coefficient in front of any omitted variables. This acts as a placeholder so you will subtract only like terms.

EXAMPLE 3   Divide $6x^2 + 13x - 5$ by $3x - 1$.

Solution   Since we are dividing by a polynomial, we must use long division. Each polynomial is arranged in descending powers and all powers are present.

$$
\begin{array}{r}
2x \\
3x - 1 \overline{)6x^2 + 13x - 5}
\end{array}
\qquad \text{divide } 3x - 1 \text{ into } 6x^2 + 13x
$$

$$
\begin{array}{r}
2x \\
3x - 1 \overline{)6x^2 + 13x - 5} \\
\ominus 6x^2 \oplus 2x \\
\hline
15x - 5
\end{array}
\qquad
\begin{array}{l}
\text{multiply } 3x - 1 \text{ by } 2x, \text{ subtract and bring} \\
\text{down the next term}
\end{array}
$$

$$
\begin{array}{r}
2x + 5 \\
3x - 1 \overline{)6x^2 + 13x - 5} \\
\ominus 6x^2 \oplus 2x \\
\hline
15x - 5 \\
\ominus 15x \oplus 5 \\
\hline
0
\end{array}
\qquad
\begin{array}{l}
\text{divide } 3x - 1 \text{ into } 15x - 5 \text{ using } 3x \text{ into } 15x \\
\text{as a trial divisor; multiply; and subtract}
\end{array}
$$

The quotient is $2x + 5$.   ∎

EXAMPLE 4   Divide $2x^4 + 5x + 3x^3 - 3 - 7x^2$ by $x^2 - 3 + 2x$.

Solution   Since you are dividing by a polynomial, you must use long division. Arranging the polynomials in descending powers gives

$$2x^4 + 3x^3 - 7x^2 + 5x - 3 \text{ and } x^2 + 2x - 3$$

$$
\begin{array}{r}
2x^2 - x + 1 \\
x^2 + 2x - 3 \overline{)2x^4 + 3x^3 - 7x^2 + 5x - 3} \\
2x^4 + 4x^3 - 6x^2 \\
\hline
- x^3 - x^2 + 5x \\
- x^3 - 2x^2 + 3x \\
\hline
x^2 + 2x - 3 \\
x^2 + 2x - 3 \\
\hline
0
\end{array}
$$

(1) divide $x^2 + 2x - 3$ into $2x^4 + 3x^3 - 7x^2$; multiply by $2x^2$ and subtract:

(2) divide $x^2 + 2x - 3$ into $-x^3 - x^2 + 5x$; multiply by $-x$ and subtract;

(3) divide $x^2 + 2x - 3$ into $x^2 + 2x - 3$ into $x^2 + 2x - 3$; multiply and subtract

The quotient is $2x^2 - x + 1$. ■■■

EXAMPLE 5 Divide $m^4 - 2m^2 + 6$ by $m - 3$.

Solution Long division is necessary. In arranging the terms in descending powers, the m^3 and m terms are omitted. Therefore, these terms are filled in with a zero numerical coefficient.

$$
\begin{array}{r}
m^3 + 3m^2 + 7m + 21 + \dfrac{69}{(m-3)} \\[4pt]
m - 3{\overline{\smash{\big)}\,m^4 + 0m^3 - 2m^2 + 0m + 6}} \\
\underline{m^4 - 3m^3} \\
3m^3 - 2m^2 \\
\underline{3m^3 - 9m^2} \\
7m^2 + 0m \\
\underline{7m^2 - 21m} \\
21m + 6 \\
\underline{21m - 63} \\
69
\end{array}
$$

■■■

A–9 EXERCISES

Find each of the following quotients.

1. $\dfrac{21a^2}{7a}$

2. $\dfrac{20x^3y^4}{-4xy^3}$

3. $\dfrac{-6a^3c^3d^2}{24a^3c}$

4. $\dfrac{5st}{-25s^3t^4}$

5. $\dfrac{15xy^6}{3xy^4}$

6. $\dfrac{p^{12}q^{10}}{p^8q^{12}}$

7. $\dfrac{(2x^2)^2}{8xy}$

8. $\dfrac{(-3x^2y)^3}{9xy^2}$

9. $\dfrac{(xy)^3(2a)^2}{-6x^3ya}$

10. $\dfrac{7x^2y - 21xy^2}{7xy}$

11. $\dfrac{12a^3b - 4ab^6}{2ab}$

12. $\dfrac{3a^2b - 9ab^2 + 6b^3}{3b}$

13. $\dfrac{32s^2t - 16st + 5}{8st^2}$

14. $\dfrac{x^2 + 11x + 30}{x + 5}$

15. $\dfrac{x^2 + 4x - 12}{x - 2}$

16. $\dfrac{4y^3 - 17y + 20}{2y + 5}$

17. $\dfrac{4x^3 + 4x^2 - 5x + 1}{2x - 1}$

18. $\dfrac{x^3 + 27}{x + 3}$

19. $\dfrac{3x^3 - 8x^2 - 9}{x - 3}$

20. $\dfrac{6x^2 + 5x - 5}{2x + 3}$

21. $\dfrac{x^3 + 11}{x + 2}$

22. $\dfrac{30x^2 - 49x + 20}{5x - 4}$

23. $\dfrac{6x^4 - 3x^2 - 63}{2x^2 - 7}$

24. $\dfrac{x^3 - 1}{x^2 - 1}$

25. $\dfrac{2x^3 - 3x^2 + 8x - 2}{x^2 - x + 2}$

26. $(3x^4 - 15x^3 - 19x^2 - 25x - 40) \div (x^2 - 5x - 8)$

27. $(15x^2 - 76x + 5) \div (x - 5)$

28. Resistance is equal to the applied voltage divided by the current. Find the resistance if the voltage is represented by $16r^2 - 22r$ and the current is represented by $2r$.

29. A free-falling body travels a distance of $-6gt^2 + 4t + 2$. Find an algebraic expression for one-half the distance.

30. The volume of a sphere is $4/3\pi r^3$ and its surface area is $4\pi r^2$. Find the relation between volume and the area. (Find $V \div A$.)

A–10

PERCENTAGES

Many technical applications require a thorough knowledge of percentages. In this section, applied percentage problems will be presented in depth.

The concept of percentages is difficult because there are many applications of it and each one may appear to be different. Actually, once the basics are learned all percentage problems can be solved using a procedural approach. **Percent,** denoted %, always means "per hundred" or "divided by 100." If 36% of the employees in a company received a raise, this simply is another way of saying 36 out of every 100 employees received a raise. Now, if the company employs 3,000 people, how many people in all received a raise? "36 out of every 100" and 3,000 divided by 100 equals 30. Thus, $30 \times 36 = 1,080$ employees.

It would be time-consuming to do all percentage problems in this fashion. To simplify matters, a formula Base × Rate = Percentage is used. In the previous example the base was the total number of employees, 3,000; the rate was the percent value, 36%; and the percentage was the number of employees receiving a raise, 1,080. Using the formula, the problem would be set up this way.

$$\text{Base} \times \text{Rate} = \text{Percentage}$$
$$(3,000)(36\%) = 1,080$$

NOTE ✦ Most calculators have a percent $\boxed{\%}$ key but some do not. If the percent key is not used, then the percent value must be changed to a decimal form. Remember "36 out of every 100" is equal to the fraction 36/100 and this fraction is equal to the decimal value 0.36. A quick rule for this procedure is to drop the percent symbol and move the decimal point two places to the left.

EXAMPLE 1 Change the following percents to decimal form:

(a) 45%   (b) 9%   (c) 104%   (d) 0.5%   (e) $\frac{1}{4}\%$

Solution

(a) $\dfrac{45}{100} = 0.45$ or $45\% = 0.45$

(b) $\dfrac{9}{100} = 0.09$ or $9\% = 0.09$

(c) $\dfrac{104}{100} = 1.04$ or $104\% = 1.04$

(d) $\dfrac{.5}{100} = 0.005$ or $0.5\% = 0.005$

(e) $\dfrac{.25}{100} = 0.0025$ or $1/4\% = 0.25\% = 0.0025$ ■ ■ ■

Many times it is also necessary to rewrite a decimal value as a percent value, such as when writing the results taken from a calculator. A decimal value may be changed to a percent value by moving the decimal two places to the right and adding the percent symbol.

EXAMPLE 2 Change the following decimals to percents:

(a) 0.356   (b) 0.03   (c) 1.67   (d) 0.0075

Solution

(a) $0.356 = 35.6\%$

(b) $0.03 = 3\%$

(c) $1.67 = 167\%$

(d) $0.0075 = 0.75\%$ ■ ■ ■

When setting up applied percentage problems, always remember to identify what quantities in the formula are given and what is the unknown quantity. Review the following examples carefully.

EXAMPLE 3 There are 45 accidents within 1 year for a manufacturing company with 450 employees. What is the accident rate for the company for the year?

Solution

Step 1   Identify the given quantities.
  Base: 450   total
  Percentage: 45   part of base

Step 2   Identify the unknown quantity.
  Rate: R Number of accidents per 100 employees

Step 3   Substitute known quantities into the formula and solve for the unknown value.

$$\text{Base} \times \text{Rate} = \text{Percentage}$$
$$450R = 45$$

$$R = \frac{45}{450}$$

$$= 0.10$$
$$= 10\% \qquad \text{always express in percent form} \qquad ■ ■ ■$$

EXAMPLE 4 A solution consists of 3 parts water and 5 parts alcohol. How many ounces of water are there in a solution of 60 oz?

Solution

Step 1:   Identify the known quantities.
  Base: 6 oz
  Rate for water: 3/8 or .375 or 37.5%

Step 2:   Identify the unknown quantity.
  Percentage: P (ounces of water in total solution)

Step 3:   Apply the formula and solve.

$$B \times R = P$$
$$60(37.5\%) = P$$
$$22.5 \text{ oz} = P \qquad\qquad ■ ■ ■$$

Example 4 is one of the cases that is often confusing. Remember always to add the parts together to obtain the denominator of the fraction that must be converted to a percent or rate. These types of percentage problems are often called ratio problems.

As stated earlier, there are many ways to approach a percentage problem. Many basic problems encountered by a technician might simply be stated as

45% of 180 is what number?

68 is what percent of 88?

23 is 12% of what number?

To solve each of these, the formula Base × Rate = Percentage applies. Again, use the procedure for solving for the unknown quantity.

EXAMPLE 5 Solve each of the following:

(a) 85% of 235 is what number?
(b) 35 is what percent of 56?
(c) 78 is 25% of what number?

Solution

(a) $R = 85\%$ and $B = 235$; find P.

$$(235)(0.85) = P$$
$$P = 199.75$$

(b) $B = 56$ and $P = 35$; find R.

$$56R = 35$$
$$R = \frac{35}{56}$$
$$R = 0.625$$
$$R = 62.5\%$$

(c) $R = 25\%$ and $P = 78$; find B. Remember to convert 25% to a decimal before substituting into the formula.

$$B(0.25) = 78$$
$$B = \frac{78}{0.25}$$
$$B = 312$$ ■ ■ ■

In summary, when solving percentage problems, remember always to identify the unknown quantity in the formula, Base × Rate = Percentage ($BR = P$), and use algebra to solve for the unknown value.

A–10 EXERCISES

Change to percents.

1. 0.32

2. 0.045

5. $\dfrac{1}{5}$

6. 1.34

3. $\dfrac{3}{8}$

4. $\dfrac{5}{1000}$

7. 0.00625

8. 0.89

9. 0.05

10. 5.23

Change to decimals.

11. 56%

12. 130%

13. 0.25%

14. 12.3%

15. 18%

16. 12.75%

17. 255%

18. 0.01%

19. 0.03%

20. 5½%

Solve the following problems.

21. 125 is what percent of 600?

22. 346 is what percent of 890?

23. 35% of 780 is what number?

24. 150% of 456 is what number?

25. 56 is 60% of what number?

26. 904 is 28% of what number?

27. If a company employs 560 people and 23.5% of these are females, how many females work for the company?

28. If 34% of a technician's salary is deducted for taxes and other items, find his take home pay for a salary of $34,560.

29. A company has an accident rate of 3% for a certain period. During this period 23 employees were injured. How many employees does the company employ in all?

30. An electrical engineer for a particular company must spend 12 hours of his 40-hour work week "on call." What percent of his work week is spent "on call"?

31. A solution consists of 2 parts alcohol and 4 parts water. How many ounces of alcohol are there in 40 oz of the solution?

32. A school has 1,250 males and 680 females enrolled. What percent of the student body is male? What percent is female?

A–11

SIGNIFICANT DIGITS AND ROUNDING

In science and technology, calculations involving numbers must take into consideration the accuracy of the numbers because accuracy affects the results. In general, a number obtained from counting is considered an **exact number,** while any number obtained from a measurement is considered an **approximate number.** The accuracy of approximate numbers is limited by the precision of the measuring device.

In working with an approximate number, we discuss the number of significant digits it contains. A **significant digit** is any digit that denotes an actual physical amount. The rules for determining if a digit is significant are given below:

■ All nonzero numbers are significant.

■ All zeros between significant digits are significant.

■ All zeros, except those used as placeholders to position the decimal, are significant.

EXAMPLE 1 Determine the number of significant digits for each of the following approximate numbers:

(a) 307　　(b) 0.07603　　(c) 40　　(d) 17.00　　(e) 34.063

Solution

(a) 3　　(b) 4　　(c) 1　　(d) 4　　(e) 5　　　　　　　■ ■ ■

In computations involving approximate numbers, we must consider the accuracy and precision of the numbers. The **accuracy** of a number is the number of significant digits it contains. The **precision** of a number refers to the decimal position of the last significant digit.

EXAMPLE 2 Discuss the accuracy and precision of the following:

(a) 26.80　　(b) 0.146

Solution

(a) The numbers 26.80 is accurate to 4 significant digits and precise to 2 decimal places (hundredths).
(b) The number 0.146 is accurate to 3 significant digits and precise to 3 decimal places (thousandths).
■ ■ ■

The last significant digit in an approximate number is understood to contain a degree of uncertainty. In mathematical operations involving approximate numbers, it is often necessary to drop digits that might give an erroneous impression of the meaningfulness of the number. To achieve the desired result, we **round off** the number using the following procedure.

■ Locate the last digit to be retained.
■ Determine the next number to the right.
■ If that number is less than 5, drop the digit.
■ If that number is 5 or greater, increase the last digit retained by 1.

EXAMPLE 3 Complete the following:

(a) Round 618.27 to 4 significant digits.
(b) Round 217,360 to the nearest hundred thousand.
(c) Round 0.00863 to 4 decimal places.
(d) Round 3,286,400 to 3 significant digits.

Solution

round here

(a) 618.27

greater than 5; round up

The rounded result is 618.3.

round here

(b) 217,360

less than 5; drop

The rounded result is 200,000.

round here

(c) 0.00863

less than 5; drop

The rounded result is 0.0086.

round here

(d) 3,286.400

more than 5; round up

The rounded result is 3,290,000.
■ ■ ■

When performing calculations with approximate numbers, be careful not to express the result to a greater degree of accuracy or precision than that of the original numbers. The following discussion summarizes how to perform basic arithmetic operations with approximate numbers.

- Operations involving addition or subtraction of approximate numbers should have the result rounded to the precision of the least precise number.

- Operations involving multiplication or division of approximate numbers should have the result rounded to the accuracy of the least accurate number.

- Operations involving a power or root of approximate numbers should have the result rounded to the same accuracy as the number.

EXAMPLE 4 Calculate the following:

(a) $73.8 + 16.23 - 19.178$ (b) 2.6×1.75 (c) $\dfrac{14.8732}{2.56}$ (d) $3.6 \times (1.56)^2$

Solution

(a) $73.8 \boxed{+} 16.23 \boxed{-} 19.178 \boxed{=} \longrightarrow 70.852000$

The result should be rounded to tenths since 73.8 is the least precise number. The rounded result is 70.9.

(b) $2.6 \boxed{\times} 1.75 \boxed{=} \longrightarrow 4.550000$

The result should be rounded to 2 significant digits to give 4.6.

(c) $14.8732 \boxed{\div} 2.56 \boxed{=} \longrightarrow 5.809844$

The result should be rounded to 3 significant digits to give 5.81.

(d) $3.6 \boxed{\times} 1.56 \boxed{x^2} \boxed{=} \longrightarrow 8.760960$

The result should be rounded to 2 significant digits to give 8.8.   ■ ■ ■

CAUTION ✦ Addition and subtraction are rounded according to precision; multiplication, division, roots, and powers are rounded according to accuracy.

In applications involving trigonometric functions, both the sides and angles of the triangle are considered approximate numbers. The result from a trigonometric calculation must be rounded to reflect the accuracy of the side lengths and the precision of the angle measure. The correlation between side length and angle measure follows.

Side Length Measured To	Corresponding Angle Measured To
1 or 2 significant digits	1°
3 significant digits	0.1 or 10^1
4 significant digits	0.01 or 1^1

EXAMPLE 5 In solving a particular right triangle, we must calculate the expression $a = 4.73 \cos 26°$.

Solution

$$4.73 \; \boxed{\times} \; 26 \; \boxed{\cos} \; \boxed{=} \; \longrightarrow \; 4.251296$$

In rounding the result, we must consider the accuracy of the side and angle. The angle has 3 significant digits, but the angle corresponds to 1 or 2 significant digits. Therefore, the result should be rounded to 2 significant digits: 4.3. ∎∎∎

NOTE ✦ Exact numbers have no limit to the number of decimal places. Therefore, rounding is determined by the approximate numbers in the calculation.

A–11 EXERCISES

Perform the indicated operations. Assume all numbers are approximate.

1. $6.73 + 2.468 + 1.59$

2. $16.8 - 3.4 + 7.631$

3. $256 + 76.81 - 13.2$

4. $17.83 + 4.27 + 19.4 + 16.756$

5. $2.7(16.45) \div 5.1$

6. $(43.78)(0.708)$

7. $(9.1)(16.734)$

8. $\dfrac{17.286}{128}$

9. $\dfrac{(4.9)(1.36)}{(0.068)}$

10. $2.1(14.5) - 10 + 5.3$

11. $\dfrac{(6.7)(4.13)}{(5.1)} + 128.5$

12. $(6.8)^2$

13. $\sqrt{138.7}$

14. $\sqrt{14.5} + 75.0 - 9.63$

15. $(1,023)(16.7) - \sqrt{97.6}$

16. $(7.5)^3(127)$

17. $\dfrac{(1,486)(12.6)^2}{3.5} + (7.36)^2(11.34)$

18. $\dfrac{\cos 17.8°}{14.9}$

19. $23.76 \tan 43.5°$

20. $145.6 \sin 37.4°$

21. $\cos 16.7° + 137.65$

22. $\dfrac{487.23}{\tan 16.73°}$

23. $\sin 78.23° + 1,768.1$

A–12

UNITS AND UNIT CONVERSION

There are two commonly used systems of measurement, the metric system and the British or English system. In addition, there is a standard system based on the metric system called the International System of Units (SI), adopted in 1960 by the Eleventh General Conference of Weights and Measures. The SI has seven base units of measure. These units are given below.

SI Base Units

Quantity	Unit	Symbol
Length	meter	m
Mass	kilogram	kg
Time	second	s
Electric Current	ampere	A
Temperature	Kelvin	K
Luminous Intensity	candela	cd
Amount of Substance	mole	mol

The SI also has derived units that are based on previously defined SI units. A list of the derived SI units follows.

SI Derived Units		
Quantity	Unit	Symbol
Angle	radian	rad
Capacitance	farad	F
Conductance	siemens	S
Electric Charge	coulomb	C
Electromotive Force	volt	V
Work	joule	J
Force	Newton	N
Frequency	hertz	Hz
Inductance	henry	H
Power	watt	W
Resistance	ohm	Ω

The metric system is a system of weights and measures developed in France in 1793. It is widely used in Europe and in scientific work in the United States. The metric system is based on powers of ten. For example, the basic unit of length is the meter. The centimeter is 10^{-2} meters, and the millimeter is 10^{-3} meters. The larger or smaller units are denoted by placing a **prefix** before the basic unit. The most commonly used prefixes are given below.

Metric Prefixes		
Prefix	Factor	Symbol
tera	10^{12}	T
giga	10^{9}	G
mega	10^{6}	M
kilo	10^{3}	k
centi	10^{-2}	c
milli	10^{-3}	m
micro	10^{-6}	μ
nano	10^{-9}	n
pico	10^{-12}	p

EXAMPLE 1   Change 3.1 μs to ms.

Solution   If we look at the prefixes on a number line, they would look like the following:

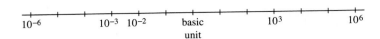

In converting μs to ms, we are going from 10^{-6} to 10^{-3}. Therefore, we move the decimal 3 places to the left.

$$3.1 \ \mu s = .0031 \ ms$$

∎∎∎

EXAMPLE 2   Change 160 km to cm.

Solution   Using the number line approach km is 10^3 and cm is 10^{-2}.

Therefore, we move the decimal 5 places to the right.

$$160 \ km = 16,000,000 \ cm$$

∎∎∎

In converting units within the metric system, we just move the decimal the appropriate number of places. However, we often have to convert between the British and metric systems. Some basic conversion factors are given below.

Conversion Factors

1 in. = 2.54 cm	1 kg = 2.205 lb
1 km = 0.6214 mi	1 lb = 4.448 N
1 lb = 454 g	1 ft^3 = 28.32 L
1 L = 1.057 qt	1 hp = 550 ft·lb/s
1 btu = 778 ft-lb	1 hp = 746 W
1 ft-lb = 1.356 J	1 slug = 14.59 kg

EXAMPLE 3   Convert 30 lb to newtons.

Solution   To perform the conversion from lb to newtons, we take the conversion factor 1 lb = 4.448 N and form it into a fraction. We could have the following fractions:

$$\frac{1 \ lb}{4.448 \ N} \quad or \quad \frac{4.448 \ N}{1 \ lb}$$

To determine which fraction to use, we look at the units we want to eliminate. We can write 30 lb as a fraction by placing one in the denominator.

$$\frac{30 \ lb}{1}$$

In order to eliminate lb, which is in the numerator, it must appear in the denominator of the conversion factor. This means we use

$$\frac{4.448 \ N}{1 \ lb}$$

Performing the conversion gives

$$\frac{30 \cancel{lb}}{1} \cdot \frac{4.448 \text{ N}}{1 \cancel{lb}} = 133.44 \text{ N}$$ ∎∎∎

Unit Conversion

To convert units of measurement

1. Perform algebraic operations with the units of measure as you would perform them with algebraic expressions.
2. Convert units of measure using conversion fractions to obtain the desired units. You may need to use more than one conversion factor.

EXAMPLE 4   Convert 55 miles/h to km/min.

Solution   In analyzing the units, we need to convert miles to kilometers and hours to minutes. The conversion factors needed are

$$1 \text{ km} = 0.6214 \text{ mi}$$

$$1 \text{ h} = 60 \text{ min}$$

Then we arrange the conversion factors in a fractional form so the desired units remain.

$$55 \frac{\cancel{mi}}{\cancel{h}} \cdot \frac{1 \text{ km}}{0.6214 \cancel{mi}} \cdot \frac{1 \cancel{h}}{60 \text{ min}}$$

Notice that mi and h cancel leaving km and min. Last, we multiply and divide the numbers to obtain the result.

$$55 \frac{\text{mi}}{\text{h}} = 1.4 \frac{\text{km}}{\text{min}}$$

$$55 \boxed{\div} \ 0.6214 \ \boxed{\div} \ 60 \ \boxed{=} \longrightarrow 1.4752$$ ∎∎∎

EXAMPLE 5   Convert 16 lb/in^2 to N/m^2.

Solution   We begin by analyzing the units that must be converted and the conversion factors required. We must convert lb to N and in^2 to m^2. We can use the conversion factor 1 lb = 4.448 N. Then, we must also convert in^2 to m^2. However, we don't have a conversion factor for in^2 to m^2. Therefore, we will need to create one. We know 1 in. = 2.54 cm. To obtain in^2, we must square both sides of this conversion factor. The conversion factor we use is

$$(1 \text{ in.})^2 = (2.54 \text{ cm})^2$$

However, we want to convert to m^2. To convert from cm to m, we know 1 m = 100 cm. Again, we must square both sides of the conversion factor.

$$(1 \text{ m})^2 = (100 \text{ cm})^2$$

Now that we have all of the conversion factors, we arrange them as fractions in an order such that the desired units remain. This gives

$$16 \frac{lb}{in^2} = 16 \frac{lb}{in^2} \cdot \frac{4.448 \text{ N}}{1 \text{ lb}} \cdot \frac{(1 \text{ in})^2}{(2.54 \text{ cm})^2} \cdot \frac{(100 \text{ cm})^2}{(1 \text{ m})^2}$$

$$16 \frac{lb}{in^2} = 16 \frac{\cancel{lb}}{\cancel{in^2}} \cdot \frac{4.448 \text{ N}}{1 \cancel{lb}} \cdot \frac{1 \cancel{in^2}}{6.4516 \cancel{cm^2}} \cdot \frac{10,000 \cancel{cm^2}}{1 \text{ m}^2}$$

$$16 \frac{lb}{in^2} = 110,311 \frac{N}{m^2}$$

$$16 \boxed{\times} 4.448 \boxed{\times} 10000 \boxed{\div} 6.4516 \boxed{=} \longrightarrow 110310.621 \qquad \blacksquare\blacksquare\blacksquare$$

A–12 EXERCISES

Convert the units of measure as required.

1. 15 in. to yd
2. 168 cm to m
3. 6.8 km to cm
4. 38.4 in. to m
5. 86 m to ft
6. 7.3 ft^3 to L
7. 0.52 ms to μs
8. 2.5 yd^3 to ft^3
9. 19 lb to g
10. 73 lb to N
11. 215 km to mi
12. 275 mL to qt
13. 36 m to in.
14. 210 km to mi
15. 47 in^2 to cm^2
16. 10.5 in^2 to cm^2
17. 39 lb/ft to N/cm
18. 285 in^3 to L
19. 66 mi/h to km/h
20. 85.3 N/m^2 to lb/in^2
21. 17.4 m^3 to ft^3
22. 1,350 J to btu
23. 168 in · lb/h to hp
24. 18.2 m^3 to ft^3
25. 3.4 hp to N·cm/s
26. 53 kg to slugs
27. 70 mi/h to ft/s
28. 6.7 N/cm^3 to lb/ft^3
29. 3,500 W to hp
30. 20 J to ft-lb

A–13

CALCULATOR USAGE

Electric calculators have increasingly become a tool to help in scientific and engineering calculations. Since you will probably use a calculator in this text, this is an introduction to the scientific calculator. A scientific calculator performs some additional calculations beyond basic arithmetic operations, such as $\boxed{\sin}$, $\boxed{\cos}$, $\boxed{\tan}$, $\boxed{y^x}$, $\boxed{\log}$, and $\boxed{\ln x}$. The information and keystrokes in this text are given for calculations using algebraic notation. Other calculators use reverse Polish notation. Refer to your owner's manual for instructions. You may also have a graphing calculator. This calculator has a larger display and usually extensive graphing and mathematical capabilities. Instructions on how to use a graphing calculator are given throughout the text in sections where its use is appropriate.

Order of Operations

When entering information into your calculator, you must remember the order in which operations must be performed. For example, multiplication and division must be performed before addition and subtraction. Consider the problem $8 + 3 \times 5$. The correct answer is $8 + (3 \times 5) = 8 + 15 = 23$. Use the following keystrokes on your calculator and verify that your display shows 23.

$$8 \boxed{+} 3 \boxed{\times} 5 \boxed{=}$$

If your calculator displays 55, you will have to enter parentheses to group your calculations or the calculator memory so the order of operations is performed correctly.

Arithmetic Functions

The basic arithmetic functions are used to perform addition, subtraction, multiplication, and division with positive and negative numbers. The keys are $\boxed{+}$, $\boxed{-}$, $\boxed{\times}$, and $\boxed{\div}$. To enter a negative number most calculators have a $\boxed{+/-}$ or $\boxed{\text{CHS}}$ key. Once you have entered your data and are ready to obtain the answer, press $\boxed{=}$.

EXAMPLE 1 Perform the following calculations on a calculator.

(a) $16.7 + 14.9$ (b) $17.48 - 3.6$ (c) -10.3×2.8 (d) $286.45 \div 10.258$

Solution The keystrokes are given below.

(a) $16.7 \boxed{+} 14.9 \boxed{=} \longrightarrow 31.6$
(b) $17.48 \boxed{-} 3.6 \boxed{=} \longrightarrow 13.88$ Rounded to 13.8
(c) $10.3 \boxed{+/-} \boxed{\times} 2.8 \boxed{=} \longrightarrow -28.84$ Rounded to -29
(d) $286.45 \boxed{\div} 10.258 \boxed{=} \longrightarrow 27.924547$ Rounded to 27.925

Calculators will generally show you 6–10 digits after the decimal. You must round according to the number of significant digits. (Refer to Appendix A–11.)   ■ ■ ■

Dual Function Key

Most calculators have keys that represent two functions. The first function is written on the key, and the second function is written above the key. In order to perform the second function for any given key, press the $\boxed{\text{2nd}}$ key. For example, the $\boxed{\sqrt{x}}$ key has x^2 placed above it. In order to perform $(25)^2$, press $25 \boxed{\text{2nd}} \boxed{x^2} \longrightarrow 625$.

EXAMPLE 2 Use a calculator to find

$$3.8 + (6.7)^2 - \frac{1}{7.6}$$

Solution Use the following keystrokes:

$3.6 \boxed{+} 6.7 \boxed{\text{2nd}} \boxed{x^2} \boxed{-} 7.6 \boxed{\text{2nd}} \boxed{1/x} \boxed{=} \longrightarrow 48.358421$

The solution is 48.   ■ ■ ■

Parentheses or Memory Functions

Under some circumstances, a problem may call for calculations to be performed contrary to the order of operations. For example, in the expression $2 \times 6 + 7$, the order of operations will multiply 2 and 6 and then add 7. However, if you want to add 6 and 7 and then multiply by 2, you must use parentheses or the memory feature. You could use either of the following keystrokes.

$2 \boxed{\times} \boxed{(} 6 \boxed{+} 7 \boxed{)} \boxed{=} \longrightarrow 26$

$2 \boxed{\text{STO}} 16 \boxed{+} 7 \boxed{=} \boxed{\times} \boxed{\text{RCL}} 1 \boxed{=} \longrightarrow 26$

Note that $\boxed{\text{STO}}$ 1 and $\boxed{\text{RCL}}$ 1 stores and recalls the data in register 1.

EXAMPLE 3 Perform the following calculations using a calculator.

$$\frac{3.68 - (7.42)^2}{(1.68)^2 + 11.8}$$

Solution There are several ways to perform this calculation. The following keystrokes use the memory key.

3.68 $\boxed{-}$ 7.42 $\boxed{2^{nd}}$ $\boxed{x^2}$ $\boxed{=}$ $\boxed{STO}$ 1

1.68 $\boxed{2^{nd}}$ $\boxed{x^2}$ $\boxed{+}$ 11.8 $\boxed{=}$ $\boxed{\div}$ $\boxed{RCL}$ 1 $\boxed{x \leftrightarrows y}$ $\boxed{=}$ ⟶ -3.513409

The $\boxed{x \leftrightarrows y}$ key interchanges the data in the x and y registers.

Another set of keystrokes using parentheses follows.

$\boxed{(}$ 3.68 $\boxed{-}$ 7.42 $\boxed{2^{nd}}$ $\boxed{x^2}$ $\boxed{)}$ $\boxed{\div}$ $\boxed{(}$ 1.68 $\boxed{2^{nd}}$ $\boxed{x^2}$ $\boxed{+}$ 11.8 $\boxed{)}$ $\boxed{=}$ ⟶ -3.513409

The solution rounded correctly is -3.51 ■■■

Note: The $\boxed{y^x}$ key can be used to raise a number to any power. For example, 6 $\boxed{x^2}$ 4 $\boxed{=}$ 1296 calculates 6^4.

Scientific Notation   Most calculators will perform calculations with numbers in scientific notation. You may want to review the information on scientific notation given in Appendix A–5. To perform calculations using numbers in scientific notation, use the $\boxed{EE}$ key (consult your user's manual since some calculators use different keys to perform this function). For example, to enter 2.3×10^{12}, follow these keystrokes:

$$2.3 \boxed{EE} 12$$

EXAMPLE 4 Use a calculator to evaluate

$$\frac{3.5 \times 10^{-4} \times 6.9 \times 10^3}{2.9 \times 10^{-5}}$$

Solution Perform the following keystrokes:

3.5 $\boxed{EE}$ 4 $\boxed{+/-}$ $\boxed{\times}$ 6.9 $\boxed{EE}$ 3 $\boxed{\div}$ 2.9 $\boxed{EE}$ 5 $\boxed{+/-}$ $\boxed{=}$ ⟶ 8.3275862×10^4

The solution is 8.3×10^4.   ■■■

A–13 EXERCISES

Use a calculator to evaluate the following:

1. $6.9 + 7.8 \times 1.4 - 9.3 \div 3.1$ **2.** $7.3 + 3.4 \times 16 \div 2.1$ **5.** 14^2 **6.** 8^4

3. $\sqrt{186}$ **4.** $\sqrt[3]{216}$ **7.** $\dfrac{1}{1.78}$ **8.** $\dfrac{1}{5.9}$

9. $\dfrac{3 + (8 \times 9)}{5 - 7}$

10. $\dfrac{6.3 + (7.5 \times 4.2)}{16.1}$

11. $\sqrt[4]{130}$

12. $\sqrt{286}$

13. $\dfrac{4.8 + 9.5}{7.2} - \dfrac{5.7(2.9 - 6.8)}{3.5}$

14. $\dfrac{1.8 + 2.9}{3} \div \dfrac{16 + 11.4}{2}$

15. $\dfrac{2.3 \times 10^3 + 7.5 \times 10^{-6}}{9.8 \times 10^4}$

16. $\dfrac{8.3 \times 10^{-4} + 9.2 \times 10^5}{3.9 \times 10^{-8}}$

17. $\log 48.7$

18. $\log 187.4$

19. $8\pi - 16 \div 14$

20. $2\pi(6)(0.012) - 14.7 \div 3$

21. 12^3

22. 121^2

23. $16.7 \times 10^9 \div 8.7 \times 10^{23}$

24. $(7.3 \times 10^4)(2.4 \times 10^{-3})$

25. $(6 + 9) \times 14 \div (8.3 \times 2.9)$

26. $(17.3 + 8.12) \times 4.5 \div (-30.6 \div 3)$

A–14

COMPUTER NUMBER SYSTEMS

This appendix will discuss the number systems commonly used in computer systems. We are all familiar with the decimal number system that we use every day. The decimal system is called base 10 because it uses 10 digits, 0, 1, 2, 3, 4, 5, 6, 7, 8, and 9. Beyond these digits numbers are assigned values dependent on their position relative to the decimal.

Three additional number systems are used in computers, digital logic trainers, and digital devices. These number systems are binary, octal, and hexadecimal systems. The **binary** number system is base 2, consisting of the two digits 0 and 1, called bits. The **octal** number system is base 8, consisting of the eight digits 0, 1, 2, 3, 4, 5, 6, and 7. The third system is the **hexadecimal** system, which is base 16 consisting of the 16 digits 0, 1, 2, 3, 4, 5, 6, 7, 8, 9, A, B, C, D, E, and F. These three additional number systems also use positional values. The digit to the far left is called the **most significant digit,** MSD, and the digit to the far right is called the **least significant digit,** LSD. To denote the number system a subscript of the base is appended to each number, such as 101_2, 70_8, $1B40_{16}$.

Converting Numbers to Decimal

To determine the decimal value of a number in one of the other number systems follow the steps below.

Decimal Value of Digits

---

1. Determine its position relative to the decimal point.

2. For a digit n places to the left of the decimal point, the value is given by

$$\text{value} = \text{digit} \times R^{n-1}$$

3. For a digit m places to the right of the decimal point, the value is given by

$$\text{value} = \text{digit} \times R^{-m}$$

EXAMPLE 1 Find the decimal value of

(a) 1 in 100_2 (b) 4 in 743_8 (c) 7 in $47F_{16}$

Solution   (a) The 1 is 3 places to the left of the implied decimal point.
The base is 2. Therefore,

$$\text{value} = 1 \times 2^{3-1} = 1 \times 2^2 = 4_{10}$$

(b) The 4 is 2 places to the left of the decimal with a base of 8.

$$\text{value} = 4 \times 8^{2-1} = 4 \times 8 = 32_{10}$$

(c) The 7 is 2 places to the left of the decimal with a base of 16.

$$\text{value} = 7 \times 16^{2-1} = 7 \times 16 = 112_{10}$$ ■ ■ ■

To find the decimal value of a number in one of the other number systems, find the value of each digit and total the values. This process is illustrated in the next example.

EXAMPLE 2   Find the decimal value of 25.34_8

Solution   We start with the LSD and work to the left. The LSD of 4 is 2 places to the right of the decimal, $m = -2$.

$$\text{value} = 4 \times 8^{-2} = 4 \times \frac{1}{64} = 0.0625_{10}$$

The 3 is 1 place to the right of the decimal, $m = -1$.

$$\text{value} = 3 \times 8^{-1} = 3 \times \frac{1}{8} = 0.3750_{10}$$

The 5 is 1 place to the left of the decimal, $n = 1$.

$$\text{value} = 5 \times 8^{1-1} = 5 \times 8^0 = 5 \times 1 = 5_{10}$$

The 2 is 2 places to the left of the decimal, $n = 2$.

$$\text{value} = 2 \times 8^{2-1} = 2 \times 8 = 16_{10}$$

$$25.34_8 = 16 + 5 + 0.3750 + 0.0625$$

$$25.34_8 = 21.4_{10}$$ ■ ■ ■

Converting Decimal Numbers to Other Systems

To convert a whole number from decimal to binary, octal, or hexadecimal, use repeated division by the appropriate base. The remainder on the first division is the LSD, and the remainder on the last division is the MSD. This process is illustrated in the next example.

EXAMPLE 3   Convert 486_{10} to octal.

Solution   We repeatedly divide by 8.

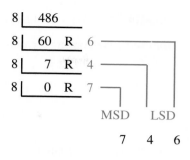

$$486_{10} = 746_8$$

You can check your answer by converting 746_8 to decimal. ■■■

To convert decimal fractions to binary, octal, or hexadecimal, multiply the fraction by the new base. Repeat this process until the remaining fraction is zero or until the desired digits have been found. The next example illustrates this process.

EXAMPLE 4   Convert 0.15625_{10} to hexadecimal.

Solution

$$
\begin{array}{r}
0.15625 \\
\times\ 16 \\
\hline
②.50000 \\
\times\ 16 \\
\hline
\boxed{8}.0000
\end{array}
$$

MSD   LSD
 2   8

Therefore, $0.15625_{10} = 0.28_{16}$ ■■■

Converting Between Octal, Binary, Hexadecimal

The octal and hexadecimal systems were chosen because of their ease of conversion to and from binary. To convert an octal number to binary, replace each digit with the appropriate set of three binary digits from the list that follows.

Octal to Binary Conversion

Octal	Binary	Octal	Binary
0	000	4	100
1	001	5	101
2	010	6	110
3	011	7	111

EXAMPLE 5   Convert 74.63_8 to binary.

Solution   We convert each digit using the information above.

	7	4	.	6	3
	111	100	.	110	011

$$74.63_8 = 111\ 100.110\ 011_2$$

∎∎∎

To convert a binary number to octal, start at the decimal and group the binary digits into groups of three. If necessary, add leading and trailing zeros to complete the groups. Then, use the information in the listed above to convert each group of binary digits to octal.

EXAMPLE 6   Convert 11101001111_2 to octal.

Solution   We start by grouping in threes starting at the implied decimal and working left.

11	101	001	111

We add a leading 0 to complete that group of three digits.

011	101	001	111

Last, we convert each group to its octal equivalent.

011	101	001	111
↓	↓	↓	↓
3	5	1	7

Therefore, $11101001111_2 = 3517_8$

∎∎∎

The conversion of hexadecimal to binary is done the same way. The hexadecimal to binary conversion is in the following list.

Hexadecimal to Binary Conversion

Hexadecimal	Binary	Hexadecimal	Binary
0	0000	8	1000
1	0001	9	1001
2	0010	A	1010
3	0011	B	1011
4	0100	C	1100
5	0101	D	1101
6	0110	E	1110
7	0111	F	1111

EXAMPLE 7   Convert $F2D8_{16}$ to binary.

Solution   We look up each digit in the conversion table.

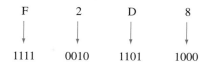

F	2	D	8
↓	↓	↓	↓
1111	0010	1101	1000

Therefore, $F2D8_{16} = 1\ 111\ 001\ 011\ 011\ 000_2$.   ■■■

NOTE ✦ To convert between octal and hexadecimal, first convert to binary and then to the number system desired.

EXAMPLE 8   Convert 235_8 to hexadecimal.

Solution   We begin by converting 235_8 to binary.

2	3	5
↓	↓	↓
010	011	101

Therefore, $235_8 = 010011101_2$. Then, we convert from the binary number to hexadecimal. Regroup in sets of four and convert.

1001	1101
↓	↓
9	D

Therefore, $235_8 = 9D_{16}$ ■■■

A–14 EXERCISES

Find the decimal value of the following numbers.

1. 321_8
2. 11001_2
3. $F37.A2_{16}$
4. 71.35_8
5. 1110.01_2
6. 916_{16}
7. 4321.5_8
8. 10011011_2
9. $6B3.2D_{16}$

Convert the following decimal numbers to binary, octal, and hexadecimal.

10. 148_{10}
11. 786_{10}
12. 938_{10}
13. 39_{10}
14. 839_{10}
15. 456_{10}

16. 1024_{10}
17. 271_{10}
18. 49201_{10}
19. 605_{10}
20. 304_{10}
21. 583_{10}

Convert the following numbers to binary.

22. 471.6_8
23. $A94_{16}$
24. $2CFA_{16}$
25. 6312_8
26. 1073_{16}
27. 1374_8
28. $3987.A2_{16}$
29. 314.068_8
30. 38.24_{16}
31. 0.147_8
32. $F5CB1_{16}$
33. 24.35_8

Appendix B

Geometry Review

Geometric figures, facts, and formulas are essential to many technical fields. This appendix will serve not only as a review of basic applied geometry but as a reference for the many geometric problems found in the text. For our study of geometry, we will assume no formal definitions are necessary for these geometric terms: point, line, and plane.

Most geometric figures are made up of a combination of intersecting and nonintersecting lines or line segments. A **straight line** (Figure B–1) extends infinitely in both directions, but a **line segment** (Figure B–2) is a finite portion of a line bounded by and including two endpoints. Sometimes it is necessary to speak of a half-line or ray. A **half-line** or **ray** (Figure B–3) is a line extending infinitely to one side of a given endpoint.

FIGURE B–1

A ——————————— B

FIGURE B–2

FIGURE B–3

Two lines that intersect are assumed to have one point in common (Figure B–4). Two lines that do not intersect and have no points in common are said to be **parallel lines** (Figure B–5).

An angle is formed by rotating a half-line about its endpoint (Figure B–6). The endpoint, *V*, is called the vertex of the angle. Angles are usually measured in degree or radian units. In this review, we will state all angles in degree units. Although angle measurement is discussed in detail in the text, it is sufficient to know for this material that 1 degree (1°) is equal to 1/360 of a complete revolution (Figure B–7).

Point of intersection

FIGURE B–4

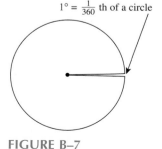

$1° = \frac{1}{360}$ th of a circle

FIGURE B–5

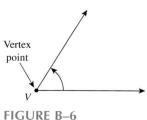

Vertex point

V

FIGURE B–6

FIGURE B–7

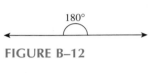

FIGURE B–8

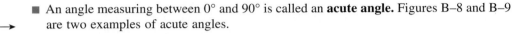

FIGURE B–9

FIGURE B–10

FIGURE B–11

Recognition of the characteristics and names of the following angles is an essential ingredient when solving many technical problems:

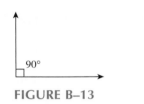

FIGURE B–12

- An angle measuring between 0° and 90° is called an **acute angle.** Figures B–8 and B–9 are two examples of acute angles.
- An angle measuring between 90° and 180° is called an **obtuse angle.** Figures B–10 and B–11 are two examples of obtuse angles.
- A **straight angle** measures exactly 180° (Figure B–12).
- A **right angle** measures exactly 90° (Figure B–13).

FIGURE B–13

If two lines intersect to form a 90° or right angle, they are called **perpendicular.** The symbol, ⊥, is used to denote this property.

The comparison of one angle to another is often important in setting up a problem. The following facts pertaining to angles are invaluable problem solving aids:

- If the sum of two angles is 90°, the angles are said to be **complementary.** For example, ∠A and ∠B in Figure B–14 are complements of each other.
- If the sum of two angles is 180°, the angles are said to be **supplementary.** For example, ∠C and ∠D in Figure B–15 are supplements of each other.

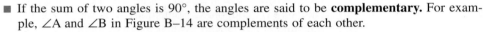

FIGURE B–14

FIGURE B–15

- **Adjacent** angles have a common vertex point, share a common side, and have no interior points in common. In Figure B–16, ∠1 is adjacent to ∠2.
- **Vertical** angles are formed when two lines intersect. Vertical angles are opposite each other and are always equal to each other. In Figure B–17, ∠1 and ∠3 are vertical angles, as are ∠2 and ∠4.

FIGURE B–16

FIGURE B–17

Parallel lines intersected by a third line, called a **transversal,** form several pairs of special angles:

- **Alternate interior angles** are equal to each other. In Figure B–18, $\angle 1$ and $\angle 7$ are alternate interior angles, as are $\angle 4$ and $\angle 6$.

- **Alternate exterior angles** are equal to each other. In Figure B–18, $\angle 3$ and $\angle 5$, and $\angle 2$ and $\angle 8$ are two pairs of alternate exterior angles.

- **Corresponding angles** are equal to each other. The pairs of corresponding angles in Figure B–18 are $\angle 3$ and $\angle 7$, $\angle 2$ and $\angle 6$, $\angle 8$ and $\angle 4$, and $\angle 5$ and $\angle 1$.

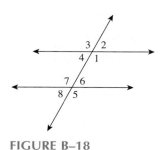

FIGURE B–18

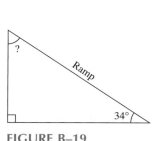

FIGURE B–19

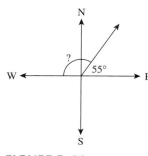

FIGURE B–20

EXAMPLE 1   Solve each of the following:

(a) A ramp makes an angle of 34° with the horizontal (see Figure B–19). What angle does it make with the vertical? Name these two angles.

(b) A plane takes off at an angle of 55° with the east as shown (see Figure B–20). What angle does it make with the west? Name these two angles.

Solution

(a) $90° - 34° = 56°$ complementary angles

(b) $180° - 55° = 125°$ supplementary angles   ■ ■ ■

Any closed figure in a plane with three or more straight sides is called a **polygon.** Polygons are usually named according to the number of sides they contain. The most familiar polygons are triangles, three-sided figures, and quadrilaterals, four-sided figures. A six-sided polygon is called a hexagon; a five-sided polygon is a pentagon; an eight-sided polygon is an octagon; and so on.

If the sides of a polygon are all equal, and in addition, the interior angles are equal, the polygon is said to be **regular.**

An important feature of every polygon is the sum of the angles it contains. If a polygon has n sides, the angles add up to $(n - 2)(180°)$. Try this formula on the polygons we stated above. You will find that the angles of a triangle ($n = 3$) add up to 180°, a quadrilateral's ($n = 4$) add up to 360°, a pentagon's ($n = 5$) add up to 540°, and an octagon's ($n = 8$) add up to 1,080°.

EXAMPLE 2   What is a six-sided polygon called? Find the sum of the angles in this figure.

Solution   A six-sided polygon is called a hexagon.

$$(6 - 2)(180°) = 4 \cdot 180° = 720°$$ ■ ■ ■

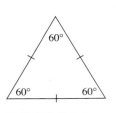

FIGURE B–21

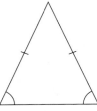

FIGURE B–22

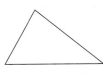

FIGURE B–23

FIGURE B–24

The polygons used most often in technical work, the triangle and the quadrilateral, have many features and properties that you should be able to recognize and use.

Triangles are classified by the relations between the sides (or angles):

- If all three sides (or angles) are equal, the triangle is called **equilateral** (Figure B–21).
- If only two sides (two angles) are equal, the triangle is called **isosceles** (Figure B–22).
- If all the sides (angles) are of different sizes, the triangle is called **scalene** (Figure B–23).
- If the triangle has a right angle, it is called a **right** triangle (Figure B–24).

Right triangles are of special importance because they are the basis of trigonometry. Because all right triangles have one right angle, the side opposite this angle has been given a name, the hypotenuse. The other two sides are called legs. All right triangles share a common relationship, the square of the hypotenuse is equal to the sum of the squares of the legs. This relationship is called the Pythagorean theorem and is expressed as the formula

$$a^2 + b^2 = c^2$$

where a and b represent the two legs and c represents the hypotenuse of the right triangle. This formula can be used to find any missing side of a right triangle, provided the two other sides are known.

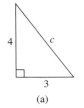

FIGURE B–25

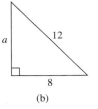

FIGURE B–26

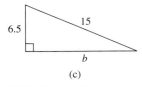

FIGURE B–27

EXAMPLE 3 Solve for the missing side in each of the following right triangles (Figures B–25, B–26, B–27).

Solution

(a) $c = \sqrt{3^2 + 4^2} = \sqrt{25} = 5$

(b) $a = \sqrt{12^2 - 8^2} = \sqrt{80} = 4\sqrt{5}$

(c) $b = \sqrt{15^2 - 6.5^2} = \sqrt{182.75} = 13.5$ (rounded) ■■■

Similar triangles are another classification important to the study of trigonometry. If two triangles are similar, their corresponding angles must be equal and their corresponding sides must be proportional. Figure B–28 is an example of two similar triangles.

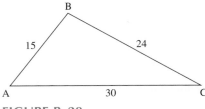

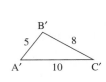

$$\angle A = \angle A'$$
$$\angle C = \angle C'$$
$$\angle B = \angle B'$$

$$\frac{AB}{A'B'} = \frac{AC}{A'C'} = \frac{BC}{B'C'}$$

FIGURE B–28

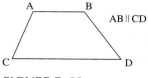

FIGURE B–29

Quadrilaterals are the second most-often-used polygon. There are several categories of quadrilaterals.

- **Trapezoids** require one pair of parallel sides and one pair of nonparallel sides. Figure B–29 is an example of a trapezoid.

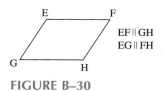

FIGURE B–30

- **Parallelograms** require that each pair of sides be parallel. In a parallelogram, opposite sides are equal and opposite angles are equal. Figure B–30 is an example of a parallelogram.
- A **rhombus** is a parallelogram with 4 equal sides. Figure B–31 is an example of a rhombus. Although a rhombus has 4 equal sides, it does not necessarily have 4 equal angles.
- The **rectangle,** the most familiar parallelogram, has 4 right angles. Figure B–32 is an example of a rectangle.
- A **square** is a rectangle with equal sides. Figure B–33 is an example of a square.

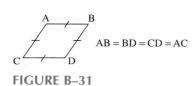

FIGURE B–31

FIGURE B–32

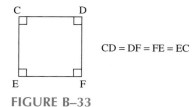

FIGURE B–33

When a quadrilateral is encountered in a problem, remember to examine its characteristics carefully for identification purposes.

The last plane figure we will describe in detail is the circle. A **circle** is a closed plane figure defined to be the set (or group) of points that are the same distance from a fixed point (Figure B–34). The fixed point is called the center of the circle, and the fixed distance is called the radius, r (plural—radii). The diameter is a line segment passing through the center of the circle and having its endpoints on the circle (Figure B–34). Thus, the diameter is twice the length of the radius. All diameters in a given circle are the same length.

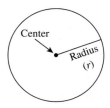

FIGURE B–34

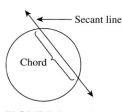

FIGURE B–35

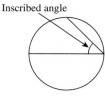

FIGURE B–36

Many facts and terms relating to circles are essential to solving technical problems. Some important ones follow.

- A line that intersects a circle at two points is called a **secant** line (Figure B–35).
- The portion of the secant line that lies within the circle is called a **chord** (Figure B–35).
- Two chords that have one common endpoint form an **inscribed angle** (Figure B–36).
- A **tangent line** intersects a circle at one and only one point. A radius of a circle is perpendicular to a tangent line at the point of tangency (Figure B–37).

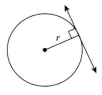

FIGURE B–37

- The **circumference** of a circle is the distance around the circle, or its perimeter. A portion of a circle or part of its circumference is called an **arc** (Figure B–38).
- An inscribed angle is equal to one-half of the intercepted arc (Figure B–39).
- Any angle formed by two radii of a circle is called a **central angle.** Arcs are measured by the central angle that intercepts them (Figure B–40).
- A **sector** is the region formed by two radii and the intercepted arc. A sector can also be identified by its central angle (Figure B–41).

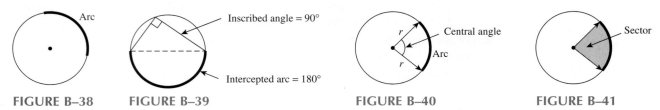

FIGURE B–38   FIGURE B–39   FIGURE B–40   FIGURE B–41

The perimeter and the area can be found for each of the geometric figures we have described. The **perimeter** is the distance around the outside of a figure, and the **area** is the region bounded by the sides (or circumference) of a figure. When using the formulas stated below for computing the area and perimeter of many plane figures, it is important to understand the term **altitude.** The altitude of a figure is the line drawn from any vertex perpendicular to the opposite side (extended if necessary). The letter h is used to denote the altitude in the formulas summarized in the following box.

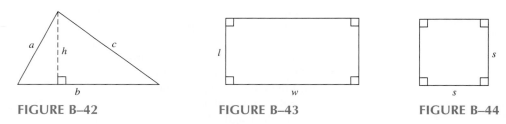

FIGURE B–42   FIGURE B–43   FIGURE B–44

Geometric Formulas for Area and Perimeter

Figure	*Perimeter, P*	*Area, A*
Triangle (Figure B–42)	$P = a + b + c$	$A = .5bh$
Rectangle (Figure B–43)	$P = 2l + 2w$	$A = lw$
Square (Figure B–44)	$P = 4s$	$A = s^2$
Parallelogram (Figure B–45)	$P = 2a + 2b$	$A = bh$
Trapezoid (Figure B–46)	$P = a + b + c + d$	$A = .5(a + b)h$
Circle (Figure B–47)	$C = \pi d = 2\pi r$	$A = \pi r^2$

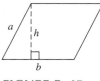

FIGURE B–45

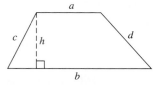

FIGURE B–46

FIGURE B–47

In plane geometry (two-dimensional), figures have dimensions of length and width. Technical problems also require a knowledge of solid geometry (three-dimensional). In solid geometry the third dimension of height is added. Many solid figures you encounter every day, such as the sphere (ball), the cone, and the cylinder, are not part of the group of solids called polyhedrons. A *polyhedron* is a solid figure that is bounded by a finite number of polygonal regions. Thus a cube, a pyramid, and a prism are polyhedrons.

Formulas have been derived to simplify computations involving solid figures. The ones listed in the following box are for finding the volume and surface area of the most common solids. The following letters are used: B = area of base; L = lateral surface area [area of all surfaces other than the base(s)]; S = total surface area [includes area of base(s)]; V = volume.

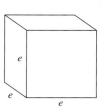

FIGURE B–48

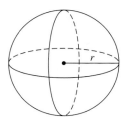

FIGURE B–49

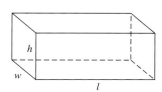

FIGURE B–50

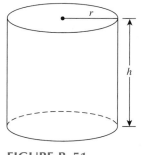

FIGURE B–51

Formulas for Surface Area and Volume

Solid	Surface Area, S	Volume, V
Cube (Figure B–48)	$S = 6e^2$	$V = e^3$
Sphere (Figure B–49)	$S = 4\pi r^2$	$V = \frac{4}{3}\pi r^3$
Rectangular prism (Figure B–50)	$S = 2(lw + wh + lh)$	$V = lwh$
Right circular cylinder (Figure B–51)	$S = 2\pi rh + 2\pi r^2$	$V = \pi r^2 h$
Right circular cone (Figure B–52)	$S = \pi rs + \pi r^2$	$V = \frac{1}{3}\pi r^2 h$
General cone or pyramid (Figure B–53)		$V = \frac{1}{3}Bh$

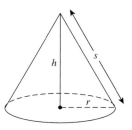

FIGURE B–52

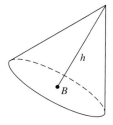

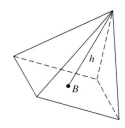

FIGURE B–53

APPENDIX B EXERCISES

Find the complement of each angle. Find the supplement of each angle.

1. 18°   **2.** 35°   **3.** 56.4°   **4.** 83.5°   **5.** 103°   **6.** 65.2°   **7.** 120°   **8.** 95.5°

Use Figure B–54 for Problems 9–13.

9. State all pairs of alternate interior angles.

10. State all pairs of alternate exterior angles.

11. State all pairs of corresponding angles.

12. Find the measurement of ∠4 and ∠8.

13. Find the measurement of ∠2 and ∠5.

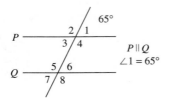

FIGURE B–54

Use Figure B–55 for Problems 14–16.

14. Find ∠x.

15. Find side BC.

16. Name the triangle.

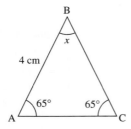

FIGURE B–55

In problems 17–20, using the Pythagorean theorem and given side measurements, find the length of the missing side of the right triangle (Figure B–56). (Round to tenths place.)

17. Given $a = 12$ cm and $b = 15$ cm, find c.

18. Given $b = 8$ in. and $c = 14$ in., find a.

19. Given $a = 34.6$ ft and $c = 54.8$ ft, find b.

20. Given $a = 125$ m and $b = 75$ m, find c.

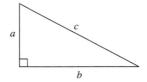

FIGURE B–56

For Figures B–57 through B–63, (a) name the figure; (b) find the perimeter of the figure; and (c) find the area of the figure (Problems 21–27).

21. **FIGURE B–57**

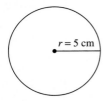

22. **FIGURE B–58**

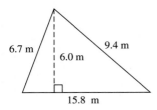

23. **FIGURE B–59**

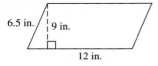

24. **FIGURE B–60**

25. **FIGURE B–61**

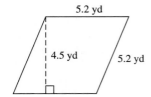

26. **FIGURE B–62**

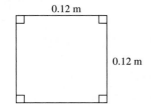

27. FIGURE B–63

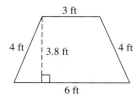

3 ft

4 ft

3.8 ft

4 ft

6 ft

For Figures B–64 through B–66, (a) name the figure and (b) find the volume of the figure (Problems 28–30).

28. FIGURE B–64

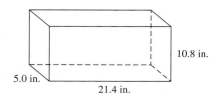

10.8 in.

5.0 in.

21.4 in.

29. FIGURE B–65

11 ft

120 ft

30. FIGURE B–66

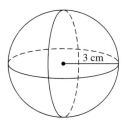

3 cm

For Figures B–67 through B–69, (a) name the figure and (b) find the total surface area of the figure (Problems 31–33).

31. FIGURE B–67

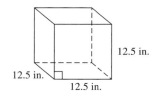

12.5 in.

12.5 in.

12.5 in.

32. FIGURE B–68

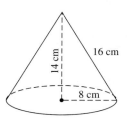

16 cm

14 cm

8 cm

33. FIGURE B–69

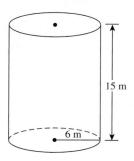

15 m

6 m

Solve the following problems.

34. A right triangle with legs 18.3 cm and 6.5 cm is rotated about the 18.3 cm leg. Find the volume of the generated cone (Figure B–70).

FIGURE B–70

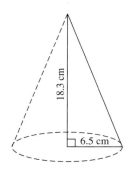

18.3 cm

6.5 cm

35. The triangle shown in Figure B–71 is known as an impedance triangle and is used in many ac circuit problems. The side Z is called the impedance, the side X is the reactance, and the side R is the resistance. All three sides are measured in units of ohms (Ω). The angle θ is known as the phase angle. Use the impedance triangle and solve for the unknown.

(a) Given $X = 9\ \Omega$, $R = 12\ \Omega$, find Z.
(b) Given $Z = 26\ \Omega$, $R = 10\ \Omega$, find X.
(c) Given $X = 90\ \Omega$, $Z = 150\ \Omega$, find R.

FIGURE B–71

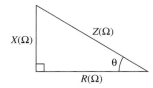

$X(\Omega)$

$Z(\Omega)$

θ

$R(\Omega)$

36. A rectangular tank is 12 ft long, 10 ft wide, and 8 ft deep. How many gallons of water will it hold? (1 gal = 231 in³)

37. A hollow pipe has an outside circumference of 10.5 in. and an inner diameter of 2.8 in. In Figure B–72, find the shaded cross-sectional area of the pipe.

FIGURE B–72

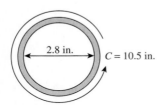

38. A wheel with a diameter of 36 cm makes 100 turns. How far does it roll?

39. An A-frame structure has a base of 18 m, while a scale drawing of the structure has a base of 20 cm (0.20 m). If the sides of the structure in the drawing are 40 cm (0.40 m) each, what are the lengths of the real sides (Figure B–73)?

FIGURE B–73

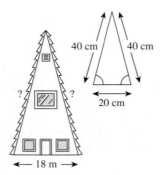

40. In the truss shown in Figure B–74, $\overrightarrow{BC}\|\overrightarrow{AD}$, $\overrightarrow{AB}\|\overrightarrow{CE}$, $\overrightarrow{BE}\|\overrightarrow{CD}$. $BAE = 45°$, $BEA = 25°$. Find angles 1, 2, 3, 4, 5, 6, and 7.

FIGURE B–74

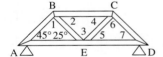

41. In Figure B–75, find the area of the shaded region.

FIGURE B–75

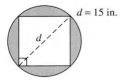

42. Find the total area of the I-beam shown in Figure B–76. All thicknesses are 1.5 in.

FIGURE B–76

9.0 in. 7.0 in. 9.0 in.

43. How many liters of gasoline can be stored in a cylindrical tank with a diameter of 5 m and a height of 12 m? (1,000 L = 1 m³)

44. Find the total surface area of the cylindrical tank described in Problem 43.

Appendix C

Tables

TABLE C–1
Table of Integrals

Basic Integrals

1. $\int u^n \, du = \dfrac{u^{n+1}}{n+1} + C \quad [n \neq -1]$

2. $\int \dfrac{du}{u} = \ln |u| + C$

3. $\int e^u \, du = e^u + C$

4. $\int a^u \, du = \dfrac{a^u}{\ln a} + C \quad [a > 0]$

5. $\int \sin u \, du = -\cos u + C$

6. $\int \cos u \, du = \sin u + C$

7. $\int \tan u \, du = -\ln |\cos u| + C$

8. $\int \cot u \, du = \ln |\sin u| + C$

9. $\int \sec u \, du = \ln |\sec u + \tan u| + C$

10. $\int \csc u \, du = \ln |\csc u - \cot u| + C$

11. $\int \sec^2 u \, du = \tan u + C$

12. $\int \csc^2 u \, du = -\cot u + C$

13. $\int \sec u \tan u \, du = \sec u + C$

14. $\int \csc u \cot u \, du = -\csc u + C$

15. $\int \dfrac{du}{\sqrt{a^2 - u^2}} = \mathrm{Arcsin} \left(\dfrac{u}{a} \right) + C$

16. $\int \dfrac{du}{a^2 + u^2} = \dfrac{1}{a} \mathrm{Arctan} \left(\dfrac{u}{a} \right) + C$

17. $\int \dfrac{du}{\sqrt{u^2 \pm a^2}} = \ln u + \left| \sqrt{u^2 \pm a^2} \right| + C$

18. $\int \dfrac{du}{u^2 - a^2} = \dfrac{1}{2a} \ln \left| \dfrac{u - a}{u + a} \right| + C$

Integrals Involving $a + bu$

19. $\int (a + bu)^n \, du = \dfrac{(a + bu)^{n+1}}{b(n+1)} + C \quad [n \neq -1]$

20. $\int u \, (a + bu)^n \, du = \dfrac{(a + bu)^{n+1}}{b^2} \left(\dfrac{a + bu}{n + 2} - \dfrac{a}{n + 1} \right) + C \quad [n \neq -1, -2]$

21. $\int \dfrac{du}{a + bu} = \dfrac{1}{b} \ln |a + bu| + C$

22. $\int \dfrac{du}{u(a + bu)} = -\dfrac{1}{a} \ln \left| \dfrac{a + bu}{u} \right| + C$

23. $\int \dfrac{du}{u^2(a + bu)} = -\dfrac{1}{au} + \dfrac{b}{a^2} \ln \left| \dfrac{a + bu}{u} \right| + C$

24. $\int \dfrac{u \, du}{a + bu} = \dfrac{1}{b^2} [a + bu - a \ln |a + bu|] + C$

25. $\displaystyle \int \frac{u^2\,du}{a + bu} = \frac{1}{b^3}\left[\frac{1}{2}(a + bu)^2 - 2a(a + bu) + a^2\ln|a + bu|\right] + C$

26. $\displaystyle \int \frac{du}{u(a + bu)^2} = \frac{1}{a(a + bu)} - \frac{1}{a^2}\ln\left|\frac{a + bu}{u}\right| + C$

27. $\displaystyle \int \frac{du}{u^2(a + bu)^2} = -\frac{a + 2bu}{a^2u(a + bu)} + \frac{2b}{a^3}\ln\left|\frac{a + bu}{u}\right| + C$

28. $\displaystyle \int \frac{u\,du}{(a + bu)^2} = \frac{1}{b^2}\left[\ln|a + bu| + \frac{a}{a + bu}\right] + C$

Integrals Involving $\sqrt{a + bu}$

29. $\displaystyle \int \sqrt{a + bu}\,du = \frac{2}{3b}(a + bu)^{3/2} + C$

30. $\displaystyle \int (\sqrt{a + bu})^n\,du = \frac{2}{b}\frac{(\sqrt{a + bu})^{n+2}}{n + 2} + C \quad [n \ne -2]$

31. $\displaystyle \int u\sqrt{a + bu}\,du = \frac{2(3bu - 2a)}{15b^2}(a + bu)^{3/2} + C$

32. $\displaystyle \int u^2\sqrt{a + bu}\,du = \frac{2(15b^2u^2 - 12abu + 8a^2)}{105b^3}(a + bu)^{3/2} + C$

33. $\displaystyle \int u^n\sqrt{a + bu}\,du = \frac{2u^n(a + bu)^{3/2}}{(2n + 3)b} - \frac{2an}{(2n + 3)b}\int u^{n-1}\sqrt{a + bu}\,du \quad [2n \ne -3]$

34. $\displaystyle \int \frac{\sqrt{a + bu}}{u}\,du = 2\sqrt{a + bu} + a\int \frac{du}{u\sqrt{a + bu}}$

35. $\displaystyle \int \frac{du}{\sqrt{a + bu}} = \frac{2}{b}\sqrt{a + bu} + C$

36. $\displaystyle \int \frac{u\,du}{\sqrt{a + bu}} = \frac{2(bu - 2a)}{3b^2}\sqrt{a + bu} + C$

37. $\displaystyle \int \frac{u^n\,du}{\sqrt{a + bu}} = \frac{2u^n\sqrt{a + bu}}{(2n + 1)b} - \frac{2an}{(2n + 1)b}\int \frac{u^{n-1}\,du}{\sqrt{a + bu}} \quad [2n \ne -1]$

38. $\displaystyle \int \frac{u\,du}{\sqrt{a + bu}} = \frac{2(bu - 2a)}{3b^2}\sqrt{a + bu} + C$

39. $\displaystyle \int \frac{du}{u\sqrt{a + bu}} = \frac{1}{\sqrt{a}}\ln\left|\frac{\sqrt{a + bu} - \sqrt{a}}{\sqrt{a + bu} + \sqrt{a}}\right| + C \quad [a > 0]$

40. $\displaystyle \int \frac{du}{u\sqrt{a + bu}} = \frac{2}{\sqrt{-a}}\text{Arctan}\sqrt{\frac{a + bu}{-a}} + C \quad [a < 0]$

41. $\displaystyle \int \frac{du}{u^n\sqrt{a + bu}} = -\frac{\sqrt{a + bu}}{(n - 1)au^{n-1}} - \frac{(2n - 3)b}{2(n - 1)a}\int \frac{du}{u^{n-1}\sqrt{a + bu}} \quad [n \ne 1]$

Integrals Involving $\sqrt{a^2 - u^2}$

42. $\displaystyle \int \sqrt{a^2 - u^2}\,du = \frac{u}{2}\sqrt{a^2 - u^2} + \frac{a^2}{2}\text{Arcsin}\left(\frac{u}{a}\right) + C$

43. $\displaystyle \int u\sqrt{a^2 - u^2}\,du = -\frac{1}{3}(a^2 - u^2)^{3/2} + C$

44. $\int u^n \sqrt{a^2 - u^2}\, du = -\dfrac{u^{n-1}(a^2 - u^2)^{3/2}}{n + 2} + \dfrac{(n-1)a^2}{n+2} \int u^{n-2}\sqrt{a^2 - u^2}\, du \quad [n \neq -2]$

45. $\int \dfrac{\sqrt{a^2 - u^2}}{u}\, du = \sqrt{a^2 - u^2} - a \ln \left| \dfrac{a + \sqrt{a^2 - u^2}}{u} \right| + C$

46. $\int \dfrac{\sqrt{a^2 - u^2}}{u^2}\, du = -\dfrac{\sqrt{a^2 - u^2}}{u} - \text{Arcsin}\left(\dfrac{u}{a}\right) + C$

47. $\int \dfrac{du}{u\sqrt{a^2 - u^2}} = -\dfrac{1}{a} \ln \left| \dfrac{a + \sqrt{a^2 - u^2}}{u} \right| + C$

48. $\int \dfrac{du}{u^2\sqrt{a^2 - u^2}} = -\dfrac{\sqrt{a^2 - u^2}}{a^2 u} + C$

49. $\int \dfrac{u^2\, du}{\sqrt{a^2 - u^2}} = -\dfrac{u}{2}\sqrt{a^2 - u^2} + \dfrac{a^2}{2} \text{Arcsin}\left(\dfrac{u}{a}\right) + C$

50. $\int \dfrac{u^n\, du}{\sqrt{a^2 - u^2}} = -\dfrac{u^{n-1}\sqrt{a^2 - u^2}}{n} + \dfrac{(n-1)a^2}{n} \int \dfrac{u^{n-2}\, du}{\sqrt{a^2 - u^2}} \quad [n \neq 0]$

51. $\int \dfrac{du}{u^n \sqrt{a^2 - u^2}} = \dfrac{-\sqrt{a^2 - u^2}}{(n-1)a^2 u^{n-1}} + \dfrac{n-2}{(n-1)a^2} \int \dfrac{du}{u^{n-2}\sqrt{a^2 - u^2}} \quad [n \neq 1]$

52. $\int (a^2 - u^2)^{3/2}\, du = \dfrac{u}{4}(a^2 - u^2)^{3/2} + \dfrac{3a^2 u}{8}\sqrt{a^2 - u^2} + \dfrac{3a^4}{8} \text{Arcsin}\left(\dfrac{u}{a}\right) + C$

53. $\int \dfrac{(a^2 - u^2)^{3/2}}{u}\, du = \dfrac{1}{3}(a^2 - u^2)^{3/2} - a^2\sqrt{a^2 - u^2} + a^3 \ln \left| \dfrac{a + \sqrt{a^2 - u^2}}{u} \right| + C$

54. $\int \dfrac{du}{(a^2 - u^2)^{3/2}} = \dfrac{u}{a^2\sqrt{a^2 - u^2}} + C$

55. $\int \dfrac{u^2\, du}{(a^2 - u^2)^{3/2}} = \dfrac{u}{\sqrt{a^2 - u^2}} - \text{Arcsin}\left(\dfrac{u}{a}\right) + C$

56. $\int \dfrac{du}{u(a^2 - u^2)^{3/2}} = \dfrac{1}{a^2\sqrt{a^2 - u^2}} - \dfrac{1}{a^3} \ln \left| \dfrac{a + \sqrt{a^2 - u^2}}{u} \right| + C$

Integrals Involving $\sqrt{u^2 + a^2}$ or $\sqrt{u^2 - a^2}$

57. $\int \sqrt{u^2 \pm a^2}\, du = \dfrac{1}{2}\left[u\sqrt{u^2 \pm a^2} \pm a^2 \ln |u + \sqrt{u^2 \pm a^2}| \right] + C$

58. $\int \dfrac{\sqrt{u^2 + a^2}}{u}\, du = \sqrt{u^2 + a^2} - a \ln \left| \dfrac{a + \sqrt{u^2 + a^2}}{u} \right| + C$

59. $\int \dfrac{\sqrt{u^2 - a^2}}{u}\, du = \sqrt{u^2 - a^2} - a \, \text{Arccos}\left(\dfrac{a}{u}\right) + C$

60. $\int \dfrac{\sqrt{u^2 \pm a^2}}{u^2}\, du = -\dfrac{\sqrt{u^2 \pm a^2}}{u} + \ln |u + \sqrt{u^2 \pm a^2}| + C$

61. $\int u^2\sqrt{u^2 \pm a^2}\, du = \dfrac{u}{8}(2u^2 \pm a^2)\sqrt{u^2 \pm a^2} - \dfrac{a^4}{8} \ln |u + \sqrt{u^2 \pm a^2}| + C$

62. $\int \dfrac{du}{u\sqrt{u^2 + a^2}} = \dfrac{1}{a} \ln \left| \dfrac{u}{a + \sqrt{u^2 + a^2}} \right| + C$

63. $\displaystyle \int \frac{du}{u\sqrt{u^2 - a^2}} = \frac{1}{a}\,\text{Arccos}\left(\frac{a}{u}\right) + C$

64. $\displaystyle \int \frac{u^2\,du}{\sqrt{u^2 \pm a^2}} = \frac{u}{2}\sqrt{u^2 \pm a^2} \mp \frac{a^2}{2}\ln\left|u + \sqrt{u^2 \pm a^2}\right| + C$

65. $\displaystyle \int \frac{du}{u^2\sqrt{u^2 \pm a^2}} = \frac{\mp\sqrt{u^2 \pm a^2}}{a^2 u} + C$

66. $\displaystyle \int \frac{du}{(u^2 \pm a^2)^{3/2}} = \frac{\pm u}{a^2\sqrt{u^2 \pm a^2}} + C$

67. $\displaystyle \int \frac{u^2\,du}{(u^2 \pm a^2)^{3/2}} = -\frac{u}{\sqrt{u^2 \pm a^2}} + \ln\left|u + \sqrt{u^2 \pm a^2}\right| + C$

68. $\displaystyle \int (u^2 \pm a^2)^{3/2}\,du = \frac{u}{8}(2u^2 \pm 5a^2)\sqrt{u^2 \pm a^2} + \frac{3a^4}{8}\ln\left|u + \sqrt{u^2 \pm a^2}\right| + C$

Integrals Involving Trigonometric Expressions

69. $\displaystyle \int \sin^2 u\,du = \frac{1}{2}u - \frac{1}{4}\sin 2u + C$

70. $\displaystyle \int \sin^n u\,du = -\frac{1}{n}\sin^{n-1}u\cos u + \frac{n-1}{n}\int \sin^{n-2}u\,du$

71. $\displaystyle \int \cos^2 u\,du = \frac{1}{2}u + \frac{1}{4}\sin 2u + C$

72. $\displaystyle \int \cos^n u\,du = \frac{1}{n}\cos^{n-1}u\sin u + \frac{n-1}{n}\int \cos^{n-2}u\,du$

73. $\displaystyle \int \tan^2 u\,du = \tan u - u + C$

74. $\displaystyle \int \tan^n u\,du = \frac{\tan^{n-1}u}{n-1} - \int \tan^{n-2}u\,du$

75. $\displaystyle \int \cot^2 u\,du = -\cot u - u + C$

76. $\displaystyle \int \cot^n u\,du = -\frac{\cot^{n-1}u}{n-1} - \int \cot^{n-2}u\,du$

77. $\displaystyle \int \sec^n u\,du = \frac{\sec^{n-2}u\tan u}{n-1} + \frac{n-2}{n-1}\int \sec^{n-2}u\,du$

78. $\displaystyle \int \csc^n u\,du = -\frac{\csc^{n-2}u\cot u}{n-1} + \frac{n-2}{n-1}\int \csc^{n-2}u\,du$

79. $\displaystyle \int u\sin u\,du = \sin u - u\cos u + C$

80. $\displaystyle \int u\cos u\,du = \cos u + u\sin u + C$

81. $\displaystyle \int u^n\sin au\,du = -\frac{u^n}{a}\cos au + \frac{n}{a}\int u^{n-1}\cos au\,du \quad [n \text{ a positive integer}]$

82. $\displaystyle \int u^n\cos au\,du = \frac{u^n}{a}\sin au - \frac{n}{a}\int u^{n-1}\sin au\,du \quad [n \text{ a positive integer}]$

83. $\displaystyle \int \frac{\sin u}{u^n}\,du = -\frac{\sin u}{(n-1)u^{n-1}} + \frac{1}{n-1}\int \frac{\cos u}{u^{n-1}}\,du \quad [n \neq 1]$

84. $\displaystyle \int \frac{\cos u}{u^n}\,du = -\frac{\cos u}{(n-1)u^{n-1}} - \frac{1}{n-1}\int \frac{\sin u}{u^{n-1}}\,du \quad [n \neq 1]$

85. $\displaystyle\int \frac{du}{\sin u \cos u} = \ln |\tan u| + C$

86. $\displaystyle\int \sin au \sin bu \, du = -\frac{\sin (a + b)u}{2(a + b)} + \frac{\sin (a - b)u}{2(a - b)} + C$

87. $\displaystyle\int \cos au \cos bu \, du = \frac{\sin (a + b)u}{2(a + b)} + \frac{\sin (a - b)u}{2(a - b)} + C$

88. $\displaystyle\int \sin au \cos bu \, du = -\frac{\cos (a + b)u}{2(a + b)} - \frac{\cos (a - b)u}{2(a - b)} + C$

89. $\displaystyle\int \sin^m u \cos^n u \, du = \frac{\sin^{m+1} u \cos^{n-1} u}{m + n} + \frac{n - 1}{m + n} \int \sin^m u \cos^{n-2} u \, du \quad [m, n > 0]$

$\displaystyle\quad\quad = -\frac{\sin^{m-1} u \cos^{n+1} u}{m + n} + \frac{m - 1}{m + n} \int \sin^{m-2} u \cos^n u \, du \quad [m, n > 0]$

Integrals Involving Exponential Functions

90. $\displaystyle\int u e^{au} \, du = \left(\frac{au - 1}{a^2}\right) e^{au} + C$

91. $\displaystyle\int u^2 e^{au} \, du = \frac{e^{au}}{a^3}(a^2 u^2 - 2au + 2) + C$

92. $\displaystyle\int u^n e^{au} \, du = \frac{1}{a} u^n e^{au} - \frac{n}{a} \int u^{n-1} e^{au} \, du \quad [n > 0]$

93. $\displaystyle\int e^{au} \ln u \, du = \frac{e^{au} \ln u}{a} - \frac{1}{a} \int \frac{e^{au}}{u} \, du$

94. $\displaystyle\int e^{au} \sin bu \, du = \frac{e^{au}(a \sin bu - b \cos bu)}{a^2 + b^2} + C$

95. $\displaystyle\int e^{au} \cos bu \, du = \frac{e^{au}(a \cos bu + b \sin bu)}{a^2 + b^2} + C$

96. $\displaystyle\int u^n a^u \, du = \frac{a^u u^n}{\ln a} - \frac{n}{\ln a} \int u^{n-1} a^u \, du$

97. $\displaystyle\int \frac{e^{au}}{u^n} \, du = \frac{1}{n - 1}\left(-\frac{e^{au}}{u^{n-1}} + a \int \frac{e^{au}}{u^{n-1}} du\right) \quad [n \text{ an integer} > 1]$

98. $\displaystyle\int \frac{du}{a + be^u} = \frac{u - \ln |a + be^u|}{a} + C$

Integrals Involving Logarithmic functions

99. $\displaystyle\int \ln u \, du = u \ln u - u + C$

100. $\displaystyle\int u^n \ln u \, du = \frac{u^{n+1} \ln u}{n + 1} - \frac{u^{n+1}}{(n + 1)^2} + C \quad [n \neq -1]$

101. $\displaystyle\int \frac{\ln u}{u} \, du = \frac{1}{2}(\ln u)^2 + C$

102. $\displaystyle\int \frac{du}{u \ln u} = \ln |\ln u| + C$

Integrals Involving Inverse Trigonometric Functions

103. $\int \text{Arcsin } u \; du = u \text{ Arcsin } u + \sqrt{1 - u^2} + C$

104. $\int \text{Arccos } u \; du = u \text{ Arccos } u - \sqrt{1 - u^2} + C$

105. $\int \text{Arctan } u \; du = u \text{ Arctan } u - \frac{1}{2} \ln(1 + u^2) + C$

106. $\int u^n \text{ Arcsin } u \; du = \frac{u^{n+1} \text{ Arcsin } u}{n + 1} - \frac{1}{n + 1} \int \frac{u^{n+1} \; du}{\sqrt{1 - u^2}}$ $[n \neq -1]$

107. $\int u^n \text{ Arccos } u \; du = \frac{u^{n+1} \text{ Arccos } u}{n + 1} + \frac{1}{n + 1} \int \frac{u^{n+1} \; du}{\sqrt{1 - u^2}}$ $[n \neq -1]$

108. $\int u^n \text{ Arctan } u \; du = \frac{u^{n+1} \text{ Arctan } u}{n + 1} - \frac{1}{n + 1} \int \frac{u^{n+1} \; du}{\sqrt{1 + u^2}}$ $[n \neq -1]$

Miscellaneous Integrals

109. $\int \frac{du}{au^2 + bu + c} = \frac{2}{\sqrt{4ac - b^2}} \text{ Arctan} \left(\frac{2au + b}{\sqrt{4ac - b^2}} \right) + C$ $[4ac - b^2 > 0]$

110. $\int \frac{u \; du}{au^2 + bu + c} = \frac{1}{2a} \ln |au^2 + bu + c| - \frac{b}{a\sqrt{4ac - b^2}} \text{ Arctan} \left(\frac{2au + b}{\sqrt{4ac - b^2}} \right) + C$ $[4ac - b^2 > 0]$

111. $\int \frac{du}{(au^2 + bu + c)^n} = \frac{2au + b}{(n - 1)(4ac - b^2)(au^2 + bu + c)^{n-1}} + \frac{2a(2n - 3)}{(n - 1)(4ac - b^2)}$

$\int \frac{du}{(au^2 + bu + c)^{n-1}}$ $[n \neq 1, 4ac - b^2 > 0]$

112. $\int \frac{du}{\sqrt{au^2 + bu + c}} = \frac{1}{\sqrt{a}} \ln |2au + b + 2\sqrt{a}\sqrt{au^2 + bu + c}|$ $[a > 0]$

$= \frac{1}{\sqrt{-a}} \text{Arcsin} \left(-\frac{2au + b}{\sqrt{b^2 - 4ac}} \right)$ $[a < 0, b^2 - 4ac > 0]$

113. $\int \sqrt{au^2 + bu + c} \; du = \frac{2au + b}{4a} \sqrt{au^2 + bu + c} + \frac{4ac - b}{8a} \int \frac{du}{\sqrt{au^2 + bu + c}}$

114. $\int \sqrt{2au - u^2} \; du = \frac{u - a}{2} \sqrt{2au - u^2} + \frac{a^2}{2} \text{ Arcsin} \left(\frac{u - a}{a} \right) + C$

115. $\int \frac{du}{\sqrt{2au - u^2}} = \text{Arccos} \left(\frac{a - u}{a} \right) + C$

TABLE C–2

Comparison of English and Metric Systems of Units

English	Metric		
	MKS	CGS	SI
Length:			
Yard (yd)	Meter (m)	Centimeter (cm)	Meter (m)
(0.914 m)	(39.37 in.)	(2.54 cm = 1 in.)	
	(100 cm)		
Mass:			
Slug	Kilogram (kg)	Gram (g)	Kilogram (kg)
(14.6 kg)	(1,000 g)		
Force:			
Pound (lb)	Newton (N)	Dyne	Newton (N)
(4.45 N)	(100,000 dynes)		
Temperature:			
Fahrenheit (°F)	Celsius or centigrade (°C)	Centigrade (°C)	Kelvin (K)
$\left(= \dfrac{9}{5}°C + 32\right)$	$\left(= \dfrac{5}{9}(°F - 32)\right)$		K = 273.15 + °C
Energy:			
Foot-pound (ft-lb)	Newton-meter (N-m) or joule (J)	Dyne-centimeter or erg	Joule (J)
(1.356 joules)	(0.7378 ft-lb)	(1 joule = 10^7 ergs)	
Time:			
Second (s)	Second (s)	Second (s)	Second (s)

Answers to Odd-Numbered Exercises

CHAPTER 1

1–1 Exercises, p. 8

1. -17 **3.** -4 **5.** 2 **7.** 0.51 **9.** 9 **11.** 18 **13.** 16 **15.** -4.6 **17.** $\dfrac{-23}{6}$ **19.** 19 **21.** -13

23. 2 **25.** $\dfrac{42}{31}$ **27.** -20 **29.** 10 **31.** -8 **33.** 4.8 **35.** $\dfrac{-57}{11}$ **37.** $\dfrac{27}{5}$ **39.** $\dfrac{141}{68}$ **41.** 4 or -1

43. $\dfrac{6}{5}$ or $\dfrac{-1}{5}$ **45.** $\dfrac{17}{2}$ or $\dfrac{3}{2}$ **47.** $352°\text{K}$ **49.** 1.59% **51.** $t = \dfrac{51}{8}$ **53.** \$17,500 **55.** 423.7

57. $\approx .717$ **59.** 35,156.25 **61.** 122,500 **63.** $95°$

1–2 Exercises, p. 14

1. $\dfrac{A}{Lw}$ **3.** $\dfrac{S - 2\pi r^2}{2\pi r}$ **5.** $\dfrac{E}{gh}$ **7.** $\dfrac{kLT - I}{kL}$ **9.** $\dfrac{V^2 - V_0^2}{2s}$ **11.** $2T + d$ **13.** $\dfrac{2E}{v^2}$ **15.** $\dfrac{L}{2\pi h}$

17. $\dfrac{5F - 160}{9}$ or $\dfrac{5}{9}(F - 32)$ **19.** $v - at$ **21.** $2v_{av} - v_0$ **23.** $\dfrac{L - 3.14r_2 - 2d}{3.14}$ **25.** $at + w_0$ **27.** $\dfrac{Fr^2}{kq_2}$

29. $\dfrac{y - b}{x}$ **31.** $\dfrac{fu + fv_s}{u}$ **33.** $\dfrac{2T + agd_2}{ag}$ **35.** $\dfrac{E - IR}{I}$ **37.** $\dfrac{PgJ + V^2_1}{V_1}$ **39.** $\dfrac{ab - ra - rb}{r}$

41. 1,254.4 m **43.** \$11.60 **45.** $\dfrac{45}{8}\,\Omega$ **47.** 2 yrs. **49.** \$1,500 **51.** 12 ft

1–3 Exercises, p. 22

1. $x \geq -13$

3. $x \leq 7$

5. $x < \dfrac{11}{3}$

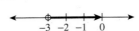

7. $x > -12$

9. $x > -3$

11. $x \leq -5$

13. $x > 6$

15. $x > 2$

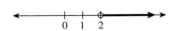

17. $x > \dfrac{-21}{2}$

19. $x \leq 4$

21. $x < \dfrac{1}{2}$

23. $x > \dfrac{38}{15}$

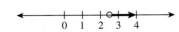

25. $x \geq \dfrac{21}{11}$

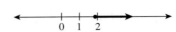

27. $x \leq \dfrac{1}{6}$

29. $x < 0$

31. $x > \dfrac{31}{58}$

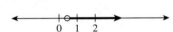

33. $x \leq 2$

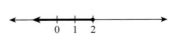

35. $x < \dfrac{31}{7}$

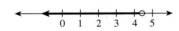

37. $x \geq 2$

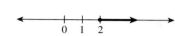

39. $x > \dfrac{1}{2}$

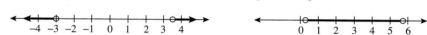

41. $-4 < x < \dfrac{4}{3}$

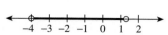

43. $x > \dfrac{7}{2}$ or $x < -3$

45. $\dfrac{3}{8} < x < \dfrac{45}{8}$

47. more than 375 miles

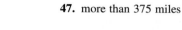

49. from 0 to $\dfrac{9}{2}$ sec   **51.** $x > 750$ mi   **53.** 47 hrs   **55.** 98.0 V $< V <$ 149 V

1–4 Exercises, p. 28

1. $x = kz$ **3.** $a = kbc$ **5.** $x = \dfrac{ky}{s}$ **7.** $m = \dfrac{knp^3}{r}$ **9.** $s = kt^2v$ **11.** $\dfrac{5}{2}$ **13.** 6   **15.** 4   **17.** 56

19. $\dfrac{63}{2}$ **21.** 90   **23.** 2   **25.** $\dfrac{25}{8}$ **27.** 1.35 m   **29.** 78.5 cm^2 **31.** 6,250 lb   **33.** $\dfrac{3}{7}$ Ω   **35.** 1.7 m

37. 2.5 s   **39.** 176.4 m   **41.** 818 rpm   **43.** 19   **45.** $t = 697$ days   **47.** \$96,750

1–5 Exercises, p. 36

1. 125 rods   **3.** 121 ft   **5.** 4 to 1   **7.** 56.56 in.   **9.** 175 sq. ft   **11.** $\dfrac{24}{7}$ hr or $3\dfrac{3}{7}$ **13.** 15 dimes 45 quarters

15. 300 lines/min   **17.** \$570   **19.** 351   **21.** 7.5 ft/min   **23.** 24, 25, 26   **25.** 84   **27.** 12 hrs

29. 62 in. = 5⅙ ft   **31.** 6 H, 4 H   **33.** \$345   **35.** 9, 9, 12 in.   **37.** 55 mph   **39.** 9.0 A, 6.5 A, 2.5 A

41. 39 V, 15 V, 24 V   **43.** 7.5 V, 1.3 V, 3.8 V, 2.5 V   **45.** \$.0833

Chapter Review, p. 39

1. 11   **3.** $\dfrac{1}{2}$ **5.** 3.7   **7.** 3   **9.** $\dfrac{5}{3}$ or $\dfrac{-2}{3}$ **11.** $at + v_0$ **13.** $\dfrac{y - y_0 + mx_0}{m}$ **15.** $IR + Ir$ **17.** $\dfrac{QJ}{I_2t}$

19. $\dfrac{L - a + d}{d}$ **21.** 64 N   **23.** 8,000 ft^2 **25.** $x > \dfrac{3}{7}$

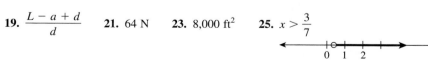

27. $x \geq \dfrac{90}{91}$

29. $x > \dfrac{7}{2}$ or $x < -5$

31. $x \geq \dfrac{16}{5}$ or $x \leq -4$ **33.** $x \geq \dfrac{1}{10}$ **35.** $x = \dfrac{k\sqrt{y}}{z^2}$ **37.** $\dfrac{3}{2}$

39. 96   **41.** 11 men   **43.** 8.68 h   **45.** 1,200 lb   **47.** 15 A, 39 A   **49.** \$2,000, \$2,300   **51.** \$80   **53.** 8 s

Chapter Test, p. 40

1. $\dfrac{6}{5}$ **3.** \$2,200 @ 7.3%, \$3,800 @ 9.5%   **5.** $b = 14$ ft   **7.** $\dfrac{-7}{6}$ or $\dfrac{-47}{6}$ **9.** 25 V

11. $x \geq 2$ or $x \leq \dfrac{-10}{7}$ **13.** 6 h @ 55 mph, 2 h @ 60 mph   **15.** $x = -14$ **17.** $\dfrac{Vk - a}{V^2}$ **19.** 1,800 lb

CHAPTER 2

2–1 Exercises, p. 52

1. $3a^2(b + 3a)$ **3.** $2a^2c(3ab - ac^2 + 2b)$ **5.** $(7x + 5y)(7x - 5y)$ **7.** $10x^2(3x^3 - 4 + 10x)$

9. $(2p + 3)(2p - 3)$ **11.** $y^3(9 + 7y^3 + 10yx^2)$ **13.** $(6m + 11)(6m - 11)$ **15.** $8st(4s - 5s^2t^2 + 1)$

17. $2xy(3x - y)(3x + y)$ **19.** $5x^2(5x^3 + 6 - 3x^4)$ **21.** $(5s + 7t)(5s - 7t)$ **23.** $19m^3(2mn^3 + 3n^4 - 1)$

25. $3xy(5x - y)(5x + y)$ **27.** $16m^4(4m^4 + 2 + 3m^{11})$ **29.** $(7c + 6d)(7c - 6d)$ **31.** $7a^5c^{10}(9a^2b^5 - 7)$

33. $(2x + 7y)(2x - 7y)$ **35.** $9x^3z(6y^2 - 7xy^5z^2 + 9)$ **37.** $(a^2 + 6b^2)(a^2 - 6b^2)$ **39.** $23s^8t^4(2st^2 + s^4 + 3t^5)$

41. $(x - 3)(2x + 3)$ **43.** $(x - 5)(3 - 5x)$ **45.** $4\pi(r_2 + r_1)(r_2 - r_1)$ **47.** $125(p + 3)(p - 3) = 0$

49. $m(v_2 - v_1)$ **51.** $\pi r^2(h_1 - h_2)$ **53.** $P(1 + rt)$

2–2 Exercises, p. 58

1. $(x - 3)(x - 2)$ **3.** $(y - 5)(y + 4)$ **5.** $(x + 8)(x - 4)$ **7.** $(a + 11b)(a - 3b)$ **9.** $(5x - 4)^2$

11. $(x + 6y)(x - 5y)$ **13.** $(3x - 2y)^2$ **15.** $(5x - 3y)^2$ **17.** $(m - 6n)(m + 4n)$ **19.** $2(2n + 1)(n - 3)$

21. $(3 - 4y)(2 + 3y)$ **23.** $(12a + 7)^2$ **25.** $(7k + 6g)(5k - 3g)$ **27.** $(8n - 7p)^2$ **29.** $(5x^2 - 9y^2)(x^2 + 2y^2)$

31. $(2m^2 + 7n^2)^2$ **33.** $3xy(2x - y)(x + y)$ **35.** $(7x^2 - 9y^2)(3x^2 + 2y^2)$ **37.** $(8a - 3b)^2$ **39.** $(9n + 5p)^2$

41. $(6x^2 - 5y^2)^2$ **43.** $8xy(x^2y - 8x^3 - 10y^3)$ **45.** $y(7x + 5y)^2$ **47.** $z(17xy + 34x^2 + 16y^3)$ **49.** $9(x - 3z)^2$

51. $4z(2x - 9y)^2$ **53.** $5(x + 3y)$ **55.** $6(2x + 5y)(x - 3y)$ **57.** $(12x + 7y)(12x - 7y)$ **59.** $2(x + 6)(x - 5)$

61. $(2t + 3)(t - 4)$ **63.** $4(x - 6)(x - 2)$ **65.** $(x - 14)(x + 12)$

2–3 Exercises, p. 63

1. $3xyz$ **3.** $4xyz^2$ **5.** $4a^2 + 4ab$ **7.** $18m^2$ **9.** $3xy - 12y$ **11.** $6x^2 + 3xy$ **13.** $\dfrac{x^4}{3y^3}$ **15.** $\dfrac{3a^4c^8}{b^4}$

17. $\dfrac{x^2}{9z^3}$ **19.** $\dfrac{4n^3p^2}{5m^5}$ **21.** $\dfrac{7a^2d^3}{c^7}$ **23.** $\dfrac{2b}{a}$ **25.** $\dfrac{7x(x - 1)}{8}$ **27.** $\dfrac{y - 2x}{x + 2y}$ **29.** $\dfrac{x + 5}{5 - x}$ **31.** $\dfrac{2a^3}{b^2}$

33. $\dfrac{x}{2}$ **35.** $\dfrac{a}{3b^3}$ **37.** $\dfrac{7x - y}{2x + 3y}$

2–4 Exercises, p. 68

1. $4x^3$ **3.** $\dfrac{1}{7x^4}$ **5.** $\dfrac{2x + 3}{2x}$ **7.** $\dfrac{5x(7x - 1)}{7x + 3}$ **9.** $3x^2y$ **11.** $\dfrac{a^4c^4}{15b}$ **13.** $\dfrac{m^3(a + 2b)}{a - 2b}$ **15.** $\dfrac{-3}{x}$

17. $\dfrac{3x + 2}{4x^2}$ **19.** $2(3x + 1)$ **21.** $\dfrac{(a - b)(2a - b)}{a^2 + b^2}$ **23.** $\dfrac{(3m + 8)(m - 1)}{(2m - 5)(2m + 3)}$ **25.** 1 **27.** $\dfrac{(7a - b)(5a - b)}{(a - 4b)}$

29. $\dfrac{3x + 5y}{4x + y}$ **31.** $\dfrac{(x - 3y)^2(2x + y)}{(x + 2y)^2(x - y)}$ **33.** $\dfrac{(5x + 6t)(2s - 7t)}{(s - t)^2}$ **35.** $\dfrac{3m + 2n}{3m - 2n}$ **37.** $\dfrac{(3 - x)(x + 4)}{x + 3}$

39. $\dfrac{wv^2}{gr}$ **41.** s **43.** 366.7 ft/s   **45.** 7.5%

2–5 Exercises, p. 74

1. $\dfrac{x^2 - 2x + 2}{4x^2}$ **3.** $\dfrac{10x^3 + 6x^2 - 1}{4x^2}$ **5.** $\dfrac{10m^2 - 15m - 6}{30m^3}$ **7.** $\dfrac{4x^2 - 7x - 6}{x(x + 1)}$ **9.** $\dfrac{4y^2 - 17y - 4}{y(7y + 2)}$

11. $\dfrac{2x^2 + (7x - 1)}{3x(2x - 1)}$ **13.** $\dfrac{-14a + 1}{3(a - 2)}$ **15.** $\dfrac{4x^2 - 8x + 1}{5x(x - 3)}$ **17.** $\dfrac{17y^2 + 2y}{(2y - 1)(3y + 2)}$ **19.** $\dfrac{2x - 23}{(x + 2)(2x - 5)}$

21. $\dfrac{7z^3 - 15z^2 + 25z + 3}{7z(z - 2)(2z + 3)}$ **23.** $\dfrac{14x^2 - 17x - 7}{(8x + 7)(3x - 8)}$ **25.** $\dfrac{b^2 - 32b - 18}{(2b + 1)(5b + 3)}$ **27.** $\dfrac{57x^2 + 49x - 30}{(7x - 4)(4x + 3)}$

29. $\dfrac{4n - 2}{n - 4}$ **31.** $\dfrac{9k^2 + 10k - 1}{(3k - 1)(k + 2)(2k + 1)}$ **33.** $\dfrac{63x + 33}{4(3x - 1)(x - 3)(x + 3)}$ **35.** $\dfrac{6y^2 + 21y + 19}{(y + 3)(y - 3)}$

37. $\dfrac{5x^2 + 8x}{(x - 4)(x + 1)(3x + 2)}$ **39.** $\dfrac{5x^3 + 4x^2 - 23x - 4}{(x - 3)(x + 2)(x + 4)(x - 1)}$ **41.** $\dfrac{x(5y - 3x^2)}{y(6 + 5x^2)}$ **43.** $\dfrac{4x^2 - 6x - 2}{6x^2 - 3x + 3}$

45. $\dfrac{-6x^2 + 4x - 7}{4x + 5}$ **47.** $\dfrac{3x(2x^2 + 4x + 18)}{(x - 4)^2(2x + 3)^2}$ **49.** $\dfrac{3a(8a^2 + 16ab - 6b^2)}{(2a - b)(4ab - 18a - 9b)}$ **51.** $\dfrac{2x^3 - 6x^2 + x - 4}{16x^3(x - 4)}$

53. 9.1% **55.** $\dfrac{2vt + at^2}{2}$ **57.** $\dfrac{v_1v_2(d_1 + d_2)}{d_1v_2 + d_2v_1}$ **59.** $\dfrac{2R_1R_2 + R_1R_3}{2R_1 + 2R_2 + R_3}$

2–6 Exercises, p. 81

1. $\dfrac{-21}{10}$ **3.** $\dfrac{16}{3}$ **5.** -12 **7.** $\dfrac{-16}{3}$ **9.** ≈ 6.1 **11.** ≈ 3.1 **13.** $\dfrac{2}{3}$ **15.** 6 **17.** $\dfrac{1}{6}$ **19.** -3

21. no solution   **23.** $\dfrac{-1}{4}$ **25.** $\dfrac{-25}{36}$ **27.** -2 **29.** 1 **31.** no solution   **33.** no solution   **35.** $\dfrac{mx}{2x - m}$

37. $M - \dfrac{m^2 d}{2\pi^2 r^2}$ **39.** $0.42m^3$ **41.** 6 A   **43.** $0.735 \ \Omega$ **45.** $15.625 \ \Omega$ **47.** $0.023 \ \mu F$ **49.** 5   **51.** 4, 6, 8

53. $\dfrac{6}{5}$ h   **55.** $\dfrac{10}{3}$ h   **57.** 0.3 m or 30 cm   **59.** 4.8 min

Chapter Review, p. 83

1. $x^2(7x + 38x^6 - 28)$ **3.** $5(3a^2 b + 8b^2 c - 5ac^3)$ **5.** $8m(4m^7 - 1 - 3m^3)$ **7.** $18x^2(xy^2 + 2z - 4x^3 y^3 z^2)$

9. $(2a + b)(7x - 6y)$ **11.** $6(m - 3n)(2x^2 + 1)$ **13.** $3r^2(-5s - 6t)$ **15.** $4r^2(6rx - 2ry + y + 4z)$

17. $(4x + 5y)(4x - 5y)$ **19.** prime   **21.** $(13s + 11t)(13s - 11t)$ **23.** $(5c^2 + 6d^2)(5c^2 - 6d^2)$

25. $\dfrac{1}{3} \pi(r_1 + r_2)(r_1 - r_2)$ **27.** $8p(50 - p)$ **29.** prime   **31.** $(13a + 5)^2$ **33.** $(7m - 8n)^2$

35. $(10s^2 - 13t^2)^2$ **37.** $(5x - 6y)(2x + 7y)$ **39.** $(2a + 11b)(a - 9b)$ **41.** $(m + 13)(m + 3)$

43. $(8s - 7t)(4s + 5t)$ **45.** $(8x + 5v)(9x - 8v)$ **47.** $(9s^2 - 5y^2)(2s^2 - 7y^2)$ **49.** $16(t^2 - 3t + 4)$

51. $\dfrac{m^3}{2n^2}$ **53.** $\dfrac{2x}{y(x^2 y^2 - 3)}$ **55.** $\dfrac{8(x + 1)}{5x + 6}$ **57.** $-\dfrac{x + 1}{5 + 4x}$ **59.** $\dfrac{7}{12x^3 y^2}$ **61.** $\dfrac{5a^7 c^7}{9b^3}$ **63.** $\dfrac{7x - 3}{3x(3x - 4)}$

65. $\dfrac{(5x - y)(x - y)}{(3x + y)(x + 4y)}$ **67.** $\dfrac{54x^3}{(3x + 4y)^2}$ **69.** $\dfrac{28xy^3 + 30y^2 - 25x^2}{10x^2 y^3}$ **71.** $\dfrac{3x + 7}{2(x - 4)}$ **73.** $\dfrac{10x^2 - 26x}{(x - 2)(x + 1)(x - 3)}$

75. $\dfrac{4x^2 - 24x + 31}{(3x + 4)(3x - 4)(2x - 1)}$ **77.** $\dfrac{15x^2 - 33x + 22}{(3x - 4)(x + 5)(x - 1)}$ **79.** $\dfrac{5}{12y^2}$ **81.** $\dfrac{2(3x - 1)}{3xy^2}$ **83.** $\dfrac{3x^2(x + 2)}{4}$

85. $\dfrac{21x^2(18x^2 + 3x + 18)}{(63x - 4)(3x - 1)(x + 2)}$ **87.** $\dfrac{(5x + 4)(6x - 5)(3x^2 + 10x + 3)}{(x - 3)(x + 2)(17x^2 - x + 4)}$ **89.** -18 **91.** no solution   **93.** $\dfrac{3}{2}$

95. -1 **97.** 10

Chapter Test, p. 85

1. $\dfrac{-3y^3}{8x}$ **3.** $\dfrac{-6x + 15}{(x - 3)(2x - 3)}$ **5.** 5   **7.** $4m(2m + 3n)(m - 5n)$ **9.** $(7x + 3y)(4x - 9y)$

11. $e\pi(T_1^2 + T_1^2)(T_1 + T_2)(T_1 - T_2)$ **13.** $\dfrac{2}{11}$ **15.** $8(2x + 5)$, 120 when $x = 5$ **17.** $6\dfrac{1}{4}$ ft

CHAPTER 3

3–1 Exercises, p. 95

1. b^5 **3.** $\dfrac{1}{m^4}$ **5.** x **7.** $\dfrac{1}{p^4}$ **9.** $\dfrac{a^6}{b^3}$ **11.** a **13.** $x^{2/3}$ **15.** $\dfrac{x^6}{27y^{12}}$ **17.** $\dfrac{4}{x^2 y^2}$ **19.** $\dfrac{a^8}{36c^6}$

21. 1   **23.** 2   **25.** $3 + y$ **27.** 1   **29.** 1   **31.** $\dfrac{y^7 z^8}{x^{10}}$ **33.** $a^3 c^3$ **35.** $\dfrac{p^6}{m^3 n^2}$ **37.** $\dfrac{a^8}{b^8 c^6}$ **39.** $\dfrac{y^{24}}{x^{12} z^{12}}$

41. $\dfrac{8y^{12}}{x^{21}}$ **43.** $\dfrac{5x^2 + 37x + 68}{(2x + 5)^2} = \dfrac{(5x + 17)(x + 4)}{(2x + 5)^2}$ **45.** $\dfrac{ab^2 - a + 1}{a}$ **47.** $\dfrac{x^2y^3 + 1}{y^3}$

49. $H = \dfrac{4mLr}{(r - L)^2(r + L)^2}$ **51.** $F = \$6,785$ **53.** $A = \dfrac{m}{(1 + n)^x}$ **55.** 256 bits   **57.** 5%

3–2 Exercises, p. 100

1. 9   **3.** 4   **5.** 2   **7.** -2 **9.** 6   **11.** $\dfrac{15\sqrt[3]{3}}{4}$ **13.** 11,520   **15.** $\dfrac{7}{2}$ **17.** 27.35   **19.** 0.14

21. 0.52   **23.** 0.78   **25.** $\dfrac{1}{x^{5/6}}$ **27.** $a^{41/35}$ **29.** $\dfrac{m^{1/2}}{n^{3/2}}$ **31.** $\dfrac{1}{y^{5/8}}$ **33.** $b^{15/4}$ **35.** $x^{9/8}$ **37.** $\dfrac{1}{a^{47/21}}$

39. $\dfrac{y^{7/4}}{x^{5/12}}$ **41.** 81   **43.** 16   **45.** 6   **47.** $\dfrac{3x + 4}{(x - 1)^{1/4}}$ **49.** $\dfrac{-5y - 12}{(y + 2)^{3/5}}$ **51.** $\dfrac{2mL}{\sqrt{(r^2 + L^2)^3}}$ **53.** \$10,473.05

55. 0.82   **57.** 5%

3–3 Exercises p. 105

1. $6\sqrt{5}$ **3.** $2\sqrt[3]{3}$ **5.** $\sqrt[4]{90}$ (simplified)   **7.** $4a^2b\sqrt{5b}$ **9.** $5xy^2\sqrt{6y}$ **11.** $2y\sqrt[3]{4x^2y}$

13. $9mn^2y^4\sqrt{2mn}$ **15.** $11b^2c^6\sqrt{3ab}$ **17.** $12y^2z^4\sqrt{2xy}$ **19.** $\sqrt[6]{y}$ **21.** $\sqrt[6]{4x}$ **23.** $\dfrac{2a^3\sqrt{2a}}{b^3}$

25. $4xz^2\sqrt[3]{3x^2y^2z^2}$ **27.** $2mnp^3\sqrt[4]{5n^3p^3}$ **29.** $2km^2\sqrt[5]{2m^2}$ **31.** $\dfrac{\sqrt{30}}{4}$ **33.** $\dfrac{\sqrt{105}}{5}$ **35.** $\dfrac{\sqrt{138ab}}{3}$

37. $\dfrac{a\sqrt{14a}}{4}$ **39.** $\dfrac{9m^2\sqrt{7n}}{7n}$ **41.** $\dfrac{\sqrt{a}}{2}$ **43.** $\dfrac{\sqrt{2b}}{2}$ **45.** $\dfrac{2m\sqrt{70n}}{7n^2}$ **47.** $\dfrac{m\sqrt[3]{12}}{2}$ **49.** $\dfrac{2m\sqrt[3]{12m^2n^2}}{3n^2}$

51. $\dfrac{3\sqrt{2ab}}{2b}$ **53.** $\dfrac{6xy\sqrt{10yz}}{5z^2}$ **55.** $\dfrac{6m^3\sqrt[3]{n}}{n^2}$ **57.** already simplified   **59.** $x + 4$ **61.** $y + \dfrac{5}{2}$

63. $\dfrac{(4m - 1)\sqrt{2}}{2}$ **65.** $6\sqrt{30}$ rad/s   **67.** 433.01 m^2 **69.** $\dfrac{\sqrt{8x^2 + 12x + 11}}{2}$

3–4 Exercises, p. 111

1. $10\sqrt{3}$ **3.** $\sqrt{5}$ **5.** $15\sqrt{3} - 12\sqrt{5}$ **7.** $-6\sqrt{6}$ **9.** $5a^2\sqrt{10a} + 18a\sqrt{10}$ **11.** $(3ab^2 + 4a^2b)\sqrt{3a}$

13. $-\sqrt[3]{2}$ **15.** $17\sqrt[3]{3}$ **17.** $60x\sqrt{2}$ **19.** $\dfrac{11\sqrt{2}}{a}$ **21.** $\dfrac{(3 - 2a^2)\sqrt{2a} + 12a\sqrt{a}}{a^2}$ **23.** $\sqrt{15} + 6\sqrt{5}$

25. $8\sqrt{3} - 8\sqrt{30}$ **27.** $\sqrt{6} + 2\sqrt{3} - \sqrt{15} - \sqrt{30}$ **29.** $\sqrt[3]{20x^2} + 7\sqrt[3]{10x} + 6\sqrt[3]{2x} + 42$

31. $8\sqrt{15} - 24\sqrt{35} + 2\sqrt{21} - 42$ **33.** $86 + 13\sqrt{3}$ **35.** $6\sqrt[4]{40} + 14\sqrt[4]{30} - 24\sqrt[4]{6} - 28\sqrt[4]{72}$

37. $56m + 22\sqrt{70mn} + 120n$ **39.** $240 - 80\sqrt{5}$ **41.** $157 - 84\sqrt{3}$ **43.** 74 **45.** $\dfrac{\sqrt{7}}{7}$ **47.** $\dfrac{2\sqrt{30}}{5}$

49. $\dfrac{2\sqrt{7} + 3\sqrt{14}}{-42}$ **51.** $\dfrac{12a + 8\sqrt{ab} - 3b\sqrt{a} - 2b\sqrt{b}}{16a - b^2}$ **53.** $\dfrac{20(d - \sqrt{3d + 400})}{d^2 - 3d - 400}$ **55.** $\dfrac{\sqrt{w^2 w_1^2 - c^2 g^2}}{w_1}$

3–5 Exercises, p. 115

1. 8 **3.** 4 **5.** $\dfrac{17}{5}$ **7.** no solution   **9.** 15 **11.** 4 **13.** -5 **15.** 5 **17.** $\dfrac{2}{3}$ **19.** $\dfrac{5}{2}$

21. 12 **23.** no solution   **25.** no solution   **27.** 3 **29.** $\dfrac{13}{5}$ **31.** $\dfrac{-8}{3}$ **33.** 0 **35.** $\dfrac{17}{8}$

37. $\dfrac{1}{2}$ **39.** $0, -5$ **41.** $-\dfrac{1}{2}$ **43.** -2 **45.** no solution   **47.** $\dfrac{v^2 m}{3k}$ **49.** $\dfrac{4}{3}\pi r^3$ **51.** $\dfrac{v_0^2 - v^2}{2a}$

53. $3{,}598{,}500$

Chapter Review, p. 117

1. $\dfrac{8y^8}{x^3}$ **3.** $\dfrac{2n^{11}}{3m^2}$ **5.** $\dfrac{3p^9}{m^2 n}$ **7.** $\dfrac{a^{12}}{b^9 c^{15}}$ **9.** $\dfrac{y^2 + 3x}{xy^2}$ **11.** 27 **13.** $\dfrac{1}{2}$ **15.** $\dfrac{-19}{4}$ **17.** $6^{5/6}$ or 4.45

19. 2.83 **21.** $\dfrac{m^{11/6}}{n^{8/5}}$ **23.** $\dfrac{y^{2/3} z^2}{x^{13/4}}$ **25.** $6xy^2\sqrt{5y}$ **27.** $3x^2 y^4 \sqrt{6x}$ **29.** $\dfrac{2xy\sqrt{2y}}{z^3}$ **31.** $\dfrac{3xy^2\sqrt{3xyz}}{4z^3}$

33. $\dfrac{a^2 b^4 \sqrt{abc}}{2c}$ **35.** $\dfrac{(2x + 5)\sqrt{2}}{2}$ **37.** $-3\sqrt{2}$ **39.** $-9\sqrt{3} + 12\sqrt{2}$ **41.** $8 - 3\sqrt{6} + 24\sqrt{2}$

43. $3\sqrt{10} + 12\sqrt{6}$ **45.** $3 + 2\sqrt{6} - 3\sqrt{15} - 6\sqrt{10}$ **47.** -84 **49.** $\dfrac{2}{x}$ **51.** $\dfrac{3\sqrt{21}}{7}$ **53.** $\dfrac{3x - 10\sqrt{x} + 8}{x - 4}$

55. $\dfrac{2m - 3\sqrt{mn} + n}{m - n}$ **57.** 9 **59.** 11 **61.** $\dfrac{8}{3}$ **63.** no solution   **65.** $\dfrac{45}{7}$

67. no solution   **69.** no solution

Chapter Test, p. 118

1. $11\sqrt{3} - 14\sqrt{6}$ **3.** $\dfrac{33}{4}$ **5.** $\dfrac{M}{LT^2}$ **7.** $\dfrac{3y^5}{5x^5 z^4}$ **9.** $6x^2 y\sqrt{6xyz}$ **11.** $\dfrac{x^4 z\sqrt{2xy}}{6y^3}$ **13.** $\dfrac{81x^8}{4y^{14}}$

15. $n = 144.53$ **17.** no solution

CHAPTER 4

4–1 Exercises, p. 128

1. $P(s) = 4s$ **3.** $A(w) = 5w$ **5.** $C(m) = 0.60 + 0.40m$ **7.** $I(V) = \dfrac{V}{10}$ **9. a.** 26 **b.** 51 **c.** 11

11. a. -19 **b.** 1 **c.** -19 **13. a.** 6.66714 **b.** -7 **c.** -5.6 **15. a.** -3 **b.** 2.27075 **c.** 0.98

17. a. 48 **b.** 8 **c.** 93   **19.** $3x - 38, 6s + 1$ **21.** $\dfrac{6n + 8}{2n - 1}, \dfrac{-11}{m(n - 1)}$ **23.** $30y^2 - 34y + 10$

25. All reals ≥ -7 **27.** All reals ≥ 0 except $\dfrac{1}{2}$ **29.** All reals except 0 and -6 **31.** 5 mi/h^2

33. 61.5%   **35.** 84 m^3

4–2 Exercises, p. 132

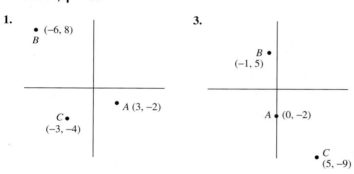

5. A: $(8, -4)$ B: $(-1, 2)$ C: $(-4, -5)$ **7.** II   **9.** III   **11.** on the y axis   **13.** 0   **15.** I and III

17. $(0, 5)$ **19.** $\dfrac{-1}{2}$ **21.** IV

4–3 Exercises, p. 139

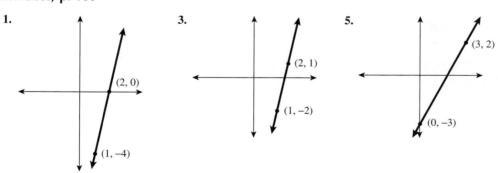

7.

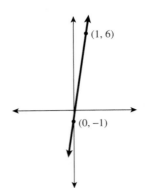

9.

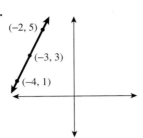

11.

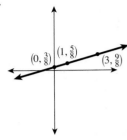

13.

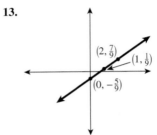

15.

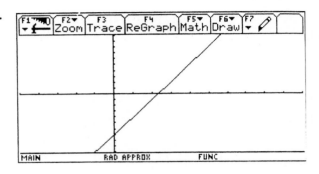

17.

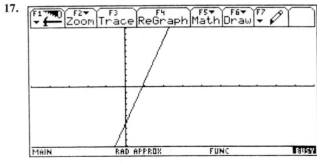

19.

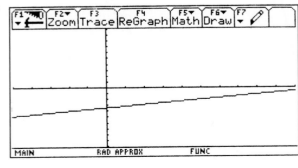

21.

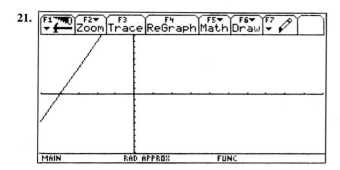

23.

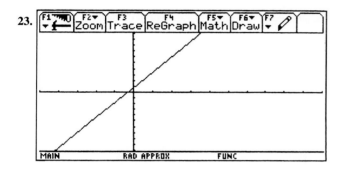

25.

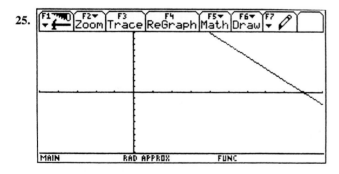

27.

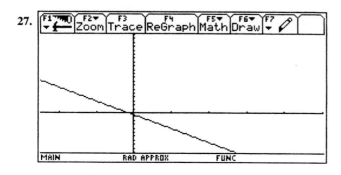

29.

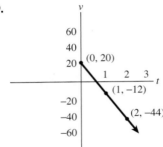

31.

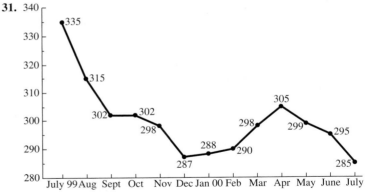

The stockpile is falling below the normal range.

33.

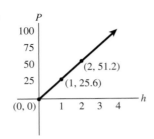

4–4 Exercises, p. 147

1. 5.4　　**3.** 4.24　　**5.** 9.2　　**7.** 3　　**9.** 6　　**11.** (5.5, 1)　　**13.** $\left(\dfrac{-5}{2}, 6\right)$　　**15.** (2.35, −3.65)

17. (−3, −10)　　**19.** $\left(\dfrac{23}{6}, \dfrac{29}{16}\right)$　　**21.** $\dfrac{2}{7}$　　**23.** undef　　**25.** $\dfrac{15}{2}$　　**27.** −1.3　　**29.** $\dfrac{-4}{7}$　　**31.** neither

33. perpendicular　　**35.** neither　　**37.** $\overline{AC}$ is perpendicular to $\overline{CB}$; therefore it is a right triangle.　　**39.** square

4–5 Exercises, p. 153

1.

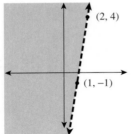

3.

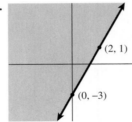

5.

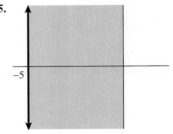

7.

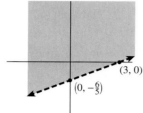

$(3, 0)$
$(0, -\frac{6}{5})$

9.

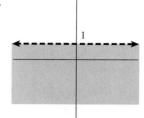

$(\frac{9}{4}, 0)$
$(0, -3)$

11.

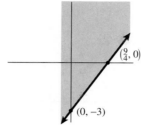

1

13.

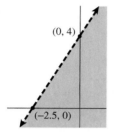

$(0, 4)$
$(-2.5, 0)$

15.

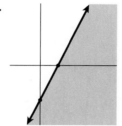

17.

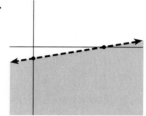

19.

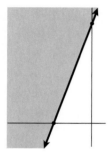

21.

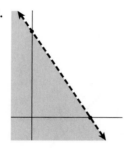

23.

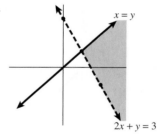

$x = y$
$2x + y = 3$

25.

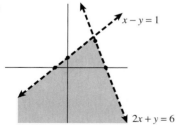

$x - y = 1$
$2x + y = 6$

27.

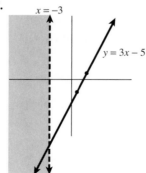

$x = -3$
$y = 3x - 5$

29. max profix at $(6.9, 3.3)$

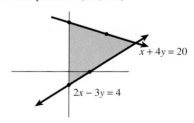

$x + 4y = 20$
$2x - 3y = 4$

31. 0 economy and 6 deluxe   **33.** 270 of type one and 4,300 of type two   **35.** 0 of first computer and 13 of second

4–6 Exercises, p. 159

1.

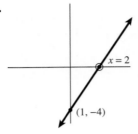

3.

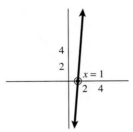

5.

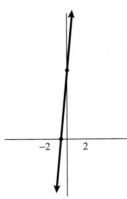

7.

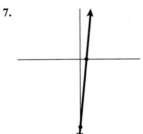

9.

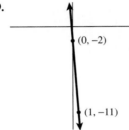

11.

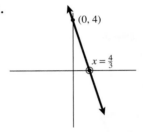

13.

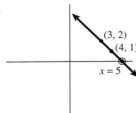

15.

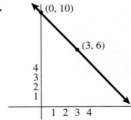

17. 4,000 units @ $12,000   **19.** 3,000 units at $42,000

21. Since he must produce 3,000 units to break even, he suffers a loss at 1,000 units.

23.

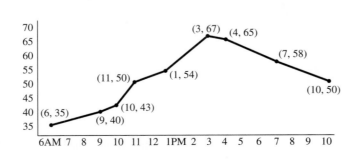

25. approx. 0.76 V

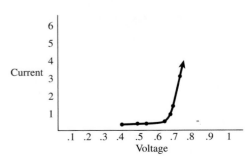

27.

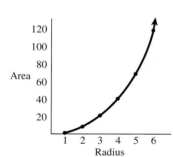

29. approx. 265 mA

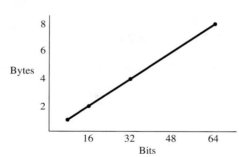

Chapter Review, p. 161

1. $P(s) = 3s$ **3.** $A(b) = \dfrac{1}{2} b(b + 3)$ **5.** $-6, 12$ **7.** 31, 3   **9.** $24y - 26$ **11.** $x \geq \dfrac{-8}{3}$ **13.** $t \geq 4$

15. All reals except 0 and 3

17.

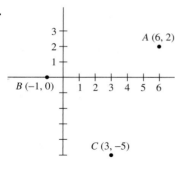

19.

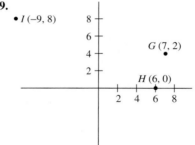

21.

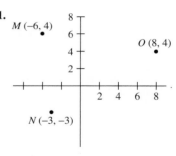

23.

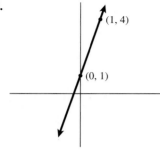

25.   **27.**   **29.**

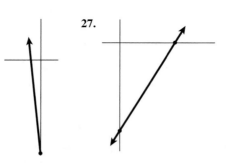

31. $11.18; \dfrac{-1}{2}; \left(3, \dfrac{7}{2}\right)$ **33.** $8.6; \dfrac{-5}{7}; \left(\dfrac{-5}{2}, \dfrac{-15}{2}\right)$ **35.** $15; 0; \left(\dfrac{-1}{2}, -3\right)$ **37.** $7.06; \dfrac{-105}{4}; \left(\dfrac{4}{15}, \dfrac{-1}{2}\right)$

39. $11.76; \dfrac{39}{4}; (2.3, 2.25)$

41.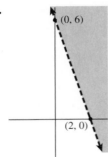
(0, 6)
(2, 0)

43.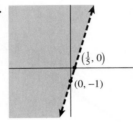
$\left(\frac{1}{5}, 0\right)$
(0, −1)

45.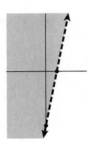

47. no solution

49.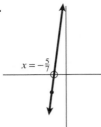
$x = -\frac{5}{7}$

51.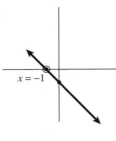
$x = -1$

53.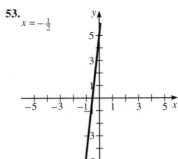
$x = -\frac{1}{2}$

55.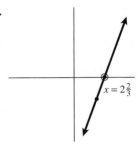
$x = 2\frac{2}{3}$

57. 2,000 units at $16,000   **59.** Yes

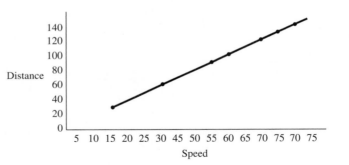

Distance

Speed

Chapter Test, p. 162

1. 14   **3.**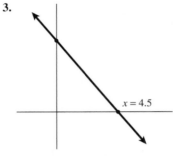
$x = 4.5$

5.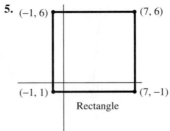
(−1, 6)   (7, 6)
(−1, 1)   (7, −1)
Rectangle

7.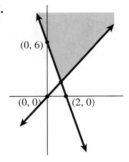
(0, 6)
(0, 0)   (2, 0)

9.

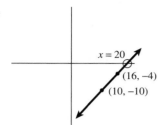

13. $19.72; \left(-1, \dfrac{-3}{2}\right)$ **15.** $16 \, m^2$ **17.** $r(P, t) = \dfrac{I}{Pt}$

CHAPTER 5

5–1 Exercises, p. 175

1. $5, 4$ **3.** $7, -9$ **5.** $\dfrac{1}{2}, \dfrac{-5}{3}$ **7.** $-3/2$ **9.** $-\dfrac{1}{5}, -\dfrac{2}{3}$ **11.** $-\dfrac{3}{7}, \dfrac{5}{4}$ **13.** $0, 4$ **15.** $\dfrac{5}{4}, -\dfrac{5}{4}$

17. $-1, -4$ **19.** -6 **21.** $6, -4$ **23.** $2 \pm 2\sqrt{6}$ **25.** $-3 \pm \sqrt{23}$ **27.** $3, -1$ **29.** $\dfrac{9 \pm \sqrt{113}}{2}$

31. $\dfrac{-4 \pm \sqrt{42}}{2}$ **33.** $\dfrac{9 \pm \sqrt{249}}{6}$ **35.** $\dfrac{-10 \pm 2\sqrt{55}}{5}$ **37.** $\dfrac{21 \pm j\sqrt{35}}{14}$ **39.** $\dfrac{9 \pm \sqrt{65}}{4}$ **41.** $\dfrac{17 \pm j\sqrt{31}}{8}$

43. $0.03 \, A, -0.63 \, A$ **45.** $12, 16, 20$ **47.** $750, 250$ **49.** $24 \, h$ **51.** $\dfrac{7}{5} \, A$ **53.** $5 \, A$ **55.** $1.65 \, ft$ **57.** $19 \, ft$

5–2 Exercises, p. 185

1. $\dfrac{3}{4}, -2$ **3.** $2 \pm j$ **5.** $\dfrac{5 \pm \sqrt{457}}{18}$ **7.** $2^{2/3}, -2$ **9.** $\dfrac{1 \pm j\sqrt{439}}{20}$ **11.** $\dfrac{1 \pm \sqrt{113}}{8}$ **13.** $6 \pm \sqrt{66}$

15. $\dfrac{\pm 3\sqrt{2}}{4}$ **17.** $0, \dfrac{15}{8}$ **19.** $\dfrac{3 \pm j\sqrt{31}}{5}$ **21.** $\dfrac{-9 \pm \sqrt{67}}{2}$ **23.** $\dfrac{27 \pm \sqrt{985}}{16}$ **25.** $\dfrac{11 \pm \sqrt{109}}{2}$

27. $\dfrac{1 \pm \sqrt{1,697}}{16}$ **29.** $\dfrac{-1 \pm \sqrt{6}}{3}$ **31.** $-2 \pm \sqrt{7}$ **33.** $8 \pm 5\sqrt{2}$ **35.** $\dfrac{9 \pm \sqrt{65}}{2}$ **37.** no real solution

39. $\dfrac{9 - \sqrt{193}}{8}$ is only solution   **41.** 2 real   **43.** 2 real   **45.** 2 real   **47.** 2 real   **49.** 2 complex   **51.** 2 real

53. 2 real   **55.** 2 complex   **57.** $\dfrac{-\pi h \pm \sqrt{\pi^2 h^2 + 2\pi s}}{2\pi}$ **59.** 496.64 **61.** $8.7 \, ft, 0.3 \, ft$ **63.** $9.1 \, s$

65. $12 \, in \times 14 \, in$

5–3 Exercises, p. 194

1. $x = \pm 2$ **3.** $x = \pm 1.73$ **5.** $x = 0, \pm 3$ **7.** $x = \pm 1.53$ **9.** $x = 3, -5$ **11.** 81 **13.** $2.04, 2.13$

15. no real solution   **17.** $6,561, 8$ **19.** $19,683, 512$ **21.** $-2, -0.0067$ **23.** $-3, 1$ **25.** $1,666 \, rad/s$

27. $45 \, hrs$

5–4 Exercises, p. 194

1. $x < -3$ or $x > 0$

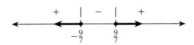

3. $-6 \leq x \leq -\frac{1}{3}$

5. $x \leq 0$ or $x \geq 3$

7. $x < -\frac{5}{3}$ or $x > \frac{2}{7}$

9. $x \leq -\frac{9}{7}$ or $x \geq \frac{9}{7}$

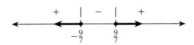

11. $-8 < x < \frac{5}{6}$

13. $-7 \leq x \leq -4$

15. $-\frac{1}{2} \leq x \leq 0$ or $x \geq 5$

17. $x \leq -2$ or $0 \leq x \leq 2$

19. $x < -4$ or $x > \frac{1}{2}$

21. $-9 \leq x < -3$ or $x \geq 1$

23. $x < -2$ or $0 < x < 3$

25. $x \leq -6$ or $-1 < x \leq \frac{4}{3}$

27. $v_0 > 149.9$ **29.** $1 < 8$ m

31. $t < 13\ s$ **33.** week 15

35. 19°C

5–5 Exercises, p. 201

1. $\frac{5}{2}$ s   **3.** 3 ft   **5.** -4 or -2 **7.** 17.64 d   **9.** \$4.16 or \$4.40   **11.** 0.56 cm   **13.** 1 cm   **15.** $-3, -1; 1, 3$

17. 20.84 ft   **19.** 2   **21.** $6\frac{1}{4}$ units   **23.** 10 and 10   **25.** 9.65 m   **27.** $\approx$3 ft   **29.** 600 kbs   **31.** 4 s

Chapter Review, p. 203

1. $\frac{2}{3}, -6$ **3.** $-4, 4$ **5.** $-\frac{1}{2}, -5$ **7.** $0, \frac{5}{7}$ **9.** $-\frac{5}{4}, 4$ **11.** $-10, 2$ **13.** $\frac{1}{2}$ **15.** $\frac{5}{3}, -\frac{3}{8}$

17. $-3 \pm \sqrt{21}$ **19.** $5, 0$ **21.** $\frac{3 \pm \sqrt{37}}{2}$ **23.** $\frac{-7 \pm j\sqrt{155}}{6}$ **25.** $\frac{-5 \pm 7\sqrt{5}}{10}$ **27.** $\frac{-1 \pm \sqrt{3}}{3}$ **29.** $\frac{1 \pm j}{2}$

31. $\frac{\pm 2\sqrt{7}}{7}$ **33.** $\frac{6 \pm j\sqrt{29}}{5}$ **35.** $0, -\frac{8}{9}$ **37.** $\frac{-1 \pm \sqrt{229}}{6}$ **39.** $\frac{-3 \pm j\sqrt{19}}{2}$ **41.** $\pm 3\sqrt{2}$ **43.** $\frac{6 \pm 2j}{5}$

45. $\frac{3 \pm \sqrt{33}}{8}$ **47.** 2 real   **49.** 2 complex   **51.** 2 real   **53.** 2 real   **55.** $\pm 5, \pm 2$ **57.** $\frac{4}{9}, 1$ **59.** $\frac{1}{8}, 512$

61. $-3, \frac{1}{2}$ **63.** $\pm 3, \pm 2$ **65.** $x \le -1$ or $x \ge \frac{3}{4}$ **67.** $x < -5$ or $0 < x < \frac{3}{2}$ **69.** $-1 < x < 3$

71. $-3 \le x \le 1$ or $x > 2$ **73.** 18, 19 or $-19, -18$ **75.** $R_1 = 20\ \Omega, R_2 = 5\ \Omega; R_1 = 5\ \Omega, R_2 = 20\ \Omega$

77. 17.64 h, 14.64 h

Chapter Test, p. 204

1. $-\frac{4}{7}, 3$ **3.** $\frac{-13 \pm \sqrt{163}}{2}$ **5.** $\frac{7}{5}, \frac{3}{2}$ **7.** $\frac{-13 \pm \sqrt{357}}{8}$ **9.** $\frac{\pm \sqrt{14}}{3}$ **11.** $\frac{5}{3}, 1$ **13.** $\frac{6}{7}, -\frac{4}{3}$

15. $\approx \pm .8, \pm 1$ **17.** $\frac{3}{4}, -\frac{7}{2}$ **19.** $-8 \pm \sqrt{82}$

CHAPTER 6

6–1 Exercises, p. 217

1. not a solution   **3.** is a solution   **5.** not a solution   **7.** is a solution   **9.** not a solution

11.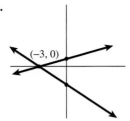

13.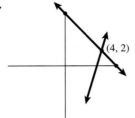

15.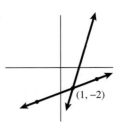

17.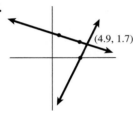

19.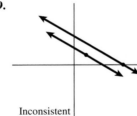

Inconsistent

21.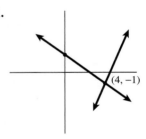

23.

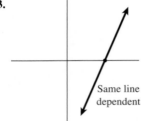

Same line
dependent

25.

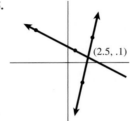

(2.5, .1)

27. $A = 103.5$ tons, $B = 113.5$ tons

29.

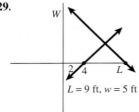

$L = 9$ ft, $w = 5$ ft

31. $I_1 \approx 3, I_2 \approx 3$ **33.** 30, 46   **35.** $S_1 = 12.3, S_2 = 8.6$

6–2 Exercises, p. 228

1. $x = 3, y = -1$ **3.** $x = -\dfrac{5}{2}, y = 2$ **5.** $x = -3, y = -\dfrac{1}{2}$ **7.** $x = \dfrac{71}{17}, y = \dfrac{30}{17}$ **9.** $x = -1, y = 4$

11. $x = \dfrac{59}{11}, y = \dfrac{69}{11}$ **13.** $y = -2, x = 2$ **15.** $x = \dfrac{10}{3}, y = 1$ **17.** $x = \dfrac{38}{17}, y = \dfrac{14}{17}$ **19.** $y = 6, x = 2$

21. $x = -3, y = -7$ **23.** $x = 2, y = 3$ **25.** $x = \dfrac{1}{2}, y = -1$ **27.** $x = 3, y = -2$ **29.** $x = -4, y = -1$

31. 10   **33.** 62   **35.** -1 **37.** -17.02 **39.** -21 **41.** 18, 12   **43.** 54, 58 mph   **45.** 23, 9

47. 1.0 A, 1.5 A   **49.** 100 hrs   **51.** 25, 22   **53.** 3,000, 1,500   **55.** 1.2 h, 2.8 h   **57.** 640 L of 1.4%, 86 L of 2.1%

59. .49, 1.5

6–3 Exercises, p. 239

1. $x = 0, y = 1, z = -2$ **3.** $x = 3, y = -2, z = 0$ **5.** $x = 2, y = 1, z = -2$ **7.** $x = 3, y = 0, z = 1$

9. $x = -1, y = 3, z = -2$ **11.** $x = -1, y = 0, z = -2$ **13.** $x = 2, y = 3, z = -2$ **15.** $x = 0, y = 1, z = 3$

17. $x = -2, y = 1, z = 2$ **19.** $x = -4, y = 1, z = -3$ **21.** $x = -3, y = 6, z = 0$ **23.** $x = -2, y = -1, z = 0$

25. $x = \dfrac{1}{2}, y = -\dfrac{1}{3}, z = 0$ **27.** $x = 0, y = \dfrac{1}{5}, z = -2$ **29.** -10 **31.** 0   **33.** 13   **35.** 16   **37.** 77   **39.** 60

41. 102   **43.** -675 **45.** (14, 20, 38)   **47.** (20n 12d, 6q)   **49.** $-2x^2 + x + 6$ **51.** \$2,015, \$2,310

53. -0.3 A, 0.3 A, 0.6 A   **55.** \$1,030   **57.** 8, 6, 4

6–4 Exercises, p. 248

1. $x = \frac{4}{3}h,\ y = \frac{2}{3}h$ **3.** 160 ml of 5%, 240 ml of 30% **5.** $\dfrac{16x + 3}{(2x - 2)(x + 3)} = \dfrac{19/4}{2x - 2} + \dfrac{45/8}{x + 3}$ **7.** 27 yd × 34 yd

9. $B = 117.5$ N, $A = 207.5$ N **11.** $33\frac{1}{3}yd^3$ of 25%, $66\frac{2}{3}yd^3$ of 40% **13.** Rate of plane = 350 mph, Wind = 50 mph

15. fixed = $200, variable = $1.80 **17.** $\dfrac{-2x - 42}{(x + 3)(x - 1)} = \dfrac{9}{x + 3} - \dfrac{11}{x - 1}$ **19.** $I_1 = 0.58$ A, $I_2 = 0.45$ A, $I_3 = 0.13$ A

21. $I_2 = 0.4$ A, $I_1 = 1$ A, $I_3 = 0.6$ A **23.** 12 dimes, 18 quarters **25.** 425, 425, 50 **27.** $-0.30, -0.45, 0.75$

29. 512, 1,024

6–5 Exercises, p. 257

1. $\begin{bmatrix} 4 & -4 \\ 11 & 8 \end{bmatrix}$ **3.** $\begin{bmatrix} 6 & 14 \\ 0 & 8 \end{bmatrix}$ **5.** $\begin{bmatrix} 5 & -1 \\ 4 & 4 \end{bmatrix}$ **7.** $\begin{bmatrix} -6 & 14 & 0 \\ 10 & -5 & 0 \end{bmatrix}$ **9.** $\begin{bmatrix} 1 & -4 & 19 \\ 9 & 1 & 5 \end{bmatrix}$ **11.** $\begin{bmatrix} 8 & 4 & 7 \\ 6 & 4 & 2 \end{bmatrix}$ **13.** $\begin{bmatrix} 19 & -1 & 5 \\ 1 & 9 & 9 \\ 13 & 2 & -2 \end{bmatrix}$

15. $\begin{bmatrix} -6 & 11 & -2 \\ -7 & 13 & 4 \\ 7 & 3 & -4 \end{bmatrix}$ **17.** $\begin{bmatrix} 18 & 11 \\ -3 & -45 \end{bmatrix}$ **19.** $\begin{bmatrix} 6 & -18 \\ -27 & -31 \end{bmatrix}$ **21.** $\begin{bmatrix} 17 & 47 \\ -19 & 79 \end{bmatrix}$ **23.** $\begin{bmatrix} 54 & 21 \\ 33 & -40 \end{bmatrix}$ **25.** $\begin{bmatrix} 47 & -43 \\ -3 & -18 \end{bmatrix}$

27. $\begin{bmatrix} -109 & -6 \\ -64 & 35 \end{bmatrix}$ **29.** $\begin{bmatrix} -10 & -74 \\ 31 & 4 \end{bmatrix}$ **31.** $-\frac{1}{2}\begin{bmatrix} -4 & -2 \\ -3 & -1 \end{bmatrix}$ **33.** $\frac{1}{34}\begin{bmatrix} 1 & 5 \\ -6 & 4 \end{bmatrix}$ **35.** $\frac{1}{23}\begin{bmatrix} 1 & 3 \\ -7 & 2 \end{bmatrix}$

37. $\begin{bmatrix} 0.186 & -0.058 & 0.093 \\ -0.628 & 0.384 & 0.186 \\ -0.047 & 0.140 & -0.023 \end{bmatrix}$ **39.** $\begin{bmatrix} -0.286 & 0.714 & 1 \\ -0.429 & 1.57 & 3 \\ 0.857 & -2.14 & -4 \end{bmatrix}$

Chapter Review, p. 259

1. No, it does not satisfy the second equation.

3.

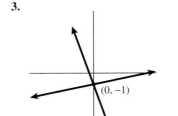

(0, −1)

5.

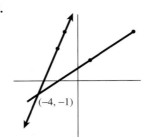

(−4, −1)

7.

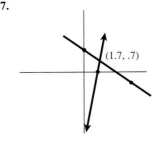

(1.7, .7)

9. $x = -3, y = 6$ **11.** $x = \dfrac{-8}{19}, y = \dfrac{7}{19}$ **13.** $x = 2, y = 2$ **15.** no solution, dependent **17.** -2

19. $6,000 @ $8\frac{1}{2}$%, $9,000 @ 9.75% **21.** $x = -2, y = 1, z = 3$ **23.** $x = 0, y = 4, z = 0$

25. $x = -1, y = 2, z = 3$ **27.** $x = 0, y = -3, z = 1$ **29.** 118 **31.** 8 **33.** 145 **35.** 32°, 105°, 43°

37. $x = 18, y = 37$ **39.** $I_1 = 1.06$ A, $I_2 = 0.46$ A, $I_3 = 0.6$ A   **41.** $\dfrac{34x - 18}{(4x - 3)(2x + 1)} = \dfrac{3}{4x - 3} + \dfrac{7}{2x + 1}$

43. $\begin{bmatrix} 15 & -7 \\ 3 & -5 \end{bmatrix}$ **45.** $\begin{bmatrix} 21 & -16 \\ 5 & 17 \end{bmatrix}$ **47.** $\begin{bmatrix} 9 & 20 & 2 \\ 0 & -1 & 13 \end{bmatrix}$ **49.** $\begin{bmatrix} 11 & 20 \\ 23 & 17 \end{bmatrix}$ **51.** $\begin{bmatrix} -10 & -62 \\ 5 & 35 \end{bmatrix}$ **53.** $\begin{bmatrix} -44 & 19 \\ 152 & -62 \end{bmatrix}$

55. $-\dfrac{1}{14}\begin{bmatrix} 2 & -3 \\ 0 & -7 \end{bmatrix}$ **57.** $\dfrac{1}{21}\begin{bmatrix} 9 & 3 \\ 8 & 5 \end{bmatrix}$ **59.** $-\dfrac{1}{16}\begin{bmatrix} -5 & -6 \\ -11 & -10 \end{bmatrix}$

Chapter Test, p. 262

1. $x = 6, y = -7$ **3.** It is a solution, it satisfies both equations.   **5.**

7. $\begin{bmatrix} 7 & -1 & 0 \\ 5 & 13 & 0 \end{bmatrix}$ **9.** 23 lb of $1.99 and 27 lb of $3.49   **11.** 137

13. $A = 240, B = 135$ **15.** no solution

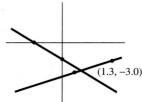
$(1.3, -3.0)$

CHAPTER 7

7–1 Exercises, p. 270

1. $f(2)=6$ **3.** $f(-1)=-28$ **5.** $f\left(\dfrac{1}{3}\right)=17$ **7.** $f(-2)=-30$ **9.** $f(-3)=79$ **11.** $f(-2)=71$ **13.** $f\left(-\dfrac{2}{3}\right)=-2$

15. $f(3) = 257$ **17.** $f(2) = 8$ **19.** $f(4) = 37$ **21.** $f(3) = 270$ **23.** $f(-2) = 7$ **25.** $f(-2) = 65$

27. $f\left(\dfrac{1}{2}\right) = -\dfrac{57}{8}$ **29.** $f\left(\dfrac{2}{3}\right) = -\dfrac{34}{9}$ **31.** not a factor   **33.** is a factor   **35.** is a factor   **37.** not a factor

39. not a factor   **41.** not a factor   **43.** is a factor   **45.** not a zero   **47.** is a zero   **49.** is a zero

51. is a zero   **53.** not a zero   **55.** is a zero   **57.** not a zero   **59.** is a root   **61.** is a root   **63.** not a root

65. not a root   **67.** not a root   **69.** not a root

7–2 Exercises, p. 275

1. $x^2 + x - 20$ **3.** $2x^2 - 7x + 10 - \dfrac{2}{x + 1}$ **5.** $9x^2 + 10x + 25 + \dfrac{60}{x - 2}$ **7.** $4x^2 + 3x + 9 + \dfrac{2}{x - 1}$

9. $x^2 + 5x + 10 + \dfrac{10}{x - 2}$ **11.** $6x^2 - 29x + 116 - \dfrac{448}{x + 4}$ **13.** $3x^3 + 13x^2 + 72x + 359 + \dfrac{1,804}{x - 5}$

15. $x^3 - 2x^2 - 3x + 6 - \dfrac{4}{x + 2}$ **17.** $x^3 - 2x^2 + 3x - 7 + \dfrac{15}{x + 3}$ **19.** $7x^4 + 14x^3 + 28x^2 + 60x + 120 + \dfrac{224}{x - 2}$

21. $x^4 - 3x^3 + 12x^2 - 56x + 224 - \dfrac{890}{x + 4}$ **23.** $3x^5 + 6x^4 + 12x^3 + 21x^2 + 42x + 84 + \dfrac{176}{x - 2}$ **25.** is a factor

27. is a factor   **29.** not a factor   **31.** not a factor   **33.** not a factor   **35.** not a factor   **37.** is a factor

39. not a factor   **41.** is a factor   **43.** not a zero   **45.** is a zero   **47.** is a zero   **49.** not a zero

7–3 Exercises, p. 280

1. $-3 \pm j\sqrt{5}, 4$ **3.** $\frac{2}{3}, \frac{5}{2}, -6$ **5.** $\frac{-3 \pm \sqrt{137}}{8}, 1$ **7.** $\frac{5 \pm j\sqrt{87}}{4}, 1$ **9.** $\frac{7 \pm \sqrt{13}}{2}, 5$

11. $\frac{5}{4}, -\frac{9}{2}, -7$ **13.** $-4, 3, -1, -2$ **15.** $4, -2, -2, -2$ **17.** $-7, -1, 3, 3$ **19.** $-2, -2, 5, 3$

21. $\frac{1}{3}, \frac{1}{5}, -2, 5$ **23.** $-1, -1, -1, 4$ **25.** $\frac{-3 \pm \sqrt{13}}{2}, -2, 3$ **27.** $\frac{-3 \pm 6\sqrt{2}}{7}, 6$ **29.** $5, -5, 3$

7–4 Exercises, p. 284

1. $\pm 1, \pm 2, \pm 4, \pm 5, \pm 20$ **3.** $\pm 1, \pm 2, \pm 3, \pm 4, \pm 6, \pm 8, \pm 12, \pm 24$ **5.** $\pm 1, \pm 2, \pm 3, \pm 4, \pm 6, \pm 9, \pm 12, \pm 18, \pm 36$

7. $\pm 1, \pm 2, \pm 3, \pm 6, \pm 9, \pm 18, \pm\frac{1}{3}, \pm\frac{2}{3}$ **9.** $\pm 1, \pm\frac{7}{2}, \pm 7, \pm\frac{1}{2}, \pm 14, \pm 2$ **11.** $\pm 1, \pm 2, \pm 3, \pm 4, \pm 6, \pm 12, \pm\frac{1}{2}, \pm\frac{3}{2},$

$\pm\frac{3}{5}, \pm\frac{4}{5}, \pm\frac{6}{5}, \pm\frac{12}{5}$ **13.** $\pm 1, \pm 2, \pm 3, \pm 6, \pm 9, \pm 18, \pm\frac{1}{2}, \pm\frac{3}{2}, \pm\frac{9}{2}, \pm\frac{1}{3}, \pm\frac{2}{3}, \pm\frac{1}{6}$ **15.** $\pm 1, \pm 3, \pm 5, \pm 15, \pm\frac{1}{2}, \pm\frac{3}{2},$

$\pm\frac{5}{2}, \pm\frac{15}{2}, \pm\frac{1}{7}, \pm\frac{3}{7}, \pm\frac{5}{7}, \pm\frac{15}{7}, \pm\frac{1}{14}, \pm\frac{3}{14}, \pm\frac{5}{14}, \pm\frac{15}{14}$ **17.** 3 positive, 1 negative   **19.** 3 positive, 0 negative

21. 3 positive, 0 negative   **23.** 4 positive, 3 negative   **25.** 3 positive, 1 negative   **27.** 3 positive, 4 negative

29. $1, \frac{1}{2}, -3$ **31.** $-2, -\frac{2}{3}, 4$ **33.** $-3, \frac{-7 \pm \sqrt{17}}{2}$ **35.** $3, -6, 1$ **37.** $-2, -7, 1$ **39.** $1, -\frac{3}{2}, \pm j$

41. $4, -3, -5, 1$ **43.** $1, -\frac{3}{2}, -1, -7$ **45.** $1, -1, 5, \frac{1 \pm j\sqrt{11}}{2}$ **47.** \$380,000 for advertising

49. $x = 10.2$ Dimensions: 4 cm $\times$ 4 cm $\times$ 9 cm or 10.2 $\times$ 10.2 $\times$ 1.4

51. \$39   **53.** 6 cm $\times$ 3 cm $\times$ 2 cm or 0.788 cm $\times$ 8.424 cm $\times$ 5.424 cm

7–5 Exercises, p. 289

1. 4.2   **3.** 1.9   **5.** 0.9   **7.** 2.2   **9.** $-5.8, -1.6$ **11.** $-1.6, -5.7$ **13.** $0.9, -1$ **15.** $0.8, -1.2$

17. -1.4 **19.** -1.3 **21.** increase by 3 cm   **23.** $t = 5$ s   **25.** $x = 22$

Chapter Review, p. 290

1. -198 **3.** 10   **5.** not a factor   **7.** not a factor   **9.** not a factor   **11.** $6x^2 - 11x + 14 - \frac{15}{x + 1}$

13. $x^3 - 5x^2 + 10x - 13 + \frac{18}{x + 2}$ **15.** $4x^3 + 11x^2 + 35x + 105 + \frac{306}{x - 3}$ **17.** is a zero   **19.** not a zero

21. is a zero   **23.** $-3, \frac{1}{2}, 1$ **25.** $-1, \frac{5 \pm j\sqrt{47}}{4}$ **27.** $\frac{-1}{2}, 3, j\sqrt{3}, -j\sqrt{3}$ **29.** $5, 1, 1 \pm j\sqrt{7}$

31. $\pm 1, \pm 2, \pm\frac{1}{2}, \pm\frac{1}{4}$ **33.** $\pm 1, \pm 2, \pm 3, \pm 4, \pm 6, \pm 12, \pm\frac{1}{3}, \pm\frac{2}{3}, \pm\frac{4}{3}$ **35.** $\pm 1, \pm 3, \pm 5, \pm 15, \pm\frac{1}{5}, \pm\frac{3}{5}$

37. 2 positive, 1 negative   **39.** 3 positive, 1 negative   **41.** 2 positive, 3 negative   **43.** $2, \frac{-3}{2}, 4$ **45.** $9, \frac{-3}{2}, \frac{-1}{2}$

47. $-3, 4, -2$ **49.** $-7, -2, \frac{3}{2}, 1$ **51.** 1.1   **53.** -1.6

Chapter Test, p. 291

1. The remainder is zero, so -2 is a zero of the equation.   **3.** $5, \dfrac{4}{3}, \dfrac{-1}{2}$ **5.** It is not a factor.   **7.** $6, \dfrac{-1}{2}, -2$

9. 3 positive, 1 negative   **11.** $7x^3 - 6x^2 + 6x - 12 + \dfrac{34}{x + 2}$ **13.** 2 positive, 3 negative

15. $x^4 + 8x^3 + 16x^2 + 25x + 50 + \dfrac{108}{x - 2}$

CHAPTER 8

8–1 Exercises, p. 302

1.
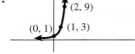

3.
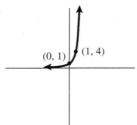

5.
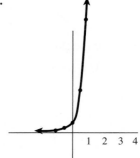

7.
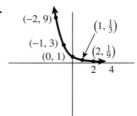

9.
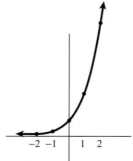

11.
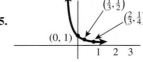

13.
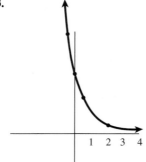

15.
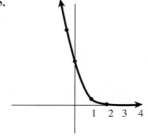

17. 8,100 at 4 h

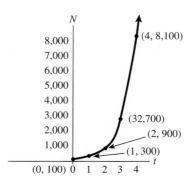

19. $13,345.74   **21.** $17,470.14   **23.** $38.59   **25.** $V_c = 60.6$ volts   **27.** $482,665.37   **31.** $27,644.81

8–2 Exercises, p. 310

1. $3^3 = 27$ **3.** $2^4 = 16$ **5.** $5^3 = 125$ **7.** $e^3 = 20.1$ **9.** $6^{-2} = \dfrac{1}{36}$ **11.** $e^2 = 7.39$ **13.** $\left(\dfrac{8}{27}\right)^{-2/3} = \dfrac{9}{4}$

15. $\log_6 216 = 3$ **17.** $\log_2 32 = 5$ **19.** $\log_3\left(\dfrac{1}{81}\right) = -4$ **21.** $\log_{1/3}\left(\dfrac{1}{27}\right) = 3$ **23.** $\ln 7.39 = 2$

25. $\log_{16}\left(\dfrac{1}{8}\right) = -\dfrac{3}{4}$ **27.** $\ln 0.37 = -1$ **29.** $\log_8\left(\dfrac{1}{4}\right) = -\dfrac{2}{3}$ **31.** 6 **33.** 512 **35.** $\dfrac{3}{4}$

37. 8 **39.** 6 **41.** 1 **43.** -1

45.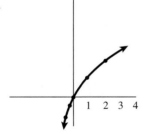

47.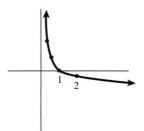

49.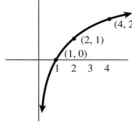

51.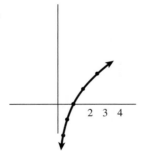

53.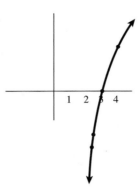

55.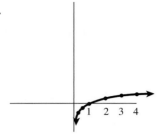

57.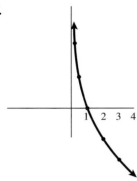

59.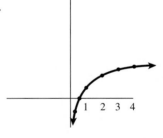

61. $t = 8.2$ yrs **63.** $x = \log_{1/2}\left(\dfrac{N}{N_0}\right)$

65. $x = -\dfrac{1}{k}\ln\left(\dfrac{I}{I_0}\right)$ **67.** $t = 5.8$ yrs

69. 1.9 dB

8–3 Exercises, p. 316

1. 7 **3.** 8 **5.** 4 **7.** 24 **9.** 4 **11.** $\dfrac{2}{3}$ **13.** $\log\dfrac{x}{(x-3)^2}$ **15.** $\log[(x+1)^2(x-3)]$ **17.** $\log\left(\dfrac{xy^3}{\sqrt{x-5}}\right)$

19. $\log\left(\dfrac{5\sqrt[3]{2y+5}}{y}\right)$ **21.** $\log[\sqrt[3]{x^2}(x+1)]$ **23.** $\log\left[\dfrac{(x-1)^3}{x^9}\right]$ **25.** $\log\left(\dfrac{x^4}{yz^3}\right)$ **27.** 0.7781 **29.** 0.3980

31. 0.5881   **33.** 1.3801   **35.** 1.8751   **37.** 1.9542   **39.** 1.2731   **41.** 2.965   **43.** 5.196   **45.** 3.0566

47.

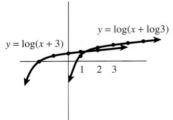

49. $14.7e^{-0.21h}$

8–4 Exercises, p. 324

1. 2.74   **3.** −2.41   **5.** 0.876   **7.** −4.75   **9.** −6.59   **11.** 0.536   **13.** 1.079   **15.** 0.924   **17.** $\dfrac{1}{17}$

19. 0.601   **21.** 1,002   **23.** 7   **25.** no solution   **27.** $\dfrac{1}{10}$ **29.** no solution   **31.** 1.58   **33.** no solution

35. $\Delta T_m = 148.26°$ **37.** 11.05 y   **39.** 1.875 dB   **41.** $x = 1,704$ units   **43.** $x = 2.136 \times 10^{96}$ **45.** 0.28 s

Chapter Review, p. 325

1.

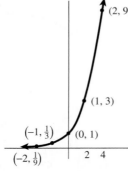

3.

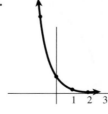

5.

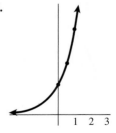

7.

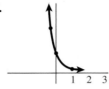

9.

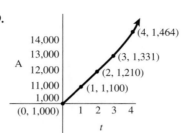

11. $450.51

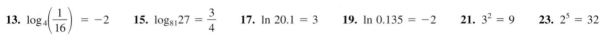

13. $\log_4\left(\dfrac{1}{16}\right) = -2$ **15.** $\log_{81}27 = \dfrac{3}{4}$ **17.** $\ln 20.1 = 3$ **19.** $\ln 0.135 = -2$ **21.** $3^2 = 9$ **23.** $2^5 = 32$

25. $e^{-1.4} \approx 0.25$ **27.** $e^3 \approx 20.1$ **29.** $\dfrac{1}{3}$ **31.** 6   **33.** −4   **35.** 5   **37.** 6   **39.** 14   **41.** $\log(x + 1)^2$

43. $\log[x^3(x+1)^6]$ **45.** $\log\left(\dfrac{xy}{z^2}\right)$ **47.** 0.79 **49.** 1.52 **51.** 2.25 **53.** $-\dfrac{3}{7}$ **55.** 2 is the only root

57. 0.056 **59.** 6.38 y

Chapter Test, p. 326

1. 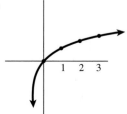 **3.** 8.5 y **5.** $C(1-r)^n$ **7.** $\log_{1/8}\left(\dfrac{1}{2}\right) = \dfrac{1}{3}$ **9.** 136 lumens **11.** $\dfrac{3}{11}$

13. 3.7818 **15.** 3 **17.** 1.862×10^{-6}

CHAPTER 9

9–1 Exercises, p. 335

1. 63.28° **3.** 120.31° **5.** 148.29° **7.** 81.73° **9.** 256.43° **11.** 221°46′ **13.** 58°23′ **15.** 173°11′

17. 330°17′ **19.** 395.8°, −324.2° **21.** 678.6°, −41.4°

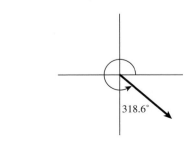

23. 416.50, −303.50 **25.** 378.9°, −341.1° **27.** 613.6°, −106.4°

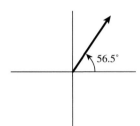 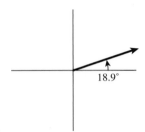 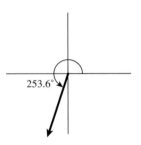

29. II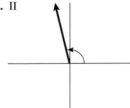

31. I

33. II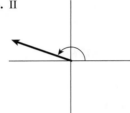

35. III

37.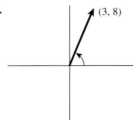

39.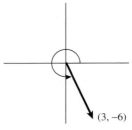

41.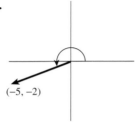

43.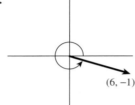

45.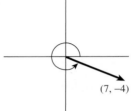

9–2 Exercises, p. 341

1. sin θ = 0.8638, cos θ = 0.5039, tan θ = 1.7143, csc θ = 1.1577, sec θ = 1.9846, cot θ = 0.5833

3. sin θ = 0.9080, cos θ = 0.4191, tan θ = 2.1667, csc θ = 1.1014, sec θ = 2.3863, cot θ = 0.4615

5. sin θ = 0.6644, cos θ = 0.7474, tan θ = 0.8889, csc θ = 1.5052, sec θ = 1.3380, cot θ = 1.1250

7. sin θ = 0.5000, cos θ = 0.8660, tan θ = 0.5774, csc θ = 2.000, sec θ = 1.1547, cot θ = 1.7321

9. sin θ = 0.7809, cos θ = 0.6247, tan θ = 1.2500, csc θ = 1.2806, sec θ = 1.6008, cot θ = 0.800

11. sin θ = 0.6000, cos θ = 0.8000, tan θ = 0.7500, csc θ = 1.6667, sec θ = 1.2500, cot θ = 1.3333

13. $\cos θ = \dfrac{\sqrt{105}}{13} = 0.7882$, $\csc θ = \dfrac{13}{8} = 1.6250$ **15.** sin θ = 0.7714, cot θ = 0.8250

17. cos θ = 0.3162, csc θ = 1.0541   **19.** sec β = 2.6000, cot β = 0.4167   **21.** sin α = 0.8824, cot α = 0.5333

23. sin θ = 0.3846, csc θ = 2.6000

9–3 Exercises, p. 347

1. 1.6390   **3.** 0.1045   **5.** 0.9587   **7.** 2.2148   **9.** 1.1736   **11.** 1.8115   **13.** 0.3134   **15.** 1.0515

17. 1.5003   **19.** 0.8785   **21.** 0.8059   **23.** 0.9954   **25.** 1.1363   **27.** 4.2273   **29.** θ = 52.6°

31. θ = 67.8°   **33.** θ = 14.9°   **35.** θ = 68.4°   **37.** θ = 69.6°   **39.** θ = 31.2°   **41.** θ = 74.5°

43. θ = 12.0°   **45.** θ = 81.6°   **47.** θ = 14.8°   **49.** θ = 49.9°   **51.** θ = 53.0°   **53.** θ = 46.4°

55. θ = 57.3°   **57.** 59.5 in.-lb   **59.** 358.2 V   **61.** 6°

9–4 Exercises, p. 354

1. $A = 46.2°, b = 6.5$ m, $c = 9.4$ m   **3.** $B = 53°17', a = 23.6$ cm, $b = 31.7$ cm   **5.** $A = 75.5°, b = 4.1$ m, $c = 16.5$ m

7. $B = 24°23', a = 107.29$ cm, $c = 117.79$ cm   **9.** $b = 51.5$ ft, $A = 36.4°, B = 53.6°$ **11.** $c = 52.6$ ft, $A = 34.5°$,

$B = 55.5°$ **13.** 50 ft   **15.** 3.4 mi   **17.** 10.5 ft   **19.** 79.3 yd   **21.** 260 ft   **23.** 35.87°, 13.00 cm

25. 16°   **27.** 80.2°, 320.3 ft   **29.** 142 ft   **31.** 27.0 ft   **33.** $x ≈ 102.0$ ft   **35.** pole ≈ 33 ft   **37.** α ≈ 79°

39. 48 ft

Chapter Review, p. 359

1. 67.7°   **3.** 274.7°   **5.** 218.8°   **7.** 16°47′   **9.** 318°9′   **11.** 108°26′

13. quad II, 507.8°, −212.2°   **15.** quad IV, 650.3°, −69.7°   **17.** quad IV, 674.63°, −45.37°

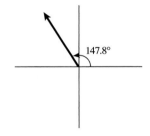
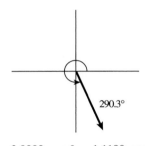
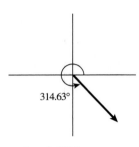

19. sin θ = 0.8944, cos θ = 0.4472, tan θ = 2.0000, csc θ = 1.1180, sec θ = 2.2361, cot θ = 0.5000

21. sin θ = 0.8319, cos θ = 0.5546, tan θ = 1.5000, csc θ = 1.2020, sec θ = 1.8030, cot θ = 0.6667

23. sin θ = 0.7559, cos θ = 0.6547, tan θ = 1.1547, csc θ = 1.3229, sec θ = 1.5275, cot θ = 0.8660

25. tan θ = 2.0647, sec θ = 2.2942   **27.** sin θ = 0.9402, cos θ = 0.3406   **29.** cos θ = 0.5598, tan θ = 1.4803

31. 0.957   **33.** 1.080   **35.** 3.378   **37.** θ = 75.2°   **39.** θ = 52.2°   **41.** θ = 21.5°

43. $B = 33.3°, a = 11.1$ yd, $c = 13.3$ yd   **45.** $a = 44.6$ m, $A = 38.4°, B = 51.6°$ **47.** $B = 36°12', c = 121.8$ km,

$b = 71.9$ km   **49.** 66.4 ft

Chapter Test, p. 359

1. 1.0202   **3.** cos A = 0.4061, cot A = 0.4444   **5.** 23.8°   **7.** 516°, −204°

9. $A = 1.8868$ **11.** 124°50′   **13.** 1.0976   **15.** 29.4°

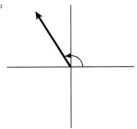

CHAPTER 10

10–1 Exercises, p. 370

1. + **3.** − **5.** − **7.** − **9.** − **11.** − **13.** − **15.** − **17.** − **19.** + **21.** + **23.** −

25. − **27.** − **29.** − **31.** I and IV **33.** II and IV **35.** II and IV **37.** IV **39.** III **41.** IV

43. I and II **45.** III **47.** sin θ = 0.9138, cos θ = −0.4061, tan θ = −2.25, csc θ = 1.0943, sec θ = −2.4622, cot θ = −0.4444 **49.** sin θ = −0.8320, cos θ = 0.5547, tan θ = −1.5000, csc θ = −1.2019, sec θ = 1.8028, cot θ = −0.6667

51. sin θ = −0.2747, cos θ = −0.9615, tan θ = −2.2000, csc θ = 1.0984, sec θ = −2.4166, cot θ = −0.4545

53. sin θ = −0.4061, cos θ = −0.4138, tan θ = 0.4444, csc θ = −2.4622, sec θ = −1.0943, cot θ = 2.2500

55. sin θ = 0.9104, cos θ = −0.7894, tan θ = −0.7778, csc θ = −1.6288, sec θ = 1.2669, cot θ = −1.2857

57. sin θ = −0.5812, cos θ = 0.8137, tan θ = −0.7143, csc θ = −1.7205, sec θ = 1.2289, cot θ = −1.4000

59. sin θ = −0.2425, cos θ = 0.9701, tan θ = −0.2500, csc θ = −4.1231, sec θ = 1.0308, cot θ = −4.0000

61. tan θ = 0.577

10–2 Exercises, p. 376

1. 0.766 **3.** −1.004 **5.** −0.848 **7.** −0.205 **9.** −3.487 **11.** 0.279 **13.** −0.111 **15.** 13.431

17. −0.967 **19.** 0.869 **21.** 0.445 **23.** 0.854 **25.** −1.788 **27.** 0.532 **29.** 66.4°, II **31.** 82.6°, II

33. 33.27°, IV **35.** 83.8°, I **37.** 78.4°, IV **39.** 42.57°, II **41.** 64°, IV **43.** 5.9°, II **45.** 50.4°, I

47. 20.6°, III **49.** 54.60°, 125.40° **51.** 137.90°, 317.90° **53.** 65.70°, 245.70° **55.** 174.62°, 185.38°

57. 136.40°, 223.60° **59.** 258.39°, 281.61° **61.** 253°36′, 286°24′ **63.** 47°6′, 312°54′ **65.** 56°6′, 236°6′

67. 148°18′, 211°42′ **69.** −0.9622 **71.** −30.1371 **73.** 3.1189 **75.** 189.5 **77.** 336.8° **79.** 126.9°

81. 308.7° **83.** 5.83 in. **85.** 40.8° **87.** 78.14 W

10–3 Exercises, p. 381

1. II **3.** III **5.** II **7.** IV **9.** $\dfrac{17\pi}{45}$ **11.** $\dfrac{53\pi}{30}$ **13.** $\dfrac{11\pi}{6}$ **15.** $-\dfrac{14\pi}{9}$ **17.** 0.82 **19.** 1.71

21. 3.40 **23.** −5.67 **25.** 93.97° **27.** 322.58° **29.** 420° **31.** −330° **33.** 0.278 **35.** 1.410

37. 0.651 **39.** 1.342 **41.** −1.161 **43.** 1 **45.** 0.383 **47.** −2 **49.** 0.58, 5.70 **51.** 3.51, 5.91

53. 0.41, 2.73 **55.** 2.23, 4.05 **57.** 1.91, 4.37 **59.** 2.88, 6.02 **61.** 55.82 m^2

63. 0; the weight is at the starting position. **65.** Tan 90°: undefined (π/2 radians converts to 90°)

10–4 Exercises, p. 386

1. 8.7 ft **3.** 0.88 **5.** 8.3 yd **7.** 1,016.87 cm^2 **9.** 491.5 m **11.** 3 **13.** r = 17.2 in.

15. $0.61\left(\dfrac{180}{\pi}\right) = 35°$ **17.** 8.02 ft/s **19.** $1.17°\left(\dfrac{180}{\pi}\right) = 67°$ **21.** 1.73 ft/s **23.** 19,699 mm^2 **25.** 18.85 ft/s

27. 65.4 ft **29.** 41,871 ft/min

Chapter Review, p. 388

1. − **3.** + **5.** + **7.** + **9.** + **11.** I **13.** II **15.** IV **17.** $\sin \theta = -0.6508$,
$\cos \theta = 0.7593$, $\tan \theta = -0.8571$, $\csc \theta = -1.5365$, $\sec \theta = 1.3171$, $\cot \theta = -1.1667$

19. $\sin \theta = -0.5215$, $\cos \theta = 0.8533$, $\tan \theta = -0.6111$, $\csc \theta = -1.9177$, $\sec \theta = 1.1719$, $\cot \theta = -1.6364$

21. $\sin \theta = 0.4856$, $\cos \theta = -0.8742$, $\tan \theta = -0.5556$, $\csc \theta = 2.0591$, $\sec \theta = -1.1440$, $\cot \theta = -1.8000$

23. undefined **25.** 21.3557 **27.** −3.5687 **29.** 67.3° **31.** 43.3° **33.** 65.3° **35.** 69.5° **37.** 256.4°

39. 330° **41.** $\dfrac{2\pi}{3}$ **43.** $\dfrac{11\pi}{6}$ **45.** $\dfrac{31\pi}{18}$ **47.** 2.23 **49.** 5.56 **51.** 3.74 **53.** 157.9° **55.** 315°

57. −294.9° **59.** −1.0023 **61.** −1 **63.** −0.1506 **65.** $\theta = 2.63$, $\theta = 5.77$ **67.** $\theta = 1.65$ **69.** 8.46 m

71. 2.21° **73.** 15.4 in. **75.** 1,233.7 yd^2 **77.** 451.56 m^2 **79.** 25,132.7 m/min.

Chapter Test, p. 389

1. −1.2020 **3.** 7.96 ft **4.** −2 **5.** 303.3° **7.** 85.3° **9.** 221.7° **11.** 140 m^2 **13.** 0.367 or 21.05°

15. $\dfrac{7\pi}{9}$ **17.** 230 ft^2

CHAPTER 11

11–1 Exercises, p. 401

1. 37 mi east **3.** 65 m, 270° **5.** 66.4 ft/s, 126.5° **7.** 43 lb, 124° **9.** 377.6 N, 144.9° **11.** 35.6 mi, 328.9°
13. 33 N, 29° **15.** 39.1 ft/s^2, 236.6° **17.** $x = 33.5$ ft/s, $y = 113.1$ ft/s **19.** $x = -74.5$ yd, $y = -11.9$ yd
21. $x = 32.7$ yd, $y = -119.4$ yd **23.** $x = -64.4$ m, $y = -18.2$ m **25.** $x = -384.2$ mi, $y = 165.5$ mi
27. 24.3 ft/s, 148.7° **29.** 109.2 yd, 97.0° **31.** 60 N at 256° **33.** 144.3 m, 59.4° **35.** 79.1 N, 287.9°
37. 555.2 mi, 187.7° **39.** 932.4 N, 214.3° **41.** $A \approx 140$ in^2 **43.** $x = 59$ ft/s, $y = 46$ ft/s

11–2 Exercises, p. 410

1. 48° north of east **3.** 468 mph at 30.2° **5.** 231 N **7.** 14 lb **9.** 37.4° or 217.4° **11.** 51.3°
13. $R = 77$ lb at 55.5° below horizontal **15.** 30.43 N, 22.74 N **17.** 461 mph, 328.8° **19.** 121.6 lb, 179.3 lb
21. 432 lb **23.** $T \approx 192$ lbs, $F \approx 107$ lb. **25.** $F_A = 6443.3165$, $F_B = 1765.416$, $\alpha = 83.351°$ **27.** $F = 269.316$
29. 110 mi/hr at 8°N of E **31.** 240 ft-lb

11–3 Exercises, p. 420

1. $A = 27°, b = 32.5, c = 26.4$ **3.** $b = 18.4, C = 31.7°, B = 105.3°$ **5.** $a = 16.7, b = 22.5, A = 43°$
7. $A = 45°, C = 77°, c = 10.3$ **9.** $A = 79.3°, b = 29.4, a = 42.6$ **11.** $B = 45°, a = 21.1, c = 13.6$
13. $A = 61°, b = 26.1, c = 43.0$ **15.** $B = 43.1°$ or $B = 136.9°, A = 106.7°$ or $A = 12.9°, a = 28.2$ or $a = 6.6$

17. $A = 62.2°$, $a = 52.1$, $c = 41.9$ **19.** no solution   **21.** The pilot should fly at 41.5° north of east.   **23.** 67.5 m

25. 85.9 m   **27.** 56.8 km   **29.** 33.2°   **31.** $x \approx 39$, $y \approx 41$ **33.** $F \approx 470$ lb   **35.** $AC \approx 28$ cm

11–4 Exercises, p. 429

1. $A = 40.3°$, $B = 48.2°$, $C = 91.5°$ **3.** $c = 23.2$, $A = 88.2°$, $B = 48.8°$ **5.** $a = 29.5$, $B = 33.2°$, $C = 73°$

7. $A = 55.4°$, $B = 75.3°$, $C = 49.3°$ **9.** $A = 111.3°$, $B = 41.3°$, $C = 27.4°$ **11.** $C = 142.5°$, $A = 21.8°$, $B = 15.7°$

13. $b = 91.7$, $A = 46.5°$, $C = 31.5°$ **15.** $c = 126.7$, $A = 48.1°$, $B = 56.6°$ **17.** $B = 148.9°$, $A = 20.3°$, $C = 10.8°$

19. $a = 106.3$, $B = 31.3°$, $C = 40.7°$ **21.** Any 2 sides of a triangle must have a total length longer than the third side.

23. 80.7° and 99.3°   **25.** 38.5°   **27.** 9.57 knots at 32.6° S of E   **29.** $c \approx 94$ ft   **31.** $x \approx 21$ ft

Chapter Review, p. 430

1. 323.8 lb, 215.1°   **3.** 40.9 m/s, 84.2°   **5.** 310.7 N, 133.3°   **7.** 17.3 lb, 334.3°   **9.** $x = -219.6$ lb, $y = -190.91$ lb

11. $x = 255.0$ km, $y = 414.5$ km   **13.** 16 blocks 10° S of E   **15.** 150.2 lb, 90.4 lb   **17.** 240.6 lb, 78.2 lb

19. 454.6 mph at 26° N of W   **21.** $A = 70.7°$, $b = 78.1$, $c = 21.8$ **23.** $B = 26.8°$, $A = 96.2°$, $a = 30.8$

25. $B = 22.6°$, $C = 90.3°$, $c = 61.7$ **27.** Heading should be 62° S of W.   **29.** 57.8°   **31.** $b = 195.8$, $A = 55.7°$,

$C = 85.3°$ **33.** $C = 87.6°$, $A = 57.5°$, $B = 34.9°$ **35.** $c = 196.1$, $A = 76.6°$, $B = 49.8°$ **37.** 26 ft

39. $B = 81.6°$, $C = 98.4°$

Chapter Test, p. 432

1. 99.3 lb at 72.8° S of W   **3.** 7.2   **5.** 41.6 N   **7.** 41.4 at 26.9° S of W   **9.** $c = 28.4$ **11.** 280.5

13. 8.7 mph at 60° relative to shore   **15.** $b = 40.3$, $C = 41°$, $c = 26.4$

CHAPTER 12

12–1 Exercises, p. 443

1.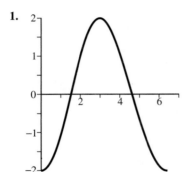

3.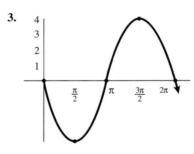

5.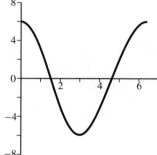

7.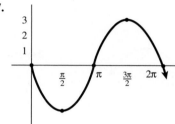

9.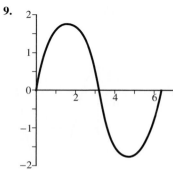

11.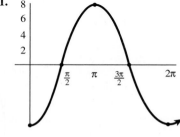

13.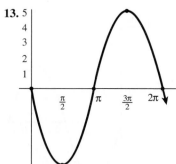

15.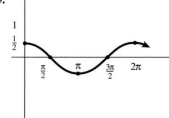

17.

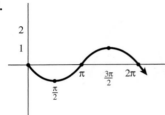

19.

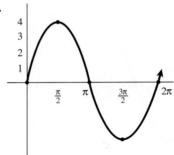

21. $y = -3 \sin x$ **23.** $y = -5 \cos x$ **25.** $y = -1.5 \sin x$

12–2 Exercises, p. 448

1. $1, \pi$

3. $1, \dfrac{2}{3}$

5. $1, \dfrac{\pi}{2}$

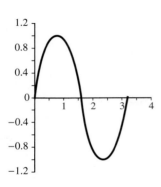

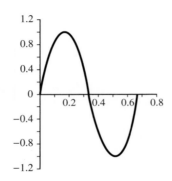

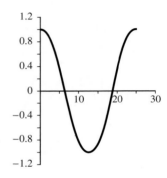

7. $2, \dfrac{2}{3}$

9. $3, \dfrac{\pi}{2}$

11. $2, \dfrac{2}{3}$

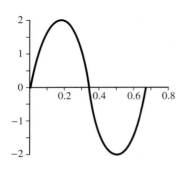

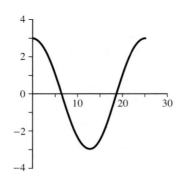

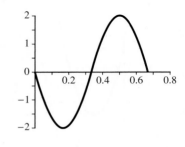

13. $3, \dfrac{\pi}{2}$

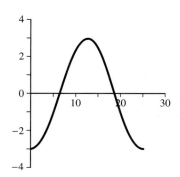

15. $\dfrac{1}{2}, 4\pi$

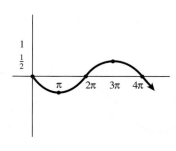

17. $4, \dfrac{2}{5}$

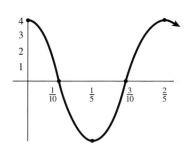

19. $3, \dfrac{4\pi}{3}$

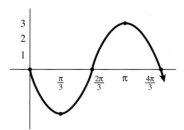

21. $y = 2 \cos 3\pi x$ **23.** $y = -3 \sin \pi x$

25. $3, 1$

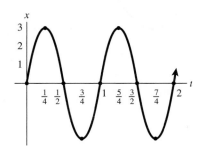

27. $8, \dfrac{1}{3}$

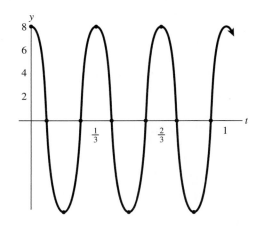

12–3 Exercises, p. 454

1. 1, 2π, $\dfrac{\pi}{4}$ left

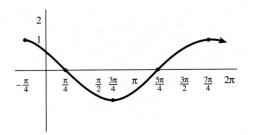

3. 1, 2π, $\dfrac{\pi}{3}$ right

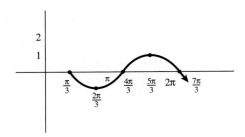

5. 2, 2π, 1 left

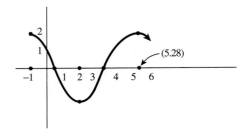

7. 1, π, $\dfrac{\pi}{2}$ right

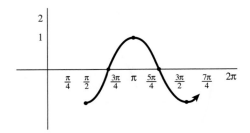

9. 1, $\dfrac{2\pi}{3}$, $\dfrac{2\pi}{3}$ left

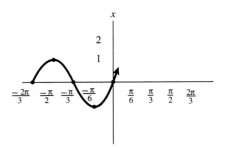

11. 2, 2π, $\dfrac{\pi}{4}$ left

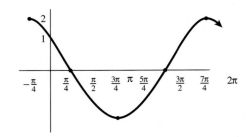

13. 3, 2π, 3 left

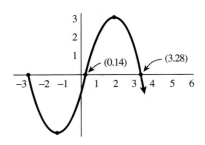

15. 2, 1, $\dfrac{1}{2\pi}$ left (0.16)

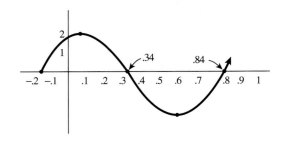

17. $1, \dfrac{2}{3}, \dfrac{2}{3\pi}$ right (0.2)

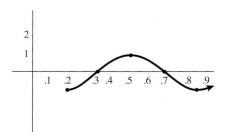

19. $4, \dfrac{1}{3}, \dfrac{1}{12\pi}$ left (0.026)

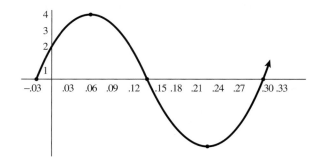

21. $2.5, \dfrac{2\pi}{3}, \dfrac{2\pi}{3}$ right

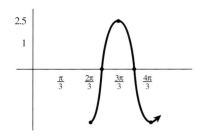

23. $1, 1, \dfrac{5}{2}$ right

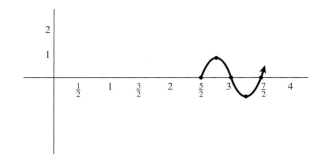

25. $y = 3\cos\left(\dfrac{\pi}{2}x\right)$ **27.** $y = -\pi\sin\left(\dfrac{\pi}{3}x + \dfrac{\pi^2}{9}\right)$ **29.** $y = 4\cos\left(2x - \dfrac{2\pi}{3}\right)$

31.

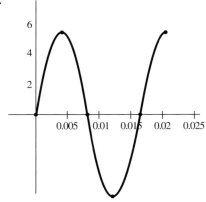

33.

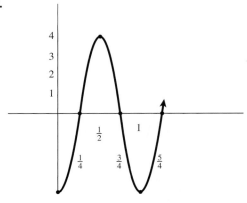

35. $V = 17$ V, $I = 30$ mA

37. $V = 9.5$ V, $I = 3.6$ mA

12–4 Exercises, p. 460

1. (a) $1.4, \dfrac{1}{3}, \dfrac{-\pi}{2}$ **(b)** 1.4 at 1.5 s   **(c)** 1/6 s   **(d)**

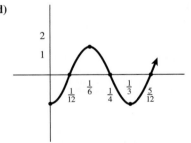

3. (a) $2, \dfrac{2}{3}, \dfrac{\pi}{2}$

(b) 0 at 1.5 s   **(c)** 0 s or $\dfrac{2}{3}$ s in 2nd cycle   **(d)**

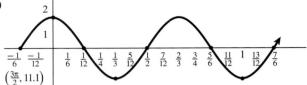

5. (a) $3.1, 2, -\pi$ **(b)** 3.1 at 1.5 s   **(c)** $\dfrac{3}{2}$ s   **(d)**

7. $y = 4 \sin\left(\pi t + \dfrac{\pi}{3}\right)$ **9.** $y = 6 \sin\left(\dfrac{20\pi}{13}t - \dfrac{\pi}{4}\right)$ **11.** $y = 2 \sin\left(\dfrac{\pi}{2}t + \pi\right)$

13.

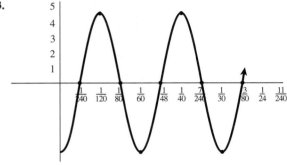

15.

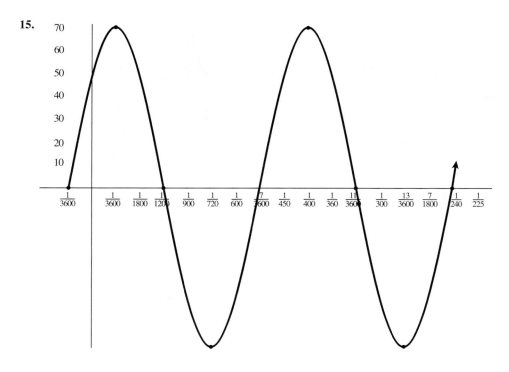

17. $y = 18 \sin\left(\dfrac{4\pi}{3}t\right)$, 7.32 cm at 0.65 s

19. $V = 135 \sin 100\pi t$, amp $= 135$, per $= \dfrac{1}{50}$

21. frequency $= 264$

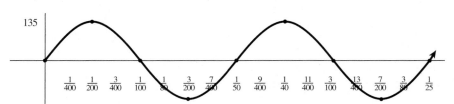

12–5 Exercises, p. 466

1.

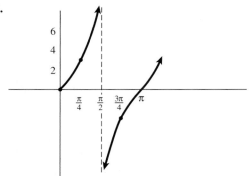

3.

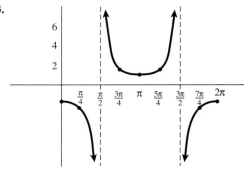

5.

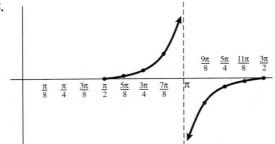

7.

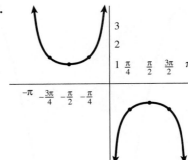

9.

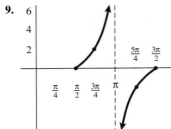

11.

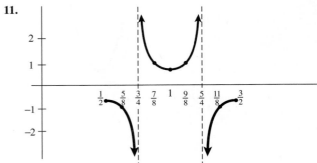

13.

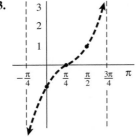

15.

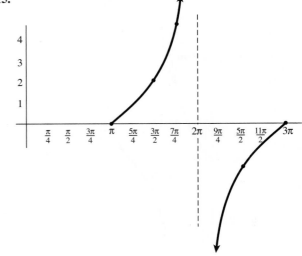

17.

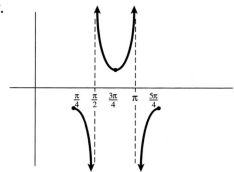

19.

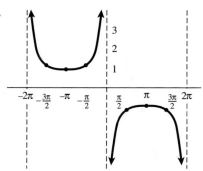

12–6 Exercises, p. 471

1.

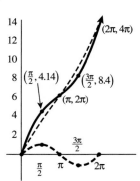

3.

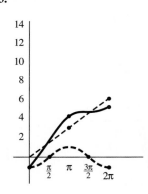

5.

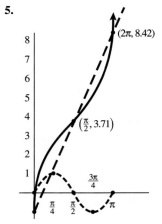

7.

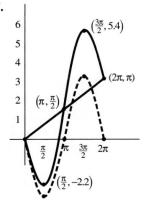

9.

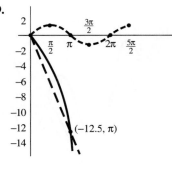

11.

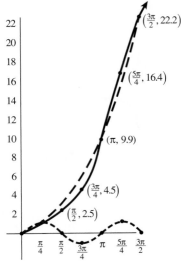

13.

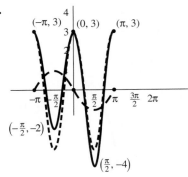

15.

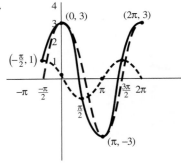

17.

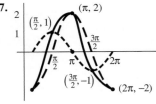

19.

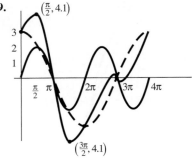

21.

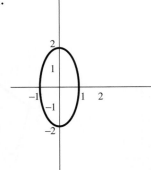

23.

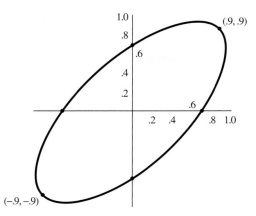

25.

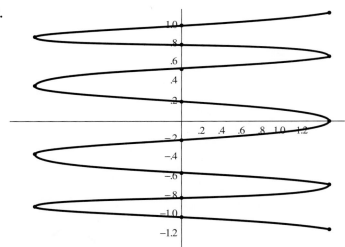

27.

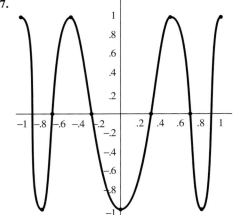

29.

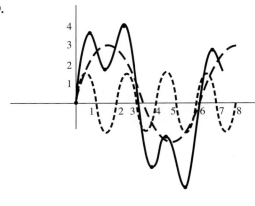

31.

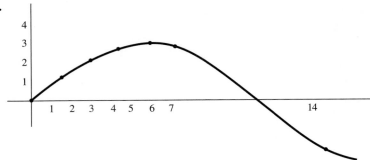

33.

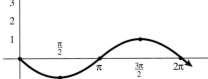

Chapter Review, p. 472

1.

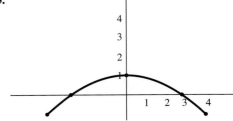

3.

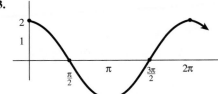

5.

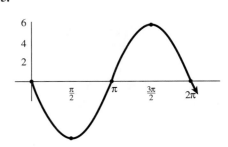

7. 1, π

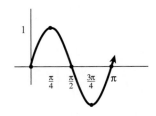

9. 3, 8π

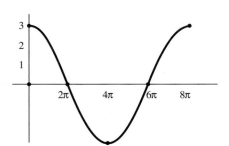

11. 6, $\dfrac{\pi}{2}$

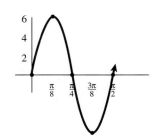

13. 1.6, $\dfrac{2}{3}$

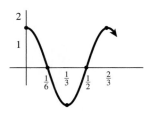

15. 4, 2

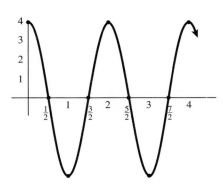

17. 2, 2π, $\dfrac{\pi}{4}$

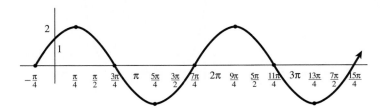

19. 1, π, $\dfrac{-\pi}{2}$

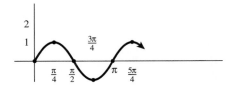

21. 3, 1, −3

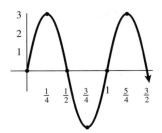

23. $4, 6\pi, \dfrac{\pi}{2}$

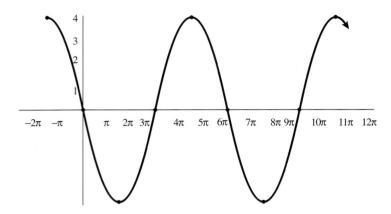

25. $y = 2\cos(2\pi^2 x - \pi^3)$

27. (a) $8, 2\pi, \dfrac{\pi}{4}$ right   **(b)** 8 at 2.45   **(c)** $\dfrac{3\pi}{4}$

(d)

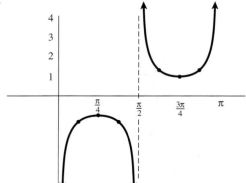

29. (a) $3.1, \pi, \dfrac{\pi}{2}$ left   **(b)** 3.1 at 2.45

(c) $\dfrac{-\pi}{4}$ or $\dfrac{3\pi}{4}$ in 2nd cycle

(d)

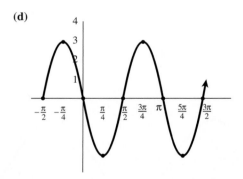

31.

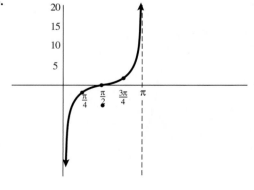

33.

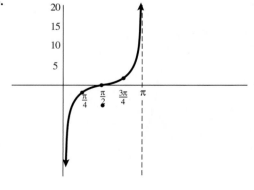

35.

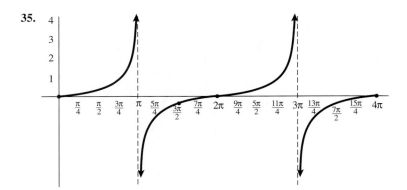

37.

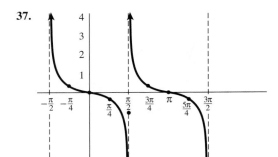

39.

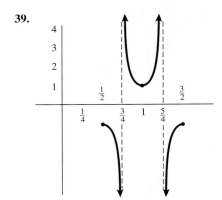

41. $60, \dfrac{\pi}{15}, -\pi$

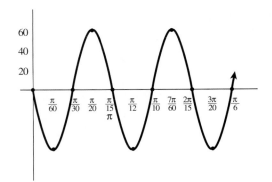

43.

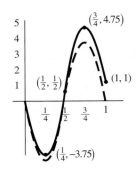

45.

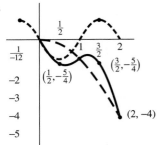

47.

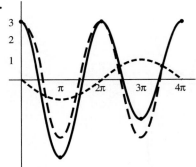

49.

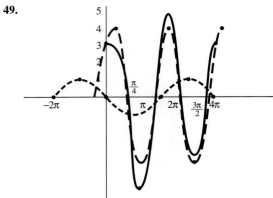

51.

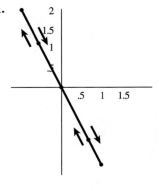

53.

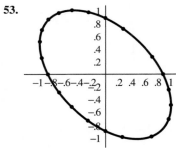

Chapter Test, p. 473

1.

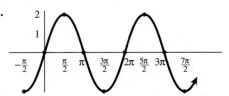

3. $y = 2 \sin\left(\dfrac{5}{2}\pi t + \dfrac{\pi}{3}\right)$

5.

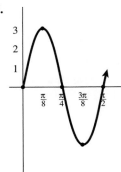

7.

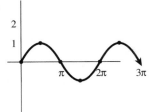

9.

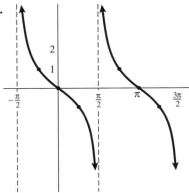

11. $3, 2, \dfrac{1}{\pi}$ right

13. $y = 115 \sin\left(100\pi t - \dfrac{\pi}{6}\right)$

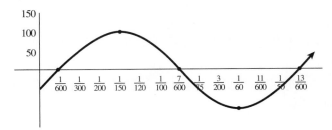

CHAPTER 13

13–1 Exercises, p. 484

The problems in this section can all be solved.

13–2 Exercises, p. 490

1. $\dfrac{\sqrt{6} + \sqrt{2}}{4}$ **3.** $\dfrac{\sqrt{6} + \sqrt{2}}{4}$ **5.** $\dfrac{-\sqrt{2} - \sqrt{6}}{4}$ **7.** $\dfrac{-\sqrt{2} - \sqrt{6}}{4}$ **9. (a)** $\dfrac{-171}{221}$ **(b)** $\dfrac{-140}{171}$ **(c)** quad II

11. (a) $\dfrac{-171}{221}$ **(b)** $\dfrac{-140}{221}$ **(c)** quad III   **13.** $\sin 35°$ **15.** $\cos 6A$ **17.** $-\cos 4A$ **19.** $\tan B$

21–31. These identities can be verified.   **33.** $a \sin(2\pi ft + B)$ **35.** $y = 12 \cos\left(\dfrac{\pi t}{3}\right)$ **37.** $y = 2B \cos kx \cos \omega t$

13–3 Exercises, p. 497

1. $\dfrac{\sqrt{2 - \sqrt{3}}}{2}$ **3.** $\dfrac{-\sqrt{2 - \sqrt{3}}}{2}$ **5.** $-\sqrt{3}$ **7.** $\cos 2A = \dfrac{161}{289}$, $\tan \dfrac{A}{2} = \dfrac{1}{4}$ **9.** $\tan \dfrac{A}{2} = -7$, $\sin \dfrac{A}{2} = \dfrac{7\sqrt{2}}{10}$

11. $2 \sin 4A$ **13.** $\sin^2 2x$ **15.** $\cos 105°$ **17–31.** These problems can be proven.

33. $n = \cos \dfrac{D}{2} + \cot \dfrac{A}{2} \sin \dfrac{D}{2}$ (others are possible)   **35.** This problem can be proven.

13–4 Exercises, p. 505

1. $0.62, 2.53, \dfrac{2\pi}{3}, \dfrac{4\pi}{3}$ **3.** $x = \dfrac{3\pi}{2}$ **5.** $\dfrac{7\pi}{6}, \dfrac{11\pi}{6}, \dfrac{\pi}{6}, \dfrac{5\pi}{6}$ **7.** $2.14, 5.28, 1.79, 4.93$ **9.** $\dfrac{\pi}{4}, \dfrac{5\pi}{4}$ **11.** $\dfrac{\pi}{2}, \dfrac{3\pi}{2}, 0$

13. $\dfrac{\pi}{6}, \dfrac{5\pi}{6}, \dfrac{3\pi}{2}$ **15.** no solution   **17.** $0, \pi, \dfrac{\pi}{6}, \dfrac{5\pi}{6}$ **19.** $1.77, 4.51, 0$ **21.** $\dfrac{3\pi}{10}, \dfrac{7\pi}{10}, \dfrac{11\pi}{10}, \dfrac{15\pi}{10}, \dfrac{19\pi}{10}$

23. $\dfrac{\pi}{3}, \dfrac{5\pi}{3}, \pi$ **25.** $\dfrac{\pi}{9}, \dfrac{5\pi}{9}, \dfrac{7\pi}{9}, \dfrac{11\pi}{9}, \dfrac{13\pi}{9}, \dfrac{17\pi}{9}$ **27.** $\dfrac{\pi}{2}, \dfrac{3\pi}{2}, \dfrac{\pi}{6}, \dfrac{5\pi}{6}$ **29.** $\dfrac{\pi}{2}, \dfrac{3\pi}{2}, \dfrac{\pi}{4}, \dfrac{5\pi}{4}$

31. $20.75°, 69.25°$ **33.** 3.97 s, 5.43 s   **35.** February   **37.** $35°$

13–5 Exercises, p. 513

1. $\dfrac{\pi}{3}$ **3.** $\dfrac{\pi}{3}$ **5.** 0 **7.** $\dfrac{\pi}{4}$ **9.** $\dfrac{\pi}{6}$ **11.** $-\dfrac{\pi}{2}$ **13.** 0.845 **15.** 2.305 **17.** -1.064 **19.** 0.810

21. 0.5 **23.** 0.5 **25.** 1 **27.** undefined   **29.** $\dfrac{3}{4}$ **31.** $-\dfrac{12}{13}$ **33.** $-\dfrac{15}{17}$ **35.** $\dfrac{1}{\sqrt{1 - x^2}}$

37. $2x\sqrt{1 - x^2}$ **39.** $\dfrac{1 - \sqrt{1 - x^2}}{x}$ **41.** $\sqrt{1 - 10x^2 + 9x^4} - 3x^2$ **43.** $\text{Arcsin } \dfrac{C}{HmL}$

45. $4 \arcsin \pm 2\sqrt{\dfrac{P - T}{T}}$ **47.** 1.2

Chapter Review, p. 515

1–11. These problems can be proven.   **13.** (a) $\dfrac{435}{533}$ (b) $\dfrac{92}{525}$ (c) $\dfrac{308}{533}$ (d) quad I   **15.** $\sin(C - D + B)$

17. $\tan 5x$ **19.** can be proven   **21.** $\sin\left(x + \dfrac{\pi}{2}\right) - \cos\left(x + \dfrac{\pi}{2}\right) = \cos x + \sin x$; $\sin x \cos \dfrac{\pi}{2} + \cos x \sin \dfrac{\pi}{2} -$

$\left(\cos x \cos \dfrac{\pi}{2} - \sin x \sin \dfrac{\pi}{2}\right) = \cos x + \sin x$; $0 + \cos x - 0 + \sin x = \cos x + \sin x$

23. can be proven   **25.** can be proven   **27.** $8 \cos 2B$ **29.** $\tan \dfrac{5B}{2}$ **31.** $\sin \dfrac{B}{2} = \dfrac{4\sqrt{17}}{17}$, $\cos 2B = \dfrac{161}{289}$

33–37. These can be proven.   **39.** $0.61, 2.53$ **41.** $2.3, 3.98, \dfrac{\pi}{3}, \dfrac{5\pi}{3}$ **43.** $\pi, 0, \dfrac{\pi}{4}, \dfrac{5\pi}{4}$

45. $\dfrac{\pi}{2}$ **47.** $1.15, 1.98, 4.29, 5.13$ **49.** $\dfrac{8}{17}$ **51.** $-\dfrac{40}{9}$ **53.** $\dfrac{x}{\sqrt{1 + x^2}}$ **55.** $\sqrt{1 - x^2}$

57. $\sqrt{\dfrac{1 + \sqrt{1 - x^2}}{2}}$

Chapter Test, p. 516

Note: Problems 3, 6, 8, 10, 13, 14 can be proven.   **1.** $\tan\dfrac{A}{2} = \dfrac{5}{4}$, $\sin 2A = -\dfrac{720}{1,681}$ **5.** $\cos(A + B + C - D)$

7. $\dfrac{13}{5}$ **9.** 3.47, 5.94, $\dfrac{\pi}{6}$, $\dfrac{5\pi}{6}$ **11.** 39.26°   **15.** $\dfrac{\sqrt{2}}{2}A_1(\sin x + \cos x) + \dfrac{\sqrt{2}}{2}A_2(\cos x - \sin x)$

CHAPTER 14

14–1 Exercises, p. 525

1. $8j$ **3.** $2j\sqrt{2}$ **5.** $6j\sqrt{2}$ **7.** $-2j\sqrt{10}$ **9.** $\dfrac{2}{7}j\sqrt{2}$ **11.** -6 **13.** $3j\sqrt{6}$ **15.** -54 **17.** $-8j$

19. 1 **21.** $-j$ **23.** -1 **25.** -1 **27.** 8 **29.** $-3j$ **31.** $x = \dfrac{1}{2}$, $y = -7$ **33.** $x = 5$, $y = \dfrac{14}{3}$

35. $x = -2$, $y = -3$ **37.** $x = -8$, $y = \dfrac{1}{3}$ **39.** $x = -8$, $y = 11$ **41.** $8 + 5j$ **43.** $-7 - 9j$ **45.** $2j$

47. -10 **49.** $Z = 44 - 14j$

14–2 Exercises, p. 531

1. $3 - 7j$ **3.** $-14 + 16j$ **5.** $-3 + 8j$ **7.** $6 + j$ **9.** $-23 + 9j$ **11.** $32 + j$ **13.** $-8 + 3j$

15. $22 - 2j$ **17.** -33 **19.** $30 + 15j$ **21.** $10 + 37j$ **23.** $-7 + 103j$ **25.** $68 + 11j$ **27.** $-21 + 20j$

29. 53 **31.** $27 - 36j$ **33.** $-46 - 9j$ **35.** $\dfrac{7 - 6j}{2}$ **37.** $\dfrac{2 + 3j}{-3}$ **39.** $\dfrac{6 - 2j}{5}$ **41.** $-5 - 8j$

43. $\dfrac{-3 - 11j}{10}$ **45.** $\dfrac{-51 + 3j}{29}$ **47.** $\dfrac{43 - 18j}{53}$ **49.** $\dfrac{-53 + 22j}{37}$

51.
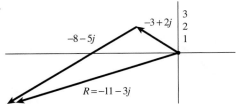

53.
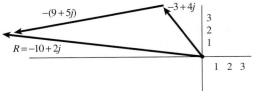

55.
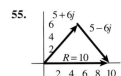

57. $\dfrac{174 - 100j}{10,069}$ **59.** $\dfrac{858j}{12,100}$

14–3 Exercises, p. 538

1. $\sqrt{13}(\cos 123.6° + j \sin 123.6°), \sqrt{13}e^{2.16j}$

$-2+3j$

3. $\sqrt{149}(\cos 234.9° + j \sin 234.9°), \sqrt{149}e^{4.1j}$

$-7-10j$

5. $3\sqrt{13}(\cos 56.3° + j \sin 56.3°), 3\sqrt{13}e^{0.98j}$

$6+9j$

7. $\sqrt{205}(\cos 294.6° + j \sin 294.6°), \sqrt{205}e^{5.14j}$

$6-13j$

9. $12.58(\cos 126.5° + j \sin 126.5°), 12.58e^{2.21j}$

$-7.5+10.1j$

11. $10.9(\cos 238.4° + j \sin 238.4°), 10.9e^{4.16j}$

$-5.7-9.3j$

13. $8(\cos 90° + j \sin 90°), 8e^{1.57j}$

$8j$

15. $15(\cos 0° + j \sin 0°), 15e^{0j}$

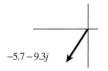

15

17. $3 - 5.2j$ **19.** $-44.8 + 254j$ **21.** $4.02 - 5.98j$ **23.** -137 **25.** $-1.59 + 6.8j$ **27.** $-139.7 + 84j$

29. $7.35 + 27.6j$ **31.** $0.34 + 0.66j$ **33.** $6.75(\cos 223.5° + j \sin 223.5°), -4.9 - 4.6j$ **35.** $175(\cos 103.1° + j$ $\sin 103.1°), -39.7 + 170.4j$ **37.** $26.9(\cos 143.2° + j \sin 143.2°), -21.5 + 16.1j$ **39.** $6.4(\cos 304° + j \sin 304°), 6.4e^{5.3j}$

41. Let z be the complex number $a + bj$. The absolute value of z is the distance from (a, b) to the origin. The conjugate of z is the mirror image of z in the real axis.

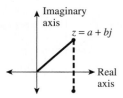

43. $Z = 20 - 18j$,
$26.9(\cos (-42°) + j \sin (-42°))$,
$26.9e^{-0.73j}$

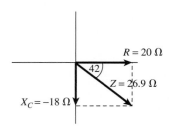

14–4 Exercises, p. 544

1. $28e^{5j}$ **3.** $50.02e^{6.1j}$ **5.** $18.4e^{0.78j}$ **7.** $4e^{5j}$ **9.** $4.16e^{5.18j}$ **11.** $2.16e^{1.52j}$ **13.** $156e^{2.71j}$

15. $112(\cos 74.2° + j \sin 74.2°)$ **17.** $192(\cos 264.5° + j \sin 264.5°)$ **19.** $0.3(\cos 45° + j \sin 45°)$

21. $2.44(\cos 159.5° + j \sin 159.5°)$ **23.** $216(\cos 81° + j \sin 81°)$ **25.** $4{,}220(\cos 241° + j \sin 241°)$

27. $1.12(\cos 105° + j \sin 105°)$, $1.12(\cos 225° + j \sin 225°)$, $1.12(\cos 345° + j \sin 345°)$

29. $2.45(\cos 156° + j \sin 156°)$, $2.45(\cos 336° + j \sin 336°)$ **31.** $1(\cos 22.5° + j \sin 22.5°)$,

$1(\cos 112.5° + j \sin 112.5°)$, $1(\cos 202.5° + j \sin 202.5°)$, $1(\cos 292.5° + j \sin 292.5°)$ **33.** $1(\cos 0° + j \sin 0°)$,

$1(\cos 60° + j \sin 60°)$, $1(\cos 120° + j \sin 120°)$, $1(\cos 180° + j \sin 180°)$, $1(\cos 240° + j \sin 240°)$, $1(\cos 300° + j \sin 300°)$

35. $2.1(\cos 31.7° + j \sin 31.7°)$, $2.1(\cos 211.7° + j \sin 211.7°)$

37. $24(\cos 115° + j \sin 115°)$ V **39.** $\dfrac{31}{20} + \dfrac{38j}{20}$ **41.** When x and z are both zero in each rule

14–5 Exercises, p. 551

1. **(a)** $331.6 \ \Omega$ **(b)** $13.57 \ \Omega$ **(c)** $318.12 \ \Omega$ **(d)** $\theta = -88.65°$ voltage lags current **(e)** 0.0283 A
(f) 0.2122 V **(g)** 9.381 V **(h)** 0.384 V **3.** 393.61 Hz **5.** $8.6 \ \Omega, -35.54°$ voltage lags current **7.** 49.65 V
9. 1.06 A **11.** 265.25 Hz **13.** 2.64×10^{-12} F or 2.64 pF **15.** 0.87 W

17. $V = IR$ can be represented by $V = IZ$ where Z is impedance. $Z = R + j(X_L - X_C)$, which implies that if $X_L \neq X_C$, we will get an imaginary number as a result.

19. $\dfrac{12 + 6j}{180}$

Chapter Review, p. 553

1. $2j\sqrt{6}$ **3.** -180 **5.** $-63 + \dfrac{2}{5}j$ **7.** $\dfrac{5}{3}j$ **9.** $-j$ **11.** 1 **13.** -4 **15.** $-4j$ **17.** $14 - 22j$

19. $5 + j\sqrt{114} - 7j$ **21.** 29 **23.** $84 + 23j$ **25.** $-198 - 10j$ **27.** $7 + 10j$ **29.** $\dfrac{-11 + 17j}{82}$

31. $10.6(\cos 131.2° + j \sin 131.2°)$, $10.6e^{2.29j}$ **33.** $16.1(\cos 209.7° + j \sin 209.7°)$, $16.1e^{3.66j}$

35. $8(\cos 90° + j \sin 90°)$, $8e^{1.57j}$ **37.** $14(\cos 0° + j \sin 0°)$, $14e^{0j}$ **39.** $1.46 - 13.9j$ **41.** $-1.4 - 0.95j$

43. $19.2 - 12.7j$ **45.** $-7.4 + 16.2j$ **47.** $2e^{11j}$ or $2e^{4.72j}$ **49.** $15e^{2.4j}$ **51.** $3.3e^{-0.7j}$ or $3.3e^{5.6j}$

53. $60.45(\cos 70° + j \sin 70°)$ **55.** $0.53(\cos 296.9° + j \sin 296.9°)$ **57.** $60.84(\cos 259.6° + j \sin 259.6°)$

59. $1.75(\cos 97.3° + j \sin 97.3°)$, $1.75(\cos 217.3° + j \sin 217.3°)$, $1.75(\cos 337.3° + j \sin 337.3°)$

61. $1.22(\cos 15.85° + j \sin 15.85°)$, $1.22(\cos 105.85° + j \sin 105.85°)$, $1.22(\cos 195.85° + j \sin 195.85°)$,

$1.22(\cos 285.85° + j \sin 285.85°)$ **63.** $1(\cos 15° + j \sin 15°)$, $1(\cos 75° + j \sin 75°)$, $1(\cos 135° + j \sin 135°)$,

$1(\cos 195° + j \sin 195°)$, $1(\cos 255° + j \sin 255°)$ $1(\cos 315° + j \sin 315°)$ **65.** $1(\cos 36° + j \sin 36°)$,

$1(\cos 108° + j \sin 108°)$, $1(\cos 180° + j \sin 180°)$, $1(\cos 252° + j \sin 252°)$, $1(\cos 324° + j \sin 324°)$ **67.** $530.5 \ \Omega$

69. $517.1 \ \Omega$ **71.** 23 mA **73.** 12.20 V **75.** 127 V **77.** 41.23 pF

Chapter Test, p. 554

1. $16j$ **3.** $6.631 \times 10^2 \ \Omega$ **5.** $1.75(\cos 22.7° + j \sin 22.7°)$, $1.75(\cos 142.7° + j \sin 142.7°)$,

$1.75(\cos 262.7° + j \sin 262.7°)$ **7.** $0.21 + 0.77j$ **9.** $-12\sqrt{5}$ **11.** -1 **13.** $142.86 \ \Omega$

15. $6.46e^{3j}$ **17.** 5.33 A

Chapter 15

15–1 Exercises, p. 566

1. $2x + y + 1 = 0$ **3.** $4x - y + 6 = 0$ **5.** $3x + 5y - 14 = 0$ **7.** $2x + y - 5 = 0$ **9.** $x - 4 = 0$

11. $8x + 7y - 27 = 0$ **13.** $y + 3 = 0$ **15.** $x - 3 = 0$ **17.** $3x - y + 9 = 0$ **19.** $x + 2y + 9 = 0$

21. $m = \dfrac{3}{4}$, $b = -3$, x int $= 4$

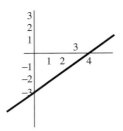

23. $m = $ undef, no y int, x int $= -8$

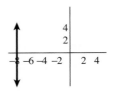

25. $m = \dfrac{2}{3}$, $b = \dfrac{5}{2}$, x int $= \dfrac{-15}{4}$

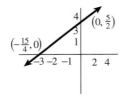

27. perpendicular   **29.** neither   **31.** parallel

33. $c = 3x + 800$ **35.** $P = \dfrac{4}{9}D + 13.6$

37. $s = 20t + 15$ **39.** $R = 0.075t + 3.5$

41. $C = 125 + 35n$

15–2 Exercises, p. 571

1. $(0, 0)$, 3

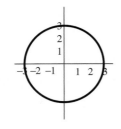

3. $(0, 0)$, 7

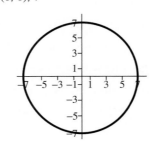

5. $(1, 0)$, $\sqrt[4]{3}$ or 1.3

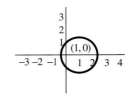

7. $(3, -2)$, $\sqrt{8} \approx 2.8$

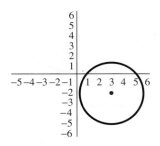

9. $(6, -1)$, 3

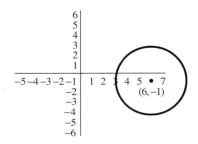

11. $\left(\dfrac{-5}{2}, 1\right)$, $\sqrt{10.25}$ or 3.2

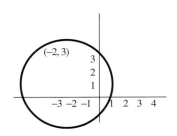

13. $\left(-1, \dfrac{4}{3}\right)$, $\sqrt{\dfrac{10}{9}}$ or 1.05

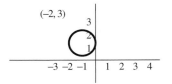

15. $x^2 + y^2 = 81$ **17.** $(x + 3)^2 + (y + 1)^2 = 9$ **19.** $x^2 + (y + 4)^2 = 6$ **21.** $(x + 8)^2 + (y - 5)^2 = 104$

23. $(x - 6)^2 + (y + 4)^2 = 45$ **25.** $(x - 3)^2 + (y - 4)^2 = 16$ **27.** $\left(x + \dfrac{3}{2}\right)^2 + \left(y - \dfrac{1}{2}\right)^2 = 40.5$

29. $(x + 4)^2 + (y - 6)^2 = 1.03 \times 10^{-11}$

31.

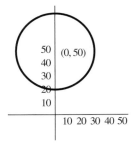

15–3 Exercises, p. 577

1. $V(0, 0)$; $F(2, 0)$; $x = -2$

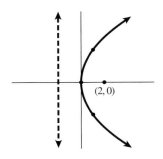

3. $V(0, 0)$; $F(0, 3)$; $y = -3$

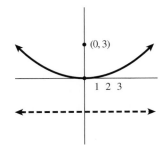

5. $V(0, 0)$; $F(-4, 0)$; $x = 4$

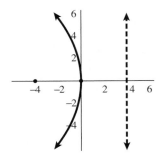

7. $V(0, 0)$; $F\left(0, \frac{-5}{2}\right)$; $y = \frac{5}{2}$

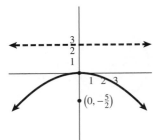

9. $V(0, 3)$; $F(-2, 3)$; $x = 2$

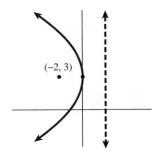

11. $V(-2, 3)$; $F(-2, 6)$; $y = 0$

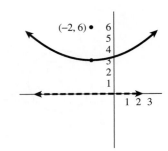

13. $V(-4, 6)$; $F(-4.5, 6)$; $x = -\frac{7}{2}$

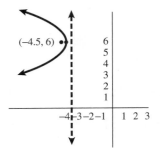

15. $V(7, -3)$; $F(7, -7)$; $y = 1$

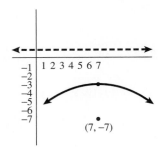

17. $V(-1, 4)$; $F\left(-1, 3\frac{3}{4}\right)$; $y = 4\frac{1}{4}$

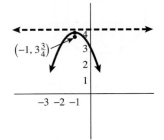

19. $V(-5, -5)$; $F(-4.5, -5)$; $x = -5.5$

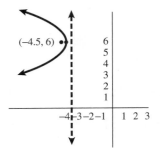

21. $V(-1, 3)$; $F\left(-1, \frac{10}{3}\right)$; $y = 2\frac{2}{3}$

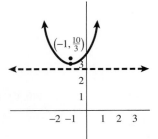

23. $x^2 = -16y$ **25.** $(x - 3)^2 = 12(y - 2)$ **27.** $(y - 4)^2 = 8(x + 1)$ **29.** $x^2 = \frac{8}{5}(y + 3)$

31. $(x - 3)^2 = 16(y + 6)$ **33.** $(t - 2.6)^2 = -.2(s - 31.9)$

35. $(v + 10)^2 = 20(y + 5)$

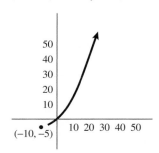

37.

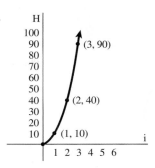

39. Approximately 1 ft from the base along the axis of symmetry

41. Approximately 1 ft from the base along the axis of symmetry

43. 12,308 inches from the center of the mirror

45. 2.6 ft

15–4 Exercises, p. 584

1. $C(0, 0)$; $V(5, 0)$ $(-5, 0)$; $MA(0, 3)$ $(0, -3)$; $F(4, 0)$ $(-4, 0)$

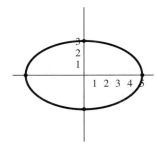

3. $C(0, 0)$; $V(0, 6)$ $(0, -6)$; $MA(4, 0)$ $(-4, 0)$; $F(0, 4.5)$ $(0, -4.5)$

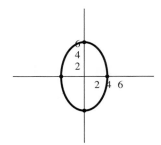

5. $C(0, 0)$; $V(3.2, 0)$ $(-3.2, 0)$; $MA(0, 2.4)$ $(0, -2.4)$; $F(2, 0)$ $(-2, 0)$

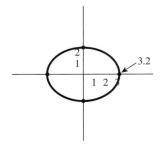

7. $C(-2, 6)$; $V(-2, 11)$ $(-2, 1)$; $MA(0, 6)$ $(-4, 6)$; $F(-2, 10.6)$ $(-2, 1.4)$

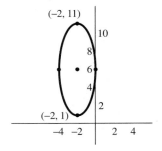

9. $C(-1, 0)$; $V(-1, 9)$ $(-1, -9)$; $MA(5, 0)$ $(-7, 0)$; $F(-1, 6.7)$ $(-1, -6.7)$

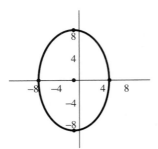

11. $C(3, 1)$; $V(6.5, 1)$ $(-0.5, 1)$; $MA(3, 3)$ $(3, -1)$; $F(5.8, 1)$ $(0.2, 1)$

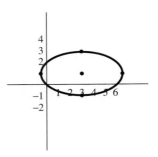

13. $C(-8, -4)$; $V(-2.5, -4)$ $(-13.5, -4)$; $MA(-8, -0.5)$ $(-8, -7.5)$; $F(-3.8, -4)$ $(-12.2, 4)$

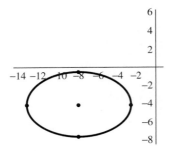

15. $C(0, 0)$; $V(3, 0)$ $(-3, 0)$; $MA(0, 2)$ $(0, -2)$; $F(2.2, 0)$ $(-2.2, 0)$

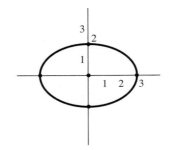

17. $C(-4, 2)$; $V(1.4, 2)$ $(-9.4, 2)$; $MA(-4, 5.1)$ $(-4, -1.1)$; $F(0.4, 2)$ $(-8.4, 2)$

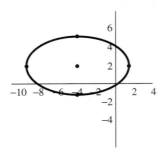

19. $C(-3, -1)$; $V(2.5, -1)$ $(-8.4, -1)$; $MA(-3, 3.1)$ $(-3, -5.1)$; $F(0.6, -1)$ $(-6.6, -1)$

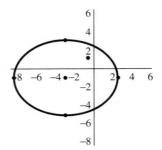

21. $C(-2, 0)$; $V(-2, 6)$ $(-2, -6)$; $MA(0.4, 0)$ $(-4.4, 0)$; $F(-2, 5.5)$ $(-2, -5.5)$

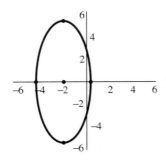

23. $\dfrac{y^2}{49} + \dfrac{x^2}{40} = 1$ **25.** $\dfrac{(x-4)^2}{5} + \dfrac{(y+3)^2}{4} = 1$ **27.** $\dfrac{(x+3)^2}{25} + \dfrac{(y+3)^2}{9} = 1$ **29.** $\dfrac{(y+2)^2}{36} + \dfrac{(x+3)^2}{25} = 1$

31. $\dfrac{x^2}{(4{,}187.5)^2} + \dfrac{y^2}{(4{,}186.4)^2} = 1$ **33.** $\dfrac{x^2}{5{,}625} + \dfrac{y^2}{2{,}500} = 1$

15–5 Exercises, p. 592

1. $C(0, 0)$; $V(4.9, 0)$ $(-4.9, 0)$; $F(5.3, 0)$ $(-5.3, 0)$;
$CA(0, 2)$ $(0, -2)$; $\pm\dfrac{\sqrt{6}}{6}$

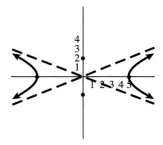

3. $C(0, 0)$; $V(0, 7)$ $(0, -7)$; $F(0, 8.1)$ $(0, -8.1)$;
$CA(4, 0)$ $(-4, 0)$; $\pm\dfrac{7}{4}$

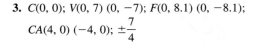
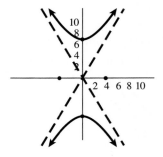

5. $C(0, 0)$; $V(8, 0)$ $(-8, 0)$; $F(9.4, 0)$ $(-9.4, 0)$;
$CA(0, 5)$ $(0, -5)$; $\pm\dfrac{5}{8}$

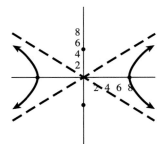

7. $C(2, -1)$; $V(2, 3)$ $(2, -5)$; $F(2, 8.8)$ $(2, -10.8)$;
$CA(11, -1)$ $(-7, -1)$; $\pm\dfrac{4}{9}$

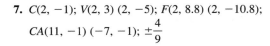
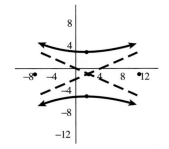

9. $C(-2, 3)$; $V(8, 3)$ $(-12, 3)$; $F(10.2, 3)$ $(-14.2, 3)$;
$CA(-2, 10)$ $(-2, -4)$; $\pm\dfrac{7}{10}$

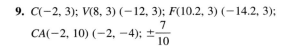
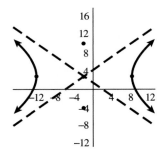

11. $C(3, 7)$; $V(3, 11.5)$ $(3, 2.5)$;
$F(3, 13)$ $(3, 1)$; $CA(7, 7)$ $(-1, 7)$; $\pm\dfrac{4.5}{4}$

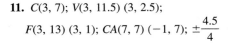
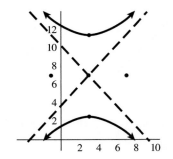

13. $C(2, 5)$; $V(0.3, 5)$ $(3.7, 5)$; $F(-1.5, 5)$ $(5.5, 5)$; $CA(2, 8)$ $(2, 2)$; $\pm\dfrac{\sqrt{3}}{1}$

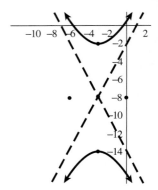

15. $C(-7, -3)$; $V(-4.6, -3)$ $(-9.4, -3)$; $F(-4, -3)$ $(-10, -3)$; $CA(-7, -1.3)$ $(-7, -4.7)$; $\pm\dfrac{\sqrt{2}}{2}$

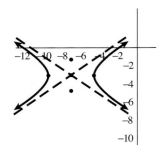

17. $C(-3, -8)$; $V(-3, -2)$ $(-3, -14)$; $F(-3, -13)$ $(-3, -14.7)$; $CA(0, -8)$ $(-6, -8)$; ± 2

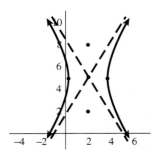

19. $C(2, 2)$; $V(0.3, 2)$ $(3.7, 2)$; $F(-0.8, 2)$ $(4.8, 2)$; $CA(2, 4.2)$ $(2, -0.2)$; $\pm\dfrac{2.2}{1.7}$

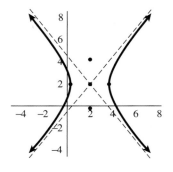

21. $\dfrac{x^2}{16} - \dfrac{y^2}{9} = 1$ **23.** $\dfrac{(x-5)^2}{3.9} - \dfrac{(y-3)^2}{21.1} = 1$ **25.** $\dfrac{(x-3)^2}{9} - \dfrac{(y-4)^2}{64} = 1$

27. $\dfrac{(y+5)^2}{9} - \dfrac{(x+1)^2}{25} = 1$

29.

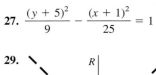

31.

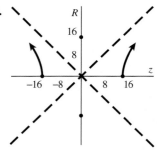

33. Approximately 63 miles from the Master Station

15–6 Exercises, p. 598

1. parabola **3.** hyperbola **5.** circle **7.** parabola **9.** ellipse **11.** hyperbola

13. circle **15.** ellipse

17. parabola; $V(3, -4)$; $F(3, -4.5)$; $y = -3.5$

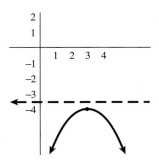

19. ellipse; $C(1, -3)$; $V(3.8 -3) (-4.7, -3)$;
$MA(1, -0.7) (1, -5.3)$; $F(2.6, -3) (-0.6, -3)$

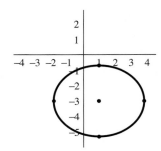

21. circle; $(-4, -1)$; $r = 2$

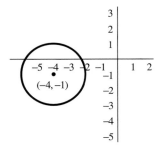

23. hyperbola; $C(0, 0)$; asymptotes are axes;
$V(\sqrt{3}, \sqrt{3}) (-\sqrt{3}, -\sqrt{3})$

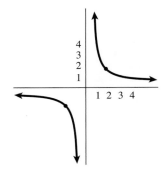

25. hyperbola; $C(0, -3)$; $V(3, -3) (-3, -3)$;
$CA(0, 2) (0, -8)$; $F(5.8, -3) (-5.8, -3)$; $\pm\dfrac{5}{3}$

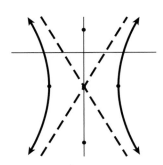

27. $v = 8L^2 - 72L$; parabola   **29.** $x^2 = 2y + 1$; parabola

Chapter Review, p. 599

1. $6x + 8y - 45 = 0$ **3.** $2x + 13y + 47 = 0$ **5.** $x = -3$ **7.** $y = -8$ **9.** $3x - 2y - 1 = 0$

11. $5x + y - 31 = 0$

13. $m = \dfrac{7}{4}$, x int $= \dfrac{10}{7}$, y int $= -2.5$

15. $m = \dfrac{-2}{3}$, x int $= -3$, y int $= -2$

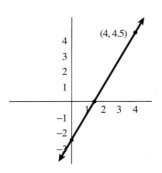

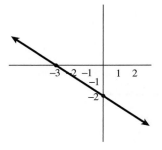

17. $m = \dfrac{1}{3}$, x int $= 9$, y int $= -3$

19. $2T = 100R - 230$

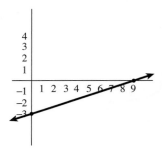

21. $v = 10t + 3.5$

23. $(-4, -3)$; $r = 2$

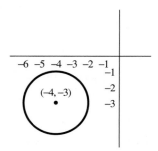

25. $(-1, 6)$; $r = \sqrt[4]{6} = 1.6$

27. $\left(\dfrac{3}{2}, \dfrac{-5}{2}\right)$; $r = \sqrt{14.5} = 3.8$

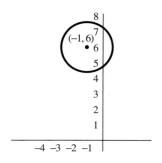

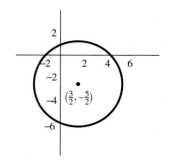

29. $(x + 2)^2 + (y + 4)^2 = 3$ **31.** $(x - 6)^2 + (y + 5)^2 = 25$ **33.** $(x + 2)^2 + (y + 1)^2 = 50$

35. $(40, 0)$; $r = 25$

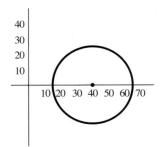

37. $V(3, 0)$; $F(2.5, 0)$; $x = 3.5$

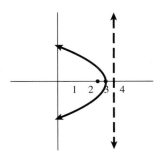

39. $V(2, -3)$; $F(2, -3.75)$; $y = -2.25$

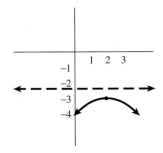

41. $V(-4, 3)$; $F(-4, 2.5)$; $y = 3.5$

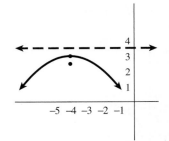

43. $y^2 = 12(x + 2)$ **45.** $(x - 4)^2 = 6(y + 1.5)$ **47.** $x^2 = 285.7y$

49. $C(-1, 3)$; $V(-1, 10)$ $(-1, -4)$; $MA(5, 3)$ $(-7, 3)$; $F(-1, 6.6)(-1, -0.6)$

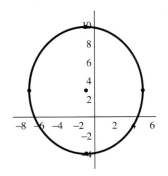

51. $C(0, 6)$; $V(0, 16.4)$ $(0, -4.4)$; $MA(5.8, 6)$ $(-5.8, 6)$; $F(0, 14.6)$ $(0, -2.6)$

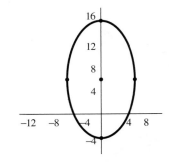

53. $C(-3, -1)$; $V(-2.5, -1)$ $(-3.5, -1)$; $MA\left(-3, \dfrac{-2}{3}\right)\left(-3, \dfrac{-4}{3}\right)$; $F(-2.6, -1)$ $(-3.4, -1)$

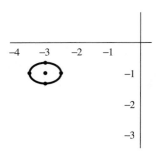

55. $\dfrac{(y-1)^2}{34} + \dfrac{(x-5)^2}{9} = 1$ **57.** $\dfrac{x^2}{16} + \dfrac{(y-2)^2}{9} = 1$ **59.** $\dfrac{x^2}{(4{,}242.5)^2} + \dfrac{(y-112.5)^2}{(4{,}241)^2} = 1$

61. $C(-3, 1)$; $V(1, 1)$ $(-7, 1)$; $CA(-3, 3)$ $(-3, -1)$; $F(1.5, 1)(-7.5, 1)$

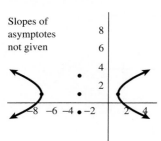

63. $C(-3, -4)$; $V(-3, -9)$ $(-3, 1)$; $CA(3, -4)$ $(-9, -4)$; $F(-3, -11.8)$ $(-3, 3.8)$

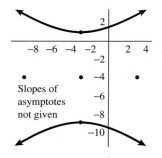

65. $C(0, 4)$; $V(0, 6.8)$ $(0, 1.2)$; $CA(3.2, 4)$ $(-3.2, 4)$; $F(0, 8.2)$ $(0, -0.2)$

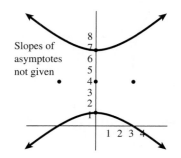

67. $\dfrac{y^2}{9} - \dfrac{x^2}{16} = 1$ **69.** $\dfrac{(x-2)^2}{9} - \dfrac{(y-7)^2}{25} = 1$

71.

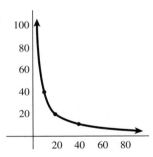

73. circle; $(1, -6)$; $r = 1.6$

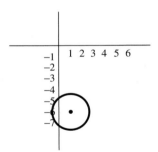

75. parabola; $V(4, -7)$; $F(4, -8)$; $y = -6$

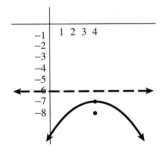

77. ellipse; $C(3, -1)$; $V(3, -5.2)$ $(3, 3.2)$; $MA(5.4, -1)$ $(0.6, -1)$; $F(3, -4.5)$ $(3, 2.5)$

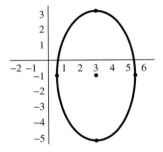

79. hyperbola; $C(3, -1)$; $V(5.8, -1)$ $(0.2, -1)$; $CA(3, 4)$ $(3, -6)$; $F(8.7, -1)$ $(-2.7, -1)$

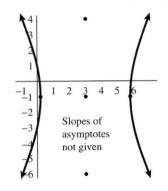

Slopes of asymptotes not given

81. ellipse; $C(-3, -1)$; $V(-3, 4)$ $(-3, -6)$; $MA(1, -1)$ $(-7, -1)$; $F(-3, 2)$ $(-3, -4)$

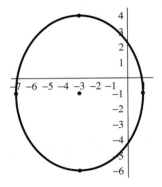

Chapter Test, p. 601

1. $m = \dfrac{1}{5}$, x int $= \dfrac{9}{2}$, y int $= -\dfrac{9}{10}$

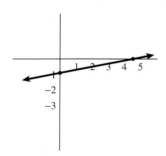

3. $C(1, -3)$; $F(-3.5, -3)$ $(4.5, -3)$;

$V(-3, -3)$ $(5, -3)$; $CA(1, -1)$ $(1, -5)$; $\pm\dfrac{1}{2}$

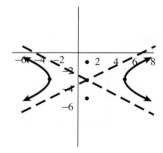

5. $(4, -8)$; $r = 5$ **7.** $\dfrac{(x - 45)^2}{16,933,225} + \dfrac{y^2}{16,931,200} = 1$ **9.** $(y + 5)^2 = 12(x - 1)$

11. $3x + y + 15 = 0$ **13.** $C(-3, 2)$; $V(1, 2)$ $(-7, 2)$; $F(-0.4, 2)$ $(-5.6, 2)$; $MA(-3, 5)$ $(-3, -1)$

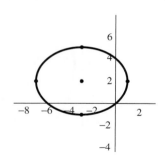

CHAPTER 16

16–1 Exercises, p. 613

1.

grade	65	69	72	74	81	86	88	93	97
frequency	2	1	2	3	1	2	2	1	1

3.

hours	35	38	40	41	42	43	46	48	50	56
frequency	2	1	3	1	1	2	2	2	2	1

5.

IQ	80	83	84	86	89	95	97	98	100	110	120	134	147
frequency	2	1	1	1	1	1	1	2	3	3	1	2	1

7.
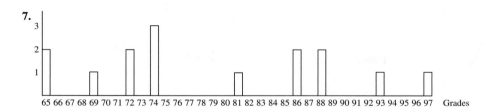

9.
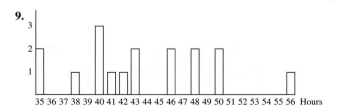

11.
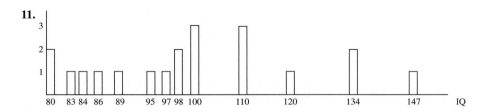

13.
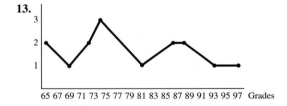

15.

17.
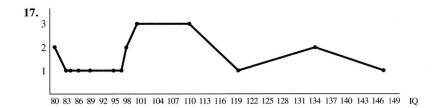

19.

21. 34,200 employees

23.

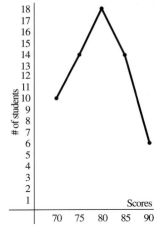

25. 38 students

27.

Math scores	Cumulative frequency
less than 70	10
less than 75	24
less than 80	42
less than 85	56
less than 90	62

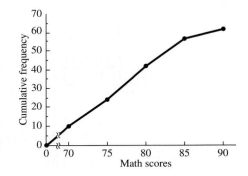

16–2 Exercises, p. 620

1. 21.1, 19, 16 **3.** 81, 83, 73 **5.** 8; 8; 3, 10 **7.** 77.8, 75, 75 **9.** 856.5; 570; 1, 50, 360, 780, 1,840, 2,108

11. 8.125, 7, 6 **13.** 79.2, 82, 85 **15.** 75 **17.** \$181.30 **19.** \$90.43; \$91.09; \$74.12; \$96.25

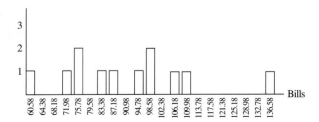

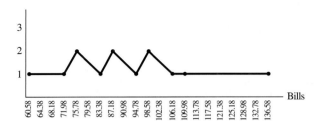

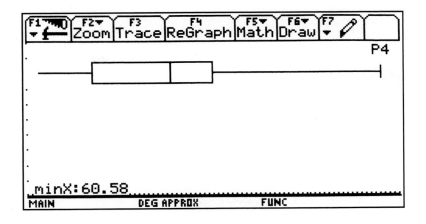

16–3 Exercises, p. 625

1. 57, 19.83   **3.** 38, 11.49   **5.** 54, 14.16   **7.** 21.6 to 49.2, 60%   **9.** 19.4 to 28.2, 58.5%

11. 81.1, 85, 85, 10.9   **13.** 101.25   **15.** 8.24

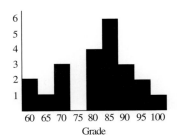

16–4 Exercises, p. 637

1. $y = 3.17x + 4.7$ **3.** $y = 2.71x + 4.67$ **5.** $y = 3.7x^2 - 1.7$

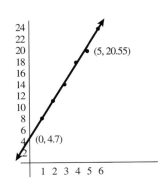

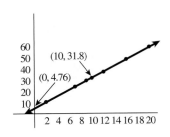

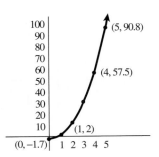

7. $y = 2.53(10)^x + 6.88$

9. $s = 0.17t + 14.3$; at $t = 12$, $s = 16.34$

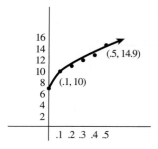

11. $s = 37t + 19$; at $t = 12$, $s = 463$

13. $R = 18.65T - 2.53$; 1,489 Ω

Chapter Review, p. 639

1.

Age	18	27	35	36	45	57	63
Frequency	3	5	2	1	1	1	1

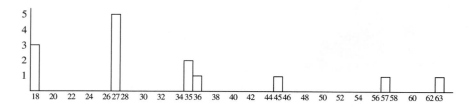

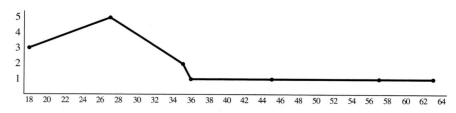

3.

salary	18	19	21	23	24	27	30	32
frequency	2	2	2	2	3	1	2	1

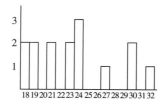

5. 32.86, 27, 27   **7.** 23.53, 23, 24   **9.** $134.00   **11.** $84.33   **13.** 45, 13.3   **15.** 14, 4.35

17. 19.56 to 46.16, 64.2%   **19.** 19.18 to 27.88, 53.3%   **21.** 3.4 (thousand)

23. $y = 7.9x + 14$ **25.** $y = 3.87x + 15.1$

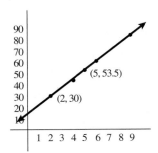

(5, 53.5)

(2, 30)

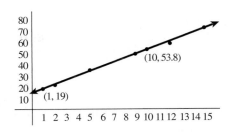

(10, 53.8)

(1, 19)

27. $y = 2.1x^2 + 10.04$ **29.** $y = 3.1(10)^x + 10$

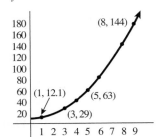

(8, 144)

(1, 12.1)

(5, 63)

(3, 29)

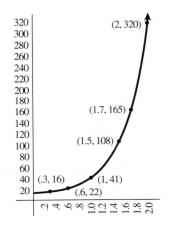

(2, 320)

(1.7, 165)

(1.5, 108)

(.3, 16)   (1, 41)

(.6, 22)

Chapter Test, p. 640

1. 22.2, 21, 18   **3.** 7.08

5.

data	2	3	6	7	8	9	11	14	18
frequency	3	3	1	2	3	1	1	2	1

7. $33,522; $7,563.88

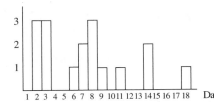

CHAPTER 17

17–1 Exercises, p. 652

1. 27 **3.** $\dfrac{7}{2}$ **5.** 165 **7.** -17 **9.** 73 **11.** 252 **13.** 180 **15.** 225 **17.** $n = 8$, $S_8 = -32$

19. $d = 3$, $S_{18} = 549$ **21.** $a_{10} = 31$, $S_{10} = 130$ **23.** $d = 3$, $n = 8$ **25.** $a_6 = \$30,686$, $S_6 = \$161,616$

27. $C = \$9.90$ **29.** 7,150 **31.** Total sales is $1,722,000 **33.** $140,000 **35.** $9,600

17–2 Exercises, p. 657

1. 1,536 **3.** 0.03125 **5.** $-4,374$ **7.** 11,585.6 **9.** 820,160 **11.** -11.953 **13.** 79.125 **15.** -2.65625

17. $-2,730.5$ **19.** $a_{10} = 0.00024$, $S_{10} = 0.25$ **21.** $n = 4$, $a_4 = \dfrac{1}{3}$ **23.** $n = 7$, $S_7 = 8,744$ **25.** $\dfrac{1}{512}$

27. $31,944 **29.** $18,271.19 **31.** Total value = $100,934 **33.** $8,874.11 **35.** 103 ml **37.** 2 billion kWh

17–3 Exercises, p. 661

1. $\dfrac{-16}{3}$ **3.** 18 **5.** 294 **7.** 118.49 **9.** 108 **11.** $\dfrac{256}{3}$ **13.** $14(2 - \sqrt{2})$ **15.** $\dfrac{12x^3}{2x - 1}$ **17.** 1

19. $\dfrac{83}{99}$ **21.** $\dfrac{419}{990}$ **23.** $\dfrac{415}{111}$ **25.** $a_6 = 21.25764$ in. $s = 168.68$ in. **27.** $S = 75$ ft

17–4 Exercises, p. 666

1. $a^4 + 20a^3 + 150a^2 + 500a + 625$ **3.** $1m^7 - 7m^6 + 21m^5 - 35m^4 + 35m^3 - 21m^2 + 7m - 1$

5. $8x^3 - 60x^2y + 150xy^2 - 125y^3$ **7.** $729y^6 + 1,458xy^5 + 1,215x^2y^4 + 540x^3y^3 + 135x^4y^2 + 18x^5y + x^6$

9. $m^5 + 15m^4 + 90m^3 + 270m^2 + 405m + 243$

11. $b^7 - 42b^6 + 756b^5 - 7,560b^4 + 45,360b^3 - 163,296b^2 + 326,592b - 279,936$

13. $81x^4 - 108x^3 + 54x^2 - 12x + 1$

15. $15,625m^6 - 75,000m^5 + 150,000m^4 - 160,000m^3 + 96,000m^2 - 30,720m + 4,096$

17. $x^7 - 28x^6y + 336x^5y^2 - 2,240x^4y^3 + 8,960x^3y^4 - 21,504x^2y^5 + 28,672xy^6 - 16,384y^7$

19. $16a^8 - 96a^6b + 216a^4b^2 - 216a^2b^3 + 81b^4$

21. $1 - 8x + 28x^2 - 56x^3 \cdots$ **23.** $1 - 3x + 6x^2 - 10x^3 \cdots$ **25.** $55,050,240y^4$ **27.** $-1,792x^3$

Chapter Review, p. 667

1. 38 **3.** -84 **5.** 162 **7.** -60 **9.** 1,364 **11.** -243 **13.** 0.0016 **15.** 6,560 **17.** 2,728

19. $21,578.40 **21.** $\dfrac{4}{9}$ **23.** 27 **25.** $\dfrac{34}{99}$ **27.** $\dfrac{3,121}{999}$ **29.** $\dfrac{1,000}{11}$ ft **31.** $x^3 - 3x^2z + 3xz^2 - z^3$

33. $16x^4 - 96x^3y + 216x^2y^2 - 216xy^3 + 81y^4$ **35.** $a^4 + 8a^3b + 24a^2b^2 + 32ab^3 + 16b^4$

37. $y^6 - 15y^4z + 75y^2z^2 - 125z^3$ **39.** $90,720x^4$ **41.** $945x^4$

Chapter Test, p. 668

1. 160   **3.** $28,051.04   **5.** $16x^4 - 32x^3y + 24x^2y^2 - 8xy^3 + y^4$ **7.** $m^4 - 4m^3n + 6m^2n^2 - 4mn^3 + n^4$

9. 2,509   **11.** 1,577   **13.** $5,670x^4$ **15.** $S_{11} = -44,287$ **17.** $23,224,320y^4$

Appendix A

A–1 Exercises, p. 676

1. rational   **3.** rational   **5.** integer and rational   **7.** integer and rational   **9.** rational   **11.** F   **13.** T

15. T   **17.** F   **19.** T

21.

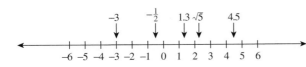

23.

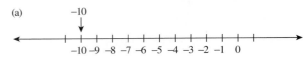

(a)

(b)

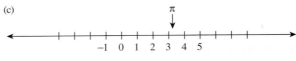

(c)

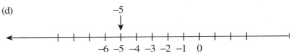

(d)

A–2 Exercises, p. 679

1. -6 **3.** -17 **5.** 18   **7.** -14 **9.** -54 **11.** 4   **13.** 13.95   **15.** -13 **17.** 19   **19.** 48.28

21. distributive   **23.** commutative   **25.** 17°   **27. (a)** 300   **(b)** 360   **(c)** 44   **(d)** 425

29. (a) -8.6 **(b)** -529 **(c)** 24.8   **(d)** -123

A–3 Exercises, p. 682

1. 23   **3.** -12 **5.** 8   **7.** undefined   **9.** -4 **11.** 14   **13.** -6 **15.** 3/19   **17.** $\dfrac{-5}{11.5}$

19. -56.6 **21. (a)** 96 ft   **(b)** 250 m   **23.** 61 ft/s   **25.** 122.2 ft^2

A–4 Exercises, p. 687

1. -32 **3.** x^4 **5.** 10   **7.** $\dfrac{4}{25}$ **9.** $\dfrac{2}{x^3}$ **11.** $\dfrac{1,000}{y^6}$ **13.** $\dfrac{1}{ab^4}$ **15.** $\dfrac{-2}{s^3}$ **17.** $16x^2$ **19.** $\dfrac{1}{10^4}$

21. 198   **23.** 19   **25.** 175.929 m^2 **27.** 14.137 m^3 **29.** $(3 + 2)^2 = 5^2 = 25$, $3^2 + 2^2 = 9 + 4 = 13$, $25 \neq 13$

A–5 Exercises, p. 689

1. 8.956×10^6 **3.** -4.50×10^5 **5.** 6.9×10^9 **7.** 5.0×10^6 mm^3 **9.** 12,300,000

11. 5,400,000,000 **13.** 0.002 **15.** 105,000 **17.** 1.8066×10^{-8} **19.** 1.6×10^9 **21.** 7.6549×10^2

23. 3.76×10^{40} joules

A–6 Exercises, p. 692

1. 9 **3.** 2 **5.** 2 **7.** $\dfrac{4}{5}$ **9.** $\dfrac{1}{2}$ **11.** $4\sqrt{2}$ **13.** 50 **15.** $2\sqrt{6}$ **17.** 5 **19.** 0.2 **21.** 182

23. 0.71 **25.** 2.19 **27.** 2.29 **29.** 13.39 **31.** 14.76 cm **33.** 4 in. **35.** 4.830

A–7 Exercises, p. 695

1. $5x$ **3.** $2a$ **5.** 0 **7.** $11xz$ **9.** $3a - 7b + 5$ **11.** $7x + 5y$ **13.** $-a + 5b$ **15.** $3x - 3y + 12$
17. $x^2 + 3y^2 - 2y$ **19.** $-18m + 23n$ **21.** $4m - 8$ **23.** $0.0406x + 206.8$ **25.** $-2I_1 + 3I_2$

A–8 Exercises, p. 698

1. $12a^3b$ **3.** $-24s^6$ **5.** $2x^4 - 6x^2y$ **7.** $-m^2n^2 + m^2n - mn^2$ **9.** $x^9 + x^7 - x^5$ **11.** $x^2 + x - 12$
13. $2a^2 - 6ab + 4b^2$ **15.** $15s^4 + 2s^2t - t^2$ **17.** $108x^5y^2$ **19.** $4x^2 + 12xy + 9y^2$ **21.** $x^3 - 2xy^2 + y^3$
23. $x^3 - 5x^2 + 4x + 6$ **25.** $12y^3 - 8y^2 - 28y - 8$ **27.** $R^2 + 2Rr + 3r^2$

A–9 Exercises, p. 701

1. $3a$ **3.** $-\dfrac{c^2d^2}{4}$ **5.** $5y^2$ **7.** $\dfrac{x^3}{2y}$ **9.** $-\dfrac{2y^2a}{3}$ **11.** $6a^2 - 2b^5$ **13.** $\dfrac{4s}{t} - \dfrac{2}{t} + \dfrac{5}{8st^2}$ **15.** $x + 6$

17. $2x^2 + 3x - 1$ **19.** $3x^2 + x + 3$ **21.** $x^2 - 2x + 4 + \dfrac{3}{x + 2}$ **23.** $3x^2 + 9$

25. $2x - 1 + \dfrac{3x}{x^2 - x + 2}$ **27.** $15x - 1$ **29.** $-3gt^2 + 2t + 1$

A–10 Exercises, p. 704

1. 32% **3.** 37.5% **5.** 20% **7.** 0.625% **9.** 5% **11.** 0.56 **13.** 0.0025 **15.** 0.18 **17.** 2.55
19. 0.0003 **21.** 20.8% **23.** 273 **25.** 93.3 **27.** 132 **29.** 767 **31.** 13.3 oz

A–11 Exercises, p. 708

1. 10.79 **3.** 320 **5.** 8.7 **7.** 150 **9.** 98 **11.** 133.9 **13.** 11.8 **15.** 171,000 **17.** 68,000
19. 22.5 **21.** 139 **23.** 1,769

A–12 Exercises, p. 712

1. 0.42 yd **3.** 6.8×10^5 cm **5.** 1,820 ft **7.** 520 μs **9.** 8,626 g **11.** 133.6 mi **13.** 1,417.3 in
15. 303.2 cm^2 **17.** 5.7 N/cm **19.** 106.2 km/h **21.** 614.4 ft^3 **23.** 7.1×10^{-6} hp **25.** 253,525 $\dfrac{\text{N} \cdot \text{cm}}{\text{s}}$
27. 102.7 ft/s **29.** 4.7 hp

A–13 Exercises, p. 714

1. 14.8 **3.** 13.6 **5.** 196 **7.** 0.562 **9.** −37.5 **11.** 3.38 **13.** 8.3 **15.** 2.3×10^{-2} **17.** 1.68

19. 24 **21.** 1,728 **23.** 1.9×10^{-14} **25.** 8.7

A–14 Exercises, p. 719

1. 209 **3.** 3,895.633 **5.** 14.25 **7.** 2,257.625 **9.** 1,715.1758 **11.** 1100010010, 1,422, 312

13. 10111, 47, 27 **15.** 111001000, 710, 1C8 **17.** 100001111, 417, 10F **19.** 1001011101, 1,135, 25d

21. 1001000111, 1,107, 247 **23.** 1010 1001 0100 **25.** 110 011 001 010 **27.** 001 011 111 100

29. 011 001 100. 000 110 **31.** 000. 001 100 111 **33.** 010 100. 011 101

Appendix B

Exercises, p. 727

1. 72° **3.** 33.6° **5.** 77° **7.** 60° **9.** ∠3 and ∠6; ∠5 and ∠4 **11.** ∠2 and ∠5; ∠3 and ∠7; ∠4 and ∠8; ∠1 and ∠6 **13.** 115° **15.** 4 cm **17.** 19.2 cm **19.** 42.5 ft **21. (a)** circle **(b)** 31.4 cm **(c)** 78.5 cm^2 **23. (a)** parallelogram **(b)** 37 in. **(c)** 108 in^2 **25. (a)** rhombus **(b)** 20.8 yd **(c)** 23.4 yd^2 **27. (a)** trapezoid **(b)** 17 ft **(c)** 17.1 ft^2 **29. (a)** right circular cylinder **(b)** 45,592.8 ft^3 **31. (a)** cube **(b)** 937.5 in^2 **33. (a)** right circular cylinder **(b)** 791.3 m^2 **35. (a)** 15 Ω **(b)** 24 Ω **(c)** 120 Ω **37.** 2.6 in^2 **39.** 36 m **41.** 64.1 in^2 **43.** 235,500 H

Index